Study and Solutions Guide for

ALGEBRA AND TRIGONOMETRY

SEVENTH EDITION

Larson / Hostetler

Dianna L. Zook

Indiana University – Purdue University
Fort Wayne, Indiana

Houghton Mifflin Company   Boston   New York

Publisher: Richard Stratton
Sponsoring Editor: Cathy Cantin
Development Manager: Maureen Ross
Development Editor: Lisa Collette
Supervising Editor: Karen Carter
Senior Project Editor: Patty Bergin
Art and Design Manager: Gary Crespo
Executive Marketing Manager: Brenda Bravener-Greville
Senior Marketing Manager: Jennifer Jones
Director of Manufacturing: Priscilla Manchester

Printed in the United States of America

ISBN 10: 0-618-64323-0
ISBN 13: 978-0-618-64323-3

23456789-CRS-09 08 07

TO THE STUDENT

The *Study and Solutions Guide* for *Algebra and Trigonometry*, Seventh Edition, is a supplement to the textbook by Ron Larson and Robert Hostetler.

As a mathematics instructor, I often have students come to me with questions about assigned homework. When I ask to see their work, the reply often is "I didn't know where to start." The purpose of the study guide is to provide brief summaries of the topics covered in the textbook and enough detailed solutions to problems so that you will be able to work the remaining exercises.

I have made every effort to see that the solutions are correct. However, I would appreciate hearing about any errors or other suggestions for improvement.

I would like to thank the staff at Larson Texts, Inc., for their help in the production of this guide.

Dianna L. Zook
Indiana University – Purdue University
Fort Wayne, IN 46805
(zook@ipfw.edu)

Study Strategies

- Attend all classes and come prepared. Have your homework completed. Bring the text, paper, pen or pencil, and a calculator (scientific or graphing) to each class.

- Read the section in the text that is to be covered before class. Make notes about any questions that you have and if they are not answered during the lecture, ask them at the appropriate time.

- Participate in class. As mentioned above ask questions.

- Take notes on all definitions, concepts, rules, formulas, and examples. After class, read your notes and fill in any gaps, or make notations of questions that you have.

- DO THE HOMEWORK!!! You learn mathematics by doing it yourself. Allow at least two hours outside of each class for homework. Do not fall behind.

- Seek help when needed. Visit your instructor during office hours and come prepared with specific questions; check with your school's tutoring service; find a study partner in class; check additional books in the library for more examples – just do something before the problem becomes insurmountable.

- Do not cram for exams. each chapter in the text contains a review chapter and a chapter test, and this Study Guide contains a practice test at the end of each chapter. (The answers are at the end of Part I.) Work these problems a few days before the exam and review any areas of weakness.

CONTENTS

PART I

C H A P T E R   P
Prerequisites

CHAPTER P
Prerequisites

Section P.1   Review of Real Numbers and Their Properties

■ You should know the following sets.

(a) The set of real numbers includes the rational numbers and the irrational numbers.

(b) The set of rational numbers includes all real numbers that can be written as the ratio p/q of two integers, where $q \neq 0$.

(c) The set of irrational numbers includes all real numbers which are not rational.

(d) The set of integers: $\{\ldots, -3, -2, -1, 0, 1, 2, 3, \ldots\}$

(e) The set of whole numbers: $\{0, 1, 2, 3, 4, \ldots\}$

(f) The set of natural numbers: $\{1, 2, 3, 4, \ldots\}$

■ The real number line is used to represent the real numbers.

■ Know the inequality symbols.

(a) $a < b$ means a is less than b.

(b) $a \leq b$ means a is less than or equal to b.

(c) $a > b$ means a is greater than b.

(d) $a \geq b$ means a is greater than or equal to b.

■

Interval Notation	Inequality Notation	Graph	Type
$[a, b]$	$a \leq x \leq b$		Bounded and Closed
(a, b)	$a < x < b$		Bounded and Open
$[a, b)$	$a \leq x < b$		Bounded
$(a, b]$	$a < x \leq b$		Bounded
$[a, \infty)$	$x \geq a$		Unbounded
(a, ∞)	$x > a$		Unbounded
$(-\infty, b]$	$x \leq b$		Unbounded
$(-\infty, b)$	$x < b$		Unbounded
$(-\infty, \infty)$	$-\infty < x < \infty$		Unbounded

■ You should know that $|a| = \begin{cases} a, & \text{if } a \geq 0 \\ -a, & \text{if } a < 0 \end{cases}$

■ Know the properties of absolute value.

(a) $|a| \geq 0$

(b) $|-a| = |a|$

(c) $|ab| = |a|\,|b|$

(d) $\left|\dfrac{a}{b}\right| = \dfrac{|a|}{|b|}, \ b \neq 0$

—CONTINUED—

■ The distance between a and b on the real line is $d(a, b) = |b - a| = |a - b|$.

■ You should be able to identify the terms in an algebraic expression.

■ You should know and be able to use the basic rules of algebra.

■ Commutative Property

(a) Addition: $a + b = b + a$ (b) Multiplication: $a \cdot b = b \cdot a$

■ Associative Property

(a) Addition: $(a + b) + c = a + (b + c)$ (b) Multiplication: $(ab)c = a(bc)$

■ Identity Property

(a) Addition: 0 is the identity; $a + 0 = 0 + a = a$.   (b) Multiplication: 1 is the identity; $a \cdot 1 = 1 \cdot a = a$.

■ Inverse Property

(a) Addition: $-a$ is the additive inverse of a; $a + (-a) = -a + a = 0$.

(b) Multiplication: $1/a$ is the multiplicative inverse of a, $a \neq 0$; $a(1/a) = (1/a)a = 1$.

■ Distributive Property

(a) $a(b + c) = ab + ac$ (b) $(a + b)c = ac + bc$

■ Properties of Negation

(a) $(-1)a = -a$ (b) $-(-a) = a$ (c) $(-a)b = a(-b) = -ab$

(d) $(-a)(-b) = ab$ (e) $-(a + b) = (-a) + (-b) = -a - b$

■ Properties of Equality

(a) If $a = b$, then $a \pm c = b \pm c$.   (b) If $a = b$, then $ac = bc$.

(c) If $a \pm c = b \pm c$, then $a = b$.   (d) If $ac = bc$ and $c \neq 0$, then $a = b$.

■ Properties of Zero

(a) $a \pm 0 = a$ (b) $a \cdot 0 = 0$ (c) $0 \div a = 0/a = 0, a \neq 0$

(d) $a/0$ is undefined.   (e) If $ab = 0$, then $a = 0$ or $b = 0$.

■ Properties of Fractions ($b \neq 0, d \neq 0$)

(a) Equivalent Fractions: $a/b = c/d$ if and only if $ad = bc$.

(b) Rule of Signs: $-a/b = a/-b = -(a/b)$ and $-a/-b = a/b$

(c) Equivalent Fractions: $a/b = ac/bc, c \neq 0$

(d) Addition and Subtraction

1. Like Denominators: $(a/b) \pm (c/b) = (a \pm c)/b$ 2. Unlike Denominators: $(a/b) \pm (c/d) = (ad \pm bc)/bd$

(e) Multiplication: $(a/b) \cdot (c/d) = (ac)/(bd)$

(f) Division: $(a/b) \div (c/d) = (a/b) \cdot (d/c) = (ad)/(bc)$ if $c \neq 0$.

Vocabulary Check

1. rational   2. irrational   3. absolute value

4. composite   5. prime   6. variables; constants

7. terms   8. coefficient   9. zero-factor property

1. $-9, -\frac{7}{2}, 5, \frac{2}{3}, \sqrt{2}, 0, 1, -4, 2, -11$

   (a) Natural numbers: 5, 1, 2

   (b) Whole numbers: 0, 5, 1, 2

   (c) Integers: $-9, 5, 0, 1, -4, 2, -11$

   (d) Rational numbers: $-9, -\frac{7}{2}, 5, \frac{2}{3}, 0, 1, -4, 2, -11$

   (e) Irrational numbers: $\sqrt{2}$

3. $2.01, 0.666\ldots, -13, 0.010110111\ldots, 1, -6$

   (a) Natural numbers: 1

   (b) Whole numbers: 1

   (c) Integers: $-13, 1, -6$

   (d) Rational numbers: $2.01, 0.666\ldots, -13, 1, -6$

   (e) Irrational numbers: $0.010110111\ldots$

5. $-\pi, -\frac{1}{3}, \frac{6}{3}, \frac{1}{2}\sqrt{2}, -7.5, -1, 8, -22$

   (a) Natural numbers: $\frac{6}{3}$ (since it equals 2), 8

   (b) Whole numbers: $\frac{6}{3}, 8$

   (c) Integers: $\frac{6}{3}, -1, 8, -22$

   (d) Rational numbers: $-\frac{1}{3}, \frac{6}{3}, -7.5, -1, 8, -22$

   (e) Irrational numbers: $-\pi, \frac{1}{2}\sqrt{2}$

7. $\frac{5}{8} = 0.625$

9. $\frac{41}{333} = 0.\overline{123}$

11. $-1 < 2.5$

13. $-4 > -8$

15. $\frac{3}{2} < 7$

17. $\frac{5}{6} > \frac{2}{3}$

19. (a) The inequality $x \le 5$ denotes the set of all real numbers less than or equal to 5.

    (b)

    (c) The interval is unbounded.

21. (a) The inequality $x < 0$ denotes the set of all negative real numbers.

    (b)

    (c) The interval is unbounded.

23. (a) The interval $[4, \infty)$ denotes the set of all real numbers greater than or equal to 4.

    (b)

    (c) The interval is unbounded.

25. (a) The inequality $-2 < x < 2$ denotes the set of all real numbers greater than -2 and less than 2.

    (b)

    (c) The interval is bounded.

27. (a) The inequality $-1 \le x < 0$ denotes the set of all negative real numbers greater than or equal to -1.

    (b)

    (c) The interval is bounded.

29. (a) The interval $[-2, 5)$ denotes the set of all real numbers greater than or equal to -2 and less than 5.

    (b)

    (c) The interval is bounded.

31. $-2 < x \le 4$

33. $y \ge 0$

35. $10 \le t \le 22$

37. $W > 65$

39. $|-10| = -(-10) = 10$

41. $|3 - 8| = |-5| = -(-5) = 5$

43. $|-1| - |-2| = 1 - 2 = -1$

45. $\dfrac{-5}{|-5|} = \dfrac{-5}{-(-5)} = \dfrac{-5}{5} = -1$

47. If $x < -2$, then $x + 2$ is negative.

Thus $\dfrac{|x + 2|}{x + 2} = \dfrac{-(x + 2)}{x + 2} = -1$.

49. $|-3| > -|-3|$ since $3 > -3$.

51. $-5 = -|5|$ since $-5 = -5$.

53. $-|-2| = -|2|$ since $-2 = -2$.

55. $d(126, 75) = |75 - 126| = 51$

57. $d\left(-\frac{5}{2}, 0\right) = \left|0 - \left(-\frac{5}{2}\right)\right| = \frac{5}{2}$

59. $d\left(\frac{16}{5}, \frac{112}{75}\right) = \left|\frac{112}{75} - \frac{16}{5}\right| = \frac{128}{75}$

61.

Budgeted Expense, b	Actual Expense, a	$\|a - b\|$	0.05b
$112,700	$113,356	$656	$0.05(112,700) = \$5635$

Since $656 < $5635 but $656 > $500, the actual expense does not pass the "budget variance test."

63.

Budgeted Expense, b	Actual Expense, a	$\|a - b\|$	0.05b
$37,640	$37,335	$305	$0.05(37,640) = \$1882$

Since $305 < $500 and $305 < $1882, the actual expense passes the "budget variance test."

65. (a)

Year	Expenditures (in billions)	Surplus or Deficit (in billions)
1960	$92.2	$\|92.5 - 92.2\| = \$0.3$ surplus
1970	$195.6	$\|192.8 - 195.6\| = \$2.8$ deficit
1980	$590.9	$\|517.1 - 590.9\| = \$73.8$ deficit
1990	$1253.2	$\|1032.0 - 1253.2\| = \$221.2$ deficit
2000	$1788.8	$\|2025.2 - 1788.8\| = \$236.4$ surplus

(b)

67. $d(x, 5) = |x - 5|$ and $d(x, 5) \le 3$, thus $|x - 5| \le 3$.

69. $d(y, 0) = |y - 0| = |y|$ and $d(y, 0) \ge 6$, thus $|y| \ge 6$.

71. $d(326, 351) = |351 - 326| = 25$ miles

73. $7x + 4$

Terms: $7x, 4$

Coefficient: 7

75. $\sqrt{3}x^2 - 8x - 11$

Terms: $\sqrt{3}x^2, -8x, -11$

Coefficients: $\sqrt{3}, -8$

77. $4x^3 + \dfrac{x}{2} - 5$

Terms: $4x^3, \dfrac{x}{2}, -5$

Coefficients: $4, \dfrac{1}{2}$

79. $4x - 6$

(a) $4(-1) - 6 = -4 - 6 = -10$

(b) $4(0) - 6 = 0 - 6 = -6$

81. $x^2 - 3x + 4$

(a) $(-2)^2 - 3(-2) + 4 = 4 + 6 + 4 = 14$

(b) $(2)^2 - 3(2) + 4 = 4 - 6 + 4 = 2$

83. $\dfrac{x+1}{x-1}$

(a) $\dfrac{1+1}{1-1} = \dfrac{2}{0}$

Division by zero is undefined.

(b) $\dfrac{-1+1}{-1-1} = \dfrac{0}{-2} = 0$

85. $x + 9 = 9 + x$

Commutative Property of Addition

87. $\dfrac{1}{(h+6)}(h+6) = 1, h \neq -6$

Multiplicative Inverse Property

89. $2(x+3) = 2x + 6$

Distributive Property

91. $1 \cdot (1+x) = 1 + x$

Multiplicative Identity Property

93. $x + (y + 10) = (x + y) + 10$

Associative Property of Addition

95. $3(t-4) = 3 \cdot t - 3 \cdot 4$

Distributive Property

97. $\dfrac{3}{16} + \dfrac{5}{16} = \dfrac{8}{16} = \dfrac{1}{2}$

99. $\dfrac{5}{8} - \dfrac{5}{12} + \dfrac{1}{6} = \dfrac{15}{24} - \dfrac{10}{24} + \dfrac{4}{24} = \dfrac{9}{24} = \dfrac{3}{8}$

101. $12 \div \dfrac{1}{4} = 12 \cdot \dfrac{4}{1} = 12 \cdot 4 = 48$

103. $\dfrac{2x}{3} - \dfrac{x}{4} = \dfrac{8x}{12} - \dfrac{3x}{12} = \dfrac{5x}{12}$

105. (a)

n	1	0.5	0.01	0.0001	0.000001
$5/n$	5	10	500	50,000	5,000,000

(b) The value of $5/n$ approaches infinity as n approaches 0.

107. False. If $a < b$, then $\dfrac{1}{a} > \dfrac{1}{b}$, where $a \neq b \neq 0$.

109. (a) $|u+v| \neq |u| + |v|$ if u is positive and v is negative or vice versa.

(b) $|u+v| \leq |u| + |v|$

They are equal when u and v have the same sign. If they differ in sign, $|u+v|$ is less than $|u| + |v|$.

111. The only even prime number is 2, because its factors are itself and 1.

113. (a) Since $A > 0$, $-A < 0$. The expression is negative.

(b) Since $B < A$, $B - A < 0$. The expression is negative.

115. Yes, if a is a negative number, then $-a$ is positive. Thus, $|a| = -a$ if a is negative.

Section P.2   Exponents and Radicals

- You should know the properties of exponents.

    (a) $a^1 = a$

    (b) $a^0 = 1, a \neq 0$

    (c) $a^m a^n = a^{m+n}$

    (d) $a^m/a^n = a^{m-n}, a \neq 0$

    (e) $a^{-n} = 1/a^n = (1/a)^n, a \neq 0$

    (f) $(a^m)^n = a^{mn}$

    (g) $(ab)^n = a^n b^n$

    (h) $(a/b)^n = a^n/b^n, b \neq 0$

    (i) $(a/b)^{-n} = (b/a)^n, a \neq 0, b \neq 0$

    (j) $|a^2| = |a|^2 = a^2$

- You should be able to write numbers in scientific notation, $c \times 10^n$, where $1 \leq c < 10$ and n is an integer.

- You should be able to use your calculator to evaluate expressions involving exponents.

- You should know the properties of radicals.

    (a) $\sqrt[n]{a^m} = \left(\sqrt[n]{a}\right)^m, a > 0$

    (b) $\sqrt[n]{a} \cdot \sqrt[n]{b} = \sqrt[n]{ab}$

    (c) $\dfrac{\sqrt[n]{a}}{\sqrt[n]{b}} = \sqrt[n]{\dfrac{a}{b}}, b \neq 0$

    (d) $\sqrt[m]{\sqrt[n]{a}} = \sqrt[mn]{a}$

    (e) $\left(\sqrt[n]{a}\right)^n = a$

    (f) For n even, $\sqrt[n]{a^n} = |a|$.
    For n odd, $\sqrt[n]{a^n} = a$.

    (g) $a^{1/n} = \sqrt[n]{a}$

    (h) $a^{m/n} = \left(\sqrt[n]{a}\right)^m = \sqrt[n]{a^m}, a \geq 0$

- You should be able to simplify radicals.

    (a) All possible factors have been removed from the radical sign.

    (b) All fractions have radical-free denominators.

    (c) The index for the radical has been reduced as far as possible.

- You should be able to use your calculator to evaluate radicals.

Vocabulary Check

1. exponent; base
2. scientific notation
3. square root
4. principal nth root
5. index; radicand
6. simplest form
7. conjugates
8. rationalizing
9. power; index

1. $8^5 = 8 \times 8 \times 8 \times 8 \times 8$

3. $(4.9)(4.9)(4.9)(4.9)(4.9)(4.9) = 4.9^6$

5. (a) $3^2 \cdot 3 = 3^3 = 27$

   (b) $3 \cdot 3^3 = 3^4 = 81$

7. (a) $(3^3)^0 = 1$

   (b) $-3^2 = -9$

9. (a) $\dfrac{3 \cdot 4^{-4}}{3^{-4} \cdot 4^{-1}} = 3^{1-(-4)} \cdot 4^{-4-(-1)} = 3^5 \cdot 4^{-3}$

   $= \dfrac{3^5}{4^3} = \dfrac{243}{64}$

   (b) $32(-2)^{-5} = \dfrac{32}{(-2)^5} = \dfrac{32}{-32} = -1$

11. (a) $2^{-1} + 3^{-1} = \dfrac{1}{2} + \dfrac{1}{3} = \dfrac{3}{6} + \dfrac{2}{6} = \dfrac{5}{6}$

    (b) $(2^{-1})^{-2} = 2^{(-1)(-2)} = 2^2 = 4$

13. $(-4)^3(5^2) = (-64)(25) = -1600$

15. $\dfrac{3^6}{7^3} = \dfrac{729}{343} \approx 2.125$

17. When $x = 2$,
$$-3x^3 = -3(2)^3 = -24.$$

19. When $x = 10$,
$$6x^0 = 6(10)^0 = 6(1) = 6.$$

21. When $x = -3$,
$$2x^3 = 2(-3)^3 = 2(-27) = -54.$$

23. When $x = -\tfrac{1}{2}$,
$$4x^2 = 4\left(-\tfrac{1}{2}\right)^2 = 4\left(\tfrac{1}{4}\right) = 1.$$

25. (a) $(-5z)^3 = (-5)^3 z^3 = -125z^3$

(b) $5x^4(x^2) = 5x^{4+2} = 5x^6$

27. (a) $6y^2(2y^0)^2 = 6y^2(2 \cdot 1)^2 = 6y^2(4) = 24y^2$

(b) $\dfrac{3x^5}{x^3} = 3x^{5-3} = 3x^2$

29. (a) $\dfrac{7x^2}{x^3} = 7x^{2-3} = 7x^{-1} = \dfrac{7}{x}$

(b) $\dfrac{12(x+y)^3}{9(x+y)} = \dfrac{4}{3}(x+y)^{3-1} = \dfrac{4}{3}(x+y)^2$

31. (a) $(x+5)^0 = 1,\ x \neq -5$

(b) $(2x^2)^{-2} = \dfrac{1}{(2x^2)^2} = \dfrac{1}{4x^4}$

33. (a) $(-2x^2)^3(4x^3)^{-1} = \dfrac{-8x^6}{4x^3} = -2x^3$

(b) $\left(\dfrac{x}{10}\right)^{-1} = \dfrac{10}{x}$

35. (a) $3^n \cdot 3^{2n} = 3^{n+2n} = 3^{3n}$

(b) $\left(\dfrac{a^{-2}}{b^{-2}}\right)\left(\dfrac{b}{a}\right)^3 = \left(\dfrac{b^2}{a^2}\right)\left(\dfrac{b^3}{a^3}\right) = \dfrac{b^5}{a^5}$

37. $57{,}300{,}000 = 5.73 \times 10^7$ square miles

39. $0.0000899 = 8.99 \times 10^{-5}$ gram per cubic centimeter

41. $4.568 \times 10^9 = 4{,}568{,}000{,}000$ ounces

43. $1.6022 \times 10^{-19} = 0.00000000000000000016022$ coulomb

45. (a) $\sqrt{25 \times 10^8} = 5 \times 10^4 = 50{,}000$

(b) $\sqrt[3]{8 \times 10^{15}} = 2 \times 10^5 = 200{,}000$

47. (a) $750\left(1 + \dfrac{0.11}{365}\right)^{800} \approx 954.448$

(b) $\dfrac{67{,}000{,}000 + 93{,}000{,}000}{0.0052} = 30{,}769{,}230{,}769.2$
$$\approx 3.077 \times 10^{10}$$

49. (a) $\sqrt{4.5 \times 10^9} \approx 67{,}082.039$

(b) $\sqrt[3]{6.3 \times 10^4} \approx 39.791$

51. (a) $\sqrt{9} = 3$

(b) $\sqrt[3]{\dfrac{27}{8}} = \dfrac{\sqrt[3]{27}}{\sqrt[3]{8}} = \dfrac{3}{2}$

53. (a) $32^{-3/5} = \dfrac{1}{32^{3/5}} = \dfrac{1}{(\sqrt[5]{32})^3} = \dfrac{1}{(2)^3} = \dfrac{1}{8}$

(b) $\left(\dfrac{16}{81}\right)^{-3/4} = \left(\dfrac{81}{16}\right)^{3/4} = \left(\sqrt[4]{\dfrac{81}{16}}\right)^3 = \left(\dfrac{3}{2}\right)^3 = \dfrac{27}{8}$

55. (a) $\left(-\dfrac{1}{64}\right)^{-1/3} = (-64)^{1/3} = \sqrt[3]{-64} = -4$

(b) $\left(\dfrac{1}{\sqrt{32}}\right)^{-2/5} = \left(\sqrt{32}\right)^{2/5} = \sqrt[5]{\left(\sqrt{32}\right)^2} = \sqrt[5]{32} = 2$

57. (a) $\sqrt{57} \approx 7.550$

(b) $\sqrt[5]{-27^3} = (-27)^{3/5}$
$$\approx -7.225$$

59. (a) $(-12.4)^{-1.8} \approx -0.011$

(b) $\left(5\sqrt{3}\right)^{-2.5} \approx 0.005$

61. (a) $\left(\sqrt[3]{4}\right)^3 = 4^{3/3} = 4^1 = 4$

(b) $\sqrt[5]{96x^5} = \sqrt[5]{32x^5 \cdot 3}$
$$= 2x\sqrt[5]{3}$$
$$= 2 \cdot 3^{1/5} \cdot x$$

63. (a) $\sqrt{8} = \sqrt{4 \cdot 2}$

$= \sqrt{4}\sqrt{2} = 2\sqrt{2}$

(b) $\sqrt[3]{24} = \sqrt[3]{8 \cdot 3}$

$= \sqrt[3]{8}\sqrt[3]{3} = 2\sqrt[3]{3}$

65. (a) $\sqrt{72x^3} = \sqrt{36x^2 \cdot 2x}$

$= 6x\sqrt{2x}$

(b) $\sqrt{\dfrac{18^2}{z^3}} = \dfrac{\sqrt{18^2}}{\sqrt{z^2 \cdot z}} = \dfrac{18}{z\sqrt{z}}$

67. (a) $\sqrt[3]{16x^5} = \sqrt[3]{8x^3 \cdot 2x^2}$

$= 2x\sqrt[3]{2x^2}$

(b) $\sqrt{75x^2y^{-4}} = \sqrt{\dfrac{75x^2}{y^4}}$

$= \dfrac{\sqrt{25x^2 \cdot 3}}{\sqrt{y^4}}$

$= \dfrac{5|x|\sqrt{3}}{y^2}$

69. (a) $2\sqrt{50} + 12\sqrt{8} = 2\sqrt{25 \cdot 2} + 12\sqrt{4 \cdot 2} = 2(5\sqrt{2}) + 12(2\sqrt{2}) = 10\sqrt{2} + 24\sqrt{2} = 34\sqrt{2}$

(b) $10\sqrt{32} - 6\sqrt{18} = 10\sqrt{16 \cdot 2} - 6\sqrt{9 \cdot 2} = 10(4\sqrt{2}) - 6(3\sqrt{2}) = 40\sqrt{2} - 18\sqrt{2} = 22\sqrt{2}$

71. (a) $5\sqrt{x} - 3\sqrt{x} = 2\sqrt{x}$

(b) $-2\sqrt{9y} + 10\sqrt{y} = -2(3\sqrt{y}) + 10\sqrt{y} = -6\sqrt{y} + 10\sqrt{y} = 4\sqrt{y}$

73. (a) $3\sqrt{x+1} + 10\sqrt{x+1} = 13\sqrt{x+1}$

(b) $7\sqrt{80x} - 2\sqrt{125x} = 7\sqrt{16 \cdot 5x} - 2\sqrt{25 \cdot 5x} = 7(4\sqrt{5x}) - 2(5\sqrt{5x}) = 28\sqrt{5x} - 10\sqrt{5x} = 18\sqrt{5x}$

75. $\sqrt{5} + \sqrt{3} \approx 3.968$ and

$\sqrt{5+3} = \sqrt{8} \approx 2.828$

Thus, $\sqrt{5} + \sqrt{3} > \sqrt{5+3}$.

77. $\sqrt{3^2 + 2^2} = \sqrt{9+4}$

$= \sqrt{13} \approx 3.606$

Thus, $5 > \sqrt{3^2 + 2^2}$.

79. $\dfrac{1}{\sqrt{3}} = \dfrac{1}{\sqrt{3}} \cdot \dfrac{\sqrt{3}}{\sqrt{3}} = \dfrac{\sqrt{3}}{3}$

81. $\dfrac{2}{5 - \sqrt{3}} = \dfrac{2}{5 - \sqrt{3}} \cdot \dfrac{5 + \sqrt{3}}{5 + \sqrt{3}} = \dfrac{2(5 + \sqrt{3})}{5^2 - (\sqrt{3})^2} = \dfrac{2(5 + \sqrt{3})}{25 - 3} = \dfrac{2(5 + \sqrt{3})}{22} = \dfrac{5 + \sqrt{3}}{11}$

83. $\dfrac{\sqrt{8}}{2} = \dfrac{\sqrt{4 \cdot 2}}{2} = \dfrac{2\sqrt{2}}{2} = \dfrac{\sqrt{2}}{1} \cdot \dfrac{\sqrt{2}}{\sqrt{2}} = \dfrac{2}{\sqrt{2}}$

85. $\dfrac{\sqrt{5} + \sqrt{3}}{3} = \dfrac{\sqrt{5} + \sqrt{3}}{3} \cdot \dfrac{\sqrt{5} - \sqrt{3}}{\sqrt{5} - \sqrt{3}} = \dfrac{5 - 3}{3(\sqrt{5} - \sqrt{3})} = \dfrac{2}{3(\sqrt{5} - \sqrt{3})}$

Radical Form	*Rational Exponent Form*
87. $\sqrt{9} = 3$, Given	$9^{1/2} = 3$, Answer
89. $\sqrt[5]{32} = 2$, Answer	$32^{1/5} = 2$, Given
91. $\sqrt[3]{-216} = -6$, Given	$(-216)^{1/3} = -6$, Answer
93. $\sqrt[4]{81^3} = 27$, Given	$81^{3/4} = 27$, Answer

95. $\dfrac{(2x^2)^{3/2}}{2^{1/2}x^4} = \dfrac{2^{3/2}(x^2)^{3/2}}{2^{1/2}x^4}$

$= \dfrac{2^{3/2}x^3}{2^{1/2}x^4} = 2^{3/2 - 1/2}x^{3-4} = 2^1 x^{-1} = \dfrac{2}{x}$

97. $\dfrac{x^{-3} \cdot x^{1/2}}{x^{3/2} \cdot x^{-1}} = \dfrac{x^{1/2} \cdot x^1}{x^{3/2} \cdot x^3}$

$= x^{1/2 + 1 - 3/2 - 3} = x^{-3} = \dfrac{1}{x^3}, \; x > 0$

99. (a) $\sqrt[4]{3^2} = 3^{2/4} = 3^{1/2} = \sqrt{3}$

(b) $\sqrt[6]{(x+1)^4} = (x+1)^{4/6} = (x+1)^{2/3} = \sqrt[3]{(x+1)^2}$

101. (a) $\sqrt{\sqrt{32}} = (32^{1/2})^{1/2}$

$= 32^{1/4} = \sqrt[4]{32} = \sqrt[4]{16 \cdot 2} = 2\sqrt[4]{2}$

(b) $\sqrt{\sqrt[4]{2x}} = ((2x)^{1/4})^{1/2} = (2x)^{1/8} = \sqrt[8]{2x}$

103. $T = 2\pi\sqrt{\dfrac{2}{32}}$

$= 2\pi\sqrt{\dfrac{1}{16}}$

$= 2\pi\left(\dfrac{1}{4}\right)$

$= \dfrac{\pi}{2} \approx 1.57$ seconds

105. $t = 0.03[12^{5/2} - (12 - h)^{5/2}], 0 \le h \le 12$

(a)

h (in centimeters)	t (in seconds)
0	0
1	2.93
2	5.48
3	7.67
4	9.53
5	11.08
6	12.32
7	13.29
8	14.00
9	14.50
10	14.80
11	14.93
12	14.96

(b) As h approaches 12, t approaches

$0.03(12^{5/2}) = 8.64\sqrt{3} \approx 14.96$ seconds.

107. True. When dividing variables, you subtract exponents.

109. $1 = \dfrac{a^m}{a^m} = a^{m-m} = a^0, a \neq 0$

111. When any positive integer is squared, the units digit is 0, 1, 4, 5, 6, or 9. Therefore, $\sqrt{5233}$ is not an integer.

Section P.3   Polynomials and Special Products

- Given a polynomial in x, $a_n x^n + a_{n-1}x^{n-1} + \cdots + a_1 x + a_0$, where $a_n \neq 0$, and n is a nonnegative integer, you should be able to identify the following.

  (a) Degree: n (b) Terms: $a_n x^n, a_{n-1}x^{n-1}, \ldots, a_1 x, a_0$

  (c) Coefficients: $a_n, a_{n-1}, \ldots, a_1$ (d) Leading coefficient: a_n

  (e) Constant term: a_0

- You should be able to add and subtract polynomials.

- You should be able to multiply polynomials by either

  (a) The Distributive Properties   (b) The Vertical Method.

- You should know the special binomial products.

  (a) $(ax+b)(cx+d) = acx^2 + adx + bcx + bd$ FOIL   (b) $(u+v)^2 = u^2 + 2uv + v^2$
  $= acx^2 + (ad+bc)x + bd$ $(u-v)^2 = u^2 - 2uv + v^2$

  (c) $(u+v)(u-v) = u^2 - v^2$ (d) $(u+v)^3 = u^3 + 3u^2v + 3uv^2 + v^3$
  $(u-v)^3 = u^3 - 3u^2v + 3uv^2 - v^3$

Vocabulary Check

1. n; a_n; a_0 2. descending

3. monomial; binomial; trinomial   4. like terms

5. First terms; Outer terms; Inner terms; Last terms   6. (c) the sum and difference of same terms

7. (a) a binomial sum squared   8. (b) a binomial difference squared

---

1. (d) 12 is a polynomial of degree zero.

3. (b) $1 - 2x^3 = -2x^3 + 1$ is a binomial with leading coefficient -2.

5. (f) $\frac{2}{3}x^4 + x^2 + 10$ is a trinomial with leading coefficient $\frac{2}{3}$.

7. $-2x^3$; $-2x^3 + 5$; $-2x^3 + 4x^2 - 3x + 20$, etc. (Answers will vary.)

9. $-15x^4 + 1$; $-3x^4 + 7x^2$; $-5x^4 - 6x$, etc. (Answers will vary.)

11. (a) Standard form: $-\frac{1}{2}x^5 + 14x$
    (b) Degree: 5
    Leading coefficient: $-\frac{1}{2}$
    (c) Binomial

13. (a) Standard form: $-3x^4 + 2x^2 - 5$
    (b) Degree: 4
    Leading coefficient: -3
    (c) Trinomial

15. (a) Standard form: $x^5 - 1$
    (b) Degree: 5
    Leading coefficient: 1
    (c) Binomial

17. (a) Standard form: 3
    (b) Degree: 0
    Leading coefficient: 3
    (c) Monomial

19. (a) Standard form: $-4x^5 + 6x^4 + 1$
    (b) Degree: 5
    Leading coefficient: -4
    (c) Trinomial

21. (a) Standard form: $4x^3y$
    (b) Degree: 4 (add the exponents on x and y)
    Leading coefficient: 4
    (c) Monomial

23. $2x - 3x^3 + 8$ *is* a polynomial.

Standard form: $-3x^3 + 2x + 8$

25. $\dfrac{3x + 4}{x} = 3 + \dfrac{4}{x} = 3 + 4x^{-1}$ is *not* a polynomial because it includes a term with a negative exponent.

27. $y^2 - y^4 + y^3$ *is* a polynomial.

Standard form: $-y^4 + y^3 + y^2$

29. $(6x + 5) - (8x + 15) = 6x + 5 - 8x - 15$

$\qquad = (6x - 8x) + (5 - 15)$

$\qquad = -2x - 10$

31. $-(x^3 - 2) + (4x^3 - 2x) = -x^3 + 2 + 4x^3 - 2x$

$\qquad = (4x^3 - x^3) - 2x + 2$

$\qquad = 3x^3 - 2x + 2$

33. $(15x^2 - 6) - (-8.3x^3 - 14.7x^2 - 17) = 15x^2 - 6 + 8.3x^3 + 14.7x^2 + 17$

$\qquad = 8.3x^3 + (15x^2 + 14.7x^2) + (-6 + 17)$

$\qquad = 8.3x^3 + 29.7x^2 + 11$

35. $5z - [3z - (10z + 8)] = 5z - (3z - 10z - 8)$

$\qquad = 5z - 3z + 10z + 8$

$\qquad = (5z - 3z + 10z) + 8$

$\qquad = 12z + 8$

37. $3x(x^2 - 2x + 1) = 3x(x^2) + 3x(-2x) + 3x(1)$

$\qquad = 3x^3 - 6x^2 + 3x$

39. $-5z(3z - 1) = -5z(3z) + (-5z)(-1)$

$\qquad = -15z^2 + 5z$

41. $(1 - x^3)(4x) = 1(4x) - x^3(4x)$

$\qquad = 4x - 4x^4$

$\qquad = -4x^4 + 4x$

43. $(2.5x^2 + 3)(3x) = (2.5x^2)(3x) + (3)(3x)$

$\qquad = 7.5x^3 + 9x$

45. $-4x\left(\tfrac{1}{8}x + 3\right) = (-4x)\left(\tfrac{1}{8}x\right) + (-4x)(3)$

$\qquad = -\tfrac{1}{2}x^2 - 12x$

47. $(7x^3 - 2x^2 + 8) + (-3x^3 - 4) = (7x^3 - 3x^3) + (-2x^2) + (8 - 4)$

$\qquad = 4x^3 - 2x^2 + 4$

49. $(5x^2 - 3x + 8) - (x - 3) = 5x^2 - 3x + 8 - x + 3$

$\qquad = 5x^2 + (-3x - x) + (8 + 3)$

$\qquad = 5x^2 - 4x + 11$

51. Multiply:

$$
\begin{array}{r}
-6x^2 + 15x - 4 \\
5x + 3 \\
\hline
-30x^3 + 75x^2 - 20x \\
-18x^2 + 45x - 12 \\
\hline
-30x^3 + 57x^2 + 25x - 12
\end{array}
$$

53. Multiply:

$$
\begin{array}{r}
x^2 - x - 4 \\
x^2 + 9 \\
\hline
x^4 - x^3 - 4x^2 \\
9x^2 - 9x - 36 \\
\hline
x^4 - x^3 + 5x^2 - 9x - 36
\end{array}
$$

55. $(x + 3)(x + 4) = x^2 + 4x + 3x + 12$ FOIL

$\qquad = x^2 + 7x + 12$

57. $(3x - 5)(2x + 1) = 6x^2 + 3x - 10x - 5$ FOIL

$\qquad = 6x^2 - 7x - 5$

59. Multiply:

$$\begin{array}{r} x^2 - x + 1 \\ \underline{x^2 + x + 1} \\ x^4 - x^3 + x^2 \\ x^3 - x^2 + x \\ \underline{x^2 - x + 1} \\ x^4 - 0x^3 + x^2 + 0x + 1 = x^4 + x^2 + 1 \end{array}$$

61. $(x + 10)(x - 10) = x^2 - 10^2 = x^2 - 100$

63. $(x + 2y)(x - 2y) = x^2 - (2y)^2 = x^2 - 4y^2$

65. $(2x + 3)^2 = (2x)^2 + 2(2x)(3) + 3^2$
$$= 4x^2 + 12x + 9$$

67. $(2x - 5y)^2 = (2x)^2 - 2(2x)(5y) + (5y)^2$
$$= 4x^2 - 20xy + 25y^2$$

69. $(x + 1)^3 = x^3 + 3x^2(1) + 3x(1^2) + 1^3$
$$= x^3 + 3x^2 + 3x + 1$$

71. $(2x - y)^3 = (2x)^3 - 3(2x)^2y + 3(2x)y^2 - y^3$
$$= 8x^3 - 12x^2y + 6xy^2 - y^3$$

73. $(4x^3 - 3)^2 = (4x^3)^2 - 2(4x^3)(3) + (3)^2$
$$= 16x^6 - 24x^3 + 9$$

75. $[(m - 3) + n][(m - 3) - n] = (m - 3)^2 - n^2$
$$= m^2 - 6m + 9 - n^2$$
$$= m^2 - n^2 - 6m + 9$$

77. $[(x - 3) + y]^2 = (x - 3)^2 + 2y(x - 3) + y^2$
$$= x^2 - 6x + 9 + 2xy - 6y + y^2$$
$$= x^2 + 2xy + y^2 - 6x - 6y + 9$$

79. $(2r^2 - 5)(2r^2 + 5) = (2r^2)^2 - 5^2 = 4r^4 - 25$

81. $\left(\tfrac{1}{2}x - 3\right)^2 = \left(\tfrac{1}{2}x\right)^2 - 2\left(\tfrac{1}{2}x\right)(3) + 3^2$
$$= \tfrac{1}{4}x^2 - 3x + 9$$

83. $\left(\tfrac{1}{3}x - 2\right)\left(\tfrac{1}{3}x + 2\right) = \left(\tfrac{1}{3}x\right)^2 - (2)^2$
$$= \tfrac{1}{9}x^2 - 4$$

85. $(1.2x + 3)^2 = (1.2x)^2 + 2(1.2x)(3) + 3^2$
$$= 1.44x^2 + 7.2x + 9$$

87. $(1.5x - 4)(1.5x + 4) = (1.5x)^2 - 4^2$
$$= 2.25x^2 - 16$$

89. $5x(x + 1) - 3x(x + 1) = 2x(x + 1)$
$$= 2x^2 + 2x$$

91. $(u + 2)(u - 2)(u^2 + 4) = (u^2 - 4)(u^2 + 4)$
$$= u^4 - 16$$

93. $\left(\sqrt{x} + \sqrt{y}\right)\left(\sqrt{x} - \sqrt{y}\right) = \left(\sqrt{x}\right)^2 - \left(\sqrt{y}\right)^2$
$$= x - y$$

95. $\left(x - \sqrt{5}\right)^2 = x^2 - 2(x)\left(\sqrt{5}\right) + \left(\sqrt{5}\right)^2$
$$= x^2 - 2\sqrt{5}x + 5$$

97. (a) Profit = Revenue − Cost
$$\text{Profit} = 95x - (73x + 25{,}000)$$
$$= 95x - 73x - 25{,}000 = 22x - 25{,}000$$

(b) For $x = 5000$:
$$\text{Profit} = 22(5000) - 25{,}000$$
$$= 110{,}000 - 25{,}000 = \$85{,}000$$

99. (a) $500(1 + r)^2 = 500(r + 1)^2 = 500(r^2 + 2r + 1)$

$$= 500r^2 + 1000r + 500$$

(b)

r	$2\frac{1}{2}\%$	3%	4%	$4\frac{1}{2}\%$	5%
$500(1 + r)^2$	\$525.31	\$530.45	\$540.80	\$546.01	\$551.25

(c) As r increases, the amount increases.

101. (a) $V = l \cdot w \cdot h = (26 - 2x)(18 - 2x)(x)$

$$= 2(13 - x)(2)(9 - x)(x)$$

$$= 4x(-1)(x - 13)(-1)(x - 9)$$

$$= 4x(x - 13)(x - 9)$$

$$= 4x^3 - 88x^2 + 468x$$

(b)

x (cm)	1	2	3
V (cm³)	384	616	720

103. (a) Area of shaded region = Area of outer rectangle − Area of inner rectangle

$$A = 2x(2x + 6) - x(x + 4)$$

$$= 4x^2 + 12x - x^2 - 4x$$

$$= 3x^2 + 8x$$

(b) Area of shaded region = Area of outer triangle − Area of inner triangle

$$A = \frac{1}{2}(9x)(12x) - \frac{1}{2}(6x)(8x)$$

$$= 54x^2 - 24x^2$$

$$= 30x^2$$

(c) Area of shaded region = Area of outer triangle − Area of inner triangle

$$A = \frac{1}{2}(x + 2)(5x) - \frac{1}{2}(x + 1)(3x)$$

$$= \frac{5}{2}x^2 + 5x - \frac{3}{2}x^2 - \frac{3}{2}x$$

$$= x^2 + \frac{7}{2}x$$

OR

Area of shaded region = Area of a trapezoid

$$A = \frac{1}{2}(2x)[(x + 2) + (x + 1)]$$

$$= x(2x + 3)$$

$$= 2x^2 + 3x$$

Note: $x = \frac{1}{2}$ is the only x-value yielding a triangle with the given dimensions.

(d) Area of shaded region = Area of rectangle − Area of triangle

$$A = (3x + 10)(x + 6) - \frac{1}{2}(3x + 10)(x + 6)$$

$$= \frac{1}{2}(3x + 10)(x + 6)$$

$$= \frac{1}{2}(3x^2 + 28x + 60)$$

$$= \frac{3}{2}x^2 + 14x + 30$$

105. Area = length × width

$$= (2x + 14)(22)$$

$$= (2x)(22) + (14)(22)$$

$$= 44x + 308$$

107. (a) Estimates will vary. Actual safe loads for $x = 12$:

$S_6 = (0.06(12)^2 - 2.42(12) + 38.71)^2 = 335.2561$ (using a calculator)

$S_8 = (0.08(12)^2 - 3.30(12) + 51.93)^2 = 568.8225$ (using a calculator)

Difference in safe loads $= 568.8225 - 335.2561 = 233.5664$ pounds

(b) The difference in safe loads decreases in magnitude as the span increases.

109. $(x + 1)(x + 4) = x^2 + 4x + x + 4$

This illustrates the Distributive Property.

111. False.

$(4x^2 + 1)(3x + 1) = 12x^3 + 4x^2 + 3x + 1$

113. Since $x^m x^n = x^{m+n}$, the degree of the product is $m + n$.

115. $(x - 3)^2 \neq x^2 + 9$

The student did not remember the middle term when squaring the binomial.
The correct method for squaring this binomial is:

$(x - 3)^2 = (x)^2 - 2(x)(3) + (3)^2 = x^2 - 6x + 9$

117. No; $(x^2 + 1) + (-x^2 + 3) = 4$,
which is not a second-degree polynomial.

119. $(x + y)^2 \neq x^2 + y^2$

Let $x = 3$ and $y = 4$.

$(3 + 4)^2 = (7)^2 = 49$

$3^2 + 4^2 = 9 + 16 = 25$ $\rangle$ Not Equal

If either x or y is zero, then $(x + y)^2$ would equal $x^2 + y^2$.

Section P.4   Factoring Polynomials

- You should be able to factor out all common factors, the first step in factoring.
- You should be able to factor the following special polynomial forms.

  (a) $u^2 - v^2 = (u + v)(u - v)$

  (b) $u^2 + 2uv + v^2 = (u + v)^2$
     $u^2 - 2uv + v^2 = (u - v)^2$

  (c) $u^3 + v^3 = (u + v)(u^2 - uv + v^2)$
     $u^3 - v^3 = (u - v)(u^2 + uv + v^2)$

- You should be able to factor trinomials with binomial factors.
- You should be able to factor by grouping.
- You should be able to factor some trinomials by grouping.

Vocabulary Check

1. factoring

2. irreducible

3. completely factored

4. factoring by grouping

5. (a) **ii** Difference of two squares

(b) **iii** Difference of two cubes

(c) **i** Perfect square trinomial

1. $90 = 2 \cdot 3 \cdot 3 \cdot 5$

  $300 = 2 \cdot 2 \cdot 3 \cdot 5 \cdot 5$

  Greatest common factor: $2 \cdot 3 \cdot 5 = 30$

3. $12x^2y^3 = 2 \cdot 2 \cdot 3 \cdot x \cdot x \cdot y \cdot y \cdot y$

  $18x^2y = 2 \cdot 3 \cdot 3 \cdot x \cdot x \cdot y$

  $24x^3y^2 = 2 \cdot 2 \cdot 2 \cdot 3 \cdot x \cdot x \cdot x \cdot y \cdot y$

  Greatest common factor: $2 \cdot 3 \cdot x \cdot x \cdot y = 6x^2y$

5. $3x + 6 = 3(x + 2)$

7. $2x^3 - 6x = 2x(x^2 - 3)$

9. $x(x - 1) + 6(x - 1) = (x - 1)(x + 6)$

11. $(x + 3)^2 - 4(x + 3) = (x + 3)[(x + 3) - 4]$

  $= (x + 3)(x - 1)$

13. $\frac{1}{2}x + 4 = \frac{1}{2}x + \frac{8}{2}$

  $= \frac{1}{2}(x + 8)$

15. $\frac{1}{2}x^3 + 2x^2 - 5x = \frac{1}{2}x^3 + \frac{4}{2}x^2 - \frac{10}{2}x$

  $= \frac{1}{2}x(x^2 + 4x - 10)$

17. $\frac{2}{3}x(x - 3) - 4(x - 3) = \frac{2}{3}x(x - 3) - \frac{12}{3}(x - 3)$

  $= \frac{2}{3}(x - 3)(x - 6)$

19. $x^2 - 81 = x^2 - 9^2$

  $= (x + 9)(x - 9)$

21. $32y^2 - 18 = 2(16y^2 - 9)$

  $= 2[(4y)^2 - 3^2]$

  $= 2(4y + 3)(4y - 3)$

23. $16x^2 - \frac{1}{9} = (4x)^2 - \left(\frac{1}{3}\right)^2$

  $= \left(4x + \frac{1}{3}\right)\left(4x - \frac{1}{3}\right)$

25. $(x - 1)^2 - 4 = (x - 1)^2 - (2)^2$

  $= [(x - 1) + 2][(x - 1) - 2]$

  $= (x + 1)(x - 3)$

27. $9u^2 - 4v^2 = (3u)^2 - (2v)^2$

  $= (3u + 2v)(3u - 2v)$

29. $x^2 - 4x + 4 = x^2 - 2(2)x + 2^2$

  $= (x - 2)^2$

31. $4t^2 + 4t + 1 = (2t)^2 + 2(2t)(1) + 1^2$

  $= (2t + 1)^2$

33. $25y^2 - 10y + 1 = (5y)^2 - 2(5y)(1) + 1^2$

  $= (5y - 1)^2$

35. $9u^2 + 24uv + 16v^2 = (3u)^2 + 2(3u)(4v) + (4v)^2$

  $= (3u + 4v)^2$

37. $x^2 - \frac{4}{3}x + \frac{4}{9} = x^2 - 2(x)\left(\frac{2}{3}\right) + \left(\frac{2}{3}\right)^2$

  $= \left(x - \frac{2}{3}\right)^2$

39. $x^3 - 8 = x^3 - 2^3$

  $= (x - 2)(x^2 + 2x + 4)$

41. $y^3 + 64 = y^3 + 4^3$

  $= (y + 4)(y^2 - 4y + 16)$

43. $8t^3 - 1 = (2t)^3 - 1^3$

  $= (2t - 1)(4t^2 + 2t + 1)$

45. $u^3 + 27v^3 = u^3 + (3v)^3$

  $= (u + 3v)(u^2 - 3uv + 9v^2)$

47. $25 - 5x^2 = -5(-5 + x^2)$

  $= -5(x^2 - 5)$

49. $-2t^3 + 4t + 6 = -2(t^3 - 2t - 3)$

51. $x^2 + x - 2 = (x + 2)(x - 1)$

53. $s^2 - 5s + 6 = (s - 3)(s - 2)$

55. $20 - y - y^2 = -(y^2 + y - 20)$

  $= -(y + 5)(y - 4)$

57. $x^2 - 30x + 200 = (x - 20)(x - 10)$

59. $3x^2 - 5x + 2 = (3x - 2)(x - 1)$

61. $5x^2 + 26x + 5 = (5x + 1)(x + 5)$

63. $-9z^2 + 3z + 2 = -(9z^2 - 3z - 2)$
$$= -(3z - 2)(3z + 1)$$

65. $x^3 - x^2 + 2x - 2 = x^2(x - 1) + 2(x - 1)$
$$= (x - 1)(x^2 + 2)$$

67. $2x^3 - x^2 - 6x + 3 = x^2(2x - 1) - 3(2x - 1)$
$$= (2x - 1)(x^2 - 3)$$

69. $6 + 2x - 3x^3 - x^4 = 2(3 + x) - x^3(3 + x)$
$$= (3 + x)(2 - x^3)$$

71. $6x^3 - 2x + 3x^2 - 1 = 2x(3x^2 - 1) + 1(3x^2 - 1)$
$$= (3x^2 - 1)(2x + 1)$$

73. $a \cdot c = (3)(8) = 24.$ Rewrite the middle term,
$10x = 6x + 4x,$ since $(6)(4) = 24$ and $6 + 4 = 10.$
$$3x^2 + 10x + 8 = 3x^2 + 6x + 4x + 8$$
$$= 3x(x + 2) + 4(x + 2)$$
$$= (x + 2)(3x + 4)$$

75. $a \cdot c = (6)(-2) = -12.$ Rewrite the middle term,
$x = 4x - 3x,$ since $4(-3) = -12$ and $4 + (-3) = 1.$
$$6x^2 + x - 2 = 6x^2 + 4x - 3x - 2$$
$$= 2x(3x + 2) - 1(3x + 2)$$
$$= (2x - 1)(3x + 2)$$

77. $a \cdot c = (15)(2) = 30.$ Rewrite the middle term,
$-11x = -6x - 5x,$ since $(-6)(-5) = 30$ and
$(-6) + (-5) = -11.$
$$15x^2 - 11x + 2 = 15x^2 - 6x - 5x + 2$$
$$= 3x(5x - 2) - 1(5x - 2)$$
$$= (3x - 1)(5x - 2)$$

79. $6x^2 - 54 = 6(x^2 - 9)$
$$= 6(x + 3)(x - 3)$$

81. $x^3 - 4x^2 = x^2(x - 4)$

83. $x^2 - 2x + 1 = (x - 1)^2$

85. $1 - 4x + 4x^2 = (1 - 2x)^2$

87. $2x^2 + 4x - 2x^3 = -2x(-x - 2 + x^2)$
$$= -2x(x^2 - x - 2)$$
$$= -2x(x + 1)(x - 2)$$

89. $9x^2 + 10x + 1 = (9x + 1)(x + 1)$

91. $\frac{1}{81}x^2 + \frac{2}{9}x - 8 = \frac{1}{81}x^2 + \frac{18}{81}x - \frac{648}{81}$
$$= \frac{1}{81}(x^2 + 18x - 648)$$
$$= \frac{1}{81}(x + 36)(x - 18)$$

93. $3x^3 + x^2 + 15x + 5 = x^2(3x + 1) + 5(3x + 1)$
$$= (3x + 1)(x^2 + 5)$$

95. $x^4 - 4x^3 + x^2 - 4x = x(x^3 - 4x^2 + x - 4)$
$$= x[x^2(x - 4) + (x - 4)]$$
$$= x(x - 4)(x^2 + 1)$$

97. $\frac{1}{4}x^3 + 3x^2 + \frac{3}{4}x + 9 = \frac{1}{4}x^3 + \frac{12}{4}x^2 + \frac{3}{4}x + \frac{36}{4}$
$$= \frac{1}{4}(x^3 + 12x^2 + 3x + 36)$$
$$= \frac{1}{4}[x^2(x + 12) + 3(x + 12)]$$
$$= \frac{1}{4}(x + 12)(x^2 + 3)$$

99. $(t - 1)^2 - 49 = (t - 1)^2 - (7)^2$
$$= [(t - 1) + 7][(t - 1) - 7]$$
$$= (t + 6)(t - 8)$$

101. $(x^2 + 8)^2 - 36x^2 = (x^2 + 8)^2 - (6x)^2$
$$= [(x^2 + 8) - 6x][(x^2 + 8) + 6x]$$
$$= (x^2 - 6x + 8)(x^2 + 6x + 8)$$
$$= (x - 4)(x - 2)(x + 4)(x + 2)$$

103. $5x^3 + 40 = 5(x^3 + 8)$

$\qquad\qquad = 5(x^3 + 2^3)$

$\qquad\qquad = 5(x + 2)(x^2 - 2x + 4)$

105. $5(3 - 4x)^2 - 8(3 - 4x)(5x - 1) = (3 - 4x)[5(3 - 4x) - 8(5x - 1)]$

$\qquad\qquad\qquad\qquad\qquad\qquad = (3 - 4x)[15 - 20x - 40x + 8]$

$\qquad\qquad\qquad\qquad\qquad\qquad = (3 - 4x)(23 - 60x)$

107. $7(3x + 2)^2(1 - x)^2 + (3x + 2)(1 - x)^3 = (3x + 2)(1 - x)^2[7(3x + 2) + (1 - x)]$

$\qquad\qquad\qquad\qquad\qquad\qquad\qquad = (3x + 2)(1 - x)^2(21x + 14 + 1 - x)$

$\qquad\qquad\qquad\qquad\qquad\qquad\qquad = (3x + 2)(1 - x)^2(20x + 15)$

$\qquad\qquad\qquad\qquad\qquad\qquad\qquad = 5(3x + 2)(1 - x)^2(4x + 3)$

109. $3(x - 2)^2(x + 1)^4 + (x - 2)^3(4)(x + 1)^3 = (x - 2)^2(x + 1)^3[3(x + 1) + 4(x - 2)]$

$\qquad\qquad\qquad\qquad\qquad\qquad\qquad = (x - 2)^2(x + 1)^3(3x + 3 + 4x - 8)$

$\qquad\qquad\qquad\qquad\qquad\qquad\qquad = (x - 2)^2(x + 1)^3(7x - 5)$

111. $5(x^6 + 1)^4(6x^5)(3x + 2)^3 + 3(3x + 2)^2(3)(x^6 + 1)^5 = 3(x^6 + 1)^4(3x + 2)^2[10x^5(3x + 2) + 3(x^6 + 1)]$

$\qquad\qquad\qquad\qquad\qquad\qquad\qquad\qquad = 3(x^6 + 1)^4(3x + 2)^2(30x^6 + 20x^5 + 3x^6 + 3)$

$\qquad\qquad\qquad\qquad\qquad\qquad\qquad\qquad = 3(x^6 + 1)^4(3x + 2)^2(33x^6 + 20x^5 + 3)$

$\qquad\qquad\qquad\qquad\qquad\qquad\qquad\qquad = 3[(x^2)^3 + 1]^4(3x + 2)^2(33x^6 + 20x^5 + 3)$

$\qquad\qquad\qquad\qquad\qquad\qquad\qquad\qquad = 3[(x^2 + 1)(x^4 - x^2 + 1)]^4(3x + 2)^2(33x^6 + 20x^5 + 3)$

$\qquad\qquad\qquad\qquad\qquad\qquad\qquad\qquad = 3(x^2 + 1)^4(x^4 - x^2 + 1)^4(3x + 2)^2(33x^6 + 20x^5 + 3)$

113. $a^2 - b^2 = (a + b)(a - b)$

Matches model (b).

115. $a^2 + 2a + 1 = (a + 1)^2$

Matches model (a).

117. $3x^2 + 7x + 2 = (3x + 1)(x + 2)$

119. $2x^2 + 7x + 3 = (2x + 1)(x + 3)$

121. $A = \pi(r + 2)^2 - \pi r^2$

$\qquad = \pi[(r + 2)^2 - r^2]$

$\qquad = \pi[r^2 + 4r + 4 - r^2]$

$\qquad = \pi(4r + 4)$

$\qquad = 4\pi(r + 1)$

123. $A = 8(18) - 4x^2$

$\qquad = 4(36 - x^2)$

$\qquad = 4(6 - x)(6 + x)$

125. For $x^2 + bx - 15$ to be factorable, b must equal $m + n$ where $mn = -15$.

Factors of -15	Sum of factors
$(15)(-1)$	$15 + (-1) = 14$
$(-15)(1)$	$-15 + 1 = -14$
$(3)(-5)$	$3 + (-5) = -2$
$(-3)(5)$	$-3 + 5 = 2$

The possible b-values are $14, -14, -2,$ or 2.

127. For $x^2 + bx - 12$ to be factorable, b must equal $m + n$ where $mn = -12$.

Factors of -12	Sum of factors
$(12)(-1)$	$12 + (-1) = 11$
$(-12)(1)$	$-12 + 1 = -11$
$(2)(-6)$	$2 + (-6) = -4$
$(-2)(6)$	$-2 + 6 = 4$
$(3)(-4)$	$3 + (-4) = -1$
$(-3)(4)$	$-3 + 4 = 1$

The possible b-values are $11, -11, -4, 4, -1, 1$.

129. For $2x^2 + 5x + c$ to be factorable, the factors of $2c$ must add up to 5.

Possible c-values	$2c$	Factors of $2c$ that add up to 5
2	4	$(1)(4) = 4$ and $1 + 4 = 5$
3	6	$(2)(3) = 6$ and $2 + 3 = 5$
-3	-6	$(6)(-1) = -6$ and $6 + (-1) = 5$
-7	-14	$(7)(-2) = -14$ and $7 + (-2) = 5$
-12	-24	$(8)(-3) = -24$ and $8 + (-3) = 5$

These are a few possible c-values. There are *many* correct answers.

If $c = \quad 2 : 2x^2 + 5x + 2 = (2x + 1)(x + 2)$

If $c = \quad 3 : 2x^2 + 5x + 3 = (2x + 3)(x + 1)$

If $c = \quad -3 : 2x^2 + 5x - 3 = (2x - 1)(x + 3)$

If $c = \quad -7 : 2x^2 + 5x - 7 = (2x + 7)(x - 1)$

If $c = -12 : 2x^2 + 5x - 12 = (2x - 3)(x + 4)$

131. For $3x^2 - x + c$ to be factorable, the factors of $3c$ must add up to -1.

Possible c-values	$3c$	Factors of $3c$ must add up to -1
-2	-6	$(2)(-3) = -6$ and $2 + (-3) = -1$
-4	-12	$(3)(-4) = -12$ and $3 + (-4) = -1$
-10	-30	$(5)(-6) = -30$ and $5 + (-6) = -1$

These are a few possible c-values. There are *many* correct answers.

If $c = \quad -2 : 3x^2 - x - 2 = (3x + 2)(x - 1)$

If $c = \quad -4 : 3x^2 - x - 4 = (3x - 4)(x + 1)$

If $c = -10 : 3x^2 - x - 10 = (3x + 5)(x - 2)$

133. $9x^2 - 9x - 54 = 9(x^2 - x - 6) = 9(x + 2)(x - 3)$

The error in the problem in the book was that 3 was factored out of the first binomial but not out of the second binomial.

$(3x + 6)(3x - 9) = 3(x + 2)(3)(x - 3) = 9(x + 2)(x - 3)$

135. $kQx - kx^2 = kx(Q - x)$

137. True. $a^2 - b^2 = (a + b)(a - b)$

139. $x^{2n} - y^{2n} = (x^n)^2 - (y^n)^2$
$$= (x^n + y^n)(x^n - y^n)$$

141. $x^{3n} - y^{2n} = (x^n)^3 - (y^n)^2 = x^{3n} - y^{2n}$ is completely factored.

143. Answers will vary. Some examples:
$$x^2 - 3; \ x^2 + x + 1; \ x^2 + 16$$

Section P.5   Rational Expressions

- You should be able to find the domain of a rational expression.

- You should know that a rational expression is the quotient of two polynomials.

- You should be able to simplify rational expressions by reducing them to lowest terms. This may involve factoring both the numerator and the denominator.

- You should be able to add, subtract, multiply, and divide rational expressions.

- You should be able to simplify complex fractions.

- You should be able to simplify expressions with negative or fraction exponents.

Vocabulary Check

1. domain

2. rational expression

3. complex

4. smaller

5. equivalent

6. difference quotient

---

1. The domain of the polynomial $3x^2 - 4x + 7$ is the set of all real numbers.

3. The domain of the polynomial $4x^3 + 3$, $x \geq 0$ is the set of non-negative real numbers, since the polynomial is restricted to that set.

5. The domain of $1/(x - 2)$ is the set of all real numbers x such that $x \neq 2$.

7. The domain of $\sqrt{x + 1}$ is the set of all real numbers x such that $x \geq -1$.

9. $\dfrac{5}{2x} = \dfrac{5(3x)}{(2x)(3x)} = \dfrac{5(3x)}{6x^2}, \quad x \neq 0$

The missing factor is $3x$, $x \neq 0$.

11. $\dfrac{15x^2}{10x} = \dfrac{5x(3x)}{5x(2)} = \dfrac{3x}{2}, \quad x \neq 0$

13. $\dfrac{3xy}{xy + x} = \dfrac{x(3y)}{x(y + 1)} = \dfrac{3y}{y + 1}, x \neq 0$

15. $\dfrac{4y - 8y^2}{10y - 5} = \dfrac{-4y(2y - 1)}{5(2y - 1)}$
$$= -\dfrac{4y}{5}, \ y \neq \dfrac{1}{2}$$

17. $\dfrac{x - 5}{10 - 2x} = \dfrac{x - 5}{-2(x - 5)}$
$$= -\dfrac{1}{2}, \ x \neq 5$$

19. $\dfrac{y^2 - 16}{y + 4} = \dfrac{(y + 4)(y - 4)}{y + 4}$
$$= y - 4, \ y \neq -4$$

21. $\dfrac{x^3 + 5x^2 + 6x}{x^2 - 4} = \dfrac{x(x + 2)(x + 3)}{(x + 2)(x - 2)} = \dfrac{x(x + 3)}{x - 2}, \quad x \neq -2$

23. $\dfrac{y^2 - 7y + 12}{y^2 + 3y - 18} = \dfrac{(y - 3)(y - 4)}{(y + 6)(y - 3)} = \dfrac{y - 4}{y + 6}, \quad y \neq 3$

25. $\dfrac{2 - x + 2x^2 - x^3}{x^2 - 4} = \dfrac{(2 - x) + x^2(2 - x)}{(x + 2)(x - 2)}$

$\qquad\qquad\quad = \dfrac{(2 - x)(1 + x^2)}{(x + 2)(x - 2)}$

$\qquad\qquad\quad = \dfrac{-(x - 2)(x^2 + 1)}{(x + 2)(x - 2)}$

$\qquad\qquad\quad = -\dfrac{x^2 + 1}{x + 2}, \quad x \neq 2$

27. $\dfrac{z^3 - 8}{z^2 + 2z + 4} = \dfrac{(z - 2)(z^2 + 2z + 4)}{z^2 + 2z + 4} = z - 2$

29.

x	0	1	2	3	4	5	6
$\dfrac{x^2 - 2x - 3}{x - 3}$	1	2	3	Undef.	5	6	7
$x + 1$	1	2	3	4	5	6	7

The expressions are equivalent except at $x = 3$.

31. $\dfrac{5x^3}{2x^3 + 4} = \dfrac{5x^3}{2(x^3 + 2)}$

There are no common factors so this expression cannot be simplified. In this case factors of terms were incorrectly cancelled.

33. $\dfrac{\pi r^2}{(2r)^2} = \dfrac{\pi r^2}{4r^2} = \dfrac{\pi}{4}, \; r \neq 0$

35. $\dfrac{5}{x - 1} \cdot \dfrac{x - 1}{25(x - 2)} = \dfrac{1}{5(x - 2)}, \; x \neq 1$

37. $\dfrac{r}{r - 1} \cdot \dfrac{r^2 - 1}{r^2} = \dfrac{r(r + 1)(r - 1)}{r^2(r - 1)} = \dfrac{r + 1}{r}, \; r \neq 1, r \neq 0$

39. $\dfrac{t^2 - t - 6}{t^2 + 6t + 9} \cdot \dfrac{t + 3}{t^2 - 4} = \dfrac{(t - 3)(t + 2)(t + 3)}{(t + 3)^2(t + 2)(t - 2)} = \dfrac{t - 3}{(t + 3)(t - 2)}, \quad t \neq -2$

41. $\dfrac{x^2 - 36}{x} \div \dfrac{x^3 - 6x^2}{x^2 + x} = \dfrac{x^2 - 36}{x} \cdot \dfrac{x^2 + x}{x^3 - 6x^2}$

$\qquad\qquad = \dfrac{(x + 6)(x - 6)}{x} \cdot \dfrac{x(x + 1)}{x^2(x - 6)}$

$\qquad\qquad = \dfrac{(x + 6)(x + 1)}{x^2}, \quad x \neq 6$

43. $\dfrac{5}{x - 1} + \dfrac{x}{x - 1} = \dfrac{5 + x}{x - 1} = \dfrac{x + 5}{x - 1}$

45. $6 - \dfrac{5}{x + 3} = \dfrac{6(x + 3)}{(x + 3)} - \dfrac{5}{x + 3}$

$\qquad = \dfrac{6(x + 3) - 5}{x + 3}$

$\qquad = \dfrac{6x + 18 - 5}{x + 3}$

$\qquad = \dfrac{6x + 13}{x + 3}$

47. $\dfrac{3}{x - 2} + \dfrac{5}{2 - x} = \dfrac{3}{x - 2} - \dfrac{5}{x - 2} = -\dfrac{2}{x - 2}$

49. $\dfrac{1}{x^2 - x - 2} - \dfrac{x}{x^2 - 5x + 6} = \dfrac{1}{(x - 2)(x + 1)} - \dfrac{x}{(x - 2)(x - 3)}$

$\qquad\qquad = \dfrac{(x - 3) - x(x + 1)}{(x + 1)(x - 2)(x - 3)} = \dfrac{x - 3 - x^2 - x}{(x + 1)(x - 2)(x - 3)}$

$\qquad\qquad = \dfrac{-x^2 - 3}{(x + 1)(x - 2)(x - 3)} = -\dfrac{x^2 + 3}{(x + 1)(x - 2)(x - 3)}$

51. $-\dfrac{1}{x} + \dfrac{2}{x^2 + 1} + \dfrac{1}{x^3 + x} = \dfrac{-(x^2 + 1)}{x(x^2 + 1)} + \dfrac{2x}{x(x^2 + 1)} + \dfrac{1}{x(x^2 + 1)}$

$$= \dfrac{-x^2 - 1 + 2x + 1}{x(x^2 + 1)} = \dfrac{-x^2 + 2x}{x(x^2 + 1)} = \dfrac{-x(x - 2)}{x(x^2 + 1)}$$

$$= -\dfrac{x - 2}{x^2 + 1} = \dfrac{2 - x}{x^2 + 1}, \quad x \neq 0$$

53. $\dfrac{x + 4}{x + 2} - \dfrac{3x - 8}{x + 2} = \dfrac{(x + 4) - (3x - 8)}{x + 2}$

$$= \dfrac{x + 4 - 3x + 8}{x + 2} = \dfrac{-2x + 12}{x + 2} = \dfrac{-2(x - 6)}{x + 2}$$

The error was incorrect subtraction in the numerator.

55. $\dfrac{\left(\dfrac{x}{2} - 1\right)}{(x - 2)} = \dfrac{\left(\dfrac{x}{2} - \dfrac{2}{2}\right)}{\left(\dfrac{x - 2}{1}\right)}$

$$= \dfrac{x - 2}{2} \cdot \dfrac{1}{x - 2}$$

$$= \dfrac{1}{2}, \quad x \neq 2$$

57. $\dfrac{\left[\dfrac{x^2}{(x + 1)^2}\right]}{\left[\dfrac{x}{(x + 1)^3}\right]} = \dfrac{x^2}{(x + 1)^2} \cdot \dfrac{(x + 1)^3}{x}$

$$= x(x + 1), \quad x \neq -1, 0$$

59. $\dfrac{\left(\sqrt{x} - \dfrac{1}{2\sqrt{x}}\right)}{\sqrt{x}} = \dfrac{\left(\sqrt{x} - \dfrac{1}{2\sqrt{x}}\right)}{\sqrt{x}} \cdot \dfrac{2\sqrt{x}}{2\sqrt{x}} = \dfrac{2x - 1}{2x}, \quad x > 0$

61. $x^5 - 2x^{-2} = x^{-2}(x^7 - 2) = \dfrac{x^7 - 2}{x^2}$

63. $x^2(x^2 + 1)^{-5} - (x^2 + 1)^{-4} = (x^2 + 1)^{-5}\left[x^2 - (x^2 + 1)\right] = -\dfrac{1}{(x^2 + 1)^5}$

65. $2x^2(x - 1)^{1/2} - 5(x - 1)^{-1/2} = (x - 1)^{-1/2}\left[2x^2(x - 1)^1 - 5\right] = \dfrac{2x^3 - 2x^2 - 5}{(x - 1)^{1/2}}$

67. $\dfrac{3x^{1/3} - x^{-2/3}}{3x^{-2/3}} = \dfrac{3x^{1/3} - x^{-2/3}}{3x^{-2/3}} \cdot \dfrac{x^{2/3}}{x^{2/3}} = \dfrac{3x^1 - x^0}{3x^0} = \dfrac{3x - 1}{3}, \quad x \neq 0$

69. $\dfrac{\left(\dfrac{1}{x + h} - \dfrac{1}{x}\right)}{h} = \dfrac{\left(\dfrac{1}{x + h} - \dfrac{1}{x}\right)}{h} \cdot \dfrac{x(x + h)}{x(x + h)}$

$$= \dfrac{x - (x + h)}{hx(x + h)}$$

$$= \dfrac{-h}{hx(x + h)}$$

$$= -\dfrac{1}{x(x + h)}, \quad h \neq 0$$

71. $\dfrac{\left(\dfrac{1}{x+h-4}-\dfrac{1}{x-4}\right)}{h} = \dfrac{\left(\dfrac{1}{x+h-4}-\dfrac{1}{x-4}\right)}{h} \cdot \dfrac{(x-4)(x+h-4)}{(x-4)(x+h-4)}$

$\qquad = \dfrac{(x-4)-(x+h-4)}{h(x-4)(x+h-4)}$

$\qquad = \dfrac{-h}{h(x-4)(x+h-4)}$

$\qquad = -\dfrac{1}{(x-4)(x+h-4)}, \quad h \neq 0$

73. $\dfrac{\sqrt{x+2}-\sqrt{x}}{2} = \dfrac{\sqrt{x+2}-\sqrt{x}}{2} \cdot \dfrac{\sqrt{x+2}+\sqrt{x}}{\sqrt{x+2}+\sqrt{x}}$

$\qquad = \dfrac{(x+2)-x}{2(\sqrt{x+2}+\sqrt{x})}$

$\qquad = \dfrac{2}{2(\sqrt{x+2}+\sqrt{x})}$

$\qquad = \dfrac{1}{\sqrt{x+2}+\sqrt{x}}$

75. $\dfrac{\sqrt{x+h+1}-\sqrt{x+1}}{h} = \dfrac{\sqrt{x+h+1}-\sqrt{x+1}}{h} \cdot \dfrac{\sqrt{x+h+1}+\sqrt{x+1}}{\sqrt{x+h+1}+\sqrt{x+1}}$

$\qquad = \dfrac{(x+h+1)-(x+1)}{h(\sqrt{x+h+1}+\sqrt{x+1})}$

$\qquad = \dfrac{h}{h(\sqrt{x+h+1}+\sqrt{x+1})}$

$\qquad = \dfrac{1}{\sqrt{x+h+1}+\sqrt{x+1}}, \quad h \neq 0$

77. Probability $= \dfrac{\text{Shaded area}}{\text{Total area}} = \dfrac{x(x/2)}{x(2x+1)} = \dfrac{x/2}{2x+1} \cdot \dfrac{2}{2} = \dfrac{x}{2(2x+1)}$

79. (a) $\dfrac{1}{16}$ minute

(b) $x\left(\dfrac{1}{16}\right) = \dfrac{x}{16}$ minutes

(c) $\dfrac{60}{16} = \dfrac{15}{4}$ minutes

81. (a) $r = \dfrac{\left(\dfrac{24[48(400)-16,000]}{48}\right)}{\left[16,000+\dfrac{48(400)}{12}\right]} \approx 0.0909 = 9.09\%$

(b) $r = \dfrac{\left[\dfrac{24(NM-P)}{N}\right]}{\left(P+\dfrac{NM}{12}\right)} = \dfrac{24(NM-P)}{N} \cdot \dfrac{12}{12P+NM} = \dfrac{288(NM-P)}{N(12P+NM)}$

$r = \dfrac{288[48(400)-16,000]}{48[12(16,000)+48(400)]} \approx 0.0909 = 9.09\%$

83. $T = 10\left(\dfrac{4t^2 + 16t + 75}{t^2 + 4t + 10}\right)$

(a)

t	0	2	4	6	8	10	12	14	16	18	20	22
T	75°	55.9°	48.3°	45°	43.3°	42.3°	41.7°	41.3°	41.1°	40.9°	40.7°	40.6°

(b) T is approaching 40°.

85. False. In order for the simplified expression to be equivalent to the original expression, the domain of the simplified expression needs to be restricted. If n is even, $x \neq \pm 1$. If n is odd, $x \neq 1$.

87. Completely factor the numerator and the denominator. A rational expression is in **simplest** form if there are no common factors in the numerator and the denominator other than ± 1.

Section P.6   Errors and the Algebra of Calculus

■ You should be able to recognize and avoid the common algebraic errors involving parentheses, fractions, exponents, radicals, and cancellation.

■ You should be able to "unsimplify" algebraic expressions by the following methods.

(a) Unusual Factoring

(b) Rewriting with Negative Exponents

(c) Writing a Fraction as a Sum of Terms

(d) Inserting Factors or Terms

Vocabulary Check

1. numerator

2. reciprocal

1. $2x - (3y + 4) \neq 2x - 3y + 4$

Change all signs when distributing the minus sign.

$2x - (3y + 4) = 2x - 3y - 4$

3. $\dfrac{4}{16x - (2x + 1)} \neq \dfrac{4}{14x + 1}$

Change all signs when distributing the minus sign.

$\dfrac{4}{16x - (2x + 1)} = \dfrac{4}{16x - 2x - 1} = \dfrac{4}{14x - 1}$

5. $(5z)(6z) \neq 30z$

z occurs twice as a factor.

$(5z)(6z) = 30z^2$

7. $a\left(\dfrac{x}{y}\right) \neq \dfrac{ax}{ay}$

The fraction as a whole is multiplied by a, not the numerator and denominator separately.

$a\left(\dfrac{x}{y}\right) = \dfrac{a}{1} \cdot \dfrac{x}{y} = \dfrac{ax}{y}$

9. $\sqrt{x + 9} \neq \sqrt{x} + 3$

Do not apply the radical to the terms.

$\sqrt{x + 9}$ does not simplify.

11. $\dfrac{2x^2 + 1}{5x} \neq \dfrac{2x + 1}{5}$

Divide out common factors not common terms.

$\dfrac{2x^2 + 1}{5x}$ cannot be simplified.

13. $\dfrac{1}{a^{-1} + b^{-1}} \neq \left(\dfrac{1}{a + b}\right)^{-1}$

To get rid of negative exponents:

$$\frac{1}{a^{-1} + b^{-1}} = \frac{1}{a^{-1} + b^{-1}} \cdot \frac{ab}{ab} = \frac{ab}{b + a}$$

15. $(x^2 + 5x)^{1/2} \neq x(x + 5)^{1/2}$

Factor within grouping symbols before applying the exponent to each factor.

$$(x^2 + 5x)^{1/2} = [x(x + 5)]^{1/2} = x^{1/2}(x + 5)^{1/2}$$

17. $\dfrac{3}{x} + \dfrac{4}{y} = \dfrac{3}{x} \cdot \dfrac{y}{y} + \dfrac{4}{y} \cdot \dfrac{x}{x} = \dfrac{3y + 4x}{xy}$

To add fractions, they must have a common denominator.

19. $\dfrac{3x + 2}{5} = \dfrac{1}{5}(3x + 2)$

The required factor is $3x + 2$.

21. $\dfrac{2}{3}x^2 + \dfrac{1}{3}x + 5 = \dfrac{2}{3}x^2 + \dfrac{1}{3}x + \dfrac{15}{3} = \dfrac{1}{3}(2x^2 + x + 15)$

The required factor is $2x^2 + x + 15$.

23. $x^2(x^3 - 1)^4 = \dfrac{1}{3}(x^3 - 1)^4(3x^2)$

The required factor is $\dfrac{1}{3}$.

25. $\dfrac{4x + 6}{(x^2 + 3x + 7)^3} = \dfrac{2(2x + 3)}{(x^2 + 3x + 7)^3} = \dfrac{2}{1} \cdot \dfrac{(2x + 3)}{1} \cdot \dfrac{1}{(x^2 + 3x + 7)^3} = (2)\dfrac{1}{(x^2 + 3x + 7)^3}(2x + 3)$

The required factor is 2.

27. $\dfrac{3}{x} + \dfrac{5}{2x^2} - \dfrac{3}{2}x = \dfrac{6x}{2x^2} + \dfrac{5}{2x^2} - \dfrac{3x^3}{2x^2}$

$$= \left(\dfrac{1}{2x^2}\right)(6x + 5 - 3x^3)$$

The required factor is $\dfrac{1}{2x^2}$.

29. $\dfrac{9x^2}{25} + \dfrac{16y^2}{49} = \dfrac{9}{25} \cdot \dfrac{x^2}{1} + \dfrac{16}{49} \cdot \dfrac{y^2}{1}$

$$= \dfrac{1}{25/9} \cdot \dfrac{x^2}{1} + \dfrac{1}{49/16} \cdot \dfrac{y^2}{1}$$

$$= \dfrac{x^2}{(25/9)} + \dfrac{y^2}{(49/16)}$$

The required factors are $\dfrac{25}{9}$ and $\dfrac{49}{16}$.

31. $\dfrac{x^2}{1/12} - \dfrac{y^2}{2/3} = x^2\left(\dfrac{12}{1}\right) - y^2\left(\dfrac{3}{2}\right) = \dfrac{12x^2}{1} - \dfrac{3y^2}{2}$

The required factors are 1 and 2.

33. $x^{1/3} - 5x^{4/3} = x^{1/3}(1 - 5x^{3/3}) = x^{1/3}(1 - 5x)$

The required factor is $1 - 5x$.

35. $(1 - 3x)^{4/3} - 4x(1 - 3x)^{1/3} = (1 - 3x)^{1/3}[(1 - 3x)^1 - 4x]$

$$= (1 - 3x)^{1/3}(1 - 7x)$$

The required factor is $1 - 7x$.

37. $\dfrac{1}{10}(2x + 1)^{5/2} - \dfrac{1}{6}(2x + 1)^{3/2} = \dfrac{3}{30}(2x + 1)^{3/2}(2x + 1)^1 - \dfrac{5}{30}(2x + 1)^{3/2}$

$$= \dfrac{1}{30}(2x + 1)^{3/2}[3(2x + 1) - 5]$$

$$= \dfrac{1}{30}(2x + 1)^{3/2}(6x - 2)$$

$$= \dfrac{1}{30}(2x + 1)^{3/2}2(3x - 1)$$

$$= \dfrac{1}{15}(2x + 1)^{3/2}(3x - 1)$$

The required factor is $3x - 1$.

39. $\dfrac{3x^2}{(2x - 1)^3} = 3x^2(2x - 1)^{-3}$

41. $\dfrac{4}{3x} + \dfrac{4}{x^4} - \dfrac{7x}{\sqrt[3]{2x}} = 4(3x)^{-1} + 4x^{-4} - 7x(2x)^{-1/3}$

43. $\dfrac{16 - 5x - x^2}{x} = \dfrac{16}{x} - \dfrac{5x}{x} - \dfrac{x^2}{x} = \dfrac{16}{x} - 5 - x$

45. $\dfrac{4x^3 - 7x^2 + 1}{x^{1/3}} = \dfrac{4x^3}{x^{1/3}} - \dfrac{7x^2}{x^{1/3}} + \dfrac{1}{x^{1/3}}$

$\qquad\qquad\qquad\qquad = 4x^{3-1/3} - 7x^{2-1/3} + \dfrac{1}{x^{1/3}}$

$\qquad\qquad\qquad\qquad = 4x^{8/3} - 7x^{5/3} + \dfrac{1}{x^{1/3}}$

47. $\dfrac{3 - 5x^2 - x^4}{\sqrt{x}} = \dfrac{3}{\sqrt{x}} - \dfrac{5x^2}{\sqrt{x}} - \dfrac{x^4}{\sqrt{x}}$

$\qquad\qquad\qquad = \dfrac{3}{\sqrt{x}} - 5x^{2-1/2} - x^{4-1/2}$

$\qquad\qquad\qquad = \dfrac{3}{x^{1/2}} - 5x^{3/2} - x^{7/2}$

49. $\dfrac{-2(x^2 - 3)^{-3}(2x)(x + 1)^3 - 3(x + 1)^2(x^2 - 3)^{-2}}{[(x + 1)^3]^2} = \dfrac{(x^2 - 3)^{-3}(x + 1)^2[-4x(x + 1) - 3(x^2 - 3)]}{(x + 1)^6}$

$\qquad\qquad\qquad\qquad\qquad\qquad\qquad\qquad\qquad = \dfrac{-4x^2 - 4x - 3x^2 + 9}{(x^2 - 3)^3(x + 1)^4}$

$\qquad\qquad\qquad\qquad\qquad\qquad\qquad\qquad\qquad = \dfrac{-7x^2 - 4x + 9}{(x^2 - 3)^3(x + 1)^4}$

51. $\dfrac{(6x + 1)^3(27x^2 + 2) - (9x^3 + 2x)(3)(6x + 1)^2(6)}{[(6x + 1)^3]^2} = \dfrac{(6x + 1)^2[(6x + 1)(27x^2 + 2) - 18(9x^3 + 2x)]}{(6x + 1)^6}$

$\qquad\qquad\qquad\qquad\qquad\qquad\qquad\qquad\qquad\qquad = \dfrac{162x^3 + 12x + 27x^2 + 2 - 162x^3 - 36x}{(6x + 1)^4}$

$\qquad\qquad\qquad\qquad\qquad\qquad\qquad\qquad\qquad\qquad = \dfrac{27x^2 - 24x + 2}{(6x + 1)^4}$

53. $\dfrac{(x + 2)^{3/4}(x + 3)^{-2/3} - (x + 3)^{1/3}(x + 2)^{-1/4}}{[(x + 2)^{3/4}]^2} = \dfrac{(x + 2)^{-1/4}(x + 3)^{-2/3}[(x + 2) - (x + 3)]}{(x + 2)^{6/4}}$

$\qquad\qquad\qquad\qquad\qquad\qquad\qquad\qquad\qquad = \dfrac{x + 2 - x - 3}{(x + 2)^{1/4}(x + 3)^{2/3}(x + 2)^{6/4}}$

$\qquad\qquad\qquad\qquad\qquad\qquad\qquad\qquad\qquad = -\dfrac{1}{(x + 3)^{2/3}(x + 2)^{7/4}}$

55. $\dfrac{2(3x - 1)^{1/3} - (2x + 1)(1/3)(3x - 1)^{-2/3}(3)}{(3x - 1)^{2/3}} = \dfrac{(3x - 1)^{-2/3}[2(3x - 1) - (2x + 1)]}{(3x - 1)^{2/3}}$

$\qquad\qquad\qquad\qquad\qquad\qquad\qquad\qquad\qquad = \dfrac{6x - 2 - 2x - 1}{(3x - 1)^{2/3}(3x - 1)^{2/3}}$

$\qquad\qquad\qquad\qquad\qquad\qquad\qquad\qquad\qquad = \dfrac{4x - 3}{(3x - 1)^{4/3}}$

57. $\dfrac{1}{(x^2 + 4)^{1/2}} \cdot \dfrac{1}{2}(x^2 + 4)^{-1/2}(2x) = \dfrac{1}{(x^2 + 4)^{1/2}} \cdot \dfrac{1}{(x^2 + 4)^{1/2}} \cdot \dfrac{1}{2}(2x)$

$\qquad\qquad\qquad\qquad\qquad\qquad\qquad = \dfrac{1}{(x^2 + 4)^1}(x)$

$\qquad\qquad\qquad\qquad\qquad\qquad\qquad = \dfrac{x}{x^2 + 4}$

59. $(x^2 + 5)^{1/2}\left(\dfrac{3}{2}\right)(3x - 2)^{1/2}(3) + (3x - 2)^{3/2}\left(\dfrac{1}{2}\right)(x^2 + 5)^{-1/2}(2x) = \dfrac{9}{2}(x^2 + 5)^{1/2}(3x - 2)^{1/2} + x(x^2 + 5)^{-1/2}(3x - 2)^{3/2}$

$$= \frac{9}{2}(x^2 + 5)^{1/2}(3x - 2)^{1/2} + \frac{2}{2}x(x^2 + 5)^{-1/2}(3x - 2)^{3/2}$$

$$= \frac{1}{2}(x^2 + 5)^{-1/2}(3x - 2)^{1/2}[9(x^2 + 5)^1 + 2x(3x - 2)^1]$$

$$= \frac{1}{2}(x^2 + 5)^{-1/2}(3x - 2)^{1/2}(9x^2 + 45 + 6x^2 - 4x)$$

$$= \frac{(3x - 2)^{1/2}(15x^2 - 4x + 45)}{2(x^2 + 5)^{1/2}}$$

61. $t = \dfrac{\sqrt{x^2 + 4}}{2} + \dfrac{\sqrt{(4 - x)^2 + 4}}{6}$

(a)

x	t
0.5	1.70
1.0	1.72
1.5	1.78
2.0	1.89
2.5	2.02
3.0	2.18
3.5	2.36
4.0	2.57

(b) She should swim to a point about $\frac{1}{2}$ mile down the coast to minimize the time required to reach the finish line.

(c) $\dfrac{1}{2}x(x^2 + 4)^{-1/2} + \dfrac{1}{6}(x - 4)(x^2 - 8x + 20)^{-1/2} = \dfrac{3}{6}x(x^2 + 4)^{-1/2} + \dfrac{1}{6}(x - 4)(x^2 - 8x + 20)^{-1/2}$

$$= \frac{1}{6}[3x(x^2 + 4)^{-1/2} + (x - 4)(x^2 - 8x + 20)^{-1/2}]$$

$$= \frac{1}{6}\left[\frac{3x}{(x^2 + 4)^{1/2}} + \frac{x - 4}{(x^2 - 8x + 20)^{1/2}}\right]$$

$$= \frac{3x\sqrt{x^2 - 8x + 20} + (x - 4)\sqrt{x^2 + 4}}{6\sqrt{x^2 + 4}\,\sqrt{x^2 - 8x + 20}}$$

63. True.

$$x^{-1} + y^{-2} = \frac{1}{x} + \frac{1}{y^2} = \frac{y^2 + x}{xy^2}$$

65. True.

$$\frac{1}{\sqrt{x} + 4} = \frac{1}{\sqrt{x} + 4} \cdot \frac{\sqrt{x} - 4}{\sqrt{x} - 4} = \frac{\sqrt{x} - 4}{x - 16}$$

67. $x^n \cdot x^{3n} \neq x^{3n^2}$

Add exponents when multiplying powers with like bases.

$x^n \cdot x^{3n} = x^{4n}$

69. $x^{2n} + y^{2n} \neq (x^n + y^n)^2$

When squaring binomials, there is also a middle term.

$(x^n + y^n)^2 = x^{2n} + 2x^ny^n + y^{2n}$

71. The two answers are equivalent and can be obtained by factoring.

$\dfrac{1}{10}(2x - 1)^{5/2} + \dfrac{1}{6}(2x - 1)^{3/2} = \dfrac{1}{60}(2x - 1)^{3/2}[6(2x - 1) + 10]$

$$= \frac{1}{60}(2x - 1)^{3/2}(12x + 4)$$

$$= \frac{4}{60}(2x - 1)^{3/2}(3x + 1)$$

$$= \frac{1}{15}(2x - 1)^{3/2}(3x + 1)$$

(a) $\dfrac{2}{3}x(2x - 3)^{3/2} - \dfrac{2}{15}(2x - 3)^{5/2} = \dfrac{2}{15}(2x - 3)^{3/2}[5x - (2x - 3)]$

$$= \frac{2}{15}(2x - 3)^{3/2}(3x + 3)$$

$$= \frac{2}{15}(2x - 3)^{3/2}3(x + 1)$$

$$= \frac{2}{5}(2x - 3)^{3/2}(x + 1)$$

—**CONTINUED**—

71. **—CONTINUED—**

(b) $\frac{2}{3}x(4 + x)^{3/2} - \frac{2}{15}(4 + x)^{5/2} = \frac{2}{15}(4 + x)^{3/2}[5x - (4 + x)]$

$$= \frac{2}{15}(4 + x)^{3/2}(4x - 4) = \frac{2}{15}(4 + x)^{3/2}4(x - 1) = \frac{8}{15}(4 + x)^{3/2}(x - 1)$$

Section P.7   The Rectangular Coordinate System and Graphs

■ You should be able to plot points and sketch scatter plots.

■ You should know that the distance between (x_1, y_1) and (x_2, y_2) in the plane is $d = \sqrt{(x_2 - x_1)^2 + (y_2 - y_1)^2}$.

■ You should know that the midpoint of the line segment joining (x_1, y_1) and (x_2, y_2) is $\left(\dfrac{x_1 + x_2}{2}, \dfrac{y_1 + y_2}{2}\right)$.

Vocabulary Check

1. (a) **v**   horizontal real number line

(b) **vi**   vertical real number line

(c) **i**   point of intersection of vertical axis and horizontal axis

(d) **iv**   four regions of the coordinate plane

(e) **iii**   directed distance from the y-axis

(f) **ii**   directed distance from the x-axis

2. Cartesian

3. Distance Formula

4. Midpoint Formula

1. A: $(2, 6)$, B: $(-6, -2)$, C: $(4, -4)$, D: $(-3, 2)$

3.

5.

7. $(-3, 4)$

9. $(-5, -5)$

11. $x > 0$ and $y < 0$ in Quadrant IV.

13. $x = -4$ and $y > 0$ in Quadrant II.

15. $y < -5$ in Quadrants III and IV.

17. $(x, -y)$ is in the second Quadrant means that (x, y) is in Quadrant III.

19. (x, y), $xy > 0$ means x and y have the same signs. This occurs in Quadrants I and III.

21.

Year, x	Number of stores, y
1996	3054
1997	3406
1998	3599
1999	3985
2000	4189
2001	4414
2002	4688
2003	4906

Year (6 ↔ 1996)

23. $d = |5 - (-3)| = 8$

25. $d = |2 - (-3)| = 5$

27. (a) The distance between $(0, 2)$ and $(4, 2)$ is 4.

The distance between $(4, 2)$ and $(4, 5)$ is 3.

The distance between $(0, 2)$ and $(4, 5)$ is

$$\sqrt{(4 - 0)^2 + (5 - 2)^2} = \sqrt{16 + 9} = \sqrt{25} = 5.$$

(b) $4^2 + 3^2 = 16 + 9 = 25 = 5^2$

29. (a) The distance between $(-1, 1)$ and $(9, 1)$ is 10.

The distance between $(9, 1)$ and $(9, 4)$ is 3.

The distance between $(-1, 1)$ and $(9, 4)$ is

$$\sqrt{(9 - (-1))^2 + (4 - 1)^2} = \sqrt{100 + 9} = \sqrt{109}.$$

(b) $10^2 + 3^2 = 109 = \left(\sqrt{109}\right)^2$

31. (a)

(b) $d = \sqrt{(9 - 1)^2 + (7 - 1)^2}$

$= \sqrt{64 + 36} = 10$

(c) $\left(\dfrac{9 + 1}{2}, \dfrac{7 + 1}{2}\right) = (5, 4)$

33. (a)

(b) $d = \sqrt{(4 + 4)^2 + (-5 - 10)^2}$

$= \sqrt{64 + 225} = 17$

(c) $\left(\dfrac{4 - 4}{2}, \dfrac{-5 + 10}{2}\right) = \left(0, \dfrac{5}{2}\right)$

35. (a)

(b) $d = \sqrt{(5 + 1)^2 + (4 - 2)^2}$

$= \sqrt{36 + 4} = 2\sqrt{10}$

(c) $\left(\dfrac{-1 + 5}{2}, \dfrac{2 + 4}{2}\right) = (2, 3)$

37. (a)

(b) $d = \sqrt{\left(\dfrac{1}{2} + \dfrac{5}{2}\right)^2 + \left(1 - \dfrac{4}{3}\right)^2}$

$= \sqrt{9 + \dfrac{1}{9}} = \dfrac{\sqrt{82}}{3}$

(c) $\left(\dfrac{-(5/2) + (1/2)}{2}, \dfrac{(4/3) + 1}{2}\right) = \left(-1, \dfrac{7}{6}\right)$

39. (a)

(b) $d = \sqrt{(6.2 + 3.7)^2 + (5.4 - 1.8)^2}$

$= \sqrt{98.01 + 12.96} = \sqrt{110.97}$

(c) $\left(\dfrac{6.2 - 3.7}{2}, \dfrac{5.4 + 1.8}{2}\right) = (1.25, 3.6)$

41. $d_1 = \sqrt{(4-2)^2 + (0-1)^2} = \sqrt{5}$

$d_2 = \sqrt{(4+1)^2 + (0+5)^2} = \sqrt{50}$

$d_3 = \sqrt{(2+1)^2 + (1+5)^2} = \sqrt{45}$

$\left(\sqrt{5}\right)^2 + \left(\sqrt{45}\right)^2 = \left(\sqrt{50}\right)^2$

43. Since $x_m = \dfrac{x_1 + x_2}{2}$ and $y_m = \dfrac{y_1 + y_2}{2}$ we have:

$$2x_m = x_1 + x_2 \qquad\qquad 2y_m = y_1 + y_2$$

$$2x_m - x_1 = x_2 \qquad\qquad 2y_m - y_1 = y_2$$

Thus, $(x_2, y_2) = (2x_m - x_1, 2y_m - y_1)$.

45. The midpoint of the given line segment is $\left(\dfrac{x_1 + x_2}{2}, \dfrac{y_1 + y_2}{2}\right)$.

The midpoint between (x_1, y_1) and $\left(\dfrac{x_1 + x_2}{2}, \dfrac{y_1 + y_2}{2}\right)$ is $\left(\dfrac{x_1 + \dfrac{x_1 + x_2}{2}}{2}, \dfrac{y_1 + \dfrac{y_1 + y_2}{2}}{2}\right) = \left(\dfrac{3x_1 + x_2}{4}, \dfrac{3y_1 + y_2}{4}\right)$.

The midpoint between $\left(\dfrac{x_1 + x_2}{2}, \dfrac{y_1 + y_2}{2}\right)$ and (x_2, y_2) is $\left(\dfrac{\dfrac{x_1 + x_2}{2} + x_2}{2}, \dfrac{\dfrac{y_1 + y_2}{2} + y_2}{2}\right) = \left(\dfrac{x_1 + 3x_2}{4}, \dfrac{y_1 + 3y_2}{4}\right)$.

Thus, the three points are

$$\left(\dfrac{3x_1 + x_2}{4}, \dfrac{3y_1 + y_2}{4}\right), \left(\dfrac{x_1 + x_2}{2}, \dfrac{y_1 + y_2}{2}\right), \text{ and } \left(\dfrac{x_1 + 3x_2}{4}, \dfrac{y_1 + 3y_2}{4}\right).$$

47. $d = \sqrt{(42-18)^2 + (50-12)^2}$

$= \sqrt{24^2 + 38^2}$

$= \sqrt{2020}$

$= 2\sqrt{505}$

≈ 45 yards

49. $\left(\dfrac{2001 + 2003}{2}, \dfrac{3433 + 4174}{2}\right) = (2002, 3803.5)$

In 2002, the sales for Big Lots was approximately $3803.5 million.

51. $(-2 + 2, -4 + 5) = (0, 1)$

$(2 + 2, -3 + 5) = (4, 2)$

$(-1 + 2, -1 + 5) = (1, 4)$

53. $(-7 + 4, -2 + 8) = (-3, 6)$

$(-2 + 4, 2 + 8) = (2, 10)$

$(-2 + 4, -4 + 8) = (2, 4)$

$(-7 + 4, -4 + 8) = (-3, 4)$

55. The highest price of butter is approximately $3.31 per pound. This occurred in 2001.

57. $\left[\dfrac{2400 - 700}{700}\right](100) \approx 242.9\%$ increase

59.

(a) The point is reflected through the y-axis.

(b) The point is reflected through the x-axis.

(c) The point is reflected through the origin.

61. (a) The minimum wage had the greatest increase in the 1990s.

(b) Minimum wage in 1990: $3.80

Minimum wage in 1995: $4.25

Percent increase: $\left(\dfrac{\$4.25 - \$3.80}{\$3.80}\right)(100) \approx 11.8\%$

Minimum wage in 1995: $4.25

Minimum wage in 2000: $5.15

Percent increase: $\left(\dfrac{\$5.15 - \$4.25}{\$4.25}\right)(100) \approx 21.2\%$

(c) $\$5.15 + 0.212(\$5.15) \approx \$6.24$

(d) The political nature of the minimum wage makes it difficult to predict, but this does seem like a reasonable value.

63. (1996, 18,546), (2004, 21,900)

By Exercise 45 we have the following:

$$\left(\frac{3(1996) + 2004}{4}, \frac{3(18,546) + 21,900}{4} \right) = (1998, 19,384.5)$$

$$\left(\frac{1996 + 2004}{2}, \frac{18,546 + 21,900}{2} \right) = (2000, 20,223)$$

$$\left(\frac{1996 + 3(2004)}{4}, \frac{18,546 + 3(21,900)}{4} \right) = (2002, 21,061.5)$$

Year	Sales for Pepsi Bottling Group, Inc.
1998	$19,384.5 million
2000	$20,223 million
2002	$21,061.5 million

65. (a)

Year, x	Pieces of mail, y (in billions)
1996	183
1997	191
1998	197
1999	202
2000	208
2001	207
2002	203
2003	202

(b) The greatest decrease occurred in 2002.

(c) Answers will vary. Technology now enables us to transport information in ways other than by mail. The Internet is one example.

67. (a)

Year, x	Men's teams, M	Women's teams, W
1994	858	859
1995	868	864
1996	866	874
1997	865	879
1998	895	911
1999	926	940
2000	932	956
2001	937	958
2002	936	975
2003	967	1009

(b) In 1994, the number of men's and women's teams were nearly equal.

(c) In 2003, the difference between the number of teams was greatest: $1009 - 967 = 42$ teams

69. False, you would have to use the Midpoint Formula 15 times.

71. On the x-axis: $y = 0$; on the y-axis: $x = 0$

73. Since (x_0, y_0) lies in Quadrant II, $(x_0, -y_0)$ must lie in Quadrant III. Matches (b).

75. Since (x_0, y_0) lies in Quadrant II, $\left(x_0, \frac{1}{2}y_0\right)$ must lie in Quadrant II. Matches (d).

77. Use the Midpoint Formula to prove the diagonals of the parallelogram bisect each other.

$$\left(\frac{b+a}{2}, \frac{c+0}{2}\right) = \left(\frac{a+b}{2}, \frac{c}{2}\right)$$

$$\left(\frac{a+b+0}{2}, \frac{c+0}{2}\right) = \left(\frac{a+b}{2}, \frac{c}{2}\right)$$

Review Exercises for Chapter P

1. $\{11, -14, -\frac{8}{9}, \frac{5}{2}, \sqrt{6}, 0.4\}$

   (a) Natural numbers: 11

   (b) Whole numbers: 0, 11

   (c) Integers: 11, −14

   (d) Rational numbers: 11, −14, −$\frac{8}{9}$, $\frac{5}{2}$, 0.4

   (e) Irrational numbers: $\sqrt{6}$

3. (a) $\frac{5}{6} = 0.8\overline{3}$

   (b) $\frac{7}{8} = 0.875$

   $\frac{5}{6} < \frac{7}{8}$

5. $x \le 7$

   The set consists of all real numbers less than or equal to 7.

7. $d(-92, 63) = |63 - (-92)|$

              $= 155$

9. $d(x, 7) = |x - 7|$ and $d(x, 7) \ge 4$, thus $|x - 7| \ge 4$.

11. $d(y, -30) = |y - (-30)|$

             $= |y + 30|$

   and $d(y, -30) < 5$, thus $|y + 30| < 5$.

13. $12x - 7$

   (a) $12(0) - 7 = -7$

   (b) $12(-1) - 7 = -19$

15. $-x^2 + x - 1$

   (a) $-(1)^2 + 1 - 1 = -1$

   (b) $-(-1)^2 + (-1) - 1 = -3$

17. $2x + (3x - 10) = (2x + 3x) - 10$

   Illustrates the Associative Property of Addition

19. $0 + (a - 5) = a - 5$

   Illustrates the Additive Identity Property

21. $|-3| + 4(-2) - 6 = 3 - 8 - 6$

                $= -11$

23. $\frac{5}{18} \div \frac{10}{3} = \frac{\cancel{5}}{\cancel{18}_6} \cdot \frac{\cancel{3}}{\cancel{10}_2} = \frac{1}{12}$

25. $6[4 - 2(6 + 8)] = 6[4 - 2(14)]$

                      $= 6[4 - 28]$

                      $= 6(-24)$

                      $= -144$

27. (a) $3x^2(4x^3)^3 = 3x^2(64x^9) = 192x^{11}$

   (b) $\frac{5y^6}{10y} = \frac{y^{6-1}}{2} = \frac{y^5}{2}$, $y \ne 0$

29. (a) $\frac{6^2 u^3 v^{-3}}{12 u^{-2} v} = \frac{36 u^{3-(-2)} v^{-3-1}}{12} = 3u^5 v^{-4} = \frac{3u^5}{v^4}$

   (b) $\frac{3^{-4} m^{-1} n^{-3}}{9^{-2} mn^{-3}} = \frac{9^2 n^3}{3^4 mmn^3} = \frac{81}{81m^2} = \frac{1}{m^2} = m^{-2}$

31. $2,134,300,000 = 2.1343 \times 10^9$

33. $4.837 \times 10^8 = 483,700,000$

35. (a) $\sqrt[3]{27^2} = \left(\sqrt[3]{27}\right)^2 = (3)^2 = 9$

   (b) $\sqrt{49^3} = \left(\sqrt{49}\right)^3 = (7)^3 = 343$

37. (a) $\left(\sqrt[3]{216}\right)^3 = \left(\sqrt[3]{6^3}\right)^3 = (6)^3 = 216$

   (b) $\sqrt[4]{32^4} = \left(\sqrt[4]{32}\right)^4 = 32$

39. (a) $\sqrt{50} - \sqrt{18} = \sqrt{25 \cdot 2} - \sqrt{9 \cdot 2} = 5\sqrt{2} - 3\sqrt{2} = 2\sqrt{2}$

(b) $2\sqrt{32} + 3\sqrt{72} = 2\sqrt{16 \cdot 2} + 3\sqrt{36 \cdot 2} = 2(4\sqrt{2}) + 3(6\sqrt{2}) = 8\sqrt{2} + 18\sqrt{2} = 26\sqrt{2}$

41. These are not like terms. Radicals cannot be combined by addition or subtraction unless the index and the radicand are the same.

43. $\dfrac{1}{2 - \sqrt{3}} = \dfrac{1}{2 - \sqrt{3}} \cdot \dfrac{2 + \sqrt{3}}{2 + \sqrt{3}} = \dfrac{2 + \sqrt{3}}{2^2 - (\sqrt{3})^2}$

$= \dfrac{2 + \sqrt{3}}{4 - 3} = \dfrac{2 + \sqrt{3}}{1} = 2 + \sqrt{3}$

45. $\dfrac{\sqrt{7} + 1}{2} = \dfrac{\sqrt{7} + 1}{2} \cdot \dfrac{\sqrt{7} - 1}{\sqrt{7} - 1} = \dfrac{(\sqrt{7})^2 - 1^2}{2(\sqrt{7} - 1)}$

$= \dfrac{7 - 1}{2(\sqrt{7} - 1)} = \dfrac{6}{2(\sqrt{7} - 1)} = \dfrac{3}{\sqrt{7} - 1}$

47. $(16)^{3/2} = \sqrt{16^3} = (\sqrt{16})^3 = (4)^3 = 64$

49. $(3x^{2/5})(2x^{1/2}) = 6x^{2/5 + 1/2} = 6x^{9/10}$

51. Standard form: $-11x^2 + 3$

Degree: 2

Leading coefficient: -11

53. Standard form: $-12x^2 - 4$

Degree: 2

Leading coefficient: -12

55. $-(3x^2 + 2x) + (1 - 5x) = -3x^2 - 2x + 1 - 5x$

$= -3x^2 - 7x + 1$

57. $2x(x^2 - 5x + 6) = (2x)(x^2) + (2x)(-5x) + (2x)(6)$

$= 2x^3 - 10x^2 + 12x$

59. $(3x - 6)(5x + 1) = 15x^2 + 3x - 30x - 6$

$= 15x^2 - 27x - 6$

61. $(2x - 3)^2 = (2x)^2 - 2(2x)(3) + 3^2$

$= 4x^2 - 12x + 9$

63. $(3\sqrt{5} + 2)(3\sqrt{5} - 2) = (3\sqrt{5})^2 - 2^2$

$= 9(5) - 4$

$= 41$

65. (a) The surface is the sum of the area of the side, $2\pi rh$, and the areas of the top and bottom which are each πr^2.

$S = 2\pi rh + \pi r^2 + \pi r^2 = 2\pi rh + 2\pi r^2$

(b) $S = 2\pi(6)(8) + 2\pi(6)^2$

$= 96\pi + 72\pi$

$= 168\pi \approx 527.79 \text{ in}^2$

67. $x^3 - x = x(x^2 - 1) = x(x + 1)(x - 1)$

69. $25x^2 - 49 = (5x)^2 - 7^2 = (5x + 7)(5x - 7)$

71. $x^3 - 64 = x^3 - 4^3 = (x - 4)(x^2 + 4x + 16)$

73. $2x^2 + 21x + 10 = (2x + 1)(x + 10)$

75. $x^3 - x^2 + 2x - 2 = x^2(x - 1) + 2(x - 1)$

$= (x - 1)(x^2 + 2)$

77. The domain of $\dfrac{1}{x + 6}$ is the set of all real numbers except $x = -6$.

79. $\dfrac{x^2 - 64}{5(3x + 24)} = \dfrac{(x + 8)(x - 8)}{5 \cdot 3(x + 8)} = \dfrac{x - 8}{15}, \quad x \neq -8$

81. $\dfrac{x^2 - 4}{x^4 - 2x^2 - 8} \cdot \dfrac{x^2 + 2}{x^2} = \dfrac{(x^2 - 4)(x^2 + 2)}{(x^2 - 4)(x^2 + 2)x^2}$

$$= \dfrac{1}{x^2}, \quad x \neq \pm 2$$

83. $\dfrac{1}{x - 1} + \dfrac{1 - x}{x^2 + x + 1} = \dfrac{x^2 + x + 1 + (1 - x)(x - 1)}{(x - 1)(x^2 + x + 1)}$

$$= \dfrac{x^2 + x + 1 + x - 1 - x^2 + x}{(x - 1)(x^2 + x + 1)}$$

$$= \dfrac{3x}{(x - 1)(x^2 + x + 1)}$$

85. $\dfrac{\left(\dfrac{3a}{\dfrac{x}{a^2} - 1}\right)}{\left(\dfrac{a}{x} - 1\right)} = \dfrac{\left(\dfrac{\dfrac{3a}{a^2 - x}}{x}\right)}{\left(\dfrac{a - x}{x}\right)}$

$$= \dfrac{3a}{1} \cdot \dfrac{x}{a^2 - x} \cdot \dfrac{x}{a - x}$$

$$= \dfrac{3ax^2}{(a^2 - x)(a - x)}$$

87. $\dfrac{\dfrac{1}{2(x + h)} - \dfrac{1}{2x}}{h} = \dfrac{\dfrac{x - (x + h)}{2x(x + h)}}{h}$

$$= \dfrac{-h}{2x(x + h)} \cdot \dfrac{1}{h}$$

$$= \dfrac{-1}{2x(x + h)}, \quad h \neq 0$$

89. $10(4 \cdot 7) = 10(28) = 280$

The multiplication in parentheses comes first.

91. $(3^4)^4 = 3^{4 \cdot 4} = 3^{16}$

Multiply exponents when raising a power to a power.

93. $(5 + 8)^2 = 13^2 \neq 5^2 + 8^2$

Add the numbers in parentheses before squaring.

95. $\dfrac{2}{3}x^4 - \dfrac{3}{8}x^3 + \dfrac{5}{6}x^2 = \dfrac{16}{24}x^4 - \dfrac{9}{24}x^3 + \dfrac{20}{24}x^2$

$$= \dfrac{1}{24}x^2(16x^2 - 9x + 20)$$

The missing factor is $16x^2 - 9x + 20$.

97. $2x(x^2 - 3)^{1/3} - 5(x^2 - 3)^{4/3} = (x^2 - 3)^{1/3}[2x - 5(x^2 - 3)]$

$$= (x^2 - 3)^{1/3}(-5x^2 + 2x + 15)$$

The missing factor is $-5x^2 + 2x + 15$.

99. $\dfrac{x^3 + 5x^2 + 7}{x} = \dfrac{x^3}{x} + \dfrac{5x^2}{x} + \dfrac{7}{x} = x^2 + 5x + \dfrac{7}{x}$

101. $\dfrac{x(x + 2)^{-1/2} + (x + 2)^{1/2}}{(x + 2)^{3/2}} = \dfrac{(x + 2)^{-1/2}[x + (x + 2)]}{(x + 2)^{3/2}}$

$$= \dfrac{2x + 2}{(x + 2)^{1/2}(x + 2)^{3/2}}$$

$$= \dfrac{2(x + 1)}{(x + 2)^2}$$

103.

105. $x > 0$ and $y = -2$ in Quadrant IV.

107. (a)

(b) $d = \sqrt{(-3 - 1)^2 + (8 - 5)^2} = \sqrt{16 + 9} = 5$

(c) Midpoint: $\left(\dfrac{-3 + 1}{2}, \dfrac{8 + 5}{2}\right) = \left(-1, \dfrac{13}{2}\right)$

109. (a)

(b) $d = \sqrt{(5.6 - 0)^2 + (0 - 8.2)^2}$

$= \sqrt{31.36 + 67.24} = \sqrt{98.6} \approx 9.9$

(c) Midpoint: $\left(\dfrac{0 + 5.6}{2}, \dfrac{8.2 + 0}{2}\right) = (2.8, 4.1)$

111. $(4 - 2, 8 - 3) = (2, 5)$

$(6 - 2, 8 - 3) = (4, 5)$

$(4 - 2, 3 - 3) = (2, 0)$

$(6 - 2, 3 - 3) = (4, 0)$

113. $(2001, 539.1), (2003, 773.8)$

$\left(\dfrac{2001 + 2003}{2}, \dfrac{539.1 + 773.8}{2}\right) = (2002, 656.45)$

In 2002, the sales were approximately \$656.45 million.

115. False, $(a + b)^2 = a^2 + 2ab + b^2 \neq a^2 + b^2$

There is also a cross-product term when a binomial sum is squared.

117. $\dfrac{ax - b}{b - ax} = -1$ for all real nonzero numbers a and b

except for $x = \dfrac{b}{a}$. This makes the expression undefined.

Problem Solving for Chapter P

1. (a) Men's

Minimum Volume: $V = \dfrac{4}{3}\pi(55)^3 \approx 696{,}910$ mm³

Maximum Volume: $V = \dfrac{4}{3}\pi(65)^3 \approx 1{,}150{,}347$ mm³

Women's

Minimum Volume: $V = \dfrac{4}{3}\pi\left(\dfrac{95}{2}\right)^3 \approx 448{,}921$ mm³

Maximum Volume: $V = \dfrac{4}{3}\pi(55)^3 \approx 696{,}910$ mm³

(b) Men's

Minimum density: $\dfrac{7.26}{1{,}150{,}347} \approx 6.31 \times 10^{-6}$ kg/mm³

Maximum density: $\dfrac{7.26}{696{,}910} \approx 1.04 \times 10^{-5}$ kg/mm³

Women's

Minimum density: $\dfrac{4.00}{696{,}910} \approx 5.74 \times 10^{-6}$ kg/mm³

Maximum density: $\dfrac{4.00}{448{,}921} \approx 8.91 \times 10^{-6}$ kg/mm³

(c) No. The weight would be different. Cork is much lighter than iron so it would have a much smaller density.

3. One golf ball: $\dfrac{1.6 \times 10^7}{1.58 \times 10^8} \approx 0.101$ pound

$$0.101(16) = 1.62 \text{ ounces}$$

5. To say that a number has n significant digits means that the number has n digits with the left most non-zero digit and ending with the right most non-zero digit. For example; 28,000, 1.400, 0.00079 each have two significant digits.

7. $r = 1 - \left(\dfrac{3225}{12,000}\right)^{1/4} \approx 0.280$ or 28%

9.

Perimeter: $P = 2l + w + \pi r$

$$13.14 = 2l + 2 + \pi(1)$$

$$l = \dfrac{13.14 - 2 - \pi}{2} \approx 4 \text{ feet}$$

Amount of glass = Area of window

$$A = lw + \dfrac{1}{2}\pi r^2$$

$$= (4)(2) + \dfrac{1}{2}\pi(1)^2$$

$$\approx 9.57 \text{ square feet}$$

11. $y_1 = 2x\sqrt{1 - x^2} - \dfrac{x^3}{\sqrt{1 - x^2}}$ $\qquad y_2 = \dfrac{2 - 3x^2}{\sqrt{1 - x^2}}$

When $x = 0$, $y_1 = 0$. When $x = 0$, $y_2 = 2$.

Thus, $y_1 \neq y_2$.

$$y_1 = \dfrac{2x\sqrt{1 - x^2}}{1} - \dfrac{x^3}{\sqrt{1 - x^2}}$$

$$= \dfrac{2x\sqrt{1 - x^2}}{1} \cdot \dfrac{\sqrt{1 - x^2}}{\sqrt{1 - x^2}} - \dfrac{x^3}{\sqrt{1 - x^2}}$$

$$= \dfrac{2x(1 - x^2) - x^3}{\sqrt{1 - x^2}} = \dfrac{2x - 2x^3 - x^3}{\sqrt{1 - x^2}}$$

$$= \dfrac{2x - 3x^3}{\sqrt{1 - x^2}}$$

$$= \dfrac{x(2 - 3x^2)}{\sqrt{1 - x^2}}$$

Let $y_2 = \dfrac{x(2 - 3x^2)}{\sqrt{1 - x^2}}$. Then $y_1 = y_2$.

13. (a) $(1, -2)$ and $(4, 1)$

The points of trisection are:

$$\left(\dfrac{2(1) + 4}{3}, \dfrac{2(-2) + 1}{3}\right) = (2, -1)$$

$$\left(\dfrac{1 + 2(4)}{3}, \dfrac{-2 + 2(1)}{3}\right) = (3, 0)$$

(b) $(-2, -3)$ and $(0, 0)$

The points of trisection are:

$$\left(\dfrac{2(-2) + 0}{3}, \dfrac{2(-3) + 0}{3}\right) = \left(-\dfrac{4}{3}, -2\right)$$

$$\left(\dfrac{-2 + 2(0)}{3}, \dfrac{-3 + 2(0)}{3}\right) = \left(-\dfrac{2}{3}, -1\right)$$

Practice Test for Chapter P

1. Evaluate $\dfrac{|-42| - 20}{15 - |-4|}$.

2. Simplify $\dfrac{x}{z} - \dfrac{z}{y}$.

3. The distance between x and 7 is no more than 4. Use absolute value notation to describe this expression.

4. Evaluate $10(-x)^3$ for $x = 5$.

5. Simplify $(-4x^3)(-2x^{-5})(\frac{1}{16}x)$.

6. Change 0.0000412 to scientific notation.

7. Evaluate $125^{2/3}$.

8. Simplify $\sqrt[4]{64x^7y^9}$.

9. Rationalize the denominator and simplify $\dfrac{6}{\sqrt{12}}$.

10. Simplify $3\sqrt{80} - 7\sqrt{500}$.

11. Simplify $(8x^4 - 9x^2 + 2x - 1) - (3x^3 + 5x + 4)$.

12. Multiply $(x - 3)(x^2 + x - 7)$.

13. Multiply $[(x - 2) - y]^2$.

14. Factor $16x^4 - 1$.

15. Factor $6x^2 + 5x - 4$.

16. Factor $x^3 - 64$.

17. Combine and simplify $-\dfrac{3}{x} + \dfrac{x}{x^2 + 2}$.

18. Combine and simplify $\dfrac{x - 3}{4x} \div \dfrac{x^2 - 9}{x^2}$.

19. Simplify $\dfrac{1 - \left(\dfrac{1}{x}\right)}{1 - \dfrac{1}{1 - \left(\dfrac{1}{x}\right)}}$.

20. (a) Plot the points $(-3, 7)$ and $(5, -1)$,

    (b) find the distance between the points, and

    (c) find the midpoint of the line segment joining the points.

CHAPTER 1
Equations, Inequalities, and Mathematical Modeling

C H A P T E R   1
Equations, Inequalities, and Mathematical Modeling

Section 1.1   Graphs of Equations

■ You should be able to use the point-plotting method of graphing.

■ You should be able to find x- and y-intercepts.

    (a) To find the x-intercepts, let $y = 0$ and solve for x.

    (b) To find the y-intercepts, let $x = 0$ and solve for y.

■ You should be able to test for symmetry.

    (a) To test for x-axis symmetry, replace y with $-y$.

    (b) To test for y-axis symmetry, replace x with $-x$.

    (c) To test for origin symmetry, replace x with $-x$ and y with $-y$.

■ You should know the standard equation of a circle with center (h, k) and radius r:

$$(x - h)^2 + (y - k)^2 = r^2$$

Vocabulary Check

1. solution or solution point   **2.** graph   **3.** intercepts

4. y-axis   **5.** circle; (h, k); r **6.** point-plotting

---

1. $y = \sqrt{x + 4}$

    (a) $(0, 2)$: $2 \overset{?}{=} \sqrt{0 + 4}$

                $2 = 2$

    Yes, the point *is* on the graph.

    (b) $(5, 3)$: $3 \overset{?}{=} \sqrt{5 + 4}$

                $3 = \sqrt{9}$

    Yes, the point *is* on the graph.

3. $y = 4 - |x - 2|$

    (a) $(1, 5)$: $5 \overset{?}{=} 4 - |1 - 2|$

                $5 \neq 4 - 1$

    No, the point *is not* on the graph.

    (b) $(6, 0)$: $0 \overset{?}{=} 4 - |6 - 2|$

                $0 = 4 - 4$

    Yes, the point *is* on the graph.

5. $y = -2x + 5$

x	-1	0	1	2	$\frac{5}{2}$
y	7	5	3	1	0
(x, y)	$(-1, 7)$	$(0, 5)$	$(1, 3)$	$(2, 1)$	$\left(\frac{5}{2}, 0\right)$

7. $y = x^2 - 3x$

x	-1	0	1	2	3
y	4	0	-2	-2	0
(x, y)	$(-1, 4)$	$(0, 0)$	$(1, -2)$	$(2, -2)$	$(3, 0)$

9. $y = 16 - 4x^2$

  x-intercepts: $0 = 16 - 4x^2$

$\qquad\qquad\qquad 4x^2 = 16$

$\qquad\qquad\qquad\; x^2 = 4$

$\qquad\qquad\qquad\;\; x = \pm 2$

$\qquad\qquad\quad (-2, 0), (2, 0)$

  y-intercept: $y = 16 - 4(0)^2 = 16$

$\qquad\qquad\quad (0, 16)$

11. $y = 2x^3 - 4x^2$

  x-intercepts: $0 = 2x^3 - 4x^2$

$\qquad\qquad\quad 0 = 2x^2(x - 2)$

$\qquad\qquad\quad x = 0 \;$ or $\; x = 2$

$\qquad\qquad\quad (0, 0), (2, 0)$

  y-intercept: $y = 2(0)^3 - 4(0)^2$

$\qquad\qquad\qquad y = 0$

$\qquad\qquad\qquad (0, 0)$

13. y-axis symmetry

15. Origin symmetry

17. $x^2 - y = 0$

$\quad (-x)^2 - y = 0 \implies x^2 - y = 0 \implies y$-axis symmetry

$\quad x^2 - (-y) = 0 \implies x^2 + y = 0 \implies$ No x-axis symmetry

$\quad (-x)^2 - (-y) = 0 \implies x^2 + y = 0 \implies$ No origin symmetry

19. $y = x^3$

$\quad y = (-x)^3 \implies y = -x^3 \implies$ No y-axis symmetry

$\quad -y = x^3 \implies y = -x^3 \implies$ No x-axis symmetry

$\quad -y = (-x)^3 \implies -y = -x^3 \implies y = x^3 \implies$ Origin symmetry

21. $y = \dfrac{x}{x^2 + 1}$

$\quad y = \dfrac{-x}{(-x)^2 + 1} \implies y = \dfrac{-x}{x^2 + 1} \implies$ No y-axis symmetry

$\quad -y = \dfrac{x}{x^2 + 1} \implies y = \dfrac{-x}{x^2 + 1} \implies$ No x-axis symmetry

$\quad -y = \dfrac{-x}{(-x)^2 + 1} \implies -y = \dfrac{-x}{x^2 + 1} \implies y = \dfrac{x}{x^2 + 1} \implies$ Origin symmetry

23. $xy^2 + 10 = 0$

$\quad (-x)y^2 + 10 = 0 \implies -xy^2 + 10 = 0 \implies$ No y-axis symmetry

$\quad x(-y)^2 + 10 = 0 \implies xy^2 + 10 = 0 \implies x$-axis symmetry

$\quad (-x)(-y)^2 + 10 = 0 \implies -xy^2 + 10 = 0 \implies$ No origin symmetry

25. $y = -3x + 1$

x-intercept: $\left(\frac{1}{3}, 0\right)$

y-intercept: $(0, 1)$

No axis or origin symmetry

27. $y = x^2 - 2x$

Intercepts: $(0, 0)$, $(2, 0)$

No axis or origin symmetry

x	-1	0	1	2	3
y	3	0	-1	0	3

29. $y = x^3 + 3$

Intercepts: $(0, 3)$, $\left(\sqrt[3]{-3}, 0\right)$

No axis or origin symmetry

x	-2	-1	0	1	2
y	-5	2	3	4	11

31. $y = \sqrt{x - 3}$

Domain: $[3, \infty)$

Intercept: $(3, 0)$

No axis or origin symmetry

x	3	4	7	12
y	0	1	2	3

33. $y = |x - 6|$

Intercepts: $(0, 6)$, $(6, 0)$

No axis or origin symmetry

x	-2	0	2	4	6	8	10
y	8	6	4	2	0	2	4

35. $x = y^2 - 1$

Intercepts: $(0, -1)$, $(0, 1)$, $(-1, 0)$

x-axis symmetry

x	-1	0	3
y	0	± 1	± 2

37. $y = 3 - \frac{1}{2}x$

Intercepts: $(6, 0)$, $(0, 3)$

39. $y = x^2 - 4x + 3$

Intercepts: $(3, 0)$, $(1, 0)$, $(0, 3)$

41. $y = \dfrac{2x}{x - 1}$

Intercept: $(0, 0)$

43. $y = \sqrt[3]{x}$

Intercept: $(0, 0)$

45. $y = x\sqrt{x + 6}$

Intercepts: $(0, 0), (-6, 0)$

47. $y = |x + 3|$

Intercepts: $(-3, 0), (0, 3)$

49. Center: $(0, 0)$; radius: 4

Standard form:

$$(x - 0)^2 + (y - 0)^2 = 4^2$$

$$x^2 + y^2 = 16$$

51. Center: $(2, -1)$; radius: 4

Standard form:

$$(x - 2)^2 + (y - (-1))^2 = 4^2$$

$$(x - 2)^2 + (y + 1)^2 = 16$$

53. Center: $(-1, 2)$; solution point: $(0, 0)$

$$(x - (-1))^2 + (y - 2)^2 = r^2$$

$$(0 + 1)^2 + (0 - 2)^2 = r^2 \Longrightarrow 5 = r^2$$

Standard form: $(x + 1)^2 + (y - 2)^2 = 5$

55. Endpoints of a diameter: $(0, 0), (6, 8)$

Center: $\left(\dfrac{0 + 6}{2}, \dfrac{0 + 8}{2}\right) = (3, 4)$

$$(x - 3)^2 + (y - 4)^2 = r^2$$

$$(0 - 3)^2 + (0 - 4)^2 = r^2 \Longrightarrow 25 = r^2$$

Standard form: $(x - 3)^2 + (y - 4)^2 = 25$

57. $x^2 + y^2 = 25$

Center: $(0, 0)$, radius: 5

59. $(x - 1)^2 + (y + 3)^2 = 9$

Center: $(1, -3)$, radius: 3

61. $\left(x - \tfrac{1}{2}\right)^2 + \left(y - \tfrac{1}{2}\right)^2 = \dfrac{9}{4}$

Center: $\left(\tfrac{1}{2}, \tfrac{1}{2}\right)$, radius: $\dfrac{3}{2}$

63. $y = 225{,}000 - 20{,}000t, \ 0 \le t \le 8$

65. (a)

(c)

(b) $2x + 2y = \dfrac{1040}{3}$

$2y = \dfrac{1040}{3} - 2x$

$y = \dfrac{520}{3} - x$

$A = xy = x\left(\dfrac{520}{3} - x\right)$

(d) When $x = y = 86\tfrac{2}{3}$ yards, the area is a maximum of $7511\tfrac{1}{9}$ square yards.

(e) A regulation NFL playing field is 120 yards long and $53\tfrac{1}{3}$ yards wide. The actual area is 6400 square yards.

67. $y = -0.0025t^2 + 0.574t + 44.25, \ 20 \le t \le 100$

(a) and (b)

Year $(20 \leftrightarrow 1920)$

(c) For the year 1948, let $t = 48$: $y \approx 66.0$ years.

(d) For the year 2005, let $t = 105$: $y \approx 77.0$ years.

For the year 2010, let $t = 110$: $y \approx 77.1$ years.

(e) No. The graph reaches a maximum of $y \approx 77.2$ years when $t \approx 114.8$, or during the year 2014. After this time, the model has life expectancy decreasing, which is not realistic.

69. False. A graph is symmetric with respect to the x-axis if, whenever (x, y) is on the graph, $(x, -y)$ is also on the graph.

71. The viewing window is incorrect. Change the viewing window. Examples will vary. For example, $y = x^2 + 20$ will not appear in the standard window setting.

73. $9x^5 + 4x^3 - 7$

Terms: $9x^5, 4x^3, -7$

75. $\sqrt{18x} - \sqrt{2x} = 3\sqrt{2x} - \sqrt{2x} = 2\sqrt{2x}$

77. $\sqrt[6]{t^2} = t^{2/6} = |t|^{1/3} = \sqrt[3]{|t|}$

Section 1.2   Linear Equations in One Variable

- You should know how to solve linear equations.

  $ax + b = 0, a \ne 0$

- An identity is an equation whose solution consists of every real number in its domain.

- To solve an equation you can:

  (a) Add or subtract the same quantity from both sides.

  (b) Multiply or divide both sides by the same nonzero quantity.

  (c) Remove all symbols of grouping and all fractions.

  (d) Combine like terms.

  (e) Interchange the two sides.

- Check the answer!

- A "solution" that does not satisfy the original equation is called an extraneous solution.

- Be able to find intercepts algebraically.

Vocabulary Check

1. equation

2. solve

3. identities; conditional

4. $ax + b = 0$

5. extraneous

1. $5x - 3 = 3x + 5$

   (a) $5(0) - 3 \stackrel{?}{=} 3(0) + 5$

          $-3 \neq 5$

      $x = 0$ *is not* a solution.

   (c) $5(4) - 3 \stackrel{?}{=} 3(4) + 5$

          $17 = 17$

      $x = 4$ *is* a solution.

   (b) $5(-5) - 3 \stackrel{?}{=} 3(-5) + 5$

          $-28 \neq -10$

      $x = -5$ *is not* a solution.

   (d) $5(10) - 3 \stackrel{?}{=} 3(10) + 5$

          $47 \neq 35$

      $x = 10$ *is not* a solution.

3. $3x^2 + 2x - 5 = 2x^2 - 2$

   (a) $3(-3)^2 + 2(-3) - 5 \stackrel{?}{=} 2(-3)^2 - 2$

       $27 - 6 - 5 \stackrel{?}{=} 18 - 2$

            $16 = 16$

      $x = -3$ *is* a solution.

   (c) $3(4)^2 + 2(4) - 5 \stackrel{?}{=} 2(4)^2 - 2$

      $48 + 8 - 5 \stackrel{?}{=} 32 - 2$

           $51 \neq 30$

      $x = 4$ *is not* a solution.

   (b) $3(1)^2 + 2(1) - 5 \stackrel{?}{=} 2(1)^2 - 2$

       $3 + 2 - 5 \stackrel{?}{=} 2 - 2$

           $0 = 0$

      $x = 1$ *is* a solution.

   (d) $3(-5)^2 + 2(-5) - 5 \stackrel{?}{=} 2(-5)^2 - 2$

       $75 - 10 - 5 \stackrel{?}{=} 50 - 2$

           $60 \neq 48$

      $x = -5$ *is not* a solution.

5. $\dfrac{5}{2x} - \dfrac{4}{x} = 3$

   (a) $\dfrac{5}{2(-1/2)} - \dfrac{4}{(-1/2)} \stackrel{?}{=} 3$

          $-5 + 8 \stackrel{?}{=} 3$

              $3 = 3$

      $x = -\frac{1}{2}$ *is* a solution.

   (c) $\dfrac{5}{2(0)} - \dfrac{4}{0}$ is undefined.

      $x = 0$ *is not* a solution.

   (b) $\dfrac{5}{2(4)} - \dfrac{4}{4} \stackrel{?}{=} 3$

          $\dfrac{5}{8} - 1 \stackrel{?}{=} 3$

           $-\dfrac{3}{8} \neq 3$

      $x = 4$ *is not* a solution.

   (d) $\dfrac{5}{2(1/4)} - \dfrac{4}{1/4} \stackrel{?}{=} 3$

        $10 - 16 \stackrel{?}{=} 3$

           $-6 \neq 3$

      $x = \frac{1}{4}$ *is not* a solution.

7. $\sqrt{3x - 2} = 4$

   (a) $\sqrt{3(3) - 2} \stackrel{?}{=} 4$

       $\sqrt{7} \neq 4$

    $x = 3$ *is not* a solution.

   (b) $\sqrt{3(2) - 2} \stackrel{?}{=} 4$

       $\sqrt{4} \neq 4$

    $x = 2$ *is not* a solution.

   (c) $\sqrt{3(9) - 2} \stackrel{?}{=} 4$

       $\sqrt{25} \neq 4$

    $x = 9$ *is not* a solution.

   (d) $\sqrt{3(-6) - 2} \stackrel{?}{=} 4$

       $\sqrt{-20} \neq 4$

    $x = -6$ *is not* a solution.

9. $6x^2 - 11x - 35 = 0$

(a) $6\left(-\frac{5}{3}\right)^2 - 11\left(-\frac{5}{3}\right) - 35 \overset{?}{=} 0$

$\quad\quad \frac{50}{3} + \frac{55}{3} - \frac{105}{3} \overset{?}{=} 0$

$\quad\quad\quad\quad\quad\quad\quad\quad 0 = 0$

$x = -\frac{5}{3}$ *is* a solution.

(c) $6\left(\frac{7}{2}\right)^2 - 11\left(\frac{7}{2}\right) - 35 \overset{?}{=} 0$

$\quad\quad \frac{147}{2} - \frac{77}{2} - \frac{70}{2} \overset{?}{=} 0$

$\quad\quad\quad\quad\quad\quad\quad 0 = 0$

$x = \frac{7}{2}$ *is* a solution.

(b) $6\left(-\frac{2}{7}\right)^2 - 11\left(-\frac{2}{7}\right) - 35 \overset{?}{=} 0$

$\quad\quad \frac{24}{49} + \frac{154}{49} - \frac{1715}{49} \overset{?}{=} 0$

$\quad\quad\quad\quad\quad\quad\quad -\frac{1537}{49} \neq 0$

$x = -\frac{2}{7}$ *is not* a solution.

(d) $6\left(\frac{5}{3}\right)^2 - 11\left(\frac{5}{3}\right) - 35 \overset{?}{=} 0$

$\quad\quad \frac{50}{3} - \frac{55}{3} - \frac{105}{3} \overset{?}{=} 0$

$\quad\quad\quad\quad\quad\quad -\frac{110}{3} \neq 0$

$x = \frac{5}{3}$ *is not* a solution.

11. $2(x - 1) = 2x - 2$ is an *identity* by the Distributive Property. It is true for all real values of x.

13. $-6(x - 3) + 5 = -2x + 10$ is *conditional*. There are real values of x for which the equation is not true.

15. $4(x + 1) - 2x = 4x + 4 - 2x = 2x + 4 = 2(x + 2)$

This is an *identity* by simplification. It is true for all real values of x.

17. $(x - 4)^2 - 11 = x^2 - 8x + 16 - 11 = x^2 - 8x + 5$

Thus, $x^2 - 8x + 5 = (x - 4)^2 - 11$ is an *identity* by simplification. It is true for all real values of x.

19. $3 + \dfrac{1}{x + 1} = \dfrac{4x}{x + 1}$ is *conditional*. There are real values of x for which the equation is not true.

21.

$4x + 32 = 83$	Original equation
$4x + 32 - 32 = 83 - 32$	Subtract 32 from both sides.
$4x = 51$	Simplify.
$\dfrac{4x}{4} = \dfrac{51}{4}$	Divide both sides by 4.
$x = \dfrac{51}{4}$	Simplify.

23.
$$x + 11 = 15$$
$$x + 11 - 11 = 15 - 11$$
$$x = 4$$

25.
$$7 - 2x = 25$$
$$7 - 7 - 2x = 25 - 7$$
$$-2x = 18$$
$$\frac{-2x}{-2} = \frac{18}{-2}$$
$$x = -9$$

27.
$$8x - 5 = 3x + 20$$
$$8x - 3x - 5 = 3x - 3x + 20$$
$$5x - 5 = 20$$
$$5x - 5 + 5 = 20 + 5$$
$$5x = 25$$
$$\frac{5x}{5} = \frac{25}{5}$$
$$x = 5$$

29. $2(x + 5) - 7 = 3(x - 2)$
$$2x + 10 - 7 = 3x - 6$$
$$2x + 3 = 3x - 6$$
$$2x - 3x + 3 = 3x - 3x - 6$$
$$-x + 3 = -6$$
$$-x + 3 - 3 = -6 - 3$$
$$-x = -9$$
$$x = 9$$

31. $x - 3(2x + 3) = 8 - 5x$
$$x - 6x - 9 = 8 - 5x$$
$$-5x - 9 = 8 - 5x$$
$$-5x + 5x - 9 = 8 - 5x + 5x$$
$$-9 \neq 8$$

No solution

33.
$$\frac{5x}{4} + \frac{1}{2} = x - \frac{1}{2}$$
$$4\left(\frac{5x}{4} + \frac{1}{2}\right) = 4\left(x - \frac{1}{2}\right)$$
$$4\left(\frac{5x}{4}\right) + 4\left(\frac{1}{2}\right) = 4(x) - 4\left(\frac{1}{2}\right)$$
$$5x + 2 = 4x - 2$$
$$x = -4$$

35.
$$\tfrac{3}{2}(z + 5) - \tfrac{1}{4}(z + 24) = 0$$
$$4\left[\tfrac{3}{2}(z + 5) - \tfrac{1}{4}(z + 24)\right] = 4(0)$$
$$4\left(\tfrac{3}{2}\right)(z + 5) - 4\left(\tfrac{1}{4}\right)(z + 24) = 4(0)$$
$$6(z + 5) - (z + 24) = 0$$
$$6z + 30 - z - 24 = 0$$
$$5z = -6$$
$$z = -\tfrac{6}{5}$$

37.
$$0.25x + 0.75(10 - x) = 3$$
$$0.25x + 7.5 - 0.75x = 3$$
$$-0.50x + 7.5 = 3$$
$$-0.50x = -4.5$$
$$x = 9$$

39.

$3(x - 1) = 4$	or	$3(x - 1) = 4$
$x - 1 = \tfrac{4}{3}$		$3x - 3 = 4$
$x = \tfrac{4}{3} + 1$		$3x = 7$
$x = \tfrac{7}{3}$		$x = \tfrac{7}{3}$

The second way is easier since you are not working with fractions until the end of the solution.

41.

$\tfrac{1}{3}(x + 2) = 5$	or	$\tfrac{1}{3}(x + 2) = 5$
$3\left(\tfrac{1}{3}\right)(x + 2) = 3(5)$		$\tfrac{1}{3}x + \tfrac{2}{3} = 5$
$x + 2 = 15$		$\tfrac{1}{3}x = 5 - \tfrac{2}{3}$
$x = 13$		$\tfrac{1}{3}x = \tfrac{13}{3}$
		$x = 3\left(\tfrac{13}{3}\right)$
		$x = 13$

The first way is easier. The fraction is eliminated in the first step.

43.
$$x + 8 = 2(x - 2) - x$$
$$x + 8 = 2x - 4 - x$$
$$x + 8 = x - 4$$
$$8 \neq -4$$

Contradiction; no solution

45.
$$\frac{100 - 4x}{3} = \frac{5x + 6}{4} + 6$$
$$12\left(\frac{100 - 4x}{3}\right) = 12\left(\frac{5x + 6}{4}\right) + 12(6)$$
$$4(100 - 4x) = 3(5x + 6) + 72$$
$$400 - 16x = 15x + 18 + 72$$
$$-31x = -310$$
$$x = 10$$

47.
$$\frac{5x - 4}{5x + 4} = \frac{2}{3}$$
$$3(5x - 4) = 2(5x + 4)$$
$$15x - 12 = 10x + 8$$
$$5x = 20$$
$$x = 4$$

49.
$$10 - \frac{13}{x} = 4 + \frac{5}{x}$$
$$\frac{10x - 13}{x} = \frac{4x + 5}{x}$$
$$10x - 13 = 4x + 5$$
$$6x = 18$$
$$x = 3$$

51.
$$3 = 2 + \frac{2}{z + 2}$$
$$3(z + 2) = \left(2 + \frac{2}{z + 2}\right)(z + 2)$$
$$3z + 6 = 2z + 4 + 2$$
$$z = 0$$

53.
$$\frac{x}{x + 4} + \frac{4}{x + 4} + 2 = 0$$
$$\frac{x + 4}{x + 4} + 2 = 0$$
$$1 + 2 = 0$$
$$3 \neq 0$$

Contradiction; no solution

55. $\dfrac{2}{(x-4)(x-2)} = \dfrac{1}{x-4} + \dfrac{2}{x-2}$ Multiply both sides by $(x-4)(x-2)$.

$$2 = 1(x-2) + 2(x-4)$$

$$2 = x - 2 + 2x - 8$$

$$2 = 3x - 10$$

$$12 = 3x$$

$$4 = x$$

A check reveals that $x = 4$ is an extraneous solution—it makes the denominator zero. There is no real solution.

57. $\dfrac{1}{x-3} + \dfrac{1}{x+3} = \dfrac{10}{x^2-9}$

$\dfrac{1}{x-3} + \dfrac{1}{x+3} = \dfrac{10}{(x+3)(x-3)}$ Multiply both sides by $(x+3)(x-3)$.

$$1(x+3) + 1(x-3) = 10$$

$$2x = 10$$

$$x = 5$$

59. $\dfrac{3}{x^2-3x} + \dfrac{4}{x} = \dfrac{1}{x-3}$

$\dfrac{3}{x(x-3)} + \dfrac{4}{x} = \dfrac{1}{x-3}$ Multiply both sides by $x(x-3)$.

$$3 + 4(x-3) = x$$

$$3 + 4x - 12 = x$$

$$3x = 9$$

$$x = 3$$

A check reveals that $x = 3$ is an extraneous solution since it makes the denominator zero, so there is no solution.

61. $(x+2)^2 + 5 = (x+3)^2$

$$x^2 + 4x + 4 + 5 = x^2 + 6x + 9$$

$$4x + 9 = 6x + 9$$

$$-2x = 0$$

$$x = 0$$

63. $(x+2)^2 - x^2 = 4(x+1)$

$$x^2 + 4x + 4 - x^2 = 4x + 4$$

$$4 = 4$$

The equation is an identity; every real number is a solution.

65. $y = 2(x-1) - 4$

$$0 = 2(x-1) - 4$$

$$0 = 2x - 2 - 4$$

$$0 = 2x - 6$$

$$6 = 2x$$

$$3 = x$$

$$x = 3$$

The x-intercept is at 3. The solution to $0 = 2(x-1) - 4$ and the x-intercept of $y = 2(x-1) - 4$ are the same. They are both $x = 3$. The x-intercept is $(3, 0)$.

67. $y = 20 - (3x - 10)$

$$0 = 20 - (3x - 10)$$

$$0 = 20 - 3x + 10$$

$$0 = 30 - 3x$$

$$3x = 30$$

$$x = 10$$

The x-intercept is at 10. The solution to $0 = 20 - (3x - 10)$ and the x-intercept of $y = 20 - (3x - 10)$ are the same. They are both $x = 10$. The x-intercept is $(10, 0)$.

69. $y = -38 + 5(9 - x)$

$$0 = -38 + 5(9 - x)$$
$$0 = -38 + 45 - 5x$$
$$0 = 7 - 5x$$
$$5x = 7$$
$$x = \frac{7}{5}$$

The x-intercept is at $\frac{7}{5}$. The solution to $0 = -38 + 5(9 - x)$ and the x-intercept of $y = -38 + 5(9 - x)$ are the same. They are both $x = \frac{7}{5}$. The x-intercept is $\left(\frac{7}{5}, 0\right)$.

71. $y = 12 - 5x$

x-intercept: $0 = 12 - 5x \implies 5x = 12 \implies x = \frac{12}{5}$

y-intercept: $y = 12 - 5(0) \implies y = 12$

The x-intercept is $\left(\frac{12}{5}, 0\right)$ and the y-intercept is $(0, 12)$.

73. $y = -3(2x + 1)$

x-intercept: $0 = -3(2x + 1) \implies 0 = 2x + 1 \implies x = -\frac{1}{2}$

y-intercept: $y = -3(2(0) + 1) \implies y = -3$

The x-intercept is $\left(-\frac{1}{2}, 0\right)$ and the y-intercept is $(0, -3)$.

75. $2x + 3y = 10$

x-intercept: $2x + 3(0) = 10 \implies 2x = 10 \implies x = 5$

y-intercept: $2(0) + 3y = 10 \implies 3y = 10 \implies y = \frac{10}{3}$

The x-intercept is $(5, 0)$ and the y-intercept is $\left(0, \frac{10}{3}\right)$.

77. $\frac{2x}{5} + 8 - 3y = 0 \implies 2x + 40 - 15y = 0$ Multiply both sides by 5.

x-intercept: $2x + 40 - 15(0) = 0 \implies 2x + 40 = 0 \implies x = -20$

y-intercept: $2(0) + 40 - 15y = 0 \implies 40 - 15y = 0 \implies y = \frac{40}{15} = \frac{8}{3}$

The x-intercept is $(-20, 0)$ and the y-intercept is $\left(0, \frac{8}{3}\right)$.

79. $4y - 0.75x + 1.2 = 0$

x-intercept: $4(0) - 0.75x + 1.2 = 0 \implies -0.75x + 1.2 = 0 \implies x = \frac{1.2}{0.75} = 1.6$

y-intercept: $4y - 0.75(0) + 1.2 = 0 \implies 4y + 1.2 = 0 \implies y = \frac{-1.2}{4} = -0.3$

The x-intercept is $(1.6, 0)$ and the y-intercept is $(0, -0.3)$.

81. $4(x + 1) - ax = x + 5$

$4x + 4 - ax = x + 5$

$3x - ax = 1$

$x(3 - a) = 1$

$x = \frac{1}{3 - a}, a \neq 3$

83. $6x + ax = 2x + 5$

$4x + ax = 5$

$x(4 + a) = 5$

$x = \frac{5}{4 + a}, a \neq -4$

85. $19x + \frac{1}{2}ax = x + 9$

$18x + \frac{1}{2}ax = 9$ Multiply both sides by 2.

$36x + ax = 18$

$x(36 + a) = 18$

$x = \frac{18}{36 + a}, a \neq -36$

87. $-2ax + 6(x + 3) = -4x + 1$

$-2ax + 6x + 18 = -4x + 1$

$-2ax + 10x + 18 = 1$

$-2ax + 10x = -17$

$x(-2a + 10) = -17$

$x = \frac{-17}{-2a + 10} = \frac{-17}{10 - 2a}, a \neq 5$

89. $0.275x + 0.725(500 - x) = 300$

$0.275x + 362.5 - 0.725x = 300$

$-0.45x = -62.5$

$x = \dfrac{62.5}{0.45}$

≈ 138.889

91. $\dfrac{2}{7.398} - \dfrac{4.405}{x} = \dfrac{1}{x}$ Multiply both sides by $7.398x$.

$2x - (4.405)(7.398) = 7.398$

$2x = (4.405)(7.398) + 7.398$

$2x = (5.405)(7.398)$

$x = \dfrac{(5.405)(7.398)}{2} \approx 19.993$

93. $471 = 2\pi(25) + 2\pi(5h)$

$471 = 50\pi + 10\pi h$

$471 - 50\pi = 10\pi h$

$h = \dfrac{471 - 50\pi}{10\pi} = \dfrac{471 - 50(3.14)}{10(3.14)} = 10$

$h = 10$ feet

95. (a) Female: $y = 0.432x - 10.44$

For $y = 16$: $16 = 0.432x - 10.44$

$26.44 = 0.432x$

$\dfrac{26.44}{0.432} = x$

$x \approx 61.2$ inches

(b) Male: $y = 0.449x - 12.15$

For $y = 19$: $19 = 0.449x - 12.15$

$31.15 = 0.449x$

$69.4 \approx x$

Yes, it is likely that both bones came from the same person because the estimated height of a male with a 19-inch thigh bone is 69.4 inches.

(d) $0.432x - 10.44 = 0.449x - 12.15$

$1.71 = 0.017x$

$x \approx 100.59$ inches

It is unlikely that a female would be over 8 feet tall, so if a femur of this length was found, it most likely belonged to a very tall male.

(c)

Height x	Female Femur Length	Male Femur Length
60	15.48	14.79
70	19.80	19.28
80	24.12	23.77
90	28.44	28.26
100	32.76	32.75
110	37.08	37.24

The lengths of the male and female femurs are approximately equal when the lengths are 100 inches.

97. $y = 1.64t + 36.8, \ -1 \le t \le 12$

(a)

The y-intercept is $(0, 36.8)$.

(b) Let $t = 0$: $y = 1.64(0) + 36.8 = 36.8$

y-intercept: $(0, 36.8)$

(c) $65 = 1.64t + 36.8$

$28.2 = 1.64t$

$t = \dfrac{28.2}{1.64} \approx 17.2$

This corresponds with the year 2017. Explanations will vary.

99. $10,000 = 0.32m + 2500$

$7500 = 0.32m$

$\dfrac{7500}{0.32} = m$

$m = 23,437.5$ miles

101. False. $x(3 - x) = 10 \implies 3x - x^2 = 10$

This is a quadratic equation. The equation cannot be written in the form $ax + b = 0$.

103. Equivalent equations are derived from the substitution principle and simplification techniques. They have the same solution(s).

$2x + 3 = 8$ and $2x = 5$ are equivalent equations.

105. (a)

x	-1	0	1	2	3	4
$3.2x - 5.8$	-9	-5.8	-2.6	0.6	3.8	7

(b) Since the sign changes from negative at 1 to positive at 2, the root is somewhere between 1 and 2.

$1 < x < 2$

(c)

x	1.5	1.6	1.7	1.8	1.9	2
$3.2x - 5.8$	-1	-0.68	-0.36	-0.04	0.28	0.6

(d) Since the sign changes from negative at 1.8 to positive at 1.9, the root is somewhere between 1.8 and 1.9.

$1.8 < x < 1.9$

To improve accuracy, evaluate the expression at subintervals within this interval and determine where the sign changes.

107. $\dfrac{x^2 + 5x - 36}{2x^2 + 17x - 9} = \dfrac{(x + 9)(x - 4)}{(2x - 1)(x + 9)} = \dfrac{x - 4}{2x - 1}, \ x \neq -9$

109. $y = 3x - 5$

Intercepts: $(0, -5), \left(\dfrac{5}{3}, 0\right)$

111. $y = -x^2 - 5x = -x(x + 5)$

Intercepts: $(0, 0), (-5, 0)$

Section 1.3   Modeling with Linear Equations

- ■ You should be able to set up mathematical models to solve problems.

- ■ You should be able to translate key words and phrases.

    (a) Equality:

    Equals, equal to, is, are, was, will be, represents

    (b) Addition:

    Sum, plus, greater, increased by, more than, exceeds, total of

    (c) Subtraction:

    Difference, minus, less than, decreased by, subtracted from, reduced by, the remainder

    (d) Multiplication:

    Product, multiplied by, twice, times, percent of

    (e) Division:

    Quotient, divided by, ratio, per

    (f) Consecutive:

    Next, subsequent

- ■ You should know the following formulas:

    (a) Perimeter:

    1. Square: $P = 4s$

    2. Rectangle: $P = 2l + 2w$

    3. Circle: $C = 2\pi r$

    4. Triangle: $P = a + b + c$

    (b) Area:

    1. Square: $A = s^2$

    2. Rectangle: $A = lw$

    3. Circle: $A = \pi r^2$

    4. Triangle: $A = \left(\frac{1}{2}\right)bh$

    (c) Volume

    1. Cube: $V = s^3$

    2. Rectangular solid: $V = lwh$

    3. Cylinder: $V = \pi r^2 h$

    4. Sphere: $V = \left(\frac{4}{3}\right)\pi r^3$

    (d) Simple Interest: $I = Prt$

    (e) Compound Interest: $A = P\left(1 + \frac{r}{n}\right)^{nt}$

    (f) Distance: $d = rt$

    (g) Temperature: $F = \frac{9}{5}C + 32$

- ■ You should be able to solve word problems. Study the examples in the text carefully.

Vocabulary Check

1. mathematical modeling

2. verbal model; algebraic equation

3. $A = \pi r^2$

4. $P = 2l + 2w$

5. $V = s^3$

6. $V = \pi r^2 h$

7. $A = P\left(1 + \frac{r}{12}\right)^{12t}$

8. $I = Prt$

1. $x + 4$

The sum of a number and 4
A number increased by 4

3. $\frac{u}{5}$

The ratio of a number and 5
The quotient of a number and 5
A number divided by 5

5. $\frac{y - 4}{5}$

The difference of a number and 4 is divided by 5.
A number decreased by 4 is divided by 5.

7. $-3(b + 2)$

The product of -3 and the sum of a number and 2.
Negative 3 is multiplied by a number increased by 2.

9. $12x(x - 5)$

The difference of a number and 5 is multiplied by 12 times the number. 12 is multiplied by a number and that product is multiplied by the number decreased by 5.

11. *Verbal Model:* (Sum) = (first number) + (second number)

Labels: Sum = S, first number = n, second number = $n + 1$

Expression: $S = n + (n + 1) = 2n + 1$

13. *Verbal Model:* Product = (first odd integer)(second odd integer)

Labels: Product = P, first odd integer = $2n - 1$, second odd integer = $2n - 1 + 2 = 2n + 1$

Expression: $P = (2n - 1)(2n + 1) = 4n^2 - 1$

15. *Verbal Model:* (Distance) = (rate) $\cdot$ (time)

Labels: Distance = d, rate = 50 mph, time = t

Expression: $d = 50t$

17. *Verbal Model:* (Amount of acid) = 20% $\cdot$ (amount of solution)

Labels: Amount of acid (in gallons) = A, amount of solution (in gallons) = x

Expression: $A = 0.20x$

19. *Verbal Model:* Perimeter = 2(width) + 2(length)

Labels: Perimeter = P, width = x,
               length = 2(width) = $2x$

Expression: $P = 2x + 2(2x) = 6x$

21. *Verbal Model:* (Total cost) = (unit cost)(number of units) + (fixed cost)

Labels: Total cost = C, fixed cost = \$1200, unit cost = \$25, number of units = x

Expression: $C = 25x + 1200$

23. *Verbal Model:* Thirty percent of the list price L

Expression: $0.30L$

25. *Verbal Model:* percent of 500 that is represented by the number N

Equation: $N = p(500)$, p is in decimal form

27.

Area = Area of top rectangle + Area of bottom rectangle

$A = 4x + 8x = 12x$

29. *Verbal Model:* Sum = (first number) + (second number)

   Labels: Sum = 525, first number = n, second number = $n + 1$

   Equations: $525 = n + (n + 1)$

                    $525 = 2n + 1$

                    $524 = 2n$

                    $n = 262$

Answer: First number = n = 262, second number = $n + 1$ = 263

31. *Verbal Model:* Difference = (one number) − (another number)

   Labels: Difference = 148, one number = $5x$, another number = x

   Equation: $148 = 5x - x$

                    $148 = 4x$

                    $x = 37$

                    $5x = 185$

Answer: The two numbers are 37 and 185.

33. *Verbal Model:* Product = (smaller number) · (larger number) = (smaller number)² − 5

   Labels: Smaller number = n, larger number = $n + 1$

   Equation: $n(n + 1) = n^2 - 5$

                    $n^2 + n = n^2 - 5$

                    $n = -5$

Answer: Smaller number = $n = -5$, larger number = $n + 1 = -4$

35. x = percent · number

   = (30%)(45)

   = 0.30(45)

   = 13.5

37. 432 = percent · 1600

   $432 = p(1600)$

   $\frac{432}{1600} = p$

   $p = 0.27 = 27\%$

39. $12 = \frac{1}{2}\% \cdot$ number

   $12 = 0.005x$

   $\frac{12}{0.005} = x$

   $x = 2400$

41. *Verbal Model:* Loan payments = 58.6% · Annual income

   Labels: Loan payments = 13,077.75

                    Annual income = I

   Equation: $13,077.75 = 0.586I$

                    $I \approx 22,316.98$

The family's annual income is $22,316.98.

43. *Verbal Model:* (2002 price for a gallon of unleaded gasoline) = (percentage increase)(1990 price for a gallon of unleaded gasoline)

+ (1990 price for a gallon of unleaded gasoline)

     Labels: 2002 price for a gallon of unleaded gasoline: $1.36

                    1990 price for a gallon of unleaded gasoline: $1.16

                    Percentage increase: p

     Equation: $1.36 = 1.16p + 1.16$

                    $0.20 = 1.16p$

                    $0.172 \approx p$

    Answer: Percentage increase $\approx 17.2\%$

45. *Verbal Model:* (2002 price for a pound of tomatoes) = (percentage increase)(1990 price for a pound of tomatoes)

+ (1990 price for a pound of tomatoes)

     Labels: 2002 price for a pound of tomatoes: $1.66

                    1990 price for a pound of tomatoes: $0.86

                    Percentage increase: p

     Equation: $1.66 = 0.86p + 0.86$

                    $0.80 = 0.86p$

                    $0.930 \approx p$

    Answer: Percentage increase $\approx 93\%$

47. *Verbal Model:* (Sale price) = (list price) − (discount)

     Labels: Sale price = $1210.75, list price = L, discount = $0.165L$

     Equation: $1210.75 = L - 0.165L$

                    $1210.75 = 0.835L$

                    $1450 = L$

    Answer: The list price of the pool is $1450.

49. (a)

    (b) $l = 1.5w$

       $P = 2l + 2w$

          $= 2(1.5w) + 2w$

          $= 5w$

    (c) $25 = 5w$

       $5 = w$

      Width: $w = 5$ meters

      Length: $l = 1.5w = 7.5$ meters

      Dimensions: 7.5 meters $\times$ 5 meters

51. *Verbal Model:* Average $= \dfrac{(\text{test \#1}) + (\text{test \#2}) + (\text{test \#3}) + (\text{test \#4})}{4}$

     Labels: Average = 90, test #1 = 87, test #2 = 92, test #3 = 84, test #4 = x

     Equation: $90 = \dfrac{87 + 92 + 84 + x}{4}$

                    $360 = 263 + x$

                    $x = 97$

    Answer: You must score 97 (or better) on test #4 to earn an A for the course.

53. Rate = $\dfrac{\text{distance}}{\text{time}} = \dfrac{50 \text{ kilometers}}{\frac{1}{2} \text{ hour}} = 100$ kilometers/hour

Total time = $\dfrac{\text{total distance}}{\text{rate}} = \dfrac{300 \text{ kilometers}}{100 \text{ kilometers/hour}} = 3$ hours

55. (time on first part) + (time on second part) = (Total time)

$$t_1 + t_2 = T$$

$$\frac{d_1}{r_1} + \frac{d_2}{r_2} = T$$

$$\frac{x}{58} + \frac{317 - x}{52} = 5.75 \qquad \text{Multiply both sides by } 58(52).$$

$$52x + 58(317 - x) = 5.75(58)(52)$$

$$52x + 18{,}386 - 58x = 17{,}342$$

$$-6x = -1044$$

$$x = 174 \text{ miles}$$

$$t_1 = \frac{174}{58} = 3 \text{ hours}$$

$$t_2 = \frac{317 - 174}{52} = 2.75 \text{ hours}$$

The salesman averaged 58 miles per hour for 3 hours and 52 miles per hour for 2 hours and 45 minutes.

57. (a) Time for the first family: $t_1 = \dfrac{d}{r_1} = \dfrac{160}{42} \approx 3.81$ hours

Time for the other family: $t_2 = \dfrac{d}{r_2} = \dfrac{160}{50} = 3.2$ hours

(b) $t = \dfrac{d}{r} = \dfrac{100}{42 + 50} = \dfrac{100}{92} \approx 1.08$ hours

(c) $d = rt = 42\left(\dfrac{160}{42} - \dfrac{160}{50}\right) = 25.6$ miles

59. *Verbal Model:* time = $\dfrac{\text{distance}}{\text{rate}}$

Labels: Let x = wind speed, then the rate to the city = $600 + x$, the rate from the city = $600 - x$, the distance to the city = 1500 kilometers, the distance traveled so far in the return trip = $1500 - 300 = 1200$ kilometers.

Equation: $\dfrac{1500}{600 + x} = \dfrac{1200}{600 - x}$

$$1500(600 - x) = 1200(600 + x)$$

$$900{,}000 - 1500x = 720{,}000 + 1200x$$

$$180{,}000 = 2700x$$

$$66\frac{2}{3} = x$$

Wind speed: $66\frac{2}{3}$ kilometers per hour

61. *Verbal Model:* time = $\dfrac{\text{distance}}{\text{rate}}$

Equation: $t = \dfrac{3.84 \times 10^8 \text{ meters}}{3.0 \times 10^8 \text{ meters per second}}$

$t = 1.28$ seconds

The radio wave travels from Mission Control to the moon in 1.28 seconds.

63. *Verbal Model:* $\dfrac{\text{(height of building)}}{\text{(length of building's shadow)}} = \dfrac{\text{(height of stake)}}{\text{(length of stake's shadow)}}$

   Label: Let h = height of the building in feet.

   Equation: $\dfrac{h \text{ feet}}{87 \text{ feet}} = \dfrac{4 \text{ feet}}{1/3 \text{ foot}}$

   $\dfrac{h}{87} = \dfrac{4}{1/3}$

   $\dfrac{1}{3}h = 348$

   $h = 1044 \text{ feet}$

The Chrysler building is 1044 feet tall.

65. *Verbal Model:* $\dfrac{\text{(height of silo)}}{\text{(length of silo's shadow)}} = \dfrac{\text{(height of person)}}{\text{(height of person's shadow)}}$

   Label: Let x = length of person's shadow.

   Equation: $\dfrac{50}{32 + x} = \dfrac{6}{x}$

   $50x = 6(32 + x)$

   $50x = 192 + 6x$

   $44x = 192$

   $x \approx 4.36 \text{ feet}$

67. *Verbal Model:*   (Interest in 3% fund) + $\left(\text{interest in } 4\tfrac{1}{2}\% \text{ fund}\right)$ = (total interest)

   Labels: Let x = amount in the 3% fund. Then $25,000 - x$ = amount in the $4\tfrac{1}{2}\%$ fund.

   Equation: $1000 = 0.03x + 0.045(25,000 - x)$

   $1000 = 0.03x + 1125 - 0.045x$

   $-125 = -0.015x$

   $x \approx \$8333.33 \text{ at } 3\%$

   $25,000 - x \approx \$16,666.67 \text{ at } 4\tfrac{1}{2}\%$

69. *Verbal Model:*   (profit on minivans) + (profit on SUVs) = (total profit)

   Labels: Let x = amount invested in minivans. Then, $600,000 - x$ = amount invested in SUVs.

   Equation: $0.24x + 0.28(600,000 - x) = 0.25(600,000)$

   $0.24x + 168,000 - 0.28x = 150,000$

   $-0.04x = -18,000$

   $x = 450,000$

$\$450,000$ is invested in minivans and $600,000 - x = \$150,000$ is invested in SUVs.

71. *Verbal Model:*   (Final concentration)(Amount) = (Solution 1 concentration)(Amount) + (Solution 2 concentration)(Amount)

   Label: Let x = amount of 100% concentrate

   Equation: $0.75(55) = 0.40(55 - x) + 1.00x$

   $41.25 = 22 - 0.40x + 1.00x$

   $41.25 = 0.60x + 22$

   $x \approx 32.1 \text{ gallons}$

Approximately 32.1 gallons of the 100% concentrate will be needed.

73. *Verbal Model:* (price per pound of peanuts)(number of pounds of peanuts) + (price per pound of walnuts)(number of pounds of walnuts) = (price per pound of nut mixture)(number of pounds of nut mixture)

Labels: Let x = number of pounds of \$2.49 peanuts. Then $100 - x$ = number of pounds of \$3.89 walnuts.

Equation: $2.49x + 3.89(100 - x) = 3.19(100)$

$$2.49x + 389 - 3.89x = 319$$

$$-1.40x = -70$$

$$x = \frac{-70}{-1.40}$$

$$x = 50 \text{ pounds of \$2.49 peanuts}$$

$$100 - x = 50 \text{ pounds of \$3.89 walnuts}$$

Use 50 pounds of peanuts and 50 pounds of walnuts.

75. *Verbal Model:* Total cost = (Fixed cost) + (Variable cost per unit) · (Number of units)

Labels: Total cost = \$85,000, variable costs = \$9.30x, fixed costs = \$10,000

Equation: $\$85,000 = \$10,000 + \$9.30x$

$$x = \frac{75,000}{9.3} \approx 8064.52 \text{ units}$$

At most the company can manufacture 8064 units.

77. $A = \frac{1}{2}bh$

$2A = bh$

$\frac{2A}{b} = h$

79. $S = C + RC$

$S = C(1 + R)$

$\frac{S}{1 + R} = C$

81. $V = \frac{4}{3}\pi a^2 b$

$\frac{3}{4}V = \pi a^2 b$

$\frac{(3/4)V}{\pi a^2} = b$

$\frac{3V}{4\pi a^2} = b$

83. $h = v_0 t + \frac{1}{2}at^2$

$h - v_0 t = \frac{1}{2}at^2$

$2(h - v_0 t) = at^2$

$\frac{2(h - v_0 t)}{t^2} = a$

85. $C = \dfrac{1}{\dfrac{1}{C_1} + \dfrac{1}{C_2}}$

$\frac{1}{C} = \frac{1}{C_1} + \frac{1}{C_2}$

$\frac{1}{C} - \frac{1}{C_2} = \frac{1}{C_1}$

$\frac{C_2 - C}{CC_2} = \frac{1}{C_1}$

$\frac{CC_2}{C_2 - C} = C_1$

87. $L = a + (n - 1)d$

$L - a = nd - d$

$L - a + d = nd$

$\frac{L - a + d}{d} = n$

89. $W_1 x = W_2(L - x)$

$50x = 75(10 - x)$

$50x = 750 - 75x$

$125x = 750$

$x = 6$ feet from 50-pound child.

91. $V = \dfrac{4}{3}\pi r^3$

$5.96 = \dfrac{4}{3}\pi r^3$

$17.88 = 4\pi r^3$

$\dfrac{17.88}{4\pi} = r^3$

$r = \sqrt[3]{\dfrac{4.47}{\pi}} \approx 1.12$ inches

93. $C = \frac{5}{9}(F - 32)$

When $F = 64.4°$,

$\frac{5}{9}(64.4 - 32) = 18°C.$

95. $F = \frac{9}{5}C + 32$

When $C = 50°$,

$\frac{9}{5}(50) + 32 = 122°F.$

97. False, it should be written as

$\dfrac{z^3 - 8}{z^2 - 9}.$

99. $ax + b = 0 \implies x = -\dfrac{b}{a}$

(a) If $ab > 0$, then a and b have the same sign and $x = -b/a$ is negative.

(b) If $ab < 0$, then a and b have opposite signs and $x = -b/a$ is positive.

101. $(5x^4)(25x^2)^{-1} = \dfrac{5x^4}{25x^2} = \dfrac{x^2}{5}, \ x \neq 0$

103. $\dfrac{3}{x - 5} + \dfrac{2}{5 - x} = \dfrac{3}{x - 5} + \dfrac{(-1)(2)}{x - 5} = \dfrac{3 - 2}{x - 5} = \dfrac{1}{x - 5}$

105. $\dfrac{10}{7\sqrt{3}} = \dfrac{10}{7\sqrt{3}} \cdot \dfrac{\sqrt{3}}{\sqrt{3}} = \dfrac{10\sqrt{3}}{21}$

107. $\dfrac{5}{\sqrt{6} + \sqrt{11}} = \dfrac{5}{\sqrt{6} + \sqrt{11}} \cdot \dfrac{\sqrt{6} - \sqrt{11}}{\sqrt{6} - \sqrt{11}}$

$= \dfrac{5\sqrt{6} - 5\sqrt{11}}{-5}$

$= -\left(\sqrt{6} - \sqrt{11}\right)$

$= \sqrt{11} - \sqrt{6}$

Section 1.4   Quadratic Equations and Applications

■ You should be able to solve a quadratic equation by factoring, if possible.

■ You should be able to solve a quadratic equation of the form $u^2 = d$ by extracting square roots.

■ You should be able to solve a quadratic equation by completing the square.

■ You should know and be able to use the Quadratic Formula: For $ax^2 + bx + c = 0$, $a \neq 0$,

$x = \dfrac{-b \pm \sqrt{b^2 - 4ac}}{2a}.$

■ You should be able to determine the types of solutions of a quadratic equation by checking the discriminant $b^2 - 4ac$.

(a) If $b^2 - 4ac > 0$, there are two distinct real solutions. The graph has two x-intercepts.

(b) If $b^2 - 4ac = 0$, there is one repeated real solution. The graph has one x-intercept.

(c) If $b^2 - 4ac < 0$, there is no real solution. The graph has no x-intercepts.

■ You should be able to use your calculator to solve quadratic equations.

■ You should be able to solve applications involving quadratic equations. Study the examples in the text carefully.

Vocabulary Check

1. quadratic equation

2. factoring, extracting square roots, completing the square, and the Quadratic Formula

3. discriminant

4. position equation; $-16t^2 + v_0 t + s_0$; velocity of the object; initial height of the object

5. Pythagorean Theorem

1. $2x^2 = 3 - 8x$

General form: $2x^2 + 8x - 3 = 0$

3. $(x - 3)^2 = 3$

$x^2 - 6x + 9 = 3$

General form: $x^2 - 6x + 6 = 0$

5. $\frac{1}{5}(3x^2 - 10) = 18x$

$3x^2 - 10 = 90x$

General form: $3x^2 - 90x - 10 = 0$

7. $6x^2 + 3x = 0$

$3x(2x + 1) = 0$

$3x = 0$ or $2x + 1 = 0$

$x = 0$ or $\qquad x = -\frac{1}{2}$

9. $x^2 - 2x - 8 = 0$

$(x - 4)(x + 2) = 0$

$x - 4 = 0$ or $x + 2 = 0$

$x = 4$ or $\qquad x = -2$

11. $x^2 + 10x + 25 = 0$

$(x + 5)^2 = 0$

$x + 5 = 0$

$x = -5$

13. $3 + 5x - 2x^2 = 0$

$(3 - x)(1 + 2x) = 0$

$3 - x = 0$ or $1 + 2x = 0$

$x = 3$ or $\qquad x = -\frac{1}{2}$

15. $x^2 + 4x = 12$

$x^2 + 4x - 12 = 0$

$(x + 6)(x - 2) = 0$

$x + 6 = 0$ or $x - 2 = 0$

$x = -6$ or $\qquad x = 2$

17. $\frac{3}{4}x^2 + 8x + 20 = 0$

$4\left(\frac{3}{4}x^2 + 8x + 20\right) = 4(0)$

$3x^2 + 32x + 80 = 0$

$(3x + 20)(x + 4) = 0$

$3x + 20 = 0$ or $x + 4 = 0$

$x = -\frac{20}{3}$ or $\qquad x = -4$

19. $x^2 + 2ax + a^2 = 0$

$(x + a)^2 = 0$

$x + a = 0$

$x = -a$

21. $x^2 = 49$

$x = \pm 7$

23. $x^2 = 11$

$x = \pm\sqrt{11}$

25. $3x^2 = 81$

$x^2 = 27$

$x = \pm 3\sqrt{3}$

27. $(x - 12)^2 = 16$

$x - 12 = \pm 4$

$x = 12 \pm 4$

$x = 16$ or $x = 8$

29. $(x + 2)^2 = 14$

$x + 2 = \pm\sqrt{14}$

$x = -2 \pm \sqrt{14}$

31. $(2x - 1)^2 = 18$

$2x - 1 = \pm\sqrt{18}$

$2x = 1 \pm 3\sqrt{2}$

$x = \dfrac{1 \pm 3\sqrt{2}}{2}$

33. $(x - 7)^2 = (x + 3)^2$

$x - 7 = \pm(x + 3)$

$x - 7 = x + 3$ or $x - 7 = -x - 3$

$-7 \neq 3$ or $2x = 4$

$x = 2$

The only solution to the equation is $x = 2$.

35. $x^2 + 4x - 32 = 0$

$\qquad x^2 + 4x = 32$

$\qquad x^2 + 4x + 2^2 = 32 + 2^2$

$\qquad (x + 2)^2 = 36$

$\qquad x + 2 = \pm 6$

$\qquad x = -2 \pm 6$

$\qquad x = 4 \quad \text{or} \quad x = -8$

37. $x^2 + 12x + 25 = 0$

$\qquad x^2 + 12x = -25$

$\qquad x^2 + 12x + 6^2 = -25 + 6^2$

$\qquad (x + 6)^2 = 11$

$\qquad x + 6 = \pm\sqrt{11}$

$\qquad x = -6 \pm \sqrt{11}$

39. $\quad 9x^2 - 18x = -3$

$\qquad x^2 - 2x = -\dfrac{1}{3}$

$\qquad x^2 - 2x + 1^2 = -\dfrac{1}{3} + 1^2$

$\qquad (x - 1)^2 = \dfrac{2}{3}$

$\qquad x - 1 = \pm\sqrt{\dfrac{2}{3}}$

$\qquad x = 1 \pm \sqrt{\dfrac{2}{3}}$

$\qquad x = 1 \pm \dfrac{\sqrt{6}}{3}$

41. $\quad 8 + 4x - x^2 = 0$

$\qquad -x^2 + 4x + 8 = 0$

$\qquad x^2 - 4x - 8 = 0$

$\qquad x^2 - 4x = 8$

$\qquad x^2 - 4x + 2^2 = 8 + 2^2$

$\qquad (x - 2)^2 = 12$

$\qquad x - 2 = \pm\sqrt{12}$

$\qquad x = 2 \pm 2\sqrt{3}$

43. $\quad 2x^2 + 5x - 8 = 0$

$\qquad 2x^2 + 5x = 8$

$\qquad x^2 + \dfrac{5}{2}x = 4$

$\qquad x^2 + \dfrac{5}{2}x + \left(\dfrac{5}{4}\right)^2 = 4 + \left(\dfrac{5}{4}\right)^2$

$\qquad \left(x + \dfrac{5}{4}\right)^2 = \dfrac{89}{16}$

$\qquad x + \dfrac{5}{4} = \pm\dfrac{\sqrt{89}}{4}$

$\qquad x = -\dfrac{5}{4} \pm \dfrac{\sqrt{89}}{4}$

$\qquad x = \dfrac{-5 \pm \sqrt{89}}{4}$

45. $\dfrac{1}{x^2 + 2x + 5} = \dfrac{1}{x^2 + 2x + 1^2 - 1^2 + 5}$

$\qquad\qquad = \dfrac{1}{(x + 1)^2 + 4}$

47. $\dfrac{4}{x^2 + 4x - 3} = \dfrac{4}{x^2 + 4x + 4 - 4 - 3}$

$\qquad\qquad = \dfrac{4}{(x + 2)^2 - 7}$

49. $\dfrac{1}{\sqrt{6x - x^2}} = \dfrac{1}{\sqrt{-1(x^2 - 6x + 3^2 - 3^2)}}$

$\qquad\qquad = \dfrac{1}{\sqrt{-1[(x - 3)^2 - 9]}}$

$\qquad\qquad = \dfrac{1}{\sqrt{-(x - 3)^2 + 9}} = \dfrac{1}{\sqrt{9 - (x - 3)^2}}$

51. (a) $y = (x + 3)^2 - 4$

(b) The x-intercepts are $(-1, 0)$ and $(-5, 0)$.

(c) $\qquad 0 = (x + 3)^2 - 4$

$\qquad\quad 4 = (x + 3)^2$

$\qquad \pm\sqrt{4} = x + 3$

$\qquad -3 \pm 2 = x$

$\qquad\qquad x = -1 \quad \text{or} \quad x = -5$

(d) The x-intercepts of the graph are solutions to the equation $0 = (x + 3)^2 - 4$.

53. (a) $y = 1 - (x - 2)^2$

(b) The x-intercepts are $(1, 0)$ and $(3, 0)$.

(c)
$$0 = 1 - (x - 2)^2$$
$$(x - 2)^2 = 1$$
$$x - 2 = \pm 1$$
$$x = 2 \pm 1$$
$$x = 3 \quad \text{or} \quad x = 1$$

(d) The x-intercepts of the graph are solutions to the equation $0 = 1 - (x - 2)^2$.

55. (a) $y = -4x^2 + 4x + 3$

(b) The x-intercepts are $\left(-\frac{1}{2}, 0\right)$ and $\left(\frac{3}{2}, 0\right)$.

(d) The x-intercepts of the graph are solutions to the equation $0 = -4x^2 + 4x + 3$.

(c)
$$0 = -4x^2 + 4x + 3$$
$$4x^2 - 4x = 3$$
$$4(x^2 - x) = 3$$
$$x^2 - x = \frac{3}{4}$$
$$x^2 - x + \left(\frac{1}{2}\right)^2 = \frac{3}{4} + \left(\frac{1}{2}\right)^2$$
$$\left(x - \frac{1}{2}\right)^2 = 1$$
$$x - \frac{1}{2} = \pm\sqrt{1}$$
$$x = \frac{1}{2} \pm 1$$
$$x = \frac{3}{2} \quad \text{or} \quad x = -\frac{1}{2}$$

57. (a) $y = x^2 + 3x - 4$

(b) The x-intercepts are $(-4, 0)$ and $(1, 0)$.

(c)
$$0 = x^2 + 3x - 4$$
$$0 = (x + 4)(x - 1)$$
$$x + 4 = 0 \quad \text{or} \quad x - 1 = 0$$
$$x = -4 \quad \text{or} \quad x = 1$$

(d) The x-intercepts of the graph are solutions to the equation $0 = x^2 + 3x - 4$.

59. $2x^2 - 5x + 5 = 0$

$b^2 - 4ac = (-5)^2 - 4(2)(5) = -15 < 0$

No real solution

61. $2x^2 - x - 1 = 0$

$b^2 - 4ac = (-1)^2 - 4(2)(-1) = 9 > 0$

Two real solutions

63. $\frac{1}{3}x^2 - 5x + 25 = 0$

$b^2 - 4ac = (-5)^2 - 4\left(\frac{1}{3}\right)(25) = -\frac{25}{3} < 0$

No real solution

65. $0.2x^2 + 1.2x - 8 = 0$

$b^2 - 4ac = (1.2)^2 - 4(0.2)(-8) = 7.84 > 0$

Two real solutions

67. $2x^2 + x - 1 = 0$

$$x = \frac{-b \pm \sqrt{b^2 - 4ac}}{2a}$$
$$= \frac{-1 \pm \sqrt{1^2 - 4(2)(-1)}}{2(2)}$$
$$= \frac{-1 \pm 3}{4} = \frac{1}{2}, -1$$

69. $16x^2 + 8x - 3 = 0$

$$x = \frac{-b \pm \sqrt{b^2 - 4ac}}{2a}$$
$$= \frac{-8 \pm \sqrt{8^2 - 4(16)(-3)}}{2(16)}$$
$$= \frac{-8 \pm 16}{32} = \frac{1}{4}, -\frac{3}{4}$$

71. $2 + 2x - x^2 = 0$

$$-x^2 + 2x + 2 = 0$$
$$x = \frac{-b \pm \sqrt{b^2 - 4ac}}{2a}$$
$$= \frac{-2 \pm \sqrt{2^2 - 4(-1)(2)}}{2(-1)}$$
$$= \frac{-2 \pm 2\sqrt{3}}{-2} = 1 \pm \sqrt{3}$$

73. $x^2 + 14x + 44 = 0$

$$x = \frac{-b \pm \sqrt{b^2 - 4ac}}{2a}$$

$$= \frac{-14 \pm \sqrt{14^2 - 4(1)(44)}}{2(1)}$$

$$= \frac{-14 \pm 2\sqrt{5}}{2} = -7 \pm \sqrt{5}$$

75. $x^2 + 8x - 4 = 0$

$$x = \frac{-b \pm \sqrt{b^2 - 4ac}}{2a}$$

$$= \frac{-8 \pm \sqrt{8^2 - 4(1)(-4)}}{2(1)}$$

$$= \frac{-8 \pm 4\sqrt{5}}{2} = -4 \pm 2\sqrt{5}$$

77. $12x - 9x^2 = -3$

$$-9x^2 + 12x + 3 = 0$$

$$x = \frac{-b \pm \sqrt{b^2 - 4ac}}{2a}$$

$$= \frac{-12 \pm \sqrt{12^2 - 4(-9)(3)}}{2(-9)}$$

$$= \frac{-12 \pm 6\sqrt{7}}{-18} = \frac{2}{3} \pm \frac{\sqrt{7}}{3}$$

79. $9x^2 + 24x + 16 = 0$

$$x = \frac{-b \pm \sqrt{b^2 - 4ac}}{2a}$$

$$= \frac{-24 \pm \sqrt{24^2 - 4(9)(16)}}{2(9)}$$

$$= \frac{-24 \pm 0}{18}$$

$$= -\frac{4}{3}$$

81. $4x^2 + 4x = 7$

$$4x^2 + 4x - 7 = 0$$

$$x = \frac{-b \pm \sqrt{b^2 - 4ac}}{2a}$$

$$= \frac{-4 \pm \sqrt{4^2 - 4(4)(-7)}}{2(4)}$$

$$= \frac{-4 \pm 8\sqrt{2}}{8} = -\frac{1}{2} \pm \sqrt{2}$$

83. $28x - 49x^2 = 4$

$$-49x^2 + 28x - 4 = 0$$

$$x = \frac{-b \pm \sqrt{b^2 - 4ac}}{2a}$$

$$= \frac{-28 \pm \sqrt{28^2 - 4(-49)(-4)}}{2(-49)}$$

$$= \frac{-28 \pm 0}{-98} = \frac{2}{7}$$

85. $8t = 5 + 2t^2$

$$-2t^2 + 8t - 5 = 0$$

$$t = \frac{-b \pm \sqrt{b^2 - 4ac}}{2a}$$

$$= \frac{-8 \pm \sqrt{8^2 - 4(-2)(-5)}}{2(-2)}$$

$$= \frac{-8 \pm 2\sqrt{6}}{-4} = 2 \pm \frac{\sqrt{6}}{2}$$

87. $(y - 5)^2 = 2y$

$$y^2 - 12y + 25 = 0$$

$$y = \frac{-b \pm \sqrt{b^2 - 4ac}}{2a}$$

$$= \frac{-(-12) \pm \sqrt{(-12)^2 - 4(1)(25)}}{2(1)}$$

$$= \frac{12 \pm 2\sqrt{11}}{2} = 6 \pm \sqrt{11}$$

89. $\frac{1}{2}x^2 + \frac{3}{8}x = 2$

$$4x^2 + 3x = 16$$

$$4x^2 + 3x - 16 = 0$$

$$x = \frac{-b \pm \sqrt{b^2 - 4ac}}{2a}$$

$$= \frac{-3 \pm \sqrt{3^2 - 4(4)(-16)}}{2(4)}$$

$$= \frac{-3 \pm \sqrt{265}}{8}$$

$$= -\frac{3}{8} \pm \frac{\sqrt{265}}{8}$$

91. $5.1x^2 - 1.7x - 3.2 = 0$

$$x = \frac{1.7 \pm \sqrt{(-1.7)^2 - 4(5.1)(-3.2)}}{2(5.1)}$$

$$x \approx 0.976, -0.643$$

93. $-0.067x^2 - 0.852x + 1.277 = 0$

$$x = \frac{-(-0.852) \pm \sqrt{(-0.852)^2 - 4(-0.067)(1.277)}}{2(-0.067)}$$

$$x \approx -14.071, 1.355$$

95. $422x^2 - 506x - 347 = 0$

$$x = \frac{506 \pm \sqrt{(-506)^2 - 4(422)(-347)}}{2(422)}$$

$$x \approx 1.687, -0.488$$

97. $12.67x^2 + 31.55x + 8.09 = 0$

$$x = \frac{-31.55 \pm \sqrt{(31.55)^2 - 4(12.67)(8.09)}}{2(12.67)}$$

$$x \approx -2.200, -0.290$$

99. $x^2 - 2x - 1 = 0$ Complete the square.

$$x^2 - 2x = 1$$
$$x^2 - 2x + 1^2 = 1 + 1^2$$
$$(x - 1)^2 = 2$$
$$x - 1 = \pm\sqrt{2}$$
$$x = 1 \pm \sqrt{2}$$

101. $(x + 3)^2 = 81$ Extract square roots.

$$x + 3 = \pm 9$$
$$x + 3 = 9 \quad \text{or} \quad x + 3 = -9$$
$$x = 6 \quad \text{or} \qquad x = -12$$

103. $x^2 - x - \frac{11}{4} = 0$ Complete the square.

$$x^2 - x = \frac{11}{4}$$
$$x^2 - x + \left(\tfrac{1}{2}\right)^2 = \frac{11}{4} + \left(\tfrac{1}{2}\right)^2$$
$$\left(x - \tfrac{1}{2}\right)^2 = \frac{12}{4}$$
$$x - \tfrac{1}{2} = \pm\sqrt{\tfrac{12}{4}}$$
$$x = \tfrac{1}{2} \pm \sqrt{3}$$

105. $(x + 1)^2 = x^2$ Extract square roots.

$$x^2 = (x + 1)^2$$
$$x = \pm(x + 1)$$

For $x = +(x + 1)$:

$$0 \neq 1 \quad \text{No solution}$$

For $x = -(x + 1)$:

$$2x = -1$$
$$x = -\tfrac{1}{2}$$

107. $3x + 4 = 2x^2 - 7$ Quadratic Formula

$$0 = 2x^2 - 3x - 11$$
$$x = \frac{-(-3) \pm \sqrt{(-3)^2 - 4(2)(-11)}}{2(2)}$$
$$= \frac{3 \pm \sqrt{97}}{4}$$
$$= \frac{3}{4} \pm \frac{\sqrt{97}}{4}$$

109. (a)

(b) $w(w + 14) = 1632$

(c) $w^2 + 14w - 1632 = 0$

$$(w + 48)(w - 34) = 0$$
$$w = -48 \quad \text{or} \quad w = 34$$

Since w must be greater than zero, we have $w = 34$ feet and the length is $w + 14 = 48$ feet.

111. $S = x^2 + 4xh$

$$84 = x^2 + 4x(2)$$
$$0 = x^2 + 8x - 84$$
$$0 = (x + 14)(x - 6)$$
$$x = -14 \quad \text{or} \quad x = 6$$

Since x must be positive, we have $x = 6$ inches. The dimensions of the box are 6 inches × 6 inches × 2 inches.

113. $(200 - 2x)(100 - 2x) = \dfrac{1}{2}(100)(200)$

$$20,000 - 600x + 4x^2 = 10,000$$
$$4x^2 - 600x + 10,000 = 0$$
$$4(x^2 - 150x + 2500) = 0$$

Thus, $a = 1$, $b = -150$, and $c = 2500$.

—CONTINUED—

113. —CONTINUED—

$$x = \frac{150 \pm \sqrt{(-150)^2 - 4(1)(2500)}}{2(1)} \approx \frac{150 \pm 111.8034}{2}$$

$$x \approx \frac{150 + 111.8034}{2} \approx 130.902 \text{ feet, not possible since the lot is only 100 feet wide.}$$

$$x \approx \frac{150 - 111.8034}{2} \approx 19.098 \text{ feet}$$

The person must go around the lot $\dfrac{19.098 \text{ feet}}{24 \text{ inches}} = \dfrac{19.098 \text{ feet}}{2 \text{ feet}} \approx 9.5$ times.

115. $s = -16t^2 + 32,000$

(a) $-16t^2 + 32,000 = 0$

$$t^2 = 2000$$
$$t = \sqrt{2000}$$
$$= 20\sqrt{5} \text{ seconds}$$
$$\approx 44.72 \text{ seconds}$$

(b) *Model:* (Rate) · (time) = (distance)

Labels: Rate = 500 miles per hour

Time $= \dfrac{20\sqrt{5}}{3600} \approx 0.0124$ hour

Distance $= d$

Equation: $d = 500(0.0124) \approx 6.2$ miles

The bomb will travel approximately 6.2 miles horizontally.

117. $s = -16t^2 + v_0 t + s_0$

(a) $v_0 = 100 \text{ mph} = \dfrac{(100)(5280)}{3600} = 146\frac{2}{3}$ ft/sec

$s_0 = 6\frac{1}{4}$ feet

$s = -16t^2 + 146\frac{2}{3}t + 6\frac{1}{4}$

(b) When $t = 3$: $s(3) = 302.25$ feet

When $t = 4$: $s(4) \approx 336.92$ feet

When $t = 5$: $s(5) \approx 339.58$ feet

During the interval $3 \le t \le 5$, the baseball's speed decreased due to gravity.

(c) The ball hits the ground when $s = 0$.

$-16t^2 + 146\frac{2}{3}t + 6\frac{1}{4} = 0$

By the Quadratic Formula, $t \approx -0.042$ or $t \approx 9.209$. Assuming that the ball is not caught and drops to the ground, it will be in the air for approximately 9.209 seconds.

119. $P = -0.0081t^2 + 0.417t + 1.99$, $7 \le t \le 13$

(a)

t	7	8	9	10	11	12	13
P	\$4.51	\$4.81	\$5.09	\$5.35	\$5.60	\$5.83	\$6.04

The average admission price reached or surpassed \$5.00 in 1999.

(b) $-0.0081t^2 + 0.417t + 1.99 = 5.00$

$-0.0081t^2 + 0.417t - 3.01 = 0$

By the Quadratic Formula, $t \approx 8.68$ or 42.80. Since $7 \le t \le 13$, we choose $t \approx 8.68 \approx 9$ which corresponds to 1999.

(c) For 2008, let $t = 18$: $P(18) \approx \$6.87$. Answers will vary.

121. $x^2 + x^2 = 5^2$ Pythagorean Theorem

$\qquad 2x^2 = 25$

$\qquad x^2 = \dfrac{25}{2}$

$\qquad x = \sqrt{\dfrac{25}{2}}$

$\qquad\quad = \dfrac{5}{\sqrt{2}}$

$\qquad\quad = \dfrac{5\sqrt{2}}{2} \approx 3.54$ centimeters

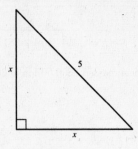

Each leg in the right triangle is approximately 3.54 centimeters.

123. $\qquad\qquad d_N = (3 \text{ hours})(r + 50 \text{ mph})$

$\qquad\qquad d_E = (3 \text{ hours})(r \text{ mph})$

$\qquad\quad d_N{}^2 + d_E{}^2 = 2440^2$

$\qquad 9(r + 50)^2 + 9r^2 = 2440^2$

$18r^2 + 900r - 5{,}931{,}100 = 0$

$\qquad\qquad r = \dfrac{-900 \pm \sqrt{900^2 - 4(18)(-5{,}931{,}100)}}{2(18)} = \dfrac{-900 \pm 60\sqrt{118{,}847}}{36}$

Using the positive value for r, we have one plane moving northbound at $r + 50 \approx 600$ miles per hour and one plane moving eastbound at $r \approx 550$ miles per hour.

125. $x(20 - 0.0002x) = 500{,}000$

$\qquad 0 = 0.0002x^2 - 20x + 500{,}000$

$\qquad 0 = x^2 - 100{,}000x + 2{,}500{,}000{,}000$

$\qquad 0 = (x - 50{,}000)^2$

$\qquad x = 50{,}000$ units

127. $\qquad 0.125x^2 + 20x + 500 = 14{,}000$

$\qquad 0.125x^2 + 20x - 13{,}500 = 0$

$\qquad x = \dfrac{-20 \pm \sqrt{20^2 - 4(0.125)(-13{,}500)}}{2(0.125)}$

Using the positive value for x, we have

$\qquad x = \dfrac{-20 + \sqrt{7150}}{0.25} \approx 258$ units.

129. $800 + 0.04x + 0.002x^2 = 1680$

$\qquad 0.002x^2 + 0.04x - 880 = 0$

$\qquad x = \dfrac{-0.04 \pm \sqrt{(0.04)^2 - 4(0.002)(-880)}}{2(0.002)}$

$\qquad\quad = \dfrac{-0.04 \pm \sqrt{7.0416}}{0.004}$

Choosing the positive value for x, we have

$\qquad x = \dfrac{-0.04 + \sqrt{7.0416}}{0.004} \approx 653$ units.

131. $M = 1.835t^2 + 3.58t + 333.0, \ 5 \le t \le 13$

(a)

t	5	6	7	8	9	10	11	12	13
M (in billions)	\$396.78	\$420.54	\$447.98	\$479.08	\$513.86	\$552.30	\$594.42	\$640.20	\$689.66

The total money in circulation reached or surpassed \$600 billion in 2001.

(b) $1.835t^2 + 3.58t + 333.0 = 600$

$1.835t^2 + 3.58t - 267.0 = 0$

By the Quadratic Formula, $t \approx 11.1$ or -13.1. Since $5 \le t \le 13$, we choose $t \approx 11.1$ which corresponds to 2001.

(c) For 2008, let $t = 18$: $M(18) \approx \$991.98$ billion
Answers will vary.

133. $\left(\begin{array}{c}\text{Distance from Oklahoma} \\ \text{City to New Orleans}\end{array}\right)^2 = \left(\begin{array}{c}\text{Distance from Oklahoma} \\ \text{City to Austin}\end{array}\right)^2 + \left(\begin{array}{c}\text{Distance from} \\ \text{New Orleans to Austin}\end{array}\right)^2$

Distance from Oklahoma City to New Orleans = 560

Distance from Oklahoma City to Austin = x

Distance from New Orleans to Austin = $1348 - 560 - x = 788 - x$

$560^2 = x^2 + (788 - x)^2$

$313{,}600 = x^2 + 620{,}944 - 1576x + x^2$

$0 = 2x^2 - 1576x + 307{,}344$

$x = \dfrac{1576 \pm \sqrt{1576^2 - 4(2)(307{,}344)}}{2(2)} \approx 354.5 \text{ or } 433.5$

The other two distances are 354.5 miles and 433.5 miles.

135. False

$b^2 - 4ac = (-1)^2 - 4(-3)(-10) < 0,$

so, the quadratic equation has no real solutions.

137. The student should have subtracted $15x$ from both sides so that the equation is equal to zero. By factoring out an x, there are two solutions, $x = 0$ and $x = 6$.

139. $3(x + 4)^2 + (x + 4) - 2 = 0$

(a) Let $u = x + 4$

$3u^2 + u - 2 = 0$

$(3u - 2)(u + 1) = 0$

$3u - 2 = 0 \quad \text{or} \quad u + 1 = 0$

$u = \frac{2}{3} \qquad\qquad u = -1$

$x + 4 = \frac{2}{3} \qquad\quad x + 4 = -1$

$x = -\frac{10}{3} \quad \text{or} \qquad x = -5$

(b) $3(x^2 + 8x + 16) + (x + 4) - 2 = 0$

$3x^2 + 24x + 48 + x + 4 - 2 = 0$

$3x^2 + 25x + 50 = 0$

$(3x + 10)(x + 5) = 0$

$3x + 10 = 0 \quad \text{or} \quad x + 5 = 0$

$x = -\frac{10}{3} \qquad\qquad x = -5$

(c) The method of part (a) reduces the number of algebraic steps.

141. -3 and 6

One possible equation is:

$(x - (-3))(x - 6) = 0$

$(x + 3)(x - 6) = 0$

$x^2 - 3x - 18 = 0$

Any non-zero multiple of this equation would also have these solutions.

143. 8 and 14

One possible equation is:

$(x - 8)(x - 14) = 0$

$x^2 - 22x + 112 = 0$

Any non-zero multiple of this equation would also have these solutions.

145. $1 + \sqrt{2}$ and $1 - \sqrt{2}$

One possible equation is:

$$\left[x - \left(1 + \sqrt{2}\right)\right]\left[x - \left(1 - \sqrt{2}\right)\right] = 0$$
$$\left[(x - 1) - \sqrt{2}\right]\left[(x - 1) + \sqrt{2}\right] = 0$$
$$(x - 1)^2 - \left(\sqrt{2}\right)^2 = 0$$
$$x^2 - 2x + 1 - 2 = 0$$
$$x^2 - 2x - 1 = 0$$

Any non-zero multiple of this equation would also have these solutions.

147. $(10x)y = 10(xy)$ by the Associative Property of Multiplication.

149. $7x^4 + (-7x^4) = 0$ by the Additive Inverse Property.

151.
$$(x + 3)(x - 6) = x^2 - 6x + 3x - 18$$
$$= x^2 - 3x - 18$$

153.
$$(x + 4)(x^2 - x + 2) = x(x^2 - x + 2) + 4(x^2 - x + 2)$$
$$= x^3 - x^2 + 2x + 4x^2 - 4x + 8$$
$$= x^3 + 3x^2 - 2x + 8$$

155.
$$x^5 - 27x^2 = x^2(x^3 - 27)$$
$$= x^2(x^3 - 3^3)$$
$$= x^2(x - 3)(x^2 + 3x + 9)$$

157.
$$x^3 + 5x^2 - 2x - 10 = x^2(x + 5) - 2(x + 5)$$
$$= (x + 5)(x^2 - 2)$$

159. Answers will vary.

Section 1.5   Complex Numbers

- Standard form: $a + bi$.

  If $b = 0$, then $a + bi$ is a real number.

  If $a = 0$ and $b \neq 0$, then $a + bi$ is a pure imaginary number.

- Equality of Complex Numbers: $a + bi = c + di$ if and only if $a = c$ and $b = d$

- Operations on complex numbers

  (a) Addition: $(a + bi) + (c + di) = (a + c) + (b + d)i$

  (b) Subtraction: $(a + bi) - (c + di) = (a - c) + (b - d)i$

  (c) Multiplication: $(a + bi)(c + di) = (ac - bd) + (ad + bc)i$

  (d) Division: $\dfrac{a + bi}{c + di} = \dfrac{a + bi}{c + di} \cdot \dfrac{c - di}{c - di} = \dfrac{ac + bd}{c^2 + d^2} + \dfrac{bc - ad}{c^2 + d^2}i$

- The complex conjugate of $a + bi$ is $a - bi$:

  $(a + bi)(a - bi) = a^2 + b^2$

- The additive inverse of $a + bi$ is $-a - bi$.

- $\sqrt{-a} = \sqrt{a}\,i$ for $a > 0$.

Vocabulary Check

1. (a) iii   (b) i   (c) ii

2. $\sqrt{-1}$; -1

3. principal square

4. complex conjugates

---

1. $a + bi = -10 + 6i$
$a = -10$
$b = 6$

3. $(a - 1) + (b + 3)i = 5 + 8i$
$a - 1 = 5 \implies a = 6$
$b + 3 = 8 \implies b = 5$

5. $4 + \sqrt{-9} = 4 + 3i$

7. $2 - \sqrt{-27} = 2 - \sqrt{27}\,i$
$= 2 - 3\sqrt{3}\,i$

9. $\sqrt{-75} = \sqrt{75}\,i = 5\sqrt{3}\,i$

11. $8 = 8 + 0i = 8$

13. $-6i + i^2 = -6i - 1$
$= -1 - 6i$

15. $\sqrt{-0.09} = \sqrt{0.09}\,i$
$= 0.3i$

17. $(5 + i) + (6 - 2i) = 11 - i$

19. $(8 - i) - (4 - i) = 8 - i - 4 + i$
$= 4$

21. $\left(-2 + \sqrt{-8}\right) + \left(5 - \sqrt{-50}\right) = -2 + 2\sqrt{2}i + 5 - 5\sqrt{2}i$
$= 3 - 3\sqrt{2}i$

23. $13i - (14 - 7i) = 13i - 14 + 7i$
$= -14 + 20i$

25. $-\left(\frac{3}{2} + \frac{5}{2}i\right) + \left(\frac{5}{3} + \frac{11}{3}i\right) = -\frac{3}{2} - \frac{5}{2}i + \frac{5}{3} + \frac{11}{3}i$
$= -\frac{9}{6} - \frac{15}{6}i + \frac{10}{6} + \frac{22}{6}i$
$= \frac{1}{6} + \frac{7}{6}i$

27. $(1 + i)(3 - 2i) = 3 - 2i + 3i - 2i^2$
$= 3 + i + 2 = 5 + i$

29. $6i(5 - 2i) = 30i - 12i^2 = 30i + 12$
$= 12 + 30i$

31. $\left(\sqrt{14} + \sqrt{10}\,i\right)\left(\sqrt{14} - \sqrt{10}\,i\right) = 14 - 10i^2$
$= 14 + 10 = 24$

33. $(4 + 5i)^2 = 16 + 40i + 25i^2$
$= 16 + 40i - 25$
$= -9 + 40i$

35. $(2 + 3i)^2 + (2 - 3i)^2 = 4 + 12i + 9i^2 + 4 - 12i + 9i^2$
$= 4 + 12i - 9 + 4 - 12i - 9$
$= -10$

37. The complex conjugate of $6 + 3i$ is $6 - 3i$.
$(6 + 3i)(6 - 3i) = 36 - (3i)^2 = 36 + 9 = 45$

39. The complex conjugate of $-1 - \sqrt{5}i$ is $-1 + \sqrt{5}i$.
$\left(-1 - \sqrt{5}i\right)\left(-1 + \sqrt{5}i\right) = (-1)^2 - \left(\sqrt{5}i\right)^2$
$= 1 + 5 = 6$

41. The complex conjugate of $\sqrt{-20} = 2\sqrt{5}i$ is $-2\sqrt{5}i$.
$\left(2\sqrt{5}i\right)\left(-2\sqrt{5}i\right) = -20i^2 = 20$

43. The complex conjugate of $\sqrt{8}$ is $\sqrt{8}$.
$\left(\sqrt{8}\right)\left(\sqrt{8}\right) = 8$

45. $\dfrac{5}{i} = \dfrac{5}{i} \cdot \dfrac{-i}{-i} = \dfrac{-5i}{1} = -5i$

47. $\dfrac{2}{4-5i} = \dfrac{2}{4-5i} \cdot \dfrac{4+5i}{4+5i}$

$\quad = \dfrac{2(4+5i)}{16+25} = \dfrac{8+10i}{41} = \dfrac{8}{41} + \dfrac{10}{41}i$

49. $\dfrac{3+i}{3-i} = \dfrac{3+i}{3-i} \cdot \dfrac{3+i}{3+i}$

$\quad = \dfrac{9+6i+i^2}{9+1} = \dfrac{8+6i}{10} = \dfrac{4}{5} + \dfrac{3}{5}i$

51. $\dfrac{6-5i}{i} = \dfrac{6-5i}{i} \cdot \dfrac{-i}{-i}$

$\quad = \dfrac{-6i+5i^2}{1} = -5 - 6i$

53. $\dfrac{3i}{(4-5i)^2} = \dfrac{3i}{16-40i+25i^2} = \dfrac{3i}{-9-40i} \cdot \dfrac{-9+40i}{-9+40i}$

$\quad = \dfrac{-27i+120i^2}{81+1600} = \dfrac{-120-27i}{1681}$

$\quad = -\dfrac{120}{1681} - \dfrac{27}{1681}i$

55. $\dfrac{2}{1+i} - \dfrac{3}{1-i} = \dfrac{2(1-i)-3(1+i)}{(1+i)(1-i)}$

$\quad = \dfrac{2-2i-3-3i}{1+1}$

$\quad = \dfrac{-1-5i}{2}$

$\quad = -\dfrac{1}{2} - \dfrac{5}{2}i$

57. $\dfrac{i}{3-2i} + \dfrac{2i}{3+8i} = \dfrac{i(3+8i)+2i(3-2i)}{(3-2i)(3+8i)}$

$\quad = \dfrac{3i+8i^2+6i-4i^2}{9+24i-6i-16i^2}$

$\quad = \dfrac{4i^2+9i}{9+18i+16}$

$\quad = \dfrac{-4+9i}{25+18i} \cdot \dfrac{25-18i}{25-18i}$

$\quad = \dfrac{-100+72i+225i-162i^2}{625+324}$

$\quad = \dfrac{-100+297i+162}{949}$

$\quad = \dfrac{62+297i}{949} = \dfrac{62}{949} + \dfrac{297}{949}i$

59. $\sqrt{-6} \cdot \sqrt{-2} = \left(\sqrt{6}i\right)\left(\sqrt{2}i\right) = \sqrt{12}i^2 = \left(2\sqrt{3}\right)(-1)$

$\qquad\qquad = -2\sqrt{3}$

61. $\left(\sqrt{-10}\right)^2 = \left(\sqrt{10}i\right)^2 = 10i^2 = -10$

63. $\left(3+\sqrt{-5}\right)\left(7-\sqrt{-10}\right) = \left(3+\sqrt{5}i\right)\left(7-\sqrt{10}i\right)$

$\qquad\qquad = 21 - 3\sqrt{10}i + 7\sqrt{5}i - \sqrt{50}i^2$

$\qquad\qquad = \left(21+\sqrt{50}\right) + \left(7\sqrt{5}-3\sqrt{10}\right)i$

$\qquad\qquad = \left(21+5\sqrt{2}\right) + \left(7\sqrt{5}-3\sqrt{10}\right)i$

65. $x^2 - 2x + 2 = 0;\ a = 1,\ b = -2,\ c = 2$

$\quad x = \dfrac{-(-2) \pm \sqrt{(-2)^2 - 4(1)(2)}}{2(1)}$

$\quad = \dfrac{2 \pm \sqrt{-4}}{2}$

$\quad = \dfrac{2 \pm 2i}{2}$

$\quad = 1 \pm i$

67. $4x^2 + 16x + 17 = 0;\ a = 4,\ b = 16,\ c = 17$

$\quad x = \dfrac{-16 \pm \sqrt{(16)^2 - 4(4)(17)}}{2(4)}$

$\quad = \dfrac{-16 \pm \sqrt{-16}}{8}$

$\quad = \dfrac{-16 \pm 4i}{8} = -2 \pm \dfrac{1}{2}i$

69. $4x^2 + 16x + 15 = 0$; $a = 4$, $b = 16$, $c = 15$

$$x = \frac{-16 \pm \sqrt{(16)^2 - 4(4)(15)}}{2(4)}$$

$$= \frac{-16 \pm \sqrt{16}}{8} = \frac{-16 \pm 4}{8}$$

$$x = -\frac{12}{8} = -\frac{3}{2} \quad \text{or} \quad x = -\frac{20}{8} = -\frac{5}{2}$$

71. $\frac{3}{2}x^2 - 6x + 9 = 0$ Multiply both sides by 2.

$$3x^2 - 12x + 18 = 0$$

$$x = \frac{-(-12) \pm \sqrt{(-12)^2 - 4(3)(18)}}{2(3)}$$

$$= \frac{12 \pm \sqrt{-72}}{6}$$

$$= \frac{12 \pm 6\sqrt{2}i}{6} = 2 \pm \sqrt{2}i$$

73. $1.4x^2 - 2x - 10 = 0$ Multiply both sides by 5.

$$7x^2 - 10x - 50 = 0$$

$$x = \frac{-(-10) \pm \sqrt{(-10)^2 - 4(7)(-50)}}{2(7)}$$

$$= \frac{10 \pm \sqrt{1500}}{14} = \frac{10 \pm 10\sqrt{15}}{14}$$

$$= \frac{5 \pm 5\sqrt{15}}{7} = \frac{5}{7} \pm \frac{5\sqrt{15}}{7}$$

75. $-6i^3 + i^2 = -6i^2i + i^2$

$$= -6(-1)i + (-1)$$

$$= 6i - 1$$

$$= -1 + 6i$$

77. $-5i^5 = -5i^2i^2i$

$$= -5(-1)(-1)i$$

$$= -5i$$

79. $\left(\sqrt{-75}\right)^3 = \left(5\sqrt{3}i\right)^3$

$$= 5^3\left(\sqrt{3}\right)^3 i^3$$

$$= 125\left(3\sqrt{3}\right)(-1)i$$

$$= -375\sqrt{3}i$$

81. $\dfrac{1}{i^3} = \dfrac{1}{-i} = \dfrac{1}{-i} \cdot \dfrac{i}{i} = \dfrac{i}{-i^2} = \dfrac{i}{1} = i$

83. (a) $z_1 = 9 + 16i$, $z_2 = 20 - 10i$

(b) $\dfrac{1}{z} = \dfrac{1}{z_1} + \dfrac{1}{z_2} = \dfrac{1}{9 + 16i} + \dfrac{1}{20 - 10i} = \dfrac{20 - 10i + 9 + 16i}{(9 + 16i)(20 - 10i)} = \dfrac{29 + 6i}{340 + 230i}$

$z = \left(\dfrac{340 + 230i}{29 + 6i}\right)\left(\dfrac{29 - 6i}{29 - 6i}\right) = \dfrac{11{,}240 + 4630i}{877} = \dfrac{11{,}240}{877} + \dfrac{4630}{877}i$

85. (a) $2^4 = 16$

(b) $(-2)^4 = 16$

(c) $(2i)^4 = 2^4 i^4 = 16(1) = 16$

(d) $(-2i)^4 = (-2)^4 i^4 = 16(1) = 16$

87. False, if $b = 0$ then $a + bi = a - bi = a$.

That is, if the complex number is real, the number equals its conjugate.

89. False

$$i^{44} + i^{150} - i^{74} - i^{109} + i^{61} = (i^4)^{11} + (i^4)^{37}(i^2) - (i^4)^{18}(i^2) - (i^4)^{27}(i) + (i^4)^{15}(i)$$

$$= (1)^{11} + (1)^{37}(-1) - (1)^{18}(-1) - (1)^{27}(i) + (1)^{15}(i)$$

$$= 1 + (-1) + 1 - i + i = 1$$

91. $(a_1 + b_1 i)(a_2 + b_2 i) = a_1 a_2 + a_1 b_2 i + a_2 b_1 i + b_1 b_2 i^2$

$$= (a_1 a_2 - b_1 b_2) + (a_1 b_2 + a_2 b_1)i$$

The complex conjugate of this product is $(a_1 a_2 - b_1 b_2) - (a_1 b_2 + a_2 b_1)i$.

The product of the complex conjugates is:

$(a_1 - b_1 i)(a_2 - b_2 i) = a_1 a_2 - a_1 b_2 i - a_2 b_1 i + b_1 b_2 i^2$

$$= (a_1 a_2 - b_1 b_2) - (a_1 b_2 + a_2 b_1)i$$

Thus, the complex conjugate of the product of two complex numbers is the product of their complex conjugates.

93. $(4 + 3x) + (8 - 6x - x^2) = -x^2 - 3x + 12$

95. $\left(3x - \frac{1}{2}\right)(x + 4) = 3x^2 + 12x - \frac{1}{2}x - 2$

$$= 3x^2 + \frac{23}{2}x - 2$$

97. $-x - 12 = 19$

$-x = 31$

$x = -31$

99. $4(5x - 6) - 3(6x + 1) = 0$

$20x - 24 - 18x - 3 = 0$

$2x - 27 = 0$

$2x = 27$

$x = \frac{27}{2}$

101. $V = \frac{4}{3}\pi a^2 b$

$3V = 4\pi a^2 b$

$\frac{3V}{4\pi b} = a^2$

$\sqrt{\frac{3V}{4\pi b}} = a$

$a = \frac{1}{2}\sqrt{\frac{3V}{\pi b}} = \frac{\sqrt{3V\pi b}}{2\pi b}$

103. Let x = # liters withdrawn and replaced.

$0.50(5 - x) + 1.00x = 0.60(5)$

$2.50 - 0.50x + 1.00x = 3.00$

$0.50x = 0.50$

$x = 1$ liter

Section 1.6   Other Types of Equations

- You should be able to solve certain types of nonlinear or nonquadratic equations by rewriting them in a form in which you can factor, extract square roots, complete the square, or use the Quadratic Formula.

- For equations involving radicals or rational exponents, raise both sides to the same power.

- For equations that are of the quadratic type, $au^2 + bu + c = 0, a \neq 0$, use either factoring, the Quadratic Formula, or completing the square.

- For equations with fractions, multiply both sides by the least common denominator to clear the fractions.

- For equations involving absolute value, remember that the expression inside the absolute value can be positive or negative.

- Always check for extraneous solutions.

Vocabulary Check

1. polynomial

2. extraneous

3. quadratic type

1. $4x^4 - 18x^2 = 0$

$2x^2(2x^2 - 9) = 0$

$2x^2 = 0 \implies x = 0$

$2x^2 - 9 = 0 \implies x = \pm\dfrac{3\sqrt{2}}{2}$

3. $x^4 - 81 = 0$

$(x^2 + 9)(x + 3)(x - 3) = 0$

$x^2 + 9 = 0 \implies x = \pm 3i$

$x + 3 = 0 \implies x = -3$

$x - 3 = 0 \implies x = 3$

5. $x^3 + 216 = 0$

$x^3 + 6^3 = 0$

$(x + 6)(x^2 - 6x + 36) = 0$

$x + 6 = 0 \implies x = -6$

$x^2 - 6x + 36 = 0 \implies x = 3 \pm 3\sqrt{3}i$

7. $5x^3 + 30x^2 + 45x = 0$

$5x(x^2 + 6x + 9) = 0$

$5x(x + 3)^2 = 0$

$5x = 0 \implies x = 0$

$x + 3 = 0 \implies x = -3$

9. $x^3 - 3x^2 - x + 3 = 0$

$x^2(x - 3) - (x - 3) = 0$

$(x - 3)(x^2 - 1) = 0$

$(x - 3)(x + 1)(x - 1) = 0$

$x - 3 = 0 \implies x = 3$

$x + 1 = 0 \implies x = -1$

$x - 1 = 0 \implies x = 1$

11. $x^4 - x^3 + x - 1 = 0$

$x^3(x - 1) + (x - 1) = 0$

$(x - 1)(x^3 + 1) = 0$

$(x - 1)(x + 1)(x^2 - x + 1) = 0$

$x - 1 = 0 \implies x = 1$

$x + 1 = 0 \implies x = -1$

$x^2 - x + 1 = 0 \implies x = \dfrac{1}{2} \pm \dfrac{\sqrt{3}}{2}i$

13. $x^4 - 4x^2 + 3 = 0$

$(x^2 - 3)(x^2 - 1) = 0$

$\left(x + \sqrt{3}\right)\left(x - \sqrt{3}\right)(x + 1)(x - 1) = 0$

$x + \sqrt{3} = 0 \implies x = -\sqrt{3}$

$x - \sqrt{3} = 0 \implies x = \sqrt{3}$

$x + 1 = 0 \implies x = -1$

$x - 1 = 0 \implies x = 1$

15. $4x^4 - 65x^2 + 16 = 0$

$(4x^2 - 1)(x^2 - 16) = 0$

$(2x + 1)(2x - 1)(x + 4)(x - 4) = 0$

$2x + 1 = 0 \implies x = -\tfrac{1}{2}$

$2x - 1 = 0 \implies x = \tfrac{1}{2}$

$x + 4 = 0 \implies x = -4$

$x - 4 = 0 \implies x = 4$

17.
$$x^6 + 7x^3 - 8 = 0$$
$$(x^3 + 8)(x^3 - 1) = 0$$
$$(x + 2)(x^2 - 2x + 4)(x - 1)(x^2 + x + 1) = 0$$
$$x + 2 = 0 \implies x = -2$$
$$x^2 - 2x + 4 = 0 \implies x = 1 \pm \sqrt{3}i$$
$$x - 1 = 0 \implies x = 1$$
$$x^2 + x + 1 = 0 \implies x = -\frac{1}{2} \pm \frac{\sqrt{3}}{2}i$$

19.
$$\frac{1}{x^2} + \frac{8}{x} + 15 = 0$$
$$1 + 8x + 15x^2 = 0$$
$$(1 + 3x)(1 + 5x) = 0$$
$$1 + 3x = 0 \implies x = -\frac{1}{3}$$
$$1 + 5x = 0 \implies x = -\frac{1}{5}$$

21.
$$2x + 9\sqrt{x} = 5$$
$$2x + 9\sqrt{x} - 5 = 0$$
$$2(\sqrt{x})^2 + 9\sqrt{x} - 5 = 0$$
$$(2\sqrt{x} - 1)(\sqrt{x} + 5) = 0$$
$$\sqrt{x} = \tfrac{1}{2} \implies x = \tfrac{1}{4}$$
$$\left(\sqrt{x} = -5 \text{ is not a solution.}\right)$$

23.
$$3x^{1/3} + 2x^{2/3} = 5$$
$$2x^{2/3} + 3x^{1/3} - 5 = 0$$
$$2(x^{1/3})^2 + 3x^{1/3} - 5 = 0$$
$$(2x^{1/3} + 5)(x^{1/3} - 1) = 0$$
$$2x^{1/3} + 5 = 0 \implies x^{1/3} = -\tfrac{5}{2} \implies x = \left(-\tfrac{5}{2}\right)^3 = -\tfrac{125}{8}$$
$$x^{1/3} - 1 = 0 \implies x^{1/3} = 1 \implies x = (1)^3 = 1$$

25. $y = x^3 - 2x^2 - 3x$

(a)

(b) x-intercepts: $(-1, 0), (0, 0), (3, 0)$

(c)
$$0 = x^3 - 2x^2 - 3x$$
$$0 = x(x + 1)(x - 3)$$
$$x = 0$$
$$x + 1 = 0 \implies x = -1$$
$$x - 3 = 0 \implies x = 3$$

(d) The x-intercepts of the graph are the same as the solutions to the equation.

27. $y = x^4 - 10x^2 + 9$

(a)

(b) x-intercepts: $(\pm 1, 0), (\pm 3, 0)$

(c)
$$0 = x^4 - 10x^2 + 9$$
$$0 = (x^2 - 1)(x^2 - 9)$$
$$0 = (x + 1)(x - 1)(x + 3)(x - 3)$$
$$x + 1 = 0 \implies x = -1$$
$$x - 1 = 0 \implies x = 1$$
$$x + 3 = 0 \implies x = -3$$
$$x - 3 = 0 \implies x = 3$$

(d) The x-intercepts of the graph are the same as the solutions to the equation.

29. $\sqrt{2x} - 10 = 0$

$\qquad \sqrt{2x} = 10$

$\qquad 2x = 100$

$\qquad x = 50$

31. $\sqrt{x - 10} - 4 = 0$

$\qquad \sqrt{x - 10} = 4$

$\qquad x - 10 = 16$

$\qquad x = 26$

33. $\sqrt[3]{2x + 5} + 3 = 0$

$\qquad \sqrt[3]{2x + 5} = -3$

$\qquad 2x + 5 = -27$

$\qquad 2x = -32$

$\qquad x = -16$

35. $-\sqrt{26 - 11x} + 4 = x$

$\qquad 4 - x = \sqrt{26 - 11x}$

$\qquad 16 - 8x + x^2 = 26 - 11x$

$\qquad x^2 + 3x - 10 = 0$

$\qquad (x + 5)(x - 2) = 0$

$\qquad x + 5 = 0 \Rightarrow x = -5$

$\qquad x - 2 = 0 \Rightarrow x = 2$

37. $\sqrt{x + 1} = \sqrt{3x + 1}$

$\qquad x + 1 = 3x + 1$

$\qquad -2x = 0$

$\qquad x = 0$

39. $\sqrt{x} - \sqrt{x - 5} = 1$

$\qquad \sqrt{x} = 1 + \sqrt{x - 5}$

$\qquad \left(\sqrt{x}\right)^2 = \left(1 + \sqrt{x - 5}\right)^2$

$\qquad x = 1 + 2\sqrt{x - 5} + x - 5$

$\qquad 4 = 2\sqrt{x - 5}$

$\qquad 2 = \sqrt{x - 5}$

$\qquad 4 = x - 5$

$\qquad 9 = x$

41. $\sqrt{x + 5} + \sqrt{x - 5} = 10$

$\qquad \sqrt{x + 5} = 10 - \sqrt{x - 5}$

$\qquad \left(\sqrt{x + 5}\right)^2 = \left(10 - \sqrt{x - 5}\right)^2$

$\qquad x + 5 = 100 - 20\sqrt{x - 5} + x - 5$

$\qquad -90 = -20\sqrt{x - 5}$

$\qquad 9 = 2\sqrt{x - 5}$

$\qquad 81 = 4(x - 5)$

$\qquad 81 = 4x - 20$

$\qquad 101 = 4x$

$\qquad \frac{101}{4} = x$

43. $\sqrt{x + 2} - \sqrt{2x - 3} = -1$

$\qquad \sqrt{x + 2} = \sqrt{2x - 3} - 1$

$\qquad \left(\sqrt{x + 2}\right)^2 = \left(\sqrt{2x - 3} - 1\right)^2$

$\qquad x + 2 = 2x - 3 - 2\sqrt{2x - 3} + 1$

$\qquad x + 2 = 2x - 2 - 2\sqrt{2x - 3}$

$\qquad -x + 4 = -2\sqrt{2x - 3}$

$\qquad x - 4 = 2\sqrt{2x - 3}$

$\qquad (x - 4)^2 = \left(2\sqrt{2x - 3}\right)^2$

$\qquad x^2 - 8x + 16 = 4(2x - 3)$

$\qquad x^2 - 8x + 16 = 8x - 12$

$\qquad x^2 - 16x + 28 = 0$

$\qquad (x - 2)(x - 14) = 0$

$\qquad x - 2 = 0 \Rightarrow x = 2, \text{extraneous}$

$\qquad x - 14 = 0 \Rightarrow x = 14$

45. $(x - 5)^{3/2} = 8$

$\qquad (x - 5)^3 = 8^2$

$\qquad x - 5 = \sqrt[3]{64}$

$\qquad x = 5 + 4 = 9$

47. $(x + 3)^{2/3} = 8$

$(x + 3)^2 = 8^3$

$x + 3 = \pm\sqrt{8^3}$

$x + 3 = \pm\sqrt{512}$

$x = -3 \pm 16\sqrt{2}$

49. $(x^2 - 5)^{3/2} = 27$

$(x^2 - 5)^3 = 27^2$

$x^2 - 5 = \sqrt[3]{27^2}$

$x^2 = 5 + 9$

$x^2 = 14$

$x = \pm\sqrt{14}$

51. $3x(x - 1)^{1/2} + 2(x - 1)^{3/2} = 0$

$(x - 1)^{1/2}[3x + 2(x - 1)] = 0$

$(x - 1)^{1/2}(5x - 2) = 0$

$(x - 1)^{1/2} = 0 \implies x - 1 = 0 \implies x = 1$

$5x - 2 = 0 \implies x = \frac{2}{5}$, extraneous

53. $y = \sqrt{11x - 30} - x$

(a)

(b) *x*-intercepts: $(5, 0)$, $(6, 0)$

(c) $0 = \sqrt{11x - 30} - x$

$x = \sqrt{11x - 30}$

$x^2 = 11x - 30$

$x^2 - 11x + 30 = 0$

$(x - 5)(x - 6) = 0$

$x - 5 = 0 \implies x = 5$

$x - 6 = 0 \implies x = 6$

(d) The *x*-intercepts of the graph are the same as the solutions to the equation.

55. $y = \sqrt{7x + 36} - \sqrt{5x + 16} - 2$

(a)

(b) *x*-intercepts: $(0, 0)$, $(4, 0)$

(c) $0 = \sqrt{7x + 36} - \sqrt{5x + 16} - 2$

$-\sqrt{7x + 36} = -\sqrt{5x + 16} - 2$

$\sqrt{7x + 36} = 2 + \sqrt{5x + 16}$

$\left(\sqrt{7x + 36}\right)^2 = \left(2 + \sqrt{5x + 16}\right)^2$

$7x + 36 = 4 + 4\sqrt{5x + 16} + 5x + 16$

$7x + 36 = 5x + 20 + 4\sqrt{5x + 16}$

$2x + 16 = 4\sqrt{5x + 16}$

$x + 8 = 2\sqrt{5x + 16}$

$(x + 8)^2 = \left(2\sqrt{5x + 16}\right)^2$

$x^2 + 16x + 64 = 4(5x + 16)$

$x^2 + 16x + 64 = 20x + 64$

$x^2 - 4x = 0$

$x(x - 4) = 0$

$x = 0$

$x - 4 = 0 \implies x = 4$

(d) The *x*-intercepts of the graph are the same as the solutions to the equation.

57.
$$x = \frac{3}{x} + \frac{1}{2}$$
$$(2x)(x) = (2x)\left(\frac{3}{x}\right) + (2x)\left(\frac{1}{2}\right)$$
$$2x^2 = 6 + x$$
$$2x^2 - x - 6 = 0$$
$$(2x + 3)(x - 2) = 0$$
$$2x + 3 = 0 \implies x = -\frac{3}{2}$$
$$x - 2 = 0 \implies x = 2$$

59.
$$\frac{1}{x} - \frac{1}{x + 1} = 3$$
$$x(x + 1)\frac{1}{x} - x(x + 1)\frac{1}{x + 1} = x(x + 1)(3)$$
$$x + 1 - x = 3x(x + 1)$$
$$1 = 3x^2 + 3x$$
$$0 = 3x^2 + 3x - 1$$
$$a = 3, \quad b = 3, \quad c = -1$$
$$x = \frac{-3 \pm \sqrt{(3)^2 - 4(3)(-1)}}{2(3)} = \frac{-3 \pm \sqrt{21}}{6}$$

61. $\dfrac{20 - x}{x} = x$
$$20 - x = x^2$$
$$0 = x^2 + x - 20$$
$$0 = (x + 5)(x - 4)$$
$$x + 5 = 0 \implies x = -5$$
$$x - 4 = 0 \implies x = 4$$

63.
$$\frac{x}{x^2 - 4} + \frac{1}{x + 2} = 3$$
$$(x + 2)(x - 2)\frac{x}{x^2 - 4} + (x + 2)(x - 2)\frac{1}{x + 2} = 3(x + 2)(x - 2)$$
$$x + x - 2 = 3x^2 - 12$$
$$3x^2 - 2x - 10 = 0$$
$$a = 3, \quad b = -2, \quad c = -10$$
$$x = \frac{-(-2) \pm \sqrt{(-2)^2 - 4(3)(-10)}}{2(3)}$$
$$= \frac{2 \pm \sqrt{124}}{6} = \frac{2 \pm 2\sqrt{31}}{6} = \frac{1 \pm \sqrt{31}}{3}$$

65. $|2x - 1| = 5$
$$2x - 1 = 5 \implies x = 3$$
$$-(2x - 1) = 5 \implies x = -2$$

67. $|x| = x^2 + x - 3$

First equation:
$$x = x^2 + x - 3$$
$$x^2 - 3 = 0$$
$$x = \pm\sqrt{3}$$

Second equation:
$$-x = x^2 + x - 3$$
$$x^2 + 2x - 3 = 0$$
$$(x - 1)(x + 3) = 0$$
$$x - 1 = 0 \implies x = 1$$
$$x + 3 = 0 \implies x = -3$$

Only $x = \sqrt{3}$ and $x = -3$ are solutions to the original equation. $x = -\sqrt{3}$ and $x = 1$ are extraneous.

69. $|x + 1| = x^2 - 5$

First equation:

$$x + 1 = x^2 - 5$$

$$x^2 - x - 6 = 0$$

$$(x - 3)(x + 2) = 0$$

$$x - 3 = 0 \implies x = 3$$

$$x + 2 = 0 \implies x = -2$$

Second equation:

$$-(x + 1) = x^2 - 5$$

$$-x - 1 = x^2 - 5$$

$$x^2 + x - 4 = 0$$

$$x = \frac{-1 \pm \sqrt{17}}{2}$$

Only $x = 3$ and $x = \dfrac{-1 - \sqrt{17}}{2}$ are solutions to the original equation. $x = -2$ and $x = \dfrac{-1 + \sqrt{17}}{2}$ are extraneous.

71. $y = \dfrac{1}{x} - \dfrac{4}{x - 1} - 1$

(a)

(b) x-intercept: $(-1, 0)$

(c)

$$0 = \frac{1}{x} - \frac{4}{x - 1} - 1$$

$$0 = (x - 1) - 4x - x(x - 1)$$

$$0 = x - 1 - 4x - x^2 + x$$

$$0 = -x^2 - 2x - 1$$

$$0 = x^2 + 2x + 1$$

$$0 = (x + 1)^2$$

$$x + 1 = 0 \implies x = -1$$

(d) The x-intercept of the graph is the same as the solution to the equation.

73. $y = |x + 1| - 2$

(a)

(b) x-intercepts: $(1, 0)$, $(-3, 0)$

(c)

$$0 = |x + 1| - 2$$

$$2 = |x + 1|$$

$$x + 1 = 2 \qquad \text{OR} \qquad -(x + 1) = 2$$

$$x = 1 \qquad \text{OR} \qquad -x - 1 = 2$$

$$-x = 3$$

$$x = -3$$

(d) The x-intercepts of the graph are the same as the solutions to the equation.

75. $3.2x^4 - 1.5x^2 - 2.1 = 0$

$$x^2 = \frac{1.5 \pm \sqrt{1.5^2 - 4(3.2)(-2.1)}}{2(3.2)}$$

Using the positive value for x^2, we have $x = \pm \sqrt{\dfrac{1.5 + \sqrt{29.13}}{6.4}} \approx \pm 1.038$.

77. $1.8x - 6\sqrt{x} - 5.6 = 0$ Given equation

$1.8\left(\sqrt{x}\right)^2 - 6\sqrt{x} - 5.6 = 0$

Use the Quadratic Formula with $a = 1.8$, $b = -6$, and $c = -5.6$.

$$\sqrt{x} = \frac{6 \pm \sqrt{36 - 4(1.8)(-5.6)}}{2(1.8)} \approx \frac{6 \pm 8.7361}{3.6}$$

Considering only the positive value for $\sqrt{x}$, we have:

$$\sqrt{x} \approx 4.0934$$

$$x \approx 16.756.$$

79. -2 and 5

One possible equation is:

$$(x - (-2))(x - 5) = 0$$

$$(x + 2)(x - 5) = 0$$

$$x^2 - 3x - 10 = 0$$

Any non-zero multiple of this equation would also have these solutions.

81. $-\frac{7}{3}$ and $\frac{6}{7}$

One possible equation is:

$x = -\frac{7}{3} \implies 3x = -7 \implies 3x + 7$ is a factor.

$x = \frac{6}{7} \implies 7x = 6 \implies 7x - 6$ is a factor.

$(3x + 7)(7x - 6) = 0$

$21x^2 + 31x - 42 = 0$

Any non-zero multiple of this equation would also have these solutions.

83. $\sqrt{3}, -\sqrt{3},$ and 4

One possible equation is:

$(x - \sqrt{3})(x - (-\sqrt{3}))(x - 4) = 0$

$(x - \sqrt{3})(x + \sqrt{3})(x - 4) = 0$

$(x^2 - 3)(x - 4) = 0$

$x^3 - 4x^2 - 3x + 12 = 0$

Any non-zero multiple of this equation would also have these solutions.

85. $-1, 1, i,$ and $-i$

One possible equation is:

$(x - (-1))(x - 1)(x - i)(x - (-i)) = 0$

$(x + 1)(x - 1)(x - i)(x + i) = 0$

$(x^2 - 1)(x^2 + 1) = 0$

$x^4 - 1 = 0$

Any non-zero multiple of this equation would also have these solutions.

87. Let $x =$ the number of students in the original group. Then, $\dfrac{1700}{x} =$ the original cost per student.

When six more students join the group, the cost per student becomes $\dfrac{1700}{x} - 7.50$.

Model: (Cost per student) · (Number of students) = Total cost

$$\left(\frac{1700}{x} - 7.5\right)(x + 6) = 1700$$

$(3400 - 15x)(x + 6) = 3400x$ Multiply both sides by $2x$ to clear fractions.

$-15x^2 - 90x + 20,400 = 0$

$$x = \frac{90 \pm \sqrt{(-90)^2 - 4(-15)(20,400)}}{2(-15)} = \frac{90 \pm 1110}{-30}$$

Using the positive value for x we conclude that the original number was $x = 34$ students.

89. *Model:* Time $= \dfrac{\text{Distance}}{\text{Rate}}$

Labels: Let $x =$ average speed of the plane. Then we have a travel time of $t = 145/x$. If the average speed is increased by 40 mph, then

$$t - \frac{12}{60} = \frac{145}{x + 40}$$

$$t = \frac{145}{x + 40} + \frac{1}{5}$$

Now, we equate these two equations and solve for x.

Equation:

$$\frac{145}{x} = \frac{145}{x + 40} + \frac{1}{5}$$

$145(5)(x + 40) = 145(5)x + x(x + 40)$

$725x + 29,000 = 725x + x^2 + 40x$

$0 = x^2 + 40x - 29,000$

Using the positive value for x found by the Quadratic Formula, we have $x \approx 151.5$ mph and $x + 40 = 191.5$ mph. The airspeed required to obtain the decrease in travel time is 191.5 miles per hour.

91.

$$A = P\left(1 + \frac{r}{n}\right)^{nt}$$

$$3052.49 = 2500\left(1 + \frac{r}{12}\right)^{(12)(5)}$$

$$1.220996 = \left(1 + \frac{r}{12}\right)^{60}$$

$$(1.220996)^{1/60} = 1 + \frac{r}{12}$$

$$[(1.220996)^{1/60} - 1](12) = r$$

$$r \approx 0.04 = 4\%$$

93. $M = 463.97 + 111.6\sqrt{t},\ 4 \le t \le 12$

(a) $463.97 + 111.6\sqrt{t} = 816$

$$111.6\sqrt{t} = 352.03$$

$$t = \left(\frac{352.03}{111.6}\right)^2$$

$$t \approx 9.95$$

The number of medical doctors reached 816,000 late during the year 1999.

(b) $463.97 + 111.6\sqrt{t} = 900$

$$111.6\sqrt{t} = 436.03$$

$$t = \left(\frac{436.03}{111.6}\right)^2$$

$$t \approx 15.27$$

The model predicts the number of medical doctors will reach 900,000 during the year 2005.

The actual number of medical doctors in 2005 is about 800,000.

95. $T = 75.82 - 2.11x + 43.51\sqrt{x},\ 5 \le x \le 40$

(a)

Absolute Pressure, x	Temperature T
5	162.56
10	192.31
15	212.68
20	228.20
25	240.62
30	250.83
35	259.38
40	266.60

(b) $T = 212°$ when $x \approx 15$ pounds per square inch.

(c) $212 = 75.82 - 2.11x + 43.51\sqrt{x}$

$$0 = -2.11x + 43.51\sqrt{x} - 136.18$$

By the Quadratic Formula, we have

$$\sqrt{x} \approx 16.77298 \implies x \approx 281.333$$

$$\sqrt{x} \approx 3.84787 \implies x \approx 14.806$$

Since x is restricted to $5 \le x \le 40$, let $x = 14.806$ pounds per square inch.

(d)

97.

$$37.55 = 40 - \sqrt{0.01x + 1}$$

$$\sqrt{0.01x + 1} = 2.45$$

$$0.01x + 1 = 6.0025$$

$$0.01x = 5.0025$$

$$x = 500.25$$

Rounding x to the nearest whole unit yields $x \approx 500$ units.

99. *Model:* $\left(\genfrac{}{}{0pt}{}{\text{Distance from}}{\text{home to 1st}}\right)^2 + \left(\genfrac{}{}{0pt}{}{\text{distance from}}{\text{1st to 2nd}}\right)^2 = \left(\genfrac{}{}{0pt}{}{\text{distance from}}{\text{home to 2nd}}\right)^2$

Labels: Distance from home to 1st = x, distance from 1st to 2nd = x, distance from home to 2nd = 127.5

Equation: $x^2 + x^2 = (127.5)^2$

$$2x^2 = 16{,}256.25$$

$$x^2 = \frac{16{,}256.25}{2}$$

$$x = \pm\sqrt{8128.125} \approx \pm 90$$

The distance between bases is approximately 90 feet.

101. $S = 8\pi\sqrt{64 + h^2}$

(a)

When $S = 350$, $h \approx 11.4$.

(b)

h	8	9	10	11	12	13
S	284.3	302.6	321.9	341.8	362.5	383.6

$S = 350$ when h is between 11 and 12 inches.

(c) $\qquad 350 = 8\pi\sqrt{64 + h^2}$

$$(350)^2 = \left(8\pi\sqrt{64 + h^2}\right)^2$$

$$122{,}500 = 64\pi^2(64 + h^2)$$

$$64 + h^2 \approx 193.935$$

$$h^2 \approx 129.935$$

$$h \approx 11.4$$

(d) Solving graphically or numerically yields an approximate solution. An exact solution is obtained algebraically.

103. *Model:* $\left(\genfrac{}{}{0pt}{}{\text{Portion done}}{\text{by first person}}\right) + \left(\genfrac{}{}{0pt}{}{\text{portion done}}{\text{by second person}}\right) = \left(\genfrac{}{}{0pt}{}{\text{work}}{\text{done}}\right)$

Labels: Work done = 1, rate of first person = $\dfrac{1}{r}$, time worked by first person = 12,

rate of second person = $\dfrac{1}{r+3}$, time worked by second person = 12

Equation: $\qquad\qquad \dfrac{12}{r} + \dfrac{12}{r+3} = 1$

$$r(r+3)\frac{12}{r} + r(r+3)\frac{12}{r+3} = r(r+3)$$

$$12r + 36 + 12r = r^2 + 3r$$

$$0 = r^2 - 21r - 36$$

$$r = \frac{-(-21) \pm \sqrt{(-21)^2 - 4(1)(-36)}}{2(1)} = \frac{21 \pm \sqrt{585}}{2}$$

$$r \approx 23 \qquad \text{(Choose the positive value for } r.\text{)}$$

It would take approximately 23 hours and 26 hours individually.

105. $\qquad v = \sqrt{\dfrac{gR}{\mu s}}$

$$v^2 = \frac{gR}{\mu s}$$

$$v^2 \mu s = gR$$

$$\frac{v^2 \mu s}{R} = g$$

107. False—See Example 7 on page 137.

109. The distance between $(1, 2)$ and $(x, -10)$ is 13.

$$\sqrt{(x - 1)^2 + (-10 - 2)^2} = 13$$

$$(x - 1)^2 + (-12)^2 = 13^2$$

$$x^2 - 2x + 1 + 144 = 169$$

$$x^2 - 2x - 24 = 0$$

$$(x + 4)(x - 6) = 0$$

$$x + 4 = 0 \implies x = -4$$

$$x - 6 = 0 \implies x = 6$$

Both $(-4, -10)$ and $(6, -10)$ are a distance of 13 from $(1, 2)$.

111. The distance between $(0, 0)$ and $(8, y)$ is 17.

$$\sqrt{(8 - 0)^2 + (y - 0)^2} = 17$$

$$(8)^2 + (y)^2 = 17^2$$

$$64 + y^2 = 289$$

$$y^2 = 225$$

$$y = \pm\sqrt{225}$$

$$= \pm 15$$

Both $(8, 15)$ and $(8, -15)$ are a distance of 17 from $(0, 0)$.

113. $9 + |9 - a| = b$

$$|9 - a| = b - 9$$

$$9 - a = b - 9 \quad \text{OR} \quad 9 - a = -(b - 9)$$

$$-a = b - 18 \qquad\qquad 9 - a = -b + 9$$

$$a = 18 - b \qquad\qquad\quad -a = -b$$

$$a = b$$

Thus, $a = 18 - b$ or $a = b$. From the original equation we know that $b \geq 9$.

Some possibilities are: $b = 9, a = 9$

$$b = 10, a = 8 \text{ or } a = 10$$

$$b = 11, a = 7 \text{ or } a = 11$$

$$b = 12, a = 6 \text{ or } a = 12$$

$$b = 13, a = 5 \text{ or } a = 13$$

$$b = 14, a = 4 \text{ or } a = 14$$

115. $20 + \sqrt{20 - a} = b$

$$\sqrt{20 - a} = b - 20$$

$$20 - a = b^2 - 40b + 400$$

$$-a = b^2 - 40b + 380$$

$$a = -b^2 + 40b - 380$$

This formula gives the relationship between a and b. From the original equation we know that $a \leq 20$ and $b \geq 20$. Choose a b value, where $b \geq 20$ and then solve for a, keeping in mind that $a \leq 20$.

Some possibilities are: $b = 20, \quad a = 20$

$$b = 21, \ a = 19$$

$$b = 22, \ a = 16$$

$$b = 23, \ a = 11$$

$$b = 24, \ a = 4$$

$$b = 25, \ a = -5$$

117. $\dfrac{8}{3x} + \dfrac{3}{2x} = \dfrac{16}{6x} + \dfrac{9}{6x} = \dfrac{25}{6x}$

119. $\dfrac{2}{z + 2} - \left(3 - \dfrac{2}{z}\right) = \dfrac{2}{z + 2} - 3 + \dfrac{2}{z}$

$$= \frac{2z - 3z(z + 2) + 2(z + 2)}{z(z + 2)}$$

$$= \frac{2z - 3z^2 - 6z + 2z + 4}{z(z + 2)}$$

$$= \frac{-3z^2 - 2z + 4}{z(z + 2)}$$

121. $x^2 - 22x + 121 = 0$

$$(x - 11)^2 = 0$$

$$x - 11 = 0$$

$$x = 11$$

Section 1.7   Linear Inequalities in One Variable

- ■ You should know the properties of inequalities.

  (a) Transitive: $a < b$ and $b < c$ implies $a < c$.

  (b) Addition: $a < b$ and $c < d$ implies $a + c < b + d$.

  (c) Adding or Subtracting a Constant: $a \pm c < b \pm c$ if $a < b$.

  (d) Multiplying or Dividing a Constant: For $a < b$,

    1. If $c > 0$, then $ac < bc$ and $\dfrac{a}{c} < \dfrac{b}{c}$.

    2. If $c < 0$, then $ac > bc$ and $\dfrac{a}{c} > \dfrac{b}{c}$.

- ■ You should be able to solve absolute value inequalities.

  (a) $|x| < a$ if and only if $-a < x < a$.

  (b) $|x| > a$ if and only if $x < -a$ or $x > a$.

Vocabulary Check

1. solution set

2. graph

3. negative

4. equivalent

5. double

6. union

1. Interval: $[-1, 5]$

  (a) Inequality: $-1 \le x \le 5$

  (b) The interval is bounded.

3. Interval: $(11, \infty)$

  (a) Inequality: $x > 11$

  (b) The interval is unbounded.

5. Interval: $(-\infty, -2)$

  (a) Inequality: $x < -2$

  (b) The interval is unbounded.

7. $x < 3$

Matches (b).

9. $-3 < x \le 4$

Matches (d).

11. $|x| < 3 \implies -3 < x < 3$

Matches (e).

13. $5x - 12 > 0$

  (a) $x = 3$

  $5(3) - 12 \overset{?}{>} 0$

  $3 > 0$

  Yes, $x = 3$ *is* a solution.

  (b) $x = -3$

  $5(-3) - 12 \overset{?}{>} 0$

  $-27 \not> 0$

  No, $x = -3$ *is not* a solution.

  (c) $x = \dfrac{5}{2}$

  $5\left(\dfrac{5}{2}\right) - 12 \overset{?}{>} 0$

  $\dfrac{1}{2} > 0$

  Yes, $x = \dfrac{5}{2}$ *is* a solution.

  (d) $x = \dfrac{3}{2}$

  $5\left(\dfrac{3}{2}\right) - 12 \overset{?}{>} 0$

  $-\dfrac{9}{2} \not> 0$

  No, $x = \dfrac{3}{2}$ *is not* a solution.

15. $0 < \dfrac{x - 2}{4} < 2$

  (a) $x = 4$

  $0 \overset{?}{<} \dfrac{4 - 2}{4} \overset{?}{<} 2$

  $0 < \dfrac{1}{2} < 2$

  Yes, $x = 4$ *is* a solution.

  (b) $x = 10$

  $0 \overset{?}{<} \dfrac{10 - 2}{4} \overset{?}{<} 2$

  $0 < 2 \not< 2$

  No, $x = 10$ *is not* a solution.

  (c) $x = 0$

  $0 \overset{?}{<} \dfrac{0 - 2}{4} \overset{?}{<} 2$

  $0 \not< -\dfrac{1}{2} < 2$

  No, $x = 0$ *is not* a solution.

  (d) $x = \dfrac{7}{2}$

  $0 \overset{?}{<} \dfrac{(7/2) - 2}{4} \overset{?}{<} 2$

  $0 < \dfrac{3}{8} < 2$

  Yes, $x = \dfrac{7}{2}$ *is a solution.*

17. $|x - 10| \geq 3$

(a) $x = 13$

$|13 - 10| \overset{?}{\geq} 3$

$3 \geq 3$

Yes, $x = 13$ *is* a solution.

(b) $x = -1$

$|-1 - 10| \overset{?}{\geq} 3$

$11 \geq 3$

Yes, $x = -1$ *is* a solution.

(c) $x = 14$

$|14 - 10| \overset{?}{\geq} 3$

$4 \geq 3$

Yes, $x = 14$ *is* a solution.

(d) $x = 9$

$|9 - 10| \overset{?}{\geq} 3$

$1 \not\geq 3$

No, $x = 9$ *is not* a solution.

19. $4x < 12$

$\frac{1}{4}(4x) < \frac{1}{4}(12)$

$x < 3$

21. $-2x > -3$

$-\frac{1}{2}(-2x) < \left(-\frac{1}{2}\right)(-3)$

$x < \frac{3}{2}$

23. $x - 5 \geq 7$

$x \geq 12$

25. $2x + 7 < 3 + 4x$

$-2x < -4$

$x > 2$

27. $2x - 1 \geq 1 - 5x$

$7x \geq 2$

$x \geq \frac{2}{7}$

29. $4 - 2x < 3(3 - x)$

$4 - 2x < 9 - 3x$

$x < 5$

31. $\frac{3}{4}x - 6 \leq x - 7$

$-\frac{1}{4}x \leq -1$

$x \geq 4$

33. $\frac{1}{2}(8x + 1) \geq 3x + \frac{5}{2}$

$4x + \frac{1}{2} \geq 3x + \frac{5}{2}$

$x \geq 2$

35. $3.6x + 11 \geq -3.4$

$3.6x \geq -14.4$

$x \geq -4$

37. $1 < 2x + 3 < 9$

$-2 < 2x < 6$

$-1 < x < 3$

39. $-4 < \dfrac{2x - 3}{3} < 4$

$-12 < 2x - 3 < 12$

$-9 < 2x < 15$

$-\frac{9}{2} < x < \frac{15}{2}$

41. $\dfrac{3}{4} > x + 1 > \dfrac{1}{4}$

$-\frac{1}{4} > x > -\frac{3}{4}$

$-\frac{3}{4} < x < -\frac{1}{4}$

43. $3.2 \leq 0.4x - 1 \leq 4.4$

$4.2 \leq 0.4x \leq 5.4$

$10.5 \leq x \leq 13.5$

45. $|x| < 6$

$-6 < x < 6$

47. $\left|\dfrac{x}{2}\right| > 1$

$\dfrac{x}{2} < -1$ or $\dfrac{x}{2} > 1$

$x < -2 \qquad x > 2$

49. $|x - 5| < -1$

No solution. The absolute value of a number cannot be less than a negative number.

51. $|x - 20| \leq 6$

$$-6 \leq x - 20 \leq 6$$

$$14 \leq x \leq 26$$

53. $|3 - 4x| \geq 9$

$$3 - 4x \leq -9 \quad \text{or} \quad 3 - 4x \geq 9$$

$$-4x \leq -12 \qquad\qquad -4x \geq 6$$

$$x \geq 3 \qquad\qquad\qquad x \leq -\tfrac{3}{2}$$

55. $\left|\dfrac{x - 3}{2}\right| \geq 4$

$$\frac{x - 3}{2} \leq -4 \quad \text{or} \quad \frac{x - 3}{2} \geq 4$$

$$x - 3 \leq -8 \qquad\qquad x - 3 \geq 8$$

$$x \leq -5 \qquad\qquad\qquad x \geq 11$$

57. $|9 - 2x| - 2 < -1$

$$|9 - 2x| < 1$$

$$-1 < 9 - 2x < 1$$

$$-10 < -2x < -8$$

$$5 > x > 4$$

$$4 < x < 5$$

59. $2|x + 10| \geq 9$

$$|x + 10| \geq \tfrac{9}{2}$$

$$x + 10 \leq -\tfrac{9}{2} \quad \text{or} \quad x + 10 \geq \tfrac{9}{2}$$

$$x \leq -\tfrac{29}{2} \qquad\qquad x \geq -\tfrac{11}{2}$$

61. $6x > 12$

$$x > 2$$

63. $5 - 2x \geq 1$

$$-2x \geq -4$$

$$x \leq 2$$

65. $|x - 8| \leq 14$

$$-14 \leq x - 8 \leq 14$$

$$-6 \leq x \leq 22$$

67. $2|x + 7| \geq 13$

$$|x + 7| \geq \tfrac{13}{2}$$

$$x + 7 \leq -\tfrac{13}{2} \quad \text{or} \quad x + 7 \geq \tfrac{13}{2}$$

$$x \leq -\tfrac{27}{2} \qquad\qquad x \geq -\tfrac{1}{2}$$

69. $y = 2x - 3$

(a) $\qquad y \geq 1$

$$2x - 3 \geq 1$$

$$2x \geq 4$$

$$x \geq 2$$

(b) $\qquad y \leq 0$

$$2x - 3 \leq 0$$

$$2x \leq 3$$

$$x \leq \tfrac{3}{2}$$

71. $y = -\frac{1}{2}x + 2$

(a) $0 \le y \le 3$

$0 \le -\frac{1}{2}x + 2 \le 3$

$-2 \le -\frac{1}{2}x \le 1$

$4 \ge x \ge -2$

(b) $y \ge 0$

$-\frac{1}{2}x + 2 \ge 0$

$-\frac{1}{2}x \ge -2$

$x \le 4$

73. $y = |x - 3|$

(a) $y \le 2$

$|x - 3| \le 2$

$-2 \le x - 3 \le 2$

$1 \le x \le 5$

(b) $y \ge 4$

$|x - 3| \ge 4$

$x - 3 \le -4$ or $x - 3 \ge 4$

$x \le -1$ or $x \ge 7$

75. $x - 5 \ge 0$

$x \ge 5$

$[5, \infty)$

77. $x + 3 \ge 0$

$x \ge -3$

$[-3, \infty)$

79. $7 - 2x \ge 0$

$-2x \ge -7$

$x \le \frac{7}{2}$

$\left(-\infty, \frac{7}{2}\right]$

81. $|x - 10| < 8$

All real numbers within 8 units of 10.

83. The midpoint of the interval $[-3, 3]$ is 0. The interval represents all real numbers x no more than 3 units from 0.

$|x - 0| \le 3$

$|x| \le 3$

85. The graph shows all real numbers at least 3 units from 7.

$|x - 7| \ge 3$

87. All real numbers within 10 units of 12

$|x - 12| < 10$

89. All real numbers more than 4 units from -3

$|x - (-3)| > 4$

$|x + 3| > 4$

91. Let $x =$ the number of checks written in a month.

Type A account charges: $6.00 + 0.25x$

Type B account charges: $4.50 + 0.50x$

$6.00 + 0.25x < 4.50 + 0.50x$

$1.50 < 0.25x$

$6 < x$

If you write more than six checks a month, then the charges for the type A account are less than the charges for the type B account.

93. $1000(1 + r(2)) > 1062.50$

$1 + 2r > 1.0625$

$2r > 0.0625$

$r > 0.03125$

$r > 3.125\%$

95. $R > C$

$115.95x > 95x + 750$

$20.95x > 750$

$x > 35.7995$

$x \ge 36$ units

97. Let x = daily sales level (in dozens) of doughnuts.

   Revenue: $R = 2.95x$

   Cost: $C = 150 + 1.45x$

   Profit: $P = R - C$

$$= 2.95x - (150 + 1.45x)$$

$$= 1.50x - 150$$

$$50 \le P \le 200$$

$$50 \le 1.50x - 150 \le 200$$

$$200 \le 1.50x \le 350$$

$$133\tfrac{1}{3} \le x \le 233\tfrac{1}{3}$$

In whole dozens, $134 \le x \le 234$.

99. (a) $y = 0.067x - 5.638$

   (b) From the graph we see that $y \ge 3$ when $x \ge 129$. Algebraically we have:

$$3 \le 0.067x - 5.638$$

$$8.638 \le 0.067x$$

$$x \ge 129$$

IQ scores are not a good predictor of GPAs. Other factors include study habits, class attendance, and attitude.

101. $S = 1.05t + 31.0, \; 0 \le t \le 12$

   (a) $32 \le 1.05t + 31 \le 42$

$$1 \le 1.05t \le 11$$

$$0.95 \le t \le 10.48$$

     Rounding to the nearest year, $1 \le t \le 10$. The average salary was at least \$32,000 but not more than \$42,000 between 1991 and 2000.

   (b) $1.05t + 31 > 48$

$$1.05t > 17$$

$$t > 16$$

     According to the model, the average salary will exceed \$48,000 in 2006.

103. $|s - 10.4| \le \tfrac{1}{16}$

$$-\tfrac{1}{16} \le s - 10.4 \le \tfrac{1}{16}$$

$$-0.0625 \le s - 10.4 \le 0.0625$$

$$10.3375 \le s \le 10.4625$$

Since $A = s^2$, we have

$$(10.3375)^2 \le \text{area} \le (10.4625)^2$$

$$106.864 \le \text{area} \le 109.464.$$

105. $|x - 15| \le \tfrac{1}{10}$

$$-\tfrac{1}{10} \le x - 15 \le \tfrac{1}{10}$$

$$14.9 \le x \le 15.1 \text{ gallons}$$

$$\tfrac{1}{10}(\$1.89) \approx \$0.19$$

You might have been undercharged or overcharged by \$0.19.

107. $\left| \dfrac{t - 15.6}{1.9} \right| < 1$

$$-1 < \frac{t - 15.6}{1.9} < 1$$

$$-1.9 < t - 15.6 < 1.9$$

$$13.7 < t < 17.5$$

Two-thirds of the workers could perform the task in the time interval between 13.7 minutes and 17.5 minutes.

109. $|h - 50| \le 30$

$$-30 \le h - 50 \le 30$$

$$20 \le h \le 80$$

The minimum relative humidity is 20 and the maximum is 80.

111. False. If c is negative, then $ac \ge bc$.

113. $|x - a| \ge 2$

$$x - a \le -2$$

$$x \le a - 2 \text{ or}$$

$$x - a \ge 2$$

$$x \ge a + 2$$

Matches (b).

115. $(-4, 2)$ and $(1, 12)$

$d = \sqrt{(1 - (-4))^2 + (12 - 2)^2} = \sqrt{5^2 + 10^2} = \sqrt{125} = 5\sqrt{5}$

Midpoint: $\left(\dfrac{-4 + 1}{2}, \dfrac{2 + 12}{2}\right) = \left(-\dfrac{3}{2}, 7\right)$

117. $(3, 6)$ and $(-5, -8)$

$d = \sqrt{(-5 - 3)^2 + (-8 - 6)^2} = \sqrt{(-8)^2 + (-14)^2} = \sqrt{260} = 2\sqrt{65}$

Midpoint: $\left(\dfrac{3 + (-5)}{2}, \dfrac{6 + (-8)}{2}\right) = (-1, -1)$

119. $-6(2 - x) - 12 = 36$

$-12 + 6x - 12 = 36$

$6x - 24 = 36$

$6x = 60$

$x = 10$

121. $14x^2 + 5x - 1 = 0$

$(7x - 1)(2x + 1) = 0$

$7x - 1 = 0$ or $2x + 1 = 0$

$7x = 1 \qquad 2x = -1$

$x = \frac{1}{7} \qquad x = -\frac{1}{2}$

123. $(-3, 10)$

125. Answers will vary.

Section 1.8   Other Types of Inequalities

- You should be able to solve inequalities.

  (a) Find the critical number.

    1. Values that make the expression zero

    2. Values that make the expression undefined

  (b) Test one value in each test interval on the real number line resulting from the critical numbers.

  (c) Determine the solution intervals.

Vocabulary Check

1. critical; test intervals

2. zeroes; undefined values

3. $P = R - C$

1. $x^2 - 3 < 0$

(a) $x = 3$

$(3)^2 - 3 \overset{?}{<} 0$

$6 \not< 0$

No, $x = 3$ *is not* a solution.

(b) $x = 0$

$(0)^2 - 3 \overset{?}{<} 0$

$-3 < 0$

Yes, $x = 0$ *is* a solution.

(c) $x = \frac{3}{2}$

$\left(\frac{3}{2}\right)^2 - 3 \overset{?}{<} 0$

$-\frac{3}{4} < 0$

Yes, $x = \frac{3}{2}$ *is* a solution.

(d) $x = -5$

$(-5)^2 - 3 \overset{?}{<} 0$

$22 \not< 0$

No, $x = -5$ *is not* a solution.

3. $\dfrac{x + 2}{x - 4} \geq 3$

(a) $x = 5$

$\dfrac{5 + 2}{5 - 4} \overset{?}{\geq} 3$

$7 \geq 3$

Yes, $x = 5$ *is* a solution.

(b) $x = 4$

$\dfrac{4 + 2}{4 - 4} \overset{?}{\geq} 3$

$\dfrac{6}{0}$ is undefined.

No, $x = 4$ *is not* a solution.

(c) $x = -\dfrac{9}{2}$

$\dfrac{-\frac{9}{2} + 2}{-\frac{9}{2} - 4} \overset{?}{\geq} 3$

$\dfrac{5}{17} \not\geq 3$

No, $x = -\dfrac{9}{2}$ *is not* a solution.

(d) $x = \dfrac{9}{2}$

$\dfrac{\frac{9}{2} + 2}{\frac{9}{2} - 4} \overset{?}{\geq} 3$

$13 \geq 3$

Yes, $x = \dfrac{9}{2}$ *is* a solution.

5. $2x^2 - x - 6 = (2x + 3)(x - 2)$

$2x + 3 = 0 \implies x = -\dfrac{3}{2}$

$x - 2 = 0 \implies x = 2$

Critical numbers: $x = -\dfrac{3}{2}, x = 2$

7. $2 + \dfrac{3}{x - 5} = \dfrac{2(x - 5) + 3}{x - 5}$

$\qquad\qquad = \dfrac{2x - 7}{x - 5}$

$2x - 7 = 0 \implies x = \dfrac{7}{2}$

$x - 5 = 0 \implies x = 5$

Critical numbers: $x = \dfrac{7}{2}, x = 5$

9. $\qquad\qquad x^2 \leq 9$

$\qquad\qquad x^2 - 9 \leq 0$

$(x + 3)(x - 3) \leq 0$

Critical numbers: $x = \pm 3$

Test intervals: $(-\infty, -3), (-3, 3), (3, \infty)$

Test: Is $(x + 3)(x - 3) \leq 0$?

Interval	x-Value	Value of $x^2 - 9$	Conclusion
$(-\infty, -3)$	$x = -4$	$16 - 9 = 7$	Positive
$(-3, 3)$	$x = 0$	$0 - 9 = -9$	Negative
$(3, \infty)$	$x = 4$	$16 - 9 = 7$	Positive

Solution set: $[-3, 3]$

11. $\qquad\qquad (x + 2)^2 < 25$

$\qquad\qquad x^2 + 4x + 4 < 25$

$\qquad\qquad x^2 + 4x - 21 < 0$

$\qquad\qquad (x + 7)(x - 3) < 0$

Critical numbers: $x = -7, x = 3$

Test intervals: $(-\infty, -7), (-7, 3), (3, \infty)$

Test: Is $(x + 7)(x - 3) < 0$?

Interval	x-Value	Value of $(x + 7)(x - 3)$	Conclusion
$(-\infty, -7)$	$x = -10$	$(-3)(-13) = 39$	Positive
$(-7, 3)$	$x = 0$	$(7)(-3) = -21$	Negative
$(3, \infty)$	$x = 5$	$(12)(2) = 24$	Positive

Solution set: $(-7, 3)$

13. $x^2 + 4x + 4 \geq 9$

  $x^2 + 4x - 5 \geq 0$

  $(x + 5)(x - 1) \geq 0$

  Critical numbers: $x = -5, x = 1$

  Test intervals: $(-\infty, -5), (-5, 1), (1, \infty)$

  Test: Is $(x + 5)(x - 1) \geq 0$?

Interval	x-Value	Value of $(x + 5)(x - 1)$	Conclusion
$(-\infty, -5)$	$x = -6$	$(-1)(-7) = 7$	Positive
$(-5, 1)$	$x = 0$	$(5)(-1) = -5$	Negative
$(1, \infty)$	$x = 2$	$(7)(1) = 7$	Positive

  Solution set: $(-\infty, -5] \cup [1, \infty)$

15. $x^2 + x < 6$

  $x^2 + x - 6 < 0$

  $(x + 3)(x - 2) < 0$

  Critical numbers: $x = -3, x = 2$

  Test intervals: $(-\infty, -3), (-3, 2), (2, \infty)$

  Test: Is $(x + 3)(x - 2) < 0$?

Interval	x-Value	Value of $(x + 3)(x - 2)$	Conclusion
$(-\infty, -3)$	$x = -4$	$(-1)(-6) = 6$	Positive
$(-3, 2)$	$x = 0$	$(3)(-2) = -6$	Negative
$(2, \infty)$	$x = 3$	$(6)(1) = 6$	Positive

  Solution set: $(-3, 2)$

17. $x^2 + 2x - 3 < 0$

  $(x + 3)(x - 1) < 0$

  Critical numbers: $x = -3, x = 1$

  Test intervals: $(-\infty, -3), (-3, 1), (1, \infty)$

  Test: Is $(x + 3)(x - 1) < 0$?

Interval	x-Value	Value of $(x + 3)(x - 1)$	Conclusion
$(-\infty, -3)$	$x = -4$	$(-1)(-5) = 5$	Positive
$(-3, 1)$	$x = 0$	$(3)(-1) = -3$	Negative
$(1, \infty)$	$x = 2$	$(5)(1) = 5$	Positive

  Solution set: $(-3, 1)$

19. $x^2 + 8x - 5 \geq 0$

  $x^2 + 8x - 5 = 0$ Complete the square.

  $x^2 + 8x + 16 = 5 + 16$

  $(x + 4)^2 = 21$

  $x + 4 = \pm\sqrt{21}$

  $x = -4 \pm \sqrt{21}$

  Critical numbers: $x = -4 \pm \sqrt{21}$

  Test intervals: $\left(-\infty, -4 - \sqrt{21}\right), \left(-4 - \sqrt{21}, -4 + \sqrt{21}\right), \left(-4 + \sqrt{21}, \infty\right)$

  Test: Is $x^2 + 8x - 5 \geq 0$?

Interval	x-Value	Value of $x^2 + 8x - 5$	Conclusion
$\left(-\infty, -4 - \sqrt{21}\right)$	$x = -10$	$100 - 80 - 5 = 15$	Positive
$\left(-4 - \sqrt{21}, -4 + \sqrt{21}\right)$	$x = 0$	$0 + 0 - 5 = -5$	Negative
$\left(-4 + \sqrt{21}, \infty\right)$	$x = 2$	$4 + 16 - 5 = 15$	Positive

  Solution set: $\left(-\infty < -4 - \sqrt{21}\right] \cup \left[-4 + \sqrt{21}, \infty\right)$

21. $x^3 - 3x^2 - x + 3 > 0$

$x^2(x - 3) - 1(x - 3) > 0$

$(x^2 - 1)(x - 3) > 0$

$(x + 1)(x - 1)(x - 3) > 0$

Critical numbers: $x = \pm 1, x = 3$

Test intervals: $(-\infty, -1), (-1, 1), (1, 3), (3, \infty)$

Test: Is $(x + 1)(x - 1)(x - 3) > 0$?

Interval	x-Value	Value of $(x + 1)(x - 1)(x - 3)$	Conclusion
$(-\infty, -1)$	$x = -2$	$(-1)(-3)(-5) = -15$	Negative
$(-1, 1)$	$x = 0$	$(1)(-1)(-3) = 3$	Positive
$(1, 3)$	$x = 2$	$(3)(1)(-1) = -3$	Negative
$(3, \infty)$	$x = 4$	$(5)(3)(1) = 15$	Positive

Solution set: $(-1, 1) \cup (3, \infty)$

23. $x^3 - 2x^2 - 9x - 2 \geq -20$

$x^3 - 2x^2 - 9x + 18 \geq 0$

$x^2(x - 2) - 9(x - 2) \geq 0$

$(x - 2)(x^2 - 9) \geq 0$

$(x - 2)(x + 3)(x - 3) \geq 0$

Critical numbers: $x = 2, x = \pm 3$

Test intervals: $(-\infty, -3), (-3, 2), (2, 3), (3, \infty)$

Test: Is $(x - 2)(x + 3)(x - 3) \geq 0$?

Interval	x-Value	Value of $(x - 2)(x + 3)(x - 3)$	Conclusion
$(-\infty, -3)$	$x = -4$	$(-6)(-1)(-7) = -42$	Negative
$(-3, 2)$	$x = 0$	$(-2)(3)(-3) = 18$	Positive
$(2, 3)$	$x = 2.5$	$(0.5)(5.5)(-0.5) = -1.375$	Negative
$(3, \infty)$	$x = 4$	$(2)(7)(1) = 14$	Positive

Solution set: $[-3, 2] \cup [3, \infty)$

25. $4x^2 - 4x + 1 \leq 0$

$(2x - 1)^2 \leq 0$

Critical number: $x = \frac{1}{2}$

Test intervals: $\left(-\infty, \frac{1}{2}\right), \left(\frac{1}{2}, \infty\right)$

Test: Is $(2x - 1)^2 \leq 0$?

Interval	x-Value	Value of $(2x - 1)^2$	Conclusion
$\left(-\infty, \frac{1}{2}\right)$	$x = 0$	$(-1)^2 = 1$	Positive
$\left(\frac{1}{2}, \infty\right)$	$x = 1$	$(1)^2 = 1$	Positive

Solution set: $x = \frac{1}{2}$

27. $4x^3 - 6x^2 < 0$

$2x^2(2x - 3) < 0$

Critical numbers: $x = 0, x = \frac{3}{2}$

Test intervals: $(-\infty, 0), \left(0, \frac{3}{2}\right), \left(\frac{3}{2}, \infty\right)$

Test: Is $2x^2(2x - 3) < 0$?

By testing an x-value in each test interval in the inequality, we see that the solution set is: $(-\infty, 0) \cup \left(0, \frac{3}{2}\right)$

29. $x^3 - 4x \geq 0$

$x(x + 2)(x - 2) \geq 0$

Critical numbers: $x = 0, x = \pm 2$

Test intervals: $(-\infty, -2), (-2, 0), (0, 2), (2, \infty)$

Test: Is $x(x + 2)(x - 2) \geq 0$?

By testing an x-value in each test interval in the inequality, we see that the solution set is: $[-2, 0] \cup [2, \infty)$

31. $(x - 1)^2(x + 2)^3 \geq 0$

Critical numbers: $x = 1, x = -2$

Test intervals: $(-\infty, -2), (-2, 1), (1, \infty)$

Test: Is $(x - 1)^2(x + 3)^3 \geq 0$?

By testing an x-value in each test interval in the inequality, we see that the solution set is: $[-2, \infty)$

33. $y = -x^2 + 2x + 3$

(a) $y \leq 0$ when $x \leq -1$ or $x \geq 3$.

(b) $y \geq 3$ when $0 \leq x \leq 2$.

35. $y = \frac{1}{8}x^3 - \frac{1}{2}x$

(a) $y \geq 0$ when $-2 \leq x \leq 0, 2 \leq x < \infty$.

(b) $y \leq 6$ when $x \leq 4$.

37. $\dfrac{1}{x} - x > 0$

$\dfrac{1 - x^2}{x} > 0$

Critical numbers: $x = 0, x = \pm 1$

Test intervals: $(-\infty, -1), (-1, 0), (0, 1), (1, \infty)$

Test: Is $\dfrac{1 - x^2}{x} > 0$?

By testing an x-value in each test interval in the inequality, we see that the solution set is: $(-\infty, -1) \cup (0, 1)$

39. $\dfrac{x + 6}{x + 1} - 2 < 0$

$\dfrac{x + 6 - 2(x + 1)}{x + 1} < 0$

$\dfrac{4 - x}{x + 1} < 0$

Critical numbers: $x = -1, x = 4$

Test intervals: $(-\infty, -1), (-1, 4), (4, \infty)$

Test: Is $\dfrac{4 - x}{x + 1} < 0$?

By testing an x-value in each test interval in the inequality, we see that the solution set is: $(-\infty, -1) \cup (4, \infty)$

41. $\dfrac{3x - 5}{x - 5} > 4$

$\dfrac{3x - 5}{x - 5} - 4 > 0$

$\dfrac{3x - 5 - 4(x - 5)}{x - 5} > 0$

$\dfrac{15 - x}{x - 5} > 0$

Critical numbers: $x = 5, x = 15$

Test intervals: $(-\infty, 5), (5, 15), (15, \infty)$

Test: Is $\dfrac{15 - x}{x - 5} > 0$?

By testing an x-value in each test interval in the inequality, we see that the solution set is: $(5, 15)$

43.
$$\frac{4}{x+5} > \frac{1}{2x+3}$$

$$\frac{4}{x+5} - \frac{1}{2x+3} > 0$$

$$\frac{4(2x+3) - (x+5)}{(x+5)(2x+3)} > 0$$

$$\frac{7x+7}{(x+5)(2x+3)} > 0$$

Critical numbers: $x = -1, x = -5, x = -\frac{3}{2}$

Test intervals: $(-\infty, -5), \left(-5, -\frac{3}{2}\right),$

$$\left(-\frac{3}{2}, -1\right), (-1, \infty)$$

Test: Is $\dfrac{7(x+1)}{(x+5)(2x+3)} > 0$?

By testing an x-value in each test interval in the inequality, we see that the solution set is: $\left(-5, -\frac{3}{2}\right) \cup (-1, \infty)$

45.
$$\frac{1}{x-3} \le \frac{9}{4x+3}$$

$$\frac{1}{x-3} - \frac{9}{4x+3} \le 0$$

$$\frac{4x+3 - 9(x-3)}{(x-3)(4x+3)} \le 0$$

$$\frac{30-5x}{(x-3)(4x+3)} \le 0$$

Critical numbers: $x = 3, x = -\frac{3}{4}, x = 6$

Test intervals: $\left(-\infty, -\frac{3}{4}\right), \left(-\frac{3}{4}, 3\right), (3, 6), (6, \infty)$

Test: Is $\dfrac{30-5x}{(x-3)(4x+3)} \le 0$?

By testing an x-value in each test interval in the inequality, we see that the solution set is: $\left(-\frac{3}{4}, 3\right) \cup [6, \infty)$

47.
$$\frac{x^2+2x}{x^2-9} \le 0$$

$$\frac{x(x+2)}{(x+3)(x-3)} \le 0$$

Critical numbers: $x = 0, x = -2, x = \pm 3$

Test intervals: $(-\infty, -3), (-3, -2), (-2, 0), (0, 3), (3, \infty)$

Test: Is $\dfrac{x(x+2)}{(x+3)(x-3)} \le 0$?

By testing an x-value in each test interval in the inequality, we see that the solution set is: $(-3, -2] \cup [0, 3)$

49.
$$\frac{5}{x-1} - \frac{2x}{x+1} < 1$$

$$\frac{5}{x-1} - \frac{2x}{x+1} - 1 < 0$$

$$\frac{5(x+1) - 2x(x-1) - (x-1)(x+1)}{(x-1)(x+1)} < 0$$

$$\frac{5x + 5 - 2x^2 + 2x - x^2 + 1}{(x-1)(x+1)} < 0$$

$$\frac{-3x^2 + 7x + 6}{(x-1)(x+1)} < 0$$

$$\frac{-(3x+2)(x-3)}{(x-1)(x+1)} < 0$$

Critical numbers: $x = -\frac{2}{3}, x = 3, x = \pm 1$

Test intervals: $(-\infty, -1), \left(-1, -\frac{2}{3}\right), \left(-\frac{2}{3}, 1\right), (1, 3), (3, \infty)$

Test: Is $\dfrac{-(3x+2)(x-3)}{(x-1)(x+1)} < 0$?

By testing an x-value in each test interval in the inequality, we see that the solution set is: $(-\infty, -1) \cup \left(-\frac{2}{3}, 1\right) \cup (3, \infty)$

51. $y = \dfrac{3x}{x-2}$

(a) $y \le 0$ when $0 \le x < 2$.

(b) $y \ge 6$ when $2 < x \le 4$.

53. $y = \dfrac{2x^2}{x^2 + 4}$

(a) $y \ge 1$ when $x \le -2$ or $x \ge 2$.

This can also be expressed as $|x| \ge 2$.

(b) $y \le 2$ for all real numbers x.

This can also be expressed as $-\infty < x < \infty$.

55. $4 - x^2 \ge 0$

$(2 + x)(2 - x) \ge 0$

Critical numbers: $x = \pm 2$

Test intervals: $(-\infty, -2), (-2, 2), (2, \infty)$

Test: Is $4 - x^2 \ge 0$?

By testing an x-value in each test interval in the inequality, we see that the domain set is: $[-2, 2]$

57. $x^2 - 7x + 12 \ge 0$

$(x - 3)(x - 4) \ge 0$

Critical numbers: $x = 3, x = 4$

Test intervals: $(-\infty, 3), (3, 4), (4, \infty)$

Test: Is $(x - 3)(x - 4) \ge 0$?

By testing an x-value in each test interval in the inequality, we see that the domain set is: $(-\infty, 3] \cup [4, \infty)$

59. $\dfrac{x}{x^2 - 2x - 35} \geq 0$

$\dfrac{x}{(x + 5)(x - 7)} \geq 0$

Critical numbers: $x = 0, x = -5, x = 7$

Test intervals: $(-\infty, -5), (-5, 0), (0, 7), (7, \infty)$

Test: Is $\dfrac{x}{(x + 5)(x - 7)} \geq 0$?

By testing an x-value in each test interval in the inequality, we see that the domain set is: $(-5, 0] \cup (7, \infty)$

61. $0.4x^2 + 5.26 < 10.2$

$0.4x^2 - 4.94 < 0$

$0.4(x^2 - 12.35) < 0$

Critical numbers: $x \approx \pm 3.51$

Test intervals: $(-\infty, -3.51), (-3.51, 3.51), (3.51, \infty)$

By testing an x-value in each test interval in the inequality, we see that the solution set is: $(-3.51, 3.51)$

63. $-0.5x^2 + 12.5x + 1.6 > 0$

The zeros are $x = \dfrac{-12.5 \pm \sqrt{(12.5)^2 - 4(-0.5)(1.6)}}{2(-0.5)}$.

Critical numbers: $x \approx -0.13, x \approx 25.13$

Test intervals: $(-\infty, -0.13), (-0.13, 25.13),$

$\qquad\qquad (25.13, \infty)$

By testing an x-value in each test interval in the inequality, we see that the solution set is: $(-0.13, 25.13)$

65. $\dfrac{1}{2.3x - 5.2} > 3.4$

$\dfrac{1}{2.3x - 5.2} - 3.4 > 0$

$\dfrac{1 - 3.4(2.3x - 5.2)}{2.3x - 5.2} > 0$

$\dfrac{-7.82x + 18.68}{2.3x - 5.2} > 0$

Critical numbers: $x \approx 2.39, x \approx 2.26$

Test intervals: $(-\infty, 2.26), (2.26, 2.39), (2.39, \infty)$

By testing an x-value in each test interval in the inequality, we see that the solution set is: $(2.26, 2.39)$

67. $s = -16t^2 + v_0 t + s_0 = -16t^2 + 160t$

(a) $-16t^2 + 160t = 0$

$-16t(t - 10) = 0$

$t = 0, t = 10$

It will be back on the ground in 10 seconds.

(b) $-16t^2 + 160t > 384$

$-16t^2 + 160t - 384 > 0$

$-16(t^2 - 10t + 24) > 0$

$t^2 - 10t + 24 < 0$

$(t - 4)(t - 6) < 0$

$\qquad 4 < t < 6$ seconds

69. $2L + 2W = 100 \implies W = 50 - L$

$LW \geq 500$

$L(50 - L) \geq 500$

$-L^2 + 50L - 500 \geq 0$

By the Quadratic Formula we have:

Critical numbers: $L = 25 \pm 5\sqrt{5}$

Test: Is $-L^2 + 50L - 500 \geq 0$?

Solution set: $25 - 5\sqrt{5} \leq L \leq 25 + 5\sqrt{5}$

13.8 meters $\leq L \leq 36.2$ meters

71. $R = x(75 - 0.0005x)$ and $C = 30x + 250,000$

$$P = R - C$$

$$= (75x - 0.0005x^2) - (30x + 250,000)$$

$$= -0.0005x^2 + 45x - 250,000$$

$$P \geq 750,000$$

$$-0.0005x^2 + 45x - 250,000 \geq 750,000$$

$$-0.0005x^2 + 45x - 1,000,000 \geq 0$$

Critical numbers: $x = 40,000$, $x = 50,000$ (These were obtained by using the Quadratic Formula.)

Test intervals: $(0, 40,000)$, $(40,000, 50,000)$, $(50,000, \infty)$

By testing x-values in each test interval in the inequality, we see that the solution set is $[40,000, 50,000]$ or $40,000 \leq x \leq 50,000$. The price per unit is

$$p = \frac{R}{x} = 75 - 0.0005x.$$

For $x = 40,000$, $p = \$55$. For $x = 50,000$, $p = \$50$. Therefore, for $40,000 \leq x \leq 50,000$, $\$50.00 \leq p \leq \$55.00$.

73. $C = 0.0031t^3 - 0.216t^2 + 5.54t + 19.1$, $0 \leq t \leq 23$

(a)

(d)

t	C
36	83.2
37	85.4
38	87.8
39	90.5
40	93.5
41	96.8
42	100.4
43	104.4

(b)

t	C
24	70.5
26	71.6
28	72.9
30	74.6
32	76.8
34	79.6

C will be greater than 75% when $t \approx 31$, which corresponds to 2011.

C will be between 85% and 100% when t is between 37 and 42. These values correspond to the years 2017 to 2022.

(c) $C = 75$ when $t \approx 30.41$.

(e) $85 \leq C \leq 100$ when $36.82 \leq t \leq 41.89$ or $37 \leq t \leq 42$.

(f) The model is a third-degree polynomial and as $t \to \infty$, $C \to \infty$.

75. $\dfrac{1}{R} = \dfrac{1}{R_1} + \dfrac{1}{2}$

$2R_1 = 2R + RR_1$

$2R_1 = R(2 + R_1)$

$\dfrac{2R_1}{2 + R_1} = R$

Since $R \geq 1$, we have

$\dfrac{2R_1}{2 + R_1} \geq 1$

$\dfrac{2R_1}{2 + R_1} - 1 \geq 0$

$\dfrac{R_1 - 2}{2 + R_1} \geq 0.$

Since $R_1 > 0$, the only critical number is $R_1 = 2$. The inequality is satisfied when $R_1 \geq 2$ ohms.

77. True

$x^3 - 2x^2 - 11x + 12 = (x + 3)(x - 1)(x - 4)$

The test intervals are $(-\infty, -3)$, $(-3, 1)$, $(1, 4)$, and $(4, \infty)$.

79. $x^2 + bx + 4 = 0$

To have at least one real solution, $b^2 - 16 \geq 0$. This occurs when $b \leq -4$ or $b \geq 4$. This can be written as $(-\infty, -4] \cup [4, \infty)$.

81. $3x^2 + bx + 10 = 0$

To have at least one real solution, $b^2 - 4(3)(10) \geq 0$.

$b^2 - 120 \geq 0$

$\left(b + \sqrt{120}\right)\left(b - \sqrt{120}\right) \geq 0$

Critical numbers: $b = \pm\sqrt{120} = \pm 2\sqrt{30}$

Test intervals:

$\left(-\infty, -2\sqrt{30}\right), \left(-2\sqrt{30}, 2\sqrt{30}\right), \left(2\sqrt{30}, \infty\right)$

Test: Is $b^2 - 120 \geq 0$?

Solution set: $\left(-\infty, -2\sqrt{30}\right] \cup \left[2\sqrt{30}, \infty\right)$

83. (a) If $a > 0$ and $c \leq 0$, then b can be any real number. If $a > 0$ and $c > 0$, then for $b^2 - 4ac$ to be greater than or equal to zero, b is restricted to $b < -2\sqrt{ac}$ or $b > 2\sqrt{ac}$.

(b) The center of the interval for b in Exercises 79–82 is 0.

85. $4x^2 + 20x + 25 = (2x + 5)^2$

87. $x^2(x + 3) - 4(x + 3) = (x^2 - 4)(x + 3)$

$= (x + 2)(x - 2)(x + 3)$

89. Area $=$ (length)(width)

$= (2x + 1)(x)$

$= 2x^2 + x$

Review Exercises for Chapter 1

1. $y = 3x - 5$

x	-2	-1	0	1	2
y	-11	-8	-5	-2	1

3. $y = x^2 - 3x$

x	-1	0	1	2	3	4
y	4	0	-2	-2	0	4

5. $y - 2x - 3 = 0$

$\qquad y = 2x + 3$

Line with x-intercept $\left(-\frac{3}{2}, 0\right)$
and y-intercept $(0, 3)$

x	-2	-1	0	1
y	-1	1	3	5

7. $y = \sqrt{5 - x}$

Domain: $(-\infty, 5]$

x	5	4	1	-4
y	0	1	2	3

9. $y + 2x^2 = 0$

$\qquad y = -2x^2$ is a parabola.

x	0	± 1	± 2
y	0	-2	-8

11. $y = (x - 3)^2 - 4$

x-intercepts: $0 = (x - 3)^2 - 4 \Rightarrow (x - 3)^2 = 4 \Rightarrow x - 3 = \pm 2 \Rightarrow x = 3 \pm 2$

$\qquad\qquad\qquad\qquad\qquad\qquad\qquad\qquad\qquad \Rightarrow x = 5 \text{ or } x = 1$

y-intercept: $y = (0 - 3)^2 - 4 \Rightarrow y = 9 - 4 \Rightarrow y = 5$

The x-intercepts are $(1, 0)$ and $(5, 0)$. The y-intercept is $(0, 5)$.

13. $y = -4x + 1$

Intercepts: $\left(\frac{1}{4}, 0\right), (0, 1)$

$y = -4(-x) + 1 \Rightarrow y = 4x + 1 \Rightarrow$ No y-axis symmetry

$-y = -4x + 1 \Rightarrow y = 4x - 1 \Rightarrow$ No x-axis symmetry

$-y = -4(-x) + 1 \Rightarrow y = -4x - 1 \Rightarrow$ No origin symmetry

15. $y = 5 - x^2$

Intercepts: $(\pm\sqrt{5}, 0), (0, 5)$

$y = 5 - (-x)^2 \implies y = 5 - x^2 \implies y\text{-axis symmetry}$

$-y = 5 - x^2 \implies y = -5 + x^2 \implies \text{No } x\text{-axis symmetry}$

$-y = 5 - (-x)^2 \implies y = -5 + x^2 \implies \text{No origin symmetry}$

17. $y = x^3 + 3$

Intercepts: $\left(-\sqrt[3]{3}, 0\right), (0, 3)$

$y = (-x)^3 + 3 \implies y = -x^3 + 3 \implies \text{No } y\text{-axis symmetry}$

$-y = x^3 + 3 \implies y = -x^3 - 3 \implies \text{No } x\text{-axis symmetry}$

$-y = (-x)^3 + 3 \implies y = x^3 - 3 \implies \text{No origin symmetry}$

19. $y = \sqrt{x + 5}$

Domain: $[-5, \infty)$

Intercepts: $(-5, 0), \left(0, \sqrt{5}\right)$

$y = \sqrt{-x + 5} \implies \text{No } y\text{-axis symmetry}$

$-y = \sqrt{x + 5} \implies y = -\sqrt{x + 5} \implies \text{No } x\text{-axis symmetry}$

$-y = \sqrt{-x + 5} \implies y = -\sqrt{-x + 5} \implies \text{No origin symmetry}$

21. $x^2 + y^2 = 9$

Center: $(0, 0)$

Radius: 3

23. $(x + 2)^2 + y^2 = 16$

$(x - (-2))^2 + (y - 0)^2 = 4^2$

Center: $(-2, 0)$

Radius: 4

25. $\left(x - \dfrac{1}{2}\right)^2 + (y + 1)^2 = 36$

$\left(x - \dfrac{1}{2}\right)^2 + (y - (-1))^2 = 6^2$

Center: $\left(\frac{1}{2}, -1\right)$

Radius: 6

27. Endpoints of a diameter: $(0, 0)$ and $(4, -6)$

Center: $\left(\dfrac{0 + 4}{2}, \dfrac{0 + (-6)}{2}\right) = (2, -3)$

Radius: $r = \sqrt{(2 - 0)^2 + (-3 - 0)^2} = \sqrt{4 + 9} = \sqrt{13}$

Standard form: $(x - 2)^2 + (y - (-3))^2 = \left(\sqrt{13}\right)^2$

$(x - 2)^2 + (y + 3)^2 = 13$

29. $F = \frac{5}{4}x,\ 0 \le x \le 20$

(a)

x	0	4	8	12	16	20
F	0	5	10	15	20	25

(b)

(c) When $x = 10$, $F = \frac{50}{4} = 12.5$ pounds.

31.
$$6 - (x - 2)^2 = 2 + 4x - x^2$$
$$6 - (x^2 - 4x + 4) = 2 + 4x - x^2$$
$$2 + 4x - x^2 = 2 + 4x - x^2$$
$$0 = 0 \quad \text{Identity}$$

All real numbers are solutions.

33.
$$-x^3 + x(7 - x) + 3 = x(-x^2 - x) + 7(x + 1) - 4$$
$$-x^3 + 7x - x^2 + 3 = -x^3 - x^2 + 7x + 7 - 4$$
$$-x^3 - x^2 + 7x + 3 = -x^3 - x^2 + 7x + 3$$
$$0 = 0 \quad \text{Identity}$$

All real numbers are solutions.

35.
$$3x - 2(x + 5) = 10$$
$$3x - 2x - 10 = 10$$
$$x = 20$$

37.
$$4(x + 3) - 3 = 2(4 - 3x) - 4$$
$$4x + 12 - 3 = 8 - 6x - 4$$
$$4x + 9 = -6x + 4$$
$$10x = -5$$
$$x = -\frac{1}{2}$$

39.
$$\frac{x}{5} - 3 = \frac{2x}{2} + 1$$
$$5\left(\frac{x}{5} - 3\right) = (x + 1)5$$
$$x - 15 = 5x + 5$$
$$-4x = 20$$
$$x = -5$$

41.
$$\frac{18}{x} = \frac{10}{x - 4}$$
$$18(x - 4) = 10x$$
$$18x - 72 = 10x$$
$$8x = 72$$
$$x = 9$$

43. $y = 3x - 1$

x-intercept: $0 = 3x - 1 \Rightarrow x = \frac{1}{3}$

y-intercept: $y = 3(0) - 1 \Rightarrow y = -1$

The x-intercept is $\left(\frac{1}{3}, 0\right)$ and the y-intercept is $(0, -1)$.

45. $y = 2(x - 4)$

x-intercept: $0 = 2(x - 4) \Rightarrow x = 4$

y-intercept: $y = 2(0 - 4) \Rightarrow y = -8$

The x-intercept is $(4, 0)$ and the y-intercept is $(0, -8)$.

47. $y = -\frac{1}{2}x + \frac{2}{3}$

x-intercept: $0 = -\frac{1}{2}x + \frac{2}{3} \Rightarrow x = \frac{2/3}{1/2} = \frac{4}{3}$

y-intercept: $y = -\frac{1}{2}(0) + \frac{2}{3} \Rightarrow y = \frac{2}{3}$

The x-intercept is $\left(\frac{4}{3}, 0\right)$ and the y-intercept is $\left(0, \frac{2}{3}\right)$.

49. $3.8y - 0.5x + 1 = 0$

x-intercept: $3.8(0) - 0.5x + 1 = 0 \Rightarrow x = \frac{1}{0.5} = 2$

y-intercept: $3.8y - 0.5(0) + 1 = 0 \Rightarrow y = \frac{-1}{3.8} = -\frac{5}{19}$

The x-intercept is $(2, 0)$ and the y-intercept is $\left(0, -\frac{5}{19}\right)$.

51. $244.92 = 2(3.14)(3)^2 + 2(3.14)(3)h$
$$244.92 = 56.52 + 18.84h$$
$$188.40 = 18.84h$$
$$10 = h$$

The height is 10 inches.

53. *Verbal Model:* September's profit + October's profit = 689,000

    Labels: Let x = September's profit. Then $x + 0.12x$ = October's profit.

    Equation: $x + (x + 0.12x) = 689,000$

$$2.12x = 689,000$$

$$x = 325,000$$

$$x + 0.12x = 364,000$$

    Answer: September profit: \$325,000, October profit: \$364,000

55. Let x = height of the streetlight.

By similar triangles we have:

$$\frac{x}{20} = \frac{6}{5}$$

$$5x = 120$$

$$x = 24$$

The streetlight is 24 feet tall.

57. Let x = the number of original investors.

Each person's share is $\dfrac{90,000}{x}$.

If three more people invest, each person's share is $\dfrac{90,000}{x + 3}$.

Since this is \$2500 less than the original cost, we have:

$$\frac{90,000}{x} - 2500 = \frac{90,000}{x + 3}$$

$$90,000(x + 3) - 2500x(x + 3) = 90,000x$$

$$90,000x + 270,000 - 2500x^2 - 7500x = 90,000x$$

$$-2500x^2 - 7500x + 270,000 = 0$$

$$-2500(x^2 + 3x - 108) = 0$$

$$-2500(x + 12)(x - 9) = 0$$

$$x = -12, \text{ extraneous or } x = 9$$

There are currently nine investors.

59. Let x = the number of liters of pure antifreeze.

$$30\% \text{ of } (10 - x) + 100\% \text{ of } x = 50\% \text{ of } 10$$

$$0.30(10 - x) + 1.00x = 0.50(10)$$

$$3 - 0.30x + 1.00x = 5$$

$$0.70x = 2$$

$$x = \frac{2}{0.70} = \frac{20}{7} = 2\frac{6}{7} \text{ liters}$$

61. $V = \dfrac{1}{3}\pi r^2 h$

$$3V = \pi r^2 h$$

$$\frac{3V}{\pi r^2} = h$$

63.

	rate	time	distance
1st car	40 mph	t	$40t$
2nd car	55 mph	t	$55t$

$$55t - 40t = 10$$

$$15t = 10$$

$$t = \tfrac{2}{3} \text{ hour or 40 minutes}$$

65. $15 + x - 2x^2 = 0$

$$0 = 2x^2 - x - 15$$

$$0 = (2x + 5)(x - 3)$$

$$2x + 5 = 0 \implies x = -\tfrac{5}{2}$$

$$x - 3 = 0 \implies x = 3$$

67.
$$6 = 3x^2$$
$$2 = x^2$$
$$\pm\sqrt{2} = x$$

69. $(x + 4)^2 = 18$
$$x + 4 = \pm\sqrt{18}$$
$$x = -4 \pm 3\sqrt{2}$$

71. $x^2 - 12x + 30 = 0$
$$x^2 - 12x = -30$$
$$x^2 - 12x + 36 = -30 + 36$$
$$(x - 6)^2 = 6$$
$$x - 6 = \pm\sqrt{6}$$
$$x = 6 \pm \sqrt{6}$$

73. $-2x^2 - 5x + 27 = 0$
$$2x^2 + 5x - 27 = 0$$
$$x = \frac{-5 \pm \sqrt{5^2 - 4(2)(-27)}}{2(2)}$$
$$= \frac{-5 \pm \sqrt{241}}{4}$$

75. $M = 500x(20 - x)$

(a) $500x(20 - x) = 0$ when $x = 0$ feet and $x = 20$ feet.

(b)

(c) The bending moment is greatest when $x = 10$ feet.

77. $6 + \sqrt{-4} = 6 + 2i$

79. $i^2 + 3i = -1 + 3i$

81. $(7 + 5i) + (-4 + 2i) = (7 - 4) + (5i + 2i) = 3 + 7i$

83. $5i(13 - 8i) = 65i - 40i^2 = 40 + 65i$

85. $(10 - 8i)(2 - 3i) = 20 - 30i - 16i + 24i^2$
$$= -4 - 46i$$

87.
$$\frac{6 + i}{4 - i} = \frac{6 + i}{4 - i} \cdot \frac{4 + i}{4 + i}$$
$$= \frac{24 + 10i + i^2}{16 + 1}$$
$$= \frac{23 + 10i}{17}$$
$$= \frac{23}{17} + \frac{10}{17}i$$

89.
$$\frac{4}{2 - 3i} + \frac{2}{1 + i} = \frac{4}{2 - 3i} \cdot \frac{2 + 3i}{2 + 3i} + \frac{2}{1 + i} \cdot \frac{1 - i}{1 - i}$$
$$= \frac{8 + 12i}{4 + 9} + \frac{2 - 2i}{1 + 1}$$
$$= \frac{8}{13} + \frac{12}{13}i + 1 - i$$
$$= \left(\frac{8}{13} + 1\right) + \left(\frac{12}{13}i - i\right)$$
$$= \frac{21}{13} - \frac{1}{13}i$$

91. $3x^2 + 1 = 0$
$$3x^2 = -1$$
$$x^2 = -\frac{1}{3}$$
$$x = \pm\sqrt{-\frac{1}{3}}$$
$$= \pm\sqrt{\frac{1}{3}}\,i = \pm\frac{\sqrt{3}}{3}i$$

93. $x^2 - 2x + 10 = 0$
$$x^2 - 2x + 1 = -10 + 1$$
$$(x - 1)^2 = -9$$
$$x - 1 = \pm\sqrt{-9}$$
$$x = 1 \pm 3i$$

95. $5x^4 - 12x^3 = 0$
$$x^3(5x - 12) = 0$$
$$x^3 = 0 \quad \text{or} \quad 5x - 12 = 0$$
$$x = 0 \quad \text{or} \quad x = \frac{12}{5}$$

97. $x^4 - 5x^2 + 6 = 0$

$(x^2 - 2)(x^2 - 3) = 0$

$x^2 - 2 = 0 \quad$ or $\quad x^2 - 3 = 0$

$\quad x^2 = 2 \qquad\qquad x^2 = 3$

$\quad x = \pm\sqrt{2} \qquad\quad x = \pm\sqrt{3}$

99. $\sqrt{x + 4} = 3$

$\left(\sqrt{x + 4}\right)^2 = (3)^2$

$x + 4 = 9$

$x = 5$

101. $\sqrt{2x + 3} + \sqrt{x - 2} = 2$

$\left(\sqrt{2x + 3}\right)^2 = \left(2 - \sqrt{x - 2}\right)^2$

$2x + 3 = 4 - 4\sqrt{x - 2} + x - 2$

$x + 1 = -4\sqrt{x - 2}$

$(x + 1)^2 = \left(-4\sqrt{x - 2}\right)^2$

$x^2 + 2x + 1 = 16(x - 2)$

$x^2 - 14x + 33 = 0$

$(x - 3)(x - 11) = 0$

$x = 3$, extraneous or $x = 11$, extraneous

No solution

103. $(x - 1)^{2/3} - 25 = 0$

$(x - 1)^{2/3} = 25$

$(x - 1)^2 = 25^3$

$x - 1 = \pm\sqrt{25^3}$

$x = 1 \pm 125$

$x = 126 \quad$ or $\quad x = -124$

105. $(x + 4)^{1/2} + 5x(x + 4)^{3/2} = 0$

$(x + 4)^{1/2}[1 + 5x(x + 4)] = 0$

$(x + 4)^{1/2}(5x^2 + 20x + 1) = 0$

$(x + 4)^{1/2} = 0 \qquad$ or $\quad 5x^2 + 20x + 1 = 0$

$\qquad\qquad x = -4$

$$x = \frac{-20 \pm \sqrt{400 - 20}}{10}$$

$$x = \frac{-20 \pm \sqrt{400 - 20}}{10}$$

$$x = \frac{-20 \pm 2\sqrt{95}}{10}$$

$$x = -2 \pm \frac{\sqrt{95}}{5}$$

107. $\dfrac{5}{x} = 1 + \dfrac{3}{x + 2}$

$5(x + 2) = 1(x)(x + 2) + 3x$

$5x + 10 = x^2 + 2x + 3x$

$10 = x^2$

$\pm\sqrt{10} = x$

109. $\dfrac{3}{x + 2} - \dfrac{1}{x} = \dfrac{1}{5x}$

$3(5x) - 5(x + 2) = x + 2$

$15x - 5x - 10 = x + 2$

$9x = 12$

$x = \dfrac{4}{3}$

111. $|x - 5| = 10$

$x - 5 = -10 \quad$ or $\quad x - 5 = 10$

$x = -5 \qquad\qquad x = 15$

113. $|x^2 - 3| = 2x$

$$x^2 - 3 = 2x \quad \text{or} \quad x^2 - 3 = -2x$$

$$x^2 - 2x - 3 = 0 \qquad x^2 + 2x - 3 = 0$$

$$(x - 3)(x + 1) = 0 \qquad (x + 3)(x - 1) = 0$$

$$x = 3 \text{ or } x = -1 \qquad x = -3 \text{ or } x = 1$$

The only solutions to the original equation are $x = 3$ or $x = 1$. ($x = -3$ and $x = -1$ are extraneous.)

115.
$$29.95 = 42 - \sqrt{0.001x + 2}$$

$$-12.05 = -\sqrt{0.001x + 2}$$

$$\sqrt{0.001x + 2} = 12.05$$

$$0.001x + 2 = 145.2025$$

$$0.001x = 143.2025$$

$$x = 143{,}202.5$$

$$\approx 143{,}203 \text{ units}$$

117. Interval: $(-7, 2]$

Inequality: $-7 < x \le 2$

The interval is bounded.

119. Interval: $(-\infty, -10]$

Inequality: $x \le -10$

The interval is unbounded.

121. $9x - 8 \le 7x + 16$

$$2x \le 24$$

$$x \le 12$$

$$(-\infty, 12]$$

123. $4(5 - 2x) \le \frac{1}{2}(8 - x)$

$$20 - 8x \le 4 - \frac{1}{2}x$$

$$-\frac{15}{2}x \le -16$$

$$x \ge \frac{32}{15}$$

$\left[\frac{32}{15}, \infty\right)$

125. $-19 < 3x - 17 \le 34$

$$-2 < 3x \le 51$$

$$-\frac{2}{3} < x \le 17$$

$\left(-\frac{2}{3}, 17\right]$

127. $|x| \le 4$

$$-4 \le x \le 4$$

$$[-4, 4]$$

129. $|x - 3| > 4$

$$x - 3 < -4 \quad \text{or} \quad x - 3 > 4$$

$$x < -1 \quad \text{or} \qquad x > 7$$

$$(-\infty, -1) \cup (7, \infty)$$

131. If the side is 19.3 cm, then with the possible error of 0.5 cm we have:

$$18.8 \le \text{side} \le 19.8$$

$$353.44 \text{ cm}^2 \le \text{area} \le 392.04 \text{ cm}^2$$

133. $x^2 - 6x - 27 < 0$

$(x + 3)(x - 9) < 0$

Critical numbers: $x = -3, x = 9$

Test intervals: $(-\infty, -3), (-3, 9), (9, \infty)$

Test: Is $(x + 3)(x - 9) < 0$?

By testing an x-value in each test interval in the inequality, we see that the solution set is: $(-3, 9)$

135.
$$6x^2 + 5x < 4$$

$$6x^2 + 5x - 4 < 0$$

$$(3x + 4)(2x - 1) < 0$$

Critical numbers: $x = -\frac{4}{3}, x = \frac{1}{2}$

Test intervals: $\left(-\infty, -\frac{4}{3}, \right), \left(-\frac{4}{3}, \frac{1}{2}\right), \left(\frac{1}{2}, \infty\right)$

Test: Is $(3x + 4)(2x - 1) < 0$?

By testing an x-value in each test interval in the inequality, we see that the solution set is: $\left(-\frac{4}{3}, \frac{1}{2}\right)$

137.
$$x^3 - 16x \ge 0$$

$x(x + 4)(x - 4) \ge 0$

Critical numbers: $x = 0, x = \pm 4$

Test intervals: $(-\infty, -4), (-4, 0), (0, 4), (4, \infty)$

Test: Is $x(x + 4)(x - 4) \ge 0$?

By testing an x-value in each test interval in the inequality, we see that the solution set is: $[-4, 0] \cup [4, \infty)$

139.
$$\frac{x + 8}{x + 5} - 2 < 0$$

$$\frac{x + 8 - 2(x + 5)}{x + 5} < 0$$

$$\frac{-x - 2}{x + 5} < 0$$

Critical numbers: $x = -2, x = -5$

Test intervals: $(-\infty, -5), (-5, -2), (-2, \infty)$

By testing an x-value in each test interval in the inequality, we see that the solution set is: $(-\infty, -5) \cup (-2, \infty)$

141.
$$\frac{2}{x + 1} \le \frac{3}{x - 1}$$

$$\frac{2(x - 1) - 3(x + 1)}{(x + 1)(x - 1)} \le 0$$

$$\frac{2x - 2 - 3x - 3}{(x + 1)(x - 1)} \le 0$$

$$\frac{-(x + 5)}{(x + 1)(x - 1)} \le 0$$

Critical numbers: $x = -5$, $x = \pm 1$

Test intervals: $(-\infty, -5), (-5, -1), (-1, 1), (1, \infty)$

Test: Is $\dfrac{-(x + 5)}{(x + 1)(x - 1)} \le 0$?

By testing an x-value in each test interval in the inequality, we see that the solution set is: $[-5, -1) \cup (1, \infty)$

143.
$$\frac{x^2 + 7x + 12}{x} \ge 0$$

$$\frac{(x + 4)(x + 3)}{x} \ge 0$$

Critical numbers: $x = -4, x = -3, x = 0$

Test intervals: $(-\infty, -4), (-4, -3), (-3, 0), (0, \infty)$

Test: Is $\dfrac{(x + 4)(x + 3)}{x} \ge 0$?

By testing an x-value in each test interval in the inequality, we see that the solution set is: $[-4, -3] \cup (0, \infty)$

145. $5000(1 + r)^2 > 5500$

$\qquad (1 + r)^2 > 1.1$

$\qquad 1 + r > 1.0488$

$\qquad r > 0.0488$

$\qquad r > 4.9\%$

147. False

$$\sqrt{-18}\sqrt{-2} = \left(\sqrt{18}i\right)\left(\sqrt{2}i\right) = \sqrt{36}i^2 = -6$$

$$\sqrt{(-18)(-2)} = \sqrt{36} = 6$$

149. Rational equations, equations involving radicals, and absolute value equations, may have "solutions" that are extraneous. So checking solutions, in the original equations, is crucial to eliminate these extraneous values.

Problem Solving for Chapter 1

1.

3. (a) $A = \pi ab$

$\qquad a + b = 20 \implies b = 20 - a$, thus:

$\qquad A = \pi a(20 - a)$

(b)

a	4	7	10	13	16
A	64π	91π	100π	91π	64π

—CONTINUED—

3. **—CONTINUED—**

(c)
$$300 = \pi a(20 - a)$$

$$300 = 20\pi a - \pi a^2$$

$$\pi a^2 - 20\pi a + 300 = 0$$

$$a = \frac{20\pi \pm \sqrt{(-20\pi)^2 - 4\pi(300)}}{2\pi}$$

$$= \frac{20\pi \pm \sqrt{400\pi^2 - 1200\pi}}{2\pi}$$

$$= \frac{20\pi \pm 20\sqrt{\pi(\pi - 3)}}{2\pi}$$

$$= 10 \pm \frac{10}{\pi}\sqrt{\pi(\pi - 3)}$$

$$a \approx 12.123 \quad \text{or} \quad a \approx 7.877$$

(d) $A = na(20 - a)$

(e) The a-intercepts occur at $a = 0$ and $a = 20$. Both yield an area of 0. When $a = 0$, $b = 20$ and you have a vertical line of length 40. Likewise when $a = 20$, $b = 0$ and you have a horizontal line of length 40. They represent the minimum and maximum values of a.

(f) The maximum value of A is $100\pi \approx 314.159$. This occurs when $a = b = 10$ and the ellipse is actually a circle.

5. $h = \left(\sqrt{h_0} - \frac{2\pi d^2\sqrt{3}}{lw}t\right)^2$

$l = 60''$, $w = 30''$, $h_0 = 25''$, $d = 2''$

$h = \left(5 - \frac{8\pi\sqrt{3}}{1800}t\right)^2 = \left(5 - \frac{\pi\sqrt{3}}{225}t\right)^2$

(a) $12.5 = \left(5 - \frac{\pi\sqrt{3}}{225}t\right)^2$

$\sqrt{12.5} = 5 - \frac{\pi\sqrt{3}}{225}t$

$t = \frac{225}{\pi\sqrt{3}}(5 - \sqrt{12.5}) \approx 60.6$ seconds

(b) $0 = \left(\sqrt{12.5} - \frac{\pi\sqrt{3}}{225}t\right)^2$

$t = \frac{225\sqrt{12.5}}{\pi\sqrt{3}} \approx 146.2$ seconds

(c) The speed at which the water drains decreases as the amount of the water in the bathtub decreases.

7. (a) 5, 12, and 13; 8, 15, and 17

7, 24, and 25

(b) $5 \cdot 12 \cdot 13 = 780$ which is divisible by 3, 4, and 5

$8 \cdot 15 \cdot 17 = 2040$ which is divisible by 3, 4, and 5

$7 \cdot 24 \cdot 25 = 4200$ which is also divisible by 3, 4, and 5

(c) Conjecture: If $a^2 + b^2 = c^2$ where a, b, and c are positive integers, then abc is divisible by 60.

9. $ax^2 + bx + c = 0$

$x^2 + \frac{b}{a}x + \frac{c}{a} = 0$

$x_1 + x_2 = -\frac{b}{a}$

$x_1 \cdot x_2 = \frac{c}{a}$

11. (a) $z_m = \dfrac{1}{z}$

$$= \frac{1}{1+i} = \frac{1}{1+i} \cdot \frac{1-i}{1-i}$$

$$= \frac{1-i}{2} = \frac{1}{2} - \frac{1}{2}i$$

(b) $z_m = \dfrac{1}{z}$

$$= \frac{1}{3-i} = \frac{1}{3-i} \cdot \frac{3+i}{3+i}$$

$$= \frac{3+i}{10} = \frac{3}{10} + \frac{1}{10}i$$

(c) $z_m = \dfrac{1}{z}$

$$= \frac{1}{-2+8i}$$

$$= \frac{1}{-2+8i} \cdot \frac{-2-8i}{-2-8i}$$

$$= \frac{-2-8i}{68} = -\frac{1}{34} - \frac{2}{17}i$$

13. (a) $c = i$

The terms are: $i, -1+i, -i, -1+i, -i, -1+i, -i, -1+i, -i, \ldots$

The sequence is bounded so $c = i$ *is* in the Mandelbrot Set.

(b) $c = 1 + i$

The terms are: $1+i, 1+3i, -7+7i, 1-97i, -9407-193i, \ldots$

The sequence is unbounded so $c = 1 + i$ is *not* in the Mandelbrot Set.

(c) $c = -2$

The terms are: $-2, 2, 2, 2, 2, \ldots$

The sequence is bounded so $c = -2$ *is* in the Mandelbrot Set.

15. $y = x^4 - x^3 - 6x^2 + 4x + 8$

$\quad = (x-2)^2(x+1)(x+2)$

From the graph we see that $x^4 - x^3 - 6x^2 + 4x + 8 > 0$ on the intervals $(-\infty, -2) \cup (-1, 2) \cup (2, \infty)$.

Practice Test for Chapter 1

1. Graph $3x - 5y = 15$.

2. Graph $y = \sqrt{9 - x}$.

3. Solve $5x + 4 = 7x - 8$.

4. Solve $\dfrac{x}{3} - 5 = \dfrac{x}{5} + 1$.

5. Solve $\dfrac{3x + 1}{6x - 7} = \dfrac{2}{5}$.

6. Solve $(x - 3)^2 + 4 = (x + 1)^2$.

7. Solve $A = \frac{1}{2}(a + b)h$ for a.

8. 301 is what percent of 4300?

9. Cindy has \$6.05 in quarters and nickels. How many of each coin does she have if there are 53 coins in all?

10. Ed has \$15,000 invested in two funds paying $9\frac{1}{2}\%$ and 11% simple interest, respectively. How much is invested in each if the yearly interest is \$1582.50?

11. Solve $28 + 5x - 3x^2 = 0$ by factoring.

12. Solve $(x - 2)^2 = 24$ by taking the square root of both sides.

13. Solve $x^2 - 4x - 9 = 0$ by completing the square.

14. Solve $x^2 + 5x - 1 = 0$ by the Quadratic Formula.

15. Solve $3x^2 - 2x + 4 = 0$ by the Quadratic Formula.

16. The perimeter of a rectangle is 1100 feet. Find the dimensions so that the enclosed area will be 60,000 square feet.

17. Find two consecutive even positive integers whose product is 624.

18. Solve $x^3 - 10x^2 + 24x = 0$ by factoring.

19. Solve $\sqrt[3]{6 - x} = 4$.

20. Solve $(x^2 - 8)^{2/5} = 4$.

21. Solve $x^4 - x^2 - 12 = 0$.

22. Solve $4 - 3x > 16$.

23. Solve $\left| \dfrac{x - 3}{2} \right| < 5$.

24. Solve $\dfrac{x + 1}{x - 3} < 2$.

25. Solve $|3x - 4| \geq 9$.

CHAPTER 2
Functions and Their Graphs

CHAPTER 2
Functions and Their Graphs

Section 2.1   Linear Equations in Two Variables

You should know the following important facts about lines.

■   The graph of $y = mx + b$ is a straight line. It is called a linear equation in two variables.

   (a) The slope (steepness) is m.

   (b) The y-intercept is $(0, b)$.

■   The slope of the line through (x_1, y_1) and (x_2, y_2) is

$$m = \frac{y_2 - y_1}{x_2 - x_1} = \frac{\text{change in } y}{\text{change in } x} = \frac{\text{rise}}{\text{run}}.$$

■   (a) If $m > 0$, the line rises from left to right.

   (b) If $m = 0$, the line is horizontal.

   (c) If $m < 0$, the line falls from left to right.

   (d) If m is undefined, the line is vertical.

■   Equations of Lines

   (a) Slope-Intercept Form: $y = mx + b$

   (b) Point-Slope Form: $y - y_1 = m(x - x_1)$

   (c) Two-Point Form: $y - y_1 = \dfrac{y_2 - y_1}{x_2 - x_1}(x - x_1)$

   (d) General Form: $Ax + By + C = 0$

   (e) Vertical Line: $x = a$

   (f) Horizontal Line: $y = b$

■   Given two distinct nonvertical lines

   $L_1: y = m_1 x + b_1$ and $L_2: y = m_2 x + b_2$

   (a) L_1 is parallel to L_2 if and only if $m_1 = m_2$ and $b_1 \neq b_2$.

   (b) L_1 is perpendicular to L_2 if and only if $m_1 = -1/m_2$.

Vocabulary Check

1. linear

2. slope

3. parallel

4. perpendicular

5. rate or rate of change

6. linear extrapolation

7. (a) $Ax + By + C = 0$ (iii) general form

   (b) $x = a$ (i) vertical line

   (c) $y = b$ (v) horizontal line

   (d) $y = mx + b$ (ii) slope-intercept form

   (e) $y - y_1 = m(x - x_1)$ (iv) point-slope form

1. (a) $m = \frac{2}{3}$. Since the slope is positive, the line rises. Matches L_2.

(b) m is undefined. The line is vertical. Matches L_3.

(c) $m = -2$. The line falls. Matches L_1.

3.

5. Two points on the line: $(0, 0)$ and $(4, 6)$

$$\text{Slope} = \frac{\text{rise}}{\text{run}} = \frac{6}{4} = \frac{3}{2}$$

7. Two points on the line: $(0, 8)$ and $(2, 0)$

$$\text{Slope} = \frac{\text{rise}}{\text{run}} = \frac{-8}{2} = -4$$

9. $y = 5x + 3$

Slope: $m = 5$

y-intercept: $(0, 3)$

11. $y = -\frac{1}{2}x + 4$

Slope: $m = -\frac{1}{2}$

y-intercept: $(0, 4)$

13. $5x - 2 = 0$

$x = \frac{2}{5}$, vertical line

Slope: undefined

No y-intercept

15. $7x + 6y = 30$

$y = -\frac{7}{6}x + 5$

Slope: $m = -\frac{7}{6}$

y-intercept: $(0, 5)$

17. $y - 3 = 0$

$y = 3$, horizontal line

Slope: $m = 0$

y-intercept: $(0, 3)$

19. $x + 5 = 0$

$x = -5$

Slope: undefined (vertical line)

No y-intercept

21. $m = \frac{6 - (-2)}{1 - (-3)} = \frac{8}{4} = 2$

23. $m = \frac{4 - (-1)}{-6 - (-6)} = \frac{5}{0}$

m is undefined.

25. $m = \frac{-\frac{1}{3} - \left(-\frac{4}{3}\right)}{-\frac{3}{2} - \frac{11}{2}} = -\frac{1}{7}$

27. $m = \dfrac{1.6 - 3.1}{-5.2 - 4.8} = \dfrac{-1.5}{-10} = 0.15$

29. Point: $(2, 1)$, Slope: $m = 0$

Since $m = 0$, y does not change. Three points are $(0, 1)$, $(3, 1)$, and $(-1, 1)$.

31. Point: $(5, -6)$, Slope: $m = 1$

Since $m = 1$, y increases by 1 for every one unit increase in x. Three points are $(6, -5)$, $(7, -4)$, and $(8, -3)$.

33. Point: $(-8, 1)$, Slope is undefined.

Since m is undefined, x does not change. Three points are $(-8, 0)$, $(-8, 2)$, and $(-8, 3)$.

35. Point: $(-5, 4)$, Slope: $m = 2$

Since $m = 2 = \frac{2}{1}$, y increases by 2 for every one unit increase in x. Three additional points are $(-4, 6)$, $(-3, 8)$, and $(-2, 10)$.

37. Point: $(7, -2)$, Slope: $m = \frac{1}{2}$

Since $m = \frac{1}{2}$, y increases by 1 unit for every two unit increase in x. Three additional points are $(9, -1)$, $(11, 0)$, and $(13, 1)$.

39. Point $(0, -2)$; $m = 3$

$$y + 2 = 3(x - 0)$$
$$y = 3x - 2$$

41. Point $(-3, 6)$; $m = -2$

$$y - 6 = -2(x + 3)$$
$$y = -2x$$

43. Point $(4, 0)$; $m = -\frac{1}{3}$

$$y - 0 = -\tfrac{1}{3}(x - 4)$$
$$y = -\tfrac{1}{3}x + \tfrac{4}{3}$$

45. Point $(6, -1)$; m is undefined.

The line is vertical.

$$x = 6$$

47. Point $\left(4, \frac{5}{2}\right)$; $m = 0$

The line is horizontal.

$$y = \frac{5}{2}$$

49. Point $(-5.1, 1.8)$; $m = 5$

$$y - 1.8 = 5(x - (-5.1))$$
$$y = 5x + 27.3$$

51. $(5, -1)$ and $(-5, 5)$

$$y + 1 = \frac{5+1}{-5-5}(x - 5)$$

$$y = -\frac{3}{5}(x - 5) - 1$$

$$y = -\frac{3}{5}x + 2$$

53. $(-8, 1)$ and $(-8, 7)$

Since both points have $x = -8$, the slope is undefined, and the line is vertical.

$x = -8$

55. $\left(2, \frac{1}{2}\right)$ and $\left(\frac{1}{2}, \frac{5}{4}\right)$

$$y - \frac{1}{2} = \frac{\frac{5}{4} - \frac{1}{2}}{\frac{1}{2} - 2}(x - 2)$$

$$y = -\frac{1}{2}(x - 2) + \frac{1}{2}$$

$$y = -\frac{1}{2}x + \frac{3}{2}$$

57. $\left(-\frac{1}{10}, -\frac{3}{5}\right)$ and $\left(\frac{9}{10}, -\frac{9}{5}\right)$

$$y - \left(-\frac{3}{5}\right) = \frac{-\frac{9}{5} - \left(-\frac{3}{5}\right)}{\frac{9}{10} - \left(-\frac{1}{10}\right)}\left(x - \left(-\frac{1}{10}\right)\right)$$

$$y = -\frac{6}{5}\left(x + \frac{1}{10}\right) - \frac{3}{5}$$

$$y = -\frac{6}{5}x - \frac{18}{25}$$

59. $(1, 0.6)$ and $(-2, -0.6)$

$$y - 0.6 = \frac{-0.6 - 0.6}{-2 - 1}(x - 1)$$

$$y = 0.4(x - 1) + 0.6$$

$$y = 0.4x + 0.2$$

61. $(2, -1)$ and $\left(\frac{1}{3}, -1\right)$

$$y + 1 = \frac{-1 - (-1)}{\frac{1}{3} - 2}(x - 2)$$

$$y + 1 = 0$$

$$y = -1$$

The line is horizontal.

63. $\left(\frac{7}{3}, -8\right)$ and $\left(\frac{7}{3}, 1\right)$

$$m = \frac{1 - (-8)}{\frac{7}{3} - \frac{7}{3}} = \frac{9}{0} \text{ and is undefined.}$$

$$x = \frac{7}{3}$$

The line is vertical.

65. L_1: $(0, -1)$, $(5, 9)$

Slope of L_1: $m = \dfrac{9 + 1}{5 - 0} = 2$

L_2: $(0, 3)$, $(4, 1)$

Slope of L_2: $m = \dfrac{1 - 3}{4 - 0} = -\dfrac{1}{2}$

L_1 and L_2 are perpendicular.

67. L_1: $(3, 6)$, $(-6, 0)$

Slope of L_1: $m = \dfrac{0 - 6}{-6 - 3} = \dfrac{2}{3}$

L_2: $(0, -1)$, $\left(5, \dfrac{7}{3}\right)$

Slope of L_2: $m = \dfrac{\frac{7}{3} + 1}{5 - 0} = \dfrac{2}{3}$

L_1 and L_2 are parallel.

69. $4x - 2y = 3$

$\qquad y = 2x - \dfrac{3}{2}$

Slope: $m = 2$

(a) $(2, 1)$, $m = 2$

$\qquad y - 1 = 2(x - 2)$

$\qquad\qquad y = 2x - 3$

(b) $(2, 1)$, $m = -\dfrac{1}{2}$

$\qquad y - 1 = -\dfrac{1}{2}(x - 2)$

$\qquad\qquad y = -\dfrac{1}{2}x + 2$

71. $3x + 4y = 7$

$\qquad y = -\dfrac{3}{4}x + \dfrac{7}{4}$

Slope: $m = -\dfrac{3}{4}$

(a) $\left(-\dfrac{2}{3}, \dfrac{7}{8}\right)$, $m = -\dfrac{3}{4}$

$\quad y - \dfrac{7}{8} = -\dfrac{3}{4}\left(x - \left(-\dfrac{2}{3}\right)\right)$

$\qquad\quad y = -\dfrac{3}{4}x + \dfrac{3}{8}$

(b) $\left(-\dfrac{2}{3}, \dfrac{7}{8}\right)$, $m = \dfrac{4}{3}$

$\quad y - \dfrac{7}{8} = \dfrac{4}{3}\left(x - \left(-\dfrac{2}{3}\right)\right)$

$\qquad\quad y = \dfrac{4}{3}x + \dfrac{127}{72}$

73. $y = -3$

$\quad m = 0$

(a) $(-1, 0)$ and $m = 0$

$\qquad y = 0$

(b) $(-1, 0)$, m is undefined.

$\qquad x = -1$

75. $x = 4$

m is undefined.

(a) $(2, 5)$, m is undefined. The line is vertical, passing through $(2, 5)$.

$\qquad x = 2$

(b) $(2, 5)$, $m = 0$

$\qquad y = 5$

77. $x - y = 4$

$\qquad y = x - 4$

Slope: $m = 1$

(a) $(2.5, 6.8)$, $m = 1$ (b) $(2.5, 6.8)$, $m = -1$

$\quad y - 6.8 = 1(x - 2.5)$ $\quad y - 6.8 = (-1)(x - 2.5)$

$\qquad\quad y = x + 4.3$ $\qquad\quad y = -x + 9.3$

79. $\qquad \dfrac{x}{2} + \dfrac{y}{3} = 1$

$\qquad 3x + 2y - 6 = 0$

81. $\dfrac{x}{-1/6} + \dfrac{y}{-2/3} = 1$

$\qquad 6x + \dfrac{3}{2}y = -1$

$\qquad 12x + 3y + 2 = 0$

83. $\qquad \dfrac{x}{c} + \dfrac{y}{c} = 1$, $c \neq 0$

$\qquad\quad x + y = c$

$\qquad\quad 1 + 2 = c$

$\qquad\qquad\quad 3 = c$

$\qquad\quad x + y = 3$

$\qquad\quad x + y - 3 = 0$

85. (a) $y = 2x$

(b) $y = -2x$

(c) $y = \dfrac{1}{2}x$

(b) and (c) are perpendicular.

87. (a) $y = -\frac{1}{2}x$

(b) $y = -\frac{1}{2}x + 3$

(c) $y = 2x - 4$

(a) and (b) are parallel. (c) is perpendicular to (a) and (b).

89. Set the distance between $(4, -1)$ and (x, y) equal to the distance between $(-2, 3)$ and (x, y).

$$\sqrt{(x-4)^2 + [y-(-1)]^2} = \sqrt{[x-(-2)]^2 + (y-3)^2}$$

$$(x-4)^2 + (y+1)^2 = (x+2)^2 + (y-3)^2$$

$$x^2 - 8x + 16 + y^2 + 2y + 1 = x^2 + 4x + 4 + y^2 - 6y + 9$$

$$-8x + 2y + 17 = 4x - 6y + 13$$

$$0 = 12x - 8y - 4$$

$$0 = 4(3x - 2y - 1)$$

$$0 = 3x - 2y - 1$$

This line is the perpendicular bisector of the line segment connecting $(4, -1)$ and $(-2, 3)$.

91. Set the distance between $\left(3, \frac{5}{2}\right)$ and (x, y) equal to the distance between $(-7, 1)$ and (x, y).

$$\sqrt{(x-3)^2 + \left(y-\frac{5}{2}\right)^2} = \sqrt{[x-(-7)]^2 + (y-1)^2}$$

$$(x-3)^2 + \left(y-\frac{5}{2}\right)^2 = (x+7)^2 + (y-1)^2$$

$$x^2 - 6x + 9 + y^2 - 5y + \frac{25}{4} = x^2 + 14x + 49 + y^2 - 2y + 1$$

$$-6x - 5y + \frac{61}{4} = 14x - 2y + 50$$

$$-24x - 20y + 61 = 56x - 8y + 200$$

$$80x + 12y + 139 = 0$$

This line is the perpendicular bisector of the line segment connecting $\left(3, \frac{5}{2}\right)$ and $(-7, 1)$.

93. (a) $m = 135$. The sales are increasing 135 units per year.

(b) $m = 0$. There is no change in sales during the year.

(c) $m = -40$. The sales are decreasing 40 units per year.

95. (a) $(0, 55{,}722), (2, 61{,}768)$: $m = \dfrac{61{,}768 - 55{,}722}{2 - 0} = 3023$ $\qquad$ $(6, 69{,}277), (8, 74{,}380)$: $m = \dfrac{74{,}380 - 69{,}277}{8 - 6} = 2551.5$

$(2, 61{,}768), (4, 64{,}993)$: $m = \dfrac{64{,}993 - 61{,}768}{4 - 2} = 1612.5$ $\qquad$ $(8, 74{,}380), (10, 79{,}839)$: $m = \dfrac{79{,}839 - 74{,}380}{10 - 8} = 2729.5$

$(4, 64{,}993), (6, 69{,}277)$: $m = \dfrac{69{,}277 - 64{,}993}{6 - 4} = 2142$ $\qquad$ $(10, 79{,}839), (12, 83{,}944)$: $m = \dfrac{83{,}944 - 79{,}839}{12 - 10} = 2052.5$

The average salary increased the most from 1990 to 1992 and the least from 1992 to 1994.

(b) $(0, 55{,}722), (12, 83{,}944)$: $m = \dfrac{83{,}944 - 55{,}722}{12 - 0} \approx \$2351.83$

(c) The average salary for senior high school principals increased by $2351.83 per year over the 12 years between 1990 and 2002.

97. $y = \frac{6}{100}x$

$y = \frac{6}{100}(200) = 12$ feet

99. $(5, 2540)$, $m = -125$

$V - 2540 = -125(t - 5)$

$V - 2540 = -125t + 625$

$V = -125t + 3165,\ 0 \le t \le 5$

101. Matches graph (b).

The slope is -20, which represents the decrease in the amount of the loan each week. The y-intercept is $(0, 200)$, which represents the original amount of the loan.

103. Matches graph (a).

The slope is 0.32, which represents the increase in travel cost for each mile driven. The y-intercept is $(0, 30)$, which represents the fixed cost of $30 per day for meals. This amount does not depend on the number of miles driven.

105. $(5, 0.18)$, $(13, 4.04)$: $m = \dfrac{4.04 - 0.18}{13 - 5} = 0.4825$

$y - 0.18 = 0.4825(t - 5)$

$y = 0.4825t - 2.2325$

For 2008, use $t = 18$: $y(18) \approx \$6.45$

For 2010, use $t = 20$: $y(20) \approx \$7.42$

107. Using the points $(0, 875)$ and $(5, 0)$, where the first coordinate represents the year t and the second coordinate represents the value V, we have

$m = \dfrac{0 - 875}{5 - 0} = -175$

$V = -175t + 875,\ 0 \le t \le 5.$

109. (a) $(0, 40{,}571)$, $(4, 41{,}289)$:

$m = \dfrac{41{,}289 - 40{,}571}{4 - 0} = 179.5$

$y = 179.5t + 40{,}571$

(b) For 2008, use $t = 8$: $y(8) = 42{,}007$ students.

For 2010, use $t = 10$: $y(10) = 42{,}366$ students.

(c) The slope is $m = 179.5$, which represents the increase in the number of students each year.

111. Sale price = List price $-$ 15% of the list price

$S = L - 0.15L$

$S = 0.85L$

113. (a) $C = 36{,}500 + 5.25t + 11.50t$

$= 16.75t + 36{,}500$

(c) $P = R - C$

$= 27t - (16.75t + 36{,}500)$

$= 10.25t - 36{,}500$

(b) $R = 27t$

(d) $\quad 0 = 10.25t - 36{,}500$

$36{,}500 = 10.25t$

$t \approx 3561$ hours

115. (a)

(b) $y = 2(15 + 2x) + 2(10 + 2x) = 8x + 50$

(c)

(d) Since $m = 8$, each 1-meter increase in x will increase y by 8 meters.

117. $C = 0.38m + 120$

119. (a) and (b)

Year (0 ↔ 1990)

(c) Answers will vary. Find two points on your line and then find the equation of the line through your points. Sample answer: $y \approx 11.72x - 14.08$

(d) Answers will vary. Sample answer: The y-intercept should represent the number of initial subscribers. In this case, since b is negative, it cannot be interpreted as such. The slope of 11.72 represents the increase in the number of subscribers per year (in millions).

(e) The model is a fairly good fit to the data.

(f) Answers will vary. Sample answer:

$y(18) \approx 11.72(18) - 14.08$

$= 196.88$ million subscribers in 2008

121. False. The slope with the greatest magnitude corresponds to the steepest line.

123. Using the Distance Formula, we have $AB = 6$, $BC = \sqrt{40}$, and $AC = 2$. Since $6^2 + 2^2 = (\sqrt{40})^2$, the triangle is a right triangle.

125. No. The slope cannot be determined without knowing the scale on the y-axis. The slopes will be the same if the scale on the y-axis of (a) is $2\frac{1}{2}$ and the scale on the y-axis of (b) is 1. Then the slope of both is $\frac{5}{4}$.

127. The V-intercept measures the initial cost and the slope measures annual depreciation.

129. $y = 8 - 3x$ is a linear equation with slope $m = -3$ and y-intercept $(0, 8)$. Matches graph (d).

131. $y = \frac{1}{2}x^2 + 2x + 1$ is a quadratic equation. Its graph is a parabola with vertex $(-2, -1)$ and y-intercept $(0, 1)$. Matches graph (a).

133. $-7(3 - x) = 14(x - 1)$

$-21 + 7x = 14x - 14$

$-7x = 7$

$x = -1$

135. $2x^2 - 21x + 49 = 0$

$(2x - 7)(x - 7) = 0$

$2x - 7 = 0$ or $x - 7 = 0$

$x = \frac{7}{2}$ or $x = 7$

137. $\sqrt{x - 9} + 15 = 0$

$\sqrt{x - 9} = -15$

No real solution

The square root of $x - 9$ cannot be negative.

139. Answers will vary.

Section 2.2   Functions

■   Given a set or an equation, you should be able to determine if it represents a function.

■   Know that functions can be represented in four ways: verbally, numerically, graphically, and algebraically.

■   Given a function, you should be able to do the following.

(a) Find the domain and range.

(b) Evaluate it at specific values.

■   You should be able to use function notation.

Vocabulary Check

1. domain; range; function

2. verbally; numerically; graphically; algebraically

3. independent; dependent

4. piecewise-defined

5. implied domain

6. difference quotient

---

1. Yes, the relationship is a function. Each domain value is matched with only one range value.

3. No, the relationship is not a function. The domain values are each matched with three range values.

5. Yes, it does represent a function. Each input value is matched with only one output value.

7. No, it does not represent a function. The input values of 10 and 7 are each matched with two output values.

9. (a) Each element of A is matched with exactly one element of B, so it does represent a function.

   (b) The element 1 in A is matched with two elements, -2 and 1 of B, so it does not represent a function.

   (c) Each element of A is matched with exactly one element of B, so it does represent a function.

   (d) The element 2 in A is not matched with an element of B, so the relation does not represent a function.

11. Each is a function. For each year there corresponds one and only one circulation.

13. $x^2 + y^2 = 4 \implies y = \pm\sqrt{4 - x^2}$

   No, y *is not* a function of x.

15. $x^2 + y = 4 \implies y = 4 - x^2$

   Yes, y *is* a function of x.

17. $2x + 3y = 4 \implies y = \frac{1}{3}(4 - 2x)$

   Yes, y *is* a function of x.

19. $y^2 = x^2 - 1 \implies y = \pm\sqrt{x^2 - 1}$

   Thus, y *is not* a function of x.

21. $y = |4 - x|$

   Yes, y *is* a function of x.

23. $x = 14$

   Thus, this *is not* a function of x.

25. $f(x) = 2x - 3$

   (a) $f(1) = 2(1) - 3 = -1$

   (b) $f(-3) = 2(-3) - 3 = -9$

   (c) $f(x - 1) = 2(x - 1) - 3 = 2x - 5$

27. $V(r) = \frac{4}{3}\pi r^3$

   (a) $V(3) = \frac{4}{3}\pi(3)^3 = \frac{4}{3}\pi(27) = 36\pi$

   (b) $V\left(\frac{3}{2}\right) = \frac{4}{3}\pi\left(\frac{3}{2}\right)^3 = \frac{4}{3}\pi\left(\frac{27}{8}\right) = \frac{9}{2}\pi$

   (c) $V(2r) = \frac{4}{3}\pi(2r)^3 = \frac{4}{3}\pi(8r^3) = \frac{32}{3}\pi r^3$

29. $f(y) = 3 - \sqrt{y}$

   (a) $f(4) = 3 - \sqrt{4} = 1$

   (b) $f(0.25) = 3 - \sqrt{0.25} = 2.5$

   (c) $f(4x^2) = 3 - \sqrt{4x^2} = 3 - 2|x|$

31. $q(x) = \dfrac{1}{x^2 - 9}$

   (a) $q(0) = \dfrac{1}{0^2 - 9} = -\dfrac{1}{9}$

   (b) $q(3) = \dfrac{1}{3^2 - 9}$ is undefined.

   (c) $q(y + 3) = \dfrac{1}{(y + 3)^2 - 9} = \dfrac{1}{y^2 + 6y}$

33. $f(x) = \dfrac{|x|}{x}$

  (a) $f(2) = \dfrac{|2|}{2} = 1$

  (b) $f(-2) = \dfrac{|-2|}{-2} = -1$

  (c) $f(x-1) = \dfrac{|x-1|}{x-1} = \begin{cases} -1, & \text{if } x < 1 \\ 1, & \text{if } x > 1 \end{cases}$

35. $f(x) = \begin{cases} 2x + 1, & x < 0 \\ 2x + 2, & x \geq 0 \end{cases}$

  (a) $f(-1) = 2(-1) + 1 = -1$

  (b) $f(0) = 2(0) + 2 = 2$

  (c) $f(2) = 2(2) + 2 = 6$

37. $f(x) = \begin{cases} 3x - 1, & x < -1 \\ 4, & -1 \leq x \leq 1 \\ x^2, & x > 1 \end{cases}$

  (a) $f(-2) = 3(-2) - 1 = -7$

  (b) $f\left(-\frac{1}{2}\right) = 4$

  (c) $f(3) = 3^2 = 9$

39. $f(x) = x^2 - 3$

  $f(-2) = (-2)^2 - 3 = 1$

  $f(-1) = (-1)^2 - 3 = -2$

  $f(0) = (0)^2 - 3 = -3$

  $f(1) = (1)^2 - 3 = -2$

  $f(2) = (2)^2 - 3 = 1$

x	-2	-1	0	1	2
$f(x)$	1	-2	-3	-2	1

41. $h(t) = \frac{1}{2}|t + 3|$

  $h(-5) = \frac{1}{2}|-5 + 3| = 1$

  $h(-4) = \frac{1}{2}|-4 + 3| = \frac{1}{2}$

  $h(-3) = \frac{1}{2}|-3 + 3| = 0$

  $h(-2) = \frac{1}{2}|-2 + 3| = \frac{1}{2}$

  $h(-1) = \frac{1}{2}|-1 + 3| = 1$

t	-5	-4	-3	-2	-1
$h(t)$	1	$\frac{1}{2}$	0	$\frac{1}{2}$	1

43. $f(x) = \begin{cases} -\frac{1}{2}x + 4, & x \leq 0 \\ (x - 2)^2, & x > 0 \end{cases}$

x	-2	-1	0	1	2
$f(x)$	5	$\frac{9}{2}$	4	1	0

  $f(-2) = -\frac{1}{2}(-2) + 4 = 5$

  $f(-1) = -\frac{1}{2}(-1) + 4 = 4\frac{1}{2} = \frac{9}{2}$

  $f(0) = -\frac{1}{2}(0) + 4 = 4$

  $f(1) = (1 - 2)^2 = 1$

  $f(2) = (2 - 2)^2 = 0$

45. $15 - 3x = 0$

  $3x = 15$

  $x = 5$

47. $\dfrac{3x - 4}{5} = 0$

  $3x - 4 = 0$

  $x = \dfrac{4}{3}$

49. $x^2 - 9 = 0$

  $x^2 = 9$

  $x = \pm 3$

51. $\quad x^3 - x = 0$

  $x(x^2 - 1) = 0$

  $x(x + 1)(x - 1) = 0$

  $x = 0,\ x = -1,\ \text{or } x = 1$

53. $\quad\quad f(x) = g(x)$

  $x^2 + 2x + 1 = 3x + 3$

  $x^2 - x - 2 = 0$

  $(x + 1)(x - 2) = 0$

  $x = -1\ \text{ or }\ x = 2$

55. $\quad\quad f(x) = g(x)$

  $\sqrt{3x} + 1 = x + 1$

  $\sqrt{3x} = x$

  $3x = x^2$

  $0 = x^2 - 3x$

  $0 = x(x - 3)$

  $x = 0\ \text{ or }\ x = 3$

57. $f(x) = 5x^2 + 2x - 1$

Since $f(x)$ is a polynomial, the domain is all real numbers x.

59. $h(t) = \dfrac{4}{t}$

Domain: All real numbers except $t = 0$

61. $g(y) = \sqrt{y - 10}$

Domain: $y - 10 \geq 0$

$$y \geq 10$$

63. $f(x) = \sqrt[4]{1 - x^2}$

Domain: $1 - x^2 \geq 0$

$$-x^2 \geq -1$$
$$x^2 \leq 1$$
$$x^2 - 1 \leq 0$$

Critical numbers: $x = \pm 1$

Test intervals: $(-\infty, -1), (-1, 1), (1, \infty)$

Test: Is $x^2 - 1 \leq 0$?

Domain: $-1 \leq x \leq 1$

65. $g(x) = \dfrac{1}{x} - \dfrac{3}{x + 2}$

Domain: All real numbers except $x = 0$, $x = -2$

67. $f(s) = \dfrac{\sqrt{s - 1}}{s - 4}$

Domain: $s - 1 \geq 0 \implies s \geq 1$ and $s \neq 4$

The domain consists of all real numbers s, such that $s \geq 1$ and $s \neq 4$.

69. $f(x) = \dfrac{x - 4}{\sqrt{x}}$

The domain is all real numbers such that $x > 0$ or $(0, \infty)$.

71. $f(x) = x^2$

$\{(-2, 4), (-1, 1), (0, 0), (1, 1), (2, 4)\}$

73. $f(x) = |x| + 2$

$\{(-2, 4), (-1, 3), (0, 2), (1, 3), (2, 4)\}$

75. By plotting the points, we have a parabola, so $g(x) = cx^2$. Since $(-4, -32)$ is on the graph, we have $-32 = c(-4)^2 \implies c = -2$. Thus, $g(x) = -2x^2$.

77. Since the function is undefined at 0, we have $r(x) = c/x$. Since $(-4, -8)$ is on the graph, we have $-8 = c/-4 \implies c = 32$. Thus, $r(x) = 32/x$.

79.
$$f(x) = x^2 - x + 1$$
$$f(2 + h) = (2 + h)^2 - (2 + h) + 1$$
$$= 4 + 4h + h^2 - 2 - h + 1$$
$$= h^2 + 3h + 3$$
$$f(2) = (2)^2 - 2 + 1 = 3$$
$$f(2 + h) - f(2) = h^2 + 3h$$
$$\frac{f(2 + h) - f(2)}{h} = \frac{h^2 + 3h}{h} = h + 3, \ h \neq 0$$

81.
$$f(x) = x^3 + 3x$$
$$f(x + h) = (x + h)^3 + 3(x + h)$$
$$= x^3 + 3x^2h + 3xh^2 + h^3 + 3x + 3h$$
$$\frac{f(x + h) - f(x)}{h} = \frac{(x^3 + 3x^2h + 3xh^2 + h^3 + 3x + 3h) - (x^3 + 3x)}{h}$$
$$= \frac{h(3x^2 + 3xh + h^2 + 3)}{h}$$
$$= 3x^2 + 3xh + h^2 + 3, \ h \neq 0$$

83. $g(x) = \dfrac{1}{x^2}$

$$\dfrac{g(x) - g(3)}{x - 3} = \dfrac{\dfrac{1}{x^2} - \dfrac{1}{9}}{x - 3}$$

$$= \dfrac{9 - x^2}{9x^2(x - 3)}$$

$$= \dfrac{-(x + 3)(x - 3)}{9x^2(x - 3)}$$

$$= -\dfrac{x + 3}{9x^2}, \ x \neq 0, 3$$

85. $f(x) = \sqrt{5x}$

$$\dfrac{f(x) - f(5)}{x - 5} = \dfrac{\sqrt{5x} - 5}{x - 5}$$

87. $A = s^2$ and $P = 4s \ \Rightarrow \ \dfrac{P}{4} = s$

$$A = \left(\dfrac{P}{4}\right)^2 = \dfrac{P^2}{16}$$

89. (a)

Height, x	Volume, V
1	484
2	800
3	972
4	1024
5	980
6	864

The volume is maximum when $x = 4$ and $V = 1024$ cubic centimeters.

(b)

V is a function of x.

(c) $V = x(24 - 2x)^2$

Domain: $0 < x < 12$

91. $A = \dfrac{1}{2}bh = \dfrac{1}{2}xy$

Since $(0, y)$, $(2, 1)$, and $(x, 0)$ all lie on the same line, the slopes between any pair are equal.

$$\dfrac{1 - y}{2 - 0} = \dfrac{0 - 1}{x - 2}$$

$$\dfrac{1 - y}{2} = \dfrac{-1}{x - 2}$$

$$y = \dfrac{2}{x - 2} + 1$$

$$y = \dfrac{x}{x - 2}$$

Therefore,

$$A = \dfrac{1}{2}x\left(\dfrac{x}{x - 2}\right) = \dfrac{x^2}{2(x - 2)}.$$

The domain of A includes x-values such that $x^2/[2(x - 2)] > 0$. Using methods of Section 1.8 we find that the domain is $x > 2$.

93. $y = -\dfrac{1}{10}x^2 + 3x + 6$

$y(30) = -\dfrac{1}{10}(30)^2 + 3(30) + 6 = 6$ feet

If the child holds a glove at a height of 5 feet, then the ball *will* be over the child's head since it will be at a height of 6 feet.

95. $p(t) = \begin{cases} 0.182t^2 + 0.57t + 27.3, & 0 \le t \le 7 \\ 2.50t + 21.3, & 8 \le t \le 12 \end{cases}$

Year	Function Value	Price
1990	$p(0) = 27.3$	$27,300
1991	$p(1) = 28.052$	$28,052
1992	$p(2) = 29.168$	$29,168
1993	$p(3) = 30.648$	$30,648
1994	$p(4) = 32.492$	$32,492
1995	$p(5) = 34.7$	$34,700
1996	$p(6) = 37.272$	$37,272
1997	$p(7) = 40.208$	$40,208
1998	$p(8) = 41.3$	$41,300
1999	$p(9) = 43.8$	$43,800
2000	$p(10) = 46.3$	$46,300
2001	$p(11) = 48.8$	$48,800
2002	$p(12) = 51.3$	$51,300

97. (a) Cost = variable costs + fixed costs

$$C = 12.30x + 98,000$$

(b) Revenue = price per unit × number of units

$$R = 17.98x$$

(c) Profit = Revenue − Cost

$$P = 17.98x - (12.30x + 98,000)$$

$$P = 5.68x - 98,000$$

99. (a) $R = n(\text{rate}) = n[8.00 - 0.05(n - 80)], \ n \ge 80$

$$R = 12.00n - 0.05n^2 = 12n - \frac{n^2}{20} = \frac{240n - n^2}{20}, \ n \ge 80$$

(b)

n	90	100	110	120	130	140	150
$R(n)$	$675	$700	$715	$720	$715	$700	$675

The revenue is maximum when 120 people take the trip.

101. (a)

d

h

|← 3000 ft →|

(b) $(3000)^2 + h^2 = d^2$

$$h = \sqrt{d^2 - (3000)^2}$$

Domain: $d \ge 3000$ (since both $d \ge 0$ and $d^2 - (3000)^2 \ge 0$)

103. False. The range is $[-1, \infty)$.

105. The domain is the set of inputs of the function, and the range is the set of outputs.

107. (a) Yes. The amount that you pay in sales tax will increase as the price of the item purchased increases.

(b) No. The length of time that you study the night before an exam does not necessarily determine your score on the exam.

109.

$$\frac{t}{3} + \frac{t}{5} = 1$$

$$15\left(\frac{t}{3} + \frac{t}{5}\right) = 15(1)$$

$$5t + 3t = 15$$

$$8t = 15$$

$$t = \frac{15}{8}$$

111.

$$\frac{3}{x(x+1)} - \frac{4}{x} = \frac{1}{x+1}$$

$$x(x+1)\left[\frac{3}{x(x+1)} - \frac{4}{x}\right] = x(x+1)\left(\frac{1}{x+1}\right)$$

$$3 - 4(x+1) = x$$

$$3 - 4x - 4 = x$$

$$-1 = 5x$$

$$-\frac{1}{5} = x$$

113. $(-2, -5)$ and $(4, -1)$

$$m = \frac{-1 - (-5)}{4 - (-2)} = \frac{4}{6} = \frac{2}{3}$$

$$y - (-5) = \frac{2}{3}(x - (-2))$$

$$y + 5 = \frac{2}{3}x + \frac{4}{3}$$

$$3y + 15 = 2x + 4$$

$$2x - 3y - 11 = 0$$

115. $(-6, 5)$ and $(3, -5)$

$$m = \frac{-5 - 5}{3 - (-6)} = -\frac{10}{9}$$

$$y - 5 = -\frac{10}{9}(x - (-6))$$

$$9y - 45 = -10x - 60$$

$$10x + 9y + 15 = 0$$

Section 2.3   Analyzing Graphs of Functions

- You should be able to determine the domain and range of a function from its graph.

- You should be able to use the vertical line test for functions.

- You should be able to find the zeros of a function.

- You should be able to determine when a function is constant, increasing, or decreasing.

- You should be able to approximate relative minimums and relative maximums from the graph of a function.

- You should know that f is

  (a) odd if $f(-x) = -f(x)$.   (b) even if $f(-x) = f(x)$.

Vocabulary Check

1. ordered pairs

2. vertical line test

3. zeros

4. decreasing

5. maximum

6. average rate of change; secant

7. odd

8. even

1. Domain: $(-\infty, -1] \cup [1, \infty)$

Range: $[0, \infty)$

3. Domain: $[-4, 4]$

Range: $[0, 4]$

5. (a) $f(-2) = 0$ (b) $f(-1) = -1$

(c) $f\left(\frac{1}{2}\right) = 0$ (d) $f(1) = -3$

7. (a) $f(-2) = -3$ (b) $f(1) = 0$

(c) $f(0) = 1$ (d) $f(2) = -3$

9. $y = \frac{1}{2}x^2$

A vertical line intersects the graph just once, so y *is* a function of x.

11. $x - y^2 = 1 \implies y = \pm\sqrt{x-1}$

y *is not* a function of x. Some vertical lines cross the graph twice.

13. $x^2 = 2xy - 1$

A vertical line intersects the graph just once, so y *is* a function of x.

15. $2x^2 - 7x - 30 = 0$

$(2x + 5)(x - 6) = 0$

$2x + 5 = 0$ or $x - 6 = 0$

$x = -\dfrac{5}{2}$ or $x = 6$

17. $\dfrac{x}{9x^2 - 4} = 0$

$x = 0$

19. $\frac{1}{2}x^3 - x = 0$

$x^3 - 2x = 2(0)$

$x(x^2 - 2) = 0$

$x = 0$ or $x^2 - 2 = 0$

$x^2 = 2$

$x = \pm\sqrt{2}$

21. $4x^3 - 24x^2 - x + 6 = 0$

$4x^2(x - 6) - 1(x - 6) = 0$

$(x - 6)(4x^2 - 1) = 0$

$(x - 6)(2x + 1)(2x - 1) = 0$

$x - 6 = 0,\quad 2x + 1 = 0,\quad 2x - 1 = 0$

$x = 6,\qquad x = -\frac{1}{2},\qquad x = \frac{1}{2}$

23. $\sqrt{2x} - 1 = 0$

$\sqrt{2x} = 1$

$2x = 1$

$x = \dfrac{1}{2}$

25. (a)

Zero: $x = -\dfrac{5}{3}$

(b) $3 + \dfrac{5}{x} = 0$

$3x + 5 = 0$

$x = -\dfrac{5}{3}$

27. (a)

Zero: $x = -\dfrac{11}{2}$

(b) $\sqrt{2x + 11} = 0$

$2x + 11 = 0$

$x = -\dfrac{11}{2}$

29. (a)

Zero: $x = \dfrac{1}{3}$

(b) $\dfrac{3x - 1}{x - 6} = 0$

$3x - 1 = 0$

$x = \dfrac{1}{3}$

31. $f(x) = \frac{3}{2}x$

f is increasing on $(-\infty, \infty)$.

33. $f(x) = x^3 - 3x^2 + 2$

f is increasing on $(-\infty, 0)$ and $(2, \infty)$.

f is decreasing on $(0, 2)$.

35. $f(x) = \begin{cases} x + 3, & x \le 0 \\ 3, & 0 < x \le 2 \\ 2x + 1, & x > 2 \end{cases}$

f is increasing on $(-\infty, 0)$ and $(2, \infty)$.

f is constant on $(0, 2)$.

37. $f(x) = |x + 1| + |x - 1|$

f is increasing on $(1, \infty)$.

f is constant on $(-1, 1)$.

f is decreasing on $(-\infty, -1)$.

39. $f(x) = 3$

(a)

Constant on $(-\infty, \infty)$

(b)

x	-2	-1	0	1	2
$f(x)$	3	3	3	3	3

41. $g(s) = \dfrac{s^2}{4}$

(a)

Decreasing on $(-\infty, 0)$; Increasing on $(0, \infty)$

(b)

s	-4	-2	0	2	4
$g(s)$	4	1	0	1	4

43. $f(t) = -t^4$

(a)

Increasing on $(-\infty, 0)$; Decreasing on $(0, \infty)$

(b)

t	-2	-1	0	1	2
$f(t)$	-16	-1	0	-1	-16

45. $f(x) = \sqrt{1 - x}$

(a)

Decreasing on $(-\infty, 1)$

(b)

x	-3	-2	-1	0	1
$f(x)$	2	$\sqrt{3}$	$\sqrt{2}$	1	0

47. $f(x) = x^{3/2}$

(a)

Increasing on $(0, \infty)$

(b)

x	0	1	2	3	4
$f(x)$	0	1	2.8	5.2	8

49. $f(x) = (x - 4)(x + 2)$

Relative minimum: $(1, -9)$

51. $f(x) = -x^2 + 3x - 2$

Relative maximum: $(1.5, 0.25)$

53. $f(x) = x(x - 2)(x + 3)$

Relative minimum: $(1.12, -4.06)$

Relative maximum: $(-1.79, 8.21)$

55. $f(x) = 4 - x$

$f(x) \geq 0$ on $(-\infty, 4]$.

57. $f(x) = x^2 + x$

$f(x) \geq 0$ on $(-\infty, -1]$ and $[0, \infty)$.

59. $f(x) = \sqrt{x - 1}$

$f(x) \geq 0$ on $[1, \infty)$.

61. $f(x) = -(1 + |x|)$

$f(x)$ is never greater than 0.
($f(x) < 0$ for all x.)

63. $f(x) = -2x + 15$

$$\frac{f(3) - f(0)}{3 - 0} = \frac{9 - 15}{3} = -2$$

The average rate of change from $x_1 = 0$ to $x_2 = 3$ is -2.

65. $f(x) = x^2 + 12x - 4$

$$\frac{f(5) - f(1)}{5 - 1} = \frac{81 - 9}{4} = 18$$

The average rate of change from $x_1 = 1$ to $x_2 = 5$ is 18.

67. $f(x) = x^3 - 3x^2 - x$

$$\frac{f(3) - f(1)}{3 - 1} = \frac{-3 - (-3)}{2} = 0$$

The average rate of change from $x_1 = 1$ to $x_2 = 3$ is 0.

69. $f(x) = -\sqrt{x - 2} + 5$

$$\frac{f(11) - f(3)}{11 - 3} = \frac{2 - 4}{8} = -\frac{1}{4}$$

The average rate of change from $x_1 = 3$ to $x_2 = 11$ is $-\frac{1}{4}$.

71. $f(x) = x^6 - 2x^2 + 3$

$f(-x) = (-x)^6 - 2(-x)^2 + 3$

$\qquad = x^6 - 2x^2 + 3$

$\qquad = f(x)$

f is even.

y-axis symmetry

73. $g(x) = x^3 - 5x$

$g(-x) = (-x)^3 - 5(-x)$

$\qquad = -x^3 + 5x$

$\qquad = -g(x)$

g is odd.
Origin symmetry

75. $f(t) = t^2 + 2t - 3$

$f(-t) = (-t)^2 + 2(-t) - 3$

$\qquad = t^2 - 2t - 3$

$\qquad \ne f(t), \ne -f(t)$

f is neither even nor odd.
No symmetry

77. $h = \text{top} - \text{bottom}$

$\qquad = (-x^2 + 4x - 1) - 2$

$\qquad = -x^2 + 4x - 3$

79. $h = \text{top} - \text{bottom}$

$\qquad = (4x - x^2) - 2x$

$\qquad = 2x - x^2$

81. $L = \text{right} - \text{left}$

$\qquad = \frac{1}{2}y^2 - 0 = \frac{1}{2}y^2$

83. $L = \text{right} - \text{left}$

$\qquad = 4 - y^2$

85. $L = -0.294x^2 + 97.744x - 664.875, \ 20 \le x \le 90$

(a)

(b) $L = 2000$ when $x \approx 29.9645 \approx 30$ watts.

87. (a) For the average salaries of college professors, a scale of \$10,000 would be appropriate.

(b) For the population of the United States, use a scale of 10,000,000.

(c) For the percent of the civilian workforce that is unemployed, use a scale of 1%.

89. $r = 15.639t^3 - 104.75t^2 + 303.5t - 301, \; 2 \le t \le 7$

(a)

(b) $\dfrac{r(7) - r(2)}{7 - 2} = \dfrac{2054.927 - 12.112}{5} = 408.563$

The average rate of change from 2002 to 2007 is $408.563 billion per year. The estimated revenue is increasing each year at a rapid pace.

91. $s_0 = 6, \; v_0 = 64$

(a) $s = -16t^2 + 64t + 6$

(b)

(c) $\dfrac{s(3) - s(0)}{3 - 0} = \dfrac{54 - 6}{3} = 16$

(d) The average rate of change of the height of the object with respect to time over the interval $t_1 = 0$ to $t_2 = 3$ is 16 feet per second.

(e) $s(0) = 6, \; m = 16$

Secant line: $y - 6 = 16(t - 0)$

$$y = 16t + 6$$

(f)

93. $v_0 = 120, \; s_0 = 0$

(a) $s = -16t^2 + 120t$

(b)

(c) $\dfrac{s(5) - s(3)}{5 - 3} = \dfrac{200 - 216}{2} = -8$

(d) The average decrease in the height of the object over the interval $t_1 = 3$ to $t_2 = 5$ is 8 feet per second.

(e) $s(5) = 200, \; m = -8$

Secant line: $y - 200 = -8(t - 5)$

$$y = -8t + 240$$

(f)

95. $v_0 = 0, \; s_0 = 120$

(a) $s = -16t^2 + 120$

(b)

(c) $\dfrac{s(2) - s(0)}{2 - 0} = \dfrac{56 - 120}{2} = -32$

(d) On the interval $t_1 = 0$ to $t_2 = 2$, the height of the object is decreasing at a rate of 32 feet per second.

(e) $s(0) = 120, \; m = -32$

Secant line: $y - 120 = -32(t - 0)$

$$y = -32t + 120$$

(f)

97. False. The function $f(x) = \sqrt{x^2 + 1}$ has a domain of all real numbers.

99. (a) Even. The graph is a reflection in the *x*-axis.

(b) Even. The graph is a reflection in the *y*-axis.

(c) Even. The graph is a vertical translation of *f*.

(d) Neither. The graph is a horizontal translation of *f*.

101. $\left(-\frac{3}{2}, 4\right)$

(a) If *f* is even, another point is $\left(\frac{3}{2}, 4\right)$.

(b) If *f* is odd, another point is $\left(\frac{3}{2}, -4\right)$.

103. $(4, 9)$

(a) If *f* is even, another point is $(-4, 9)$.

(b) If *f* is odd, another point is $(-4, -9)$.

105. (a) $y = x$ (b) $y = x^2$ (c) $y = x^3$

(d) $y = x^4$ (e) $y = x^5$ (f) $y = x^6$

All the graphs pass through the origin. The graphs of the odd powers of *x* are symmetric with respect to the origin and the graphs of the even powers are symmetric with respect to the *y*-axis. As the powers increase, the graphs become flatter in the interval $-1 < x < 1$.

107. $x^2 - 10x = 0$

$x(x - 10) = 0$

$x = 0 \quad \text{or} \quad x = 10$

109. $x^3 - x = 0$

$x(x^2 - 1) = 0$

$x = 0 \quad \text{or} \quad x^2 - 1 = 0$

$x^2 = 1$

$x = \pm 1$

111. $f(x) = 5x - 8$

(a) $f(9) = 5(9) - 8 = 37$

(b) $f(-4) = 5(-4) - 8 = -28$

(c) $f(x - 7) = 5(x - 7) - 8 = 5x - 35 - 8 = 5x - 43$

113. $f(x) = \sqrt{x - 12} - 9$

(a) $f(12) = \sqrt{12 - 12} - 9 = 0 - 9 = -9$

(b) $f(40) = \sqrt{40 - 12} - 9 = \sqrt{28} - 9 = 2\sqrt{7} - 9$

(c) $f\left(-\sqrt{36}\right)$ does not exist. The given value is not in the domain of the function.

115.

$$f(x) = x^2 - 2x + 9$$

$$f(3 + h) = (3 + h)^2 - 2(3 + h) + 9$$

$$= 9 + 6h + h^2 - 6 - 2h + 9$$

$$= h^2 + 4h + 12$$

$$f(3) = 3^2 - 2(3) + 9 = 12$$

$$\frac{f(3 + h) - f(3)}{h} = \frac{(h^2 + 4h + 12) - (12)}{h} = \frac{h^2 + 4h}{h} = \frac{h(h + 4)}{h} = h + 4, \ h \neq 0$$

Section 2.4 A Library of Parent Functions

- ■ You should be able to identify and graph the following types of functions:

  (a) Linear functions like $f(x) = ax + b$

  (b) Squaring functions like $f(x) = x^2$

  (c) Cubic functions like $f(x) = x^3$

  (d) Square root functions like $f(x) = \sqrt{x}$

  (e) Reciprocal functions like $f(x) = \dfrac{1}{x}$

  (f) Constant functions like $f(x) = c$

  (g) Absolute value functions like $f(x) = |x|$

  (h) Step and piecewise-defined functions like $f(x) = [\![x]\!]$

- ■ You should be able to determine the following about these common functions:

  (a) Domain and range

  (b) x-intercept(s) and y-intercept

  (c) Symmetries

  (d) Where it is increasing, decreasing, or constant

  (e) If it is odd, even or neither

  (f) Relative maximums and relative minimums

Vocabulary Check

1. $f(x) = [\![x]\!]$

   (g) greatest integer function

2. $f(x) = x$

   (i) identity function

3. $f(x) = \dfrac{1}{x}$

   (h) reciprocal function

4. $f(x) = x^2$

   (a) squaring function

5. $f(x) = \sqrt{x}$

   (b) square root function

6. $f(x) = c$

   (e) constant function

7. $f(x) = |x|$

   (f) absolute value function

8. $f(x) = x^3$

   (c) cubic function

9. $f(x) = ax + b$

   (d) linear function

---

1. (a) $f(1) = 4$, $f(0) = 6$

    $(1, 4)$ and $(0, 6)$

    $m = \dfrac{6 - 4}{0 - 1} = -2$

    $y - 6 = -2(x - 0)$

    $y = -2x + 6$

    $f(x) = -2x + 6$

(b)

3. (a) $f(5) = -4$, $f(-2) = 17$

$(5, -4)$ and $(-2, 17)$

$m = \dfrac{17 - (-4)}{-2 - 5} = \dfrac{21}{-7} = -3$

$y - (-4) = -3(x - 5)$

$y + 4 = -3x + 15$

$y = -3x + 11$

$f(x) = -3x + 11$

(b)

5. (a) $f(-5) = -1$, $f(5) = -1$

$(-5, -1)$ and $(5, -1)$

$m = \dfrac{-1 - (-1)}{5 - (-5)} = \dfrac{0}{10} = 0$

$y - (-1) = 0(x - (-5))$

$y + 1 = 0$

$y = -1$

$f(x) = -1$

(b)

7. (a) $f\left(\dfrac{1}{2}\right) = -6, f(4) = -3$

$\left(\dfrac{1}{2}, -6\right)$ and $(4, -3)$

$m = \dfrac{-3 - (-6)}{4 - (1/2)} = \dfrac{3}{7/2} = \dfrac{6}{7}$

$y - (-3) = \dfrac{6}{7}(x - 4)$

$y + 3 = \dfrac{6}{7}x - \dfrac{24}{7}$

$y = \dfrac{6}{7}x - \dfrac{45}{7}$

$f(x) = \dfrac{6}{7}x - \dfrac{45}{7}$

(b)

9. $f(x) = -x - \frac{3}{4}$

11. $f(x) = -\frac{1}{6}x - \frac{5}{2}$

13. $f(x) = x^2 - 2x$

15. $h(x) = -x^2 + 4x + 12$

17. $f(x) = x^3 - 1$

19. $f(x) = (x - 1)^3 + 2$

21. $f(x) = 4\sqrt{x}$

23. $g(x) = 2 - \sqrt{x + 4}$

25. $f(x) = -\dfrac{1}{x}$

27. $h(x) = \dfrac{1}{x + 2}$

29. $f(x) = [\![x]\!]$

(a) $f(2.1) = 2$

(b) $f(2.9) = 2$

(c) $f(-3.1) = -4$

(d) $f\left(\dfrac{7}{2}\right) = 3$

31. $h(x) = [\![x + 3]\!]$

(a) $h(-2) = [\![1]\!] = 1$

(b) $h\left(\dfrac{1}{2}\right) = [\![3.5]\!] = 3$

(c) $h(4.2) = [\![7.2]\!] = 7$

(d) $h(-21.6) = [\![-18.6]\!] = -19$

33. $h(x) = [\![3x - 1]\!]$

(a) $h(2.5) = [\![6.5]\!] = 6$

(b) $h(-3.2) = [\![-10.6]\!] = -11$

(c) $h\left(\dfrac{7}{3}\right) = [\![6]\!] = 6$

(d) $h\left(-\dfrac{21}{3}\right) = [\![-22]\!] = -22$

35. $g(x) = 3[\![x - 2]\!] + 5$

(a) $g(-2.7) = 3[\![-4.7]\!] + 5 = 3(-5) + 5 = -10$

(b) $g(-1) = 3[\![-3]\!] + 5 = 3(-3) + 5 = -4$

(c) $g(0.8) = 3[\![-1.2]\!] + 5 = 3(-2) + 5 = -1$

(d) $g(14.5) = 3[\![12.5]\!] + 5 = 3(12) + 5 = 41$

37. $g(x) = -[\![x]\!]$

39. $g(x) = [\![x]\!] - 2$

41. $g(x) = [\![x + 1]\!]$

43. $f(x) = \begin{cases} 2x + 3, & x < 0 \\ 3 - x, & x \geq 0 \end{cases}$

45. $f(x) = \begin{cases} \sqrt{4 + x}, & x < 0 \\ \sqrt{4 - x}, & x \geq 0 \end{cases}$

47. $f(x) = \begin{cases} x^2 + 5, & x \leq 1 \\ -x^2 + 4x + 3, & x > 1 \end{cases}$

49. $h(x) = \begin{cases} 4 - x^2, & x < -2 \\ 3 + x, & -2 \le x < 0 \\ x^2 + 1, & x \ge 0 \end{cases}$

51. $s(x) = 2\left(\frac{1}{4}x - \left[\!\left[\frac{1}{4}x\right]\!\right]\right)$

(a)

(b) Domain: $(-\infty, \infty)$

Range: $[0, 2)$

(c) Sawtooth pattern

53. (a) Common function: $f(x) = |x|$

(b) $g(x) = |x + 2| - 1$

(c)

55. (a) Common function: $f(x) = x^3$

(b) $g(x) = (x - 1)^3 - 2$

(c)

57. (a) Common function: $f(x) = c$

(b) $g(x) = 2$

(c)

59. (a) Common function: $f(x) = x$

(b) $g(x) = x - 2$

(c)

61. $C = 0.60 - 0.42[\![1 - t]\!],\ t > 0$

(a)

Time (in minutes)

(b) $C(12.5) = \$5.64$

63. $C = 10.75 - 3.95[\![x]\!],\ x > 0$

(a)

Weight (in pounds)

(b) $C(10.33) = 10.75 + 3.95(10) = \$50.25$

65. $W(h) = \begin{cases} 12h, & 0 < h \le 40 \\ 18(h - 40) + 480, & h > 40 \end{cases}$

(a) $W(30) = 12(30) = \$360$

$W(40) = 12(40) = \$480$

$W(45) = 18(5) + 480 = \$570$

$W(50) = 18(10) + 480 = \$660$

(b) $W(h) = \begin{cases} 12h, & 0 < h \le 45 \\ 18(h - 45) + 540, & h > 45 \end{cases}$

67. (a) The domain of $f(x) = -1.97x + 26.3$ is $6 < x \leq 12$. One way to see this is to notice that this is the equation of a line with negative slope, so the function values are decreasing as x increases, which matches the data for the corresponding part of the table. The domain of $f(x) = 0.505x^2 - 1.47x + 6.3$ is then $1 \leq x \leq 6$.

(b)

Month (1 ↔ January)

(c) $f(5) = 0.505(5)^2 - 1.47(5) + 6.3$

$= 0.505(25) - 7.35 + 6.3 = 11.575$

$f(11) = -1.97(11) + 26.3 = 4.63$

These values represent the income in thousands of dollars for the months of May and November, respectively.

(d) The model values are very close to the actual values.

Month, x	1	2	3	4	5	6	7	8	9	10	11	12
Revenue, y	5.2	5.6	6.6	8.3	11.5	15.8	12.8	10.1	8.6	6.9	4.5	2.7
Model, $f(x)$	5.3	5.4	6.4	8.5	11.6	15.7	12.5	10.5	8.6	6.6	4.6	2.7

69. False. A piecewise-defined function is a function that is defined by two or more equations over a specified domain. That domain may or may not include x- and y-intercepts.

71. For the line through $(0, 6)$ and $(3, 2)$: $m = \dfrac{6 - 2}{0 - 3} = -\dfrac{4}{3}$

$$y - 6 = -\frac{4}{3}(x - 0) \implies y = -\frac{4}{3}x + 6$$

For the line through $(3, 2)$ and $(8, 0)$: $m = \dfrac{2 - 0}{3 - 8} = -\dfrac{2}{5}$

$$y - 0 = -\frac{2}{5}(x - 8) \implies y = -\frac{2}{5}x + \frac{16}{5}$$

$$f(x) = \begin{cases} -\frac{4}{3}x + 6, & 0 \leq x \leq 3 \\ -\frac{2}{5}x + \frac{16}{5}, & 3 < x \leq 8 \end{cases}$$

Note that the respective domains can also be $0 \leq x < 3$ and $3 \leq x \leq 8$.

73. $3x + 4 \leq 12 - 5x$

$8x + 4 \leq 12$

$8x \leq 8$

$x \leq 1$

75. L_1: $(-2, -2)$ and $(2, 10)$

$$m_1 = \frac{10 - (-2)}{2 - (-2)} = \frac{12}{4} = 3$$

L_2: $(-1, 3)$ and $(3, 9)$

$$m_2 = \frac{9 - 3}{3 - (-1)} = \frac{6}{4} = \frac{3}{2}$$

The lines are neither parallel nor perpendicular.

Section 2.5   Transformations of Functions

■ You should know the basic types of transformations.

Let $y = f(x)$ and let c be a positive real number.

1. $h(x) = f(x) + c$	Vertical shift c units upward
2. $h(x) = f(x) - c$	Vertical shift c units downward
3. $h(x) = f(x - c)$	Horizontal shift c units to the right
4. $h(x) = f(x + c)$	Horizontal shift c units to the left
5. $h(x) = -f(x)$	Reflection in the x-axis
6. $h(x) = f(-x)$	Reflection in the y-axis
7. $h(x) = cf(x), c > 1$	Vertical stretch
8. $h(x) = cf(x), 0 < c < 1$	Vertical shrink
9. $h(x) = f(cx),\ c > 1$	Horizontal shrink
10. $h(x) = f(cx),\ 0 < c < 1$	Horizontal stretch

Vocabulary Check

1. rigid

2. $-f(x);\ f(-x)$

3. nonrigid

4. horizontal shrink; horizontal stretch

5. vertical stretch; vertical shrink

6. (a) iv   (b) ii   (c) iii   (d) i

---

1. (a) $f(x) = |x| + c$ Vertical shifts

$c = -1 : f(x) = |x| - 1$ 1 unit down

$c = 1 : f(x) = |x| + 1$ 1 unit up

$c = 3 : f(x) = |x| + 3$ 3 units up

(b) $f(x) = |x - c|$ Horizontal shifts

$c = -1 : f(x) = |x + 1|$ 1 unit left

$c = 1 : f(x) = |x - 1|$ 1 unit right

$c = 3 : f(x) = |x - 3|$ 3 units right

(c) $f(x) = |x + 4| + c$ Horizontal shift four units left and a vertical shift

$c = -1 : f(x) = |x + 4| - 1$ 1 unit down

$c = 1 : f(x) = |x + 4| + 1$ 1 unit up

$c = 3 : f(x) = |x + 4| + 3$ 3 units up

3. (a) $f(x) = [\![x]\!] + c$ Vertical shifts

$c = -2 : f(x) = [\![x]\!] - 2$ 2 units down

$c = 0 : f(x) = [\![x]\!]$ Common function

$c = 2 : f(x) = [\![x]\!] + 2$ 2 units up

(b) $f(x) = [\![x + c]\!]$ Horizontal shifts

$c = -2 : f(x) = [\![x - 2]\!]$ 2 units right

$c = 0 : f(x) = [\![x]\!]$ Common function

$c = 2 : f(x) = [\![x + 2]\!]$ 2 units left

(c) $f(x) = [\![x - 1]\!] + c$ Horizontal shift 1 unit right and a vertical shift

$c = -2 : f(x) = [\![x - 1]\!] - 2$ 2 units down

$c = 0 : f(x) = [\![x - 1]\!]$

$c = 2 : f(x) = [\![x - 1]\!] + 2$ 2 units up

5. (a) $y = f(x) + 2$

Vertical shift 2 units upward

(b) $y = f(x - 2)$

Horizontal shift 2 units to the right

(c) $y = 2f(x)$

Vertical stretch (each y-value is multiplied by 2)

(d) $y = -f(x)$

Reflection in the x-axis

(e) $y = f(x + 3)$

Horizontal shift 3 units to the left

(f) $y = f(-x)$

Reflection in the y-axis

—CONTINUED—

5. **—CONTINUED—**

(g) $y = f\left(\frac{1}{2}x\right)$

Horizontal stretch (each
x-value is multiplied by 2)

7. (a) $y = f(x) - 1$

Vertical shift 1 unit downward

(b) $y = f(x - 1)$

Horizontal shift 1 unit to the right

(c) $y = f(-x)$

Reflection about the y-axis

(d) $y = f(x + 1)$

Horizontal shift 1 unit to the
left

(e) $y = -f(x - 2)$

Reflection about the x-axis and a
horizontal shift 2 units to the right

(f) $y = \frac{1}{2}f(x)$

Vertical shrink $\left(\text{each } y\text{-value is}\right.$
multiplied by $\left.\frac{1}{2}\right)$

(g) $y = f(2x)$

Horizontal shrink $\left(\text{each}\right.$
x-value is multiplied by $\left.\frac{1}{2}\right)$

9. (a) Vertical shift 1 unit downward

$$f(x) = x^2 - 1$$

(b) Reflection about the *x*-axis, horizontal shift 1 unit to the left, and a vertical shift 1 unit upward

$$f(x) = -(x + 1)^2 + 1$$

(c) Reflection about the *x*-axis, horizontal shift 2 units to the right, and a vertical shift 6 units upward

$$f(x) = -(x - 2)^2 + 6$$

(d) Horizontal shift 5 units to the right and a vertical shift 3 units downward

$$f(x) = (x - 5)^2 - 3$$

11. (a) Vertical shift 5 units upward

$$f(x) = |x| + 5$$

(b) Reflection in the *x*-axis and a horizontal shift 3 units to the left

$$f(x) = -|x + 3|$$

(c) Horizontal shift 2 units to the right and a vertical shift 4 units downward

$$f(x) = |x - 2| - 4$$

(d) Reflection in the *x*-axis, horizontal shift 6 units to the right, and a vertical shift 1 unit downward

$$f(x) = -|x - 6| - 1$$

13. Common function: $f(x) = x^3$

Horizontal shift 2 units to the right: $y = (x - 2)^3$

15. Common function: $f(x) = x^2$

Reflection in the *x*-axis: $y = -x^2$

17. Common function: $f(x) = \sqrt{x}$

Reflection in the *x*-axis and a vertical shift 1 unit upward: $y = -\sqrt{x} + 1$

19. $g(x) = 12 - x^2$

(a) Common function: $f(x) = x^2$

(b) Reflection in the *x*-axis and a vertical shift 12 units upward

(c)

(d) $g(x) = 12 - f(x)$

21. $g(x) = x^3 + 7$

(a) Common function: $f(x) = x^3$

(b) Vertical shift 7 units upward

(c)

(d) $g(x) = f(x) + 7$

23. $g(x) = \frac{2}{3}x^2 + 4$

(a) Common function: $f(x) = x^2$

(b) Vertical shrink of two-thirds, and a vertical shift 4 units upward

(c)

(d) $g(x) = \frac{2}{3}f(x) + 4$

25. $g(x) = 2 - (x + 5)^2$

(a) Common function: $f(x) = x^2$

(b) Reflection in the *x*-axis, horizontal shift 5 units to the left, and a vertical shift 2 units upward

(c)

(d) $g(x) = 2 - f(x + 5)$

27. $g(x) = \sqrt{3x}$

(a) Common function: $f(x) = \sqrt{x}$

(b) Horizontal shrink by $\frac{1}{3}$

(c)

(d) $g(x) = f(3x)$

29. $g(x) = (x - 1)^3 + 2$

(a) Common function: $f(x) = x^3$

(b) Horizontal shift 1 unit to the right and a vertical shift 2 units upward

(c)

(d) $g(x) = f(x - 1) + 2$

31. $g(x) = -|x| - 2$

(a) Common function: $f(x) = |x|$

(b) Reflection in the x-axis; vertical shift 2 units downward

(c)

(d) $g(x) = -f(x) - 2$

33. $g(x) = -|x + 4| + 8$

(a) Common function: $f(x) = |x|$

(b) Reflection in the x-axis, horizontal shift 4 units to the left, and a vertical shift 8 units upward

(c)

(d) $g(x) = -f(x + 4) + 8$

35. $g(x) = 3 - [\![x]\!]$

(a) Common function: $f(x) = [\![x]\!]$

(b) Reflection in the x-axis and a vertical shift 3 units up

(c)

(d) $g(x) = 3 - f(x)$

37. $g(x) = \sqrt{x - 9}$

(a) Common function: $f(x) = \sqrt{x}$

(b) Horizontal shift 9 units to the right

(c)

(d) $g(x) = f(x - 9)$

39. $g(x) = \sqrt{7-x} - 2$ or $g(x) = \sqrt{-(x-7)} - 2$

(a) Common function: $f(x) = \sqrt{x}$

(b) Reflection in the y-axis, horizontal shift 7 units to the right, and a vertical shift 2 units downward

(c)   (d) $g(x) = f(7-x) - 2$

41. $g(x) = \sqrt{\frac{1}{2}x} - 4$

(a) Common function: $f(x) = \sqrt{x}$

(b) Horizontal stretch (each x-value is multiplied by 2) and a vertical shift 4 units down

(c)   (d) $g(x) = f\left(\frac{1}{2}x\right) - 4$

43. $f(x) = x^2$ moved 2 units to the right and 8 units down.

$g(x) = (x-2)^2 - 8$

45. $f(x) = x^3$ moved 13 units to the right.

$g(x) = (x - 13)^3$

47. $f(x) = |x|$ moved 10 units up and reflected about the x-axis.

$g(x) = -(|x| + 10) = -|x| - 10$

49. $f(x) = \sqrt{x}$ moved 6 units to the left and reflected in both the x- and y-axes.

$g(x) = -\sqrt{-x + 6}$

51. $f(x) = x^2$

(a) Reflection in the x-axis and a vertical stretch (each y-value is multiplied by 3)

$g(x) = -3x^2$

(b) Vertical shift 3 units upward and a vertical stretch (each y-value is multiplied by 4)

$g(x) = 4x^2 + 3$

53. $f(x) = |x|$

(a) Reflection in the x-axis and a vertical shrink (each y-value is multiplied by $\frac{1}{2}$)

$g(x) = -\frac{1}{2}|x|$

(b) Vertical stretch (each y-value is multiplied by 3) and a vertical shift 3 units downward

$g(x) = 3|x| - 3$

55. Common function: $f(x) = x^3$

Vertical stretch (each y-value is multiplied by 2)

$g(x) = 2x^3$

57. Common function: $f(x) = x^2$

Reflection in the x-axis; vertical shrink (each y-value is multiplied by $\frac{1}{2}$)

$g(x) = -\frac{1}{2}x^2$

59. Common function: $f(x) = \sqrt{x}$

Reflection in the y-axis; vertical shrink (each y-value is multiplied by $\frac{1}{2}$)

$g(x) = \frac{1}{2}\sqrt{-x}$

61. Common function: $f(x) = x^3$

Reflection in the x-axis, horizontal shift 2 units to the right and a vertical shift 2 units upward

$g(x) = -(x-2)^3 + 2$

63. Common function: $f(x) = \sqrt{x}$

Reflection in the x-axis and a vertical shift 3 units downward

$g(x) = -\sqrt{x} - 3$

65. (a) $g(x) = f(x) + 2$

Vertical shift 2 units upward

(b) $g(x) = f(x) - 1$

Vertical shift 1 unit downward

(c) $g(x) = f(-x)$

Reflection in the *y*-axis

(d) $g(x) = -2f(x)$

Reflection in the *x*-axis and a vertical stretch (each *y*-value is multiplied by 2)

(e) $g(x) = f(4x)$

Horizontal shrink $\left(\text{each } x\text{-value is multiplied by } \frac{1}{4}\right)$

(f) $g(x) = f\left(\frac{1}{2}x\right)$

Horizontal stretch (each *x*-value is multiplied by 2)

67. $F = f(t) = 20.6 + 0.035t^2, \ 0 \le t \le 22$

(a) A vertical shrink by 0.035 and a vertical shift of 20.6 units upward

Year ($0 \leftrightarrow 1980$)

(b) $\dfrac{f(22) - f(0)}{22 - 0} = \dfrac{37.54 - 20.6}{22} = 0.77$

The average increase in fuel used by trucks was 0.77 billion gallons per year between 1980 and 2002.

(c) $g(t) = 20.6 + 0.035(t + 10)^2 = f(t + 10)$

This represents a horizontal shift 10 units to the left.

(d) $g(20) = 52.1$ billion gallons

Yes. There are many factors involved here. The number of trucks on the road continues to increase but are more fuel efficient. The availability and the cost of overseas and domestic fuel also plays a role in usage.

69. True, since $|x| = |-x|$, the graphs of $f(x) = |x| + 6$ and $f(x) = |-x| + 6$ are identical.

71. (a) The profits were only $\frac{3}{4}$ as large as expected:

$g(t) = \frac{3}{4}f(t)$

(b) The profits were $10,000 greater than predicted:

$g(t) = f(t) + 10{,}000$

(c) There was a two-year delay: $g(t) = f(t - 2)$

73. $y = f(x + 2) - 1$

Horizontal shift 2 units to the left and a vertical shift 1 unit downward

$(0, 1) \rightarrow (0 - 2, 1 - 1) = (-2, 0)$

$(1, 2) \rightarrow (1 - 2, 2 - 1) = (-1, 1)$

$(2, 3) \rightarrow (2 - 2, 3 - 1) = (0, 2)$

75. $\dfrac{4}{x} + \dfrac{4}{1 - x} = \dfrac{4(1 - x) + 4x}{x(1 - x)} = \dfrac{4 - 4x + 4x}{x(1 - x)} = \dfrac{4}{x(1 - x)}$

77. $\dfrac{3}{x - 1} - \dfrac{2}{x(x - 1)} = \dfrac{3x - 2}{x(x - 1)}$

79. $(x - 4)\left(\dfrac{1}{\sqrt{x^2 - 4}}\right) = \dfrac{x - 4}{\sqrt{x^2 - 4}} = \dfrac{(x - 4)\sqrt{x^2 - 4}}{x^2 - 4}$

81. $(x^2 - 9) \div \left(\dfrac{x + 3}{5}\right) = \dfrac{(x + 3)(x - 3)}{1} \cdot \dfrac{5}{x + 3} = 5(x - 3), \; x \neq -3$

83. $f(x) = x^2 - 6x + 11$

(a) $f(-3) = (-3)^2 - 6(-3) + 11 = 38$

(b) $f\left(-\frac{1}{2}\right) = \left(-\frac{1}{2}\right)^2 - 6\left(-\frac{1}{2}\right) + 11 = \frac{1}{4} + 3 + 11 = \frac{57}{4}$

(c) $f(x - 3) = (x - 3)^2 - 6(x - 3) + 11 = x^2 - 6x + 9 - 6x + 18 + 11 = x^2 - 12x + 38$

85. $f(x) = \dfrac{2}{11 - x}$

Domain: All real numbers except $x = 11$

87. $f(x) = \sqrt{81 - x^2}$

$81 - x^2 \geq 0$

$(9 + x)(9 - x) \geq 0$

Critical numbers: $x = \pm 9$

Test intervals: $(-\infty, -9), (-9, 9), (9, \infty)$

Test: Is $81 - x^2 \geq 0$?

Solution: $[-9, 9]$

Domain of $f(x)$: $-9 \leq x \leq 9$

Section 2.6   Combinations of Functions: Composite Functions

■   Given two functions, f and g, you should be able to form the following functions (if defined):

1. Sum: $(f + g)(x) = f(x) + g(x)$

2. Difference: $(f - g)(x) = f(x) - g(x)$

3. Product: $(fg)(x) = f(x)g(x)$

4. Quotient: $(f/g)(x) = f(x)/g(x), \, g(x) \neq 0$

5. Composition of f with g: $(f \circ g)(x) = f(g(x))$

6. Composition of g with f: $(g \circ f)(x) = g(f(x))$

Vocabulary Check

1. addition, subtraction, multiplication, division

2. composition

3. $g(x)$

4. inner; outer

1.

x	0	1	2	3
f	2	3	1	2
g	-1	0	$\frac{1}{2}$	0
$f + g$	1	3	$\frac{3}{2}$	2

3.

x	-2	0	1	2	4
f	2	0	1	2	4
g	4	2	1	0	2
$f + g$	6	2	2	2	6

5. $f(x) = x + 2, g(x) = x - 2$

(a) $(f + g)(x) = f(x) + g(x) = (x + 2) + (x - 2) = 2x$

(b) $(f - g)(x) = f(x) - g(x) = (x + 2) - (x - 2) = 4$

(c) $(fg)(x) = f(x) \cdot g(x) = (x + 2)(x - 2) = x^2 - 4$

(d) $\left(\dfrac{f}{g}\right)(x) = \dfrac{f(x)}{g(x)} = \dfrac{x + 2}{x - 2}$

Domain: all real numbers except $x = 2$

7. $f(x) = x^2, g(x) = 4x - 5$

(a) $(f + g)(x) = f(x) + g(x)$

$= x^2 + (4x - 5) = x^2 + 4x - 5$

(b) $(f - g)(x) = f(x) - g(x)$

$= x^2 - (4x - 5) = x^2 - 4x + 5$

(c) $(fg)(x) = f(x) \cdot g(x) = x^2(4x - 5) = 4x^3 - 5x^2$

(d) $\left(\dfrac{f}{g}\right)(x) = \dfrac{f(x)}{g(x)} = \dfrac{x^2}{4x - 5}$

Domain: all real numbers except $x = \dfrac{5}{4}$

9. $f(x) = x^2 + 6, g(x) = \sqrt{1 - x}$

(a) $(f + g)(x) = f(x) + g(x) = (x^2 + 6) + \sqrt{1 - x}$

(b) $(f - g)(x) = f(x) - g(x) = (x^2 + 6) - \sqrt{1 - x}$

(c) $(fg)(x) = f(x) \cdot g(x) = (x^2 + 6)\sqrt{1 - x}$

(d) $\left(\dfrac{f}{g}\right)(x) = \dfrac{f(x)}{g(x)} = \dfrac{x^2 + 6}{\sqrt{1 - x}} = \dfrac{(x^2 + 6)\sqrt{1 - x}}{1 - x}$,

Domain: $x < 1$

11. $f(x) = \dfrac{1}{x}, g(x) = \dfrac{1}{x^2}$

(a) $(f + g)(x) = f(x) + g(x) = \dfrac{1}{x} + \dfrac{1}{x^2} = \dfrac{x + 1}{x^2}$

(b) $(f - g)(x) = f(x) - g(x) = \dfrac{1}{x} - \dfrac{1}{x^2} = \dfrac{x - 1}{x^2}$

(c) $(fg)(x) = f(x) \cdot g(x) = \dfrac{1}{x}\left(\dfrac{1}{x^2}\right) = \dfrac{1}{x^3}$

(d) $\left(\dfrac{f}{g}\right)(x) = \dfrac{f(x)}{g(x)} = \dfrac{1/x}{1/x^2} = \dfrac{x^2}{x} = x, \; x \neq 0$

For Exercises 13–23, $f(x) = x^2 + 1$ and $g(x) = x - 4$.

13. $(f + g)(2) = f(2) + g(2) = (2^2 + 1) + (2 - 4) = 3$

15. $(f - g)(0) = f(0) - g(0) = (0^2 + 1) - (0 - 4) = 5$

17. $(f - g)(3t) = f(3t) - g(3t) = [(3t)^2 + 1] - (3t - 4)$

$= 9t^2 - 3t + 5$

19. $(fg)(6) = f(6)g(6) = (6^2 + 1)(6 - 4) = 74$

21. $\left(\dfrac{f}{g}\right)(5) = \dfrac{f(5)}{g(5)} = \dfrac{5^2 + 1}{5 - 4} = 26$

23. $\left(\dfrac{f}{g}\right)(-1) - g(3) = \dfrac{f(-1)}{g(-1)} - g(3)$

$$= \dfrac{(-1)^2 + 1}{-1 - 4} - (3 - 4)$$

$$= -\dfrac{2}{5} + 1 = \dfrac{3}{5}$$

25. $f(x) = \frac{1}{2}x,\ g(x) = x - 1,\ (f + g)(x) = \frac{3}{2}x - 1$

27. $f(x) = x^2,\ g(x) = -2x,\ (f + g)(x) = x^2 - 2x$

29. $f(x) = 3x,\ g(x) = -\dfrac{x^3}{10},\ (f + g)(x) = 3x - \dfrac{x^3}{10}$

For $0 \leq x \leq 2$, $f(x)$ contributes most to the magnitude.

For $x > 6$, $g(x)$ contributes most to the magnitude.

31. $f(x) = x^2,\ g(x) = x - 1$

(a) $(f \circ g)(x) = f(g(x)) = f(x - 1) = (x - 1)^2$

(b) $(g \circ f)(x) = g(f(x)) = g(x^2) = x^2 - 1$

(c) $(f \circ f)(x) = f(f(x)) = f(x^2) = (x^2)^2 = x^4$

33. $f(x) = \sqrt[3]{x - 1},\ g(x) = x^3 + 1$

(a) $(f \circ g)(x) = f(g(x))$

$$= f(x^3 + 1)$$

$$= \sqrt[3]{(x^3 + 1) - 1}$$

$$= \sqrt[3]{x^3} = x$$

(b) $(g \circ f)(x) = g(f(x))$

$$= g\left(\sqrt[3]{x - 1}\right)$$

$$= \left(\sqrt[3]{x - 1}\right)^3 + 1$$

$$= (x - 1) + 1 = x$$

(c) $(f \circ f)(x) = f(f(x))$

$$= f\left(\sqrt[3]{x - 1}\right)$$

$$= \sqrt[3]{\sqrt[3]{x - 1} - 1}$$

35. $f(x) = \sqrt{x + 4}$ Domain: $x \geq -4$

$g(x) = x^2$ Domain: all real numbers

(a) $(f \circ g)(x) = f(g(x)) = f(x^2) = \sqrt{x^2 + 4}$

Domain: all real numbers

(b) $(g \circ f)(x) = g(f(x))$

$$= g\left(\sqrt{x + 4}\right) = \left(\sqrt{x + 4}\right)^2 = x + 4$$

Domain: $x \geq -4$

37. $f(x) = x^2 + 1$ Domain: all real numbers

$g(x) = \sqrt{x}$ Domain: $x \geq 0$

(a) $(f \circ g)(x) = f(g(x)) = f\left(\sqrt{x}\right) = \left(\sqrt{x}\right)^2 + 1 = x + 1$

Domain: $x \geq 0$

(b) $(g \circ f)(x) = g(f(x)) = g(x^2 + 1) = \sqrt{x^2 + 1}$

Domain: all real numbers

39. $f(x) = |x|$ Domain: all real numbers

$g(x) = x + 6$ Domain: all real numbers

(a) $(f \circ g)(x) = f(g(x)) = f(x + 6) = |x + 6|$

Domain: all real numbers

(b) $(g \circ f)(x) = g(f(x)) = g(|x|) = |x| + 6$

Domain: all real numbers

41. $f(x) = \dfrac{1}{x}$ Domain: all real numbers except $x = 0$

$g(x) = x + 3$ Domain: all real numbers

(a) $(f \circ g)(x) = f(g(x)) = f(x + 3) = \dfrac{1}{x + 3}$

Domain: all real numbers except $x = -3$

(b) $(g \circ f)(x) = g(f(x)) = g\left(\dfrac{1}{x}\right) = \dfrac{1}{x} + 3$

Domain: all real numbers except $x = 0$

43. (a) $(f + g)(3) = f(3) + g(3) = 2 + 1 = 3$

(b) $\left(\dfrac{f}{g}\right)(2) = \dfrac{f(2)}{g(2)} = \dfrac{0}{2} = 0$

45. (a) $(f \circ g)(2) = f(g(2)) = f(2) = 0$

(b) $(g \circ f)(2) = g(f(2)) = g(0) = 4$

47. Let $f(x) = x^2$ and $g(x) = 2x + 1$, then $(f \circ g)(x) = h(x)$.

This is not a unique solution.

49. Let $f(x) = \sqrt[3]{x}$ and $g(x) = x^2 - 4$, then $(f \circ g)(x) = h(x)$.

This answer is not unique.

51. Let $f(x) = 1/x$ and $g(x) = x + 2$, then $(f \circ g)(x) = h(x)$.

This is not a unique solution.

53. Let $f(x) = \dfrac{x + 3}{4 + x}$ and $g(x) = -x^2$, then $(f \circ g)(x) = h(x)$.

This answer is not unique.

55. $T(x) = R(x) + B(x) = \frac{3}{4}x + \frac{1}{15}x^2$

57. (a) $c(t) = \dfrac{p(t) + b(t) - d(t)}{p(t)} \times 100$

(b) $c(5)$ represents the percent change in the population in the year 2005.

59. $A(t) = 3.36t^2 - 59.8t + 735$, $N(t) = 1.95t^2 - 42.2t + 603$

(a) $(A + N)(t) = A(t) + N(t) = 5.31t^2 - 102.0t + 1338$

This represents the combined Army and Navy personnel (in thousands) from 1990 to 2002, where $t = 0$ corresponds to 1990.

$(A + N)(4) = 1014.96$ thousand

$(A + N)(8) = 861.84$ thousand

$(A + N)(12) = 878.64$ thousand

(b) $(A - N)(t) = A(t) - N(t) = 1.41t^2 - 17.6t + 132$

This represents the number of Army personnel (in thousands) more than the number of Navy personnel from 1990 to 2002, where $t = 0$ corresponds to 1990.

$(A - N)(4) = 84.16$ thousand

$(A - N)(8) = 81.44$ thousand

$(A - N)(12) = 123.84$ thousand

61.

Year	y_1	y_2	y_3
1995	146.2	329.1	44.8
1996	152.0	344.1	48.1
1997	162.2	359.9	52.1
1998	175.2	382.0	55.6
1999	184.4	412.1	57.8
2000	194.7	449.0	57.4
2001	205.5	496.1	57.8

(a) $y_1 \approx 10.20t + 92.7$

$y_2 \approx 3.357t^2 - 26.46t + 379.5$

$y_3 \approx -0.465t^2 + 9.71t + 7.4$

(b) $y_1 + y_2 + y_3 \approx 2.892t^2 - 6.55t + 479.6$

This sum represents the total spent on health services and supplies for the years 1995 through 2001. It includes out-of-pocket payments, insurance premiums, and other types of payments.

(c)

(d) For 2008 use $t = 18$:

$(y_1 + y_2 + y_3)(18) \approx \$1298.708$ billion

For 2010 use $t = 20$:

$(y_1 + y_2 + y_3)(20) \approx \$1505.4$ billion

63. (a) $r(x) = \dfrac{x}{2}$

(b) $A(r) = \pi r^2$

(c) $(A \circ r)(x) = A(r(x)) = A\left(\dfrac{x}{2}\right) = \pi\left(\dfrac{x}{2}\right)^2$

$(A \circ r)(x)$ represents the area of the circular base of the tank on the square foundation with side length x.

65. (a) $N(T(t)) = N(3t + 2)$

$= 10(3t + 2)^2 - 20(3t + 2) + 600$

$= 10(9t^2 + 12t + 4) - 60t - 40 + 600$

$= 90t^2 + 60t + 600$

$= 30(3t^2 + 2t + 20), \ 0 \le t \le 6$

This represents the number of bacteria in the food as a function of time.

(b) $30(3t^2 + 2t + 20) = 1500$

$3t^2 + 2t + 20 = 50$

$3t^2 + 2t - 30 = 0$

By the Quadratic Formula, $t \approx -3.513$ or 2.846. Choosing the positive value for t, we have $t \approx 2.846$ hours.

67. (a) $f(g(x)) = f(0.03x) = 0.03x - 500{,}000$

(b) $g(f(x)) = g(x - 500{,}000) = 0.03(x - 500{,}000)$

$g(f(x))$ represents your bonus of 3% of an amount over \$500,000.

69. False. $(f \circ g)(x) = 6x + 1$ and $(g \circ f)(x) = 6x + 6$.

71. Let $f(x)$ and $g(x)$ be two odd functions and define $h(x) = f(x)g(x)$. Then

$h(-x) = f(-x)g(-x)$

$= [-f(x)][-g(x)]$ since f and g are odd

$= f(x)g(x)$

$= h(x)$.

Thus, $h(x)$ is even.

Let $f(x)$ and $g(x)$ be two even functions and define $h(x) = f(x)g(x)$. Then

$h(-x) = f(-x)g(-x)$

$= f(x)g(x)$ since f and g are even

$= h(x)$.

Thus, $h(x)$ is even.

73. $f(x) = 3x - 4$

$$\frac{f(x + h) - f(x)}{h} = \frac{[3(x + h) - 4] - (3x - 4)}{h}$$

$$= \frac{3x + 3h - 4 - 3x + 4}{h}$$

$$= \frac{3h}{h}$$

$$= 3, \ h \neq 0$$

75. $f(x) = \dfrac{4}{x}$

$$\frac{f(x + h) - f(x)}{h} = \frac{\dfrac{4}{x + h} - \dfrac{4}{x}}{h} = \frac{\dfrac{4x - 4(x + h)}{x(x + h)}}{\dfrac{h}{1}}$$

$$= \frac{4x - 4x - 4h}{x(x + h)} \cdot \frac{1}{h}$$

$$= \frac{-4h}{x(x + h)} \cdot \frac{1}{h}$$

$$= \frac{-4}{x(x + h)}, \ h \neq 0$$

77. Point: $(2, -4)$

    Slope: $m = 3$

$$y - (-4) = 3(x - 2)$$

$$y + 4 = 3x - 6$$

$$3x - y - 10 = 0$$

79. Point: $(8, -1)$

    Slope: $m = -\frac{3}{2}$

$$y - (-1) = -\frac{3}{2}(x - 8)$$

$$y + 1 = -\frac{3}{2}x + 12$$

$$2y + 2 = -3x + 24$$

$$3x + 2y - 22 = 0$$

Section 2.7   Inverse Functions

- Two functions f and g are inverses of each other if $f(g(x)) = x$ for every x in the domain of g and $g(f(x)) = x$ for every x in the domain of f.

- A function f has an inverse function if and only if no **horizontal** line crosses the graph of f at more than one point.

- The graph of f^{-1} is a reflection of the graph of f about the line $y = x$.

- Be able to find the inverse of a function, if it exists.

  1. Use the Horizontal Line Test to see if f^{-1} exists.

  2. Replace $f(x)$ with y.

  3. Interchange x and y and solve for y.

  4. Replace y with $f^{-1}(x)$.

Vocabulary Check

1. inverse; f-inverse

2. range; domain

3. $y = x$

4. one-to-one

5. Horizontal

1. $f(x) = 6x$

$$f^{-1}(x) = \frac{x}{6} = \frac{1}{6}x$$

$$f(f^{-1}(x)) = f\left(\frac{x}{6}\right) = 6\left(\frac{x}{6}\right) = x$$

$$f^{-1}(f(x)) = f^{-1}(6x) = \frac{6x}{6} = x$$

3. $f(x) = x + 9$

$$f^{-1}(x) = x - 9$$

$$f(f^{-1}(x)) = f(x - 9) = (x - 9) + 9 = x$$

$$f^{-1}(f(x)) = f^{-1}(x + 9) = (x + 9) - 9 = x$$

5. $f(x) = 3x + 1$

$$f^{-1}(x) = \frac{x - 1}{3}$$

$$f(f^{-1}(x)) = f\left(\frac{x - 1}{3}\right) = 3\left(\frac{x - 1}{3}\right) + 1 = x$$

$$f^{-1}(f(x)) = f^{-1}(3x + 1) = \frac{(3x + 1) - 1}{3} = x$$

7. $f(x) = \sqrt[3]{x}$

$$f^{-1}(x) = x^3$$

$$f(f^{-1}(x)) = f(x^3) = \sqrt[3]{x^3} = x$$

$$f^{-1}(f(x)) = f^{-1}(\sqrt[3]{x}) = (\sqrt[3]{x})^3 = x$$

9. The inverse is a line through $(-1, 0)$.
Matches graph (c).

11. The inverse is half a parabola starting at $(1, 0)$.
Matches graph (a).

13. $f(x) = 2x, \quad g(x) = \frac{x}{2}$

(a) $f(g(x)) = f\left(\frac{x}{2}\right) = 2\left(\frac{x}{2}\right) = x$

$\quad g(f(x)) = g(2x) = \frac{2x}{2} = x$

(b)

15. $f(x) = 7x + 1, \quad g(x) = \frac{x - 1}{7}$

(a) $f(g(x)) = f\left(\frac{x - 1}{7}\right) = 7\left(\frac{x - 1}{7}\right) + 1 = x$

$\quad g(f(x)) = g(7x + 1) = \frac{(7x + 1) - 1}{7} = x$

(b)

17. $f(x) = \frac{x^3}{8}, \quad g(x) = \sqrt[3]{8x}$

(a) $f(g(x)) = f(\sqrt[3]{8x}) = \frac{(\sqrt[3]{8x})^3}{8} = \frac{8x}{8} = x$

$\quad g(f(x)) = g\left(\frac{x^3}{8}\right) = \sqrt[3]{8\left(\frac{x^3}{8}\right)} = \sqrt[3]{x^3} = x$

(b)

19. $f(x) = \sqrt{x - 4}, \quad g(x) = x^2 + 4, x \geq 0$

(a) $f(g(x)) = f(x^2 + 4), \quad x \geq 0$

$\quad\quad = \sqrt{(x^2 + 4) - 4} = x$

$\quad g(f(x)) = g(\sqrt{x - 4})$

$\quad\quad = (\sqrt{x - 4})^2 + 4 = x$

(b)

21. $f(x) = 9 - x^2, x \geq 0; g(x) = \sqrt{9 - x}, x \leq 9$

    (a) $f(g(x)) = f\left(\sqrt{9 - x}\right), \ x \leq 9$

          $= 9 - \left(\sqrt{9 - x}\right)^2 = x$

      $g(f(x)) = g(9 - x^2), \ x \geq 0$

          $= \sqrt{9 - (9 - x^2)} = x$

(b)

23. $f(x) = \dfrac{x - 1}{x + 5}, \ g(x) = -\dfrac{5x + 1}{x - 1}$

    (a) $f(g(x)) = f\left(-\dfrac{5x + 1}{x - 1}\right)$

$$= \frac{\left(-\dfrac{5x + 1}{x - 1} - 1\right)}{\left(-\dfrac{5x + 1}{x - 1} + 5\right)} \cdot \frac{x - 1}{x - 1} = \frac{-(5x + 1) - (x - 1)}{-(5x + 1) + 5(x - 1)} = \frac{-6x}{-6} = x$$

      $g(f(x)) = g\left(\dfrac{x - 1}{x + 5}\right)$

$$= -\frac{\left[5\left(\dfrac{x - 1}{x + 5}\right) + 1\right]}{\left[\dfrac{x - 1}{x + 5} - 1\right]} \cdot \frac{x + 5}{x + 5} = -\frac{5(x - 1) + (x + 5)}{(x - 1) - (x + 5)} = -\frac{6x}{-6} = x$$

(b)

25. No, $\{(-2, -1), (1, 0), (2, 1), (1, 2), (-2, 3), (-6, 4)\}$ does not represent a function. -2 and 1 are paired with two different values.

27.

x	-2	0	2	4	6	8
$f^{-1}(x)$	-2	-1	0	1	2	3

29. Yes, since no horizontal line crosses the graph of f at more than one point, f *has* an inverse.

31. No, since some horizontal lines cross the graph of f twice, f *does not* have an inverse.

33. $g(x) = \dfrac{4 - x}{6}$

g passes the horizontal line test, so g *has* an inverse.

35. $h(x) = |x + 4| - |x - 4|$

h does not pass the horizontal line test, so h *does not* have an inverse.

37. $f(x) = -2x\sqrt{16 - x^2}$

f does not pass the horizontal line test, so f *does not* have an inverse.

39. (a) $f(x) = 2x - 3$ (b)

$y = 2x - 3$

$x = 2y - 3$

$y = \dfrac{x + 3}{2}$

$f^{-1}(x) = \dfrac{x + 3}{2}$

(c) The graph of f^{-1} is the reflection of the graph of f about the line $y = x$.

(d) The domains and ranges of f and f^{-1} are all real numbers.

41. (a) $f(x) = x^5 - 2$ (b)

$y = x^5 - 2$

$x = y^5 - 2$

$y = \sqrt[5]{x + 2}$

$f^{-1}(x) = \sqrt[5]{x + 2}$

(c) The graph of f^{-1} is the reflection of the graph of f about the line $y = x$.

(d) The domains and ranges of f and f^{-1} are all real numbers.

43. (a) $f(x) = \sqrt{x}$ (b)

$y = \sqrt{x}$

$x = \sqrt{y}$

$y = x^2$

$f^{-1}(x) = x^2,\ x \geq 0$

(c) The graph of f^{-1} is the reflection of the graph of f about the line $y = x$.

(d) The domains and ranges of f and f^{-1} are $[0, \infty)$.

45. (a) $f(x) = \sqrt{4 - x^2},\ 0 \leq x \leq 2$

$y = \sqrt{4 - x^2}$

$x = \sqrt{4 - y^2}$

$f^{-1}(x) = \sqrt{4 - x^2},\ 0 \leq x \leq 2$

(b)

(c) The graph of f^{-1} is the same as the graph of f.

(d) The domains and ranges of f and f^{-1} are $[0, 2]$.

47. (a) $f(x) = \dfrac{4}{x}$ (b)

$y = \dfrac{4}{x}$

$x = \dfrac{4}{y}$

$xy = 4$

$y = \dfrac{4}{x}$

$f^{-1}(x) = \dfrac{4}{x}$

(c) The graph of f^{-1} is the same as the graph of f.

(d) The domains and ranges of f and f^{-1} are all real numbers except for 0.

49. (a) $f(x) = \dfrac{x + 1}{x - 2}$ (b)

$y = \dfrac{x + 1}{x - 2}$

$x = \dfrac{y + 1}{y - 2}$

$x(y - 2) = y + 1$

$xy - 2x = y + 1$

$xy - y = 2x + 1$

$y(x - 1) = 2x + 1$

$y = \dfrac{2x + 1}{x - 1}$

$f^{-1}(x) = \dfrac{2x + 1}{x - 1}$

(c) The graph of f^{-1} is the reflection of the graph of f about the line $y = x$.

(d) The domain of f and the range of f^{-1} is all real numbers except 2. The range of f and the domain of f^{-1} is all real numbers except 1.

51. (a) $f(x) = \sqrt[3]{x} - 1$ (b)

$$y = \sqrt[3]{x} - 1$$

$$x = \sqrt[3]{y} - 1$$

$$x^3 = y - 1$$

$$y = x^3 + 1$$

$$f^{-1}(x) = x^3 + 1$$

(c) The graph of f^{-1} is the reflection of the graph of f about the line $y = x$.

(d) The domains and ranges of f and f^{-1} are all real numbers.

53. (a) $f(x) = \dfrac{6x + 4}{4x + 5}$ (b)

$$y = \dfrac{6x + 4}{4x + 5}$$

$$x = \dfrac{6y + 4}{4y + 5}$$

$$x(4y + 5) = 6y + 4$$

$$4xy + 5x = 6y + 4$$

$$4xy - 6y = -5x + 4$$

$$y(4x - 6) = -5x + 4$$

$$y = \dfrac{-5x + 4}{4x - 6}$$

$$f^{-1}(x) = \dfrac{-5x + 4}{4x - 6} = \dfrac{5x - 4}{6 - 4x}$$

(c) The graph of f^{-1} is the graph of f reflected about the line $y = x$.

(d) The domain of f and the range of f^{-1} is all real numbers except $-\frac{5}{4}$. The range of f and the domain of f^{-1} is all real numbers except $\frac{3}{2}$.

55. $f(x) = x^4$

$$y = x^4$$

$$x = y^4$$

$$y = \pm\sqrt[4]{x}$$

This does not represent y as a function of x. f does not have an inverse.

57. $g(x) = \dfrac{x}{8}$

$$y = \dfrac{x}{8}$$

$$x = \dfrac{y}{8}$$

$$y = 8x$$

This is a function of x, so g has an inverse.

$$g^{-1}(x) = 8x$$

59. $p(x) = -4$

$$y = -4$$

Since $y = -4$ for all x, the graph is a horizontal line and fails the horizontal line test. p does not have an inverse.

61. $f(x) = (x + 3)^2,\ x \geq -3 \implies y \geq 0$

$$y = (x + 3)^2,\ x \geq -3,\ y \geq 0$$

$$x = (y + 3)^2,\ y \geq -3,\ x \geq 0$$

$$\sqrt{x} = y + 3,\ y \geq -3,\ x \geq 0$$

$$y = \sqrt{x} - 3,\ x \geq 0,\ y \geq -3$$

This is a function of x, so f has an inverse.

$$f^{-1}(x) = \sqrt{x} - 3,\ x \geq 0$$

63. $f(x) = \begin{cases} x + 3, & x < 0 \\ 6 - x, & x \geq 0 \end{cases}$

The graph fails the horizontal line test, so $f(x)$ does not have an inverse.

65. $h(x) = -\dfrac{4}{x^2}$

The graph fails the horizontal line test so h does not have an inverse.

67. $f(x) = \sqrt{2x+3} \implies x \geq -\dfrac{3}{2}, \; y \geq 0$

$$y = \sqrt{2x+3}, \; x \geq -\dfrac{3}{2}, \; y \geq 0$$

$$x = \sqrt{2y+3}, \; y \geq -\dfrac{3}{2}, \; x \geq 0$$

$$x^2 = 2y + 3, \; x \geq 0, \; y \geq -\dfrac{3}{2}$$

$$y = \dfrac{x^2 - 3}{2}, \; x \geq 0, \; y \geq -\dfrac{3}{2}$$

This is a function of x, so f has an inverse.

$$f^{-1}(x) = \dfrac{x^2 - 3}{2}, \; x \geq 0$$

In Exercises 69–73, $f(x) = \frac{1}{8}x - 3$, $f^{-1}(x) = 8(x+3)$, $g(x) = x^3$, $g^{-1}(x) = \sqrt[3]{x}$.

69. $(f^{-1} \circ g^{-1})(1) = f^{-1}(g^{-1}(1))$
$= f^{-1}\left(\sqrt[3]{1}\right)$
$= 8\left(\sqrt[3]{1} + 3\right) = 32$

71. $(f^{-1} \circ f^{-1})(6) = f^{-1}(f^{-1}(6))$
$= f^{-1}(8[6+3])$
$= 8[8(6+3) + 3] = 600$

73. $(f \circ g)(x) = f(g(x)) = f(x^3) = \frac{1}{8}x^3 - 3$
$y = \frac{1}{8}x^3 - 3$
$x = \frac{1}{8}y^3 - 3$
$x + 3 = \frac{1}{8}y^3$
$8(x+3) = y^3$
$\sqrt[3]{8(x+3)} = y$
$(f \circ g)^{-1}(x) = 2\sqrt[3]{x+3}$

In Exercises 75 and 77, $f(x) = x + 4$, $f^{-1}(x) = x - 4$, $g(x) = 2x - 5$, $g^{-1}(x) = \dfrac{x+5}{2}$.

75. $(g^{-1} \circ f^{-1})(x) = g^{-1}(f^{-1}(x))$
$= g^{-1}(x - 4)$
$= \dfrac{(x-4)+5}{2}$
$= \dfrac{x+1}{2}$

77. $(f \circ g)(x) = f(g(x))$
$= f(2x - 5)$
$= (2x - 5) + 4$
$= 2x - 1$
$(f \circ g)^{-1}(x) = \dfrac{x+1}{2}$

Note: Comparing Exercises 75 and 77, we see that $(f \circ g)^{-1}(x) = (g^{-1} \circ f^{-1})(x)$.

79. (a) $f^{-1}(108,209) = 11$

(b) f^{-1} represents the year for a given number of households in the United States.

(c) $y \approx 1578.68t + 90,183.63$

(d)
$$y = 1578.68t + 90,183.63$$
$$t = 1578.68y + 90,183.63$$
$$\frac{t - 90,183.63}{1578.68} = y$$
$$f^{-1}(t) = \frac{t - 90,183.63}{1578.68}$$

(e) $f^{-1}(117,022) \approx 17$

(f) $f^{-1}(108,209) \approx 11.418$

This is close to the value of 11 in the table.

81. (a) Yes. Since the values of f increase each year, no two f-values are paired with the same t-value so f does have an inverse.

(b) f^{-1} would represent the year that a given number of miles was traveled by motor vehicles.

(c) Since $f(8) = 2632, f^{-1}(2632) = 8$.

(d) No. Since the new value is the same as the value given for 2000, f would not pass the Horizontal Line Test and would not have an inverse.

83. (a)
$$y = 0.03x^2 + 245.50, \ 0 < x < 100$$
$$\Rightarrow 245.50 < y < 545.50$$
$$x = 0.03y^2 + 245.50$$
$$x - 245.50 = 0.03y^2$$
$$\frac{x - 245.50}{0.03} = y^2$$
$$\sqrt{\frac{x - 245.50}{0.03}} = y, \ 245.50 < x < 545.50$$
$$f^{-1}(x) = \sqrt{\frac{x - 245.50}{0.03}}$$

x = temperature in degrees Fahrenheit

y = percent load for a diesel engine

(b)

(c)
$$0.03x^2 + 245.50 \le 500$$
$$0.03x^2 \le 254.50$$
$$x^2 \le 8483.33$$
$$x \le 92.10$$

Thus, $0 < x \le 92.10$.

85. False. $f(x) = x^2$ is even and does not have an inverse.

87. Let $(f \circ g)(x) = y$. Then $x = (f \circ g)^{-1}(y)$. Also,
$$(f \circ g)(x) = y \Rightarrow f(g(x)) = y$$
$$g(x) = f^{-1}(y)$$
$$x = g^{-1}(f^{-1}(y))$$
$$x = (g^{-1} \circ f^{-1})(y).$$

Since f and g are both one-to-one functions,
$$(f \circ g)^{-1} = g^{-1} \circ f^{-1}.$$

89.

x	1	3	4	6
f	1	2	6	7

x	1	2	6	7
$f^{-1}(x)$	1	3	4	6

91.

x	-2	-1	3	4
f	6	0	-2	-3

x	-3	-2	0	6
$f^{-1}(x)$	4	3	-1	-2

93. If $f(x) = k(2 - x - x^3)$ has an inverse and $f^{-1}(3) = -2$, then $f(-2) = 3$. Thus,

$$f(-2) = k(2 - (-2) - (-2)^3) = 3$$

$$k(2 + 2 + 8) = 3$$

$$12k = 3$$

$$k = \frac{3}{12} = \frac{1}{4}$$

So, $k = \frac{1}{4}$.

95. $x^2 = 64$

$x = \pm\sqrt{64} = \pm 8$

97. $4x^2 - 12x + 9 = 0$

$(2x - 3)^2 = 0$

$2x - 3 = 0$

$x = \frac{3}{2}$

99. $x^2 - 6x + 4 = 0$ Complete the square.

$x^2 - 6x = -4$

$x^2 - 6x + 9 = -4 + 9$

$(x - 3)^2 = 5$

$x - 3 = \pm\sqrt{5}$

$x = 3 \pm \sqrt{5}$

101. $50 + 5x = 3x^2$

$0 = 3x^2 - 5x - 50$

$0 = (3x + 10)(x - 5)$

$3x + 10 = 0 \implies x = -\frac{10}{3}$

$x - 5 = 0 \implies x = 5$

103. Let $2n$ = first positive even integer. Then $2n + 2$ = next positive even integer.

$2n(2n + 2) = 288$

$4n^2 + 4n - 288 = 0$

$4(n^2 + n - 72) = 0$

$4(n + 9)(n - 8) = 0$

$n + 9 = 0 \implies n = -9$ Not a solution since the integers are positive.

$n - 8 = 0 \implies n = 8$

So, $2n = 16$ and $2n + 2 = 18$.

Review Exercises for Chapter 2

1. (a) $m = \frac{3}{2} > 0 \implies$ The line rises. Matches L_2.

(b) $m = 0 \implies$ The line is horizontal. Matches L_3.

(c) $m = -3 < 0 \implies$ The line falls. Matches L_1.

(d) $m = -\frac{1}{5} < 0 \implies$ The line gradually falls. Matches L_4.

3. $y = -2x - 7$

Slope: $m = -2 = -\frac{2}{1}$

y-intercept: $(0, -7)$

5. $y = 6$

Horizontal line, $m = 0$

y-intercept: $(0, 6)$

7. $y = 3x + 13$

Slope: $m = 3 = \frac{3}{1}$

y-intercept: $(0, 13)$

9. $y = -\frac{5}{2}x - 1$

Slope: $m = -\frac{5}{2}$

y-intercept: $(0, -1)$

11. $(-2, 5), (0, t), (1, 1)$ are collinear.

$$\frac{t - 5}{0 - (-2)} = \frac{1 - 5}{1 - (-2)}$$

$$\frac{t - 5}{2} = \frac{-4}{3}$$

$$3(t - 5) = -8$$

$$3t - 15 = -8$$

$$3t = 7$$

$$t = \frac{7}{3}$$

13. Point: $(2, -1)$

Slope: $m = \frac{1}{4} = \frac{\text{rise}}{\text{run}}$

$(2 + 4, -1 + 1) = (6, 0)$

$(6 + 4, 0 + 1) = (10, 1)$

$(2 - 4, -1 - 1) = (-2, -2)$

15. $(3, -4), (-7, 1)$

$$m = \frac{1 - (-4)}{-7 - 3} = \frac{5}{-10} = -\frac{1}{2}$$

17. $(-4.5, 6), (2.1, 3)$

$$m = \frac{3 - 6}{2.1 - (-4.5)}$$

$$= \frac{-3}{6.6} = -\frac{30}{66} = -\frac{5}{11}$$

19. $(0, -5), m = \frac{3}{2}$

$$y - (-5) = \frac{3}{2}(x - 0)$$

$$y + 5 = \frac{3}{2}x$$

$$y = \frac{3}{2}x - 5$$

21. $(10, -3), m = -\frac{1}{2}$

$$y - (-3) = -\frac{1}{2}(x - 10)$$

$$y + 3 = -\frac{1}{2}x + 5$$

$$y = -\frac{1}{2}x + 2$$

23. $(0, 0), (0, 10)$

$$m = \frac{10 - 0}{0 - 0} = \frac{10}{0}, \quad \text{undefined}$$

The line is vertical.

$x = 0$

25. $(-1, 4), (2, 0)$

$$m = \frac{0 - 4}{2 - (-1)} = -\frac{4}{3}$$

$$y - 4 = -\frac{4}{3}(x - (-1))$$

$$y - 4 = -\frac{4}{3}x - \frac{4}{3}$$

$$y = -\frac{4}{3}x + \frac{8}{3}$$

27. Point: $(3, -2)$

$5x - 4y = 8 \implies y = \frac{5}{4}x - 2$ and $m = \frac{5}{4}$

(a) Parallel slope: $m = \frac{5}{4}$

$y - (-2) = \frac{5}{4}(x - 3)$

$y + 2 = \frac{5}{4}x - \frac{15}{4}$

$y = \frac{5}{4}x - \frac{23}{4}$

(b) Perpendicular slope: $m = -\frac{4}{5}$

$y - (-2) = -\frac{4}{5}(x - 3)$

$y + 2 = -\frac{4}{5}x + \frac{12}{5}$

$y = -\frac{4}{5}x + \frac{2}{5}$

29. $(6, 12{,}500)$ $m = 850$

$y - 12{,}500 = 850(t - 6)$

$y - 12{,}500 = 850t - 5100$

$y = 850t + 7400, \quad 6 \leq t \leq 11$

31. $(2, 160{,}000)$, $(3, 185{,}000)$

$m = \dfrac{185{,}000 - 160{,}000}{3 - 2} = 25{,}000$

$S - 160{,}000 = 25{,}000(t - 2)$

$S = 25{,}000t + 110{,}000$

For the fourth quarter let $t = 4$. Then we have

$S = 25{,}000(4) + 110{,}000 = \$210{,}000.$

33. $A = \{10, 20, 30, 40\}$ and $B = \{0, 2, 4, 6\}$

(a) 20 is matched with two elements in the range so it is not a function.

(b) Function

(c) Function

(d) 30 is not matched with any element of B so it is not a function.

35. $16x - y^4 = 0$

$y^4 = 16x$

$y = \pm 2\sqrt[4]{x}$

y is not a function of x. Some x-values correspond to two y-values.

37. $y = \sqrt{1 - x}$

Each x-value, $x \leq 1$, corresponds to only one y-value so y is a function of x.

39. $f(x) = x^2 + 1$

(a) $f(2) = (2)^2 + 1 = 5$

(b) $f(-4) = (-4)^2 + 1 = 17$

(c) $f(t^2) = (t^2)^2 + 1 = t^4 + 1$

(d) $f(t + 1) = (t + 1)^2 + 1 = t^2 + 2t + 2$

41. $h(x) = \begin{cases} 2x + 1, & x \leq -1 \\ x^2 + 2, & x > -1 \end{cases}$

(a) $h(-2) = 2(-2) + 1 = -3$

(b) $h(-1) = 2(-1) + 1 = -1$

(c) $h(0) = 0^2 + 2 = 2$

(d) $h(2) = 2^2 + 2 = 6$

43. $f(x) = \sqrt{25 - x^2}$

Domain: $25 - x^2 \geq 0$

$(5 + x)(5 - x) \geq 0$

Critical numbers: $x = \pm 5$

Test intervals: $(-\infty, -5), \ (-5, 5), \ (5, \infty)$

Test: Is $25 - x^2 \geq 0$?

Solution set: $-5 \leq x \leq 5$

Thus, the domain is all real numbers x such that $-5 \leq x \leq 5$, or $[-5, 5]$.

45. $g(s) = \dfrac{5}{3s - 9}$

Domain: All real numbers s except $s = 3$

47. $h(x) = \dfrac{x}{x^2 - x - 6} = \dfrac{x}{(x + 2)(x - 3)}$

Domain: All real numbers x except $x = -2, 3$

49. $v(t) = -32t + 48$

(a) $v(1) = 16$ feet per second

(b) $0 = -32t + 48$

$t = \frac{48}{32} = 1.5$ seconds

(c) $v(2) = -16$ feet per second

51.

(a) $2x + 2y = 24$

$y = 12 - x$

$A = xy = x(12 - x)$

(b) Since x and y cannot be negative, we have $0 < x < 12$. The domain is $0 < x < 12$.

53. $f(x) = 2x^2 + 3x - 1$

$$\frac{f(x + h) - f(x)}{h} = \frac{[2(x + h)^2 + 3(x + h) - 1] - (2x^2 + 3x - 1)}{h}$$

$$= \frac{2x^2 + 4xh + 2h^2 + 3x + 3h - 1 - 2x^2 - 3x + 1}{h}$$

$$= \frac{h(4x + 2h + 3)}{h}$$

$$= 4x + 2h + 3, \ \ h \neq 0$$

55. $y = (x - 3)^2$

The graph passes the Vertical Line Test. *y is a function of x.*

57. $x - 4 = y^2$

The graph does not pass the Vertical Line Test. *y is not a function of x.*

59. $f(x) = 3x^2 - 16x + 21$

$3x^2 - 16x + 21 = 0$

$(3x - 7)(x - 3) = 0$

$3x - 7 = 0$ or $x - 3 = 0$

$x = \frac{7}{3}$ or $x = 3$

61. $f(x) = \dfrac{8x + 3}{11 - x}$

$\dfrac{8x + 3}{11 - x} = 0$

$8x + 3 = 0$

$x = -\dfrac{3}{8}$

63. $f(x) = |x| + |x + 1|$

f is increasing on $(0, \infty)$.

f is decreasing on $(-\infty, -1)$.

f is constant on $(-1, 0)$.

65. $f(x) = -x^2 + 2x + 1$

Relative maximum: $(1, 2)$

67. $f(x) = x^3 - 6x^4$

Relative maximum: $(0.125, 0.000488) \approx (0.13, 0.00)$

69. $f(x) = -x^2 + 8x - 4$

$\dfrac{f(4) - f(0)}{4 - 0} = \dfrac{12 - (-4)}{4} = 4$

The average rate of change of f from $x_1 = 0$ to $x_2 = 4$ is 4.

71. $f(x) = 2 - \sqrt{x + 1}$

$\dfrac{f(7) - f(3)}{7 - 3} = \dfrac{(2 - \sqrt{8}) - (2 - 2)}{4}$

$= \dfrac{2 - 2\sqrt{2}}{4} = \dfrac{1 - \sqrt{2}}{2}$

The average rate of change of f from $x_1 = 3$ to $x_2 = 7$ is $\dfrac{1 - \sqrt{2}}{2}$.

73. $f(x) = x^5 + 4x - 7$

$f(-x) = (-x)^5 + 4(-x) - 7$

$= -x^5 - 4x - 7$

$\neq f(x)$

$\neq -f(x)$

Neither even nor odd

75. $f(x) = 2x\sqrt{x^2 + 3}$

$f(-x) = 2(-x)\sqrt{(-x)^2 + 3}$

$= -2x\sqrt{x^2 + 3}$

$= -f(x)$

f is odd.

77. $f(2) = -6, f(-1) = 3$

Points: $(2, -6), (-1, 3)$

$m = \dfrac{3 - (-6)}{-1 - 2} = \dfrac{9}{-3} = -3$

$y - (-6) = -3(x - 2)$

$y + 6 = -3x + 6$

$y = -3x$

79. $f\left(-\dfrac{4}{5}\right) = 2, \, f\left(\dfrac{11}{5}\right) = 7$

Points: $\left(-\dfrac{4}{5}, 2\right), \left(\dfrac{11}{5}, 7\right)$

$m = \dfrac{7 - 2}{11/5 - (-4/5)} = \dfrac{5}{3}$

$y - 2 = \dfrac{5}{3}\left(x - \left(-\dfrac{4}{5}\right)\right)$

$y - 2 = \dfrac{5}{3}x + \dfrac{4}{3}$

$y = \dfrac{5}{3}x + \dfrac{10}{3}$

81. $f(x) = 3 - x^2$

Intercepts: $(0, 3), \left(\pm\sqrt{3}, 0\right)$

y-axis symmetry

83. $f(x) = -\sqrt{x}$

Domain: $x \geq 0$

Intercepts: $(0, 0)$

x	0	1	4	9
y	0	-1	-2	-3

85. $g(x) = \dfrac{3}{x}$

No intercepts

Origin symmetry

x	-3	-1	1	3
y	-1	-3	3	1

87. $f(x) = [\![x]\!] - 2$

89. $f(x) = \begin{cases} 5x - 3, & x \geq -1 \\ -4x + 5, & x < -1 \end{cases}$

91. Common function: $f(x) = x^3$

Horizontal shift 4 units to the left and a vertical shift 4 units upward

93. (a) $f(x) = x^2$

(b) $h(x) = x^2 - 9$
Vertical shift 9 units downward

(c)

(d) $h(x) = f(x) - 9$

95. (a) $f(x) = \sqrt{x}$

(b) $h(x) = \sqrt{x - 7}$

Horizontal shift 7 units to the right

(c)

(d) $h(x) = f(x - 7)$

97. (a) $f(x) = x^2$

(b) $h(x) = -(x + 3)^2 + 1$

Reflection in the *x*-axis, a horizontal shift 3 units to the left, and a vertical shift 1 unit upward

(c)

(d) $h(x) = -f(x + 3) + 1$

99. (a) $f(x) = [\![x]\!]$

(b) $h(x) = -[\![x]\!] + 6$

Reflection in the *x*-axis and a vertical shift 6 units upward

(c)

(d) $h(x) = -f(x) + 6$

101. (a) $f(x) = |x|$

(b) $h(x) = -|-x + 4| + 6$

Reflection in both the *x*- and *y*-axes; horizontal shift of 4 units to the right; vertical shift of 6 units upward

(c)

(d) $h(x) = -f(-(x - 4)) + 6 = -f(-x + 4) + 6$

103. (a) $f(x) = [\![x]\!]$

(b) $h(x) = 5[\![x - 9]\!]$

Horizontal shift 9 units to the right and a vertical stretch (each *y*-value is multiplied by 5)

(c)

(d) $h(x) = 5f(x - 9)$

105. (a) $f(x) = \sqrt{x}$

(b) $h(x) = -2\sqrt{x - 4}$

Reflection in the *x*-axis, a vertical stretch (each *y*-value is multiplied by 2), and a horizontal shift 4 units to the right

(c)

(d) $h(x) = -2f(x - 4)$

107. $f(x) = x^2 + 3,\ g(x) = 2x - 1$

(a) $(f + g)(x) = (x^2 + 3) + (2x - 1) = x^2 + 2x + 2$

(b) $(f - g)(x) = (x^2 + 3) - (2x - 1) = x^2 - 2x + 4$

(c) $(fg)(x) = (x^2 + 3)(2x - 1) = 2x^3 - x^2 + 6x - 3$

(d) $\left(\dfrac{f}{g}\right)(x) = \dfrac{x^2 + 3}{2x - 1}$, Domain: $x \neq \dfrac{1}{2}$

109. $f(x) = \frac{1}{3}x - 3$, $g(x) = 3x + 1$

The domains of $f(x)$ and $g(x)$ are all real numbers.

(a) $(f \circ g)(x) = f(g(x))$

$\qquad = f(3x + 1)$

$\qquad = \frac{1}{3}(3x + 1) - 3$

$\qquad = x + \frac{1}{3} - 3$

$\qquad = x - \frac{8}{3}$

Domain: all real numbers

(b) $(g \circ f)(x) = g(f(x))$

$\qquad = g\left(\frac{1}{3}x - 3\right)$

$\qquad = 3\left(\frac{1}{3}x - 3\right) + 1$

$\qquad = x - 9 + 1$

$\qquad = x - 8$

Domain: all real numbers

111. $h(x) = (6x - 5)^3$

Answer is not unique.

One possibility: Let $f(x) = x^3$ and $g(x) = 6x - 5$.

$f(g(x)) = f(6x - 5) = (6x - 5)^3 = h(x)$

113. $v(t) = -31.86t^2 + 233.6t + 2594$

$d(t) = -4.18t^2 + 571.0t - 3706$

(a) $(v + d)(t) = v(t) + d(t) = -36.04t^2 + 804.6t - 1112$

$(v + d)(t)$ represents the combined factory sales (in millions of dollars) for VCRs and DVD players from 1997 to 2003.

(b)

(c) $(v + d)(10) = \$3330$ million

115. $\qquad f(x) = x - 7$

$\qquad f^{-1}(x) = x + 7$

$f(f^{-1}(x)) = f(x + 7) = (x + 7) - 7 = x$

$f^{-1}(f(x)) = f^{-1}(x - 7) = (x - 7) + 7 = x$

117. The graph passes the Horizontal Line Test. The function has an inverse.

119. $f(x) = 4 - \frac{1}{3}x$

The graph passes the Horizontal Line Test. The function has an inverse.

121. $h(t) = \dfrac{2}{t - 3}$

The graph passes the Horizontal Line Test. The function has an inverse.

123. (a) $\qquad f(x) = \frac{1}{2}x - 3$

$\qquad\qquad y = \frac{1}{2}x - 3$

$\qquad\qquad x = \frac{1}{2}y - 3$

$\qquad x + 3 = \frac{1}{2}y$

$\qquad 2(x + 3) = y$

$\qquad f^{-1}(x) = 2x + 6$

(c) The graph of f^{-1} is the reflection of the graph of f about the line $y = x$.

(b)

(d) The domains and ranges of f and f^{-1} are the set of all real numbers.

125. (a) $f(x) = \sqrt{x+1}$

$y = \sqrt{x+1}$

$x = \sqrt{y+1}$

$x^2 = y + 1$

$x^2 - 1 = y$

$f^{-1}(x) = x^2 - 1, \; x \geq 0$

Note: The inverse must have a restricted domain.

(c) The graph of f^{-1} is the reflection of the graph of f about the line $y = x$.

(b)

(d) The domain of f and the range of f^{-1} is $[-1, \infty)$. The range of f and the domain of f^{-1} is $[0, \infty)$.

127. $f(x) = 2(x-4)^2$ is increasing on $[4, \infty)$.

Let $f(x) = 2(x-4)^2, \; x \geq 4$ and $y \geq 0$

$y = 2(x-4)^2$

$x = 2(y-4)^2, \; x \geq 0, \; y \geq 4$

$\dfrac{x}{2} = (y-4)^2$

$\sqrt{\dfrac{x}{2}} = y - 4$

$\sqrt{\dfrac{x}{2}} + 4 = y$

$f^{-1}(x) = \sqrt{\dfrac{x}{2}} + 4, \; x \geq 0$

129. False. The graph is reflected in the x-axis, shifted 9 units to the left, then shifted 13 units down.

131. A function from a Set A to a Set B is a relation that assigns to each element x in the Set A exactly one element y in the Set B.

133. The basic cubic function, $y = x^3$, has the following characteristics. Its domain and range is the set of all real numbers. It is an odd function, thus its graph is symmetric with respect to the origin. It is an increasing function and has an intercept at $(0, 0)$. The cubic function, $f(x) = x^3 - 1$, has the following characterictics. Its domain and range is the set of all real numbers. It is an increasing function with x-intercept at $(1, 0)$ and y-intercept at $(0, -1)$. The graph of $f(x)$ can be obtained by shifting the basic graph of $y = x^3$ down one unit.

Problem Solving for Chapter 2

1. (a) $W_1 = 0.07x + 2000$

(b) $W_2 = 0.05x + 2300$

(d) If you think you can sell \$20,000 per month, keep your current job with the higher commission rate. For sales over \$15,000 it pays more than the other job.

(c)

Point of intersection: $(15,000, 3050)$

Both jobs pay the same, \$3050, if you sell \$15,000 per month.

3. (a) Let $f(x)$ and $g(x)$ be two even functions. Then define $h(x) = f(x) \pm g(x)$.

$$h(-x) = f(-x) \pm g(-x)$$

$$= f(x) \pm g(x) \text{ since } f \text{ and } g \text{ are even}$$

$$= h(x)$$

So, $h(x)$ is also even.

(c) Let $f(x)$ be odd and $g(x)$ be even. Then define $h(x) = f(x) \pm g(x)$.

$$h(-x) = f(-x) \pm g(-x)$$

$$= -f(x) \pm g(x) \text{ since } f \text{ is odd and } g \text{ is even}$$

$$\neq h(x)$$

$$\neq -h(x)$$

So, $h(x)$ is neither odd nor even.

(b) Let $f(x)$ and $g(x)$ be two odd functions. Then define $h(x) = f(x) \pm g(x)$.

$$h(-x) = f(-x) \pm g(-x)$$

$$= -f(x) \mp g(x) \text{ since } f \text{ and } g \text{ are odd}$$

$$= -h(x)$$

So, $h(x)$ is also odd. (If $f(x) \neq g(x)$)

5. $f(x) = a_{2n}x^{2n} + a_{2n-2}x^{2n-2} + \cdots + a_2 x^2 + a_0$

$$f(-x) = a_{2n}(-x)^{2n} + a_{2n-2}(-x)^{2n-2} + \cdots + a_2(-x)^2 + a_0$$

$$= a_{2n}x^{2n} + a_{2n-2}x^{2n-2} + \cdots + a_2 x^2 + a_0$$

$$= f(x)$$

Therefore, $f(x)$ is even.

7. (a) April 11: 10 hours

April 12: 24 hours

April 13: 24 hours

April 14: $23\frac{2}{3}$ hours

Total: $81\frac{2}{3}$ hours

(b) Speed $= \dfrac{\text{distance}}{\text{time}} = \dfrac{2100}{81\frac{2}{3}} = \dfrac{180}{7} = 25\frac{5}{7}$ mph

(c) $D = -\dfrac{180}{7}t + 3400$

Domain: $0 \leq t \leq \dfrac{1190}{9}$

Range: $0 \leq D \leq 3400$

(d)

9. (a)–(d) Use $f(x) = 4x$ and $g(x) = x + 6$.

(a) $(fg)(x) = f(x + 6) = 4(x + 6) = 4x + 24$

(b) $(f \circ g)^{-1}(x) = \dfrac{x - 24}{4} = \dfrac{1}{4}x - 6$

(c) $f^{-1}(x) = \dfrac{1}{4}x$

$g^{-1}(x) = x - 6$

(d) $(g^{-1} \circ f^{-1})(x) = g^{-1}\left(\dfrac{1}{4}x\right) = \dfrac{1}{4}x - 6$

(e) $f(x) = x^3 + 1$ and $g(x) = 2x$

$(f \circ g)(x) = f(2x) = (2x)^3 + 1 = 8x^3 + 1$

$(f \circ g)^{-1}(x) = \sqrt[3]{\dfrac{x - 1}{8}} = \dfrac{1}{2}\sqrt[3]{x - 1}$

$f^{-1}(x) = \sqrt[3]{x - 1}$

$g^{-1}(x) = \dfrac{1}{2}x$

$(g^{-1} \circ f^{-1})(x) = g^{-1}\left(\sqrt[3]{x - 1}\right) = \dfrac{1}{2}\sqrt[3]{x - 1}$

(f) Answers will vary.

(g) Conjecture: $(f \circ g)^{-1}(x) = (g^{-1} \circ f^{-1})(x)$

11. $H(x) = \begin{cases} 1, & x \geq 0 \\ 0, & x < 0 \end{cases}$

(a) $H(x) - 2$

(b) $H(x - 2)$

(c) $-H(x)$

(d) $H(-x)$

(e) $\frac{1}{2}H(x)$

(f) $-H(x - 2) + 2$

13. $(f \circ (g \circ h))(x) = f((g \circ h)(x))$

$$= f(g(h(x)))$$

$$= (f \circ g \circ h)(x)$$

$$((f \circ g) \circ h)(x) = (f \circ g)(h(x))$$

$$= f(g(h(x)))$$

$$= (f \circ g \circ h)(x)$$

15.

x	$f(x)$	$f^{-1}(x)$
-4	—	2
-3	4	1
-2	1	0
-1	0	—
0	-2	-1
1	-3	-2
2	-4	—
3	—	—
4	—	-3

(a)

x	$f(f^{-1}(x))$
-4	$f(f^{-1}(-4)) = f(2) = -4$
-2	$f(f^{-1}(-2)) = f(0) = -2$
0	$f(f^{-1}(0)) = f(-1) = 0$
4	$f(f^{-1}(4)) = f(-3) = 4$

(b)

x	$(f + f^{-1})(x)$
-3	$f(-3) + f^{-1}(-3) = 4 + 1 = 5$
-2	$f(-2) + f^{-1}(-2) = 1 + 0 = 1$
0	$f(0) + f^{-1}(0) = -2 + (-1) = -3$
1	$f(1) + f^{-1}(1) = -3 + (-2) = -5$

(c)

x	$(f \cdot f^{-1})(x)$
-3	$f(-3)f^{-1}(-3) = (4)(1) = 4$
-2	$f(-2)f^{-1}(-2) = (1)(0) = 0$
0	$f(0)f^{-1}(0) = (-2)(-1) = 2$
1	$f(1)f^{-1}(1) = (-3)(-2) = 6$

(d)

x	$\|f^{-1}(x)\|$
-4	$\|f^{-1}(-4)\| = \|2\| = 2$
-3	$\|f^{-1}(-3)\| = \|1\| = 1$
0	$\|f^{-1}(0)\| = \|-1\| = 1$
4	$\|f^{-1}(4)\| = \|-3\| = 3$

Practice Test for Chapter 2

1. Find the equation of the line through $(2, 4)$ and $(3, -1)$.

2. Find the equation of the line with slope $m = 4/3$ and y-intercept $b = -3$.

3. Find the equation of the line through $(4, 1)$ perpendicular to the line $2x + 3y = 0$.

4. If it costs a company \$32 to produce 5 units of a product and \$44 to produce 9 units, how much does it cost to produce 20 units? (Assume that the cost function is linear.)

5. Given $f(x) = x^2 - 2x + 1$, find $f(x - 3)$.

6. Given $f(x) = 4x - 11$, find $\dfrac{f(x) - f(3)}{x - 3}$.

7. Find the domain and range of $f(x) = \sqrt{36 - x^2}$.

8. Which equations determine y as a function of x?

   (a) $6x - 5y + 4 = 0$

   (b) $x^2 + y^2 = 9$

   (c) $y^3 = x^2 + 6$

9. Sketch the graph of $f(x) = x^2 - 5$.

10. Sketch the graph of $f(x) = |x + 3|$.

11. Sketch the graph of $f(x) = \begin{cases} 2x + 1 & \text{if } x \geq 0, \\ x^2 - x & \text{if } x < 0. \end{cases}$

12. Use the graph of $f(x) = |x|$ to graph the following.

    (a) $f(x + 2)$

    (b) $-f(x) + 2$

13. Given $f(x) = 3x + 7$ and $g(x) = 2x^2 - 5$, find the following:

    (a) $(g - f)(x)$

    (b) $(fg)(x)$

14. Given $f(x) = x^2 - 2x + 16$ and $g(x) = 2x + 3$, find $f(g(x))$.

15. Given $f(x) = x^3 + 7$, find $f^{-1}(x)$.

16. Which of the following functions have inverses?

(a) $f(x) = |x - 6|$

(b) $f(x) = ax + b$, $a \neq 0$

(c) $f(x) = x^3 - 19$

17. Given $f(x) = \sqrt{\dfrac{3 - x}{x}}$, $0 < x \leq 3$, find $f^{-1}(x)$.

Exercises 18–20, true or false?

18. $y = 3x + 7$ and $y = \frac{1}{3}x - 4$ are perpendicular.

19. $(f \circ g)^{-1} = g^{-1} \circ f^{-1}$

20. If a function has an inverse, then it must pass both the Vertical Line Test and the Horizontal Line Test.

CHAPTER 3
Polynomial Functions

CHAPTER 3
Polynomial Functions

Section 3.1   Quadratic Functions and Models

You should know the following facts about parabolas.

- ■ $f(x) = ax^2 + bx + c$, $a \neq 0$, is a quadratic function, and its graph is a parabola.

- ■ If $a > 0$, the parabola opens upward and the vertex is the point with the minimum y-value.
  If $a < 0$, the parabola opens downward and the vertex is the point with the maximum y-value.

- ■ The vertex is $(-b/2a, f(-b/2a))$.

- ■ To find the x-intercepts (if any), solve
  $$ax^2 + bx + c = 0.$$

- ■ The standard form of the equation of a parabola is
  $$f(x) = a(x - h)^2 + k$$
  where $a \neq 0$.

  (a) The vertex is (h, k).

  (b) The axis is the vertical line $x = h$.

Vocabulary Check

1. nonnegative integer; real

2. quadratic; parabola

3. axis or axis of symmetry

4. positive; minimum

5. negative; maximum

---

1. $f(x) = (x - 2)^2$ opens upward and has vertex $(2, 0)$. Matches graph (g).

3. $f(x) = x^2 - 2$ opens upward and has vertex $(0, -2)$. Matches graph (b).

5. $f(x) = 4 - (x - 2)^2 = -(x - 2)^2 + 4$ opens downward and has vertex $(2, 4)$. Matches graph (f).

7. $f(x) = -(x - 3)^2 - 2$ opens downward and has vertex $(3, -2)$. Matches graph (e).

9. (a) $y = \frac{1}{2}x^2$

Vertical shrink

(b) $y = -\frac{1}{8}x^2$

Vertical shrink and reflection in the x-axis

—CONTINUED—

9. —CONTINUED—

(c) $y = \frac{3}{2}x^2$

Vertical stretch

(d) $y = -3x^2$

Vertical stretch and reflection in the x-axis

11. (a) $y = (x - 1)^2$

Horizontal translation one unit to the right

(b) $y = (3x)^2 + 1$

Horizontal shrink and a vertical translation one unit upward

(c) $y = \left(\frac{1}{3}x\right)^2 - 3$

Horizontal stretch and a vertical translation three units downward

(d) $y = (x + 3)^2$

Horizontal translation three units to the left

13. $f(x) = x^2 - 5$

Vertex: $(0, -5)$

Axis of symmetry: $x = 0$ or the y-axis

Find x-intercepts:

$x^2 - 5 = 0$

$x^2 = 5$

$x = \pm\sqrt{5}$

x-intercepts:
$(-\sqrt{5}, 0), (\sqrt{5}, 0)$

15. $f(x) = \frac{1}{2}x^2 - 4 = \frac{1}{2}(x - 0)^2 - 4$

Vertex: $(0, -4)$

Axis of symmetry: $x = 0$ or the y-axis

Find x-intercepts:

$\frac{1}{2}x^2 - 4 = 0$

$x^2 = 8$

$x = \pm\sqrt{8} = \pm2\sqrt{2}$

x-intercepts:
$(-2\sqrt{2}, 0), (2\sqrt{2}, 0)$

17. $f(x) = (x + 5)^2 - 6$

Vertex: $(-5, -6)$

Axis of symmetry: $x = -5$

Find x-intercepts:

$(x + 5)^2 - 6 = 0$

$(x + 5)^2 = 6$

$x + 5 = \pm\sqrt{6}$

$x = -5 \pm \sqrt{6}$

x-intercepts: $\left(-5 - \sqrt{6}, 0\right), \left(-5 + \sqrt{6}, 0\right)$

19. $h(x) = x^2 - 8x + 16 = (x - 4)^2$

Vertex: $(4, 0)$

Axis of symmetry: $x = 4$

x-intercept: $(4, 0)$

21. $f(x) = x^2 - x + \dfrac{5}{4}$

$= \left(x^2 - x + \dfrac{1}{4}\right) - \dfrac{1}{4} + \dfrac{5}{4}$

$= \left(x - \dfrac{1}{2}\right)^2 + 1$

Vertex: $\left(\dfrac{1}{2}, 1\right)$

Axis of symmetry: $x = \dfrac{1}{2}$

Find x-intercepts:

$x^2 - x + \dfrac{5}{4} = 0$

$x = \dfrac{1 \pm \sqrt{1 - 5}}{2}$

Not a real number

No x-intercepts

23. $f(x) = -x^2 + 2x + 5$

$= -(x^2 - 2x + 1) - (-1) + 5$

$= -(x - 1)^2 + 6$

Vertex: $(1, 6)$

Axis of symmetry: $x = 1$

Find x-intercepts:

$-x^2 + 2x + 5 = 0$

$x^2 - 2x - 5 = 0$

$x = \dfrac{2 \pm \sqrt{4 + 20}}{2}$

$= 1 \pm \sqrt{6}$

x-intercepts: $\left(1 - \sqrt{6}, 0\right), \left(1 + \sqrt{6}, 0\right)$

25. $h(x) = 4x^2 - 4x + 21$

$= 4\left(x^2 - x + \dfrac{1}{4}\right) - 4\left(\dfrac{1}{4}\right) + 21$

$= 4\left(x - \dfrac{1}{2}\right)^2 + 20$

Vertex: $\left(\dfrac{1}{2}, 20\right)$

Axis of symmetry: $x = \dfrac{1}{2}$

Find x-intercepts:

$4x^2 - 4x + 21 = 0$

$x = \dfrac{4 \pm \sqrt{16 - 336}}{2(4)}$

Not a real number $\implies$ No x-intercepts

27. $f(x) = \frac{1}{4}x^2 - 2x - 12$

$\qquad = \frac{1}{4}(x^2 - 8x + 16) - \frac{1}{4}(16) - 12$

$\qquad = \frac{1}{4}(x - 4)^2 - 16$

Vertex: $(4, -16)$

Axis of symmetry: $x = 4$

Find x-intercepts:

$\frac{1}{4}x^2 - 2x - 12 = 0$

$x^2 - 8x - 48 = 0$

$(x + 4)(x - 12) = 0$

$x = -4 \quad \text{or} \quad x = 12$

x-intercepts: $(-4, 0), (12, 0)$

29. $f(x) = -(x^2 + 2x - 3) = -(x + 1)^2 + 4$

Vertex: $(-1, 4)$

Axis of symmetry: $x = -1$

x-intercepts: $(-3, 0), (1, 0)$

31. $g(x) = x^2 + 8x + 11 = (x + 4)^2 - 5$

Vertex: $(-4, -5)$

Axis of symmetry: $x = -4$

x-intercepts: $\left(-4 \pm \sqrt{5}, 0\right)$

33. $f(x) = 2x^2 - 16x + 31$

$\qquad = 2(x - 4)^2 - 1$

Vertex: $(4, -1)$

Axis of symmetry: $x = 4$

x-intercepts: $\left(4 \pm \frac{1}{2}\sqrt{2}, 0\right)$

35. $g(x) = \frac{1}{2}(x^2 + 4x - 2) = \frac{1}{2}(x + 2)^2 - 3$

Vertex: $(-2, -3)$

Axis of symmetry: $x = -2$

x-intercepts: $\left(-2 \pm \sqrt{6}, 0\right)$

37. $(1, 0)$ is the vertex.

$y = a(x - 1)^2 + 0 = a(x - 1)^2$

Since the graph passes through the point $(0, 1)$, we have:

$1 = a(0 - 1)^2$

$1 = a$

$y = 1(x - 1)^2 = (x - 1)^2$

39. $(-1, 4)$ is the vertex.

$y = a(x + 1)^2 + 4$

Since the graph passes through the point $(1, 0)$, we have:

$0 = a(1 + 1)^2 + 4$

$-4 = 4a$

$-1 = a$

$\qquad y = -1(x + 1)^2 + 4 = -(x + 1)^2 + 4$

41. $(-2, 2)$ is the vertex.

$y = a(x + 2)^2 + 2$

Since the graph passes through the point $(-1, 0)$, we have:

$0 = a(-1 + 2)^2 + 2$

$-2 = a$

$y = -2(x + 2)^2 + 2$

43. $(-2, 5)$ is the vertex.

$f(x) = a(x + 2)^2 + 5$

Since the graph passes through the point $(0, 9)$, we have:

$9 = a(0 + 2)^2 + 5$

$4 = 4a$

$1 = a$

$f(x) = 1(x + 2)^2 + 5 = (x + 2)^2 + 5$

45. $(3, 4)$ is the vertex.

$f(x) = a(x - 3)^2 + 4$

Since the graph passes through the point $(1, 2)$, we have:

$2 = a(1 - 3)^2 + 4$

$-2 = 4a$

$-\frac{1}{2} = a$

$f(x) = -\frac{1}{2}(x - 3)^2 + 4$

47. (5, 12) is the vertex.

$$f(x) = a(x - 5)^2 + 12$$

Since the graph passes through the point (7, 15), we have:

$$15 = a(7 - 5)^2 + 12$$
$$3 = 4a \implies a = \tfrac{3}{4}$$
$$f(x) = \tfrac{3}{4}(x - 5)^2 + 12$$

49. $\left(-\tfrac{1}{4}, \tfrac{3}{2}\right)$ is the vertex.

$$f(x) = a\left(x + \tfrac{1}{4}\right)^2 + \tfrac{3}{2}$$

Since the graph passes through the point $(-2, 0)$, we have:

$$0 = a\left(-2 + \tfrac{1}{4}\right)^2 + \tfrac{3}{2}$$
$$-\tfrac{3}{2} = \tfrac{49}{16}a \implies a = -\tfrac{24}{49}$$
$$f(x) = -\tfrac{24}{49}\left(x + \tfrac{1}{4}\right)^2 + \tfrac{3}{2}$$

51. $\left(-\tfrac{5}{2}, 0\right)$ is the vertex.

$$f(x) = a\left(x + \tfrac{5}{2}\right)^2$$

Since the graph passes through the point $\left(-\tfrac{7}{2}, -\tfrac{16}{3}\right)$, we have:

$$-\tfrac{16}{3} = a\left(-\tfrac{7}{2} + \tfrac{5}{2}\right)^2$$
$$-\tfrac{16}{3} = a$$
$$f(x) = -\tfrac{16}{3}\left(x + \tfrac{5}{2}\right)^2$$

53. $y = x^2 - 16$

x-intercepts: $(\pm 4, 0)$

$$0 = x^2 - 16$$
$$x^2 = 16$$
$$x = \pm 4$$

55. $y = x^2 - 4x - 5$

x-intercepts: $(5, 0), (-1, 0)$

$$0 = x^2 - 4x - 5$$
$$0 = (x - 5)(x + 1)$$
$$x = 5 \quad \text{or} \quad x = -1$$

57. $f(x) = x^2 - 4x$

x-intercepts: $(0, 0), (4, 0)$

$$0 = x^2 - 4x$$
$$0 = x(x - 4)$$
$$x = 0 \quad \text{or} \quad x = 4$$

The x-intercepts and the solutions of $f(x) = 0$ are the same.

59. $f(x) = x^2 - 9x + 18$

x-intercepts: $(3, 0), (6, 0)$

$$0 = x^2 - 9x + 18$$
$$0 = (x - 3)(x - 6)$$
$$x = 3 \quad \text{or} \quad x = 6$$

The x-intercepts and the solutions of $f(x) = 0$ are the same.

61. $f(x) = 2x^2 - 7x - 30$

x-intercepts: $\left(-\tfrac{5}{2}, 0\right), (6, 0)$

$$0 = 2x^2 - 7x - 30$$
$$0 = (2x + 5)(x - 6)$$
$$x = -\tfrac{5}{2} \quad \text{or} \quad x = 6$$

The x-intercepts and the solutions of $f(x) = 0$ are the same.

63. $f(x) = -\tfrac{1}{2}(x^2 - 6x - 7)$

x-intercepts: $(-1, 0), (7, 0)$

$$0 = -\tfrac{1}{2}(x^2 - 6x - 7)$$
$$0 = x^2 - 6x - 7$$
$$0 = (x + 1)(x - 7)$$
$$x = -1 \quad \text{or} \quad x = 7$$

The x-intercepts and the solutions of $f(x) = 0$ are the same.

65. $f(x) = [x - (-1)](x - 3)$ opens upward

$$= (x + 1)(x - 3)$$
$$= x^2 - 2x - 3$$
$$g(x) = -[x - (-1)](x - 3) \quad \text{opens downward}$$
$$= -(x + 1)(x - 3)$$
$$= -(x^2 - 2x - 3)$$
$$= -x^2 + 2x + 3$$

Note: $f(x) = a(x + 1)(x - 3)$ has x-intercepts $(-1, 0)$ and $(3, 0)$ for all real numbers $a \neq 0$.

67. $f(x) = (x - 0)(x - 10)$ opens upward

         $= x^2 - 10x$

     $g(x) = -(x - 0)(x - 10)$ opens downward

         $= -x^2 + 10x$

     Note: $f(x) = a(x - 0)(x - 10) = ax(x - 10)$ has
         x-intercepts $(0, 0)$ and $(10, 0)$ for all real numbers
         $a \neq 0$.

69. $f(x) = [x - (-3)]\left[x - \left(-\frac{1}{2}\right)\right](2)$ opens upward

         $= (x + 3)\left(x + \frac{1}{2}\right)(2)$

         $= (x + 3)(2x + 1)$

         $= 2x^2 + 7x + 3$

     $g(x) = -(2x^2 + 7x + 3)$ opens downward

         $= -2x^2 - 7x - 3$

     Note: $f(x) = a(x + 3)(2x + 1)$ has *x*-intercepts
         $(-3, 0)$ and $\left(-\frac{1}{2}, 0\right)$ for all real numbers $a \neq 0$.

71. Let $x =$ the first number and $y =$ the second number.
Then the sum is

$$x + y = 110 \implies y = 110 - x.$$

The product is $P(x) = xy = x(110 - x) = 110x - x^2$.

$$P(x) = -x^2 + 110x$$

$$= -(x^2 - 110x + 3025 - 3025)$$

$$= -[(x - 55)^2 - 3025]$$

$$= -(x - 55)^2 + 3025$$

The maximum value of the product occurs at the vertex of
$P(x)$ and is 3025. This happens when $x = y = 55$.

73. Let $x =$ the first number and $y =$ the second number.
Then the sum is

$$x + 2y = 24 \implies y = \frac{24 - x}{2}.$$

The product is $P(x) = xy = x\left(\dfrac{24 - x}{2}\right)$.

$$P(x) = \frac{1}{2}(-x^2 + 24x)$$

$$= -\frac{1}{2}(x^2 - 24x + 144 - 144)$$

$$= -\frac{1}{2}[(x - 12)^2 - 144] = -\frac{1}{2}(x - 12)^2 + 72$$

The maximum value of the product occurs at the vertex
of $P(x)$ and is 72. This happens when $x = 12$ and
$y = (24 - 12)/2 = 6$. Thus, the numbers are 12 and 6.

75. (a)

$$4x + 3y = 200 \implies y = \frac{1}{3}(200 - 4x) = \frac{4}{3}(50 - x)$$

$$A = 2xy = 2x\left[\frac{4}{3}(50 - x)\right] = \frac{8}{3}x(50 - x) = \frac{8x(50 - x)}{3}$$

(c)

This area is maximum when $x = 25$ feet and
$y = \frac{100}{3} = 33\frac{1}{3}$ feet.

(b)

x	A
5	600
10	$1066\frac{2}{3}$
15	1400
20	1600
25	$1666\frac{2}{3}$
30	1600

This area is maximum when
$x = 25$ feet and $y = \frac{100}{3} = 33\frac{1}{3}$ feet.

—CONTINUED—

75. —CONTINUED—

(d) $A = \dfrac{8}{3}x(50 - x)$

$\quad = -\dfrac{8}{3}(x^2 - 50x)$

$\quad = -\dfrac{8}{3}(x^2 - 50x + 625 - 625)$

$\quad = -\dfrac{8}{3}[(x - 25)^2 - 625]$

$\quad = -\dfrac{8}{3}(x - 25)^2 + \dfrac{5000}{3}$

The maximum area occurs at the vertex and is 5000/3 square feet. This happens when $x = 25$ feet and $y = (200 - 4(25))/3 = 100/3$ feet. The dimensions are $2x = 50$ feet by $33\frac{1}{3}$ feet.

(e) They are all identical.

$\quad x = 25$ feet and $y = 33\frac{1}{3}$ feet

77. $y = -\dfrac{4}{9}x^2 + \dfrac{24}{9}x + 12$

The vertex occurs at $-\dfrac{b}{2a} = \dfrac{-24/9}{2(-4/9)} = 3$. The maximum height is $y(3) = -\dfrac{4}{9}(3)^2 + \dfrac{24}{9}(3) + 12 = 16$ feet.

79. $C = 800 - 10x + 0.25x^2 = 0.25x^2 - 10x + 800$

The vertex occurs at $x = -\dfrac{b}{2a} = -\dfrac{-10}{2(0.25)} = 20$. The cost is minimum when $x = 20$ fixtures.

81. $P = -0.0002x^2 + 140x - 250,000$

The vertex occurs at $x = -\dfrac{b}{2a} = -\dfrac{140}{2(-0.0002)} = 350,000$. The profit is maximum when $x = 350,000$ units.

83. $R(p) = -25p^2 + 1200p$

(a) $R(20) = \$14,000$ thousand

$\quad R(25) = \$14,375$ thousand

$\quad R(30) = \$13,500$ thousand

(b) The revenue is a maximum at the vertex

$\quad -\dfrac{b}{2a} = \dfrac{-1200}{2(-25)} = 24$

$\quad R(24) = 14,400$

The unit price that will yield a maximum revenue of \$14,400 thousand is \$24.

85. $C = 4299 - 1.8t - 1.36t^2, 0 \le t \le 43$

(a)

(b) Vertex $\approx (0, 4299)$

The vertex occurs when $y \approx 4299$ which is the maximum average annual consumption. The warnings may not have had an immediate effect, but over time they and other findings about the health risks and the increased cost of cigarettes have had an effect.

(c) $C(40) = 2051$

Annually: $\dfrac{209,128,094(2051)}{48,308,590} \approx 8879$ cigarettes

Daily: $\dfrac{8879}{366} \approx 24$ cigarettes

87. (a)

(b) $0.002s^2 + 0.005s - 0.029 = 10$

$$2s^2 + 5s - 29 = 10,000$$

$$2s^2 + 5s - 10,029 = 0$$

$$a = 2, b = 5, c = -10,029$$

$$s = \frac{-5 \pm \sqrt{5^2 - 4(2)(-10,029)}}{2(2)}$$

$$s = \frac{-5 \pm \sqrt{80,257}}{4}$$

$$s \approx -72.1, 69.6$$

The maximum speed if power is not to exceed 10 horsepower is 69.6 miles per hour.

91. $f(x) = ax^2 + bx + c$

$$= a\left(x^2 + \frac{b}{a}x\right) + c$$

$$= a\left(x^2 + \frac{b}{a}x + \frac{b^2}{4a^2} - \frac{b^2}{4a^2}\right) + c$$

$$= a\left(x + \frac{b}{2a}\right)^2 - \frac{b^2}{4a} + c$$

$$= a\left(x - \left(-\frac{b}{2a}\right)\right)^2 + \frac{4ac - b^2}{4a}$$

$$f\left(-\frac{b}{2a}\right) = a\left(\frac{b^2}{4a^2}\right) + b\left(-\frac{b}{2a}\right) + c$$

$$= \frac{b^2}{4a} - \frac{b^2}{2a} + c$$

$$= \frac{b^2 - 2b^2 + 4ac}{4a} = \frac{4ac - b^2}{4a}$$

So, the vertex occurs at $\left(-\frac{b}{2a}, \frac{4ac - b^2}{4a}\right) = \left(-\frac{b}{2a}, f\left(-\frac{b}{2a}\right)\right)$.

93. Yes. A graph of a quadratic equation whose vertex is $(0, 0)$ has only one x-intercept.

95. $(-4, 3)$ and $(2, 1)$

$$m = \frac{1 - 3}{2 - (-4)} = \frac{-2}{6} = -\frac{1}{3}$$

$$y - 1 = -\frac{1}{3}(x - 2)$$

$$y - 1 = -\frac{1}{3}x + \frac{2}{3}$$

$$y = -\frac{1}{3}x + \frac{5}{3}$$

89. True. The equation $-12x^2 - 1 = 0$ has no real solution, so the graph has no x-intercepts.

97. $4x + 5y = 10 \implies y = -\frac{4}{5}x + 2$ and $m = -\frac{4}{5}$

The slope of the perpendicular line through $(0, 3)$ is $m = \frac{5}{4}$ and the y-intercept is $b = 3$.

$$y = \frac{5}{4}x + 3$$

For Exercises 99–103, let $f(x) = 14x - 3$, **and** $g(x) = 8x^2$.

99. $(f + g)(-3) = f(-3) + g(-3)$

$$= [14(-3) - 3] + 8(-3)^2 = 27$$

101. $(fg)\left(-\dfrac{4}{7}\right) = f\left(-\dfrac{4}{7}\right)g\left(-\dfrac{4}{7}\right)$

$$= \left[14\left(-\dfrac{4}{7}\right) - 3\right]\left[8\left(-\dfrac{4}{7}\right)^2\right]$$

$$= (-11)\left(\dfrac{128}{49}\right) = -\dfrac{1408}{49}$$

103. $(f \circ g)(-1) = f(g(-1)) = f(8) = 14(8) - 3 = 109$

Section 3.2   Polynomial Functions of Higher Degree

You should know the following basic principles about polynomials.

- $f(x) = a_n x^n + a_{n-1}x^{n-1} + \cdots + a_2 x^2 + a_1 x + a_0, a_n \neq 0$, is a polynomial function of degree n.
- If f is of odd degree and
  - (a) $a_n > 0$, then
    1. $f(x) \to \infty$ as $x \to \infty$.
    2. $f(x) \to -\infty$ as $x \to -\infty$.
  - (b) $a_n < 0$, then
    1. $f(x) \to -\infty$ as $x \to \infty$.
    2. $f(x) \to \infty$ as $x \to -\infty$.
- If f is of even degree and
  - (a) $a_n > 0$, then
    1. $f(x) \to \infty$ as $x \to \infty$.
    2. $f(x) \to \infty$ as $x \to -\infty$.
  - (b) $a_n < 0$, then
    1. $f(x) \to -\infty$ as $x \to \infty$.
    2. $f(x) \to -\infty$ as $x \to -\infty$.
- The following are equivalent for a polynomial function.
  - (a) $x = a$ is a zero of a function.
  - (b) $x = a$ is a solution of the polynomial equation $f(x) = 0$.
  - (c) $(x - a)$ is a factor of the polynomial.
  - (d) $(a, 0)$ is an x-intercept of the graph of f.
- A polynomial of degree n has at most n distinct zeros and at most $n - 1$ turning points.
- A factor $(x - a)^k, k > 1$, yields a repeated zero of $x = a$ of multiplicity k.
  - (a) If k is odd, the graph crosses the x-axis at $x = a$.
  - (b) If k is even, the graph just touches the x-axis at $x = a$.
- If f is a polynomial function such that $a < b$ and $f(a) \neq f(b)$, then f takes on every value between $f(a)$ and $f(b)$ in the interval $[a, b]$.
- If you can find a value where a polynomial is positive and another value where it is negative, then there is at least one real zero between the values.

Vocabulary Check

1. continuous

2. Leading Coefficient Test

3. $n; n - 1$

4. solution; $(x - a)$; x-intercept

5. touches; crosses

6. standard

7. Intermediate Value

1. $f(x) = -2x + 3$ is a line with y-intercept $(0, 3)$. Matches graph (c).

3. $f(x) = -2x^2 - 5x$ is a parabola with x-intercepts $(0, 0)$ and $\left(-\frac{5}{2}, 0\right)$ and opens downward. Matches graph (h).

5. $f(x) = -\frac{1}{4}x^4 + 3x^2$ has intercepts $(0, 0)$ and $\left(\pm 2\sqrt{3}, 0\right)$. Matches graph (a).

7. $f(x) = x^4 + 2x^3$ has intercepts $(0, 0)$ and $(-2, 0)$. Matches graph (d).

9. $y = x^3$

   (a) $f(x) = (x - 2)^3$

   Horizontal shift two units to the right

   (b) $f(x) = x^3 - 2$

   Vertical shift two units downward

   (c) $f(x) = -\frac{1}{2}x^3$

   Reflection in the x-axis and a vertical shrink

   (d) $f(x) = (x - 2)^3 - 2$

   Horizontal shift two units to the right and a vertical shift two units downward

11. $y = x^4$

    (a) $f(x) = (x + 3)^4$

    Horizontal shift three units to the left

    (b) $f(x) = x^4 - 3$

    Vertical shift three units downward

—CONTINUED—

11. —CONTINUED—

(c) $f(x) = 4 - x^4$

Reflection in the x-axis and then a vertical shift four units upward

(e) $f(x) = (2x)^4 + 1$

Vertical shift one unit upward and a horizontal shrink (each y-value is multiplied by $\frac{1}{2}$)

(d) $f(x) = \frac{1}{2}(x - 1)^4$

Horizontal shift one unit to the right and a vertical shrink (each y-value is multiplied by $\frac{1}{2}$)

(f) $f(x) = \left(\frac{1}{2}x\right)^4 - 2$

Vertical shift two units downward and a horizontal stretch (each y-value is multiplied by $\frac{1}{2}$)

13. $f(x) = \frac{1}{3}x^3 + 5x$

Degree: 3

Leading coefficient: $\frac{1}{3}$

The degree is odd and the leading coefficient is positive. The graph falls to the left and rises to the right.

15. $g(x) = 5 - \frac{7}{2}x - 3x^2$

Degree: 2

Leading coefficient: -3

The degree is even and the leading coefficient is negative. The graph falls to the left and falls to the right.

17. $f(x) = -2.1x^5 + 4x^3 - 2$

Degree: 5

Leading coefficient: -2.1

The degree is odd and the leading coefficient is negative. The graph rises to the left and falls to the right.

19. $f(x) = 6 - 2x + 4x^2 - 5x^3$

Degree: 3

Leading coefficient: -5

The degree is odd and the leading coefficient is negative. The graph rises to the left and falls to the right.

21. $h(t) = -\frac{2}{3}(t^2 - 5t + 3)$

Degree: 2

Leading coefficient: $-\frac{2}{3}$

The degree is even and the leading coefficient is negative. The graph falls to the left and falls to the right.

23. $f(x) = 3x^3 - 9x + 1$; $g(x) = 3x^3$

25. $f(x) = -(x^4 - 4x^3 + 16x); \quad g(x) = -x^4$

27. $f(x) = x^2 - 25$

(a) $0 = x^2 - 25 = (x + 5)(x - 5)$

Zeros: $x = \pm 5$

(b) Each zero has a multiplicity of 1. (odd multiplicity)

Turning point: 1 (the vertex of the parabola)

(c)

29. $h(t) = t^2 - 6t + 9$

(a) $0 = t^2 - 6t + 9 = (t - 3)^2$

Zero: $t = 3$

(b) $t = 3$ has a multiplicity of 2 (even multiplicity).

Turning point: 1 (the vertex of the parabola)

(c)

31. $f(x) = \frac{1}{3}x^2 + \frac{1}{3}x - \frac{2}{3}$

(a) $0 = \frac{1}{3}x^2 + \frac{1}{3}x - \frac{2}{3}$

$\quad = \frac{1}{3}(x^2 + x - 2)$

$\quad = \frac{1}{3}(x + 2)(x - 1)$

Zeros: $x = -2, x = 1$

(b) Each zero has a multiplicity of 1 (odd multiplicity).

Turning point: 1 (the vertex of the parabola)

(c)

33. $f(x) = 3x^3 - 12x^2 + 3x$

(a) $0 = 3x^3 - 12x^2 + 3x = 3x(x^2 - 4x + 1)$

Zeros: $x = 0, x = 2 \pm \sqrt{3}$ (by the Quadratic Formula)

(b) Each zero has a multiplicity of 1 (odd multiplicity).

Turning points: 2

(c)

35. $f(t) = t^3 - 4t^2 + 4t$

(a) $0 = t^3 - 4t^2 + 4t = t(t^2 - 4t + 4) = t(t - 2)^2$

Zeros: $t = 0, t = 2$

(b) $t = 0$ has a multiplicity of 1 (odd multiplicity)

$t = 2$ has a multiplicity of 2 (even multiplicity)

Turning points: 2

(c)

37. $g(t) = t^5 - 6t^3 + 9t$

(a) $0 = t^5 - 6t^3 + 9t = t(t^4 - 6t^2 + 9) = t(t^2 - 3)^2$

$\quad = t(t + \sqrt{3})^2(t - \sqrt{3})^2$

Zeros: $t = 0, t = \pm\sqrt{3}$

(b) $t = 0$ has a multiplicity of 1 (odd multiplicity).

$t = \pm\sqrt{3}$ each have a multiplicity of 2 (even multiplicity)

Turning points: 4

(c)

39. $f(x) = 5x^4 + 15x^2 + 10$

(a) $0 = 5x^4 + 15x^2 + 10$

$= 5(x^4 + 3x^2 + 2)$

$= 5(x^2 + 1)(x^2 + 2)$

No real zeros

(b) Turning point: 1

(c)

43. $y = 4x^3 - 20x^2 + 25x$

(a)

(b) x-intercepts: $(0, 0), \left(\frac{5}{2}, 0\right)$

(c) $0 = 4x^3 - 20x^2 + 25x$

$0 = x(2x - 5)^2$

$x = 0$ or $x = \frac{5}{2}$

(d) The solutions are the same as the x-coordinates of the x-intercepts.

47. $f(x) = (x - 0)(x - 10)$

$f(x) = x^2 - 10x$

Note: $f(x) = a(x - 0)(x - 10) = ax(x - 10)$ has zeros 0 and 10 for all real numbers $a \neq 0$.

51. $f(x) = (x - 0)(x - (-2))(x - (-3))$

$= x(x + 2)(x + 3)$

$= x^3 + 5x^2 + 6x$

Note: $f(x) = ax(x + 2)(x + 3)$ has zeros $0, -2, -3$ for all real numbers $a \neq 0$.

55. $f(x) = \left[x - \left(1 + \sqrt{3}\right)\right]\left[x - \left(1 - \sqrt{3}\right)\right]$

$= \left[(x - 1) - \sqrt{3}\right]\left[(x - 1) + \sqrt{3}\right]$

$= (x - 1)^2 - \left(\sqrt{3}\right)^2$

$= x^2 - 2x + 1 - 3$

$= x^2 - 2x - 2$

Note: $f(x) = a(x^2 - 2x - 2)$ has these zeros for all real numbers $a \neq 0$.

41. $g(x) = x^3 + 3x^2 - 4x - 12$

(a) $0 = x^3 + 3x^2 - 4x - 12 = x^2(x + 3) - 4(x + 3)$

$= (x^2 - 4)(x + 3) = (x - 2)(x + 2)(x + 3)$

Zeros: $x = \pm 2, x = -3$

(b) Each zero has a multiplicity of 1 (odd multiplicity).

Turning points: 2

(c)

45. $y = x^5 - 5x^3 + 4x$

(a)

(b) x-intercepts: $(0, 0), (\pm 1, 0), (\pm 2, 0)$

(c) $0 = x^5 - 5x^3 + 4x$

$0 = x(x^2 - 1)(x^2 - 4)$

$0 = x(x + 1)(x - 1)(x + 2)(x - 2)$

$x = 0, \pm 1, \pm 2$

(d) The solutions are the same as the x-coordinates of the x-intercepts.

49. $f(x) = (x - 2)(x - (-6))$

$= (x - 2)(x + 6)$

$= x^2 + 4x - 12$

Note: $f(x) = a(x - 2)(x + 6)$ has zeros 2 and -6 for all real numbers $a \neq 0$.

53. $f(x) = (x - 4)(x + 3)(x - 3)(x - 0)$

$= (x - 4)(x^2 - 9)x$

$= x^4 - 4x^3 - 9x^2 + 36x$

Note: $f(x) = a(x^4 - 4x^3 - 9x^2 + 36x)$ has these zeros for all real numbers $a \neq 0$.

57. $f(x) = (x - (-2))(x - (-2))$

$= (x + 2)^2 = x^2 + 4x + 4$

Note: $f(x) = a(x^2 + 4x + 4)$, $a \neq 0$, has degree 2 and zero $x = -2$.

59. $f(x) = (x - (-3))(x - 0)(x - 1)$

$= x(x + 3)(x - 1) = x^3 + 2x^2 - 3x$

Note: $f(x) = a(x^3 + 2x^2 - 3x)$, $a \neq 0$, has degree 3 and zeros $x = -3, 0, 1$.

61. $f(x) = (x - 0)(x - \sqrt{3})(x - (-\sqrt{3}))$

$= x(x - \sqrt{3})(x + \sqrt{3}) = x^3 - 3x$

Note: $f(x) = a(x^3 - 3x)$, $a \neq 0$, has degree 3 and zeros $x = 0, \sqrt{3}, -\sqrt{3}$.

63. $f(x) = (x - (-5))^2(x - 1)(x - 2) = x^4 + 7x^3 - 3x^2 - 55x + 50$

or $f(x) = (x - (-5))(x - 1)^2(x - 2) = x^4 + x^3 - 15x^2 + 23x - 10$

or $f(x) = (x - (-5))(x - 1)(x - 2)^2 = x^4 - 17x^2 + 36x - 20$

Note: Any nonzero scalar multiple of these functions would also have degree 4 and zeros $x = -5, 1, 2$.

65. $f(x) = x^4(x + 4) = x^5 + 4x^4$

or $f(x) = x^3(x + 4)^2 = x^5 + 8x^4 + 16x^3$

or $f(x) = x^2(x + 4)^3 = x^5 + 12x^4 + 48x^3 + 64x^2$

or $f(x) = x(x + 4)^4 = x^5 + 16x^4 + 96x^3 + 256x^2 + 256x$

Note: Any nonzero scalar multiple of these functions would also have degree 5 and zeros $x = 0$ and -4.

67. $f(x) = x^3 - 9x = x(x^2 - 9) = x(x + 3)(x - 3)$

(a) Falls to the left; rises to the right

(b) Zeros: $0, -3, 3$

(c)

x	-3	-2	-1	0	1	2	3
$f(x)$	0	10	8	0	-8	-10	0

(d)

69. $f(t) = \frac{1}{4}(t^2 - 2t + 15) = \frac{1}{4}(t - 1)^2 + \frac{7}{2}$

(a) Rises to the left; rises to the right

(b) No real zero (no x-intercepts)

(c)

t	-1	0	1	2	3
$f(t)$	4.5	3.75	3.5	3.75	4.5

(d) The graph is a parabola with vertex $\left(1, \frac{7}{2}\right)$.

71. $f(x) = x^3 - 3x^2 = x^2(x - 3)$

(a) Falls to the left; rises to the right

(b) Zeros: 0 and 3

(c)

x	-1	0	1	2	3
$f(x)$	-4	0	-2	-4	0

(d)

73. $f(x) = 3x^3 - 15x^2 + 18x = 3x(x - 2)(x - 3)$

(a) Falls to the left; rises to the right

(b) Zeros: $0, 2, 3$

(c)

x	0	1	2	2.5	3	3.5
$f(x)$	0	6	0	-1.875	0	7.875

(d)

75. $f(x) = -5x^2 - x^3 = -x^2(5 + x)$

(a) Rises to the left; falls to the right

(b) Zeros: $0, -5$

(c)

x	-5	-4	-3	-2	-1	0	1
$f(x)$	0	-16	-18	-12	-4	0	-6

(d)

77. $f(x) = x^2(x - 4)$

(a) Falls to the left; rises to the right

(b) Zeros: $0, 4$

(c)

x	-1	0	1	2	3	4	5
$f(x)$	-5	0	-3	-8	-9	0	25

(d)

79. $g(t) = -\frac{1}{4}(t - 2)^2(t + 2)^2$

(a) Falls to the left; falls to the right

(b) Zeros: 2 and -2

(c)

t	-3	-2	-1	0	1	2	3
$g(t)$	$-\frac{25}{4}$	0	$-\frac{9}{4}$	-4	$-\frac{9}{4}$	0	$-\frac{25}{4}$

(d)

81. $f(x) = x^3 - 4x = x(x + 2)(x - 2)$

Zeros: $0, -2, 2$ all of multiplicity 1

83. $g(x) = \frac{1}{5}(x + 1)^2(x - 3)(2x - 9)$

Zeros: -1 of multiplicity 2, 3 of multiplicity 1, and $\frac{9}{2}$ of multiplicity 1

85. $f(x) = x^3 - 3x^2 + 3$

The function has three zeros.
They are in the intervals
$(-1, 0), (1, 2)$ and $(2, 3)$.

They are $x \approx -0.879, 1.347, 2.532$.

x	y_1
-3	-51
-2	-17
-1	-1
0	3
1	1
2	-1
3	3
4	19

87. $g(x) = 3x^4 + 4x^3 - 3$

x	y_1
-4	509
-3	132
-2	13
-1	-4
0	-3
1	4
2	77
3	348

The function has two zeros.
They are in the intervals
$(-2, -1)$ and $(0, 1)$.

They are $x \approx -1.585, 0.779$.

89. (a) Volume $= l \cdot w \cdot h$

height $= x$

length $=$ width $= 36 - 2x$

Thus, $V(x) = (36 - 2x)(36 - 2x)(x) = x(36 - 2x)^2$.

(b) Domain: $0 < x < 18$

The length and width must be positive.

(d)

The maximum point on the graph occurs at $x = 6$.
This agrees with the maximum found in part (c).

(c)

Box Height	Box Width	Box Volume, V
1	$36 - 2(1)$	$1[36 - 2(1)]^2 = 1156$
2	$36 - 2(2)$	$2[36 - 2(2)]^2 = 2048$
3	$36 - 2(3)$	$3[36 - 2(3)]^2 = 2700$
4	$36 - 2(4)$	$4[36 - 2(4)]^2 = 3136$
5	$36 - 2(5)$	$5[36 - 2(5)]^2 = 3380$
6	$36 - 2(6)$	$6[36 - 2(6)]^2 = 3456$
7	$36 - 2(7)$	$7[36 - 2(7)]^2 = 3388$

The volume is a maximum of 3456 cubic inches when
the height is 6 inches and the length and width are each
24 inches. So the dimensions are $6 \times 24 \times 24$ inches.

91. (a) $A = \ell \cdot w = (12 - 2x)(x) = -2x^2 + 12x$ square inches

(b) 16 feet $= 192$ inches

$V = l \cdot w \cdot h$

$\quad = (12 - 2x)(x)(192)$

$\quad = -384x^2 + 2304x$ cubic inches

(c) Since x and $12 - 2x$ cannot be negative, we have
$0 < x < 6$ inches for the domain.

(d)

x	V
0	0
1	1920
2	3072
3	3456
4	3072
5	1920
6	0

When $x = 3$, the volume is a
maximum with $V = 3456$ in.^{3}. The
dimensions of the gutter cross-section are
3 inches $\times$ 6 inches $\times$ 3 inches.

(e)

Maximum: (3, 3456)

The maximum value is the same.

(f) No. The volume is a product of the constant length and the
cross-sectional area. The value of x would remain the
same; only the value of V would change if the length was
changed.

93. $y_1 = 0.139t^3 - 4.42t^2 + 51.1t - 39$

The model is a good fit to the actual data.

95. Midwest: $y_1(18) = \$259.368$ thousand $= \$259{,}368$

South: $y_2(18) = \$223.472$ thousand $= \$223{,}472$

Since the models are both cubic functions with positive leading coefficients, both will increase without bound as t increases, thus should only be used for short term projections.

97. $G = -0.003t^3 + 0.137t^2 + 0.458t - 0.839,\ 2 \le t \le 34$

(a)

(b) The tree is growing most rapidly at $t \approx 15$.

(c) $\quad y = -0.009t^2 + 0.274t + 0.458$

$$-\frac{b}{2a} = \frac{-0.274}{2(-0.009)} \approx 15.222$$

$y(15.222) \approx 2.543$

Vertex $\approx (15.22, 2.54)$

(d) The x-value of the vertex in part (c) is approximately equal to the value found in part (b).

99. False. A fifth degree polynomial can have at most four turning points.

101. True. A polynomial of degree 7 with a negative leading coefficient rises to the left and falls to the right.

103. $f(x) = x^4$; $f(x)$ is even.

(a) $g(x) = f(x) + 2$

Vertical shift two units upward

$g(-x) = f(-x) + 2$

$\quad\quad = f(x) + 2$

$\quad\quad = g(x)$

Even

(b) $g(x) = f(x + 2)$

Horizontal shift two units to the left

Neither odd nor even

(c) $g(x) = f(-x) = (-x)^4 = x^4$

Reflection in the y-axis. The graph looks the same.

Even

(d) $g(x) = -f(x) = -x^4$

Reflection in the x-axis

Even

(e) $g(x) = f\left(\frac{1}{2}x\right) = \frac{1}{16}x^4$

Horizontal stretch

Even

(f) $g(x) = \frac{1}{2}f(x) = \frac{1}{2}x^4$

Vertical shrink

Even

(g) $g(x) = f(x^{3/4}) = (x^{3/4})^4 = x^3, x \ge 0$

Neither odd nor even

(h) $g(x) = (f \circ f)(x) = f(f(x)) = f(x^4) = (x^4)^4 = x^{16}$

Even

105. $5x^2 + 7x - 24 = (5x - 8)(x + 3)$

107. $4x^4 - 7x^3 - 15x^2 = x^2(4x^2 - 7x - 15)$

$\quad\quad\quad\quad\quad\quad\quad = x^2(4x + 5)(x - 3)$

109. $2x^2 - x - 28 = 0$

$(2x + 7)(x - 4) = 0$

$2x + 7 = 0 \implies x = -\frac{7}{2}$

$x - 4 = 0 \implies x = 4$

111. $12x^2 + 11x - 5 = 0$

$(3x - 1)(4x + 5) = 0$

$3x - 1 = 0 \implies x = \frac{1}{3}$

$4x + 5 = 0 \implies x = -\frac{5}{4}$

113. $x^2 - 2x - 21 = 0$

$(x^2 - 2x + (-1)^2) - 21 - 1 = 0$

$(x - 1)^2 - 22 = 0$

$(x - 1)^2 = 22$

$x - 1 = \pm\sqrt{22}$

$x = 1 \pm \sqrt{22}$

115. $2x^2 + 5x - 20 = 0$

$2\left(x^2 + \frac{5}{2}x\right) - 20 = 0$

$2\left(x^2 + \frac{5}{2}x + \left(\frac{5}{4}\right)^2\right) - 20 - \frac{25}{8} = 0$

$2\left(x + \frac{5}{4}\right)^2 - \frac{185}{8} = 0$

$\left(x + \frac{5}{4}\right)^2 = \frac{185}{16}$

$x + \frac{5}{4} = \pm\frac{\sqrt{185}}{4}$

$x = \frac{-5 \pm \sqrt{185}}{4}$

117. $f(x) = (x + 4)^2$

Common function: $y = x^2$

Transformation: Horizontal shift four units to the left

119. $f(x) = \sqrt{x + 1} - 5$

Common function: $y = \sqrt{x}$

Transformation: Horizontal shift one unit to the left and a vertical shift five units downward.

121. $f(x) = 2[\![x]\!] + 9$

Common function: $y = [\![x]\!]$

Transformation: Vertical stretch (each y-value is multiplied by 2), then a vertical shift nine units upward

Section 3.3 Polynomial and Synthetic Division

You should know the following basic techniques and principles of polynomial division.

■ The Division Algorithm (Long Division of Polynomials)

■ Synthetic Division

■ $f(k)$ is equal to the remainder of $f(x)$ divided by $(x - k)$. (The Remainder Theorem)

■ $f(k) = 0$ if and only if $(x - k)$ is a factor of $f(x)$.

Vocabulary Check

1. $f(x)$ is the dividend; $d(x)$ is the divisor; $g(x)$ is the quotient; $r(x)$ is the remainder

2. improper; proper   **3.** synthetic division   **4.** factor   **5.** remainder

---

1. $y_1 = \dfrac{x^2}{x + 2}$ and $y_2 = x - 2 + \dfrac{4}{x + 2}$

$$
\begin{array}{r}
x - 2 \\
x + 2 \overline{\smash{)}\, x^2 + 0x + 0} \\
\underline{x^2 + 2x} \\
-2x + 0 \\
\underline{-2x - 4} \\
4
\end{array}
$$

Thus, $\dfrac{x^2}{x + 2} = x - 2 + \dfrac{4}{x + 2}$ and $y_1 = y_2$.

3. $y_1 = \dfrac{x^5 - 3x^3}{x^2 + 1}$ and $y_2 = x^3 - 4x + \dfrac{4x}{x^2 + 1}$

(a) and (b)

(c)
$$
\begin{array}{r}
x^3 - 4x \\
x^2 + 0x + 1 \overline{\smash{)}\, x^5 + 0x^4 - 3x^3 + 0x^2 + 0x + 0} \\
\underline{x^5 + 0x^4 + \; x^3} \\
-4x^3 + 0x^2 + 0x \\
\underline{-4x^3 + 0x^2 - 4x} \\
4x + 0
\end{array}
$$

Thus, $\dfrac{x^5 - 3x^3}{x^2 + 1} = x^3 - 4x + \dfrac{4x}{x^2 + 1}$ and $y_1 = y_2$.

5.
$$
\begin{array}{r}
2x + 4 \\
x + 3 \overline{\smash{)}\, 2x^2 + 10x + 12} \\
\underline{2x^2 + \; 6x} \\
4x + 12 \\
\underline{4x + 12} \\
0
\end{array}
$$

$$\dfrac{2x^2 + 10x + 12}{x + 3} = 2x + 4$$

7.
$$
\begin{array}{r}
x^2 - \; 3x + 1 \\
4x + 5 \overline{\smash{)}\, 4x^3 - \; 7x^2 - 11x + 5} \\
\underline{4x^3 + \; 5x^2} \\
-12x^2 - 11x \\
\underline{-12x^2 - 15x} \\
4x + 5 \\
\underline{4x + 5} \\
0
\end{array}
$$

$$\dfrac{4x^3 - 7x^2 - 11x + 5}{4x + 5} = x^2 - 3x + 1$$

9.
$$
\begin{array}{r}
x^3 + 3x^2 \qquad\; - 1 \\
x + 2 \overline{\smash{)}\, x^4 + 5x^3 + 6x^2 - x - 2} \\
\underline{x^4 + 2x^3} \\
3x^3 + 6x^2 \\
\underline{3x^3 + 6x^2} \\
-x - 2 \\
\underline{-x - 2} \\
0
\end{array}
$$

$$\dfrac{x^4 + 5x^3 + 6x^2 - x - 2}{x + 2} = x^3 + 3x^2 - 1$$

11.
$$
\begin{array}{r}
7 \\
x + 2 \overline{\smash{)}\, 7x + \; 3} \\
\underline{7x + 14} \\
-11
\end{array}
$$

$$\dfrac{7x + 3}{x + 2} = 7 - \dfrac{11}{x + 2}$$

13.

$$
\begin{array}{r}
3x + 5 \\
2x^2 + 0x + 1\overline{\smash{\big)}6x^3 + 10x^2 + x + 8} \\
\underline{6x^3 + 0x^2 + 3x} \\
10x^2 - 2x + 8 \\
\underline{10x^2 + 0x + 5} \\
-2x + 3
\end{array}
$$

$$\frac{6x^3 + 10x^2 + x + 8}{2x^2 + 1} = 3x + 5 - \frac{2x - 3}{2x^2 + 1}$$

15.

$$
\begin{array}{r}
x^2 + 2x + 4 \\
x^2 - 2x + 3\overline{\smash{\big)}x^4 + 0x^3 + 3x^2 + 0x + 1} \\
\underline{x^4 - 2x^3 + 3x^2} \\
2x^3 + 0x^2 + 0x \\
\underline{2x^3 - 4x^2 + 6x} \\
4x^2 - 6x + 1 \\
\underline{4x^2 - 8x + 12} \\
2x - 11
\end{array}
$$

$$\Rightarrow \quad \frac{x^4 + 3x^2 + 1}{x^2 - 2x + 3} = x^2 + 2x + 4 + \frac{2x - 11}{x^2 - 2x + 3}$$

17.

$$
\begin{array}{r}
x + 3 \\
x^3 - 3x^2 + 3x - 1\overline{\smash{\big)}x^4 + 0x^3 + 0x^2 + 0x + 0} \\
\underline{x^4 - 3x^3 + 3x^2 - x} \\
3x^3 - 3x^2 + x + 0 \\
\underline{3x^3 - 9x^2 + 9x - 3} \\
6x^2 - 8x + 3
\end{array}
$$

$$\frac{x^4}{(x - 1)^3} = x + 3 + \frac{6x^2 - 8x + 3}{(x - 1)^3}$$

19.

$$
\begin{array}{r|rrrr}
5 & 3 & -17 & 15 & -25 \\
  &   & 15 & -10 & 25 \\
\hline
  & 3 & -2 & 5 & 0
\end{array}
$$

$$\frac{3x^3 - 17x^2 + 15x - 25}{x - 5} = 3x^2 - 2x + 5$$

21.

$$
\begin{array}{r|rrrr}
-2 & 4 & 8 & -9 & -18 \\
   &   & -8 & 0 & 18 \\
\hline
   & 4 & 0 & -9 & 0
\end{array}
$$

$$\frac{4x^3 + 8x^2 - 9x - 18}{x + 2} = 4x^2 - 9$$

23.

$$
\begin{array}{r|rrrr}
-10 & -1 & 0 & 75 & -250 \\
    &   & 10 & -100 & 250 \\
\hline
    & -1 & 10 & -25 & 0
\end{array}
$$

$$\frac{-x^3 + 75x - 250}{x + 10} = -x^2 + 10x - 25$$

25.

$$
\begin{array}{r|rrrr}
4 & 5 & -6 & 0 & 8 \\
  &   & 20 & 56 & 224 \\
\hline
  & 5 & 14 & 56 & 232
\end{array}
$$

$$\frac{5x^3 - 6x^2 + 8}{x - 4} = 5x^2 + 14x + 56 + \frac{232}{x - 4}$$

27.

$$
\begin{array}{r|rrrrr}
6 & 10 & -50 & 0 & 0 & -800 \\
  &   & 60 & 60 & 360 & 2160 \\
\hline
  & 10 & 10 & 60 & 360 & 1360
\end{array}
$$

$$\frac{10x^4 - 50x^3 - 800}{x - 6} = 10x^3 + 10x^2 + 60x + 360 + \frac{1360}{x - 6}$$

29.

$$
\begin{array}{r|rrrr}
-8 & 1 & 0 & 0 & 512 \\
   &   & -8 & 64 & -512 \\
\hline
   & 1 & -8 & 64 & 0
\end{array}
$$

$$\frac{x^3 + 512}{x + 8} = x^2 - 8x + 64$$

31.

$$
\begin{array}{r|rrrrr}
2 & -3 & 0 & 0 & 0 & 0 \\
  &   & -6 & -12 & -24 & -48 \\
\hline
  & -3 & -6 & -12 & -24 & -48
\end{array}
$$

$$\frac{-3x^4}{x - 2} = -3x^3 - 6x^2 - 12x - 24 - \frac{48}{x - 2}$$

33.

$$
\begin{array}{r|rrrrr}
6 & -1 & 0 & 0 & 180 & 0 \\
  &   & -6 & -36 & -216 & -216 \\
\hline
  & -1 & -6 & -36 & -36 & -216
\end{array}
$$

$$\frac{180x - x^4}{x - 6} = -x^3 - 6x^2 - 36x - 36 - \frac{216}{x - 6}$$

35.

$$
\begin{array}{r|rrrr}
-\frac{1}{2} & 4 & 16 & -23 & -15 \\
 &   & -2 & -7 & 15 \\
\hline
 & 4 & 14 & -30 & 0
\end{array}
$$

$$\frac{4x^3 + 16x^2 - 23x - 15}{x + (1/2)} = 4x^2 + 14x - 30$$

37. $f(x) = x^3 - x^2 - 14x + 11,\ k = 4$

$$
\begin{array}{r|rrrr}
4 & 1 & -1 & -14 & 11 \\
  &   & 4 & 12 & -8 \\
\hline
  & 1 & 3 & -2 & 3
\end{array}
$$

$f(x) = (x - 4)(x^2 + 3x - 2) + 3$

$f(4) = 4^3 - 4^2 - 14(4) + 11 = 3$

39. $f(x) = 15x^4 + 10x^3 - 6x^2 + 14,\ k = -\frac{2}{3}$

$$
\begin{array}{r|rrrrr}
-\frac{2}{3} & 15 & 10 & -6 & 0 & 14 \\
 &   & -10 & 0 & 4 & -\frac{8}{3} \\
\hline
 & 15 & 0 & -6 & 4 & \frac{34}{3}
\end{array}
$$

$f(x) = \left(x + \frac{2}{3}\right)(15x^3 - 6x + 4) + \frac{34}{3}$

$f\left(-\frac{2}{3}\right) = 15\left(-\frac{2}{3}\right)^4 + 10\left(-\frac{2}{3}\right)^3 - 6\left(-\frac{2}{3}\right)^2 + 14 = \frac{34}{3}$

41. $f(x) = x^3 + 3x^2 - 2x - 14,\ k = \sqrt{2}$

$$
\begin{array}{r|rrrr}
\sqrt{2} & 1 & 3 & -2 & -14 \\
 &   & \sqrt{2} & 2 + 3\sqrt{2} & 6 \\
\hline
 & 1 & 3 + \sqrt{2} & 3\sqrt{2} & -8
\end{array}
$$

$f(x) = \left(x - \sqrt{2}\right)\left[x^2 + \left(3 + \sqrt{2}\right)x + 3\sqrt{2}\right] - 8$

$f\left(\sqrt{2}\right) = \left(\sqrt{2}\right)^3 + 3\left(\sqrt{2}\right)^2 - 2\sqrt{2} - 14 = -8$

43. $f(x) = -4x^3 + 6x^2 + 12x + 4,\ k = 1 - \sqrt{3}$

$$
\begin{array}{r|rrrr}
1 - \sqrt{3} & -4 & 6 & 12 & 4 \\
 &   & -4 + 4\sqrt{3} & -10 + 2\sqrt{3} & -4 \\
\hline
 & -4 & 2 + 4\sqrt{3} & 2 + 2\sqrt{3} & 0
\end{array}
$$

$f(x) = \left[x - \left(1 - \sqrt{3}\right)\right]\left[-4x^2 + \left(2 + 4\sqrt{3}\right)x + \left(2 + 2\sqrt{3}\right)\right] + 0$

$f\left(1 - \sqrt{3}\right) = -4\left(1 - \sqrt{3}\right)^3 + 6\left(1 - \sqrt{3}\right)^2 + 12\left(1 - \sqrt{3}\right) + 4 = 0$

45. $f(x) = 4x^3 - 13x + 10$

(a)
$$
\begin{array}{r|rrrr}
1 & 4 & 0 & -13 & 10 \\
  &   & 4 & 4 & -9 \\
\hline
  & 4 & 4 & -9 & 1
\end{array}
$$
$f(1) = 1$

(b)
$$
\begin{array}{r|rrrr}
-2 & 4 & 0 & -13 & 10 \\
   &   & -8 & 16 & -6 \\
\hline
   & 4 & -8 & 3 & 4
\end{array}
$$
$f(-2) = 4$

(c)
$$
\begin{array}{r|rrrr}
\frac{1}{2} & 4 & 0 & -13 & 10 \\
 &   & 2 & 1 & -6 \\
\hline
 & 4 & 2 & -12 & 4
\end{array}
$$
$f\left(\frac{1}{2}\right) = 4$

(d)
$$
\begin{array}{r|rrrr}
8 & 4 & 0 & -13 & 10 \\
  &   & 32 & 256 & 1944 \\
\hline
  & 4 & 32 & 243 & 1954
\end{array}
$$
$f(8) = 1954$

47. $h(x) = 3x^3 + 5x^2 - 10x + 1$

(a)
$$
\begin{array}{r|rrrr}
3 & 3 & 5 & -10 & 1 \\
  &   & 9 & 42 & 96 \\
\hline
  & 3 & 14 & 32 & 97
\end{array}
$$
$h(3) = 97$

(b)
$$
\begin{array}{r|rrrr}
\frac{1}{3} & 3 & 5 & -10 & 1 \\
 &   & 1 & 2 & -\frac{8}{3} \\
\hline
 & 3 & 6 & -8 & -\frac{5}{3}
\end{array}
$$
$h\left(\frac{1}{3}\right) = -\frac{5}{3}$

(c)
$$
\begin{array}{r|rrrr}
-2 & 3 & 5 & -10 & 1 \\
   &   & -6 & 2 & 16 \\
\hline
   & 3 & -1 & -8 & 17
\end{array}
$$
$h(-2) = 17$

(d)
$$
\begin{array}{r|rrrr}
-5 & 3 & 5 & -10 & 1 \\
   &   & -15 & 50 & -200 \\
\hline
   & 3 & -10 & 40 & -199
\end{array}
$$
$h(-5) = -199$

49.
$$2 \,\rvert\, \begin{array}{rrrr} 1 & 0 & -7 & 6 \\ & 2 & 4 & -6 \\ \hline 1 & 2 & -3 & 0 \end{array}$$

$$x^3 - 7x + 6 = (x - 2)(x^2 + 2x - 3)$$
$$= (x - 2)(x + 3)(x - 1)$$

Zeros: $2, -3, 1$

51.
$$\tfrac{1}{2} \,\rvert\, \begin{array}{rrrr} 2 & -15 & 27 & -10 \\ & 1 & -7 & 10 \\ \hline 2 & -14 & 20 & 0 \end{array}$$

$$2x^3 - 15x^2 + 27x - 10$$
$$= \left(x - \tfrac{1}{2}\right)(2x^2 - 14x + 20)$$
$$= (2x - 1)(x - 2)(x - 5)$$

Zeros: $\tfrac{1}{2}, 2, 5$

53.
$$\sqrt{3} \,\rvert\, \begin{array}{rrrr} 1 & 2 & -3 & -6 \\ & \sqrt{3} & 3 + 2\sqrt{3} & 6 \\ \hline 1 & 2 + \sqrt{3} & 2\sqrt{3} & 0 \end{array}$$

$$-\sqrt{3} \,\rvert\, \begin{array}{rrr} 1 & 2 + \sqrt{3} & 2\sqrt{3} \\ & -\sqrt{3} & -2\sqrt{3} \\ \hline 1 & 2 & 0 \end{array}$$

$$x^3 + 2x^2 - 3x - 6 = (x - \sqrt{3})(x + \sqrt{3})(x + 2)$$

Zeros: $\pm\sqrt{3}, -2$

55.
$$1 + \sqrt{3} \,\rvert\, \begin{array}{rrrr} 1 & -3 & 0 & 2 \\ & 1 + \sqrt{3} & 1 - \sqrt{3} & -2 \\ \hline 1 & -2 + \sqrt{3} & 1 - \sqrt{3} & 0 \end{array}$$

$$1 - \sqrt{3} \,\rvert\, \begin{array}{rrr} 1 & -2 + \sqrt{3} & 1 - \sqrt{3} \\ & 1 - \sqrt{3} & -1 + \sqrt{3} \\ \hline 1 & -1 & 0 \end{array}$$

$$x^3 - 3x^2 + 2 = \left[x - \left(1 + \sqrt{3}\right)\right]\left[x - \left(1 - \sqrt{3}\right)\right](x - 1)$$
$$= (x - 1)(x - 1 - \sqrt{3})(x - 1 + \sqrt{3})$$

Zeros: $1, 1 \pm \sqrt{3}$

57. $f(x) = 2x^3 + x^2 - 5x + 2$; Factors: $(x + 2), (x - 1)$

(a)
$$-2 \,\rvert\, \begin{array}{rrrr} 2 & 1 & -5 & 2 \\ & -4 & 6 & -2 \\ \hline 2 & -3 & 1 & 0 \end{array}$$

$$1 \,\rvert\, \begin{array}{rrr} 2 & -3 & 1 \\ & 2 & -1 \\ \hline 2 & -1 & 0 \end{array}$$

Both are factors of $f(x)$ since the remainders are zero.

(b) The remaining factor of $f(x)$ is $(2x - 1)$.

(c) $f(x) = (2x - 1)(x + 2)(x - 1)$

(d) Zeros: $\tfrac{1}{2}, -2, 1$

(e)

59. $f(x) = x^4 - 4x^3 - 15x^2 + 58x - 40$; Factors: $(x - 5), (x + 4)$

(a)
$$5 \,\rvert\, \begin{array}{rrrrr} 1 & -4 & -15 & 58 & -40 \\ & 5 & 5 & -50 & 40 \\ \hline 1 & 1 & -10 & 8 & 0 \end{array}$$

$$-4 \,\rvert\, \begin{array}{rrrr} 1 & 1 & -10 & 8 \\ & -4 & 12 & -8 \\ \hline 1 & -3 & 2 & 0 \end{array}$$

Both are factors of $f(x)$ since the remainders are zero.

(b) $x^2 - 3x + 2 = (x - 1)(x - 2)$

The remaining factors are $(x - 1)$ and $(x - 2)$.

(c) $f(x) = (x - 1)(x - 2)(x - 5)(x + 4)$

(d) Zeros: $1, 2, 5, -4$

(e)

61. $f(x) = 6x^3 + 41x^2 - 9x - 14;$

Factors: $(2x + 1), (3x - 2)$

(a) $-\frac{1}{2}$ |
$$
\begin{array}{r r r r}
6 & 41 & -9 & -14 \\
 & -3 & -19 & 14 \\
\hline
6 & 38 & -28 & 0
\end{array}
$$

$\frac{2}{3}$ |
$$
\begin{array}{r r r}
6 & 38 & -28 \\
 & 4 & 28 \\
\hline
6 & 42 & 0
\end{array}
$$

Both are factors since the remainders are zero.

(b) $6x + 42 = 6(x + 7)$

This shows that $\dfrac{f(x)}{\left(x + \frac{1}{2}\right)\left(x - \frac{2}{3}\right)} = 6(x + 7),$

so $\dfrac{f(x)}{(2x + 1)(3x - 2)} = x + 7.$

The remaining factor is $(x + 7)$.

(c) $f(x) = (x + 7)(2x + 1)(3x - 2)$

(d) Zeros: $-7, -\dfrac{1}{2}, \dfrac{2}{3}$

(e)

63. $f(x) = 2x^3 - x^2 - 10x + 5;$

Factors: $(2x - 1), \left(x + \sqrt{5}\right)$

(a) $\frac{1}{2}$ |
$$
\begin{array}{r r r r}
2 & -1 & -10 & 5 \\
 & 1 & 0 & -5 \\
\hline
2 & 0 & -10 & 0
\end{array}
$$

$-\sqrt{5}$ |
$$
\begin{array}{r r r}
2 & 0 & -10 \\
 & -2\sqrt{5} & 10 \\
\hline
2 & -2\sqrt{5} & 0
\end{array}
$$

Both are factors since the remainders are zero.

(b) $2x - 2\sqrt{5} = 2\left(x - \sqrt{5}\right)$

This shows that $\dfrac{f(x)}{\left(x - \frac{1}{2}\right)\left(x + \sqrt{5}\right)} = 2\left(x - \sqrt{5}\right),$

so $\dfrac{f(x)}{(2x - 1)\left(x + \sqrt{5}\right)} = x - \sqrt{5}.$

The remaining factor is $\left(x - \sqrt{5}\right)$.

(c) $f(x) = \left(x + \sqrt{5}\right)\left(x - \sqrt{5}\right)(2x - 1)$

(d) Zeros: $-\sqrt{5}, \sqrt{5}, \dfrac{1}{2}$

(e)

65. $f(x) = x^3 - 2x^2 - 5x + 10$

(a) The zeros of f are 2 and $\approx \pm 2.236$.

(b) An exact zero is $x = 2$.

(c) 2 |
$$
\begin{array}{r r r r}
1 & -2 & -5 & 10 \\
 & 2 & 0 & -10 \\
\hline
1 & 0 & -5 & 0
\end{array}
$$

$f(x) = (x - 2)(x^2 - 5)$

$\quad = (x - 2)\left(x - \sqrt{5}\right)\left(x + \sqrt{5}\right)$

67. $h(t) = t^3 - 2t^2 - 7t + 2$

(a) The zeros of h are $t = -2, t \approx 3.732, t \approx 0.268.$

(b) An exact zero is $t = -2$.

(c) -2 |
$$
\begin{array}{r r r r}
1 & -2 & -7 & 2 \\
 & -2 & 8 & -2 \\
\hline
1 & -4 & 1 & 0
\end{array}
$$

$h(t) = (t + 2)(t^2 - 4t + 1)$

By the Quadratic Formula, the zeros of $t^2 - 4t + 1$ are $2 \pm \sqrt{3}$. Thus,

$h(t) = (t + 2)\left[t - \left(2 + \sqrt{3}\right)\right]\left[t - \left(2 - \sqrt{3}\right)\right]$

$\quad = (t + 2)\left(t - 2 - \sqrt{3}\right)\left(t - 2 + \sqrt{3}\right).$

69. $\dfrac{4x^3 - 8x^2 + x + 3}{2x - 3}$

$\frac{3}{2}$ |
$$
\begin{array}{r r r r}
4 & -8 & 1 & 3 \\
 & 6 & -3 & -3 \\
\hline
4 & -2 & -2 & 0
\end{array}
$$

$\dfrac{4x^3 - 8x^2 + x + 3}{x - \frac{3}{2}} = 4x^2 - 2x - 2 = 2(2x^2 - x - 1)$

Thus, $\dfrac{4x^3 - 8x^2 + x + 3}{2x - 3} = 2x^2 - x - 1, x \neq \dfrac{3}{2}.$

71. $\dfrac{x^4 + 6x^3 + 11x^2 + 6x}{x^2 + 3x + 2} = \dfrac{x^4 + 6x^3 + 11x^2 + 6x}{(x + 1)(x + 2)}$

-1 |
$$
\begin{array}{r r r r r}
1 & 6 & 11 & 6 & 0 \\
 & -1 & -5 & -6 & 0 \\
\hline
1 & 5 & 6 & 0 & 0
\end{array}
$$

-2 |
$$
\begin{array}{r r r r}
1 & 5 & 6 & 0 \\
 & -2 & -6 & 0 \\
\hline
1 & 3 & 0 & 0
\end{array}
$$

$\dfrac{x^4 + 6x^3 + 11x^2 + 6x}{(x + 1)(x + 2)} = x^2 + 3x, x \neq -2, -1$

73. (a) and (b)

(c) $M \approx -0.242t^3 + 12.43t^2 - 173.4t + 2118$

Year, t	Military Personnel	M
3	1705	1703
4	1611	1608
5	1518	1532
6	1472	1473
7	1439	1430
8	1407	1402
9	1386	1388
10	1384	1385
11	1385	1393
12	1412	1409
13	1434	1433

The model is a good fit to the actual data.

(d)

$$\begin{array}{r|rrrr} 18 & -0.242 & 12.43 & -173.4 & 2118 \\ & & -4.356 & 145.332 & -505.224 \\ \hline & -0.242 & 8.074 & -28.068 & 1612.776 \end{array}$$

$M(18) \approx 1613$ thousand

No, this model should not be used to predict the number of military personnel in the future. It predicts an increase in military personnel until 2024 and then it decreases and will approach negative infinity quickly.

75. False. If $(7x + 4)$ is a factor of f, then $-\frac{4}{7}$ is a zero of f.

77. True. The degree of the numerator is greater than the degree of the denominator.

79.

$$\begin{array}{r} x^{2n} + 6x^n + 9 \\ x^n + 3 \overline{\smash{\big)}\ x^{3n} + 9x^{2n} + 27x^n + 27} \\ \underline{x^{3n} + 3x^{2n}} \\ 6x^{2n} + 27x^n \\ \underline{6x^{2n} + 18x^n} \\ 9x^n + 27 \\ \underline{9x^n + 27} \\ 0 \end{array}$$

$$\frac{x^{3n} + 9x^{2n} + 27x^n + 27}{x^n + 3} = x^{2n} + 6x^n + 9$$

81. A divisor divides evenly into a dividend if the remainder is zero.

83.

$$\begin{array}{r|rrrr} 5 & 1 & 4 & -3 & c \\ & & 5 & 45 & 210 \\ \hline & 1 & 9 & 42 & c + 210 \end{array}$$

To divide evenly, $c + 210$ must equal zero. Thus, c must equal -210.

85. $f(x) = (x + 3)^2(x - 3)(x + 1)^3$

The remainder when $k = -3$ is zero since $(x + 3)$ is a factor of $f(x)$.

87.
$$9x^2 - 25 = 0$$
$$(3x - 5)(3x + 5) = 0$$
$$3x - 5 = 0 \implies x = \tfrac{5}{3}$$
$$3x + 5 = 0 \implies x = -\tfrac{5}{3}$$

89. $5x^2 - 3x - 14 = 0$
$$(5x + 7)(x - 2) = 0$$
$$5x + 7 = 0 \implies x = -\tfrac{7}{5}$$
$$x - 2 = 0 \implies x = 2$$

91. $2x^2 + 6x + 3 = 0$
$$x = \frac{-b \pm \sqrt{b^2 - 4ac}}{2a} = \frac{-6 \pm \sqrt{6^2 - 4(2)(3)}}{2(2)} = \frac{-6 \pm \sqrt{12}}{4}$$
$$= \frac{-3 \pm \sqrt{3}}{2}$$

93. $f(x) = (x - 0)(x - 3)(x - 4)$
$$= x(x - 3)(x - 4) = x(x^2 - 7x + 12)$$
$$= x^3 - 7x^2 + 12x$$

Note: Any nonzero scalar multiple of $f(x)$ would also have these zeros.

95. $f(x) = [x - (-3)]\big[x - (1 + \sqrt{2})\big]\big[x - (1 - \sqrt{2})\big]$
$$= (x + 3)\big[(x - 1) - \sqrt{2}\big]\big[(x - 1) + \sqrt{2}\big]$$
$$= (x + 3)\big[(x - 1)^2 - (\sqrt{2})^2\big]$$
$$= (x + 3)(x^2 - 2x - 1)$$
$$= x^3 + x^2 - 7x - 3$$

Note: Any nonzero scalar multiple of $f(x)$ would also have these zeros.

Section 3.4   Zeros of Polynomial Functions

- You should know that if f is a polynomial of degree $n > 0$, then f has at least one zero in the complex number system.
- You should know the Linear Factorization Theorem.
- You should know the Rational Zero Test.
- You should know shortcuts for the Rational Zero Test. Possible rational zeros $= \dfrac{\text{factors of constant term}}{\text{factors of leading coefficient}}$
  - (a) Use a graphing or programmable calculator.
  - (b) Sketch a graph.
  - (c) After finding a root, use synthetic division to reduce the degree of the polynomial.
- You should know that if $a + bi$ is a complex zero of a polynomial f, with real coefficients, then $a - bi$ is also a complex zero of f.
- You should know the difference between a factor that is irreducible over the rationals (such as $x^2 - 7$) and a factor that is irreducible over the reals (such as $x^2 + 9$).
- You should know Descartes's Rule of Signs. (For a polynomial with real coefficients and a non-zero constant term.)
  - (a) The number of positive real zeros of f is either equal to the number of variations of sign of f or is less than that number by an even integer.
  - (b) The number of negative real zeros of f is either equal to the number of variations in sign of $f(-x)$ or is less than that number by an even integer.
  - (c) When there is only one variation in sign, there is exactly one positive (or negative) real zero.
- You should be able to observe the last row obtained from synthetic division in order to determine upper or lower bounds.
  - (a) If the test value is positive and all of the entries in the last row are positive or zero, then the test value is an upper bound.
  - (b) If the test value is negative and the entries in the last row alternate from positive to negative, then the test value is a lower bound. (Zero entries count as positive or negative.)

Vocabulary Check

1. Fundamental Theorem of Algebra
2. Linear Factorization Theme
3. Rational Zero
4. conjugate
5. irreducible; reals
6. Descarte's Rule of Signs
7. lower; upper

---

1. $f(x) = x(x - 6)^2$

The zeros are: $x = 0, x = 6$.

3. $g(x) = (x - 2)(x + 4)^3$

The zeros are: $x = 2, x = -4$.

5. $f(x) = (x + 6)(x + i)(x - i)$

The three zeros are:

$x = -6, x = -i, x = i$.

7. $f(x) = x^3 + 3x^2 - x - 3$

Possible rational zeros: $\pm 1, \pm 3$

Zeros shown on graph: $-3, -1, 1$

9. $f(x) = 2x^4 - 17x^3 + 35x^2 + 9x - 45$

Possible rational zeros: $\pm 1, \pm 3, \pm 5, \pm 9, \pm 15, \pm 45,$
$\pm \frac{1}{2}, \pm \frac{3}{2}, \pm \frac{5}{2}, \pm \frac{9}{2}, \pm \frac{15}{2}, \pm \frac{45}{2}$

Zeros shown on graph: $-1, \frac{3}{2}, 3, 5$

11. $f(x) = x^3 - 6x^2 + 11x - 6$

Possible rational zeros: $\pm 1, \pm 2, \pm 3, \pm 6$

$$
\begin{array}{r|rrrr}
1 & 1 & -6 & 11 & -6 \\
  &   & 1 & -5 & 6 \\
\hline
  & 1 & -5 & 6 & 0
\end{array}
$$

$x^3 - 6x^2 + 11x - 6 = (x - 1)(x^2 - 5x + 6)$

$\qquad\qquad\qquad\quad = (x - 1)(x - 2)(x - 3)$

Thus, the rational zeros are 1, 2, and 3.

13. $g(x) = x^3 - 4x^2 - x + 4 = x^2(x - 4) - 1(x - 4)$

$\qquad = (x - 4)(x^2 - 1)$

$\qquad = (x - 4)(x - 1)(x + 1)$

Thus, the rational zeros of $g(x)$ are 4 and ± 1.

15. $h(t) = t^3 + 12t^2 + 21t + 10$

Possible rational zeros: $\pm 1, \pm 2, \pm 5, \pm 10$

$$
\begin{array}{r|rrrr}
-1 & 1 & 12 & 21 & 10 \\
   &   & -1 & -11 & -10 \\
\hline
   & 1 & 11 & 10 & 0
\end{array}
$$

$t^3 + 12t^2 + 21t + 10 = (t + 1)(t^2 + 11t + 10)$

$\qquad\qquad\qquad\qquad = (t + 1)(t + 1)(t + 10)$

$\qquad\qquad\qquad\qquad = (t + 1)^2(t + 10)$

Thus, the rational zeros are -1 and -10.

17. $C(x) = 2x^3 + 3x^2 - 1$

Possible rational zeros: $\pm 1, \pm \frac{1}{2}$

$$
\begin{array}{r|rrrr}
-1 & 2 & 3 & 0 & -1 \\
   &   & -2 & -1 & 1 \\
\hline
   & 2 & 1 & -1 & 0
\end{array}
$$

$2x^3 + 3x^2 - 1 = (x + 1)(2x^2 + x - 1)$

$\qquad\qquad\qquad = (x + 1)(x + 1)(2x - 1)$

$\qquad\qquad\qquad = (x + 1)^2(2x - 1)$

Thus, the rational zeros are -1 and $\frac{1}{2}$.

19. $f(x) = 9x^4 - 9x^3 - 58x^2 + 4x + 24$

Possible rational zeros: $\pm 1, \pm 2, \pm 3, \pm 4, \pm 6, \pm 8, \pm 12, \pm 24,$
$\pm\frac{1}{3}, \pm\frac{2}{3}, \pm\frac{4}{3}, \pm\frac{8}{3}, \pm\frac{1}{9}, \pm\frac{2}{9}, \pm\frac{4}{9}, \pm\frac{8}{9}$

$$
\begin{array}{r|rrrrr}
-2 & 9 & -9 & -58 & 4 & 24 \\
   &   & -18 & 54 & 8 & -24 \\
\hline
   & 9 & -27 & -4 & 12 & 0 \\
\end{array}
$$

$$
\begin{array}{r|rrrr}
3 & 9 & -27 & -4 & 12 \\
  &   & 27 & 0 & -12 \\
\hline
  & 9 & 0 & -4 & 0 \\
\end{array}
$$

$9x^4 - 9x^3 - 58x^2 + 4x - 24$

$\qquad = (x + 2)(x - 3)(9x^2 - 4)$

$\qquad = (x + 2)(x - 3)(3x - 2)(3x + 2)$

Thus, the rational zeros are -2, 3, and $\pm\frac{2}{3}$.

21. $z^4 - z^3 - 2z - 4 = 0$

Possible rational zeros: $\pm 1, \pm 2, \pm 4$

$$
\begin{array}{r|rrrrr}
-1 & 1 & -1 & 0 & -2 & -4 \\
   &   & -1 & 2 & -2 & 4 \\
\hline
   & 1 & -2 & 2 & -4 & 0 \\
\end{array}
$$

$$
\begin{array}{r|rrrr}
2 & 1 & -2 & 2 & -4 \\
  &   & 2 & 0 & 4 \\
\hline
  & 1 & 0 & 2 & 0 \\
\end{array}
$$

$z^4 - z^3 - 2z - 4 = (z + 1)(z - 2)(z^2 + 2)$

The only real zeros are -1 and 2.

23. $2y^4 + 7y^3 - 26y^2 + 23y - 6 = 0$

Possible rational zeros: $\pm 1, \pm 2, \pm 3, \pm 6, \pm\frac{1}{2}, \pm\frac{3}{2}$

$$
\begin{array}{r|rrrrr}
1 & 2 & 7 & -26 & 23 & -6 \\
  &   & 2 & 9 & -17 & 6 \\
\hline
  & 2 & 9 & -17 & 6 & 0 \\
\end{array}
$$

$$
\begin{array}{r|rrrr}
-6 & 2 & 9 & -17 & 6 \\
   &   & -12 & 18 & -6 \\
\hline
   & 2 & -3 & 1 & 0 \\
\end{array}
$$

$2y^4 + 7y^3 - 26y^2 + 23y - 6 = (y - 1)(y + 6)(2y^2 - 3y + 1) = (y - 1)(y + 6)(2y - 1)(y - 1) = (y - 1)^2(y + 6)(2y - 1)$

The only real zeros are 1, -6, and $\frac{1}{2}$.

25. $f(x) = x^3 + x^2 - 4x - 4$

(a) Possible rational zeros: $\pm 1, \pm 2, \pm 4$

(b)

(c) The zeros are: $-2, -1, 2$

27. $f(x) = -4x^3 + 15x^2 - 8x - 3$

(a) Possible rational zeros: $\pm 1, \pm 3, \pm\frac{1}{2}, \pm\frac{3}{2}, \pm\frac{1}{4}, \pm\frac{3}{4}$

(b)

(c) The zeros are: $-\frac{1}{4}, 1, 3$

29. $f(x) = -2x^4 + 13x^3 - 21x^2 + 2x + 8$

(a) Possible rational zeros: $\pm 1, \pm 2, \pm 4, \pm 8, \pm\frac{1}{2}$

(b)

(c) The zeros are: $-\frac{1}{2}, 1, 2, 4$

31. $f(x) = 32x^3 - 52x^2 + 17x + 3$

(a) Possible rational zeros: $\pm 1, \pm 3, \pm\frac{1}{2}, \pm\frac{3}{2}, \pm\frac{1}{4}, \pm\frac{3}{4},$
$\pm\frac{1}{6}, \pm\frac{3}{8}, \pm\frac{1}{16}, \pm\frac{3}{16}, \pm\frac{1}{32}, \pm\frac{3}{32}$

(b)

(c) The zeros are: $-\frac{1}{8}, \frac{3}{4}, 1$

33. $f(x) = x^4 - 3x^2 + 2$

(a) From the calculator we have $x = \pm 1$ and $x \approx \pm 1.414$.

(b) An exact zero is $x = 1$.

$$
\begin{array}{r|rrrrr}
1 & 1 & 0 & -3 & 0 & 2 \\
  &   & 1 & 1 & -2 & -2 \\
\hline
  & 1 & 1 & -2 & -2 & 0
\end{array}
$$

(c)
$$
\begin{array}{r|rrrr}
-1 & 1 & 1 & -2 & -2 \\
   &   & -1 & 0 & 2 \\
\hline
   & 1 & 0 & -2 & 0
\end{array}
$$

$$
\begin{aligned}
f(x) &= (x - 1)(x + 1)(x^2 - 2) \\
     &= (x - 1)(x + 1)(x - \sqrt{2})(x + \sqrt{2})
\end{aligned}
$$

35. $h(x) = x^5 - 7x^4 + 10x^3 + 14x^2 - 24x$

(a) $h(x) = x(x^4 - 7x^3 + 10x^2 + 14x - 24)$

From the calculator we have $x = 0, 3, 4$ and $x \approx \pm 1.414$.

(b) An exact zero is $x = 3$.

$$
\begin{array}{r|rrrrr}
3 & 1 & -7 & 10 & 14 & -24 \\
  &   & 3 & -12 & -6 & 24 \\
\hline
  & 1 & -4 & -2 & 8 & 0
\end{array}
$$

(c)
$$
\begin{array}{r|rrrr}
4 & 1 & -4 & -2 & 8 \\
  &   & 4 & 0 & -8 \\
\hline
  & 1 & 0 & -2 & 0
\end{array}
$$

$$
\begin{aligned}
h(x) &= x(x - 3)(x - 4)(x^2 - 2) \\
     &= x(x - 3)(x - 4)(x - \sqrt{2})(x + \sqrt{2})
\end{aligned}
$$

37.
$$
\begin{aligned}
f(x) &= (x - 1)(x - 5i)(x + 5i) \\
     &= (x - 1)(x^2 + 25) \\
     &= x^3 - x^2 + 25x - 25
\end{aligned}
$$

Note: $f(x) = a(x^3 - x^2 + 25x - 25)$, where a is any nonzero real number, has the zeros 1 and $\pm 5i$.

39.
$$
\begin{aligned}
f(x) &= (x - 6)[x - (-5 + 2i)][x - (-5 - 2i)] \\
     &= (x - 6)[(x + 5) - 2i][(x + 5) + 2i] \\
     &= (x - 6)[(x + 5)^2 - (2i)^2] \\
     &= (x - 6)(x^2 + 10x + 25 + 4) \\
     &= (x - 6)(x^2 + 10x + 29) \\
     &= x^3 + 4x^2 - 31x - 174
\end{aligned}
$$

Note: $f(x) = a(x^3 + 4x^2 - 31x - 174)$, where a is any nonzero real number, has the zeros 6, and $-5 \pm 2i$.

41. If $3 + \sqrt{2}i$ is a zero, so is its conjugate, $3 - \sqrt{2}i$.

$$
\begin{aligned}
f(x) &= (3x - 2)(x + 1)[x - (3 + \sqrt{2}i)][x - (3 - \sqrt{2}i)] \\
     &= (3x - 2)(x + 1)[(x - 3) - \sqrt{2}i][(x - 3) + \sqrt{2}i] \\
     &= (3x^2 + x - 2)[(x - 3)^2 - (\sqrt{2}i)^2] \\
     &= (3x^2 + x - 2)(x^2 - 6x + 9 + 2) \\
     &= (3x^2 + x - 2)(x^2 - 6x + 11) \\
     &= 3x^4 - 17x^3 + 25x^2 + 23x - 22
\end{aligned}
$$

Note: $f(x) = a(3x^4 - 17x^3 + 25x^2 + 23x - 22)$, where a is any nonzero real number, has the zeros $\frac{2}{3}, -1$, and $3 \pm \sqrt{2}i$.

43. $f(x) = x^4 + 6x^2 - 27$

(a) $f(x) = (x^2 + 9)(x^2 - 3)$

(b) $f(x) = (x^2 + 9)(x + \sqrt{3})(x - \sqrt{3})$

(c) $f(x) = (x + 3i)(x - 3i)(x + \sqrt{3})(x - \sqrt{3})$

45. $f(x) = x^4 - 4x^3 + 5x^2 - 2x - 6$

$$
\begin{array}{r}
x^2 - 2x + 3 \\
x^2 - 2x - 2 \overline{)\, x^4 - 4x^3 + 5x^2 - 2x - 6} \\
\underline{x^4 - 2x^3 - 2x^2 } \\
-2x^3 + 7x^2 - 2x \\
\underline{-2x^3 + 4x^2 + 4x } \\
3x^2 - 6x - 6 \\
\underline{3x^2 - 6x - 6} \\
0
\end{array}
$$

$f(x) = (x^2 - 2x - 2)(x^2 - 2x + 3)$

(a) $f(x) = (x^2 - 2x - 2)(x^2 - 2x + 3)$

(b) $f(x) = (x - 1 + \sqrt{3})(x - 1 - \sqrt{3})(x^2 - 2x + 3)$

(c) $f(x) = (x - 1 + \sqrt{3})(x - 1 - \sqrt{3})(x - 1 + \sqrt{2}\,i)(x - 1 - \sqrt{2}\,i)$

Note: Use the Quadratic Formula for (b) and (c).

47. $f(x) = 2x^3 + 3x^2 + 50x + 75$

Since $5i$ is a zero, so is $-5i$.

$$
\begin{array}{r|rrrr}
5i & 2 & 3 & 50 & 75 \\
   &   & 10i & -50+15i & -75 \\
\hline
   & 2 & 3+10i & 15i & 0 \\
\end{array}
$$

$$
\begin{array}{r|rrr}
-5i & 2 & 3+10i & 15i \\
    &   & -10i & -15i \\
\hline
    & 2 & 3 & 0 \\
\end{array}
$$

The zero of $2x + 3$ is $x = -\frac{3}{2}$.

The zeros of $f(x)$ are $x = -\frac{3}{2}$ and $x = \pm 5i$.

Alternate Solution

Since $x = \pm 5i$ are zeros of $f(x)$, $(x + 5i)(x - 5i) = x^2 + 25$ is a factor of $f(x)$. By long division we have:

$$
\begin{array}{r}
2x + 3 \\
x^2 + 0x + 25 \overline{)2x^3 + 3x^2 + 50x + 75} \\
\underline{2x^3 + 0x^2 + 50x } \\
3x^2 + 0x + 75 \\
\underline{3x^2 + 0x + 75} \\
0
\end{array}
$$

Thus, $f(x) = (x^2 + 25)(2x + 3)$ and the zeros of f are $x = \pm 5i$ and $x = -\frac{3}{2}$.

49. $f(x) = 2x^4 - x^3 + 7x^2 - 4x - 4$

Since $2i$ is a zero, so is $-2i$.

$$
\begin{array}{r|rrrrr}
2i & 2 & -1 & 7 & -4 & -4 \\
   &   & 4i & -8-2i & 4-2i & 4 \\
\hline
   & 2 & -1+4i & -1-2i & -2i & 0 \\
\end{array}
$$

$$
\begin{array}{r|rrrr}
-2i & 2 & -1+4i & -1-2i & -2i \\
    &   & -4i & 2i & 2i \\
\hline
    & 2 & -1 & -1 & 0 \\
\end{array}
$$

The zeros of $2x^2 - x - 1 = (2x + 1)(x - 1)$ are $x = -\frac{1}{2}$ and $x = 1$.

The zeros of $f(x)$ are $x = \pm 2i$, $x = -\frac{1}{2}$, and $x = 1$.

Alternate Solution

Since $x = \pm 2i$ are zeros of $f(x)$, $(x + 2i)(x - 2i) = x^2 + 4$ is a factor of $f(x)$. By long division we have:

$$
\begin{array}{r}
2x^2 - x - 1 \\
x^2 + 0x + 4 \overline{)2x^4 - x^3 + 7x^2 - 4x - 4} \\
\underline{2x^4 + 0x^3 + 8x^2 } \\
-x^3 - x^2 - 4x \\
\underline{-x^3 + 0x^2 - 4x } \\
-x^2 + 0x - 4 \\
\underline{-x^2 + 0x - 4} \\
0
\end{array}
$$

Thus, $f(x) = (x^2 + 4)(2x^2 - x - 1)$

$$= (x + 2i)(x - 2i)(2x + 1)(x - 1)$$

and the zeros of $f(x)$ are $x = \pm 2i$, $x = -\frac{1}{2}$, and $x = 1$.

51. $g(x) = 4x^3 + 23x^2 + 34x - 10$

Since $-3 + i$ is a zero, so is $-3 - i$.

$$
\begin{array}{r|rrrr}
-3+i & 4 & 23 & 34 & -10 \\
     &   & -12+4i & -37-i & 10 \\
\hline
     & 4 & 11+4i & -3-i & 0 \\
\end{array}
$$

$$
\begin{array}{r|rrr}
-3-i & 4 & 11+4i & -3-i \\
     &   & -12-4i & 3+i \\
\hline
     & 4 & -1 & 0 \\
\end{array}
$$

The zero of $4x - 1$ is $x = \frac{1}{4}$. The zeros of $g(x)$ are $x = -3 \pm i$ and $x = \frac{1}{4}$.

Alternate Solution

Since $-3 \pm i$ are zeros of $g(x)$,
$$[x - (-3 + i)][x - (-3 - i)] = [(x + 3) - i][(x + 3) + i]$$
$$= (x + 3)^2 - i^2$$
$$= x^2 + 6x + 10$$
is a factor of $g(x)$. By long division we have:

$$
\begin{array}{r}
4x - 1 \\
x^2 + 6x + 10 \overline{)4x^3 + 23x^2 + 34x - 10} \\
\underline{4x^3 + 24x^2 + 40x } \\
-x^2 - 6x - 10 \\
\underline{-x^2 - 6x - 10} \\
0
\end{array}
$$

Thus, $g(x) = (x^2 + 6x + 10)(4x - 1)$ and the zeros of $g(x)$ are $x = -3 \pm i$ and $x = \frac{1}{4}$.

53. $f(x) = x^4 + 3x^3 - 5x^2 - 21x + 22$

Since $-3 + \sqrt{2}\,i$ is a zero, so is $-3 - \sqrt{2}\,i$, and

$$\left[x - \left(-3 + \sqrt{2}\,i\right)\right]\left[x - \left(-3 - \sqrt{2}\,i\right)\right] = \left[(x + 3) - \sqrt{2}\,i\right]\left[(x + 3) + \sqrt{2}\,i\right]$$

$$= (x + 3)^2 - \left(\sqrt{2}\,i\right)^2$$

$$= x^2 + 6x + 11$$

is a factor of $f(x)$. By long division, we have:

$$
\begin{array}{r}
x^2 - 3x + 2 \\
x^2 + 6x + 11 \overline{\smash{\big)}\ x^4 + 3x^3 - 5x^2 - 21x + 22} \\
\underline{x^4 + 6x^3 + 11x^2} \\
-3x^3 - 16x^2 - 21x \\
\underline{-3x^3 - 18x^2 - 33x} \\
2x^2 + 12x + 22 \\
\underline{2x^2 + 12x + 22} \\
0
\end{array}
$$

Thus,

$$f(x) = (x^2 + 6x + 11)(x^2 - 3x + 2)$$

$$= (x^2 + 6x + 11)(x - 1)(x - 2)$$

and the zeros of f are $x = -3 \pm \sqrt{2}\,i$, $x = 1$, and $x = 2$.

55. $f(x) = x^2 + 25$

$$= (x + 5i)(x - 5i)$$

The zeros of $f(x)$ are $x = \pm 5i$.

57. $h(x) = x^2 - 4x + 1$

By the Quadratic Formula, the zeros of $h(x)$ are

$$x = \frac{4 \pm \sqrt{16 - 4}}{2} = 2 \pm \sqrt{3}.$$

$$h(x) = \left[x - \left(2 + \sqrt{3}\right)\right]\left[x - \left(2 - \sqrt{3}\right)\right]$$

$$= \left(x - 2 - \sqrt{3}\right)\left(x - 2 + \sqrt{3}\right)$$

59. $f(x) = x^4 - 81$

$$= (x^2 - 9)(x^2 + 9)$$

$$= (x + 3)(x - 3)(x + 3i)(x - 3i)$$

The zeros of $f(x)$ are $x = \pm 3$ and $x = \pm 3i$.

61. $f(z) = z^2 - 2z + 2$

By the Quadratic Formula, the zeros of $f(z)$ are

$$z = \frac{2 \pm \sqrt{4 - 8}}{2} = 1 \pm i.$$

$$f(z) = [z - (1 + i)][z - (1 - i)]$$

$$= (z - 1 - i)(z - 1 + i)$$

63. $g(x) = x^3 - 6x^2 + 13x - 10$

Possible rational zeros: ± 1, ± 2, ± 5, ± 10

$$
\begin{array}{r|rrrr}
2 & 1 & -6 & 13 & -10 \\
  &   & 2 & -8 & 10 \\
\hline
  & 1 & -4 & 5 & 0
\end{array}
$$

By the Quadratic Formula, the zeros of $x^2 - 4x + 5$ are

$$x = \frac{4 \pm \sqrt{16 - 20}}{2} = 2 \pm i.$$

The zeros of $g(x)$ are $x = 2$ and $x = 2 \pm i$.

$$g(x) = (x - 2)[x - (2 + i)][x - (2 - i)]$$
$$= (x - 2)(x - 2 - i)(x - 2 + i)$$

65. $h(x) = x^3 - x + 6$

Possible rational zeros: ± 1, ± 2, ± 3, ± 6

$$
\begin{array}{r|rrrr}
-2 & 1 & 0 & -1 & 6 \\
   &   & -2 & 4 & -6 \\
\hline
   & 1 & -2 & 3 & 0
\end{array}
$$

By the Quadratic Formula, the zeros of $x^2 - 2x + 3$ are

$$x = \frac{2 \pm \sqrt{4 - 12}}{2} = 1 \pm \sqrt{2}\, i.$$

The zeros of $h(x)$ are $x = -2$ and $x = 1 \pm \sqrt{2}\, i$.

$$h(x) = [x - (-2)]\left[x - \left(1 + \sqrt{2}\, i\right)\right]\left[x - \left(1 - \sqrt{2}\, i\right)\right]$$
$$= (x + 2)\left(x - 1 - \sqrt{2}\, i\right)\left(x - 1 + \sqrt{2}\, i\right)$$

67. $f(x) = 5x^3 - 9x^2 + 28x + 6$

Possible rational zeros:
± 1, ± 2, ± 3, ± 6, $\pm \frac{1}{5}$, $\pm \frac{2}{5}$, $\pm \frac{3}{5}$, $\pm \frac{6}{5}$

$$
\begin{array}{r|rrrr}
-\frac{1}{5} & 5 & -9 & 28 & 6 \\
             &   & -1 & 2 & -6 \\
\hline
             & 5 & -10 & 30 & 0
\end{array}
$$

By the Quadratic Formula, the zeros of
$5x^2 - 10x + 30 = 5(x^2 - 2x + 6)$ are

$$x = \frac{2 \pm \sqrt{4 - 24}}{2} = 1 \pm \sqrt{5}\, i.$$

The zeros of $f(x)$ are $x = -\frac{1}{5}$ and $x = 1 \pm \sqrt{5}\, i$.

$$f(x) = \left[x - \left(-\tfrac{1}{5}\right)\right](5)\left[x - \left(1 + \sqrt{5}\, i\right)\right]\left[x - \left(1 - \sqrt{5}\, i\right)\right]$$
$$= (5x + 1)\left(x - 1 - \sqrt{5}\, i\right)\left(x - 1 + \sqrt{5}\, i\right)$$

69. $g(x) = x^4 - 4x^3 + 8x^2 - 16x + 16$

Possible rational zeros: ± 1, ± 2, ± 4, ± 8, ± 16

$$
\begin{array}{r|rrrrr}
2 & 1 & -4 & 8 & -16 & 16 \\
  &   & 2 & -4 & 8 & -16 \\
\hline
  & 1 & -2 & 4 & -8 & 0
\end{array}
$$

$$
\begin{array}{r|rrrr}
2 & 1 & -2 & 4 & -8 \\
  &   & 2 & 0 & 8 \\
\hline
  & 1 & 0 & 4 & 0
\end{array}
$$

$$g(x) = (x - 2)(x - 2)(x^2 + 4) = (x - 2)^2(x + 2i)(x - 2i)$$

The zeros of $g(x)$ are 2 and $\pm 2i$.

71. $f(x) = x^4 + 10x^2 + 9$

$$= (x^2 + 1)(x^2 + 9)$$

$$= (x + i)(x - i)(x + 3i)(x - 3i)$$

The zeros of $f(x)$ are $x = \pm i$ and $x = \pm 3i$.

73. $f(x) = x^3 + 24x^2 + 214x + 740$

Possible rational zeros: $\pm 1, \pm 2, \pm 4, \pm 5, \pm 10, \pm 20, \pm 37,$
$\pm 74, \pm 148, \pm 185, \pm 370, \pm 740$

Based on the graph, try $x = -10$.

$$
\begin{array}{r|rrrr}
-10 & 1 & 24 & 214 & 740 \\
    &   & -10 & -140 & -740 \\
\hline
    & 1 & 14 & 74 & 0
\end{array}
$$

By the Quadratic Formula, the zeros of $x^2 + 14x + 74$ are

$$x = \frac{-14 \pm \sqrt{196 - 296}}{2} = -7 \pm 5i.$$

The zeros of $f(x)$ are $x = -10$ and $x = -7 \pm 5i$.

75. $f(x) = 16x^3 - 20x^2 - 4x + 15$

Possible rational zeros:

$\pm 1, \pm 3, \pm 5, \pm 15, \pm\frac{1}{2}, \pm\frac{3}{2}, \pm\frac{5}{2}, \pm\frac{15}{2}, \pm\frac{1}{4}, \pm\frac{3}{4},$
$\pm\frac{5}{4}, \pm\frac{15}{4}, \pm\frac{1}{8}, \pm\frac{3}{8}, \pm\frac{5}{8}, \pm\frac{15}{8}, \pm\frac{1}{16}, \pm\frac{3}{16}, \pm\frac{5}{16}, \pm\frac{15}{16}$

Based on the graph, try $x = -\frac{3}{4}$.

$$
\begin{array}{r|rrrr}
-\frac{3}{4} & 16 & -20 & -4 & 15 \\
 & & -12 & 24 & -15 \\
\hline
 & 16 & -32 & 20 & 0
\end{array}
$$

By the Quadratic Formula, the zeros of

$16x^2 - 32x + 20 = 4(4x^2 - 8x + 5)$ are

$x = \dfrac{8 \pm \sqrt{64 - 80}}{8} = 1 \pm \dfrac{1}{2}i.$

The zeros of $f(x)$ are $x = -\frac{3}{4}$ and $x = 1 \pm \frac{1}{2}i$.

77. $f(x) = 2x^4 + 5x^3 + 4x^2 + 5x + 2$

Possible rational zeros: $\pm 1, \pm 2, \pm\frac{1}{2}$

Based on the graph, try $x = -2$ and $x = -\frac{1}{2}$.

$$
\begin{array}{r|rrrrr}
-2 & 2 & 5 & 4 & 5 & 2 \\
 & & -4 & -2 & -4 & -2 \\
\hline
 & 2 & 1 & 2 & 1 & 0
\end{array}
$$

$$
\begin{array}{r|rrrr}
-\frac{1}{2} & 2 & 1 & 2 & 1 \\
 & & -1 & 0 & -1 \\
\hline
 & 2 & 0 & 2 & 0
\end{array}
$$

The zeros of $2x^2 + 2 = 2(x^2 + 1)$ are $x = \pm i$.

The zeros of $f(x)$ are $x = -2$, $x = -\frac{1}{2}$, and $x = \pm i$.

79. $g(x) = 5x^5 + 10x = 5x(x^4 + 2)$

Let $f(x) = x^4 + 2$.

Sign variations: 0, positive zeros: 0

$f(-x) = x^4 + 2$

Sign variations: 0, negative zeros: 0

81. $h(x) = 3x^4 + 2x^2 + 1$

Sign variations: 0, positive zeros: 0

$h(-x) = 3x^4 + 2x^2 + 1$

Sign variations: 0, negative zeros: 0

83. $g(x) = 2x^3 - 3x^2 - 3$

Sign variations: 1, positive zeros: 1

$g(-x) = -2x^3 - 3x^2 - 3$

Sign variations: 0, negative zeros: 0

85. $f(x) = -5x^3 + x^2 - x + 5$

Sign variations: 3, positive zeros: 3 or 1

$f(-x) = 5x^3 + x^2 + x + 5$

Sign variations: 0, negative zeros: 0

87. $f(x) = x^4 - 4x^3 + 15$

(a)
$$
\begin{array}{r|rrrrr}
4 & 1 & -4 & 0 & 0 & 15 \\
 & & 4 & 0 & 0 & 0 \\
\hline
 & 1 & 0 & 0 & 0 & 15
\end{array}
$$

4 is an upper bound.

(b)
$$
\begin{array}{r|rrrrr}
-1 & 1 & -4 & 0 & 0 & 15 \\
 & & -1 & 5 & -5 & 5 \\
\hline
 & 1 & -5 & 5 & -5 & 20
\end{array}
$$

-1 is a lower bound.

89. $f(x) = x^4 - 4x^3 + 16x - 16$

(a)
$$
\begin{array}{r|rrrrr}
5 & 1 & -4 & 0 & 16 & -16 \\
 & & 5 & 5 & 25 & 205 \\
\hline
 & 1 & 1 & 5 & 41 & 189
\end{array}
$$

5 is an upper bound.

(b)
$$
\begin{array}{r|rrrrr}
-3 & 1 & -4 & 0 & 16 & -16 \\
 & & -3 & 21 & -63 & 141 \\
\hline
 & 1 & -7 & 21 & -47 & 125
\end{array}
$$

-3 is a lower bound.

91. $f(x) = 4x^3 - 3x - 1$

Possible rational zeros: $\pm 1,\ \pm\frac{1}{2},\ \pm\frac{1}{4}$

$$
\begin{array}{r|rrrr}
1 & 4 & 0 & -3 & -1 \\
  &   & 4 & 4 & 1 \\
\hline
  & 4 & 4 & 1 & 0
\end{array}
$$

$4x^3 - 3x - 1 = (x - 1)(4x^2 + 4x + 1)$

$\qquad\qquad\quad = (x - 1)(2x + 1)^2$

Thus, the zeros are 1 and $-\frac{1}{2}$.

93. $f(y) = 4y^3 + 3y^2 + 8y + 6$

Possible rational zeros: $\pm 1, \pm 2, \pm 3, \pm 6, \pm\frac{1}{2}, \pm\frac{3}{2}, \pm\frac{1}{4}, \pm\frac{3}{4}$

$$
\begin{array}{r|rrrr}
-\frac{3}{4} & 4 & 3 & 8 & 6 \\
  &   & -3 & 0 & -6 \\
\hline
  & 4 & 0 & 8 & 0
\end{array}
$$

$4y^3 + 3y^2 + 8y + 6 = \left(y + \frac{3}{4}\right)(4y^2 + 8)$

$\qquad\qquad\qquad\quad = \left(y + \frac{3}{4}\right)4(y^2 + 2)$

$\qquad\qquad\qquad\quad = (4y + 3)(y^2 + 2)$

Thus, the only real zero is $-\frac{3}{4}$.

95. $P(x) = x^4 - \frac{25}{4}x^2 + 9$

$\qquad = \frac{1}{4}(4x^4 - 25x^2 + 36)$

$\qquad = \frac{1}{4}(4x^2 - 9)(x^2 - 4)$

$\qquad = \frac{1}{4}(2x + 3)(2x - 3)(x + 2)(x - 2)$

The rational zeros are $\pm\frac{3}{2}$ and ± 2.

97. $f(x) = x^3 - \frac{1}{4}x^2 - x + \frac{1}{4}$

$\qquad = \frac{1}{4}(4x^3 - x^2 - 4x + 1)$

$\qquad = \frac{1}{4}[x^2(4x - 1) - 1(4x - 1)]$

$\qquad = \frac{1}{4}(4x - 1)(x^2 - 1)$

$\qquad = \frac{1}{4}(4x - 1)(x + 1)(x - 1)$

The rational zeros are $\frac{1}{4}$ and ± 1.

99. $f(x) = x^3 - 1 = (x - 1)(x^2 + x + 1)$

Rational zeros: 1 $(x = 1)$

Irrational zeros: 0

Matches (d).

101. $f(x) = x^3 - x = x(x + 1)(x - 1)$

Rational zeros: 3 $(x = 0, \pm 1)$

Irrational zeros: 0

Matches (b).

103. (a)

(c)

The volume is maximum when $x \approx 1.82$.

The dimensions are: length $\approx 15 - 2(1.82) = 11.36$

$\qquad\qquad\qquad\quad$ width $\approx 9 - 2(1.82) = 5.36$

$\qquad\qquad\qquad\quad$ height $= x \approx 1.82$

$1.82 \text{ cm} \times 5.36 \text{ cm} \times 11.36 \text{ cm}$

(b) $V = l \cdot w \cdot h = (15 - 2x)(9 - 2x)x$

$\qquad = x(9 - 2x)(15 - 2x)$

Since length, width, and height must be positive, we have $0 < x < \frac{9}{2}$ for the domain.

(d) $56 = x(9 - 2x)(15 - 2x)$

$56 = 135x - 48x^2 + 4x^3$

$0 = 4x^3 - 48x^2 + 135x - 56$

The zeros of this polynomial are $\frac{1}{2}, \frac{7}{2}$, and 8.
x cannot equal 8 since it is not in the domain of V.
[The length cannot equal -1 and the width cannot equal -7. The product of $(8)(-1)(-7) = 56$ so it showed up as an extraneous solution.]

Thus, the volume is 56 cubic centimeters when $x = \frac{1}{2}$ centimeter or $x = \frac{7}{2}$ centimeters.

105.
$$P = -76x^3 + 4830x^2 - 320{,}000,\ 0 \le x \le 60$$

$$2{,}500{,}000 = -76x^3 + 4830x^2 - 320{,}000$$

$$76x^3 - 4830x^2 + 2{,}820{,}000 = 0$$

The zeros of this equation are $x \approx 46.1$, $x \approx 38.4$, and $x \approx -21.0$. Since $0 \le x \le 60$, we disregard $x \approx -21.0$. The smaller remaining solution is $x \approx 38.4$. The advertising expense is \$384,000.

107. (a) Current bin: $V = 2 \times 3 \times 4 = 24$ cubic feet

New bin: $V = 5(24) = 120$ cubic feet

$$V = (2 + x)(3 + x)(4 + x) = 120$$

(b) $x^3 + 9x^2 + 26x + 24 = 120$

$x^3 + 9x^2 + 26x - 96 = 0$

The only real zero of this polynomial is $x = 2$. All the dimensions should be increased by 2 feet, so the new bin will have dimensions of 4 feet by 5 feet by 6 feet.

109. $C = 100\left(\dfrac{200}{x^2} + \dfrac{x}{x + 30}\right), x \geq 1$

C is minimum when $3x^3 - 40x^2 - 2400x - 36000 = 0$.

The only real zero is $x \approx 40$ or 4000 units.

111.
$$P = R - C = xp - C$$
$$= x(140 - 0.0001x) - (80x + 150,000)$$
$$= -0.0001x^2 + 60x - 150,000$$

$9,000,000 = -0.0001x^2 + 60x - 150,000$

Thus, $0 = 0.0001x^2 - 60x + 9,150,000.$

$$x = \frac{60 \pm \sqrt{-60}}{0.0002} = 300,000 \pm 10,000\sqrt{15}i$$

Since the solutions are both complex, it is not possible to determine a price p that would yield a profit of 9 million dollars.

113. False. The most nonreal complex zeros it can have is two and the Linear Factorization Theorem guarantees that there are 3 linear factors, so one zero must be real.

115. $g(x) = -f(x)$. This function would have the same zeros as $f(x)$ so r_1, r_2, and r_3 are also zeros of $g(x)$.

117. $g(x) = f(x - 5)$. The graph of $g(x)$ is a horizontal shift of the graph of $f(x)$ five units to the right so the zeros of $g(x)$ are $5 + r_1$, $5 + r_2$, and $5 + r_3$.

119. $g(x) = 3 + f(x)$. Since $g(x)$ is a vertical shift of the graph of $f(x)$, the zeros of $g(x)$ cannot be determined.

121. $f(x) = x^4 - 4x^2 + k$

$$x^2 = \frac{-(-4) \pm \sqrt{(-4)^2 - 4(1)(k)}}{2(1)} = \frac{4 \pm 2\sqrt{4 - k}}{2} = 2 \pm \sqrt{4 - k}$$

$$x = \pm\sqrt{2 \pm \sqrt{4 - k}}$$

(a) For there to be four distinct real roots, both $4 - k$ and $2 \pm \sqrt{4 - k}$ must be positive. This occurs when $0 < k < 4$. Thus, some possible k-values are $k = 1$, $k = 2$, $k = 3$, $k = \frac{1}{2}$, $k = \sqrt{2}$, etc.

(b) For there to be two real roots, each of multiplicity 2, $4 - k$ must equal zero. Thus, $k = 4$.

(c) For there to be two real zeros and two complex zeros, $2 + \sqrt{4 - k}$ must be positive and $2 - \sqrt{4 - k}$ must be negative. This occurs when $k < 0$. Thus, some possible k-values are $k = -1$, $k = -2$, $k = -\frac{1}{2}$, etc.

(d) For there to be four complex zeros, $2 \pm \sqrt{4 - k}$ must be nonreal. This occurs when $k > 4$. Some possible k-values are $k = 5$, $k = 6$, $k = 7.4$, etc.

123. Zeros: $-2, \frac{1}{2}, 3$

$$f(x) = -(x + 2)(2x - 1)(x - 3)$$

$$= -2x^3 + 3x^2 + 11x - 6$$

Any nonzero scalar multiple of f would have the same three zeros. Let $g(x) = af(x)$, $a > 0$. There are infinitely many possible functions for f.

125. Answers will vary. Some of the factoring techniques are:

1. Factor out the greatest common factor.

2. Use special product formulas.

$$a^2 - b^2 = (a + b)(a - b)$$

$$a^2 + 2ab + b^2 = (a + b)^2$$

$$a^2 - 2ab + b^2 = (a - b)^2$$

$$a^3 + b^3 = (a + b)(a^2 - ab + b^2)$$

$$a^3 - b^3 = (a - b)(a^2 + ab + b^2)$$

3. Factor by grouping, if possible.

4. Factor general trinomials with binomial factors by "guess-and-test" or by the grouping method.

5. Use the Rational Zero Test together with synthetic division to factor a polynomial.

6. Use Descartes's Rule of Signs to determine the number of real zeros. Then find any zeros and use them to factor the polynomial.

7. Find any upper and lower bounds for the real zeros to eliminate some of the possible rational zeros. Then test the remaining candidates by synthetic division and use any zeros to factor the polynomial.

127. (a) $f(x) = \left(x - \sqrt{b}\,i\right)\left(x + \sqrt{b}\,i\right) = x^2 + b$

(b) $f(x) = [x - (a + bi)][x - (a - bi)]$

$$= [(x - a) - bi][(x - a) + bi]$$

$$= (x - a)^2 - (bi)^2$$

$$= x^2 - 2ax + a^2 + b^2$$

129. $(-3 + 6i) - (8 - 3i) = -3 + 6i - 8 + 3i = -11 + 9i$

131. $(6 - 2i)(1 + 7i) = 6 + 42i - 2i - 14i^2 = 20 + 40i$

133. $g(x) = f(x - 2)$

Horizontal shift two units to the right

135. $g(x) = 2f(x)$

Vertical stretch (each y-value is multiplied by 2)

137. $g(x) = f(2x)$

Horizontal shrink (each x-value is multiplied by $\frac{1}{2}$)

Section 3.5 Mathematical Modeling and Variation

You should know the following the following terms and formulas.

- ■ Direct variation (varies directly, directly proportional)
  - (a) $y = kx$
  - (b) $y = kx^n$ (as nth power)
- ■ Inverse variation (varies inversely, inversely proportional)
  - (a) $y = k/x$
  - (b) $y = k/(x^n)$ (as nth power)
- ■ Joint variation (varies jointly, jointly proportional)
  - (a) $z = kxy$
  - (b) $z = kx^n y^m$ (as nth power of x and mth power of y)
- ■ k is called the constant of proportionality.
- ■ Least Squares Regression Line $y = ax + b$. Use your calculator or computer to enter the data points and to find the "best-fitting" linear model.

Vocabulary Check

1. variation; regression

2. sum of square differences

3. correlation coefficient

4. directly proportional

5. constant of variation

6. directly proportional

7. inverse

8. combined

9. jointly proportional

---

1. $y = 1767.0t + 123{,}916$

Year	Actual Number (in thousands)	Model (in thousands)
1992	128,105	127,450
1993	129,200	129,217
1994	131,056	130,984
1995	132,304	132,751
1996	133,943	134,518
1997	136,297	136,285
1998	137,673	138,052
1999	139,368	139,819
2000	142,583	141,586
2001	143,734	143,353
2002	144,863	145,120

The model is a good fit for the actual data.

3. Using the points $(0, 3)$ and $(4, 4)$, we have $y = \frac{1}{4}x + 3$.

5. Using the points $(2, 2)$ and $(4, 1)$, we have $y = -\frac{1}{2}x + 3$.

7. (a) and (b)

$y \approx t + 130$

(c) $y \approx 1.03t + 130.27$

(d) The models are similar.

(e) When $t = 108$, we have:

Model in part (b): 238 feet

Model in part (c): 241.51 feet

(f) Answers will vary.

9. (a) and (c)

The model is a good fit to the actual data. ($r \approx 0.98$)

(b) $S \approx 38.4t + 224$

(d) For 2005, use $t = 15$: $S \approx \$800.4$ million

For 2007, use $t = 17$: $S \approx \$877.3$ million

(e) Each year the annual gross ticket sales for Broadway shows in New York City increase by approximately \$38.4 million.

11. The graph appears to represent $y = 4/x$, so y varies inversely as x.

13. $k = 1$

x	2	4	6	8	10
$y = kx^2$	4	16	36	64	100

15. $k = \frac{1}{2}$

x	2	4	6	8	10
$y = kx^2$	2	8	18	32	50

17. $k = 2$

x	2	4	6	8	10
$y = \dfrac{k}{x^2}$	$\dfrac{1}{2}$	$\dfrac{1}{8}$	$\dfrac{1}{18}$	$\dfrac{1}{32}$	$\dfrac{1}{50}$

19. $k = 10$

x	2	4	6	8	10
$y = \dfrac{k}{x^2}$	$\dfrac{5}{2}$	$\dfrac{5}{8}$	$\dfrac{5}{18}$	$\dfrac{5}{32}$	$\dfrac{1}{10}$

21. The table represents the equation
$y = 5/x$.

23.
$$y = kx$$
$$-7 = k(10)$$
$$-\frac{7}{10} = k$$
$$y = -\frac{7}{10}x$$

This equation checks with the other points given in the table.

25.
$$y = kx$$
$$12 = k(5)$$
$$\frac{12}{5} = k$$
$$y = \frac{12}{5}x$$

27.
$$y = kx$$
$$2050 = k(10)$$
$$205 = k$$
$$y = 205x$$

29.
$$I = kP$$
$$87.50 = k(2500)$$
$$0.035 = k$$
$$I = 0.035P$$

31.
$$y = kx$$
$$33 = k(13)$$
$$\frac{33}{13} = k$$
$$y = \frac{33}{13}x$$

When $x = 10$ inches, $y \approx 25.4$ centimeters.

When $x = 20$ inches, $y \approx 50.8$ centimeters.

33.
$$y = kx$$
$$5520 = k(150{,}000)$$
$$0.0368 = k$$
$$y = 0.0368x$$
$$y = 0.0368(200{,}000)$$
$$= \$7360$$

The property tax is \$7360.

35.
$$d = kF$$
$$0.15 = k(265)$$
$$\frac{3}{5300} = k$$
$$d = \frac{3}{5300}F$$

(a) $d = \frac{3}{5300}(90) \approx 0.05$ meter

(b) $0.1 = \frac{3}{5300}F$
$$\frac{530}{3} = F$$
$$F = 176\frac{2}{3} \text{ newtons}$$

37.
$$d = kF$$
$$1.9 = k(25) \implies k = 0.076$$
$$d = 0.076F$$

When the distance compressed is 3 inches, we have
$$3 = 0.076F$$
$$F \approx 39.47.$$

No child over 39.47 pounds should use the toy.

39. $A = kr^2$

41. $y = \dfrac{k}{x^2}$

43. $F = \dfrac{kg}{r^2}$

45. $P = \dfrac{k}{V}$

47. $F = \dfrac{km_1m_2}{r^2}$

49. $A = \dfrac{1}{2}bh$

The area of a triangle is jointly proportional to its base and height.

51. $V = \dfrac{4}{3}\pi r^3$

The volume of a sphere varies directly as the cube of its radius.

53. $r = \dfrac{d}{t}$

Average speed is directly proportional to the distance and inversely proportional to the time.

55.
$$A = kr^2$$
$$9\pi = k(3)^2$$
$$\pi = k$$
$$A = \pi r^2$$

57.
$$y = \frac{k}{x}$$
$$7 = \frac{k}{4}$$
$$28 = k$$
$$y = \frac{28}{x}$$

59. $F = krs^3$

$4158 = k(11)(3)^3$

$k = 14$

$F = 14rs^3$

61. $z = \dfrac{kx^2}{y}$

$6 = \dfrac{k(6)^2}{4}$

$\dfrac{24}{36} = k$

$\dfrac{2}{3} = k$

$z = \dfrac{2/3x^2}{y} = \dfrac{2x^2}{3y}$

63. $d = kv^2$

$0.02 = k\left(\dfrac{1}{4}\right)^2$

$k = 0.32$

$d = 0.32v^2$

$0.12 = 0.32v^2$

$v^2 = \dfrac{0.12}{0.32} = \dfrac{3}{8}$

$v = \dfrac{\sqrt{3}}{2\sqrt{2}} = \dfrac{\sqrt{6}}{4} \approx 0.61 \text{ mi/hr}$

65. $r = \dfrac{kl}{A}, \; A = \pi r^2 = \dfrac{\pi d^2}{4}$

$r = \dfrac{4kl}{\pi d^2}$

$66.17 = \dfrac{4(1000)k}{\pi\left(\frac{0.0126}{12}\right)^2}$

$k \approx 5.73 \times 10^{-8}$

$r = \dfrac{4(5.73 \times 10^{-8})l}{\pi\left(\frac{0.0126}{12}\right)^2}$

$33.5 = \dfrac{4(5.73 \times 10^{-8})l}{\pi\left(\frac{0.0126}{12}\right)^2}$

$\dfrac{33.5\pi\left(\frac{0.0126}{12}\right)^2}{4(5.73 \times 10^{-8})} = l$

$l \approx 506 \text{ feet}$

67. $W = kmh$

$2116.8 = k(120)(1.8)$

$k = \dfrac{2116.8}{(120)(1.8)} = 9.8$

$W = 9.8mh$

When $m = 100$ kilograms and $h = 1.5$ meters, we have $W = 9.8(100)(1.5) = 1470$ joules.

69. $v = \dfrac{k}{A}$

$v = \dfrac{k}{0.75A} = \dfrac{4}{3}\left(\dfrac{k}{A}\right)$

The velocity is increased by one-third.

71. (a)

(b) Yes, the data appears to be modeled (approximately) by the inverse proportion model.

$4.2 = \dfrac{k_1}{1000}$ $1.9 = \dfrac{k_2}{2000}$ $1.4 = \dfrac{k_3}{3000}$ $1.2 = \dfrac{k_4}{4000}$ $0.9 = \dfrac{k_5}{5000}$

$4200 = k_1$ $3800 = k_2$ $4200 = k_3$ $4800 = k_4$ $4500 = k_5$

—CONTINUED—

71. —CONTINUED—

(c) Mean: $k = \dfrac{4200 + 3800 + 4200 + 4800 + 4500}{5} = 4300$, Model: $C = \dfrac{4300}{d}$

(d)

(e) $3 = \dfrac{4300}{d}$

$d = \dfrac{4300}{3} = 1433\dfrac{1}{3}$ meters

73. $y = \dfrac{262.76}{x^{2.12}}$

(a)

(b) $y = \dfrac{262.76}{(25)^{2.12}}$

≈ 0.2857 microwatts per sq.cm.

75. False. y will increase if k is positive and y will decrease if k is negative.

77. False. The closer the value of $|r|$ is to 1, the better the fit.

79. The accuracy of the model in predicting prize winnings is questionable because the model is based on limited data.

81. $3x + 2 > 17$

$3x > 15$

$x > 5$

83. $|2x - 1| < 9$

$-9 < 2x - 1 < 9$

$-8 < 2x < 10$

$-4 < x < 5$

85. $f(x) = \dfrac{x^2 + 5}{x - 3}$

(a) $f(0) = \dfrac{0^2 + 5}{0 - 3} = -\dfrac{5}{3}$

(b) $f(-3) = \dfrac{(-3)^2 + 5}{-3 - 3} = \dfrac{14}{-6} = -\dfrac{7}{3}$

(c) $f(4) = \dfrac{4^2 + 5}{4 - 3} = 21$

Review Exercises for Chapter 3

1. (a) $y = 2x^2$

Vertical stretch

(b) $y = -2x^2$

Vertical stretch and a reflection in the x-axis

—CONTINUED—

1. —CONTINUED—

(c) $y = x^2 + 2$

Vertical shift two units upward

(d) $y = (x + 2)^2$

Horizontal shift two units to the left

3. $g(x) = x^2 - 2x$

$= x^2 - 2x + 1 - 1$

$= (x - 1)^2 - 1$

Vertex: $(1, -1)$

Axis of symmetry: $x = 1$

$0 = x^2 - 2x = x(x - 2)$

x-intercepts: $(0, 0), (2, 0)$

5. $f(x) = x^2 + 8x + 10$

$= x^2 + 8x + 16 - 16 + 10$

$= (x + 4)^2 - 6$

Vertex: $(-4, -6)$

Axis of symmetry: $x = -4$

$0 = (x + 4)^2 - 6$

$(x + 4)^2 = 6$

$x + 4 = \pm\sqrt{6}$

$x = -4 \pm \sqrt{6}$

x-intercepts: $\left(-4 \pm \sqrt{6}, 0\right)$

7. $f(t) = -2t^2 + 4t + 1$

$= -2(t^2 - 2t + 1 - 1) + 1$

$= -2[(t - 1)^2 - 1] + 1$

$= -2(t - 1)^2 + 3$

Vertex: $(1, 3)$

Axis of symmetry: $t = 1$

$0 = -2(t - 1)^2 + 3$

$2(t - 1)^2 = 3$

$t - 1 = \pm\sqrt{\dfrac{3}{2}}$

$t = 1 \pm \dfrac{\sqrt{3}}{6}$

t-intercepts: $\left(1 \pm \dfrac{\sqrt{6}}{2}, 0\right)$

9. $h(x) = 4x^2 + 4x + 13$

$= 4(x^2 + x) + 13$

$= 4\left(x^2 + x + \dfrac{1}{4} - \dfrac{1}{4}\right) + 13$

$= 4\left(x^2 + x + \dfrac{1}{4}\right) - 1 + 13$

$= 4\left(x + \dfrac{1}{2}\right)^2 + 12$

Vertex: $\left(-\dfrac{1}{2}, 12\right)$

Axis of symmetry: $x = -\dfrac{1}{2}$

$0 = 4\left(x + \dfrac{1}{2}\right)^2 + 12$

$\left(x + \dfrac{1}{2}\right)^2 = -3$

No real zeros

x-intercepts: none

11. $h(x) = x^2 + 5x - 4$

$$= x^2 + 5x + \frac{25}{4} - \frac{25}{4} - 4$$

$$= \left(x + \frac{5}{2}\right)^2 - \frac{25}{4} - \frac{16}{4}$$

$$= \left(x + \frac{5}{2}\right)^2 - \frac{41}{4}$$

Vertex: $\left(-\frac{5}{2}, -\frac{41}{4}\right)$

Axis of symmetry: $x = -\frac{5}{2}$

$0 = x^2 + 5x - 4$

By the Quadratic Formula, $x = \dfrac{-5 \pm \sqrt{41}}{2}$.

x-intercepts: $\left(\dfrac{-5 \pm \sqrt{41}}{2}, 0\right)$

13. $f(x) = \frac{1}{3}(x^2 + 5x - 4)$

$$= \frac{1}{3}\left(x^2 + 5x + \frac{25}{4} - \frac{25}{4} - 4\right)$$

$$= \frac{1}{3}\left[\left(x + \frac{5}{2}\right)^2 - \frac{41}{4}\right]$$

$$= \frac{1}{3}\left(x + \frac{5}{2}\right)^2 - \frac{41}{12}$$

Vertex: $\left(-\frac{5}{2}, -\frac{41}{12}\right)$

Axis of symmetry: $x = -\frac{5}{2}$

$0 = x^2 + 5x - 4$

By the Quadratic Formula, $x = \dfrac{-5 \pm \sqrt{41}}{2}$.

x-intercepts: $\left(\dfrac{-5 \pm \sqrt{41}}{2}, 0\right)$

15. Vertex: $(4, 1) \Rightarrow f(x) = a(x - 4)^2 + 1$

Point: $(2, -1) \Rightarrow -1 = a(2 - 4)^2 + 1$

$$-2 = 4a$$

$$-\frac{1}{2} = a$$

Thus, $f(x) = -\frac{1}{2}(x - 4)^2 + 1$.

17. Vertex: $(1, -4) \Rightarrow f(x) = a(x - 1)^2 - 4$

Point: $(2, -3) \Rightarrow -3 = a(2 - 1)^2 - 4$

$$1 = a$$

Thus, $f(x) = (x - 1)^2 - 4$.

19. (a) $A = xy = x\left(\dfrac{8 - x}{2}\right)$ since

$$x + 2y - 8 = 0 \implies y = \frac{8 - x}{2}.$$

(b) Since the figure is in the first quadrant and x and y must be positive, the domain of

$$A = x\left(\frac{8 - x}{2}\right) \text{ is } 0 < x < 8.$$

(c)

x	y	Area
1	$4 - \frac{1}{2}(1)$	$(1)[4 - \frac{1}{2}(1)] = \frac{7}{2}$
2	$4 - \frac{1}{2}(2)$	$(2)[4 - \frac{1}{2}(2)] = 6$
3	$4 - \frac{1}{2}(3)$	$(3)[4 - \frac{1}{2}(3)] = \frac{15}{2}$
4	$4 - \frac{1}{2}(4)$	$(4)[4 - \frac{1}{2}(4)] = 8$
5	$4 - \frac{1}{2}(5)$	$(5)[4 - \frac{1}{2}(5)] = \frac{15}{2}$
6	$4 - \frac{1}{2}(6)$	$(6)[4 - \frac{1}{2}(6)] = 6$

(d)

The maximum area of 8 occurs at the vertex when

$$x = 4 \text{ and } y = \frac{8 - 4}{2} = 2.$$

(e) $A = x\left(\dfrac{8 - x}{2}\right)$

$$= \frac{1}{8}(8x - x^2)$$

$$= -\frac{1}{2}(x^2 - 8x)$$

$$= -\frac{1}{2}(x^2 - 8x + 16 - 16)$$

$$= -\frac{1}{2}[(x - 4)^2 - 16]$$

$$= -\frac{1}{2}(x - 4)^2 + 8$$

The maximum area of 8 occurs when

$$x = 4 \text{ and } y = \frac{8 - 4}{2} = 2.$$

21. $R = -10p^2 + 800p$

(a) $R(20) = \$12,000$

$R(25) = \$13,750$

$R(30) = \$15,000$

(b) The maximum revenue occurs at the vertex of the parabola.

$$-\frac{b}{2a} = \frac{-800}{2(-10)} = \$40$$

$R(40) = \$16,000$

The revenue is maximum when the price is $40 per unit. The maximum revenue is $16,000.

23. $C = 70,000 - 120x + 0.055x^2$

The minimum cost occurs at the vertex of the parabola.

$$\text{Vertex: } -\frac{b}{2a} = -\frac{-120}{2(0.055)} \approx 1091 \text{ units}$$

Approximately 1091 units should be produced each day to yield a minimum cost.

25. $y = x^3,\ f(x) = -(x - 4)^3$

Transformation: Reflection in the x-axis and a horizontal shift four units to the right

27. $y = x^4,\ f(x) = 2 - x^4$

Transformation: Reflection in the x-axis and a vertical shift two units upward

29. $y = x^5,\ f(x) = (x - 3)^5$

Transformation: Horizontal shift three units to the right

31. $f(x) = -x^2 + 6x + 9$

The degree is even and the leading coefficient is negative. The graph falls to the left and falls to the right.

33. $g(x) = \frac{3}{4}(x^4 + 3x^2 + 2)$

The degree is even and the leading coefficient is positive. The graph rises to the left and rises to the right.

35. $f(x) = 2x^2 + 11x - 21$

$0 = 2x^2 + 11x - 21$

$= (2x - 3)(x + 7)$

Zeros: $x = \frac{3}{2}, -7$, all of multiplicity 1 (odd multiplicity)

Turning points: 1

37. $f(t) = t^3 - 3t$

$0 = t^3 - 3t$

$0 = t(t^2 - 3)$

Zeros: $t = 0, \pm\sqrt{3}$ all of multiplicity 1 (odd multiplicity)

Turning points: 2

39. $f(x) = -12x^3 + 20x^2$

$0 = -12x^3 + 20x^2$

$0 = -4x^2(3x - 5)$

Zeros: $x = 0$ of multiplicity 2 (even multiplicity)

$x = \frac{5}{3}$ of multiplicity 1 (odd multiplicity)

Turning points: 2

41. $f(x) = -x^3 + x^2 - 2$

   (a) The degree is odd and the leading coefficient is negative. The graph rises to the left and falls to the right.

   (b) Zero: $x = -1$

   (c)

x	-3	-2	-1	0	1	2
$f(x)$	34	10	0	-2	-2	-6

   (d)

43. $f(x) = x(x^3 + x^2 - 5x + 3)$

   (a) The degree is even and the leading coefficient is positive. The graph rises to the left and rises to the right.

   (b) Zeros: $x = 0, 1, -3$

   (c)

x	-4	-3	-2	-1	0	1	2	3
$f(x)$	100	0	-18	-8	0	0	10	72

   (d)

45. $f(x) = 3x^3 - x^2 + 3$

x	-3	-2	-1	0	1	2	3
$f(x)$	-87	-25	-1	3	5	23	75

The zero is in the interval $[-1, 0]$.

Zero: $x \approx -0.900$

47. $f(x) = x^4 - 5x - 1$

x	-3	-2	-1	0	1	2	3
$f(x)$	95	25	5	-1	-5	5	65

There are two zeros, one in the interval $[-1, 0]$ and one in the interval $[1, 2]$

Zeros: $x \approx -0.200, x \approx 1.772$

49.

$$
\begin{array}{r}
8x + 5 \\
3x - 2 \overline{)\,24x^2 - x - 8} \\
\underline{24x^2 - 16x} \\
15x - 8 \\
\underline{15x - 10} \\
2
\end{array}
$$

Thus, $\dfrac{24x^2 - x - 8}{3x - 2} = 8x + 5 + \dfrac{2}{3x - 2}$.

51.

$$
\begin{array}{r}
5x + 2 \\
x^2 - 3x + 1 \overline{)\,5x^3 - 13x^2 - x + 2} \\
\underline{5x^3 - 15x^2 + 5x} \\
2x^2 - 6x + 2 \\
\underline{2x^2 - 6x + 2} \\
0
\end{array}
$$

Thus, $\dfrac{5x^3 - 13x^2 - x + 2}{x^2 - 3x + 1} = 5x + 2$.

53.

$$
\begin{array}{r}
x^2 - 3x + 2 \\
x^2 + 0x + 2 \overline{)\,x^4 - 3x^3 + 4x^2 - 6x + 3} \\
\underline{x^4 + 0x^3 + 2x^2} \\
-3x^3 + 2x^2 - 6x \\
\underline{-3x^3 + 0x^2 - 6x} \\
2x^2 + 0x + 3 \\
\underline{2x^2 + 0x + 4} \\
-1
\end{array}
$$

Thus, $\dfrac{x^4 - 3x^3 + 4x^2 - 6x + 3}{x^2 + 2} = x^2 - 3x + 2 - \dfrac{1}{x^2 + 2}$.

55.

$$
\begin{array}{r|rrrrr}
2 & 6 & -4 & -27 & 18 & 0 \\
& & 12 & 16 & -22 & -8 \\
\hline
& 6 & 8 & -11 & -4 & -8
\end{array}
$$

Thus,

$$\dfrac{6x^4 - 4x^3 - 27x^2 + 18x}{x - 2} = 6x^3 + 8x^2 - 11x - 4 - \dfrac{8}{x - 2}.$$

57.

$$4 \;\big|\; \begin{array}{rrrr} 2 & -19 & 38 & 24 \\ & 8 & -44 & -24 \\ \hline 2 & -11 & -6 & 0 \end{array}$$

Thus, $\dfrac{2x^3 - 19x^2 + 38x + 24}{x - 4} = 2x^2 - 11x - 6.$

59. $f(x) = 20x^4 + 9x^3 - 14x^2 - 3x$

(a)
$$-1 \;\big|\; \begin{array}{rrrrr} 20 & 9 & -14 & -3 & 0 \\ & -20 & 11 & 3 & 0 \\ \hline 20 & -11 & -3 & 0 & 0 \end{array}$$

Yes, $x = -1$ is a zero of f.

(c)
$$0 \;\big|\; \begin{array}{rrrrr} 20 & 9 & -14 & -3 & 0 \\ & 0 & 0 & 0 & 0 \\ \hline 20 & 9 & -14 & -3 & 0 \end{array}$$

Yes, $x = 0$ is a zero of f.

(b)
$$\tfrac{3}{4} \;\big|\; \begin{array}{rrrrr} 20 & 9 & -14 & -3 & 0 \\ & 15 & 18 & 3 & 0 \\ \hline 20 & 24 & 4 & 0 & 0 \end{array}$$

Yes, $x = \tfrac{3}{4}$ is a zero of f.

(d)
$$1 \;\big|\; \begin{array}{rrrrr} 20 & 9 & -14 & -3 & 0 \\ & 20 & 29 & 15 & 12 \\ \hline 20 & 29 & 15 & 12 & 12 \end{array}$$

No, $x = 1$ is not a zero of f.

61. $f(x) = x^4 + 10x^3 - 24x^2 + 20x + 44$

(a)
$$-3 \;\big|\; \begin{array}{rrrrr} 1 & 10 & -24 & 20 & 44 \\ & -3 & -21 & 135 & -465 \\ \hline 1 & 7 & -45 & 155 & -421 \end{array}$$

Thus, $f(-3) = -421.$

(b)
$$-1 \;\big|\; \begin{array}{rrrrr} 1 & 10 & -24 & 20 & 44 \\ & -1 & -9 & 33 & -53 \\ \hline 1 & 9 & -33 & 53 & -9 \end{array}$$

$f(-1) = -9$

63. $f(x) = x^3 + 4x^2 - 25x - 28$; Factor: $(x - 4)$

(a)
$$4 \;\big|\; \begin{array}{rrrr} 1 & 4 & -25 & -28 \\ & 4 & 32 & 28 \\ \hline 1 & 8 & 7 & 0 \end{array}$$

Yes, $x - 4$ is a factor of $f(x)$.

(b) $x^2 + 8x + 7 = (x + 7)(x + 1)$

The remaining factors of f are $(x + 7)$ and $(x + 1)$.

(c) $f(x) = x^3 + 4x^2 - 25x - 28$

$= (x + 7)(x + 1)(x - 4)$

(d) Zeros: $-7, -1, 4$

(e)

65. $f(x) = x^4 - 4x^3 - 7x^2 + 22x + 24$

Factors: $(x + 2), (x - 3)$

(a)
$$-2 \;\big|\; \begin{array}{rrrrr} 1 & -4 & -7 & 22 & 24 \\ & -2 & 12 & -10 & -24 \\ \hline 1 & -6 & 5 & 12 & 0 \end{array}$$

$$3 \;\big|\; \begin{array}{rrrr} 1 & -6 & 5 & 12 \\ & 3 & -9 & -12 \\ \hline 1 & -3 & -4 & 0 \end{array}$$

Both are factors since the remainders are zero.

(b) $x^2 - 3x - 4 = (x + 1)(x - 4)$

The remaining factors are $(x + 1)$ and $(x - 4)$.

(c) $f(x) = (x + 1)(x - 4)(x + 2)(x - 3)$

(d) Zeros: $-2, -1, 3, 4$

(e)

67. $f(x) = 3x(x - 2)^2$

Zeros: $x = 0, x = 2$

69. $f(x) = x^2 - 9x + 8$

$= (x - 1)(x - 8)$

Zeros: $x = 1, x = 8$

71. $f(x) = (x + 4)(x - 6)(x - 2i)(x + 2i)$

Zeros: $x = -4, x = 6, x = 2i, x = -2i$

73. $f(x) = -4x^3 + 8x^2 - 3x + 15$

Possible rational zeros: $\pm 1, \pm 3, \pm 5, \pm 15, \pm\frac{1}{2}, \pm\frac{3}{2}, \pm\frac{5}{2}, \pm\frac{15}{2}, \pm\frac{1}{4}, \pm\frac{3}{4}, \pm\frac{5}{4}, \pm\frac{15}{4}$

75. $f(x) = x^3 - 2x^2 - 21x - 18$

Possible rational zeros: $\pm 1, \pm 2, \pm 3, \pm 6, \pm 9, \pm 18$

$$
\begin{array}{r|rrrr}
-1 & 1 & -2 & -21 & -18 \\
   &   & -1 &   3 & 18 \\
\hline
   & 1 & -3 & -18 &   0 \\
\end{array}
$$

$x^3 - 2x^2 - 21x - 18 = (x + 1)(x^2 - 3x - 18)$

$\qquad\qquad\qquad\qquad = (x + 1)(x - 6)(x + 3)$

The zeros of $f(x)$ are $x = -1, x = 6,$ and $x = -3$.

77. $f(x) = x^3 - 10x^2 + 17x - 8$

Possible rational zeros: $\pm 1, \pm 2, \pm 4, \pm 8$

$$
\begin{array}{r|rrrr}
1 & 1 & -10 & 17 & -8 \\
  &   &   1 & -9 & 8 \\
\hline
  & 1 & -9 & 8 & 0 \\
\end{array}
$$

$x^3 - 10x^2 + 17x - 8 = (x - 1)(x^2 - 9x + 8)$

$\qquad\qquad\qquad\qquad = (x - 1)(x - 1)(x - 8)$

$\qquad\qquad\qquad\qquad = (x - 1)^2(x - 8)$

The zeros of $f(x)$ are $x = 1$ and $x = 8$.

79. $f(x) = x^4 + x^3 - 11x^2 + x - 12$

Possible rational zeros: $\pm 1, \pm 2, \pm 3, \pm 4, \pm 6, \pm 12$

$$
\begin{array}{r|rrrrr}
3 & 1 & 1 & -11 & 1 & -12 \\
  &   & 3 & 12 & 3 & 12 \\
\hline
  & 1 & 4 &   1 & 4 &   0 \\
\end{array}
$$

$$
\begin{array}{r|rrrr}
-4 & 1 & 4 & 1 & 4 \\
   &   & -4 & 0 & -4 \\
\hline
   & 1 & 0 & 1 & 0 \\
\end{array}
$$

$x^4 + x^3 - 11x^2 + x - 12 = (x - 3)(x + 4)(x^2 + 1)$

The real zeros of $f(x)$ are $x = 3$, and $x = -4$.

81. $f(x) = 3\left(x - \frac{2}{3}\right)(x - 4)\left(x - \sqrt{3}i\right)\left(x + \sqrt{3}i\right)$

Multiply by 3 to clear the fraction.

$= (3x - 2)(x - 4)(x^2 + 3)$

Since $\sqrt{3}i$ is a zero, so is $-\sqrt{3}i$.

$= (3x^2 - 14x + 8)(x^2 + 3)$

$= 3x^4 - 14x^3 + 17x^2 - 42x + 24$

Note: $f(x) = a(3x^4 - 14x^3 + 17x^2 - 42x + 24)$, where a is any real nonzero number, has zeros $\frac{2}{3}$, 4, and $\pm\sqrt{3}i$.

83. $f(x) = x^3 - 4x^2 + x - 4$, Zero: i

Since i is a zero, so is $-i$.

$$
\begin{array}{r|rrrr}
i & 1 & -4 & 1 & -4 \\
  &   & i & -1 - 4i & 4 \\
\hline
  & 1 & -4 + i & -4i & 0 \\
\end{array}
$$

$$
\begin{array}{r|rrr}
-i & 1 & -4 + i & -4i \\
   &   & -i & 4i \\
\hline
   & 1 & -4 & 0 \\
\end{array}
$$

$f(x) = (x - i)(x + i)(x - 4)$, Zeros: $x = \pm i, 4$

85. $g(x) = 2x^4 - 3x^3 - 13x^2 + 37x - 15$, Zero: $2 + i$

Since $2 + i$ is a zero, so is $2 - i$

$$
\begin{array}{r|rrrrr}
2 + i & 2 & -3 & -13 & 37 & -15 \\
       &   & 4 + 2i & & 5i & -31 - 3i & 15 \\
\hline
       & 2 & 1 + 2i & -13 + 5i & 6 - 3i & 0
\end{array}
$$

$$
\begin{array}{r|rrrr}
2 - i & 2 & 1 + 2i & -13 + 5i & 6 - 3i \\
       &   & 4 - 2i & 10 - 5i & -6 + 3i \\
\hline
       & 2 & 5 & -3 & 0
\end{array}
$$

$g(x) = [x - (2 + i)][x - (2 - i)](2x^2 + 5x - 3)$

$\quad = (x - 2 - i)(x - 2 + i)(2x - 1)(x + 3)$

Zeros: $x = 2 \pm i, \frac{1}{2}, -3$

89. $g(x) = x^4 + 4x^3 - 3x^2 + 40x + 208$, Zero: $x = -4$

$$
\begin{array}{r|rrrrr}
-4 & 1 & 4 & -3 & 40 & 208 \\
   &   & -4 & 0 & 12 & -208 \\
\hline
   & 1 & 0 & -3 & 52 & 0
\end{array}
$$

$$
\begin{array}{r|rrrr}
-4 & 1 & 0 & -3 & 52 \\
   &   & -4 & 16 & -52 \\
\hline
   & 1 & -4 & 13 & 0
\end{array}
$$

$g(x) = (x + 4)^2(x^2 - 4x + 13)$

By the Quadratic Formula the zeros of $x^2 - 4x + 13$ are $x = 2 \pm 3i$. The zeros of $g(x)$ are $x = -4$ of multiplicity 2, and $x = 2 \pm 3i$.

$g(x) = (x + 4)^2[x - (2 + 3i)][x - (2 - 3i)]$

$\quad = (x + 4)^2(x - 2 - 3i)(x - 2 + 3i)$

93. $h(x) = x^3 - 6x^2 + 12x - 10$

(a)

(b) The graph has one x-intercept, so there is one real zero.

(c) $x \approx 3.26$

97. $f(x) = 4x^3 - 3x^2 + 4x - 3$

(a) $$
\begin{array}{r|rrrr}
1 & 4 & -3 & 4 & -3 \\
  &   & 4 & 1 & 5 \\
\hline
  & 4 & 1 & 5 & 2
\end{array}
$$

Since the last row has all positive entries, $x = 1$ is an upper bound.

87. $f(x) = x^3 + 4x^2 - 5x$

$\quad = x(x^2 + 4x - 5)$

$\quad = x(x + 5)(x - 1)$

Zeros: $x = 0, -5, 1$

91. $f(x) = x^4 + 2x + 1$

(a)

(b) The graph has two x-intercepts, so there are two real zeros.

(c) The zeros are $x = -1$ and $x \approx -0.54$.

95. $g(x) = 5x^3 + 3x^2 - 6x + 9$

$g(x)$ has two variations in sign, so g has either two or no positive real zeros.

$g(-x) = -5x^3 + 3x^2 + 6x + 9$

$g(-x)$ has one variation in sign, so g has one negative real zero.

(b) $$
\begin{array}{r|rrrr}
-\frac{1}{4} & 4 & -3 & 4 & -3 \\
             &   & -1 & 1 & -\frac{5}{4} \\
\hline
             & 4 & -4 & 5 & -\frac{17}{4}
\end{array}
$$

Since the last row entries alternate in sign, $x = -\frac{1}{4}$ is a lower bound.

99. $I = 2.09t + 37.2$

(a)

(b) The model is a good fit to the actual data. ($r \approx 0.992$)

101. $D = km$

$4 = k(2.5)$

$\frac{8}{5} = k$

$D = \frac{8}{5}m$

In 2 miles,
$D = \frac{8}{5}(2) = 3.2$ kilometers.

In 10 miles,
$D = \frac{8}{5}(10) = 16$ kilometers.

103. $F = ks^2$

If the speed is doubled,

$F = k(2s)^2$

$F = 4ks^2$.

The force will be changed by a factor of 4.

105. $T = \dfrac{k}{r}$

$3 = \dfrac{k}{65}$

$k = 3(65) = 195$

$T = \dfrac{195}{r}$

When $t = 80$ mph,

$T = \dfrac{195}{80} = 2.4375$ hours

≈ 2 hours, 26 minutes.

107. False. A fourth-degree polynomial can have at most four zeros and complex zeros occur in conjugate pairs.

109. The maximum (or minimum) value of a quadratic function is located at its graph's vertex. To find the vertex, either write the equation in standard form or use the formula

$$\left(-\frac{b}{2a}, f\left(-\frac{b}{2a}\right)\right).$$

If the leading coefficient is positive, the vertex is a minimum. If the leading coefficient is negative, the vertex is a maximum.

Problem Solving for Chapter 3

1. (a) (i) $g(x) = x^2 - 4x - 12$

$0 = (x - 6)(x + 2)$

Zeros: $6, -2$

(iii) $g(x) = x^2 + 3x - 10$

$0 = (x + 5)(x - 2)$

Zeros: $-5, 2$

(v) $g(x) = x^2 - 2x - 6$

$0 = x^2 - 2x - 6$

By the Quadratic Formula, $x = 1 \pm \sqrt{7}$.

Zeros: $1 \pm \sqrt{7}$

(ii) $g(x) = x^2 + 5x$

$0 = x(x + 5)$

Zeros: $0, -5$

(iv) $g(x) = x^2 - 4x + 4$

$0 = (x - 2)(x - 2)$

Zeros: $2, 2$

(vi) $g(x) = x^2 + 3x + 4$

$0 = x^2 + 3x + 4$

By the Quadratic Formula, $x = \dfrac{-3 \pm \sqrt{7}i}{2}$.

Zeros: $\dfrac{-3 \pm \sqrt{7}i}{2}$

—CONTINUED—

1. —CONTINUED—

    (b) (i) $f(x) = (x - 2)(x^2 - 4x - 12)$

    (iii) $f(x) = (x - 2)(x^2 + 3x - 10)$

    (v) $f(x) = (x - 2)(x^2 - 2x - 6)$

    (ii) $f(x) = (x - 2)(x^2 + 5x)$

    (iv) $f(x) = (x - 2)(x^2 - 4x + 4)$

    (vi) $f(x) = (x - 2)(x^2 + 3x + 4)$

    $x = 2$ is an x-intercept of $f(x)$ in all six graphs. All the graphs, except (iii), cross the x-axis at $x = 2$.

    (c) (i) other x-intercepts: $(-2, 0), (6, 0)$

      (ii) other x-intercepts: $(-5, 0), (0, 0)$

      (iii) other x-intercepts: $(-5, 0)$

      (iv) other x-intercepts: No additional x-intercepts

      (v) other x-intercepts: $(-1.6, 0), (3.6, 0)$

      (vi) other x-intercepts: No additional x-intercepts

    (d) The x-intercepts of $f(x) = (x - 2)g(x)$ are the same as the zeros of $g(x)$ plus $x = 2$.

3. $f(x) = ax^3 + bx^2 + cx + d$

$$
\require{enclose}
\begin{array}{r}
ax^2 + (ak + b)x + (ak^2 + bk + c) \\
x - k \enclose{longdiv}{ax^3 + bx^2 \qquad + cx \qquad\qquad + d} \\
\end{array}
$$

$$\underline{ax^3 - akx^2}$$
$$(ak + b)x^2 + cx$$
$$\underline{(ak + b)x^2 - (ak^2 + bk)x}$$
$$(ak^2 + bk + c)x + d$$
$$\underline{(ak^2 + bk + c)x - (ak^3 + bk^2 + ck)}$$
$$(ak^3 + bk^2 + ck + d)$$

Thus, $f(x) = ax^3 + bx^2 + cx + d = (x - k)[ax^2 + (ak + b)x + (ak^2 + bx + c)] + ak^3 + bk^2 + ck + d$ and

$f(k) = ak^3 + bk^2 + ck + d$. Since the remainder $r = ak^3 + bk^2 + ck + d$, $f(k) = r$.

5. $V = l \cdot w \cdot h = x^2(x + 3)$

$x^2(x + 3) = 20$

$x^3 + 3x^2 - 20 = 0$

Possible rational zeros: $\pm 1, \pm 2, \pm 4, \pm 5, \pm 10, \pm 20$

$$
\begin{array}{r|rrrr}
2 & 1 & 3 & 0 & -20 \\
  &   & 2 & 10 & 20 \\
\hline
  & 1 & 5 & 10 & 0
\end{array}
$$

$(x - 2)(x^2 + 5x + 10) = 0$

$x = 2$ or $x = \dfrac{-5 \pm \sqrt{15}i}{2}$

Choosing the real positive value for x we have: $x = 2$ and $x + 3 = 5$. The dimensions of the mold are 2 inches $\times$ 2 inches $\times$ 5 inches.

7. False. Since $f(x) = d(x)q(x) + r(x)$, we have

$$\frac{f(x)}{d(x)} = q(x) + \frac{r(x)}{d(x)}.$$

The statement should be corrected to read $f(-1) = 2$ since

$$\frac{f(x)}{x + 1} = q(x) + \frac{f(-1)}{x + 1}.$$

9. (a) Slope $= \dfrac{9 - 4}{3 - 2} = 5$

   Slope of tangent line is less than 5.

(b) Slope $= \dfrac{4 - 1}{2 - 1} = 3$

   Slope of tangent line is greater than 3.

(c) Slope $= \dfrac{4.41 - 4}{2.1 - 2} = 4.1$

   Slope of tangent line is less than 4.1.

(d) Slope $= \dfrac{f(2 + h) - f(2)}{(2 + h) - 2}$

$\qquad = \dfrac{(2 + h)^2 - 4}{h}$

$\qquad = \dfrac{4h + h^2}{h}$

$\qquad = 4 + h, h \neq 0$

(e) $\qquad$ Slope $= 4 + h, \quad h \neq 0$

$\qquad 4 + (-1) = 3$

$\qquad\quad 4 + 1 = 5$

$\qquad 4 + 0.1 = 4.1$

   The results are the same as in (a)–(c).

(f) Letting h get closer and closer to 0, the slope approaches 4. Hence, the slope at $(2, 4)$ is 4.

11. Let x = length of the wire used to form the square. Then $100 - x$ = length of wire used to form the circle.

(a) Let s = the side of the square. Then $4s = x \implies s = x/4$ and the area of the square is $s^2 = (x/4)^2$.
   Let r = the radius of the circle. Then

$$2\pi r = 100 - x \implies r = \frac{100 - x}{2\pi} \text{ and the area of the circle is } \pi r^2 = \pi\left(\frac{100 - x}{2\pi}\right)^2.$$

The combined area is:

$$A(x) = \left(\frac{x}{4}\right)^2 + \pi\left(\frac{100 - x}{2\pi}\right)^2$$

$$= \frac{x^2}{16} + \pi\left(\frac{10{,}000 - 200x + x^2}{4\pi^2}\right)$$

$$= \frac{x^2}{16} + \frac{2500}{\pi} - \frac{50x}{\pi} + \frac{x^2}{4\pi}$$

$$= \left(\frac{1}{16} + \frac{1}{4\pi}\right)x^2 - \frac{50x}{\pi} + \frac{2500}{\pi}$$

$$= \left(\frac{\pi + 4}{16\pi}\right)x^2 - \frac{50}{\pi}x + \frac{2500}{\pi}$$

(b) Domain: Since the wire is 100 cm, $0 \le x \le 100$.

(c) $A(x) = \left(\dfrac{\pi + 4}{16\pi}\right)x^2 - \dfrac{50}{\pi}x + \dfrac{2500}{\pi}$

$$= \left(\frac{\pi + 4}{16\pi}\right)\left(x^2 - \frac{800}{\pi + 4}x\right) + \frac{2500}{\pi}$$

$$= \left(\frac{\pi + 4}{16\pi}\right)\left[x^2 - \frac{800}{\pi + 4}x + \left(\frac{400}{\pi + 4}\right)^2 - \left(\frac{400}{\pi + 4}\right)^2\right] + \frac{2500}{\pi}$$

$$= \left(\frac{\pi + 4}{16\pi}\right)\left[x - \left(\frac{400}{\pi + 4}\right)\right]^2 - \left(\frac{\pi + 4}{16\pi}\right)\left(\frac{400}{\pi + 4}\right)^2 + \frac{2500}{\pi}$$

$$= \left(\frac{\pi + 4}{16\pi}\right)\left[x - \left(\frac{400}{\pi + 4}\right)\right]^2 - \frac{10{,}000}{\pi(\pi + 4)} + \frac{2500}{\pi}$$

$$= \left(\frac{\pi + 4}{16\pi}\right)\left[x - \left(\frac{400}{\pi + 4}\right)\right]^2 + \frac{2500}{\pi + 4}$$

The minimum occurs at the vertex when $x = 400/(\pi + 4) \approx 56$ cm and $A(x) \approx 350$ cm^2. The maximum occurs at one of the endpoints of the domain. When $x = 0$, $A(x) \approx 796$ cm^2. When $x = 100$, $A(x) \approx 625$ cm^2. Thus, the area is maximum when $x = 0$ cm.

(d) Answers will vary. Graph $A(x)$ to see where the minimum and maximum values occur.

Practice Test for Chapter 3

1. Sketch the graph of $f(x) = x^2 - 6x + 5$ and identify the vertex and the intercepts.

2. Find the number of units x that produce a minimum cost C if
   $C = 0.01x^2 - 90x + 15,000$.

3. Find the quadratic function that has a maximum at $(1, 7)$ and passes through the point $(2, 5)$.

4. Find two quadratic functions that have x-intercepts $(2, 0)$ and $\left(\frac{4}{3}, 0\right)$.

5. Use the leading coefficient test to determine the right and left end behavior of the graph of the polynomial function $f(x) = -3x^5 + 2x^3 - 17$.

6. Find all the real zeros of $f(x) = x^5 - 5x^3 + 4x$.

7. Find a polynomial function with 0, 3, and -2 as zeros.

8. Sketch $f(x) = x^3 - 12x$.

9. Divide $3x^4 - 7x^2 + 2x - 10$ by $x - 3$ using long division.

10. Divide $x^3 - 11$ by $x^2 + 2x - 1$.

11. Use synthetic division to divide $3x^5 + 13x^4 + 12x - 1$ by $x + 5$.

12. Use synthetic division to find $f(-6)$ given $f(x) = 7x^3 + 40x^2 - 12x + 15$.

13. Find the real zeros of $f(x) = x^3 - 19x - 30$.

14. Find the real zeros of $f(x) = x^4 + x^3 - 8x^2 - 9x - 9$.

15. List all possible rational zeros of the function $f(x) = 6x^3 - 5x^2 + 4x - 15$.

16. Find the rational zeros of the polynomial $f(x) = x^3 - \frac{20}{3}x^2 + 9x - \frac{10}{3}$.

17. Write $f(x) = x^4 + x^3 + 3x^2 + 5x - 10$ as a product of linear factors.

18. Find a polynomial with real coefficients that has 2, $3 + i$, and $3 - 2i$ as zeros.

19. Use synthetic division to show that $3i$ is a zero of $f(x) = x^3 + 4x^2 + 9x + 36$.

20. Find a mathematical model for the statement, "z varies directly as the square of x and inversely as the square root of y."

C H A P T E R 4
Rational Functions and Conics

CHAPTER 4
Rational Functions and Conics

Section 4.1   Rational Functions and Asymptotes

■ You should know the following basic facts about rational functions.

(a) A function of the form $f(x) = N(x)/D(x)$, $D(x) \neq 0$, where $N(x)$ and $D(x)$ are polynomials, is called a rational function.

(b) The domain of a rational function is the set of all real numbers except those which make the denominator zero.

(c) If $f(x) = N(x)/D(x)$ is in reduced form, and a is a value such that $D(a) = 0$, then the line $x = a$ is a vertical asymptote of the graph of f. $f(x) \to \infty$ or $f(x) \to -\infty$ as $x \to a$.

(d) The line $y = b$ is a horizontal asymptote of the graph of f if $f(x) \to b$ as $x \to \infty$ or $x \to -\infty$.

(e) Let $f(x) = \dfrac{N(x)}{D(x)} = \dfrac{a_n x^n + a_{n-1}x^{n-1} + \cdots + a_1 x + a_0}{b_m x^m + b_{m-1}x^{m-1} + \cdots + b_1 x + b_0}$ where $N(x)$ and $D(x)$ have no common factors.

  1. If $n < m$, then the x-axis ($y = 0$) is a horizontal asymptote.

  2. If $n = m$, then $y = \dfrac{a_n}{b_m}$ is a horizontal asymptote.

  3. If $n > m$, then there are no horizontal asymptotes.

Vocabulary Check

1. rational functions   2. vertical asymptote   3. horizontal asymptote

---

1. $f(x) = \dfrac{1}{x - 1}$

(a)

x	$f(x)$
0.5	-2
0.9	-10
0.99	-100
0.999	-1000

x	$f(x)$
1.5	2
1.1	10
1.01	100
1.001	1000

x	$f(x)$
5	0.25
10	$0.\overline{1}$
100	$0.\overline{01}$
1000	$0.\overline{001}$

(b) The zero of the denominator is $x = 1$, so $x = 1$ is a vertical asymptote. The degree of the numerator is less than the degree of the denominator so the x-axis, or $y = 0$, is a horizontal asymptote.

(c) The domain is all real numbers except $x = 1$.

---

3. $f(x) = \dfrac{3x^2}{x^2 - 1}$

(a)

x	$f(x)$
0.5	-1
0.9	-12.79
0.99	-147.8
0.999	-1498

x	$f(x)$
1.5	5.4
1.1	17.29
1.01	152.3
1.001	1502

x	$f(x)$
5	3.125
10	$3.\overline{03}$
100	$3.\overline{0003}$
1000	3

(b) The zeros of the denominator are $x = \pm 1$ so both $x = 1$ and $x = -1$ are vertical asymptotes. Since the degree of the numerator equals the degree of the denominator, $y = \frac{3}{1} = 3$ is a horizontal asymptote.

(c) The domain is all real numbers except $x = \pm 1$.

5. $f(x) = \dfrac{1}{x^2}$

Domain: all real numbers except $x = 0$

Vertical asymptote: $x = 0$

Horizontal asymptote: $y = 0$

[Degree of $N(x) <$ degree of $D(x)$]

7. $f(x) = \dfrac{2+x}{2-x} = \dfrac{x+2}{-x+2}$

Domain: all real numbers except $x = 2$

Vertical asymptote: $x = 2$

Horizontal asymptote: $y = -1$

[Degree of $N(x) =$ degree of $D(x)$]

9. $f(x) = \dfrac{x^3}{x^2-1}$

Domain: all real numbers except $x = \pm 1$

Vertical asymptotes: $x = \pm 1$

Horizontal asymptote: None

[Degree of $N(x) >$ degree of $D(x)$]

11. $f(x) = \dfrac{3x^2+1}{x^2+x+9}$

Domain: All real numbers. The denominator has no real zeros. [Try the Quadratic Formula on the denominator.]

Vertical asymptote: None

Horizontal asymptote: $y = 3$

[Degree of $N(x) =$ degree of $D(x)$]

13. $f(x) = \dfrac{2}{x+3}$

Vertical asymptote: $y = -3$

Horizontal asymptote: $y = 0$

Matches graph (d).

15. $f(x) = \dfrac{x-1}{x-4}$

Vertical asymptote: $x = 4$

Horizontal asymptote: $y = 1$

Matches graph (c).

17. $g(x) = \dfrac{x^2-1}{x+1} = \dfrac{(x-1)(x+1)}{x+1}$

The only zero of $g(x)$ is $x = 1$.

$x = -1$ makes $g(x)$ undefined.

19. $h(x) = 2 + \dfrac{5}{x^2+2}$

$$0 = 2 + \dfrac{5}{x^2+2}$$

$$-2 = \dfrac{5}{x^2+2}$$

$$-2(x^2+2) = 5$$

$$x^2 = -\dfrac{5}{2} - 2$$

No real solution, $h(x)$ has no real zeros.

21. $f(x) = 1 - \dfrac{3}{x-3}$

$$1 - \dfrac{3}{x-3} = 0$$

$$1 = \dfrac{3}{x-3}$$

$$x - 3 = 3$$

$x = 6$ is a zero of $f(x)$.

23. $g(x) = \dfrac{x^3-8}{x^2+1}$

$$\dfrac{x^3-8}{x^2+1} = 0$$

$$x^3 - 8 = 0$$

$$x^3 = 8$$

$$x = 2$$

$x = 2$ is a real zero of $g(x)$.

25. $f(x) = \dfrac{x-4}{x^2-16} = \dfrac{1}{x+4}, \ x \neq 4$

Domain: all real numbers x except $x = \pm 4$

Horizontal asymptote: $y = 0$ (Degree of $N(x) <$ degree of $D(x)$)

Vertical asymptote: $x = -4$

(Since $x - 4$ is a common factor of $N(x)$ and $D(x)$, $x = 4$ is not a vertical asymptote of $f(x)$).

27. $f(x) = \dfrac{x^2 - 1}{x^2 - 2x - 3} = \dfrac{(x + 1)(x - 1)}{(x + 1)(x - 3)} = \dfrac{x - 1}{x - 3}, \; x \neq -1$

Domain: all real numbers x except $x = -1$ and $x = 3$

Horizontal asymptote: $y = 1$ (Degree of $N(x)$ = degree of $D(x)$)

Vertical asymptote: $x = 3$
(Since $x + 1$ is a common factor of $N(x)$ and $D(x)$, $x = -1$ is not a vertical asymptote of $f(x)$).

29. $f(x) = \dfrac{x^2 - 3x - 4}{2x^2 + x - 1} = \dfrac{(x + 1)(x - 4)}{(2x - 1)(x + 1)} = \dfrac{x - 4}{2x - 1}, \; x \neq -1$

Domain: all real numbers x except $x = \dfrac{1}{2}$ and $x = -1$

Horizontal asymptote: $y = \dfrac{1}{2}$ (Degree of $N(x)$ = degree of $D(x)$)

Vertical asymptote: $x = \dfrac{1}{2}$
(Since $x + 1$ is a common factor of $N(x)$ and $D(x)$, $x = -1$ is not a vertical asymptote of $f(x)$).

31. $f(x) = \dfrac{6x^2 + 5x - 6}{3x^2 - 8x + 4} = \dfrac{(3x - 2)(2x + 3)}{(3x - 2)(x - 2)} = \dfrac{2x + 3}{x - 2}, \; x \neq \dfrac{2}{3}$

Domain: all real numbers x except $x = \frac{2}{3}$ and $x = 2$

Horizontal asymptote: $y = 2$
(Degree of $N(x)$ = degree of $D(x)$)

Vertical asymptote: $x = 2$
(Since $3x - 2$ is a common factor of $N(x)$ and $D(x)$, $x = \frac{2}{3}$ is not a vertical asymptote of $f(x)$).

33. $f(x) = \dfrac{x^2 - 4}{x + 2}, \; g(x) = x - 2$

$f(x) = \dfrac{(x + 2)(x - 2)}{x + 2}$

(a) Domain of f: all real numbers except -2
Domain of g: all real numbers

(b) Since $x + 2$ is a common factor of both the numerator and the denominator of $f(x)$, $x = -2$ is not a vertical asymptote of f. f has no vertical asymptotes.

(c)

x	-4	-3	-2.5	-2	-1.5	-1	0
$f(x)$	-6	-5	-4.5	Undef.	-3.5	-3	-2
$g(x)$	-6	-5	-4.5	-4	-3.5	-3	-2

(d) f and g differ only where f is undefined and g is defined.

35. $f(x) = \dfrac{2x - 1}{2x^2 - x}, \; g(x) = \dfrac{1}{x}$

$f(x) = \dfrac{2x - 1}{x(2x - 1)}$

(a) Domain of f: all real number except 0 and $\frac{1}{2}$
Domain of g: all real numbers except 0

(b) Since $2x - 1$ is a common factor of both the numerator and the denominator of f, $x = 0.5$ is not a vertical asymptote of f. The only vertical asymptote is $x = 0$.

(c)

x	-1	-0.5	0	0.5	2	3	4
$f(x)$	-1	-2	Undef.	Undef.	$\frac{1}{2}$	$\frac{1}{3}$	$\frac{1}{4}$
$g(x)$	-1	-2	Undef.	2	$\frac{1}{2}$	$\frac{1}{3}$	$\frac{1}{4}$

(d) They differ only at $x = 0.5$, where f is undefined and g is defined.

37. $f(x) = 4 - \dfrac{1}{x}$

(a) As $x \to \pm\infty, f(x) \to 4$.

(b) As $x \to \infty, f(x) \to 4$ but is less than 4.

(c) As $x \to -\infty, f(x) \to 4$ but is greater than 4.

39. $f(x) = \dfrac{2x - 1}{x - 3}$

(a) As $x \to \pm\infty, f(x) \to 2$.

(b) As $x \to \infty, f(x) \to 2$ but is greater than 2.

(c) As $x \to -\infty, f(x) \to 2$ but is less than 2.

41. $t = \dfrac{38M + 16{,}695}{10(M + 5000)}$

(a)

M	200	400	600	800	1000
t	0.467	0.591	0.705	0.812	0.912

M	1200	1400	1600	1800	2000
t	1.005	1.092	1.174	1.251	1.324

(b) The model is a "good fit" to the actual data. The greater the mass, the more time required per oscillation.

43. $C = \dfrac{255p}{100 - p}, \ 0 \le p < 100$

(a)

(b) $C(10) = \dfrac{255(10)}{100 - 10} \approx 28.33$ million dollars

$C(40) = \dfrac{255(40)}{100 - 40} = 170$ million dollars

$C(75) = \dfrac{255(75)}{100 - 75} = 765$ million dollars

(c) $C \to \infty$ as $x \to 100$. No, it would not be possible to remove 100% of the pollutants.

45. $N = \dfrac{20(5 + 3t)}{1 + 0.04t}, \ t \ge 0$

(a) $N(5) \approx 334$ deer

$N(10) = 500$ deer

$N(25) = 800$ deer

(b) The herd is limited by the horizontal asymptote:

$N = \dfrac{60}{0.04} = 1500$ deer

47. $P = \dfrac{0.5 + 0.9(n - 1)}{1 + 0.9(n - 1)}, \ n > 0$

(a)

n	1	2	3	4	5
P	0.50	0.74	0.82	0.86	0.89

n	6	7	8	9	10
P	0.91	0.92	0.93	0.94	0.95

P approaches 1 as n increases.

(b) $P = \dfrac{0.9n - 0.4}{0.9n + 0.1}$

The percentage of correct responses is limited by a horizontal asymptote:

$P = \dfrac{0.9}{0.9} = 1 = 100\%$

49. $D = \dfrac{1.391t^2 - 73.37}{-0.003t^2 + 1.00}, \ 8 \le t \le 13$

(a)

The model is a fairly good fit to the actual data.

(b) $D(16) \approx \$1218.65$ million

This value seems to be a little high.

(c) The function has a vertical asymptote at

$$x = \sqrt{\dfrac{1.00}{0.003}} \approx 18.26$$

and approaches infinity as x gets closer to this value, so would not be useful for forecasting after 2003.

(d) Linear model: $y_1 \approx 68.78t - 563$

Quadratic model: $y_2 \approx 9.329t^2 - 127.12t + 438.3$

(e) $y_1(16) \approx \$537.5$ million

$y_2(16) \approx \$792.6$ million

(f) The quadratic model appears to best fit the data and would be best for forecasting future values of DVDs shipped, as long as t does not vary too far from $t = 13$.

51. False. Polynomial functions do not have vertical asymptotes.

53. Vertical asymptote: None $\Rightarrow$ The denominator is not zero for any value of x (unless the numerator is also zero there).

Horizontal asymptote: $y = 2 \Rightarrow$ The degree of the numerator equals the degree of the denominator.

$f(x) = \dfrac{2x^2}{x^2 + 1}$ is one possible function. There are many correct answers.

55. Vertical asymptotes: $x = -2$, $x = 1 \Rightarrow (x + 2)(x - 1)$ are factors of the denominator.

Horizontal asymptotes: None $\Rightarrow$ The degree of the numerator is greater than the degree of the denominator.

$f(x) = \dfrac{x^3}{(x + 2)(x - 1)}$ is one possible function. There are many correct answers.

57. Domain: All real numbers

Example: $f(x) = \dfrac{1}{x^2 + 2}$

Domain: All real numbers except $x = 20$

Example: $f(x) = \dfrac{1}{x - 20}$

There are many correct answers.

59. $f(x) = 8x - 7$

$y = 8x - 7$

$x = 8y - 7$

$x + 7 = 8y$

$\dfrac{x + 7}{8} = y$

$f^{-1}(x) = \dfrac{x + 7}{8}$

61.

$$
\begin{array}{r}
x + 9 \\
x - 4 \overline{)\, x^2 + 5x + 6} \\
\underline{x^2 - 4x} \\
9x + 6 \\
\underline{9x - 36} \\
42
\end{array}
$$

Thus, $\dfrac{x^2 + 5x + 6}{x - 4} = x + 9 + \dfrac{42}{x - 4}$.

63.

$$
\begin{array}{r}
2x - 9 \\
x + 5 \overline{)\, 2x^2 + x - 11} \\
\underline{2x^2 + 10x} \\
-9x - 11 \\
\underline{-9x - 45} \\
34
\end{array}
$$

Thus, $\dfrac{2x^2 + x - 11}{x + 5} = 2x - 9 + \dfrac{34}{x + 5}$.

Section 4.2 Graphs of Rational Functions

■ You should be able to graph $f(x) = \dfrac{N(x)}{D(x)}$ where $N(x)$ and $D(x)$ are polynomials with no common factors.

(a) Find the x-and y-intercepts.

(b) Test for symmetry.

(c) Find any vertical or horizontal asymptotes.

(d) Plot additional points.

(e) If the degree of the numerator is one more than the degree of the denominator, use long division to find the slant asymptote.

Vocabulary Check

1. slant asymptote

2. vertical

1. $g(x) = \dfrac{2}{x} + 3$

Vertical shift three units upward

3. $g(x) = -\dfrac{2}{x}$

Reflection in the *x*-axis

5. $g(x) = \dfrac{2}{x^2} - 1$

Vertical shift one unit downward

7. $g(x) = \dfrac{2}{(x-1)^2}$

Horizontal shift one unit to the right

9. $g(x) = \dfrac{4}{(x+3)^3}$

Horizontal shift three units to the left

11. $g(x) = -\dfrac{4}{x^3}$

Reflection in the *x*-axis

13. $f(x) = \dfrac{1}{x+2}$

(a) Domain: all real numbers *x* except $x = -2$

(b) *y*-intercept: $\left(0, \dfrac{1}{2}\right)$

(c) Vertical asymptote: $x = -2$
    Horizontal asymptote: $y = 0$

(d)

x	-4	-3	-1	0	1
y	$-\frac{1}{2}$	-1	1	$\frac{1}{2}$	$\frac{1}{3}$

15. $h(x) = \dfrac{-1}{x+2}$

(a) Domain: all real numbers *x* except $x = -2$

(b) *y*-intercept: $\left(0, -\dfrac{1}{2}\right)$

(c) Vertical asymptote: $x = -2$
    Horizontal asymptote: $y = 0$

(d)

x	-4	-3	-1	0
y	$\frac{1}{2}$	1	-1	$-\frac{1}{2}$

Note: This is the graph of $f(x) = \dfrac{1}{x+2}$
(Exercise 13) reflected about the *x*-axis.

17. $C(x) = \dfrac{5 + 2x}{1 + x} = \dfrac{2x + 5}{x + 1}$

  (a) Domain: all real numbers x except $x = -1$

  (b) x-intercept: $\left(-\dfrac{5}{2}, 0\right)$

   y-intercept: $(0, 5)$

  (c) Vertical asymptote: $x = -1$
    Horizontal asymptote: $y = 2$

  (d)

x	-4	-3	-2	0	1	2
$C(x)$	1	$\frac{1}{2}$	-1	5	$\frac{7}{2}$	3

19. $g(x) = \dfrac{1}{x + 2} + 2 = \dfrac{2x + 5}{x + 2}$

  (a) Domain: all real numbers x except $x = -2$

  (b) x-intercept: $\left(-\dfrac{5}{2}, 0\right)$

   y-intercept: $\left(0, \dfrac{5}{2}\right)$

  (c) Vertical asymptote: $x = -2$
    Horizontal asymptote: $y = 2$

  (d)

x	-4	-3	-1	0	1
y	$\frac{3}{2}$	1	3	$\frac{5}{2}$	$\frac{7}{3}$

Note: This is the graph of $f(x) = \dfrac{1}{x + 2}$
(Exercise 13) shifted upward two units.

21. $f(x) = \dfrac{x^2}{x^2 + 9}$

  (a) Domain: all real numbers x

  (b) Intercept: $(0, 0)$

  (c) Horizontal asymptote: $y = 1$

  (d)

x	± 1	± 2	± 3
y	$\frac{1}{10}$	$\frac{4}{13}$	$\frac{1}{2}$

23. $h(x) = \dfrac{x^2}{x^2 - 9}$

  (a) Domain: all real numbers x except $x = \pm 3$

  (b) Intercept: $(0, 0)$

  (c) Vertical asymptotes: $x = \pm 3$
    Horizontal asymptote: $y = 1$

  (d)

x	± 5	± 4	± 2	± 1	0
y	$\frac{25}{16}$	$\frac{16}{7}$	$-\frac{4}{5}$	$-\frac{1}{8}$	0

25. $g(s) = \dfrac{s}{s^2 + 1}$

   (a) Domain: all real numbers s

   (b) Intercept: $(0, 0)$

   (c) Horizontal asymptote: $y = 0$

   (d)

s	-2	-1	0	1	2
$g(s)$	$-\frac{2}{5}$	$-\frac{1}{2}$	0	$\frac{1}{2}$	$\frac{2}{5}$

27. $g(x) = \dfrac{4(x + 1)}{x(x - 4)}$

   (a) Domain: all real numbers x except $x = 0$ and $x = 4$

   (b) x-intercept: $(-1, 0)$

   (c) Vertical asymptotes: $x = 0$ and $x = 4$
Horizontal asymptote: $y = 0$

   (d)

x	-2	-1	1	2	3	5	6
y	$-\frac{1}{3}$	0	$-\frac{8}{3}$	-3	$-\frac{16}{3}$	$\frac{24}{5}$	$\frac{7}{3}$

29. $f(x) = \dfrac{3x}{x^2 - x - 2} = \dfrac{3x}{(x + 1)(x - 2)}$

   (a) Domain: all real numbers x except $x = -1$ and $x = 2$

   (b) Intercept: $(0, 0)$

   (c) Vertical asymptotes: $x = -1$ and $x = 2$
Horizontal asymptote: $y = 0$

   (d)

x	-3	0	1	3	4
y	$-\frac{9}{10}$	0	$-\frac{3}{2}$	$\frac{9}{4}$	$\frac{6}{5}$

31. $h(x) = \dfrac{x^2 - 5x + 4}{x^2 - 4} = \dfrac{(x - 1)(x - 4)}{(x + 2)(x - 2)}$

   (a) Domain: all real numbers x except $x = \pm 2$

   (b) x-intercepts: $(1, 0)$, $(4, 0)$
y-intercept: $(0, -1)$

   (c) Vertical asymptotes: $x = -2$, $x = 2$
Horizontal asymptote: $y = 1$

   (d)

x	-4	-3	-1	0	1	3	4
y	$\frac{10}{3}$	$\frac{28}{5}$	$-\frac{10}{3}$	-1	0	$-\frac{2}{5}$	0

33. $f(x) = \dfrac{6x}{x^2 - 5x - 14} = \dfrac{6x}{(x + 2)(x - 7)}$

   (a) Domain: all real numbers x except $x = -2$ and $x = 7$

   (b) Intercept: $(0, 0)$

   (c) Vertical asymptotes: $x = -2$, and $x = 7$
Horizontal asymptotes: $y = 0$

   (d)

x	-6	-4	0	2	4	6	8	10
$f(x)$	$-\frac{9}{13}$	$-\frac{12}{11}$	0	$-\frac{3}{5}$	$-\frac{4}{3}$	$-\frac{9}{2}$	$\frac{24}{5}$	$\frac{5}{3}$

35. $f(x) = \dfrac{2x^2 - 5x - 3}{x^3 - 2x^2 - x + 2} = \dfrac{(2x + 1)(x - 3)}{(x - 2)(x + 1)(x - 1)}$

(a) Domain: all real numbers x except $x = 2$, $x = \pm 1$

(b) Intercepts: $\left(-\dfrac{1}{2}, 0\right)$, $(3, 0)$, $\left(0, -\dfrac{3}{2}\right)$

(c) Vertical asymptotes: $x = 2$, $x = -1$ and $x = 1$
Horizontal asymptotes: $y = 0$

(d)

x	-3	-2	0	1.5	3	4
$f(x)$	$-\dfrac{3}{4}$	$-\dfrac{5}{4}$	$-\dfrac{3}{2}$	$\dfrac{48}{5}$	0	$\dfrac{3}{10}$

37. $f(x) = \dfrac{x^2 + 3x}{x^2 + x - 6} = \dfrac{x(x + 3)}{(x + 3)(x - 2)} = \dfrac{x}{x - 2}$, $x \neq -3$

(a) Domain: all real numbers x except $x = -3$ and $x = 2$

(b) Intercept: $(0, 0)$

(c) Vertical asymptote: $x = 2$
Horizontal asymptote: $y = 1$

(d)

x	-1	0	1	3	4
y	$\dfrac{1}{3}$	0	-1	3	2

39. $f(x) = \dfrac{2x^2 - 5x + 2}{2x^2 - x - 6}$

$= \dfrac{(2x - 1)(x - 2)}{(2x + 3)(x - 2)} = \dfrac{2x - 1}{2x + 3}$, $x \neq 2$

(a) Domain: all real numbers x except $x = 2$ and

$x = -\dfrac{3}{2}$

(b) x-intercept: $\left(\dfrac{1}{2}, 0\right)$

y-intercept: $\left(0, -\dfrac{1}{3}\right)$

(c) Vertical asymptote: $x = -\dfrac{3}{2}$

Horizontal asymptote: $y = 1$

(d)

x	-3	-2	-1	0	1
y	$\dfrac{7}{3}$	5	-3	$-\dfrac{1}{3}$	$\dfrac{1}{5}$

41. $f(t) = \dfrac{t^2 - 1}{t + 1} = \dfrac{(t + 1)(t - 1)}{t + 1} = t - 1$, $t \neq -1$

(a) Domain: all real numbers t except $t = -1$

(b) t-intercept: $(1, 0)$
y-intercept: $(0, -1)$

(c) Vertical asymptote: none
Horizontal asymptote: none

(d)

t	-3	-2	0	1
y	-4	-3	-1	0

43. $f(x) = \dfrac{x^2 - 1}{x + 1}$, $g(x) = x - 1$

(a) Domain of f: all real numbers x except $x = -1$

Domain of g: all real numbers x

(b) Because $(x + 1)$ is a factor of both the numerator and the denominator of f, $x = -1$ is not a vertical asymptote. f has no vertical asymptotes.

(c)

x	-3	-2	-1.5	-1	-0.5	0	1
$f(x)$	-4	-3	-2.5	Undef.	-1.5	-1	0
$g(x)$	-4	-3	-2.5	-2	-1.5	-1	0

(d)

(e) Because there are only a finite number of pixels, the utility may not attempt to evaluate the function where it does not exist.

45. $f(x) = \dfrac{x - 2}{x^2 - 2x}$, $g(x) = \dfrac{1}{x}$

(a) Domain of f: all real numbers x except $x = 0$ and $x = 2$

Domain of g: all real numbers x except $x = 0$

(b) Because $(x - 2)$ is a factor of both the numerator and the denominator of f, $x = 2$ is not a vertical asymptote. The only vertical asymptote of f is $x = 0$.

(c)

x	-0.5	0	0.5	1	1.5	2	3
$f(x)$	-2	Undef.	2	1	$\frac{2}{3}$	Undef.	$\frac{1}{3}$
$g(x)$	-2	Undef.	2	1	$\frac{2}{3}$	$\frac{1}{2}$	$\frac{1}{3}$

(d)

(e) Because there are only a finite number of pixels, the utility may not attempt to evaluate the function where it does not exist.

47. $h(x) = \dfrac{x^2 - 4}{x} = x - \dfrac{4}{x}$

(a) Domain: all real numbers x except $x = 0$

(b) Intercepts: $(2, 0), (-2, 0)$

(c) Vertical asymptote: $x = 0$
Slant asymptote: $y = x$

(d)

x	-3	-1	1	3
y	$-\frac{5}{3}$	3	-3	$\frac{5}{3}$

49. $f(x) = \dfrac{2x^2 + 1}{x} = 2x + \dfrac{1}{x}$

(a) Domain: all real numbers x except $x = 0$

(b) No intercepts

(c) Vertical asymptote: $x = 0$
Slant asymptote: $y = 2x$

(d)

x	-4	-2	2	4	6
$f(x)$	$-\frac{33}{4}$	$-\frac{9}{2}$	$\frac{9}{2}$	$\frac{33}{4}$	$\frac{73}{6}$

51. $g(x) = \dfrac{x^2 + 1}{x} = x + \dfrac{1}{x}$

(a) Domain: all real numbers x except $x = 0$

(b) No intercepts

(c) Vertical asymptote: $x = 0$
Slant asymptote: $y = x$

(d)

x	-4	-2	2	4	6
$g(x)$	$-\frac{17}{4}$	$-\frac{5}{2}$	$\frac{5}{2}$	$\frac{17}{4}$	$\frac{37}{6}$

53. $f(t) = -\dfrac{t^2 + 1}{t + 5} = -t + 5 - \dfrac{26}{t + 5}$

(a) Domain: all real numbers t except $t = -5$

(b) Intercept: $\left(0, -\dfrac{1}{5}\right)$

(c) Vertical asymptote: $t = -5$
Slant asymptote: $y = -t + 5$

(d)

t	-7	-6	-4	-3	0
y	25	37	-17	-5	$-\frac{1}{5}$

55. $f(x) = \dfrac{x^3}{x^2 - 1} = x + \dfrac{x}{x^2 - 1}$

(a) Domain: all real numbers x except $x = \pm 1$

(b) Intercept: $(0, 0)$

(c) Vertical asymptotes: $x = \pm 1$
Slant asymptote: $y = x$

(d)

x	-4	-2	0	2	4
$f(x)$	$-\frac{64}{15}$	$-\frac{8}{3}$	0	$\frac{8}{3}$	$\frac{64}{15}$

57. $f(x) = \dfrac{x^3 - 1}{x^2 - x} = \dfrac{(x - 1)(x^2 + x + 1)}{x(x - 1)} = \dfrac{x^2 + x + 1}{x}$

$= x + 1 + \dfrac{1}{x}, x \neq 1$

(a) Domain: all real numbers except $x = 0$ and $x = 1$

(b) Intercepts: none

(c) Vertical asymptote: $x = 0$
Note: $x = 1$ is not a vertical asymptote since it also makes the numerator zero.
Slant asymptote: $y = x + 1$

(d)

x	-2	-1	$\frac{1}{2}$	2
y	$-\frac{3}{2}$	-1	$\frac{7}{2}$	$\frac{7}{2}$

59. $f(x) = \dfrac{x^2 - x + 1}{x - 1} = x + \dfrac{1}{x - 1}$

(a) Domain: all real numbers x except $x = 1$

(b) y-intercept: $(0, -1)$

(c) Vertical asymptote: $x = 1$
Slant asymptote: $y = x$

(d)

x	-4	-2	0	2	4
$f(x)$	$-\frac{21}{5}$	$-\frac{7}{3}$	-1	3	$\frac{13}{3}$

61. $f(x) = \dfrac{2x^3 - x^2 - 2x + 1}{x^2 + 3x + 2}$

$\qquad = \dfrac{(2x - 1)(x + 1)(x - 1)}{(x + 1)(x + 2)}$

$\qquad = \dfrac{(2x - 1)(x - 1)}{x + 2}, \quad x \neq -1$

$\qquad = \dfrac{2x^2 - 3x + 1}{x + 2}$

$\qquad = 2x - 7 + \dfrac{15}{x + 2}, \quad x \neq -1$

(a) Domain: all real numbers x except
$\qquad x = -1$ and $x = -2$

(b) y-intercept: $\left(0, \dfrac{1}{2}\right)$

x-intercepts: $\left(\dfrac{1}{2}, 0\right), (1, 0)$

(c) Vertical asymptote: $x = -2$
Slant asymptote: $y = 2x - 7$

(d)

x	-4	-3	$-\frac{3}{2}$	0	1
y	$-\frac{45}{2}$	-28	20	$\frac{1}{2}$	0

63. $f(x) = \dfrac{x^2 + 5x + 8}{x + 3} = x + 2 + \dfrac{2}{x + 3}$

Domain: all real numbers x except $x = -3$

y-intercept: $\left(0, \dfrac{8}{3}\right)$

Vertical asymptote: $x = -3$

Slant asymptote: $y = x + 2$

Line: $y = x + 2$

65. $g(x) = \dfrac{1 + 3x^2 - x^3}{x^2} = \dfrac{1}{x^2} + 3 - x = -x + 3 + \dfrac{1}{x^2}$

Domain: all real numbers x except $x = 0$

Vertical asymptote: $x = 0$

Slant asymptote: $y = -x + 3$

Line: $y = -x + 3$

67. $y = \dfrac{x+1}{x-3}$

(a) x-intercept: $(-1, 0)$

(b) $0 = \dfrac{x+1}{x-3}$

$0 = x + 1$

$-1 = x$

69. $y = \dfrac{1}{x} - x$

(a) x-intercepts: $(\pm 1, 0)$

(b) $0 = \dfrac{1}{x} - x$

$x = \dfrac{1}{x}$

$x^2 = 1$

$x = \pm 1$

71. $y = \dfrac{1}{x+5} + \dfrac{4}{x}$

(a)

x-intercept: $(-4, 0)$

(b) $0 = \dfrac{1}{x+5} + \dfrac{4}{x}$

$-\dfrac{4}{x} = \dfrac{1}{x+5}$

$-4(x+5) = x$

$-4x - 20 = x$

$-5x = 20$

$x = -4$

73. $y = x - \dfrac{6}{x-1}$

(a)

x-intercepts: $(-2, 0), (3, 0)$

(b) $0 = x - \dfrac{6}{x-1}$

$\dfrac{6}{x-1} = x$

$6 = x(x-1)$

$0 = x^2 - x - 6$

$0 = (x+2)(x-3)$

$x = -2, \quad x = 3$

75. (a) $0.25(50) + 0.75(x) = C(50 + x)$

$C = \dfrac{12.50 + 0.75x}{50 + x} \cdot \dfrac{4}{4}$

$C = \dfrac{50 + 3x}{4(50 + x)} = \dfrac{3x + 50}{4(x + 50)}$

(b) Domain: $x \geq 0$ and $x \leq 1000 - 50$

Thus, $0 \leq x \leq 950$. Using interval notation, the domain is $[0, 950]$.

(c)

```
C
1.0
0.8
0.6
0.4
0.2
      200 400 600 800 1000   x
```

(d) As the tank is filled, the concentration increases more slowly. It approaches the horizontal asymptote of $C = \frac{3}{4} = 0.75$.

77. (a) $A = xy$ and

$(x-4)(y-2) = 30$

$y - 2 = \dfrac{30}{x-4}$

$y = 2 + \dfrac{30}{x-4} = \dfrac{2x+22}{x-4}$

Thus, $A = xy = x\left(\dfrac{2x+22}{x-4}\right) = \dfrac{2x(x+11)}{x-4}$.

—CONTINUED—

77. —CONTINUED—

(b) Domain: Since the margins on the left and right are each 2 inches, $x > 4$. In interval notation, the domain is $(4, \infty)$.

(c)

The area is minimum when $x \approx 11.75$ inches and $y \approx 5.87$ inches.

x	5	6	7	8	9	10	11	12	13	14	15
y_1 (Area)	160	102	84	76	72	70	69.143	69	69.333	70	70.909

The area is minimum when x is approximately 12.

79. $f(x) = \dfrac{3(x + 1)}{x^2 + x + 1}$

Minimum: $(-2, -1)$

Maximum: $(0, 3)$

81. $C = 100\left(\dfrac{200}{x^2} + \dfrac{x}{x + 30}\right),\ x \geq 1$

The minimum occurs when $x \approx 40.45$, or 4045 components.

83. (a) Let $t_1 =$ time from Akron to Columbus and $t_2 =$ time from Columbus back to Akron.

$$xt_1 = 100 \implies t_1 = \frac{100}{x}$$

$$yt_2 = 100 \implies t_2 = \frac{100}{y}$$

$$50(t_1 + t_2) = 200$$

$$t_1 + t_2 = 4$$

$$\frac{100}{x} + \frac{100}{y} = 4$$

$$100y + 100x = 4xy$$

$$25y + 25x = xy$$

$$25x = xy - 25y$$

$$25x = y(x - 25)$$

Thus, $y = \dfrac{25x}{x - 25}$.

(b) Vertical asymptote: $x = 25$

Horizontal asymptote: $y = 25$

(c)

(d)

x	30	35	40	45	50	55	60
y	150	87.5	66.67	56.25	50	45.83	42.86

(e) Yes. You would expect the average speed for the round trip to be the average of the average speeds for the two parts of the trip.

(f) No. At 20 miles per hour you would use more time in one direction than is required for the round trip at an average speed of 50 miles per hour.

85. False. There are two distinct branches of the graph.

87. False. The degree of the numerator is *two* more than the degree of the denominator. To have a slant asymptote it has to be exactly one more.

89. $h(x) = \dfrac{6 - 2x}{3 - x} = \dfrac{2(3 - x)}{3 - x}$

Since $h(x)$ is not reduced and $(3 - x)$ is a factor of both the numerator and the denominator, $x = 3$ is not a vertical asymptote.

91. $C = \dfrac{3t^2 + t}{t^3 + 50}$, $t > 0$

(a) The horizontal asymptote is the t-axis, or $C = 0$. This indicates that the chemical eventually dissipates.

(b)

The time when the concentration is greatest is $t \approx 4.5$.

(c) $0 < t < 2.645$ or $t > 8.322$

93. $x^2 - 15x + 56 = (x - 8)(x - 7)$

95. $x^3 - 5x^2 + 4x - 20 = (x - 5)(x^2 + 4)$

$$= (x - 5)(x + 2i)(x - 2i)$$

97. $10 - 3x \le 0$

$3x \ge 10$

$x \ge \dfrac{10}{3}$

99. $|4(x - 2)| < 20$

$-20 < 4x - 8 < 20$

$-12 < 4x < 28$

$-3 < x < 7$

101. Answers will vary.

Section 4.3   Conics

You should know the following basic definitions of conic sections.

- ■ A parabola is the set of all points (x, y) that are equidistant from a fixed line (directrix) and a fixed point (focus) not on the line.

  (a) Standard equation with vertex $(0, 0)$ and directrix $y = -p$ (vertical axis): $x^2 = 4py$

  (b) Standard equation with vertex $(0, 0)$ and directrix $x = -p$ (horizontal axis): $y^2 = 4px$

  (c) The focus lies on the axis p units (directed distance) from the vertex.

- ■ An ellipse is the set of all points (x, y) the sum of whose distances from two distinct fixed points (foci) is constant.

  (a) Standard equation of an ellipse with center $(0, 0)$, major axis length $2a$, and minor axis length $2b$:

    1. Horizontal major axis: $\dfrac{x^2}{a^2} + \dfrac{y^2}{b^2} = 1$

    2. Vertical major axis: $\dfrac{x^2}{b^2} + \dfrac{y^2}{a^2} = 1$

  (b) The foci lie on the major axis, c units from the center, where a, b, and c are related by the equation $c^2 = a^2 - b^2$.

  (c) The vertices and endpoints of the minor axis are:

    1. Horizontal axis: $(\pm a, 0)$ and $(0, \pm b)$

    2. Vertical axis: $(0, \pm a)$ and $(\pm b, 0)$

- ■ A hyperbola is the set of all points (x, y) the difference of whose distances from two distinct fixed points (foci) is constant.

  (a) Standard equation of hyperbola with center $(0, 0)$

    1. Horizontal transverse axis: $\dfrac{x^2}{a^2} - \dfrac{y^2}{b^2} = 1$

    2. Vertical transverse axis: $\dfrac{y^2}{a^2} - \dfrac{x^2}{b^2} = 1$

  (b) The vertices and foci lie on the transverse axis and are a and c units from the center and $b^2 = c^2 - a^2$.

  (c) The asymptotes of the hyperbola are:

    1. Horizontal transverse axis: $y = \pm \dfrac{b}{a}x$

    2. Vertical transverse axis: $y = \pm \dfrac{a}{b}x$

Vocabulary Check

1. conic or conic section   **2.** circle; (h, k); r **3.** parabola; directrix; focus

4. vertex   **5.** axis   **6.** ellipse; foci

7. major axis; center   **8.** minor axis   **9.** hyperbola; foci

10. transverse axis; center

---

1. $x^2 = 2y$

Parabola opening upward

Not shown

3. $y^2 = 2x$

Parabola opening to the right

Matches (e).

5. $9x^2 + y^2 = 9$

$$\frac{x^2}{1} + \frac{y^2}{9} = 1$$

Ellipse with vertical major axis

Not shown

7. $9x^2 - y^2 = 9$

$$\frac{x^2}{1} - \frac{y^2}{9} = 1$$

Hyperbola with horizontal transverse axis

Matches (f).

9. $x^2 + y^2 = 25$

Circle with radius 5

Matches (d).

11. $y = \frac{1}{2}x^2$

$x^2 = 2y = 4\left(\frac{1}{2}\right)y$; $p = \frac{1}{2}$

Vertex: $(0, 0)$

Focus: $\left(0, \frac{1}{2}\right)$

13. $y^2 = -6x$

$y^2 = 4\left(-\frac{3}{2}\right)x$; $p = -\frac{3}{2}$

Vertex: $(0, 0)$

Focus: $\left(-\frac{3}{2}, 0\right)$

15. $x^2 + 8y = 0$

$x^2 = 4(-2)y$; $p = -2$

Vertex: $(0, 0)$

Focus: $(0, -2)$

17. Focus: $(-2, 0)$

$y^2 = 4(-2)x$

$y^2 = -8x$

19. Focus: $\left(0, -\frac{3}{2}\right)$

$x^2 = 4\left(-\frac{3}{2}\right)y$

$x^2 = -6y$

21. Directrix: $y = -1$

$x^2 = 4(1)y$

$x^2 = 4y$

23. Directrix: $x = 3$

$y^2 = 4(-3)x$

$y^2 = -12x$

25. $y^2 = 4px$

$6^2 = 4p(4)$

$36 = 16p$

$p = \frac{9}{4}$

$y^2 = 4\left(\frac{9}{4}\right)x$

$y^2 = 9x$

27. $x^2 = 4py$

$3^2 = 4p(6)$

$9 = 24p$

$\frac{3}{8} = p$

$x^2 = 4\left(\frac{3}{8}\right)y$

$x^2 = \frac{3}{2}y$

Focus: $\left(0, \frac{3}{8}\right)$

29. $y^2 = 4px$

$(-3)^2 = 4p(5)$

$9 = 20p$

$\frac{9}{20} = p$

$y^2 = 4\left(\frac{9}{20}\right)x$

$y^2 = \frac{9}{5}x$

Focus: $\left(\frac{9}{20}, 0\right)$

31. $y^2 = 4px$, $p = 1.5$

$y^2 = 4(1.5)x$

$y^2 = 6x$

33. (a)

(b) $x^2 = 4py$

$(640)^2 = 4p(152) \implies p = \dfrac{409,600}{608} = \dfrac{12,800}{19}$

$x^2 = 4\left(\dfrac{12,800}{19}\right)y$

$y = \dfrac{19x^2}{51,200}$

(c)

Distance, x	Height, y
0	0
200	$14\frac{27}{32}$
400	$59\frac{3}{8}$
500	$92\frac{99}{128}$
600	$133\frac{19}{32}$

35. $\dfrac{x^2}{25} + \dfrac{y^2}{16} = 1$

Horizontal major axis

$a = 5, \; b = 4$

Center: $(0, 0)$

Vertices: $(\pm 5, 0)$

37. $\dfrac{x^2}{25/9} + \dfrac{y^2}{16/9} = 1$

Horizontal major axis

$a = \dfrac{5}{3}, b = \dfrac{4}{3}$

Center: $(0, 0)$

Vertices: $\left(\pm \dfrac{5}{3}, 0\right)$

39. $\dfrac{x^2}{9} + \dfrac{y^2}{5} = 1$

Horizontal major axis

$a = 3, \; b = \sqrt{5}$

Center: $(0, 0)$

Vertices: $(\pm 3, 0)$

41. $4x^2 + y^2 = 1$

$\dfrac{x^2}{\frac{1}{4}} + y^2 = 1$

Vertical major axis

$a = \dfrac{1}{2}, b = 1$

Center: $(0, 0)$

Vertices: $(0, \pm 1)$

43. Vertices: $(0, \pm 2) \implies a = 2$

Minor axis of length $2 \implies b = 1$

Vertical major axis

$\dfrac{x^2}{b^2} + \dfrac{y^2}{a^2} = 1$

$\dfrac{x^2}{1} + \dfrac{y^2}{4} = 1$

45. Vertices: $(\pm 2, 0) \implies a = 2$

Minor axis of length $3 \implies b = \dfrac{3}{2}$

Horizontal major axis

$\dfrac{x^2}{a^2} + \dfrac{y^2}{b^2} = 1$

$\dfrac{x^2}{4} + \dfrac{y^2}{9/4} = 1$

47. Vertices: $(\pm 5, 0) \implies a = 5$

Foci: $(\pm 2, 0) \implies b = \sqrt{5^2 - 2^2} = \sqrt{21}$

Horizontal major axis

$\dfrac{x^2}{a^2} + \dfrac{y^2}{b^2} = 1$

$\dfrac{x^2}{25} + \dfrac{y^2}{21} = 1$

49. Foci: $(\pm 5, 0) \implies c = 5$

Major axis of length $12 \implies a = 6$

$b = \sqrt{6^2 - 5^2} = \sqrt{11}$

Horizontal major axis

$\dfrac{x^2}{a^2} + \dfrac{y^2}{b^2} = 1$

$\dfrac{x^2}{36} + \dfrac{y^2}{11} = 1$

51. Vertices: $(0, \pm 5) \implies a = 5$

Vertical major axis

$$\frac{x^2}{b^2} + \frac{y^2}{25} = 1$$

Passes through $(4, 2)$

$$\frac{(4)^2}{b^2} + \frac{(2)^2}{25} = 1$$

$$\frac{16}{b^2} = \frac{21}{25}$$

$$\frac{400}{21} = b^2$$

$$\frac{21x^2}{400} + \frac{y^2}{25} = 1$$

53. The length of string needed is $2a$ or $2(3) = 6$ feet. The positions of the tacks are $(\pm c, 0)$ where c is given by

$$c^2 = a^2 - b^2$$

$$= 3^2 - 2^2 = 9 - 4 = 5 \implies c = \pm\sqrt{5}.$$

Positions: $\left(\pm\sqrt{5}, 0\right)$

55. (a)

(b) Horizontal major axis

$a = 20, b = 15$

$$\frac{x^2}{400} + \frac{y^2}{225} = 1 \implies y^2 = 225\left(1 - \frac{x^2}{400}\right)$$

$$= \frac{225}{400}(400 - x^2)$$

$$y = \frac{15}{20}\sqrt{400 - x^2} = \frac{3}{4}\sqrt{400 - x^2}$$

(c) When $x = 5$:

$$y = \frac{3}{4}\sqrt{400 - 5^2}$$

$$y \approx 14.52 \text{ feet}$$

Yes, the truck can clear the tunnel with 0.52 foot clearance.

57. $\dfrac{x^2}{4} + \dfrac{y^2}{1} = 1$

$a = 2, \ b = 1, \ c = \sqrt{3}$

Points on the ellipse: $(\pm 2, 0), (0, \pm 1)$

Length of each latus rectum: $\dfrac{2b^2}{a} = 1$

Additional points: $\left(\sqrt{3}, \pm\dfrac{1}{2}\right), \left(-\sqrt{3}, \pm\dfrac{1}{2}\right)$

59. $9x^2 + 4y^2 = 36$

$\dfrac{x^2}{4} + \dfrac{y^2}{9} = 1$

$a = 3,\ b = 2,\ c = \sqrt{5}$

Points on the ellipse: $(\pm 2, 0),\ (0, \pm 3)$

Length of each latus rectum: $\dfrac{2b^2}{a} = \dfrac{2 \cdot (2)^2}{3} = \dfrac{8}{3}$

Additional points: $\left(\pm \dfrac{4}{3}, -\sqrt{5}\right),\ \left(\pm \dfrac{4}{3}, \sqrt{5}\right)$

61. $x^2 - y^2 = 1$

$a = 1,\ b = 1$

Center: $(0, 0)$

Vertices: $(\pm 1, 0)$

Asymptotes: $y = \pm x$

63. $\dfrac{y^2}{1} - \dfrac{x^2}{4} = 1$

$a = 1,\ b = 2$

Center: $(0, 0)$

Vertices: $(0, \pm 1)$

Asymptotes: $y = \pm \dfrac{1}{2}x$

65. $\dfrac{y^2}{25} - \dfrac{x^2}{144} = 1$

$a = 5,\ b = 12$

Center: $(0, 0)$

Vertices: $(0, \pm 5)$

Asymptotes: $y = \pm \dfrac{5}{12}x$

67. $4y^2 - x^2 = 1$

$\dfrac{y^2}{1/4} - \dfrac{x^2}{1} = 0$

$a = \dfrac{1}{2},\ b = 1$

Center: $(0, 0)$

Vertices: $\left(0, \pm \dfrac{1}{2}\right)$

Asymptotes: $y = \pm \dfrac{1}{2}x$

69. Vertices: $(0, \pm 2) \implies a = 2$

Foci: $(0, \pm 4) \implies c = 4$

$b^2 = c^2 - a^2 = 12$

Vertical transverse axis

$\dfrac{y^2}{a^2} - \dfrac{x^2}{b^2} = 1$

$\dfrac{y^2}{4} - \dfrac{x^2}{12} = 1$

71. Vertices: $(\pm 1, 0) \implies a = 1$

Asymptotes: $y = \pm 3x$

Horizontal transverse axis

$3 = \dfrac{b}{a} = \dfrac{b}{1} \implies b = 3$

$\dfrac{x^2}{a^2} - \dfrac{y^2}{b^2} = 1$

$\dfrac{x^2}{1} - \dfrac{y^2}{9} = 1$

73. Foci: $(0, \pm 8) \implies c = 8$

Asymptotes: $y = \pm 4x$

Vertical transverse axis

$$4 = \frac{a}{b} \implies a = 4b$$

$$a^2 + b^2 = c^2 \implies 16b^2 + b^2 = (8)^2$$

$$b^2 = \frac{64}{17} \implies a^2 = \frac{1024}{17}$$

$$\frac{y^2}{a^2} - \frac{x^2}{b^2} = 1$$

$$\frac{y^2}{1024/17} - \frac{x^2}{64/17} = 1$$

$$\frac{17y^2}{1024} - \frac{17x^2}{64} = 1$$

75. Vertices: $(0, \pm 3) \implies a = 3$

Vertical transverse axis

$$\frac{y^2}{9} - \frac{x^2}{b^2} = 1$$

Point on the graph: $(-2, 5)$

$$\frac{5^2}{9} - \frac{(-2)^2}{b^2} = 1$$

$$b^2 = \frac{9}{4}$$

$$\frac{y^2}{9} - \frac{x^2}{9/4} = 1$$

77. (a) Vertices: $(\pm 1, 0) \implies a = 1$

Horizontal transverse axis

$$\frac{x^2}{a^2} - \frac{y^2}{b^2} = 1$$

Point on the graph: $(2, 13)$

$$\frac{2^2}{1^2} - \frac{13^2}{b^2} = 1$$

$$4 - \frac{169}{b^2} = 1$$

$$3b^2 = 169$$

$$b^2 = \frac{169}{3} \approx 56.33$$

Thus, we have $\dfrac{x^2}{1} - \dfrac{y^2}{56.33} = 1$.

(b) When $y = 5$:

$$x^2 = 1 + \frac{5^2}{56.33}$$

$$x \approx \sqrt{1 + \frac{25}{56.33}} \approx 1.2016$$

Width: $2x \approx 2.403$ feet

79. $\dfrac{x^2}{100} - \dfrac{y^2}{4} = 1$

The shortest horizontal distance would be the distance between the center and a vertex of the hyperbola, that is, 10 miles.

81. False. The equation represents a hyperbola.

$$\frac{x^2}{144} - \frac{y^2}{144} = 1$$

83. False. If the graph crossed the directrix, there would exist points nearer the directrix than to the focus.

85. $4x^2 + 4y^2 - 16 = 0$

$$4x^2 + 4y^2 = 16$$

$$x^2 + y^2 = 4$$

Circle

87. $3y^2 - 6x = 0$

$$3y^2 = 6x$$

$$y^2 = 2x$$

Parabola

89. $4x^2 + y^2 - 16 = 0$

$$4x^2 + y^2 = 16$$

$$\frac{x^2}{4} + \frac{y^2}{16} = 1$$

Ellipse

91. (a) $a + b = 20 \implies b = 20 - a$

$A = \pi ab = \pi a(20 - a)$

(b) $264 = \pi a(20 - a)$

$\pi a^2 - 20\pi a + 264 = 0$

$a \approx 14$ or $a \approx 6$ by the Quadratic Formula

$b = 6 \qquad b = 14$

Since $a > b$ we choose $a = 14$ and $b = 6$.

$$\frac{x^2}{14^2} + \frac{y^2}{6^2} = 1, \quad \frac{x^2}{196} + \frac{y^2}{36} = 1$$

(c)

a	8	9	10	11	12	13
A	301.6	311.0	314.2	311.0	301.6	285.9

Conjecture: Area is maximum when $a = b = 10$ and the shape is a circle.

(d)

The area is maximum when $a = b = 10$ and the shape is a circle.

93. $x^2 - y^2 = 0$

$y^2 = x^2$

$y = \pm x$

95. Answers will vary. From the exercise set we have examples of hyperbolas in optics, art, navigation, and aeronautics.

97. Let (x, y) be such that the sum of the distances from $(-c, 0)$ and $(c, 0)$ is $2a$. (We are only deriving the form where the major axis is horizontal.)

$$\sqrt{(x + c)^2 + y^2} + \sqrt{(x - c)^2 + y^2} = 2a$$

$$\sqrt{(x + c)^2 + y^2} = 2a - \sqrt{(x - c)^2 + y^2}$$

$$(x + c)^2 + y^2 = 4a^2 - 4a\sqrt{(x - c)^2 + y^2} + (x - c)^2 + y^2$$

$$4a\sqrt{(x - c)^2 + y^2} = 4a^2 + (x - c)^2 - (x + c)^2$$

$$4a\sqrt{(x - c)^2 + y^2} = 4a^2 - 4cx$$

$$a\sqrt{(x - c)^2 + y^2} = a^2 - cx$$

$$a^2[(x - c)^2 + y^2] = a^4 - 2a^2cx + c^2x^2$$

$$a^2x^2 - 2a^2cx + a^2c^2 + a^2y^2 = a^4 - 2a^2cx + c^2x^2$$

$$a^2x^2 + a^2c^2 + a^2y^2 = a^4 + c^2x^2$$

$$a^2x^2 - c^2x^2 + a^2y^2 = a^4 - a^2c^2$$

$$x^2(a^2 - c^2) + a^2y^2 = a^2(a^2 - c^2)$$

Let $b^2 = a^2 - c^2$. Then $x^2b^2 + a^2y^2 = a^2b^2 \implies \dfrac{x^2}{a^2} + \dfrac{y^2}{b^2} = 1$.

99. $f(x) = x^2 - 8$

$= x^2 + 0x - 8$

$= (x^2 + 0x + 0) - 8$

$= (x - 0)^2 - 8$

x-intercepts: $\left(\pm 2\sqrt{2}, 0\right)$

y-intercept: $(0, -8)$

Vertex: $(0, -8)$

101. $f(x) = x^2 + 8x + 12$

$\qquad = (x^2 + 8x + 16 - 16) + 12$

$\qquad = (x + 4)^2 - 4$

x-intercepts:

$\qquad (x + 4)^2 - 4 = 0$

$\qquad x^2 + 8x + 12 = 0$

$\qquad (x + 6)(x + 2) = 0$

$\qquad x = -2, \ x = -6$

$\qquad (-2, 0), (-6, 0)$

y-intercept: $(0, 12)$

Vertex: $(-4, -4)$

103. $(x - 3)(x - 2 - i)(x - 2 + i) = (x - 3)(x^2 - 2x + ix - 2x + 4 - 2i - ix + 2i - i^2)$

$\qquad\qquad\qquad\qquad\qquad\qquad = (x - 3)(x^2 - 4x + 5)$

$\qquad\qquad\qquad\qquad\qquad\qquad = x^3 - 4x^2 + 5x - 3x^2 + 12x - 15$

$\qquad\qquad\qquad\qquad\qquad\qquad = x^3 - 7x^2 + 17x - 15$

105. $g(x) = 6x^4 + 7x^3 - 29x^2 - 28x + 20$

Possible rational zeros:

$\pm 1, \pm 2, \pm 4, \pm 5, \pm 10, \pm 20, \pm\frac{1}{2}, \pm\frac{5}{2} \pm\frac{1}{3}, \pm\frac{2}{3}, \pm\frac{4}{3}, \pm\frac{5}{3}, \pm\frac{10}{3}, \pm\frac{20}{3}, \pm\frac{1}{6}, \pm\frac{5}{6}$

$\left[\textit{Note:} \text{ The actual zeros are } \pm 2, \frac{1}{2}, \text{ and } -\frac{5}{3}.\right]$

Section 4.4   Translations of Conics

You should know the following basic facts about conic sections.

- Parabola with vertex (h, k)

    (a) Vertical axis

    1. Standard equation: $(x - h)^2 = 4p(y - k)$

    2. Focus: $(h, k + p)$

    3. Directrix: $y = k - p$

    (b) Horizontal axis

    1. Standard equation: $(y - k)^2 = 4p(x - h)$

    2. Focus: $(h + p, k)$

    3. Directrix: $x = h - p$

—CONTINUED—

—CONTINUED—

■ Circle with center (h, k) and radius r

Standard equation: $(x - h)^2 + (y - k)^2 = r^2$

■ Ellipse with center (h, k)

(a) Horizontal major axis:

1. Standard equation:

$$\frac{(x - h)^2}{a^2} + \frac{(y - k)^2}{b^2} = 1$$

2. Vertices: $(h \pm a, k)$

3. Foci: $(h \pm c, k)$

4. $c^2 = a^2 - b^2$

(b) Vertical major axis:

1. Standard equation:

$$\frac{(x - h)^2}{b^2} + \frac{(y - k)^2}{a^2} = 1$$

2. Vertices: $(h, k \pm a)$

3. Foci: $(h, k \pm c)$

4. $c^2 = a^2 - b^2$

■ Hyperbola with center (h, k)

(a) Horizontal transverse axis:

1. Standard equation:

$$\frac{(x - h)^2}{a^2} - \frac{(y - k)^2}{b^2} = 1$$

2. Vertices: $(h \pm a, k)$

3. Foci: $(h \pm c, k)$

4. Asymptotes: $y - k = \pm\frac{b}{a}(x - h)$

5. $c^2 = a^2 + b^2$

(b) Vertical transverse axis:

1. Standard equation:

$$\frac{(y - k)^2}{a^2} - \frac{(x - h)^2}{b^2} = 1$$

2. Vertices: $(h, k \pm a)$

3. Foci: $(h, k \pm c)$

4. Asymptotes: $y - k = \pm\frac{a}{b}(x - h)$

5. $c^2 = a^2 + b^2$

Vocabulary Check

1. Circle $\implies (x-h)^2 + (y-k)^2 = r^2$

Matches (g).

2. Ellipse with vertical major axis $\implies \dfrac{(x-h)^2}{b^2} + \dfrac{(y-k)^2}{a^2} = 1$

Matches (d).

3. Parabola with vertical axis $\implies (x-h)^2 = 4p(y-k)$

Matches (e).

4. Hyperbola with horizontal transverse axis $\implies \dfrac{(x-h)^2}{a^2} - \dfrac{(y-k)^2}{b^2} = 1$

Matches (b).

5. Ellipse with horizontal major axis $\implies \dfrac{(x-h)^2}{a^2} + \dfrac{(y-k)^2}{b^2} = 1$

Matches (a).

6. Parabola with horizontal axis $\implies (y-k)^2 = 4p(x-h)$

Matches (f).

7. Hyperbola with vertical transverse axis $\implies \dfrac{(y-k)^2}{a^2} - \dfrac{(x-h)^2}{b^2} = 1$

Matches (c).

---

1. $(x+2)^2 + (y-1)^2 = 4$

Center: $(-2, 1)$

The graph has been shifted two units to the left and one unit upward from standard position.

3. $\dfrac{(y+3)^2}{4} - (x-1)^2 = 1$

Center: $(1, -3)$

The graph has been shifted one unit to the right and three units downward from standard position.

5. $\dfrac{(x+4)^2}{9} + \dfrac{(y+2)^2}{16} = 1$

Center: $(-4, -2)$

The graph has been shifted four units to the left and two units downward from standard position.

7. $x^2 + y^2 = 49$

Center: $(0, 0)$

Radius: 7

9. $(x+3)^2 + (y-8)^2 = 16$

Center: $(-3, 8)$

Radius: 4

11. $(x-1)^2 + y^2 = 10$

Center: $(1, 0)$

Radius: $\sqrt{10}$

13.
$$x^2 + y^2 - 2x + 6y + 9 = 0$$
$$(x^2 - 2x) + (y^2 + 6y) = -9$$
$$(x^2 - 2x + 1) + (y^2 + 6y + 9) = -9 + 1 + 9$$
$$(x-1)^2 + (y+3)^2 = 1$$

Center: $(1, -3)$

Radius: 1

15. $4x^2 + 4y^2 + 12x - 24y + 41 = 0$

$$x^2 + y^2 + 3x - 6y + \tfrac{41}{4} = 0$$

$$(x^2 + 3x) + (y^2 - 6y) = -\tfrac{41}{4}$$

$$\left(x^2 + 3x + \tfrac{9}{4}\right) + (y^2 - 6y + 9) = -\tfrac{41}{4} + \tfrac{9}{4} + 9$$

$$\left(x + \tfrac{3}{2}\right)^2 + (y - 3)^2 = 1$$

Center: $\left(-\tfrac{3}{2}, 3\right)$

Radius: 1

17. $(x - 1)^2 + 8(y + 2) = 0$

$$(x - 1)^2 = 4(-2)(y + 2); p = -2$$

Vertex: $(1, -2)$

Focus: $(1, -4)$

Directrix: $y = 0$

19. $\left(y + \tfrac{1}{2}\right)^2 = 2(x - 5)$

$$\left(y + \tfrac{1}{2}\right)^2 = 4\left(\tfrac{1}{2}\right)(x - 5); p = \tfrac{1}{2}$$

Vertex: $\left(5, -\tfrac{1}{2}\right)$

Focus: $\left(\tfrac{11}{2}, -\tfrac{1}{2}\right)$

Directrix: $x = \tfrac{9}{2}$

21. $y = \tfrac{1}{4}(x^2 - 2x + 5)$

$$x^2 - 2x = 4y - 5$$

$$x^2 - 2x + 1 = 4y - 5 + 1$$

$$(x - 1)^2 = 4y - 4$$

$$(x - 1)^2 = 4(1)(y - 1); p = 1$$

Vertex: $(1, 1)$

Focus: $(1, 2)$

Directrix: $y = 0$

23. $y^2 + 6y + 8x + 25 = 0$

$$y^2 + 6y = -8x - 25$$

$$y^2 + 6y + 9 = -8x - 25 + 9$$

$$(y + 3)^2 = -8x - 16$$

$$(y + 3)^2 = 4(-2)(x + 2); p = -2$$

Vertex: $(-2, -3)$

Focus: $(-4, -3)$

Directrix: $x = 0$

25. Vertex: $(3, 2)$

Focus: $(1, 2)$

Horizontal axis

$p = 1 - 3 = -2$

$(y - 2)^2 = 4(-2)(x - 3)$

$(y - 2)^2 = -8(x - 3)$

27. Vertex: $(0, 4)$

Directrix: $y = 2$

Vertical axis

$p = 4 - 2 = 2$

$(x - 0)^2 = 4(2)(y - 4)$

$x^2 = 8(y - 4)$

29. Focus: $(2, 2)$

Directrix: $x = -2$

Horizontal axis

Vertex: $(0, 2)$

$p = 2 - 0 = 2$

$(y - 2)^2 = 4(2)(x - 0)$

$(y - 2)^2 = 8x$

31. (a) $y = -23.539t^2 + 540.232t - 2130.893$

(b) x-coordinate of vertex:

$$-\frac{b}{2a} = \frac{-540.232}{2(-23.539)} \approx 11.4753$$

y-coordinate of vertex:

$$-23.539(11.4753)^2 + 540.232(11.4753) - 2130.893 \approx 968.72$$

The vertex is about $(11.48, 969)$. This means that the model shows the maximum revenue of about \$969 million occurring during the year when $t = 11$, or 2001.

(c)

(d) 2001

(e) 2001

(f) The results are the same.

33. $x^2 = -39,204(y - 30,000)$

When $y = 0$, $x^2 = 1,176,120,000$

$x \approx 34,295$ feet ≈ 6.5 miles.

35. $\dfrac{(x - 1)^2}{9} + \dfrac{(y - 5)^2}{25} = 1$

$a = 5$, $b = 3$, $c = \sqrt{a^2 - b^2} = 4$

Center: $(1, 5)$

Foci: $(1, 1), (1, 9)$

Vertices: $(1, 0), (1, 10)$

37. $\dfrac{(x + 2)^2}{1} + \dfrac{(y + 4)^2}{1/4} = 1$

$a = 1, b = \dfrac{1}{2}, c = \sqrt{a^2 - b^2} = \sqrt{\dfrac{3}{4}} = \dfrac{\sqrt{3}}{2}$

Center: $(-2, -4)$

Foci: $\left(-2 \pm \dfrac{\sqrt{3}}{2}, -4\right)$ or $\left(\dfrac{-4 \pm \sqrt{3}}{2}, -4\right)$

Vertices: $(-3, -4), (-1, -4)$

39.
$$9x^2 + 4y^2 + 36x - 24y + 36 = 0$$
$$9(x^2 + 4x) + 4(y^2 - 6y) = -36$$
$$9(x^2 + 4x + 4) + 4(y^2 - 6y + 9) = -36 + 36 + 36$$
$$9(x + 2)^2 + 4(y - 3)^2 = 36$$
$$\frac{(x + 2)^2}{4} + \frac{(y - 3)^2}{9} = 1$$

$a = 3,\ b = 2,\ c = \sqrt{a^2 - b^2} = \sqrt{5}$

Vertical major axis

Center: $(-2, 3)$

Foci: $\left(-2, 3 - \sqrt{5}\right)\left(-2, 3 + \sqrt{5}\right)$

Vertices: $(-2, 0),\ (-2, 6)$

41.
$$16x^2 + 25y^2 - 32x + 50y + 16 = 0$$
$$16(x^2 - 2x) + 25(y^2 + 2y) = -16$$
$$16(x^2 - 2x + 1) + 25(y^2 + 2y + 1) = -16 + 16(1) + 25(1)$$
$$16(x - 1)^2 + 25(y + 1) = 25$$
$$\frac{(x - 1)^2}{25/16} + (y + 1)^2 = 1$$

$a = \dfrac{5}{4},\ b = 1,\ c = \sqrt{a^2 - b^2} = \dfrac{3}{4}$

Horizontal major axis

Center: $(1, -1)$

Foci: $\left(\dfrac{1}{4}, -1\right), \left(\dfrac{7}{4}, -1\right)$

Vertices: $\left(-\dfrac{1}{4}, -1\right), \left(\dfrac{9}{4}, -1\right)$

43. Vertices: $(4, 4),\ (4, -4)$

Minor axis of length 6

Center: $(4, 0)$

$a = 4,\ b = 3$

Vertical major axis

$$\frac{(x - 4)^2}{9} + \frac{y^2}{16} = 1$$

45. Vertices: $(0, 2),\ (4, 2)$

Minor axis of length 2

$a = \dfrac{4 - 0}{2} = 2,\ b = \dfrac{2}{2} = 1$

Center: $(2, 2)$

Horizontal major axis

$$\frac{(x - 2)^2}{4} + \frac{(y - 2)^2}{1} = 1$$

47. Foci: $(0, 0),\ (0, 8)$

Major axis of length 16

$a = 8,\ c = 4,\ b^2 = a^2 - c^2 = 48$

Center: $(0, 4)$

Vertical major axis

$$\frac{x^2}{48} + \frac{(y - 4)^2}{64} = 1$$

49. Center: $(0, 4)$

$a = 2c$

Vertices: $(-4, 4),\ (4, 4)$

$a = 4,\ c = 2,\ b^2 = a^2 - c^2 = 12$

Horizontal major axis

$$\frac{x^2}{16} + \frac{(y - 4)^2}{12} = 1$$

51. Vertices: $(0, 2),\ (4, 2)$

Endpoints of the minor axis: $(2, 3),\ (2, 1)$

Center: $(2, 2)$

$a = 2,\ b = 1$

Horizontal major axis

$$\frac{(x - 2)^2}{4} + \frac{(y - 2)^2}{1} = 1$$

53. Vertices: $(\pm 5, 0)$

$e = \dfrac{c}{a} = \dfrac{3}{5}$

$a = 5,\ c = 3,\ b = 4$

$$\frac{x^2}{25} + \frac{y^2}{16} = 1$$

55. $a = 3.67 \times 10^9$

$e = \dfrac{c}{a} = 0.249$

$c = 913,830,000$

Smallest distance: $a - c = 2,756,170,000$ miles

Greatest distance: $a + c = 4,583,830,000$ miles

57. $\dfrac{(x-1)^2}{4} - \dfrac{(y+2)^2}{1} = 1$

$a = 2$, $b = 1$, $c = \sqrt{a^2 + b^2} = \sqrt{5}$

Center: $(1, -2)$

Horizontal transverse axis

Vertices: $(-1, -2), (3, -2)$

Foci: $\left(1 \pm \sqrt{5}, -2\right)$

Asymptotes: $y = \pm \dfrac{1}{2}(x - 1) - 2$

59. $(y+6)^2 - (x-2)^2 = 1$

$a = 1$, $b = 1$, $c = \sqrt{a^2 + b^2} = \sqrt{2}$

Center: $(2, -6)$

Vertical transverse axis

Vertices: $(2, -7), (2, -5)$

Foci: $\left(2, -6 \pm \sqrt{2}\right)$

Asymptotes: $y = \pm(x - 2) - 6$

61.
$$9x^2 - y^2 - 36x - 6y + 18 = 0$$
$$9(x^2 - 4x) - (y^2 + 6y) = -18$$
$$9(x^2 - 4x + 4) - (y^2 + 6y + 9) = -18 + 9(4) - 9$$
$$9(x - 2)^2 - (y + 3)^2 = 9$$
$$(x - 2)^2 - \dfrac{(y + 3)^2}{9} = 1$$

$a = 1$, $b = 3$, $c = \sqrt{a^2 + b^2} = \sqrt{10}$

Center: $(2, -3)$

Horizontal transverse axis

Vertices: $(1, -3), (3, -3)$

Foci: $\left(2 \pm \sqrt{10}, -3\right)$

Asymptotes: $y = \pm 3(x - 2) - 3$

63.
$$x^2 - 9y^2 + 2x - 54y - 80 = 0$$
$$(x^2 + 2x) - 9(y^2 + 6y) = 80$$
$$(x^2 + 2x + 1) - 9(y^2 + 6y + 9) = 80 + 1 - 9(9)$$
$$(x + 1)^2 - 9(y + 3)^2 = 0$$
$$9(y + 3)^2 = (x + 1)^2$$
$$y = \pm \tfrac{1}{3}(x + 1) - 3$$

The graph of this equation is two lines intersecting at $(-1, -3)$.

65.
$$9y^2 - 4x^2 + 8x + 18y + 41 = 0$$
$$9(y^2 + 2y) - 4(x^2 - 2x) = -41$$
$$9(y^2 + 2y + 1) - 4(x^2 - 2x + 1) = -41 + 9 - 4$$
$$9(y + 1)^2 - 4(x - 1)^2 = -36$$
$$\frac{(x - 1)^2}{9} - \frac{(y + 1)^2}{4} = 1$$
$$a = 3, b = 2, c = \sqrt{a^2 + b^2} = \sqrt{13}$$

Center: $(1, -1)$

Horizontal transverse axis

Vertices: $(-2, -1), (4, -1)$

Foci: $\left(1 \pm \sqrt{13}, -1\right)$

Asymptotes: $y = \pm\frac{2}{3}(x - 1) - 1$

67. Vertices: $(0, 0), (0, 2)$

Foci: $(0, -1), (0, 3)$

Center: $(0, 1)$

Vertical transverse axis

$a = 1, c = 2, b^2 = c^2 - a^2 = 3$

$$\frac{(y - 1)^2}{1} - \frac{x^2}{3} = 1$$

69. Vertices: $(2, 0), (6, 0)$

Foci: $(0, 0), (8, 0)$

Center: $(4, 0)$

Horizontal transverse axis

$a = 2, c = 4, b^2 = c^2 - a^2 = 12$

$$\frac{(x - 4)^2}{4} - \frac{y^2}{12} = 1$$

71. Vertices: $(4, 1), (4, 9)$

Foci: $(4, 0), (4, 10)$

Center: $(4, 5)$

Vertical transverse axis

$a = 4, c = 5, b^2 = c^2 - a^2 = 9$

$$\frac{(y - 5)^2}{16} - \frac{(x - 4)^2}{9} = 1$$

73. Vertices: $(2, 3), (2, -3)$

Passes through the point $(0, 5)$

Center: $(2, 0)$

Vertical transverse axis

$a = 3$

$$\frac{y^2}{9} - \frac{(x - 2)^2}{b^2} = 1$$

$$\frac{5^2}{9} - \frac{(0 - 2)^2}{b^2} = 1$$

$$b^2 = \frac{9}{4}$$

$$\frac{y^2}{9} - \frac{4(x - 2)^2}{9} = 1$$

75. Vertices: $(0, 2), (6, 2)$

Asymptotes: $y = \frac{2}{3}x, \; y = 4 - \frac{2}{3}x$

Center: $(3, 2)$

Horizontal transverse axis

$a = 3, \; b = 2$

$$\frac{(x - 3)^2}{9} - \frac{(y - 2)^2}{4} = 1$$

79. $4x^2 - y^2 - 4x - 3 = 0$

$4(x^2 - x) - y^2 = 3$

$4\left(x^2 - x + \frac{1}{4}\right) - y^2 = 3 + 4\left(\frac{1}{4}\right)$

$4\left(x - \frac{1}{2}\right)^2 - y^2 = 4$

$\left(x - \frac{1}{2}\right)^2 - \frac{y^2}{4} = 1$

Hyperbola

77. $x^2 + y^2 - 6x + 4y + 9 = 0$

$(x^2 - 6x) + (y^2 + 4y) = -9$

$(x^2 - 6x + 9) + (y^2 + 4y + 4) = -9 + 9 + 4$

$(x - 3)^2 + (y + 2)^2 = 4$

Circle

81. $4x^2 + 3y^2 + 8x - 24y + 51 = 0$

$4(x^2 + 2x) + 3(y^2 - 8y) = -51$

$4(x^2 + 2x + 1) + 3(y^2 - 8y + 16) = -51 + 4(1) + 3(16)$

$4(x + 1)^2 + 3(y - 4)^2 = 1$

$\frac{(x + 1)^2}{1/4} + \frac{(y - 4)^2}{1/3} = 1$

Ellipse

83. $25x^2 - 10x - 200y - 119 = 0$

$25\left(x^2 - \frac{2}{5}x\right) = 200y + 119$

$25\left(x^2 - \frac{2}{5}x + \frac{1}{25}\right) = 200y + 119 + 1$

$25\left(x - \frac{1}{5}\right)^2 = 200\left(y + \frac{3}{5}\right)$

$\left(x - \frac{1}{5}\right)^2 = 8\left(y + \frac{3}{5}\right)$

Parabola

85. $3x^2 + 2y^2 - 18x - 16y + 58 = 0$

$3(x^2 - 6x) + 2(y^2 - 8y) = -58$

$3(x^2 - 6x + 9) + 2(y^2 - 8y + 16) = -58 + 27 + 32$

$3(x - 3)^2 + 2(y - 4)^2 = 1$

$\frac{(x - 3)^2}{1/3} + \frac{(y - 4)^2}{1/2} = 1$

True. The equation in standard form is

$\frac{(x - 3)^2}{1/3} + \frac{(y - 4)^2}{1/2} = 1.$

87. For an ellipse $e = c/a$ and $c^2 = a^2 - b^2$.

$e = \frac{c}{a} \implies ea = c$

$e^2 a^2 = c^2$

$e^2 a^2 = a^2 - b^2$

$b^2 = a^2 - e^2 a^2$

$b^2 = a^2(1 - e^2)$

Thus, by substitution, $\frac{(x - h)^2}{a^2} + \frac{(y - k)^2}{b^2} = 1$ is

equivalent to $\frac{(x - h)^2}{a^2} + \frac{(y - k)^2}{a^2(1 - e^2)} = 1.$

89.
$$f(x) = 10 - 7x$$
$$y = 10 - 7x$$
$$x = 10 - 7y$$
$$x - 10 = 7y$$
$$\frac{x - 10}{7} = y$$
$$f^{-1}(x) = \frac{x - 10}{7}$$

91.
$$f(x) = \sqrt{x + 8}, x \geq -8, y \geq 0$$
$$y = \sqrt{x + 8}$$
$$x = \sqrt{y + 8}, x \geq 0, y \geq -8$$
$$x^2 = y + 8$$
$$x^2 - 8 = y$$
$$f^{-1}(x) = x^2 - 8, x \geq 0$$

Review Exercises for Chapter 4

1. $f(x) = \dfrac{5x}{x + 12}$

Domain: all real numbers except $x = -12$

3. $f(x) = \dfrac{8}{x^2 - 10x + 24} = \dfrac{8}{(x - 4)(x - 6)}$

Domain: all real numbers except $x = 4$ and $x = 6$

5. $f(x) = \dfrac{4}{x + 3}$

Vertical asymptote: $x = -3$

Horizontal asymptote: $y = 0$

7. $g(x) = \dfrac{x^2}{x^2 - 4}$

Vertical asymptotes: $x = -2$, $x = 2$

Horizontal asymptote: $y = 1$

9. $h(x) = \dfrac{2x - 10}{x^2 - 2x - 15} = \dfrac{2(x - 5)}{(x + 3)(x - 5)} = \dfrac{2}{x + 3}$, $x \neq 5$

Vertical asymptote: $x = -3$

Horizontal asymptote: $y = 0$

11. $\overline{C} = \dfrac{C}{x} = \dfrac{0.5x + 500}{x}$, $0 < x$

Horizontal asymptote: $\overline{C} = \dfrac{0.5}{1} = 0.5$

As x increases, the average cost per unit approaches the horizontal asymptote, $\overline{C} = 0.5 = \$0.50$.

13. $f(x) = \dfrac{-5}{x^2}$

(a) Domain: all real numbers x except $x = 0$

(b) No intercepts

(c) Vertical asymptote: $x = 0$
Horizontal asymptote: $y = 0$

(d)

x	± 3	± 2	± 1
y	$-\frac{5}{9}$	$-\frac{5}{4}$	-5

15. $g(x) = \dfrac{2 + x}{1 - x} = -\dfrac{x + 2}{x - 1}$

(a) Domain: all real numbers x except $x = 1$

(b) x-intercept: $(-2, 0)$
y-intercept: $(0, 2)$

(c) Vertical asymptote: $x = 1$
Horizontal asymptote: $y = -1$

(d)

x	-1	0	2	3
y	$\frac{1}{2}$	2	-4	$-\frac{5}{2}$

17. $p(x) = \dfrac{x^2}{x^2 + 1}$

   (a) Domain: all real numbers x

   (b) Intercept: $(0, 0)$

   (c) Horizontal asymptote: $y = 1$

   (d)

x	± 3	± 2	± 1	0
y	$\frac{9}{10}$	$\frac{4}{5}$	$\frac{1}{2}$	0

19. $f(x) = \dfrac{x}{x^2 + 1}$

   (a) Domain: all real numbers x

   (b) Intercept: $(0, 0)$

   (c) Horizontal asymptote: $y = 0$

   (d)

x	-2	-1	0	1	2
y	$-\frac{2}{5}$	$-\frac{1}{2}$	0	$\frac{1}{2}$	$\frac{2}{5}$

21. $f(x) = \dfrac{-6x^2}{x^2 + 1}$

   (a) Domain: all real numbers x

   (b) Intercept: $(0, 0)$

   (c) Horizontal asymptote: $y = -6$

   (d)

x	± 3	± 2	± 1	0
y	$-\frac{27}{5}$	$-\frac{24}{5}$	-3	0

23. $f(x) = \dfrac{6x^2 - 11x + 3}{3x^2 - x}$

   $= \dfrac{(3x - 1)(2x - 3)}{x(3x - 1)} = \dfrac{2x - 3}{x}, \ x \ne \dfrac{1}{3}$

   (a) Domain: all real numbers x except $x = 0$ and $x = \dfrac{1}{3}$

   (b) x-intercept: $\left(\dfrac{3}{2}, 0\right)$
       y-intercept: none

   (c) Vertical asymptote: $x = 0$
       Horizontal asymptote: $y = 2$

   (d)

x	-2	-1	1	2	3	4
y	$\frac{7}{2}$	5	-1	$\frac{1}{2}$	1	$\frac{5}{4}$

25. $f(x) = \dfrac{2x^3}{x^2 + 1} = 2x - \dfrac{2x}{x^2 + 1}$

   (a) Domain: all real numbers x

   (b) Intercept: $(0, 0)$

   (c) Slant asymptote: $y = 2x$

   (d)

x	-2	-1	0	1	2
y	$-\frac{16}{5}$	-1	0	1	$\frac{16}{5}$

27. $f(x) = \dfrac{x^2 + 3x - 10}{x + 2} = x + 1 - \dfrac{12}{x + 2}$

(a) Domain: all real numbers x except $x = -2$

(b) Intercepts: $(-5, 0)$, $(2, 0)$, $(0, -5)$

(c) Vertical asymptote: $x = -2$
   Slant asymptote: $y = x + 1$

(d)

x	-6	-4	-3	-1	0	1	2	4
y	-2	3	10	-12	-5	-2	0	3

29. $f(x) = \dfrac{3x^3 - 2x^2 - 3x + 2}{3x^2 - x - 4}$

$\quad = \dfrac{(3x - 2)(x + 1)(x - 1)}{(3x - 4)(x + 1)}$

$\quad = \dfrac{(3x - 2)(x - 1)}{3x - 4}$

$\quad = x - \dfrac{1}{3} + \dfrac{2/3}{3x - 4}, \quad x \neq -1$

(a) Domain: all real numbers x except $x = -1, x = \dfrac{4}{3}$

(b) x-intercepts: $(1, 0)$ and $\left(\dfrac{2}{3}, 0\right)$

   y-intercept: $\left(0, -\dfrac{1}{2}\right)$

(c) Vertical asymptote: $x = \dfrac{4}{3}$

   Slant asymptote: $y = x - \dfrac{1}{3}$

(d)

x	-3	-2	0	1	2	3
y	$-\dfrac{44}{13}$	$-\dfrac{12}{5}$	$-\dfrac{1}{2}$	0	2	$\dfrac{14}{5}$

31. (a)

(b) $\overline{C} = \dfrac{100,000 + 0.9x}{x}, \, 0 < x$

When $x = 1000$, $C = \dfrac{100,000 + 900}{1000} = \$100.90.$

When $x = 10,000$, $C = \dfrac{100,000 + 9000}{10,000} = \$10.90.$

When $x = 100,000$, $C = \dfrac{100,000 + 90,000}{100,000} = \$1.90.$

(c) The average cost per unit is always greater than $\$0.90$ because $\$0.90$ is the horizontal asymptote of the function.

33. $y = \dfrac{18.47x - 2.96}{0.23x + 1}, \, 0 < x$

The limiting amount of CO_2 uptake is determined by the horizontal asymptote,

$y = \dfrac{18.47}{0.23} \approx 80.3 \text{ mg/dm}^2/\text{hr}.$

35. $y^2 = -12x$

$y^2 = 4(-3)x$

Parabola

37. $\dfrac{x^2}{9} - \dfrac{y^2}{1} = 1$

Hyperbola

39. $x^2 + 20y = 0$

$x^2 = -20y$

$x^2 = 4(-5)y$

Parabola

41. $\dfrac{y^2}{49} - \dfrac{x^2}{144} = 1$

Hyperbola

43. Vertex: $(0,0)$

Point: $(1,2)$

Horizontal axis

$y^2 = 4px$

$(2)^2 = 4p(1) \implies p = 1$

$y^2 = 4x$

45. Vertex: $(0,0)$

Focus: $(-6,0) \implies p = -6$

Horizontal axis

$y^2 = 4px$

$y^2 = 4(-6)x$

$y^2 = -24x$

47. $y = \dfrac{x^2}{200}$, $-100 \le x \le 100$

Vertex: $(0,0)$

$x^2 = 200y$

$4p = 200$

$p = 50$

Focus: $(0,50)$

49. Vertices: $(\pm 5, 0) \implies a = 5$

Minor axis of length $6 \implies b = 3$

Center: $(0,0)$

Horizontal major axis

$\dfrac{x^2}{a^2} + \dfrac{y^2}{b^2} = 1$

$\dfrac{x^2}{25} + \dfrac{y^2}{9} = 1$

51. Center: $(0,0)$

Vertices: $(0, \pm 6) \implies a = 6$

Vertical major axis

$\dfrac{x^2}{b^2} + \dfrac{y^2}{36} = 1$

Using $(2,2)$:

$\dfrac{(2)^2}{b^2} + \dfrac{(2)^2}{36} = 1$

$b^2 = \dfrac{9}{2}$

$\dfrac{2x^2}{9} + \dfrac{y^2}{36} = 1$

53. Foci: $(\pm 6, 0) \implies c = 6$

Center: $(0,0)$

Minor axis of length $10 \implies b = 5$

$c^2 = a^2 - b^2$

$36 = a^2 - 25$

$a^2 = 61$

$\dfrac{x^2}{61} + \dfrac{y^2}{25} = 1$

55. $a = 5$, $b = 4$, $c = \sqrt{a^2 - b^2} = 3$

The foci should be placed 3 feet on either side of the center and have the same height as the pillars.

57. Vertices: $(0, \pm 1)$

Foci: $(0, \pm 3)$

Vertical transverse axis

Center: $(0,0)$

$a = 1$, $c = 3$,

$b = \sqrt{9-1} = \sqrt{8}$

$\dfrac{y^2}{a^2} - \dfrac{x^2}{b^2} = 1$

$\dfrac{y^2}{1} - \dfrac{x^2}{8} = 1$

59. Center: $(0,0)$

Horizontal transverse axis

Vertices: $(\pm 1, 0) \implies a = 1$

Asymptotes: $y = \pm 2x \implies \dfrac{b}{a} = 2 \implies b = 2$

$\dfrac{x^2}{1} - \dfrac{y^2}{4} = 1$

61. $x^2 - 6x + 2y + 9 = 0$

$(x - 3)^2 = -2y$

$(x - 3)^2 = 4\left(-\frac{1}{2}\right)y \implies p = -\frac{1}{2}$

Parabola

Vertex: $(3, 0)$

Focus: $\left(3, -\frac{1}{2}\right)$

The graph of $x^2 = -2y$ has been shifted to the right three units.

63. $x^2 + y^2 - 2x - 4y + 5 = 0$

$(x^2 - 2x) + (y^2 - 4y) = -5$

$(x^2 - 2x + 1) + (y^2 - 4y + 4) = -5 + 1 + 4$

$(x - 1)^2 + (y - 2)^2 = 0$

Point: $(1, 2)$

Note: This a degenerate conic, a circle of radius zero.

The graph has been shifted to the right one unit and upward two units from standard position.

65. $x^2 + 9y^2 + 10x - 18y + 25 = 0$

$(x^2 + 10x) + 9(y^2 - 2y) = -25$

$(x^2 + 10x + 25) + 9(y^2 - 2y + 1) = -25 + 25 + 9$

$(x + 5)^2 + 9(y - 1)^2 = 9$

$\frac{(x + 5)^2}{9} + (y - 1)^2 = 1$

Ellipse

Center: $(-5, 1)$

Vertices: $(-8, 1), (-2, 1)$

The graph of $\frac{x^2}{9} + y^2 = 1$ has been shifted left five units and upward one unit.

67. $9x^2 - y^2 - 72x + 8y + 119 = 0$

$9(x^2 - 8x) - (y^2 - 8y) = -119$

$9(x^2 - 8x + 16) - (y^2 - 8y + 16) = -119 + 144 - 16$

$9(x - 4)^2 - (y - 4)^2 = 9$

$\frac{(x - 4)^2}{1} - \frac{(y - 4)^2}{9} = 1$

Hyperbola

Center: $(4, 4)$

Vertices: $(3, 4), (5, 4)$

The graph of $x^2 - \frac{y^2}{9} = 1$ has been shifted right four units and upward four units from standard position.

69. Vertex: $(-6, 4)$

Passes through $(0, 0)$

Vertical axis

$(x + 6)^2 = 4p(y - 4)$

$(0 + 6)^2 = 4p(0 - 4)$

$36 = -16p$

$-\frac{9}{4} = p$

$(x + 6)^2 = 4\left(-\frac{9}{4}\right)(y - 4)$

$(x + 6)^2 = -9(y - 4)$

71. Vertex: $(4, 2)$

Focus: $(4, 0)$

Vertical axis, $p = -2$

$(x - 4)^2 = 4(-2)(y - 2)$

$(x - 4)^2 = -8(y - 2)$

73. Vertices: $(0, 3), (10, 3)$

Passes through $(5, 0)$

Center: $(5, 3)$

Horizontal major axis

$a = 5, b = 3$

$\frac{(x - 5)^2}{25} + \frac{(y - 3)^2}{9} = 1$

75. Vertices: $(-3, 0)$, $(7, 0)$

Foci: $(0, 0)$, $(4, 0)$

Horizontal major axis

Center: $(2, 0)$

$a = 5, c = 2,$

$b = \sqrt{25 - 4} = \sqrt{21}$

$\dfrac{(x - h)^2}{a^2} + \dfrac{(y - k)^2}{b^2} = 1$

$\dfrac{(x - 2)^2}{25} + \dfrac{y^2}{21} = 1$

77. Vertices: $(\pm 6, 7) \Rightarrow$ Center: $(0, 7)$ and $a = 6$

Asymptotes: $y = \pm\dfrac{1}{2}x + 7 \Rightarrow \pm\dfrac{b}{a} = \pm\dfrac{1}{2} = \pm\dfrac{b}{6}$

$\Rightarrow b = 3$

Horizontal transverse axis

$\dfrac{x^2}{36} - \dfrac{(y - 7)^2}{9} = 1$

79. Vertices: $(-10, 3)$, $(6, 3) \Rightarrow$ Center: $(-2, 3)$ and $a = 8$

Foci: $(-12, 3)$, $(8, 3) \Rightarrow c = 10 \Rightarrow b^2 = \sqrt{c^2 - a^2} = \sqrt{100 - 64} = 6$

Horizontal transverse axis

$\dfrac{(x + 2)^2}{64} - \dfrac{(y - 3)^2}{36} = 1$

81. Foci: $(0, 0)$, $(8, 0)$

Asymptotes: $y = \pm 2(x - 4)$

Horizontal transverse axis

Center: $(4, 0) \Rightarrow c = 4$

$\dfrac{b}{a} = 2 \Rightarrow b = 2a$

$a^2 + b^2 = c^2$

$a^2 + (2a)^2 = 4^2$

$a^2 = \dfrac{16}{5}$

$b^2 = \dfrac{64}{5}$

$\dfrac{(x - h)^2}{a^2} - \dfrac{(y - k)^2}{b^2} = 1$

$\dfrac{5(x - 4)^2}{16} - \dfrac{5y^2}{64} = 1$

83. $\quad x^2 = 4p(y - 12)$

$\quad (\pm 4)^2 = 4p(10 - 12)$

$\quad\quad 16 = -8p$

$\quad\quad -2 = p$

$\quad\quad x^2 = 4(-2)(y - 12)$

$\quad\quad x^2 = -8(y - 12)$

When $y = 0$, we have:

$x^2 = 96$

$x = \pm\sqrt{96} = \pm 4\sqrt{6}$

At ground level, the width is $2x = 8\sqrt{6} \approx 19.6$ meters.

85. $x^2 + y^2 - 200x - 52{,}500 = 0$

(a) The conic is a circle. x^2 and y^2 have identical coefficients.

(c) The distance between $(0, 0)$ and $(-100, 150)$ is

$\quad d = \sqrt{(-100 - 0)^2 + (150 - 0)^2}$

$\quad = \sqrt{32{,}500}$

$\quad \approx 180.28$ meters.

(b) $x^2 - 200x + 10{,}000 + y^2 = 52{,}500 + 10{,}000$

$\quad\quad (x - 100)^2 + y^2 = 62{,}500$

Center: $(100, 0)$

Radius: 250

87. True. (See Exercise 63.)

Problem Solving for Chapter 4

1. $f(x) = \dfrac{ax + b}{cx + d}$

Vertical asymptote: $x = -\dfrac{d}{c}$

Horizontal asymptote: $y = \dfrac{a}{c}$

(i) $a > 0, b < 0, c > 0, d < 0$

   Both the vertical asymptote and the horizontal asymptote are positive. Matches graph (d).

(ii) $a > 0, b > 0, c < 0, d < 0$

   Both the vertical asymptote and the horizontal asymptote are negative. Matches graph (b).

(iii) $a < 0, b > 0, c > 0, d < 0$

   The vertical asymptote is positive and the horizontal asymptote is negative. Matches graph (a).

(iv) $a > 0, b < 0, c > 0, d > 0$

   The vertical asymptote is negative and the horizontal asymptote is positive. Matches graph (c).

3. (a)

Age, x	Near point, y
16	3.0
32	4.7
44	9.8
50	19.7
60	39.4

$y \approx 0.0313x^2 - 1.586x + 21.02$

(b) $\dfrac{1}{y} \approx -0.007x + 0.44$

$y \approx \dfrac{1}{-0.007x + 0.44}$

(c)

Age, x	Near point, y	Quadratic Model	Rational Model
16	3.0	3.66	3.05
32	4.7	2.32	4.63
44	9.8	11.83	7.58
50	19.7	19.97	11.11
60	39.4	38.54	50.00

The models are fairly good fits to the data. The quadratic model seems to be a better fit for older ages and the rational model a better fit for younger ages.

(d) For $x = 25$, the quadratic model yields $y \approx 0.9325$ inches and the rational model yields $y \approx 3.774$ inches.

(e) The reciprocal model cannot be used to predict the near point for a person who is 70 years old because it results in a negative value $(y \approx -20)$. The quadratic model yields $y \approx 63.37$ inches.

5. A hyperbola is the set of all points (x, y) in a plane, the difference of whose distances from two distinct fixed points (foci) is a positive constant.

Using the vertex $(a, 0)$, the distance to $(-c, 0)$ is $a + c$, and the distance to $(c, 0)$ is $(c - a)$.

$$d_2 - d_1 = (a + c) - (c - a) = 2a$$

By definition, $|d_2 - d_1|$ is constant for any point (x, y) on the hyperbola. Thus, $|d_2 - d_1| = 2a$.

7.

Place the vertex at $(0, 0)$. Then, $y^2 = 4px$ and $p = 1.5 \implies y^2 = 6x$. When $y = 3.5$ we have

$$x = \frac{(3.5)^2}{6} \approx 2.04.$$

The reflector is approximately 2.04 inches deep.

9. Tangent line: $y - y_1 = \dfrac{x_1}{2p}(x - x_1)$

(a) The slope is $m = \dfrac{x_1}{2p}$.

(b) For $x^2 = 4y$ or $y = x^2/4$, the endpoints of the chord are $(\pm 2, 1)$. The tangent lines are:

$$y - 1 = \frac{2}{2(1)}(x - 2) \implies y = x - 1$$

$$y - 1 = \frac{-2}{2(1)}(x + 2) \implies y = -x - 1$$

For $x^2 = 8y$ or $y = x^2/8$, the endpoints of the chord are $(\pm 4, 2)$. The tangent lines are:

$$y - 2 = \frac{4}{2(2)}(x - 4) \implies y = x - 2$$

$$y - 2 = \frac{-4}{2(2)}(x + 4) \implies y = -x - 2$$

For $x^2 = 12y$ or $y = x^2/12$, the endpoints of the chord are $(\pm 6, 3)$. The tangent lines are:

$$y - 3 = \frac{6}{2(3)}(x - 6) \implies y = x - 3$$

$$y - 3 = \frac{-6}{2(3)}(x + 6) \implies y = -x - 3$$

For $x^2 = 16y$ or $y = x^2/16$, the endpoints of the chord are $(\pm 8, 4)$. The tangent lines are:

$$y - 4 = \frac{8}{2(4)}(x - 8) \implies y = x - 4$$

$$y - 4 = \frac{-8}{2(4)}(x + 8) \implies y = -x - 4$$

11. $Ax^2 + Cy^2 + Dx + Ey + F = 0$

Assume that the conic is *not* degenerate.

(a) $A = C, A \neq 0$

$$Ax^2 + Ay^2 + Dx + Ey + F = 0$$

$$x^2 + y^2 + \frac{D}{A}x + \frac{E}{A}y + \frac{F}{A} = 0$$

$$\left(x^2 + \frac{D}{A}x + \frac{D^2}{4A^2}\right) + \left(y^2 + \frac{E}{A}y + \frac{E^2}{4A^2}\right) = -\frac{F}{A} + \frac{D^2}{4A^2} + \frac{E^2}{4A^2}$$

$$\left(x + \frac{D}{2A}\right)^2 + \left(y + \frac{E}{2A}\right)^2 = \frac{D^2 + E^2 - 4AF}{4A^2}$$

This is a circle with center $\left(-\dfrac{D}{2A}, -\dfrac{E}{2A}\right)$ and radius $\dfrac{\sqrt{D^2 + E^2 - 4AF}}{2|A|}$.

(b) $A = 0$ or $C = 0$ (but not both). Let $C = 0$.

$$Ax^2 + Dx + Ey + F = 0$$

$$x^2 + \frac{D}{A}x = -\frac{E}{A}y - \frac{F}{A}$$

$$x^2 + \frac{D}{A}x + \frac{D^2}{4A^2} = -\frac{E}{A}y - \frac{F}{A} + \frac{D^2}{4A^2}$$

$$\left(x + \frac{D}{2A}\right)^2 = -\frac{E}{A}\left(y + \frac{F}{E} - \frac{D^2}{4AE}\right)$$

This is a parabola with vertex $\left(-\dfrac{D}{2A}, \dfrac{D^2 - 4AF}{4AE}\right)$. $A = 0$ yields a similar result.

(c) $AC > 0 \implies A$ and C are either both positive or are both negative (if that is the case, move the terms to the other side of the equation so that they are both positive).

$$Ax^2 + Cy^2 + Dx + Ey + F = 0$$

$$A\left(x^2 + \frac{D}{A}x + \frac{D^2}{4A^2}\right) + C\left(y^2 + \frac{E}{C}y + \frac{E^2}{4C^2}\right) = -F + \frac{D^2}{4A} + \frac{E^2}{4C}$$

$$A\left(x + \frac{D}{2A}\right)^2 + C\left(y + \frac{E}{2C}\right)^2 = \frac{CD^2 + AE^2 - 4ACF}{4AC}$$

$$\frac{\left(x + \dfrac{D}{2A}\right)^2}{\dfrac{CD^2 + AE^2 - 4ACF}{4A^2C}} + \frac{\left(y + \dfrac{E}{2C}\right)^2}{\dfrac{CD^2 + AE^2 - 4ACF}{4AC^2}} = 1$$

Since A and C are both positive, $4A^2C$ and $4AC^2$ are both positive. $CD^2 + AE^2 - 4ACF$ must be positive or the conic is degenerate. Thus, we have an ellipse with center $(-D/2A, -E/2C)$.

(d) $AC < 0 \implies A$ and C have opposite signs. Let's assume that A is positive and C is negative. (If A is negative and C is positive, move the terms to the other side of the equation.) From part (c) we have

$$\frac{\left(x + \dfrac{D}{2A}\right)^2}{\dfrac{CD^2 + AE^2 - 4ACF}{4A^2C}} + \frac{\left(y + \dfrac{E}{2C}\right)^2}{\dfrac{CD^2 + AE^2 - 4ACF}{4AC^2}} = 1.$$

Since $A > 0$ and $C < 0$, the first denominator is positive if $CD^2 + AE^2 - 4ACF < 0$ and is negative if $CD^2 + AE^2 - 4ACF > 0$, since $4A^2C$ is negative. The second denominator would have the *opposite* sign since $4AC^2 > 0$. Thus, we have a hyperbola with center $(-D/2A, -E/2C)$.

Practice Test for Chapter 4

1. Sketch the graph of $f(x) = \dfrac{x-1}{2x}$ and label all intercepts and asymptotes.

2. Sketch the graph of $f(x) = \dfrac{3x^2 - 4}{x}$ and label all intercepts and asymptotes.

3. Find all the asymptotes of $f(x) = \dfrac{8x^2 - 9}{x^2 + 1}$.

4. Find all the asymptotes of $f(x) = \dfrac{4x^2 - 2x + 7}{x - 1}$.

5. Sketch the graph of $f(x) = \dfrac{x-5}{(x-5)^2}$.

6. Find the vertex, focus, and directrix of the parabola $x^2 = 20y$.

7. Find the equation of the parabola with vertex $(0, 0)$ and focus $(7, 0)$.

8. Find the center, foci, and vertices of the ellipse $\dfrac{x^2}{144} + \dfrac{y^2}{25} = 1$.

9. Find the equation of the ellipse with foci $(\pm 4, 0)$ and minor axis of length 6.

10. Find the center, vertices, foci, and asymptotes of the hyperbola $\dfrac{y^2}{144} - \dfrac{x^2}{169} = 1$.

11. Find the equation of the hyperbola with vertices $(\pm 4, 0)$ and asymptotes $y = \pm\tfrac{1}{2}x$.

12. Find the equation of the parabola with vertex $(6, -1)$ and focus $(6, 3)$.

13. Find the center, foci, and vertices of the ellipse $16x^2 + 9y^2 - 96x + 36y + 36 = 0$.

14. Find the equation of the ellipse with vertices $(-1, 1)$ and $(7, 1)$ and minor axis of length 2.

15. Find the center, vertices, foci, and asymptotes of the hyperbola $4(x + 3)^2 - 9(y - 1)^2 = 1$.

16. Find the equation of the hyperbola with vertices $(3, 4)$ and $(3, -4)$ and foci $(3, 7)$ and $(3, -7)$.

CHAPTER 5
Exponential and Logarithmic Functions

CHAPTER 5
Exponential and Logarithmic Functions

Section 5.1   Exponential Functions and Their Graphs

- You should know that a function of the form $f(x) = a^x$, where $a > 0$, $a \neq 1$, is called an exponential function with base a.
- You should be able to graph exponential functions.
- You should know formulas for compound interest.

  (a) For n compoundings per year: $A = P\left(1 + \dfrac{r}{n}\right)^{nt}$.

  (b) For continuous compoundings: $A = Pe^{rt}$.

Vocabulary Check

1. algebraic

2. transcendental

3. natural exponential; natural

4. $A = P\left(1 + \dfrac{r}{n}\right)^{nt}$

5. $A = Pe^{rt}$

---

1. $f(5.6) = (3.4)^{5.6} \approx 946.852$

3. $f(-\pi) = 5^{-\pi} \approx 0.006$

5. $g(x) = 5000(2^x) = 5000(2^{-1.5})$

≈ 1767.767

7. $f(x) = 2^x$

Increasing

Asymptote: $y = 0$

Intercept: $(0, 1)$

Matches graph (d).

9. $f(x) = 2^{-x}$

Decreasing

Asymptote: $y = 0$

Intercept: $(0, 1)$

Matches graph (a).

11. $f(x) = \left(\frac{1}{2}\right)^x$

x	-2	-1	0	1	2
$f(x)$	4	2	1	0.5	0.25

Asymptote: $y = 0$

13. $f(x) = 6^{-x}$

x	-2	-1	0	1	2
$f(x)$	36	6	1	0.167	0.028

Asymptote: $y = 0$

15. $f(x) = 2^{x-1}$

x	-2	-1	0	1	2
$f(x)$	0.125	0.25	0.5	1	2

Asymptote: $y = 0$

17. $f(x) = 3^x$, $g(x) = 3^{x-4}$

Because $g(x) = f(x - 4)$, the graph of g can be obtained by shifting the graph of f four units to the right.

19. $f(x) = -2^x$, $g(x) = 5 - 2^x$

Because $g(x) = 5 + f(x)$, the graph of g can be obtained by shifting the graph of f five units upward.

21. $f(x) = \left(\frac{7}{2}\right)^x$, $g(x) = -\left(\frac{7}{2}\right)^{-x+6}$

Because $g(x) = -f(-x + 6)$, the graph of g can be obtained by reflecting the graph of f in the x-axis and y-axis and shifting f six units to the right. (*Note:* This is equivalent to shifting f six units to the left and then reflecting the graph in the x-axis and y-axis.)

23. $y = 2^{-x^2}$

25. $f(x) = 3^{x-2} + 1$

27. $f\left(\frac{3}{4}\right) = e^{-3/4} \approx 0.472$ **29.** $f(10) = 2e^{-5(10)} \approx 3.857 \times 10^{-22}$ **31.** $f(6) = 5000e^{0.06(6)} \approx 7166.647$

33. $f(x) = e^x$

x	-2	-1	0	1	2
$f(x)$	0.135	0.368	1	2.718	7.389

Asymptote: $y = 0$

35. $f(x) = 3e^{x+4}$

x	-8	-7	-6	-5	-4
$f(x)$	0.055	0.149	0.406	1.104	3

Asymptote: $y = 0$

37. $f(x) = 2e^{x-2} + 4$

x	-2	-1	0	1	2
$f(x)$	4.037	4.100	4.271	4.736	6

Asymptote: $y = 4$

39. $y = 1.08^{-5x}$

41. $s(t) = 2e^{0.12t}$

43. $g(x) = 1 + e^{-x}$

45. $3^{x+1} = 27$

$3^{x+1} = 3^3$

$x + 1 = 3$

$x = 2$

47. $2^{x-2} = \frac{1}{32}$

$2^{x-2} = 2^{-5}$

$x - 2 = -5$

$x = -3$

49. $e^{3x+2} = e^3$

$3x + 2 = 3$

$3x = 1$

$x = \frac{1}{3}$

51. $e^{x^2-3} = e^{2x}$

$x^2 - 3 = 2x$

$x^2 - 2x - 3 = 0$

$(x - 3)(x + 1) = 0$

$x = 3$ or $x = -1$

53. $P = \$2500$, $r = 2.5\%$, $t = 10$ years

Compounded n times per year: $A = P\left(1 + \dfrac{r}{n}\right)^{nt} = 2500\left(1 + \dfrac{0.025}{n}\right)^{10n}$

Compounded continuously: $A = Pe^{rt} = 2500e^{0.025(10)}$

n	1	2	4	12	365	Continuous Compounding
A	\$3200.21	\$3205.09	\$3207.57	\$3209.23	\$3210.04	\$3210.06

55. $P = \$2500$, $r = 3\%$, $t = 20$ years

Compounded n times per year: $A = P\left(1 + \dfrac{r}{n}\right)^{nt} = 2500\left(1 + \dfrac{0.03}{n}\right)^{20n}$

Compounded continuously: $A = Pe^{rt} = 2500e^{0.03(20)}$

n	1	2	4	12	365	Continuous Compounding
A	\$4515.28	\$4535.05	\$4545.11	\$4551.89	\$4555.18	\$4555.30

57. $A = Pe^{rt} = 12{,}000e^{0.04t}$

t	10	20	30	40	50
A	\$17,901.90	\$26,706.49	\$39,841.40	\$59,436.39	\$88,668.67

59. $A = Pe^{rt} = 12{,}000e^{0.065t}$

t	10	20	30	40	50
A	\$22,986.49	\$44,031.56	\$84,344.25	\$161,564.86	\$309,484.08

61. $A = 25{,}000e^{(0.0875)(25)} \approx \$222{,}822.57$

63. $C(10) = 23.95(1.04)^{10} \approx \$35.45$

65. $V(t) = 100e^{4.6052t}$

(a) $V(1) \approx 10{,}000.298$ computers

(b) $V(1.5) \approx 10{,}004.472$ computers

(c) $V(2) \approx 1{,}000{,}059.63$ computers

67. $Q = 25\left(\frac{1}{2}\right)^{t/1599}$

(a) $Q(0) = 25$ grams

(b) $Q(1000) \approx 16.21$ grams

(c)

69. $y = \dfrac{100}{1 + 7e^{-0.069x}}$

(a)

(b)

x	Sample Data	Model
0	12	12.5
25	44	44.5
50	81	81.82
75	96	96.19
100	99	99.3

(c) When $x = 36$: $y = \dfrac{100}{1 + 7e^{-0.069(36)}} \approx 63.14\%$.

(d) $\dfrac{2}{3}(100) = \dfrac{100}{1 + 7e^{-0.069x}}$ when $x \approx 38$ masses.

71. True. The line $y = -2$ is a horizontal asymptote for the graph of $f(x) = 10^x - 2$.

73. $f(x) = 3^{x-2}$

$\quad\quad = 3^x 3^{-2}$

$\quad\quad = 3^x \left(\dfrac{1}{3^2}\right)$

$\quad\quad = \dfrac{1}{9}(3^x)$

$\quad\quad = h(x)$

Thus, $f(x) \neq g(x)$, but $f(x) = h(x)$.

75.

$f(x) = 16(4^{-x})$	and	$f(x) = 16(4^{-x})$
$= 4^2(4^{-x})$		$= 16(2^2)^{-x}$
$= 4^{2-x}$		$= 16(2^{-2x})$
$= \left(\dfrac{1}{4}\right)^{-(2-x)}$		$= h(x)$
$= \left(\dfrac{1}{4}\right)^{x-2}$		
$= g(x)$		

Thus, $f(x) = g(x) = h(x)$.

77. $y = 3^x$ and $y = 4^x$

x	-2	-1	0	1	2
3^x	$\frac{1}{9}$	$\frac{1}{3}$	1	3	9
4^x	$\frac{1}{16}$	$\frac{1}{4}$	1	4	16

(a) $4^x < 3^x$ when $x < 0$.

(b) $4^x > 3^x$ when $x > 0$.

79. $f(x) = \left(1 + \dfrac{0.5}{x}\right)^x$ and $g(x) = e^{0.5}$ (Horizontal line)

As $x \to \infty$, $f(x) \to g(x)$.

As $x \to -\infty$, $f(x) \to g(x)$.

81. $x^2 + y^2 = 25$

$\quad\quad y^2 = 25 - x^2$

$\quad\quad y = \pm\sqrt{25 - x^2}$

83. $f(x) = \dfrac{2}{9 + x}$

Vertical asymptote: $x = -9$

Horizontal asymptote: $y = 0$

x	-11	-10	-8	-7
$f(x)$	-1	-2	2	1

85. Answers will vary.

Section 5.2   Logarithmic Functions and Their Graphs

■ You should know that a function of the form $y = \log_a x$, where $a > 0$, $a \neq 1$, and $x > 0$, is called a logarithm of x to base a.

■ You should be able to convert from logarithmic form to exponential form and vice versa.

$y = \log_a x \iff a^y = x$

■ You should know the following properties of logarithms.

(a) $\log_a 1 = 0$ since $a^0 = 1$.

(b) $\log_a a = 1$ since $a^1 = a$.

(c) $\log_a a^x = x$ since $a^x = a^x$.

(d) $a^{\log_a x} = x$ Inverse Property

(e) If $\log_a x = \log_a y$, then $x = y$.

■ You should know the definition of the natural logarithmic function.

$\log_e x = \ln x, x > 0$

■ You should know the properties of the natural logarithmic function.

(a) $\ln 1 = 0$ since $e^0 = 1$.

(b) $\ln e = 1$ since $e^1 = e$.

(c) $\ln e^x = x$ since $e^x = e^x$.

(d) $e^{\ln x} = x$ Inverse Property

(e) If $\ln x = \ln y$, then $x = y$.

■ You should be able to graph logarithmic functions.

Vocabulary Check

1. logarithmic

2. 10

3. natural; e

4. $a^{\log_a x} = x$

5. $x = y$

---

1. $\log_4 64 = 3 \implies 4^3 = 64$

3. $\log_7 \frac{1}{49} = -2 \implies 7^{-2} = \frac{1}{49}$

5. $\log_{32} 4 = \frac{2}{5} \implies 32^{2/5} = 4$

7. $\log_{36} 6 = \frac{1}{2} \implies 36^{1/2} = 6$

9. $5^3 = 125 \implies \log_5 125 = 3$

11. $81^{1/4} = 3 \implies \log_{81} 3 = \frac{1}{4}$

13. $6^{-2} = \frac{1}{36} \implies \log_6 \frac{1}{36} = -2$

15. $7^0 = 1 \implies \log_7 1 = 0$

17. $f(x) = \log_2 x$

$f(16) = \log_2 16 = 4$ since $2^4 = 16$

19. $f(x) = \log_7 x$

$f(1) = \log_7 1 = 0$ since $7^0 = 1$

21. $g(x) = \log_a x$

$g(a^2) = \log_a a^2$

$= 2$ by the Inverse Property

23. $f(x) = \log x$

$f\left(\frac{4}{5}\right) = \log\left(\frac{4}{5}\right) \approx -0.097$

25. $f(x) = \log x$

$f(12.5) \approx 1.097$

27. $\log_3 3^4 = 4$ since $3^4 = 3^4$

29. $\log_\pi \pi = 1$ since $\pi^1 = \pi$.

31. $f(x) = \log_4 x$

Domain: $x > 0 \implies$ The domain is $(0, \infty)$.

x-intercept: $(1, 0)$

Vertical asymptote: $x = 0$

$y = \log_4 x \implies 4^y = x$

x	$\frac{1}{4}$	1	4	2
$f(x)$	-1	0	1	$\frac{1}{2}$

33. $y = -\log_3 x + 2$

Domain: $(0, \infty)$

x-intercept:

$$-\log_3 x + 2 = 0$$
$$2 = \log_3 x$$
$$3^2 = x$$
$$9 = x$$

The x-intercept is $(9, 0)$.

Vertical asymptote: $x = 0$

$y = -\log_3 x + 2$

$\log_3 x = 2 - y \implies 3^{2-y} = x$

x	27	9	3	1	$\frac{1}{3}$
y	-1	0	1	2	3

35. $f(x) = -\log_6(x + 2)$

Domain: $x + 2 > 0 \implies x > -2$

The domain is $(-2, \infty)$.

x-intercept:

$$0 = -\log_6(x + 2)$$
$$0 = \log_6(x + 2)$$
$$6^0 = x + 2$$
$$1 = x + 2$$
$$-1 = x$$

The x-intercept is $(-1, 0)$.

Vertical asymptote: $x + 2 = 0 \implies x = -2$

$$y = -\log_6(x + 2)$$
$$-y = \log_6(x + 2)$$
$$6^{-y} - 2 = x$$

x	4	-1	$-1\frac{5}{6}$	$-1\frac{35}{36}$
$f(x)$	-1	0	1	2

37. $y = \log\left(\dfrac{x}{5}\right)$

Domain: $\dfrac{x}{5} > 0 \implies x > 0$

The domain is $(0, \infty)$.

x-intercept:

$$\log\left(\frac{x}{5}\right) = 0$$
$$\frac{x}{5} = 10^0$$
$$\frac{x}{5} = 1 \implies x = 5$$

The x-intercept is $(5, 0)$.

Vertical asymptote: $\dfrac{x}{5} = 0 \implies x = 0$

The vertical asymptote is the y-axis.

x	1	2	3	4	5	6	7
y	-0.70	-0.40	-0.22	-0.10	0	0.08	0.15

39. $f(x) = \log_3 x + 2$

Asymptote: $x = 0$

Point on graph: $(1, 2)$

Matches graph (c).

The graph of $f(x)$ is obtained by shifting the graph of $g(x)$ upward two units.

41. $f(x) = -\log_3(x + 2)$

Asymptote: $x = -2$

Point on graph: $(-1, 0)$

Matches graph (d).

The graph of $f(x)$ is obtained by reflecting the graph of $g(x)$ about the x-axis and shifting the graph two units to the left.

43. $f(x) = \log_3(1 - x) = \log_3[-(x - 1)]$

Asymptote: $x = 1$

Point on graph: $(0, 0)$

Matches graph (b).

The graph of $f(x)$ is obtained by reflecting the graph of $g(x)$ about the y-axis and shifting the graph one unit to the right.

45. $\ln \frac{1}{2} = -0.693 \ldots \implies e^{-0.693 \cdots} = \frac{1}{2}$

47. $\ln 4 = 1.386 \ldots \implies e^{1.386 \cdots} = 4$

49. $\ln 250 = 5.521 \ldots \implies e^{5.521 \cdots} = 250$

51. $\ln 1 = 0 \implies e^0 = 1$

53. $e^3 = 20.0855 \ldots \implies \ln 20.0855 \ldots = 3$

55. $e^{1/2} = 1.6487 \ldots \implies \ln 1.6487 \ldots = \frac{1}{2}$

57. $e^{-0.5} = 0.6065 \ldots \implies \ln 0.6065 \ldots = -0.5$

59. $e^x = 4 \implies \ln 4 = x$

61. $f(x) = \ln x$

$f(18.42) = \ln 18.42 \approx 2.913$

63. $g(x) = 2 \ln x$

$g(0.75) = 2 \ln 0.75 \approx -0.575$

65. $g(x) = \ln x$

$g(e^3) = \ln e^3 = 3$ by the Inverse Property

67. $g(x) = \ln x$

$g(e^{-2/3}) = \ln e^{-2/3} = -\frac{2}{3}$ by the Inverse Property

69. $f(x) = \ln(x - 1)$

Domain: $x - 1 > 0 \implies x > 1$

The domain is $(1, \infty)$.

x-intercept:

$0 = \ln(x - 1)$

$e^0 = x - 1$

$2 = x$

The x-intercept is $(2, 0)$.

Vertical asymptote: $x - 1 = 0 \implies x = 1$

x	1.5	2	3	4
$f(x)$	-0.69	0	0.69	1.10

71. $g(x) = \ln(-x)$

Domain: $-x > 0 \implies x < 0$

The domain is $(-\infty, 0)$.

x-intercept:

$0 = \ln(-x)$

$e^0 = -x$

$-1 = x$

The x-intercept is $(-1, 0)$.

Vertical asymptote: $-x = 0 \implies x = 0$

x	-0.5	-1	-2	-3
$g(x)$	-0.69	0	0.69	1.10

73. $y_1 = \log(x + 1)$

75. $y_1 = \ln(x - 1)$

77. $y = \ln x + 2$

79. $\log_2(x + 1) = \log_2 4$

$x + 1 = 4$

$x = 3$

81. $\log(2x + 1) = \log 15$

$2x + 1 = 15$

$x = 7$

83. $\ln(x + 2) = \ln 6$

$x + 2 = 6$

$x = 4$

85. $\ln(x^2 - 2) = \ln 23$

$x^2 - 2 = 23$

$x^2 = 25$

$x = \pm 5$

87. $t = 12.542 \ln\left(\dfrac{x}{x - 1000}\right), x > 1000$

(a) When $x = \$1100.65$:

$$t = 12.542 \ln\left(\frac{1100.65}{1100.65 - 1000}\right) \approx 30 \text{ years}$$

When $x = \$1254.68$:

$$t = 12.542 \ln\left(\frac{1254.68}{1254.68 - 1000}\right) \approx 20 \text{ years}$$

(b) Total amounts: $(1100.65)(12)(30) = \$396,234.00$

$(1254.68)(12)(20) = \$301,123.20$

(c) Interest charges: $396,234 - 150,000 = \$246,234$

$301,123.20 - 150,000 = \$151,123.20$

(d) The vertical asymptote is $x = 1000$. The closer the payment is to \$1000 per month, the longer the length of the mortgage will be. Also, the monthly payment must be greater than \$1000.

89. $f(t) = 80 - 17 \log(t + 1), 0 \le t \le 12$

(a)

(b) $f(0) = 80 - 17 \log 1 = 80.0$

(c) $f(4) = 80 - 17 \log 5 \approx 68.1$

(d) $f(10) = 80 - 17 \log 11 \approx 62.3$

91. False. Reflecting $g(x)$ about the line $y = x$ will determine the graph of $f(x)$.

93. $f(x) = 3^x$, $g(x) = \log_3 x$

f and g are inverses. Their graphs are reflected about the line $y = x$.

95. $f(x) = e^x$, $g(x) = \ln x$

f and g are inverses. Their graphs are reflected about the line $y = x$.

97. (a) $f(x) = \ln x$, $g(x) = \sqrt{x}$

The natural log function grows at a slower rate than the square root function.

(b) $f(x) = \ln x$, $g(x) = \sqrt[4]{x}$

The natural log function grows at a slower rate than the fourth root function.

99. (a) False. If y were an exponential function of x, then $y = a^x$, but $a^1 = a$, not 0. Because one point is $(1, 0)$, y is not an exponential function of x.

(b) True. $y = \log_a x$

For $a = 2$, $y = \log_2 x$.

$x = 1, \log_2 1 = 0$

$x = 2, \log_2 2 = 1$

$x = 8, \log_2 8 = 3$

(c) True. $x = a^y$

For $a = 2$, $x = 2^y$.

$y = 0, 2^0 = 1$

$y = 1, 2^1 = 2$

$y = 3, 2^3 = 8$

(d) False. If y were a linear function of x, the slope between $(1, 0)$ and $(2, 1)$ and the slope between $(2, 1)$ and $(8, 3)$ would be the same. However,

$$m_1 = \frac{1 - 0}{2 - 1} = 1 \text{ and } m_2 = \frac{3 - 1}{8 - 2} = \frac{2}{6} = \frac{1}{3}.$$

Therefore, y is not a linear function of x.

101. $f(x) = |\ln x|$

(a)

(b) Increasing on $(1, \infty)$
Decreasing on $(0, 1)$

(c) Relative minimum: $(1, 0)$

For Exercises 103–107, use $f(x) = 3x + 2$ and $g(x) = x^3 - 1$.

103. $(f + g)(2) = f(2) + g(2)$
$= [3(2) + 2] + [(2)^3 - 1]$
$= 8 + 7$
$= 15$

105. $(fg)(6) = f(6)g(6)$
$= [3(6) + 2][(6)^3 - 1]$
$= (20)(215)$
$= 4300$

107. $(f \circ g)(7) = f(g(7))$
$= f((7)^3 - 1)$
$= f(342)$
$= 3(342) + 2$
$= 1028$

Section 5.3   Properties of Logarithms

■ You should know the following properties of logarithms.

(a) $\log_a x = \dfrac{\log_b x}{\log_b a}$ $\log_a x = \dfrac{\log_{10} x}{\log_{10} a}$ $\log_a x = \dfrac{\ln x}{\ln a}$

(b) $\log_a(uv) = \log_a u + \log_a v$ $\ln(uv) = \ln u + \ln v$

(c) $\log_a\left(\dfrac{u}{v}\right) = \log_a u - \log_a v$ $\ln\left(\dfrac{u}{v}\right) = \ln u - \ln v$

(d) $\log_a u^n = n \log_a u$ $\ln u^n = n \ln u$

■ You should be able to rewrite logarithmic expressions using these properties.

Vocabulary Check

1. change-of-base

2. $\dfrac{\log x}{\log a} = \dfrac{\ln x}{\ln a}$

3. $\log_a(uv) = \log_a u + \log_a v$
This is the Product Property. Matches (c).

4. $\ln u^n = n \ln u$
This is the Power Property. Matches (a).

5. $\log_a \dfrac{u}{v} = \log_a u - \log_a v$

This is the Quotient Property. Matches (b).

1. (a) $\log_5 x = \dfrac{\log x}{\log 5}$

    (b) $\log_5 x = \dfrac{\ln x}{\ln 5}$

3. (a) $\log_{1/5} x = \dfrac{\log x}{\log(1/5)}$

    (b) $\log_{1/5} x = \dfrac{\ln x}{\ln(1/5)}$

5. (a) $\log_x \dfrac{3}{10} = \dfrac{\log(3/10)}{\log x}$

    (b) $\log_x \dfrac{3}{10} = \dfrac{\ln(3/10)}{\ln x}$

7. (a) $\log_{2.6} x = \dfrac{\log x}{\log 2.6}$

    (b) $\log_{2.6} x = \dfrac{\ln x}{\ln 2.6}$

9. $\log_3 7 = \dfrac{\log 7}{\log 3} = \dfrac{\ln 7}{\ln 3} \approx 1.771$

11. $\log_{1/2} 4 = \dfrac{\log 4}{\log(1/2)} = \dfrac{\ln 4}{\ln(1/2)} = -2.000$

13. $\log_9(0.4) = \dfrac{\log 0.4}{\log 9} = \dfrac{\ln 0.4}{\ln 9} \approx -0.417$

15. $\log_{15} 1250 = \dfrac{\log 1250}{\log 15} = \dfrac{\ln 1250}{\ln 15} \approx 2.633$

17. $\log_4 8 = \dfrac{\log_2 8}{\log_2 4} = \dfrac{\log_2 2^3}{\log_2 2^2} = \dfrac{3}{2}$

19. $\log_5 \frac{1}{250} = \log_5\left(\frac{1}{125} \cdot \frac{1}{2}\right)$

$\qquad = \log_5 \frac{1}{125} + \log_5 \frac{1}{2}$

$\qquad = \log_5 5^{-3} + \log_5 2^{-1}$

$\qquad = -3 - \log_5 2$

21. $\ln(5e^6) = \ln 5 + \ln e^6$

$\qquad = \ln 5 + 6$

$\qquad = 6 + \ln 5$

23. $\log_3 9 = 2\log_3 3 = 2$

25. $\log_2 \sqrt[4]{8} = \frac{1}{4}\log_2 2^3 = \frac{3}{4}\log_2 2 = \frac{3}{4}(1) = \frac{3}{4}$

27. $\log_4 16^{1.2} = 1.2(\log_4 16) = 1.2\log_4 4^2 = 1.2(2) = 2.4$

29. $\log_3(-9)$ is undefined. -9 is not in the domain of $\log_3 x$.

31. $\ln e^{4.5} = 4.5$

33. $\ln \dfrac{1}{\sqrt{e}} = \ln 1 - \ln \sqrt{e}$

$\qquad = 0 - \dfrac{1}{2}\ln e$

$\qquad = 0 - \dfrac{1}{2}(1)$

$\qquad = -\dfrac{1}{2}$

35. $\ln e^2 + \ln e^5 = 2 + 5 = 7$

37. $\log_5 75 - \log_5 3 = \log_5 \dfrac{75}{3}$

$\qquad = \log_5 25$

$\qquad = \log_5 5^2$

$\qquad = 2\log_5 5$

$\qquad = 2$

39. $\log_4 5x = \log_4 5 + \log_4 x$

41. $\log_8 x^4 = 4\log_8 x$

43. $\log_5 \dfrac{5}{x} = \log_5 5 - \log_5 x$

$\qquad = 1 - \log_5 x$

45. $\ln \sqrt{z} = \ln z^{1/2} = \dfrac{1}{2}\ln z$

47. $\ln xyz^2 = \ln x + \ln y + \ln z^2$

$\qquad = \ln x + \ln y + 2\ln z$

49. $\ln z(z-1)^2 = \ln z + \ln(z-1)^2$

$\qquad = \ln z + 2\ln(z-1), \ z > 1$

51. $\log_2 \dfrac{\sqrt{a-1}}{9} = \log_2 \sqrt{a-1} - \log_2 9$

$$= \frac{1}{2}\log_2(a-1) - \log_2 3^2$$

$$= \frac{1}{2}\log_2(a-1) - 2\log_2 3, \; a > 1$$

53. $\ln \sqrt[3]{\dfrac{x}{y}} = \dfrac{1}{3}\ln\dfrac{x}{y}$

$$= \frac{1}{3}[\ln x - \ln y]$$

$$= \frac{1}{3}\ln x - \frac{1}{3}\ln y$$

55. $\ln\left(\dfrac{x^4\sqrt{y}}{z^5}\right) = \ln x^4 \sqrt{y} - \ln z^5$

$$= \ln x^4 + \ln \sqrt{y} - \ln z^5$$

$$= 4\ln x + \frac{1}{2}\ln y - 5\ln z$$

57. $\log_5\left(\dfrac{x^2}{y^2 z^3}\right) = \log_5 x^2 - \log_5 y^2 z^3$

$$= \log_5 x^2 - (\log_5 y^2 + \log_5 z^3)$$

$$= 2\log_5 x - 2\log_5 y - 3\log_5 z$$

59. $\ln \sqrt[4]{x^3(x^2+3)} = \frac{1}{4}\ln x^3(x^2+3)$

$$= \tfrac{1}{4}[\ln x^3 + \ln(x^2+3)]$$

$$= \tfrac{1}{4}[3\ln x + \ln(x^2+3)]$$

$$= \tfrac{3}{4}\ln x + \tfrac{1}{4}\ln(x^2+3)$$

61. $\ln x + \ln 3 = \ln 3x$

63. $\log_4 z - \log_4 y = \log_4 \dfrac{z}{y}$

65. $2\log_2(x+4) = \log_2(x+4)^2$

67. $\dfrac{1}{4}\log_3 5x = \log_3(5x)^{1/4} = \log_3 \sqrt[4]{5x}$

69. $\ln x - 3\ln(x+1) = \ln x - \ln(x+1)^3$

$$= \ln \frac{x}{(x+1)^3}$$

71. $\log x - 2\log y + 3\log z = \log x - \log y^2 + \log z^3$

$$= \log \frac{x}{y^2} + \log z^3$$

$$= \log \frac{xz^3}{y^2}$$

73. $\ln x - 4[\ln(x+2) + \ln(x-2)] = \ln x - 4\ln(x+2)(x-2)$

$$= \ln x - 4\ln(x^2-4)$$

$$= \ln x - \ln(x^2-4)^4$$

$$= \ln \frac{x}{(x^2-4)^4}$$

75. $\dfrac{1}{3}[2\ln(x+3) + \ln x - \ln(x^2-1)] = \dfrac{1}{3}[\ln(x+3)^2 + \ln x - \ln(x^2-1)]$

$$= \frac{1}{3}[\ln x(x+3)^2 - \ln(x^2-1)]$$

$$= \frac{1}{3}\ln \frac{x(x+3)^2}{x^2-1}$$

$$= \ln \sqrt[3]{\frac{x(x+3)^2}{x^2-1}}$$

77. $\frac{1}{3}[\log_8 y + 2\log_8(y+4)] - \log_8(y-1) = \frac{1}{3}[\log_8 y + \log_8(y+4)^2] - \log_8(y-1)$

$$= \frac{1}{3}\log_8 y(y+4)^2 - \log_8(y-1)$$

$$= \log_8 \sqrt[3]{y(y+4)^2} - \log_8(y-1)$$

$$= \log_8\left(\frac{\sqrt[3]{y(y+4)^2}}{y-1}\right)$$

79. $\log_2 \frac{32}{4} = \log_2 32 - \log_2 4 \neq \frac{\log_2 32}{\log_2 4}$

The second and third expressions are equal by Property 2.

81. $\beta = 10\log\left(\frac{I}{10^{-12}}\right)$

$= 10[\log I - \log 10^{-12}]$

$= 10[\log I + 12]$

$= 120 + 10\log I$

When $I = 10^{-6}$:

$\beta = 120 + 10\log 10^{-6}$

$= 120 + 10(-6)$

$= 60$ decibels

83. $\beta = 120 + 10\log(2I)$

$= 120 + 10(\log 2 + \log I)$

$= (120 + 10\log I) + 10\log 2$

With both stereos playing, the music is $10\log 2 \approx 3$ decibels louder.

85. By using the regression feature on a graphing calculator we obtain $y \approx 256.24 - 20.8\ln x$.

87. $f(x) = \ln x$

False, $f(0) \neq 0$ since 0 is not in the domain of $f(x)$.

$f(1) = \ln 1 = 0$

89. False. $f(x) - f(2) = \ln x - \ln 2 = \ln\frac{x}{2} \neq \ln(x-2)$

91. False.

$f(u) = 2f(v) \Rightarrow \ln u = 2\ln v \Rightarrow \ln u = \ln v^2 \Rightarrow u = v^2$

93. Let $x = \log_b u$ and $y = \log_b v$, then $b^x = u$ and $b^y = v$.

$\frac{u}{v} = \frac{b^x}{b^y} = b^{x-y}$

Then $\log_b(u/v) = \log_b(b^{x-y}) = x - y = \log_b u - \log_b v$.

95. $f(x) = \log_2 x = \frac{\log x}{\log 2} = \frac{\ln x}{\ln 2}$

97. $f(x) = \log_{1/2} x$

$= \frac{\log x}{\log(1/2)} = \frac{\ln x}{\ln(1/2)}$

99. $f(x) = \log_{11.8} x$

$= \frac{\log x}{\log 11.8} = \frac{\ln x}{\ln 11.8}$

101. $f(x) = \ln \dfrac{x}{2}$, $g(x) = \dfrac{\ln x}{\ln 2}$, $h(x) = \ln x - \ln 2$

$f(x) = h(x)$ by Property 2

103. $\dfrac{24xy^{-2}}{16x^{-3}y} = \dfrac{24xx^3}{16yy^2} = \dfrac{3x^4}{2y^3}$, $x \neq 0$

105. $(18x^3y^4)^{-3}(18x^3y^4)^3 = \dfrac{(18x^3y^4)^3}{(18x^3y^4)^3} = 1$ if $x \neq 0$, $y \neq 0$.

107. $\quad 3x^2 + 2x - 1 = 0$

$\quad (3x - 1)(x + 1) = 0$

$\quad 3x - 1 = 0 \Rightarrow x = \frac{1}{3}$

$\quad x + 1 = 0 \Rightarrow x = -1$

109. $\quad \dfrac{2}{3x + 1} = \dfrac{x}{4}$

$\quad (3x + 1)(x) = (2)(4)$

$\quad 3x^2 + x - 8 = 0$

$\quad x = \dfrac{-1 \pm \sqrt{1^2 - 4(3)(-8)}}{2(3)} = \dfrac{-1 \pm \sqrt{97}}{6}$

Section 5.4 Exponential and Logarithmic Equations

- To solve an exponential equation, isolate the exponential expression, then take the logarithm of both sides. Then solve for the variable.

  1. $\log_a a^x = x$
  2. $\ln e^x = x$

- To solve a logarithmic equation, rewrite it in exponential form. Then solve for the variable.

  1. $a^{\log_a x} = x$
  2. $e^{\ln x} = x$

- If $a > 0$ and $a \neq 1$ we have the following:

  1. $\log_a x = \log_a y \iff x = y$
  2. $a^x = a^y \iff x = y$

- Check for extraneous solutions.

Vocabulary Check

1. solve

2. (a) $x = y$ (b) $x = y$
(c) x (d) x

3. extraneous

1. $4^{2x-7} = 64$

(a) $\quad x = 5$

$\quad 4^{2(5)-7} = 4^3 = 64$

Yes, $x = 5$ *is* a solution.

(b) $\quad x = 2$

$\quad 4^{2(2)-7} = 4^{-3} = \frac{1}{64} \neq 64$

No, $x = 2$ *is not* a solution.

3. $3e^{x+2} = 75$

  (a) $x = -2 + e^{25}$

    $3e^{(-2+e^{25})+2} = 3e^{e^{25}} \neq 75$

    No, $x = -2 + e^{25}$ *is not* a solution.

  (b) $x = -2 + \ln 25$

    $3e^{(-2+\ln 25)+2} = 3e^{\ln 25} = 3(25) = 75$

    Yes, $x = -2 + \ln 25$ *is* a solution.

  (c) $x \approx 1.219$

    $3e^{1.219+2} = 3e^{3.219} \approx 75$

    Yes, $x \approx 1.219$ *is* a solution.

5. $\log_4(3x) = 3 \implies 3x = 4^3 \implies 3x = 64$

  (a) $x \approx 21.333$

    $3(21.333) \approx 64$

    Yes, 21.333 *is* an approximate solution.

  (b) $x = -4$

    $3(-4) = -12 \neq 64$

    No, $x = -4$ *is not* a solution.

  (c) $x = \frac{64}{3}$

    $3\left(\frac{64}{3}\right) = 64$

    Yes, $x = \frac{64}{3}$ *is* a solution.

7. $\ln(2x + 3) = 5.8$

  (a) $x = \frac{1}{2}(-3 + \ln 5.8)$

    $\ln\left[2\left(\frac{1}{2}\right)(-3 + \ln 5.8) + 3\right] = \ln(\ln 5.8) \neq 5.8$

    No, $x = \frac{1}{2}(-3 + \ln 5.8)$ *is not* a solution.

  (b) $x = \frac{1}{2}(-3 + e^{5.8})$

    $\ln\left[2\left(\frac{1}{2}\right)(-3 + e^{5.8}) + 3\right] = \ln(e^{5.8}) = 5.8$

    Yes, $x = \frac{1}{2}(-3 + e^{5.8})$ *is* a solution.

  (c) $x \approx 163.650$

    $\ln[2(163.650) + 3] = \ln 330.3 \approx 5.8$

    Yes, $x \approx 163.650$ *is* an approximate solution.

9. $4^x = 16$

  $4^x = 4^2$

  $x = 2$

11. $\left(\frac{1}{2}\right)^x = 32$

  $2^{-x} = 2^5$

  $-x = 5$

  $x = -5$

13. $\ln x - \ln 2 = 0$

  $\ln x = \ln 2$

  $x = 2$

15. $e^x = 2$

  $\ln e^x = \ln 2$

  $x = \ln 2$

  $x \approx 0.693$

17. $\ln x = -1$

  $e^{\ln x} = e^{-1}$

  $x = e^{-1}$

  $x \approx 0.368$

19. $\log_4 x = 3$

  $4^{\log_4 x} = 4^3$

  $x = 4^3$

  $x = 64$

21. $f(x) = g(x)$

  $2^x = 8$

  $2^x = 2^3$

  $x = 3$

  Point of intersection:
  $(3, 8)$

23. $f(x) = g(x)$

  $\log_3 x = 2$

  $x = 3^2$

  $x = 9$

  Point of intersection:
  $(9, 2)$

25. $e^x = e^{x^2-2}$

  $x = x^2 - 2$

  $0 = x^2 - x - 2$

  $0 = (x + 1)(x - 2)$

  $x = -1$ or $x = 2$

27. $e^{x^2-3} = e^{x-2}$

  $x^2 - 3 = x - 2$

  $x^2 - x - 1 = 0$

  By the Quadratic Formula
  $x \approx 1.618$ or $x \approx -0.618$.

29. $4(3^x) = 20$

  $3^x = 5$

  $\log_3 3^x = \log_3 5$

  $x = \log_3 5 = \dfrac{\log 5}{\log 3}$ or $\dfrac{\ln 5}{\ln 3}$

  $x \approx 1.465$

31. $2e^x = 10$

  $e^x = 5$

  $\ln e^x = \ln 5$

  $x = \ln 5 \approx 1.609$

33. $e^x - 9 = 19$

  $e^x = 28$

  $\ln e^x = \ln 28$

  $x = \ln 28 \approx 3.332$

35. $3^{2x} = 80$

$\ln 3^{2x} = \ln 80$

$2x \ln 3 = \ln 80$

$x = \dfrac{\ln 80}{2 \ln 3} \approx 1.994$

37. $5^{-t/2} = 0.20$

$5^{-t/2} = \dfrac{1}{5}$

$5^{-t/2} = 5^{-1}$

$-\dfrac{t}{2} = -1$

$t = 2$

39. $3^{x-1} = 27$

$3^{x-1} = 3^3$

$x - 1 = 3$

$x = 4$

41. $2^{3-x} = 565$

$\ln 2^{3-x} = \ln 565$

$(3 - x) \ln 2 = \ln 565$

$3 \ln 2 - x \ln 2 = \ln 565$

$-x \ln 2 = \ln 565 - 3 \ln 2$

$x \ln 2 = 3 \ln 2 - \ln 565$

$x = \dfrac{3 \ln 2 - \ln 565}{\ln 2}$

$= 3 - \dfrac{\ln 565}{\ln 2} \approx -6.142$

43. $8(10^{3x}) = 12$

$10^{3x} = \dfrac{12}{8}$

$\log 10^{3x} = \log\!\left(\dfrac{3}{2}\right)$

$3x = \log\!\left(\dfrac{3}{2}\right)$

$x = \dfrac{1}{3} \log\!\left(\dfrac{3}{2}\right)$

≈ 0.059

45. $3(5^{x-1}) = 21$

$5^{x-1} = 7$

$\ln 5^{x-1} = \ln 7$

$(x - 1) \ln 5 = \ln 7$

$x - 1 = \dfrac{\ln 7}{\ln 5}$

$x = 1 + \dfrac{\ln 7}{\ln 5} \approx 2.209$

47. $e^{3x} = 12$

$3x = \ln 12$

$x = \dfrac{\ln 12}{3} \approx 0.828$

49. $500e^{-x} = 300$

$e^{-x} = \dfrac{3}{5}$

$-x = \ln \dfrac{3}{5}$

$x = -\ln \dfrac{3}{5} = \ln \dfrac{5}{3} \approx 0.511$

51. $7 - 2e^x = 5$

$-2e^x = -2$

$e^x = 1$

$x = \ln 1 = 0$

53. $6(2^{3x-1}) - 7 = 9$

$6(2^{3x-1}) = 16$

$2^{3x-1} = \dfrac{8}{3}$

$\log_2 2^{3x-1} = \log_2\!\left(\dfrac{8}{3}\right)$

$3x - 1 = \log_2\!\left(\dfrac{8}{3}\right) = \dfrac{\log(8/3)}{\log 2} \ \text{ or } \ \dfrac{\ln(8/3)}{\ln 2}$

$x = \dfrac{1}{3}\left[\dfrac{\log(8/3)}{\log 2} + 1\right] \approx 0.805$

55. $e^{2x} - 4e^x - 5 = 0$

$(e^x + 1)(e^x - 5) = 0$

$e^x = -1 \qquad \text{or} \qquad e^x = 5$

(No solution) $\qquad\qquad x = \ln 5 \approx 1.609$

57. $e^{2x} - 3e^x - 4 = 0$

$(e^x + 1)(e^x - 4) = 0$

$e^x + 1 = 0 \Rightarrow e^x = -1$

Not possible since $e^x > 0$ for all x.

$e^x - 4 = 0 \Rightarrow e^x = 4 \Rightarrow x = \ln 4 \approx 1.386$

59. $\dfrac{500}{100 - e^{x/2}} = 20$

$500 = 20(100 - e^{x/2})$

$25 = 100 - e^{x/2}$

$e^{x/2} = 75$

$\dfrac{x}{2} = \ln 75$

$x = 2 \ln 75 \approx 8.635$

61. $\dfrac{3000}{2 + e^{2x}} = 2$

$3000 = 2(2 + e^{2x})$

$1500 = 2 + e^{2x}$

$1498 = e^{2x}$

$\ln 1498 = 2x$

$x = \dfrac{\ln 1498}{2} \approx 3.656$

63. $\left(1 + \dfrac{0.065}{365}\right)^{365t} = 4$

$\ln\left(1 + \dfrac{0.065}{365}\right)^{365t} = \ln 4$

$365t \ln\left(1 + \dfrac{0.065}{365}\right) = \ln 4$

$t = \dfrac{\ln 4}{365 \ln\left(1 + \frac{0.065}{365}\right)} \approx 21.330$

65. $\left(1 + \dfrac{0.10}{12}\right)^{12t} = 2$

$\ln\left(1 + \dfrac{0.10}{12}\right)^{12t} = \ln 2$

$12t \ln\left(1 + \dfrac{0.10}{12}\right) = \ln 2$

$t = \dfrac{\ln 2}{12 \ln\left(1 + \frac{0.10}{12}\right)} \approx 6.960$

67. $g(x) = 6e^{1-x} - 25$

Algebraically:

$6e^{1-x} = 25$

$e^{1-x} = \dfrac{25}{6}$

$1 - x = \ln\left(\dfrac{25}{6}\right)$

$x = 1 - \ln\left(\dfrac{25}{6}\right)$

$x \approx -0.427$

The zero is $x \approx -0.427$.

69. $f(x) = 3e^{3x/2} - 962$

Algebraically:

$3e^{3x/2} = 962$

$e^{3x/2} = \dfrac{962}{3}$

$\dfrac{3x}{2} = \ln\left(\dfrac{962}{3}\right)$

$x = \dfrac{2}{3} \ln\left(\dfrac{962}{3}\right)$

$x \approx 3.847$

The zero is $x \approx 3.847$.

71. $g(t) = e^{0.09t} - 3$

Algebraically:

$e^{0.09t} = 3$

$0.09t = \ln 3$

$t = \dfrac{\ln 3}{0.09}$

$t \approx 12.207$

The zero is $t \approx 12.207$.

73. $h(t) = e^{0.125t} - 8$

Algebraically:

$e^{0.125t} - 8 = 0$

$e^{0.125t} = 8$

$0.125t = \ln 8$

$t = \dfrac{\ln 8}{0.125}$

$t \approx 16.636$

The zero is $t \approx 16.636$.

75. $\ln x = -3$

$x = e^{-3} \approx 0.050$

77. $\ln 2x = 2.4$

$2x = e^{2.4}$

$x = \dfrac{e^{2.4}}{2} \approx 5.512$

79. $\log x = 6$

$x = 10^6 = 1{,}000{,}000.000$

81. $3 \ln 5x = 10$

$\ln 5x = \dfrac{10}{3}$

$5x = e^{10/3}$

$x = \dfrac{e^{10/3}}{5} \approx 5.606$

83. $\ln \sqrt{x + 2} = 1$

$\sqrt{x + 2} = e^1$

$x + 2 = e^2$

$x = e^2 - 2 \approx 5.389$

85. $7 + 3 \ln x = 5$

$3 \ln x = -2$

$\ln x = -\tfrac{2}{3}$

$x = e^{-2/3} \approx 0.513$

87. $6 \log_3(0.5x) = 11$

$\log_3(0.5x) = \tfrac{11}{6}$

$3^{\log_3(0.5x)} = 3^{11/6}$

$0.5x = 3^{11/6}$

$x = 2(3^{11/6}) \approx 14.988$

89. $\ln x - \ln(x + 1) = 2$

$\ln\left(\dfrac{x}{x + 1}\right) = 2$

$\dfrac{x}{x + 1} = e^2$

$x = e^2(x + 1)$

$x = e^2 x + e^2$

$x - e^2 x = e^2$

$x(1 - e^2) = e^2$

$x = \dfrac{e^2}{1 - e^2} \approx -1.157$

This negative value is extraneous. The equation has no solution.

91. $\ln x + \ln(x - 2) = 1$

$\ln[x(x - 2)] = 1$

$x(x - 2) = e^1$

$x^2 - 2x - e = 0$

$x = \dfrac{2 \pm \sqrt{4 + 4e}}{2}$

$= \dfrac{2 \pm 2\sqrt{1 + e}}{2}$

$= 1 \pm \sqrt{1 + e}$

The negative value is extraneous. The only solution is $x = 1 + \sqrt{1 + e} \approx 2.928$.

93.

$\ln(x + 5) = \ln(x - 1) - \ln(x + 1)$

$\ln(x + 5) = \ln\left(\dfrac{x - 1}{x + 1}\right)$

$x + 5 = \dfrac{x - 1}{x + 1}$

$(x + 5)(x + 1) = x - 1$

$x^2 + 6x + 5 = x - 1$

$x^2 + 5x + 6 = 0$

$(x + 2)(x + 3) = 0$

$x = -2 \text{ or } x = -3$

Both of these solutions are extraneous, so the equation has no solution.

95. $\log_2(2x - 3) = \log_2(x + 4)$

$2x - 3 = x + 4$

$x = 7$

97. $\log(x + 4) - \log x = \log(x + 2)$

$$\log\left(\frac{x + 4}{x}\right) = \log(x + 2)$$

$$\frac{x + 4}{x} = x + 2$$

$$x + 4 = x^2 + 2x$$

$$0 = x^2 + x - 4$$

$$x = \frac{-1 \pm \sqrt{17}}{2} \quad \text{Quadratic Formula}$$

Choosing the positive value of x (the negative value is extraneous), we have

$$x = \frac{-1 + \sqrt{17}}{2} \approx 1.562.$$

99. $\log_4 x - \log_4(x - 1) = \frac{1}{2}$

$$\log_4\left(\frac{x}{x - 1}\right) = \frac{1}{2}$$

$$4^{\log_4[x/(x-1)]} = 4^{1/2}$$

$$\frac{x}{x - 1} = 4^{1/2}$$

$$x = 2(x - 1)$$

$$x = 2x - 2$$

$$-x = -2$$

$$x = 2$$

101. $\log 8x - \log\left(1 + \sqrt{x}\right) = 2$

$$\log \frac{8x}{1 + \sqrt{x}} = 2$$

$$\frac{8x}{1 + \sqrt{x}} = 10^2$$

$$8x = 100\left(1 + \sqrt{x}\right)$$

$$2x = 25\left(1 + \sqrt{x}\right) = 25 + 25\sqrt{x}$$

$$2x - 25 = 25\sqrt{x}$$

$$(2x - 25)^2 = \left(25\sqrt{x}\right)^2$$

$$4x^2 - 100x + 625 = 625x$$

$$4x^2 - 725x + 625 = 0$$

$$x = \frac{725 \pm \sqrt{725^2 - 4(4)(625)}}{2(4)}$$

$$x = \frac{725 \pm \sqrt{515{,}625}}{8}$$

$$x = \frac{25\left(29 \pm 5\sqrt{33}\right)}{8}$$

$$x \approx 0.866 \text{ (extraneous) or}$$

$$x \approx 180.384$$

The only solution is $x = \dfrac{25\left(29 + 5\sqrt{33}\right)}{8} \approx 180.384.$

103. $y_1 = 7$

$y_2 = 2^x$

From the graph we have $x \approx 2.807$ when $y = 7$. Algebraically:

$$2^x = 7$$

$$\ln 2^x = \ln 7$$

$$x \ln 2 = \ln 7$$

$$x = \frac{\ln 7}{\ln 2} \approx 2.807$$

105. $y_1 = 3$

$y_2 = \ln x$

From the graph we have $x \approx 20.086$ when $y = 3$. Algebraically:

$$3 - \ln x = 0$$

$$\ln x = 3$$

$$x = e^3 \approx 20.086$$

107. (a) $\quad A = Pe^{rt}$

$$5000 = 2500e^{0.085t}$$

$$2 = e^{0.085t}$$

$$\ln 2 = 0.085t$$

$$\frac{\ln 2}{0.085} = t$$

$$t \approx 8.2 \text{ years}$$

(b) $\quad A = Pe^{rt}$

$$7500 = 2500e^{0.085t}$$

$$3 = e^{0.085t}$$

$$\ln 3 = 0.085t$$

$$\frac{\ln 3}{0.085} = t$$

$$t \approx 12.9 \text{ years}$$

109. $p = 500 - 0.5(e^{0.004x})$

(a) $\qquad p = 350$

$\qquad 350 = 500 - 0.5(e^{0.004x})$

$\qquad 300 = e^{0.004x}$

$\qquad 0.004x = \ln 300$

$\qquad\quad x \approx 1426$ units

(b) $\qquad p = 300$

$\qquad 300 = 500 - 0.5(e^{0.004x})$

$\qquad 400 = e^{0.004x}$

$\qquad 0.004x = \ln 400$

$\qquad\quad x \approx 1498$ units

111. $V = 6.7e^{-48.1/t}$, $t \geq 0$

(a)

(b) As $t \to \infty$, $V \to 6.7$.

Horizontal asymptote: $V = 6.7$

The yield will approach 6.7 million cubic feet per acre.

(c) $\qquad 1.3 = 6.7e^{-48.1/t}$

$\qquad \dfrac{1.3}{6.7} = e^{-48.1/t}$

$\qquad \ln\left(\dfrac{13}{67}\right) = \dfrac{-48.1}{t}$

$\qquad\quad t = \dfrac{-48.1}{\ln(13/67)} \approx 29.3$ years

113. $y = 7312 - 630.0 \ln t$, $5 \leq t \leq 12$

$\qquad 7312 - 630.0 \ln t = 5800$

$\qquad\qquad -630.0 \ln t = -1512$

$\qquad\qquad\qquad \ln t = 2.4$

$\qquad\qquad\qquad\quad t = e^{2.4} \approx 11$

$t \approx 11$ corresponds to the year 2001.

115. (a) From the graph shown in the textbook, we see horizontal asymptotes at $y = 0$ and $y = 100$. These represent the lower and upper percent bounds; the range falls between 0% and 100%.

(b) Males

$$50 = \frac{100}{1 + e^{-0.6114(x - 69.71)}}$$

$$1 + e^{-0.6114(x - 69.71)} = 2$$

$$e^{-0.6114(x - 69.71)} = 1$$

$$-0.6114(x - 69.71) = \ln 1$$

$$-0.6114(x - 69.71) = 0$$

$$x = 69.71 \text{ inches}$$

Females

$$50 = \frac{100}{1 + e^{-0.66607(x - 64.51)}}$$

$$1 + e^{-0.66607(x - 64.51)} = 2$$

$$e^{-0.6667(x - 64.51)} = 1$$

$$-0.66607(x - 64.51) = \ln 1$$

$$-0.66607(x - 64.51) = 0$$

$$x = 64.51 \text{ inches}$$

117. $y = -3.00 + 11.88 \ln x + \dfrac{36.94}{x}$

(a)

x	0.2	0.4	0.6	0.8	1.0
y	162.6	78.5	52.5	40.5	33.9

(b)

The model seems to fit the data well.

(c) When $y = 30$:

$$30 = -3.00 + 11.88 \ln x + \frac{36.94}{x}$$

Add the graph of $y = 30$ to the graph in part (a) and estimate the point of intersection of the two graphs. We find that $x \approx 1.20$ meters.

(d) No, it is probably not practical to lower the number of gs experienced during impact to less than 23 because the required distance traveled at $y = 23$ is $x \approx 2.27$ meters. It is probably not practical to design a car allowing a passenger to move forward 2.27 meters (or 7.45 feet) during an impact.

119. $\log_a(uv) = \log_a u + \log_a v$

True by Property 1 in Section 5.3.

121. $\log_a(u - v) = \log_a u - \log_a v$

False.

$$1.95 \approx \log(100 - 10)$$
$$\neq \log 100 - \log 10 = 1$$

123. Yes, a logarithmic equation can have more than one extraneous solution. See Exercise 93.

125. Yes.

Time to Double	Time to Quadruple
$2P = Pe^{rt}$	$4P = Pe^{rt}$
$2 = e^{rt}$	$4 = e^{rt}$
$\ln 2 = rt$	$\ln 4 = rt$
$\dfrac{\ln 2}{r} = t$	$\dfrac{2\ln 2}{r} = t$

Thus, the time to quadruple is twice as long as the time to double.

127. $\sqrt{48x^2y^5} = \sqrt{16x^2y^4 3y} = 4|x|y^2\sqrt{3y}$

129. $\sqrt[3]{25}\sqrt[3]{15} = \sqrt[3]{375} = \sqrt[3]{125 \cdot 3} = 5\sqrt[3]{3}$

131. $f(x) = |x| + 9$

Domain: all real numbers x

y-intercept: $(0, 9)$

y-axis symmetry

x	0	± 1	± 2	± 3
y	9	10	11	12

133. $g(x) = \begin{cases} 2x, & x < 0 \\ -x^2 + 4, & x \geq 0 \end{cases}$

Domain: all real numbers x

x-intercept: $(2, 0)$

y-intercept: $(0, 4)$

x	-3	-2	-1	-0.5	0	1	2	3
y	-6	-4	-2	-1	4	3	2	-5

135. $\log_6 9 = \dfrac{\log_{10} 9}{\log_{10} 6} = \dfrac{\ln 9}{\ln 6} \approx 1.226$

137. $\log_{3/4} 5 = \dfrac{\log_{10} 5}{\log_{10}(3/4)} = \dfrac{\ln 5}{\ln(3/4)} \approx -5.595$

Section 5.5 Exponential and Logarithmic Models

■ You should be able to solve growth and decay problems.

(a) Exponential growth if $b > 0$ and $y = ae^{bx}$.

(b) Exponential decay if $b > 0$ and $y = ae^{-bx}$.

■ You should be able to use the Gaussian model

$y = ae^{-(x-b)^2/c}$.

■ You should be able to use the logistic growth model

$y = \dfrac{a}{1 + be^{-rx}}$.

■ You should be able to use the logarithmic models

$y = a + b \ln x$, $y = a + b \log x$.

Vocabulary Check

1. $y = ae^{bx}$; $y = ae^{-bx}$

2. $y = a + b \ln x$; $y = a + b \log x$

3. normally distributed

4. bell; average value

5. sigmoidal

---

1. $y = 2e^{x/4}$

This is an exponential growth model. Matches graph (c).

3. $y = 6 + \log(x + 2)$

This is a logarithmic function shifted up six units and left two units. Matches graph (b).

5. $y = \ln(x + 1)$

This is a logarithmic model shifted left one unit. Matches graph (d).

7. Since $A = 1000e^{0.035t}$, the time to double is given by $2000 = 1000e^{0.035t}$ and we have

$2 = e^{0.035t}$

$\ln 2 = \ln e^{0.035t}$

$\ln 2 = 0.035t$

$t = \dfrac{\ln 2}{0.035} \approx 19.8$ years.

Amount after 10 years: $A = 1000e^{0.35} \approx \$1419.07$

9. Since $A = 750e^{rt}$ and $A = 1500$ when $t = 7.75$, we have the following.

$1500 = 750e^{7.75r}$

$2 = e^{7.75r}$

$\ln 2 = \ln e^{7.75r}$

$\ln 2 = 7.75r$

$r = \dfrac{\ln 2}{7.75} \approx 0.089438 = 8.9438\%$

Amount after 10 years: $A = 750e^{0.089438(10)} \approx \$1834.37$

11. Since $A = 500e^{rt}$ and $A = \$1505.00$ when $t = 10$, we have the following.

$1505.00 = 500e^{10r}$

$r = \dfrac{\ln(1505.00/500)}{10} \approx 0.110 = 11.0\%$

The time to double is given by

$1000 = 500e^{0.110t}$

$t = \dfrac{\ln 2}{0.110} \approx 6.3$ years.

13. Since $A = Pe^{0.045t}$ and $A = 10,000.00$ when $t = 10$, we have the following.

$10,000.00 = Pe^{0.045(10)}$

$\dfrac{10,000.00}{e^{0.045(10)}} = P \approx \$6376.28$

The time to double is given by $t = \dfrac{\ln 2}{0.045} \approx 15.40$ years.

15. $500{,}000 = P\left(1 + \dfrac{0.075}{12}\right)^{12(20)}$

$$P = \dfrac{500{,}000}{\left(1 + \dfrac{0.075}{12}\right)^{12(20)}}$$

$$= \dfrac{500{,}000}{1.00625^{240}} \approx \$112{,}087.09$$

17. $P = 1000, r = 11\%$

(a) $n = 1$

$(1 + 0.11)^t = 2$

$t \ln 1.11 = \ln 2$

$t = \dfrac{\ln 2}{\ln 1.11} \approx 6.642$ years

(b) $n = 12$

$\left(1 + \dfrac{0.11}{12}\right)^{12t} = 2$

$12t \ln\left(1 + \dfrac{0.11}{12}\right) = \ln 2$

$t = \dfrac{\ln 2}{12 \ln\left(1 + \frac{0.11}{12}\right)} \approx 6.330$ years

(c) $n = 365$

$\left(1 + \dfrac{0.11}{365}\right)^{365t} = 2$

$365t \ln\left(1 + \dfrac{0.11}{365}\right) = \ln 2$

$t = \dfrac{\ln 2}{365 \ln\left(1 + \frac{0.11}{365}\right)} \approx 6.302$ years

(d) Compounded continuously

$e^{0.11t} = 2$

$0.11t = \ln 2$

$t = \dfrac{\ln 2}{0.11} \approx 6.301$ years

19. $3P = Pe^{rt}$

$3 = e^{rt}$

$\ln 3 = rt$

$\dfrac{\ln 3}{r} = t$

r	2%	4%	6%	8%	10%	12%
$t = \dfrac{\ln 3}{r}$ (years)	54.93	27.47	18.31	13.73	10.99	9.16

21. $3P = P(1 + r)^t$

$3 = (1 + r)^t$

$\ln 3 = \ln(1 + r)^t$

$\ln 3 = t \ln(1 + r)$

$\dfrac{\ln 3}{\ln(1 + r)} = t$

r	2%	4%	6%	8%	10%	12%
$t = \dfrac{\ln 3}{\ln(1 + r)}$ (years)	55.48	28.01	18.85	14.27	11.53	9.69

23. Continuous compounding results in faster growth.

$A = 1 + 0.075\llbracket t \rrbracket$ and $A = e^{0.07t}$

Time (in years)

25. $\dfrac{1}{2}C = Ce^{k(1599)}$

$0.5 = e^{k(1599)}$

$\ln 0.5 = \ln e^{k(1599)}$

$\ln 0.5 = k(1599)$

$k = \dfrac{\ln 0.5}{1599}$

Given $C = 10$ grams after 1000 years, we have

$y = 10e^{[(\ln 0.5)/1599](1000)}$

≈ 6.48 grams.

27. $\frac{1}{2}C = Ce^{k(5715)}$

$0.5 = e^{k(5715)}$

$\ln 0.5 = \ln e^{k(5715)}$

$\ln 0.5 = k(5715)$

$k = \dfrac{\ln 0.5}{5715}$

Given $y = 2$ grams after 1000 years, we have

$2 = Ce^{[(\ln 0.5)/5715](1000)}$

$C \approx 2.26$ grams.

29. $\frac{1}{2}C = Ce^{k(24,100)}$

$0.5 = e^{k(24,100)}$

$\ln 0.5 = \ln e^{k(24,100)}$

$\ln 0.5 = k(24,100)$

$k = \dfrac{\ln 0.5}{24,100}$

Given $y = 2.1$ grams after 1000 years, we have

$2.1 = Ce^{[(\ln 0.5)/24,100](1000)}$

$C \approx 2.16$ grams.

31. $y = ae^{bx}$

$1 = ae^{b(0)} \implies 1 = a$

$10 = e^{b(3)}$

$\ln 10 = 3b$

$\dfrac{\ln 10}{3} = b \implies b \approx 0.7675$

Thus, $y = e^{0.7675x}$.

33. $y = ae^{bx}$

$5 = ae^{b(0)} \implies 5 = a$

$1 = 5e^{b(4)}$

$\dfrac{1}{5} = e^{4b}$

$\ln\left(\dfrac{1}{5}\right) = 4b$

$\dfrac{\ln(1/5)}{4} = b \implies b \approx -0.4024$

Thus, $y = 5e^{-0.4024x}$.

35. $P = 2430e^{-0.0029t}$

(a) Since the exponent is negative, this is an exponential decay model. The population is decreasing.

(b) For 2000, let $t = 0$: $P = 2430$ thousand people

For 2003, let $t = 3$: $P \approx 2408.95$ thousand people

(c) 2.3 million = 2300 thousand

$2300 = 2430e^{-0.0029t}$

$\dfrac{2300}{2430} = e^{-0.0029t}$

$\ln\left(\dfrac{2300}{2430}\right) = -0.0029t$

$t = \dfrac{\ln(2300/2430)}{-0.0029} \approx 18.96$

The population will reach 2.3 million (according to the model) during the later part of the year 2018.

37. $y = 4080e^{kt}$

When $t = 3$, $y = 10,000$:

$10,000 = 4080e^{k(3)}$

$\dfrac{10,000}{4080} = e^{3k}$

$\ln\left(\dfrac{10,000}{4080}\right) = 3k$

$k = \dfrac{\ln(10,000/4080)}{3} \approx 0.2988$

When $t = 24$: $y = 4080e^{0.2988(24)} \approx 5,309,734$ hits

39. $N = 100e^{kt}$

$300 = 100e^{5k}$

$3 = e^{5k}$

$\ln 3 = \ln e^{5k}$

$\ln 3 = 5k$

$k = \dfrac{\ln 3}{5} \approx 0.2197$

$N = 100e^{0.2197t}$

$200 = 100e^{0.2197t}$

$t = \dfrac{\ln 2}{0.2197} \approx 3.15$ hours

41. $R = \dfrac{1}{10^{12}} e^{-t/8223}$

(a)
$$R = \frac{1}{8^{14}}$$

$$\frac{1}{10^{12}} e^{-t/8223} = \frac{1}{8^{14}}$$

$$e^{-t/8223} = \frac{10^{12}}{8^{14}}$$

$$-\frac{t}{8223} = \ln\left(\frac{10^{12}}{8^{14}}\right)$$

$$t = -8223 \ln\left(\frac{10^{12}}{8^{14}}\right) \approx 12{,}180 \text{ years old}$$

(b) $\dfrac{1}{10^{12}} e^{-t/8223} = \dfrac{1}{13^{11}}$

$$e^{-t/8223} = \frac{10^{12}}{13^{11}}$$

$$-\frac{t}{8223} = \ln\left(\frac{10^{12}}{13^{11}}\right)$$

$$t = -8223 \ln\left(\frac{10^{12}}{13^{11}}\right) \approx 4797 \text{ years old}$$

43. $(0, 30{,}788)$, $(2, 18{,}000)$

(a) $m = \dfrac{18{,}000 - 30{,}788}{2 - 0} = -6394$

$b = 30{,}788$

Linear model: $V = -6394t + 30{,}788$

(b) $a = 30{,}788$

$$18{,}000 = 30{,}788 e^{k(2)}$$

$$\frac{4500}{7697} = e^{2k}$$

$$\ln\left(\frac{4500}{7697}\right) = 2k$$

$$k = \frac{1}{2} \ln\left(\frac{4500}{7697}\right) \approx -0.268$$

Exponential model: $V = 30{,}788 e^{-0.268t}$

(c)

The exponential model depreciates faster in the first two years.

(d)

t	1	3
$V = -6394t + 30{,}788$	\$24,394	\$11,606
$V = 30{,}788 e^{-0.268t}$	\$23,550	\$13,779

(e) The linear model gives a higher value for the car for the first two years, then the exponential model yields a higher value. If the car is less than two years old, the seller would most likely want to use the linear model and the buyer the exponential model. If it is more than two years old, the opposite is true.

45. $S(t) = 100(1 - e^{kt})$

(a) $15 = 100(1 - e^{k(1)})$

$$-85 = -100 e^k$$

$$\frac{85}{100} = e^k$$

$$0.85 = e^k$$

$$\ln 0.85 = \ln e^k$$

$$k = \ln 0.85$$

$$k \approx -0.1625$$

$$S(t) = 100(1 - e^{-0.1625t})$$

(b)

(c) $S(5) = 100(1 - e^{-0.1625(5)}) \approx 55.625 = 55{,}625$ units

47. $y = 0.0266e^{-(x-100)^2/450}$, $70 \leq x \leq 116$

(a)

(b) The average IQ score of an adult student is 100.

49. $p(t) = \dfrac{1000}{1 + 9e^{-0.1656t}}$

(a) $p(5) = \dfrac{1000}{1 + 9e^{-0.1656(5)}} \approx 203$ animals

(b) $500 = \dfrac{1000}{1 + 9e^{-0.1656t}}$

$1 + 9e^{-0.1656t} = 2$

$9e^{-0.1656t} = 1$

$e^{-0.1656t} = \dfrac{1}{9}$

$t = -\dfrac{\ln(1/9)}{0.1656} \approx 13$ months

(c)

The horizontal asymptotes are $p = 0$ and $p = 1000$. The asymptote with the larger p-value, $p = 1000$, indicates that the population size will approach 1000 as time increases.

51. $R = \log \dfrac{I}{I_0} = \log I$ since $I_0 = 1$.

(a) $7.9 = \log I \implies I = 10^{7.9} \approx 79,432,823$

(b) $8.3 = \log I \implies I = 10^{8.3} \approx 199,526,231$

(c) $4.2 = \log I \implies I = 10^{4.2} \approx 15,849$

53. $\beta = 10 \log \dfrac{I}{I_0}$ where $I_0 = 10^{-12}$ watt/m^2.

(a) $\beta = 10 \log \dfrac{10^{-10}}{10^{-12}} = 10 \log 10^2 = 20$ decibels

(b) $\beta = 10 \log \dfrac{10^{-5}}{10^{-12}} = 10 \log 10^7 = 70$ decibels

(c) $\beta = 10 \log \dfrac{10^{-8}}{10^{-12}} = 10 \log 10^4 = 40$ decibels

(d) $\beta = 10 \log \dfrac{1}{10^{-12}} = 10 \log 10^{12} = 120$ decibels

55. $\beta = 10 \log \dfrac{I}{I_0}$

$\dfrac{\beta}{10} = \log \dfrac{I}{I_0}$

$10^{\beta/10} = 10^{\log I/I_0}$

$10^{\beta/10} = \dfrac{I}{I_0}$

$I = I_0 10^{\beta/10}$

% decrease $= \dfrac{I_0 10^{9.3} - I_0 10^{8.0}}{I_0 10^{9.3}} \times 100 \approx 95\%$

57. pH $= -\log[H^+]$

$-\log(2.3 \times 10^{-5}) \approx 4.64$

59. $5.8 = -\log[H^+]$

$-5.8 = \log[H^+]$

$10^{-5.8} = 10^{\log[H^+]}$

$10^{-5.8} = [H^+]$

$[H^+] \approx 1.58 \times 10^{-6}$ mole per liter

61. $2.9 = -\log[H^+]$

$-2.9 = \log[H^+]$

$[H^+] = 10^{-2.9}$ for the apple juice

$8.0 = -\log[H^+]$

$-8.0 = \log[H^+]$

$[H^+] = 10^{-8}$ for the drinking water

$\dfrac{10^{-2.9}}{10^{-8}} = 10^{5.1}$ times the hydrogen ion
concentration of drinking water

63. $t = -10 \ln \dfrac{T - 70}{98.6 - 70}$

At 9:00 A.M. we have:

$t = -10 \ln \dfrac{85.7 - 70}{98.6 - 70} \approx 6$ hours

From this you can conclude that the person died at 3:00 A.M.

65. $u = 120{,}000 \left[\dfrac{0.075t}{1 - \left(\dfrac{1}{1 + 0.075/12} \right)^{12t}} - 1 \right]$

(a)

(b) From the graph, $u = \$120{,}000$ when $t \approx 21$ years. It would take approximately 37.6 years to pay \$240,000 in interest. Yes, it is possible to pay twice as much in interest charges as the size of the mortgage. It is especially likely when the interest rates are higher.

67. False. The domain can be the set of real numbers for a logistic growth function.

69. False. The graph of $f(x)$ is the graph of $g(x)$ shifted upward five units.

71. (a) Logarithmic

(b) Logistic

(c) Exponential (decay)

(d) Linear

(e) None of the above (appears to be a combination of a linear and a quadratic)

(f) Exponential (growth)

73. $(-1, 2), (0, 5)$

(a)

(b) $d = \sqrt{(0 - (-1))^2 + (5 - 2)^2} = \sqrt{1^2 + 3^2} = \sqrt{10}$

(c) Midpoint: $\left(\dfrac{-1 + 0}{2}, \dfrac{2 + 5}{2} \right) = \left(-\dfrac{1}{2}, \dfrac{7}{2} \right)$

(d) $m = \dfrac{5 - 2}{0 - (-1)} = \dfrac{3}{1} = 3$

75. $(3, 3), (14, -2)$

(a)

(b) $d = \sqrt{(14 - 3)^2 + (-2 - 3)^2}$

$= \sqrt{11^2 + (-5)^2} = \sqrt{146}$

(c) Midpoint: $\left(\dfrac{3 + 14}{2}, \dfrac{3 + (-2)}{2} \right) = \left(\dfrac{17}{2}, \dfrac{1}{2} \right)$

(d) $m = \dfrac{-2 - 3}{14 - 3} = -\dfrac{5}{11}$

77. $\left(\frac{1}{2}, -\frac{1}{4}\right), \left(\frac{3}{4}, 0\right)$

(a)

(b) $d = \sqrt{\left(\frac{3}{4} - \frac{1}{2}\right)^2 + \left(0 - \left(-\frac{1}{4}\right)\right)^2}$

$= \sqrt{\left(\frac{1}{4}\right)^2 + \left(\frac{1}{4}\right)^2} = \sqrt{\frac{1}{8}}$

(c) Midpoint: $\left(\dfrac{(1/2) + (3/4)}{2}, \dfrac{(-1/4) + 0}{2}\right) = \left(\dfrac{5}{8}, -\dfrac{1}{8}\right)$

(d) $m = \dfrac{0 - (-1/4)}{(3/4) - (1/2)} = \dfrac{1/4}{1/4} = 1$

79. $y = 10 - 3x$

Line

Slope: $m = -3$

y-intercept: $(0, 10)$

81. $y = -2x^2 - 3$

$y = -2(x - 0)^2 - 3$

Parabola

Vertex: $(0, -3)$

83. $3x^2 - 4y = 0$

$3x^2 = 4y$

$x^2 = \frac{4}{3}y$

Parabola

Vertex: $(0, 0)$

Focus: $\left(0, \frac{1}{3}\right)$

Directrix: $y = -\frac{1}{3}$

85. $y = \dfrac{4}{1 - 3x}$

Vertical asymptote: $x = \dfrac{1}{3}$

Horizontal asymptote: $y = 0$

87. $x^2 + (y - 8)^2 = 25$

Circle

Center: $(0, 8)$

Radius: 5

89. $f(x) = 2^{x-1} + 5$

Horizontal asymptote: $y = 5$

x	-5	-3	-1	0	1	3	5
$f(x)$	5.02	5.06	5.3	5.5	6	9	21

91. $f(x) = 3^x - 4$

Horizontal asymptote: $y = -4$

x	-4	-2	-1	0	1	2
$f(x)$	-3.99	-3.89	-3.67	-3	-1	5

Review Exercises for Chapter 5

1. $f(x) = 6.1^x$

$f(2.4) = 6.1^{2.4} \approx 76.699$

3. $f(x) = 2^{-0.5x}$

$f(\pi) = 2^{-0.5(\pi)} \approx 0.337$

5. $f(x) = 7(0.2^x)$

$f\left(-\sqrt{11}\right) = 7\left(0.2^{-\sqrt{11}}\right)$

≈ 1456.529

7. $f(x) = 4^x$

Intercept: $(0, 1)$

Horizontal asymptote: x-axis

Increasing on: $(-\infty, \infty)$

Matches graph (c).

9. $f(x) = -4^x$

Intercept: $(0, -1)$

Horizontal asymptote: x-axis

Decreasing on: $(-\infty, \infty)$

Matches graph (a).

11. $f(x) = 5^x$

$g(x) = 5^{x-1}$

Since $g(x) = f(x - 1)$, the graph of g can be obtained by shifting the graph of f one unit to the right.

13. $f(x) = \left(\frac{1}{2}\right)^x$

$g(x) = -\left(\frac{1}{2}\right)^{x+2}$

Since $g(x) = -f(x + 2)$, the graph of g can be obtained by reflecting the graph of f about the x-axis and shifting $-f$ two units to the left.

15. $f(x) = 4^{-x} + 4$

Horizontal asymptote: $y = 4$

x	-1	0	1	2	3
$f(x)$	8	5	4.25	4.063	4.016

17. $f(x) = -2.65^{x+1}$

Horizontal asymptote: $y = 0$

x	-2	-1	0	1	2
$f(x)$	-0.377	-1	-2.65	-7.023	-18.61

19. $f(x) = 5^{x-2} + 4$

Horizontal asymptote: $y = 4$

x	-1	0	1	2	3
$f(x)$	4.008	4.04	4.2	5	9

21. $f(x) = \left(\frac{1}{2}\right)^{-x} + 3 = 2^x + 3$

Horizontal asymptote: $y = 3$

x	-2	-1	0	1	2
$f(x)$	3.25	3.5	4	5	7

23. $3^{x+2} = \frac{1}{9}$

$3^{x+2} = 3^{-2}$

$x + 2 = -2$

$x = -4$

25. $e^{5x-7} = e^{15}$

$5x - 7 = 15$

$5x = 22$

$x = \frac{22}{5}$

27. $e^8 \approx 2980.958$

29. $e^{-1.7} \approx 0.183$

31. $h(x) = e^{-x/2}$

x	-2	-1	0	1	2
$h(x)$	2.72	1.65	1	0.61	0.37

33. $f(x) = e^{x+2}$

x	-3	-2	-1	0	1
$f(x)$	0.37	1	2.72	7.39	20.09

35. $A = 3500\left(1 + \dfrac{0.065}{n}\right)^{10n}$ or $A = 3500e^{(0.065)(10)}$

n	1	2	4	12	365	Continuous Compounding
A	\$6569.98	\$6635.43	\$6669.46	\$6692.64	\$6704.00	\$6704.39

37. $F(t) = 1 - e^{-t/3}$

(a) $F\left(\frac{1}{2}\right) \approx 0.154$ (b) $F(2) \approx 0.487$ (c) $F(5) \approx 0.811$

39. (a) $A = 50{,}000e^{(0.0875)(35)} \approx \$1{,}069{,}047.14$

(b) The doubling time is $\dfrac{\ln 2}{0.0875} \approx 7.9$ years.

41. $4^3 = 64$

$\log_4 64 = 3$

43. $e^{0.8} = 2.2255\ldots$

$\ln 2.2255\ldots = 0.8$

45. $f(x) = \log x$

$f(1000) = \log 1000$

$= \log 10^3 = 3$

47. $g(x) = \log_2 x$

$g\left(\frac{1}{8}\right) = \log_2\left(\frac{1}{8}\right) = \log_2 2^{-3} = -3$

49. $\log_4(x + 7) = \log_4 14$

$\qquad x + 7 = 14$

$\qquad\qquad x = 7$

51. $\ln(x + 9) = \ln 4$

$\qquad x + 9 = 4$

$\qquad\qquad x = -5$

53. $g(x) = \log_7 x \implies x = 7^y$

Domain: $(0, \infty)$

x-intercept: $(1, 0)$

Vertical asymptote: $x = 0$

x	$\frac{1}{7}$	1	7	49
$g(x)$	-1	0	1	2

55. $f(x) = \log\left(\dfrac{x}{3}\right) \implies \dfrac{x}{3} = 10^y \implies x = 3(10^y)$

Domain: $(0, \infty)$

x-intercept: $(3, 0)$

Vertical asymptote:
$\qquad x = 0$

x	0.03	0.3	3	30
$f(x)$	-2	-1	0	1

57. $f(x) = 4 - \log(x + 5)$

Domain: $(-5, \infty)$

x-intercept: $(9995, 0)$

Since $4 - \log(x + 5) = 0 \implies \log(x + 5) = 4$

$\qquad\qquad\qquad x + 5 = 10^4$

$\qquad\qquad x = 10^4 - 5 = 9995.$

Vertical asymptote: $x = -5$

x	-4	-3	-2	-1	0	1
$f(x)$	4	3.70	3.52	3.40	3.30	3.22

59. $\ln 22.6 \approx 3.118$

61. $\ln e^{-12} = -12$

63. $\ln\left(\sqrt{7} + 5\right) \approx 2.034$

65. $f(x) = \ln x + 3$

Domain: $(0, \infty)$

x-intercept: $\ln x + 3 = 0$

$\qquad\qquad \ln x = -3$

$\qquad\qquad\quad x = e^{-3}$

$\qquad\qquad (e^{-3}, 0)$

Vertical asymptote: $x = 0$

x	1	2	3	$\frac{1}{2}$	$\frac{1}{4}$
$f(x)$	3	3.69	4.10	2.31	1.61

67. $h(x) = \ln(x^2) = 2\ln|x|$

Domain: $(-\infty, 0) \cup (0, \infty)$

x-intercepts: $(\pm 1, 0)$

Vertical asymptote: $x = 0$

x	± 0.5	± 1	± 2	± 3	± 4
y	-1.39	0	1.39	2.20	2.77

69. $\qquad h = 116\log(a + 40) - 176$

$h(55) = 116\log(55 + 40) - 176$

$\qquad\quad \approx 53.4$ inches

71. $\log_4 9 = \dfrac{\log 9}{\log 4} \approx 1.585$

$\qquad \log_4 9 = \dfrac{\ln 9}{\ln 4} \approx 1.585$

73. $\log_{1/2} 5 = \dfrac{\log 5}{\log(1/2)} \approx -2.322$

$\qquad \log_{1/2} 5 = \dfrac{\ln 5}{\ln(1/2)} \approx -2.322$

75. $\log 18 = \log(2 \cdot 3^2)$

$\qquad\quad = \log 2 + 2\log 3$

$\qquad\quad \approx 1.255$

77. $\ln 20 = \ln(2^2 \cdot 5)$

$\qquad\quad = 2\ln 2 + \ln 5 \approx 2.996$

79. $\log_5 5x^2 = \log_5 5 + \log_5 x^2$

$\qquad\quad = 1 + 2\log_5 x$

81. $\log_3 \dfrac{6}{\sqrt[3]{x}} = \log_3 6 - \log_3 \sqrt[3]{x}$

$\qquad = \log_3(3 \cdot 2) - \log_3 x^{1/3}$

$\qquad = \log_3 3 + \log_3 2 - \dfrac{1}{3}\log_3 x$

$\qquad = 1 + \log_3 2 - \dfrac{1}{3}\log_3 x$

83. $\ln x^2 y^2 z = \ln x^2 + \ln y^2 + \ln z$

$\qquad = 2\ln x + 2\ln y + \ln z$

85. $\ln\left(\dfrac{x+3}{xy}\right) = \ln(x+3) - \ln xy$

$\qquad = \ln(x+3) - [\ln x + \ln y]$

$\qquad = \ln(x+3) - \ln x - \ln y$

87. $\log_2 5 + \log_2 x = \log_2 5x$

89. $\ln x - \dfrac{1}{4}\ln y = \ln x - \ln \sqrt[4]{y} = \ln\left(\dfrac{x}{\sqrt[4]{y}}\right)$

91. $\dfrac{1}{3}\log_8(x+4) + 7\log_8 y = \log_8 \sqrt[3]{x+4} + \log_8 y^7$

$\qquad = \log_8\left(y^7 \sqrt[3]{x+4}\right)$

93. $\dfrac{1}{2}\ln(2x-1) - 2\ln(x+1) = \ln\sqrt{2x-1} - \ln(x+1)^2$

$\qquad = \ln\dfrac{\sqrt{2x-1}}{(x+1)^2}$

95. $t = 50\log\dfrac{18{,}000}{18{,}000 - h}$

(a) Domain: $0 \le h < 18{,}000$

(b)

Vertical asymptote: $h = 18{,}000$

(c) As the plane approaches its absolute ceiling, it climbs at a slower rate, so the time required increases.

(d) $50\log\dfrac{18{,}000}{18{,}000 - 4000} \approx 5.46$ minutes

97. $8^x = 512$

$8^x = 8^3$

$x = 3$

99. $e^x = 3$

$x = \ln 3$

101. $\log_4 x = 2$

$x = 4^2 = 16$

103. $\ln x = 4$

$x = e^4$

105. $e^x = 12$

$\ln e^x = \ln 12$

$x = \ln 12 \approx 2.485$

107. $e^{4x} = e^{x^2+3}$

$4x = x^2 + 3$

$0 = x^2 - 4x + 3$

$0 = (x-1)(x-3)$

$x = 1 \text{ or } x = 3$

109. $2^x + 13 = 35$

$2^x = 22$

$x = \log_2 22$

$\qquad = \dfrac{\log 22}{\log 2} \text{ or } \dfrac{\ln 22}{\ln 2}$

$x \approx 4.459$

111. $-4(5^x) = -68$

$$5^x = 17$$

$$\ln 5^x = \ln 17$$

$$x \ln 5 = \ln 17$$

$$x = \frac{\ln 17}{\ln 5} \approx 1.760$$

113. $e^{2x} - 7e^x + 10 = 0$

$$(e^x - 2)(e^x - 5) = 0$$

$$e^x = 2 \qquad \text{or} \qquad e^x = 5$$

$$\ln e^x = \ln 2 \qquad\qquad \ln e^x = \ln 5$$

$$x = \ln 2 \approx 0.693 \qquad x = \ln 5 \approx 1.609$$

115. $2^{0.6x} - 3x = 0$

Graph $y_1 = 2^{0.6x} - 3x$.

The x-intercepts are at
$x \approx 0.392$ and at $x \approx 7.480$.

117. $25e^{-0.3x} = 12$

Graph $y_1 = 25e^{-0.3x}$ and
$y_2 = 12$.

The graphs intersect at
$x \approx 2.447$.

119. $\ln 3x = 8.2$

$$e^{\ln 3x} = e^{8.2}$$

$$3x = e^{8.2}$$

$$x = \frac{e^{8.2}}{3} \approx 1213.650$$

121. $2 \ln 4x = 15$

$$\ln 4x = \frac{15}{2}$$

$$e^{\ln 4x} = e^{7.5}$$

$$4x = e^{7.5}$$

$$x = \frac{1}{4}e^{7.5} \approx 452.011$$

123. $\ln x - \ln 3 = 2$

$$\ln \frac{x}{3} = 2$$

$$e^{\ln(x/3)} = e^2$$

$$\frac{x}{3} = e^2$$

$$x = 3e^2 \approx 22.167$$

125. $\ln \sqrt{x+1} = 2$

$$\frac{1}{2}\ln(x+1) = 2$$

$$\ln(x+1) = 4$$

$$e^{\ln(x+1)} = e^4$$

$$x + 1 = e^4$$

$$x = e^4 - 1 \approx 53.598$$

127.

$$\log_8(x-1) = \log_8(x-2) - \log_8(x+2)$$

$$\log_8(x-1) = \log_8\left(\frac{x-2}{x+2}\right)$$

$$x - 1 = \frac{x-2}{x+2}$$

$$(x-1)(x+2) = x-2$$

$$x^2 + x - 2 = x - 2$$

$$x^2 = 0$$

$$x = 0$$

Since $x = 0$ is not in the domain of $\log_8(x-1)$ or of
$\log_8(x-2)$, it is an extraneous solution. The equation
has no solution.

129. $\log(1-x) = -1$

$$1 - x = 10^{-1}$$

$$1 - \frac{1}{10} = x$$

$$x = 0.900$$

131. $2\ln(x+3) + 3x = 8$

Graph $y_1 = 2\ln(x+3) + 3x$ and $y_2 = 8$.

The graphs intersect at approximately $(1.643, 8)$.
The solution of the equation is $x \approx 1.643$.

133. $4 \ln(x + 5) - x = 10$

Graph $y_1 = 4 \ln(x + 5) - x$ and $y_2 = 10$.

The graphs do not intersect. The equation has no solution.

135. $3(7550) = 7550 e^{0.0725t}$

$$3 = e^{0.0725t}$$

$$\ln 3 = \ln e^{0.0725t}$$

$$\ln 3 = 0.0725t$$

$$t = \frac{\ln 3}{0.0725} \approx 15.2 \text{ years}$$

137. $y = 3e^{-2x/3}$

Exponential decay model

Matches graph (e).

139. $y = \ln(x + 3)$

Logarithmic model

Vertical asymptote: $x = -3$

Graph includes $(-2, 0)$

Matches graph (f).

141. $y = 2e^{-(x+4)^2/3}$

Gaussian model

Matches graph (a).

143. $y = ae^{bx}$

$2 = ae^{b(0)} \implies a = 2$

$3 = 2e^{b(4)}$

$1.5 = e^{4b}$

$\ln 1.5 = 4b \implies b \approx 0.1014$

Thus, $y \approx 2e^{0.1014x}$.

145. $P = 3499e^{0.0135t}$

$4.5 \text{ million} = 4500 \text{ thousand}$

$$4500 = 3499e^{0.0135t}$$

$$\frac{4500}{3499} = e^{0.0135t}$$

$$\ln\left(\frac{4500}{3499}\right) = 0.0135t$$

$$t = \frac{\ln(4500/3499)}{0.0135} \approx 18.6 \text{ years}$$

According to this model, the population of South Carolina will reach 4.5 million during the year 2008.

147. (a) $20,000 = 10,000 e^{r(5)}$

$$2 = e^{5r}$$

$$\ln 2 = 5r$$

$$\frac{\ln 2}{5} = r$$

$$r \approx 0.138629$$

$$= 13.8629\%$$

(b) $A = 10,000 e^{0.138629}$

$\approx \$11,486.98$

149. $y = 0.0499 e^{-(x-71)^2/128}$,

$40 \le x \le 100$

(a) Graph $y_1 = 0.0499 e^{-(x-71)^2/128}$.

(b) The average test score is 71.

151. $\beta = 10 \log\left(\dfrac{I}{10^{-16}}\right)$

$125 = 10 \log\left(\dfrac{I}{10^{-16}}\right)$

$12.5 = \log\left(\dfrac{I}{10^{-16}}\right)$

$10^{12.5} = \dfrac{I}{10^{-16}}$

$I = 10^{-3.5} \text{ watt/cm}^2$

153. True. By the inverse properties, $\log_b b^{2x} = 2x$.

155. Since graphs (b) and (d) represent exponential decay, b and d are negative.

Since graph (a) and (c) represent exponential growth, a and c are positive.

Problem Solving for Chapter 5

1. $y = a^x$

$y_1 = 0.5^x$

$y_2 = 1.2^x$

$y_3 = 2.0^x$

$y_4 = x$

The curves $y = 0.5^x$ and $y = 1.2^x$ cross the line $y = x$. From checking the graphs it appears that $y = x$ will cross $y = a^x$ for $0 \le a \le 1.44$.

3. The exponential function, $y = e^x$, increases at a faster rate than the polynomial function $y = x^n$.

5. (a) $f(u + v) = a^{u+v}$

$$= a^u \cdot a^v$$

$$= f(u) \cdot f(v)$$

(b) $f(2x) = a^{2x}$

$$= (a^x)^2$$

$$= [f(x)]^2$$

7. (a)

(b)

(c)

9.

$$f(x) = e^x - e^{-x}$$

$$y = e^x - e^{-x}$$

$$x = e^y - e^{-y}$$

$$x = \frac{e^{2y} - 1}{e^y}$$

$$xe^y = e^{2y} - 1$$

$$e^{2y} - xe^y - 1 = 0$$

$$e^y = \frac{x \pm \sqrt{x^2 + 4}}{2} \qquad \text{Quadratic Formula}$$

Choosing the positive quantity for e^y we have $y = \ln\left(\dfrac{x + \sqrt{x^2 + 4}}{2}\right)$. Thus, $f^{-1}(x) = \ln\left(\dfrac{x + \sqrt{x^2 + 4}}{2}\right)$.

11. Answer (c). $y = 6(1 - e^{-x^2/2})$

The graph passes through $(0, 0)$ and neither (a) nor (b) pass through the origin. Also, the graph has y-axis symmetry and a horizontal asymptote at $y = 6$.

13. $y_1 = c_1\left(\dfrac{1}{2}\right)^{t/k_1}$ and $y_2 = c_2\left(\dfrac{1}{2}\right)^{t/k_2}$

$$c_1\left(\dfrac{1}{2}\right)^{t/k_1} = c_2\left(\dfrac{1}{2}\right)^{t/k_2}$$

$$\dfrac{c_1}{c_2} = \left(\dfrac{1}{2}\right)^{(t/k_2 - t/k_1)}$$

$$\ln\left(\dfrac{c_1}{c_2}\right) = \left(\dfrac{t}{k_2} - \dfrac{t}{k_1}\right)\ln\left(\dfrac{1}{2}\right)$$

$$\ln c_1 - \ln c_2 = t\left(\dfrac{1}{k_2} - \dfrac{1}{k_1}\right)\ln\left(\dfrac{1}{2}\right)$$

$$t = \dfrac{\ln c_1 - \ln c_2}{[(1/k_2) - (1/k_1)]\ln(1/2)}$$

17.
$$(\ln x)^2 = \ln x^2$$

$$(\ln x)^2 - 2\ln x = 0$$

$$\ln x(\ln x - 2) = 0$$

$$\ln x = 0 \quad \text{or} \quad \ln x = 2$$

$$x = 1 \quad \text{or} \quad x = e^2$$

19. $y_4 = (x - 1) - \frac{1}{2}(x - 1)^2 + \frac{1}{3}(x - 1)^3 - \frac{1}{4}(x - 1)^4$

The pattern implies that

$$\ln x = (x - 1) - \frac{1}{2}(x - 1)^2 + \frac{1}{3}(x - 1)^3 - \frac{1}{4}(x - 1)^4 + \ldots.$$

21. $y = 80.4 - 11\ln x$

$$y(300) = 80.4 - 11\ln 300 \approx 17.7 \text{ ft}^3/\text{min}$$

23. (a)

(b) The data could best be modeled by a logarithmic model.

(c) The shape of the curve looks much more logarithmic than linear or exponential.

(d) $y \approx 2.1518 + 2.7044\ln x$

(e) The model is a good fit to the actual data.

15. (a) $y \approx 252.606(1.0310)^t$

(b) $y \approx 400.88t^2 - 1464.6t + 291{,}782$

(c)

(d) Both models appear to be "good fits" for the data, but neither would be reliable to predict the population of the United States in 2010. The exponential model approaches infinity rapidly.

25. (a)

(b) The data could best be modeled by a linear model.

(c) The shape of the curve looks much more linear than exponential or logarithmic.

(d) $y \approx -0.7884x + 8.2566$

(e) The model is a good fit to the actual data.

Practice Test for Chapter 5

1. Solve for x: $x^{3/5} = 8$.

2. Solve for x: $3^{x-1} = \frac{1}{81}$.

3. Graph $f(x) = 2^{-x}$.

4. Graph $g(x) = e^x + 1$.

5. If \$5000 is invested at 4% interest, find the amount after three years if the interest is compounded

   (a) monthly.   (b) quarterly.   (c) continuously.

6. Write the equation in logarithmic form: $7^{-2} = \frac{1}{49}$.

7. Solve for x: $x - 4 = \log_2 \frac{1}{64}$.

8. Use the properties of logarithms to expand $\ln\left(\dfrac{x^2\sqrt{y}}{z^4}\right)$.

9. Write $5 \ln x - \frac{1}{2} \ln y + 6 \ln z$ as a single logarithm.

10. Using your calculator and the change of base formula, evaluate $\log_9 28$.

11. Use your calculator to solve for N: $\log N = 0.6646$

12. Graph $y = \log_4 x$.

13. Determine the domain of $f(x) = \log_3(x^2 - 9)$.

14. Graph $y = \ln(x - 2)$.

15. True or false: $\dfrac{\ln x}{\ln y} = \ln(x - y)$

16. Solve for x: $5^x = 41$

17. Solve for x: $x - x^2 = \log_5 \frac{1}{25}$

18. Solve for x: $\log_2 x + \log_2(x - 3) = 2$

19. Find the exponential function $y = ae^{bx}$ that passes through the points $(0, 7)$ and $(5, 3)$.

20. Six thousand dollars is deposited into a fund at an annual interest rate of 6%. Find the time required for the investment to double if the interest is compounded continuously.

C H A P T E R 6
Trigonometry

C H A P T E R 6
Trigonometry

Section 6.1 Angles and Their Measures

You should know the following basic facts about angles, their measurement, and their applications.

■ Types of Angles:

   (a) Acute: Measure between $0°$ and $90°$.

   (b) Right: Measure $90°$.

   (c) Obtuse: Measure between $90°$ and $180°$.

   (d) Straight: Measure $180°$.

■ Two positive angles, α and β are complementary if $\alpha + \beta = 90°$. They are supplementary if $\alpha + \beta = 180°$.

■ Two angles in standard position that have the same terminal side are called coterminal angles.

■ To convert degrees to radians, use $1° = \pi/180$ radians.

■ To convert radians to degrees, use 1 radian $= (180/\pi)°$.

■ $1' =$ one minute $= 1/60$ of $1°$.

■ $1'' =$ one second $= 1/60$ of $1' = 1/3600$ of $1°$.

■ The length of a circular arc is $s = r\theta$ where θ is measured in radians.

■ Speed $=$ distance/time

■ Angular speed $= \theta/t = s/rt$

Vocabulary Check

1. Trigonometry

2. angle

3. coterminal

4. degree

5. acute; obtuse

6. complementary; supplementary

7. radian

8. linear

9. angular

10. $A = \frac{1}{2}r^2\theta$

---

1. The angle shown is approximately $210°$.

3. The angle shown is approximately $-60°$.

5. (a) Since $90° < 130° < 180°$; $130°$ lies in Quadrant II.

   (b) Since $270° < 285° < 360°$; $285°$ lies in Quadrant IV.

7. (a) Since $-180° < -132°50' < -90°$; $-132°50'$ lies in Quadrant III.

   (b) Since $-360° < -336° < -270°$; $-336°$ lies in Quadrant I.

9. (a)

(b)

11. (a)

(b)

13. (a) Coterminal angles for 45°

$$45° + 360° = 405°$$

$$45° - 360° = -315°$$

(b) Coterminal angles for −36°

$$-36° + 360° = 324°$$

$$-36° - 360° = -396°$$

15. (a) Coterminal angles for 300°

$$300° + 360° = 660°$$

$$300° - 360° = -60°$$

(b) Coterminal angles for 740°

$$740° - 2(360°) = 20°$$

$$20° - 360° = -340°$$

17. (a) $54°45' = 54° + \left(\frac{45}{60}\right)° = 54.75°$

(b) $-128°30' = -128° - \left(\frac{30}{60}\right)° = -128.5°$

19. (a) $85°18'30'' = \left(85 + \frac{18}{60} + \frac{30}{3600}\right)° \approx 85.308°$

(b) $330°25'' = \left(330 + \frac{25}{3600}\right)° \approx 330.007°$

21. (a) $240.6° = 240° + 0.6(60)' = 240°36'$

(b) $-145.8° = -[145° + 0.8(60')] = -145°48'$

23. (a) $2.5° = 2° + 0.5(60)' = 2°30'$

(b) $-3.58° = -[3° + 0.58(60)']$

$$= -[3°34.8']$$

$$= -[3°34' + 0.8(60)'']$$

$$= -3°34'48''$$

25. (a) Complement: $90° - 24° = 66°$

Supplement: $180° - 24° = 156°$

(b) Complement: Not possible. 126° is greater than 90°.

Supplement: $180° - 126° = 54°$

27.

The angle shown is approximately 2 radians.

29.

The angle shown is approximately −3 radians.

31. (a) Since $0 < \frac{\pi}{5} < \frac{\pi}{2}$; $\frac{\pi}{5}$ lies in Quadrant I.

(b) Since $\pi < \frac{7\pi}{5} < \frac{3\pi}{2}$; $\frac{7\pi}{5}$ lies in Quadrant III.

33. (a) Since $-\dfrac{\pi}{2} < -1 < 0$; -1 lies in Quadrant IV.

(b) Since $-\pi < -2 < -\dfrac{\pi}{2}$; -2 lies in Quadrant III.

35. (a) $\dfrac{5\pi}{4}$

(b) $-\dfrac{2\pi}{3}$

37. (a) $\dfrac{11\pi}{6}$

(b) -3

39. (a) Coterminal angles for $\dfrac{\pi}{6}$

$$\dfrac{\pi}{6} + 2\pi = \dfrac{13\pi}{6}$$

$$\dfrac{\pi}{6} - 2\pi = -\dfrac{11\pi}{6}$$

(b) Coterminal angles for $\dfrac{5\pi}{6}$

$$\dfrac{5\pi}{6} + 2\pi = \dfrac{17\pi}{6}$$

$$\dfrac{5\pi}{6} - 2\pi = -\dfrac{7\pi}{6}$$

41. (a) Coterminal angles for $-\dfrac{9\pi}{4}$

$$-\dfrac{9\pi}{4} + 4\pi = \dfrac{7\pi}{4}$$

$$-\dfrac{9\pi}{4} + 2\pi = -\dfrac{\pi}{4}$$

(b) Coterminal angles for $-\dfrac{2\pi}{15}$

$$-\dfrac{2\pi}{15} + 2\pi = \dfrac{28\pi}{15}$$

$$-\dfrac{2\pi}{15} - 2\pi = -\dfrac{32\pi}{15}$$

43. (a) Complement: $\dfrac{\pi}{2} - \dfrac{\pi}{12} = \dfrac{5\pi}{12}$

Supplement: $\pi - \dfrac{\pi}{12} = \dfrac{11\pi}{12}$

(b) Complement: Not possible; $\dfrac{11\pi}{12}$ is greater than $\dfrac{\pi}{2}$

Supplement: $\pi - \dfrac{11\pi}{12} = \dfrac{\pi}{12}$

45. (a) $30° = 30\left(\dfrac{\pi}{180}\right) = \dfrac{\pi}{6}$

(b) $150° = 150\left(\dfrac{\pi}{180}\right) = \dfrac{5\pi}{6}$

47. (a) $-20° = -20\left(\dfrac{\pi}{180}\right) = -\dfrac{\pi}{9}$

(b) $-240° = -240\left(\dfrac{\pi}{180}\right) = -\dfrac{4\pi}{3}$

49. (a) $\dfrac{3\pi}{2} = \dfrac{3\pi}{2}\left(\dfrac{180}{\pi}\right)° = 270°$

(b) $\dfrac{7\pi}{6} = \dfrac{7\pi}{6}\left(\dfrac{180}{\pi}\right)° = 210°$

51. (a) $\dfrac{7\pi}{3} = \dfrac{7\pi}{3}\left(\dfrac{180}{\pi}\right)^\circ = 420^\circ$

(b) $-\dfrac{11\pi}{30} = -\dfrac{11\pi}{30}\left(\dfrac{180}{\pi}\right)^\circ = -66^\circ$

53. $115^\circ = 115\left(\dfrac{\pi}{180}\right) \approx 2.007$ radians

55. $-216.35^\circ = -216.35\left(\dfrac{\pi}{180}\right) \approx -3.776$ radians

57. $-0.83^\circ = -0.83\left(\dfrac{\pi}{180}\right) \approx -0.014$ radian

59. $\dfrac{\pi}{7} = \dfrac{\pi}{7}\left(\dfrac{180}{\pi}\right)^\circ \approx 25.714^\circ$

61. $\dfrac{15\pi}{8} = \dfrac{15\pi}{8}\left(\dfrac{180}{\pi}\right)^\circ = 337.500^\circ$

63. $-2 = -2\left(\dfrac{180}{\pi}\right)^\circ \approx -114.592^\circ$

65. $s = r\theta$
$6 = 5\theta$
$\theta = \frac{6}{5}$ radians

67. $s = r\theta$
$32 = 7\theta$
$\theta = \frac{32}{7}$ radians

69. $s = r\theta$
$6 = 27\theta$
$\theta = \frac{6}{27} = \frac{2}{9}$ radian

71. $s = r\theta$
$25 = 14.5\theta$
$\theta = \frac{25}{14.5} = \frac{50}{29}$ radians

73. $s = r\theta$, θ in radians
$s = 15(180)\left(\dfrac{\pi}{180}\right) = 15\pi$ inches
≈ 47.12 inches

75. $s = r\theta$, θ in radians
$s = 3(1) = 3$ meters

77. $A = \dfrac{1}{2}r^2\theta$
$A = \dfrac{1}{2}(4)^2\left(\dfrac{\pi}{3}\right) = \dfrac{8\pi}{3}$ square inches
≈ 8.38 square inches

79. $A = \dfrac{1}{2}r^2\theta$
$A = \dfrac{1}{2}(2.5)^2(225)\left(\dfrac{\pi}{180}\right) \approx 12.27$ square feet

81. $\theta = 41^\circ15'50'' - 32^\circ47'39'' \approx 8.46972^\circ \approx 0.14782$ radian
$s = r\theta \approx 4000(0.14782) \approx 591$ miles

83. $\theta = \dfrac{s}{r} = \dfrac{450}{6378} \approx 0.071$ radian $\approx 4.04^\circ$

85. $\theta = \dfrac{s}{r} = \dfrac{2.5}{6} = \dfrac{25}{60} = \dfrac{5}{12}$ radian

87. (a) 65 miles per hour $= 65(5280)/60$
$= 5720$ feet per minute
The circumference of the tire is $C = 2.5\pi$ feet.
The number of revolutions per minute is
$r = 5720/2.5\pi \approx 728.3$ rev/minute.

(b) The angular speed is θ/t.
$\theta = \dfrac{5720}{2.5\pi}(2\pi) = 4576$ radians
Angular speed $= \dfrac{4576 \text{ radians}}{1 \text{ minute}} = 4576$ radians/minute

89. (a) Angular speed $= \dfrac{(5200)(2\pi) \text{ radians}}{1 \text{ minute}}$
$= 10{,}400\pi$ radians per minute

(b) Linear speed $= \dfrac{\left(\dfrac{7.25}{2} \text{ in.}\right)\left(\dfrac{1 \text{ ft}}{12 \text{ in.}}\right)(5200)(2\pi) \text{ feet}}{1 \text{ minute}}$
$= 3141\dfrac{2}{3}\pi$ feet per minute
≈ 164.5 feet per second

91. (a) $(200)(2\pi) \leq$ Angular speed $\leq (500)(2\pi)$ radians per minute

Interval: $[400\pi, 1000\pi]$ radians per minute

(b) $(6)(200)(2\pi) \leq$ Linear speed $\leq (6)(500)(2\pi)$ centimeters per minute

Interval: $[2400\pi, 6000\pi]$ centimeters per minute

93. $A = \dfrac{1}{2}r^2\theta$

$= \dfrac{1}{2}(35)^2(140°)\left(\dfrac{\pi}{180°}\right)$

$\approx 476.39\pi$ square meters

≈ 1496.62 square meters

95. False. A measurement of 4π radians corresponds to two complete revolutions from the initial to the terminal side of an angle.

97. False. The terminal side of $-1260°$ lies on the negative x-axis.

99. The speed increases, since the linear speed is proportional to the radius.

101. Since the arc length s is given by $s = r\theta$, if the central angle θ is fixed while the radius r increases, then s increases in proportion to r.

103. $\dfrac{4}{4\sqrt{2}} = \dfrac{4}{4\sqrt{2}} \cdot \dfrac{\sqrt{2}}{\sqrt{2}} = \dfrac{4\sqrt{2}}{8} = \dfrac{\sqrt{2}}{2}$

105. $\dfrac{2\sqrt{3}}{\sqrt{6}} = 2\sqrt{\dfrac{3}{6}} = 2\sqrt{\dfrac{1}{2}} = \dfrac{2}{\sqrt{2}} \cdot \dfrac{\sqrt{2}}{\sqrt{2}} = \sqrt{2}$

107. $\sqrt{2^2 + 6^2} = \sqrt{4 + 36} = \sqrt{40} = \sqrt{4 \cdot 10} = 2\sqrt{10}$

109. $\sqrt{18^2 - 6^2} = \sqrt{324 - 36} = \sqrt{288}$
$= \sqrt{144 \cdot 2} = 12\sqrt{2}$

111. Horizontal shift two units to the right

113. Reflection in the x-axis and a vertical shift two units upward

Section 6.2   Right Triangle Trigonometry

- You should know the right triangle definition of trigonometric functions.

(a) $\sin \theta = \dfrac{\text{opp}}{\text{hyp}}$ (b) $\cos \theta = \dfrac{\text{adj}}{\text{hyp}}$ (c) $\tan \theta = \dfrac{\text{opp}}{\text{adj}}$

(d) $\csc \theta = \dfrac{\text{hyp}}{\text{opp}}$ (e) $\sec \theta = \dfrac{\text{hyp}}{\text{adj}}$ (f) $\cot \theta = \dfrac{\text{adj}}{\text{opp}}$

- You should know the following identities.

(a) $\sin \theta = \dfrac{1}{\csc \theta}$ (b) $\csc \theta = \dfrac{1}{\sin \theta}$ (c) $\cos \theta = \dfrac{1}{\sec \theta}$

(d) $\sec \theta = \dfrac{1}{\cos \theta}$ (e) $\tan \theta = \dfrac{1}{\cot \theta}$ (f) $\cot \theta = \dfrac{1}{\tan \theta}$

(g) $\tan \theta = \dfrac{\sin \theta}{\cos \theta}$ (h) $\cot \theta = \dfrac{\cos \theta}{\sin \theta}$ (i) $\sin^2 \theta + \cos^2 \theta = 1$

(j) $1 + \tan^2 \theta = \sec^2 \theta$ (k) $1 + \cot^2 \theta = \csc^2 \theta$

- You should know that two acute angles α and β are complementary if $\alpha + \beta = 90°$, and that the cofunctions of complementary angles are equal.

- You should know the trigonometric function values of 30°, 45°, and 60°, or be able to construct triangles from which you can determine them.

Vocabulary Check

1.   (i) $\dfrac{\text{hypotenuse}}{\text{adjacent}} = \sec \theta$ (e)

   (ii) $\dfrac{\text{adjacent}}{\text{opposite}} = \cot \theta$ (f)

   (iii) $\dfrac{\text{hypotenuse}}{\text{opposite}} = \csc \theta$ (d)

   (iv) $\dfrac{\text{adjacent}}{\text{hypotenuse}} = \cos \theta$ (b)

   (v) $\dfrac{\text{opposite}}{\text{hypotenuse}} = \sin \theta$ (a)

   (vi) $\dfrac{\text{opposite}}{\text{adjacent}} = \tan \theta$ (c)

2. opposite; adjacent; hypotenuse

3. elevation; depression

1. $\text{hyp} = \sqrt{6^2 + 8^2} = \sqrt{36 + 64} = \sqrt{100} = 10$

$\sin \theta = \dfrac{\text{opp}}{\text{hyp}} = \dfrac{6}{10} = \dfrac{3}{5}$ $\csc \theta = \dfrac{\text{hyp}}{\text{opp}} = \dfrac{10}{6} = \dfrac{5}{3}$

$\cos \theta = \dfrac{\text{adj}}{\text{hyp}} = \dfrac{8}{10} = \dfrac{4}{5}$ $\sec \theta = \dfrac{\text{hyp}}{\text{adj}} = \dfrac{10}{8} = \dfrac{5}{4}$

$\tan \theta = \dfrac{\text{opp}}{\text{adj}} = \dfrac{6}{8} = \dfrac{3}{4}$ $\cot \theta = \dfrac{\text{adj}}{\text{opp}} = \dfrac{8}{6} = \dfrac{4}{3}$

3. $\text{adj} = \sqrt{41^2 - 9^2} = \sqrt{1681 - 81} = \sqrt{1600} = 40$

$\sin \theta = \dfrac{\text{opp}}{\text{hyp}} = \dfrac{9}{41}$ $\csc \theta = \dfrac{\text{hyp}}{\text{opp}} = \dfrac{41}{9}$

$\cos \theta = \dfrac{\text{adj}}{\text{hyp}} = \dfrac{40}{41}$ $\sec \theta = \dfrac{\text{hyp}}{\text{adj}} = \dfrac{41}{40}$

$\tan \theta = \dfrac{\text{opp}}{\text{adj}} = \dfrac{9}{40}$ $\cot \theta = \dfrac{\text{adj}}{\text{opp}} = \dfrac{40}{9}$

5. $\text{adj} = \sqrt{3^2 - 1^2} = \sqrt{8} = 2\sqrt{2}$

$\sin \theta = \dfrac{\text{opp}}{\text{hyp}} = \dfrac{1}{3}$ $\csc \theta = \dfrac{\text{hyp}}{\text{opp}} = 3$

$\cos \theta = \dfrac{\text{adj}}{\text{hyp}} = \dfrac{2\sqrt{2}}{3}$ $\sec \theta = \dfrac{\text{hyp}}{\text{adj}} = \dfrac{3}{2\sqrt{2}} = \dfrac{3\sqrt{2}}{4}$

$\tan \theta = \dfrac{\text{opp}}{\text{adj}} = \dfrac{1}{2\sqrt{2}} = \dfrac{\sqrt{2}}{4}$ $\cot \theta = \dfrac{\text{adj}}{\text{opp}} = 2\sqrt{2}$

$\text{adj} = \sqrt{6^2 - 2^2} = \sqrt{32} = 4\sqrt{2}$

$\sin \theta = \dfrac{\text{opp}}{\text{hyp}} = \dfrac{2}{6} = \dfrac{1}{3}$ $\csc \theta = \dfrac{\text{hyp}}{\text{opp}} = \dfrac{6}{2} = 3$

$\cos \theta = \dfrac{\text{adj}}{\text{hyp}} = \dfrac{4\sqrt{2}}{6} = \dfrac{2\sqrt{2}}{3}$ $\sec \theta = \dfrac{\text{hyp}}{\text{adj}} = \dfrac{6}{4\sqrt{2}} = \dfrac{3}{2\sqrt{2}} = \dfrac{3\sqrt{2}}{4}$

$\tan \theta = \dfrac{\text{opp}}{\text{adj}} = \dfrac{2}{4\sqrt{2}} = \dfrac{1}{2\sqrt{2}} = \dfrac{\sqrt{2}}{4}$ $\cot \theta = \dfrac{\text{adj}}{\text{opp}} = \dfrac{4\sqrt{2}}{2} = 2\sqrt{2}$

The function values are the same since the triangles are similar and the corresponding sides are proportional.

7. $\text{opp} = \sqrt{5^2 - 4^2} = 3$

$\sin \theta = \dfrac{\text{opp}}{\text{hyp}} = \dfrac{3}{5}$ $\csc \theta = \dfrac{\text{hyp}}{\text{opp}} = \dfrac{5}{3}$

$\cos \theta = \dfrac{\text{adj}}{\text{hyp}} = \dfrac{4}{5}$ $\sec \theta = \dfrac{\text{hyp}}{\text{adj}} = \dfrac{5}{4}$

$\tan \theta = \dfrac{\text{opp}}{\text{adj}} = \dfrac{3}{4}$ $\cot \theta = \dfrac{\text{adj}}{\text{opp}} = \dfrac{4}{3}$

$\text{opp} = \sqrt{1.25^2 - 1^2} = 0.75$

$\sin \theta = \dfrac{\text{opp}}{\text{hyp}} = \dfrac{0.75}{1.25} = \dfrac{3}{5}$ $\csc \theta = \dfrac{\text{hyp}}{\text{opp}} = \dfrac{1.25}{0.75} = \dfrac{5}{3}$

$\cos \theta = \dfrac{\text{adj}}{\text{hyp}} = \dfrac{1}{1.25} = \dfrac{4}{5}$ $\sec \theta = \dfrac{\text{hyp}}{\text{adj}} = \dfrac{1.25}{1} = \dfrac{5}{4}$

$\tan \theta = \dfrac{\text{opp}}{\text{adj}} = \dfrac{0.75}{1} = \dfrac{3}{4}$ $\cot \theta = \dfrac{\text{adj}}{\text{opp}} = \dfrac{1}{0.75} = \dfrac{4}{3}$

The function values are the same since the triangles are similar and the corresponding sides are proportional.

9. Given: $\sin \theta = \dfrac{3}{4} = \dfrac{\text{opp}}{\text{hyp}}$

$3^2 + (\text{adj})^2 = 4^2$

$\text{adj} = \sqrt{7}$

$\cos \theta = \dfrac{\text{adj}}{\text{hyp}} = \dfrac{\sqrt{7}}{4}$

$\tan \theta = \dfrac{\text{opp}}{\text{adj}} = \dfrac{3\sqrt{7}}{7}$

$\cot \theta = \dfrac{\text{adj}}{\text{opp}} = \dfrac{\sqrt{7}}{3}$

$\sec \theta = \dfrac{\text{hyp}}{\text{adj}} = \dfrac{4\sqrt{7}}{7}$

$\csc \theta = \dfrac{\text{hyp}}{\text{opp}} = \dfrac{4}{3}$

11. Given: $\sec \theta = 2 = \dfrac{2}{1} = \dfrac{\text{hyp}}{\text{adj}}$

$(\text{opp})^2 + 1^2 = 2^2$

$\text{opp} = \sqrt{3}$

$\sin \theta = \dfrac{\text{opp}}{\text{hyp}} = \dfrac{\sqrt{3}}{2}$

$\cos \theta = \dfrac{\text{adj}}{\text{hyp}} = \dfrac{1}{2}$

$\tan \theta = \dfrac{\text{opp}}{\text{adj}} = \sqrt{3}$

$\cot \theta = \dfrac{\text{adj}}{\text{opp}} = \dfrac{\sqrt{3}}{3}$

$\csc \theta = \dfrac{\text{hyp}}{\text{opp}} = \dfrac{2\sqrt{3}}{3}$

13. Given: $\tan \theta = 3 = \dfrac{3}{1} = \dfrac{\text{opp}}{\text{adj}}$

$3^2 + 1^2 = (\text{hyp})^2$

$\text{hyp} = \sqrt{10}$

$\sin \theta = \dfrac{\text{opp}}{\text{hyp}} = \dfrac{3\sqrt{10}}{10}$

$\cos \theta = \dfrac{\text{adj}}{\text{hyp}} = \dfrac{\sqrt{10}}{10}$

$\cot \theta = \dfrac{\text{adj}}{\text{opp}} = \dfrac{1}{3}$

$\sec \theta = \dfrac{\text{hyp}}{\text{adj}} = \sqrt{10}$

$\csc \theta = \dfrac{\text{hyp}}{\text{opp}} = \dfrac{\sqrt{10}}{3}$

15. Given: $\cot \theta = \dfrac{3}{2} = \dfrac{\text{adj}}{\text{opp}}$

$2^2 + 3^2 = (\text{hyp})^2$

$\text{hyp} = \sqrt{13}$

$\sin \theta = \dfrac{\text{opp}}{\text{hyp}} = \dfrac{2}{\sqrt{13}} = \dfrac{2\sqrt{13}}{13}$

$\cos \theta = \dfrac{\text{adj}}{\text{hyp}} = \dfrac{3}{\sqrt{13}} = \dfrac{3\sqrt{13}}{13}$

$\tan \theta = \dfrac{\text{opp}}{\text{adj}} = \dfrac{2}{3}$

$\csc \theta = \dfrac{\text{hyp}}{\text{opp}} = \dfrac{\sqrt{13}}{2}$

$\sec \theta = \dfrac{\text{hyp}}{\text{adj}} = \dfrac{\sqrt{13}}{3}$

17.

$30° = 30°\left(\dfrac{\pi}{180°}\right) = \dfrac{\pi}{6} \text{ radian}$

$\sin 30° = \dfrac{\text{opp}}{\text{hyp}} = \dfrac{1}{2}$

19.

$\dfrac{\pi}{3} = \dfrac{\pi}{3}\left(\dfrac{180°}{\pi}\right) = 60°$

$\tan \dfrac{\pi}{3} = \dfrac{\text{opp}}{\text{adj}} = \dfrac{\sqrt{3}}{1} = \sqrt{3}$

21.

$\cot \theta = \dfrac{\sqrt{3}}{3} = \dfrac{1}{\sqrt{3}} = \dfrac{\text{adj}}{\text{opp}}$

$\theta = 60° = \dfrac{\pi}{3} \text{ radian}$

23.

$\dfrac{\pi}{6} = \dfrac{\pi}{6}\left(\dfrac{180°}{\pi}\right) = 30°$

$\cos \dfrac{\pi}{6} = \dfrac{\text{adj}}{\text{hyp}} = \dfrac{\sqrt{3}}{2}$

25.

$$\cot \theta = 1 = \frac{1}{1} = \frac{\text{adj}}{\text{opp}}$$

$$\theta = 45° = 45°\left(\frac{\pi}{180°}\right) = \frac{\pi}{4}$$

27. $\sin 60° = \dfrac{\sqrt{3}}{2}, \ \cos 60° = \dfrac{1}{2}$

  (a) $\tan 60° = \dfrac{\sin 60°}{\cos 60°} = \sqrt{3}$

  (b) $\sin 30° = \cos 60° = \dfrac{1}{2}$

  (c) $\cos 30° = \sin 60° = \dfrac{\sqrt{3}}{2}$

  (d) $\cot 60° = \dfrac{\cos 60°}{\sin 60°} = \dfrac{1}{\sqrt{3}} = \dfrac{\sqrt{3}}{3}$

29. $\csc \theta = \dfrac{\sqrt{13}}{2}, \ \sec \theta = \dfrac{\sqrt{13}}{3}$

  (a) $\sin \theta = \dfrac{1}{\csc \theta} = \dfrac{2}{\sqrt{13}} = \dfrac{2\sqrt{13}}{13}$

  (b) $\cos \theta = \dfrac{1}{\sec \theta} = \dfrac{3}{\sqrt{13}} = \dfrac{3\sqrt{13}}{13}$

  (c) $\tan \theta = \dfrac{\sin \theta}{\cos \theta} = \dfrac{2\sqrt{13}/13}{3\sqrt{13}/13} = \dfrac{2}{3}$

  (d) $\sec(90° - \theta) = \csc \theta = \dfrac{\sqrt{13}}{2}$

31. $\cos \alpha = \dfrac{1}{3}$

  (a) $\sec \alpha = \dfrac{1}{\cos \alpha} = 3$

  (b) $\sin^2 \alpha + \cos^2 \alpha = 1$

    $\sin^2 \alpha + \left(\dfrac{1}{3}\right)^2 = 1$

    $\sin^2 \alpha = \dfrac{8}{9}$

    $\sin \alpha = \dfrac{2\sqrt{2}}{3}$

  (c) $\cot \alpha = \dfrac{\cos \alpha}{\sin \alpha} = \dfrac{\frac{1}{3}}{\frac{2\sqrt{2}}{3}} = \dfrac{1}{2\sqrt{2}} = \dfrac{\sqrt{2}}{4}$

  (d) $\sin(90° - \alpha) = \cos \alpha = \dfrac{1}{3}$

33. $\tan \theta \cot \theta = \tan \theta \left(\dfrac{1}{\tan \theta}\right) = 1$

35. $\tan \alpha \cos \alpha = \left(\dfrac{\sin \alpha}{\cos \alpha}\right) \cos \alpha = \sin \alpha$

37. $(1 + \cos \theta)(1 - \cos \theta) = 1 - \cos^2 \theta$

                       $= (\sin^2 \theta + \cos^2 \theta) - \cos^2 \theta$

                       $= \sin^2 \theta$

39. $(\sec \theta + \tan \theta)(\sec \theta - \tan \theta) = \sec^2 \theta - \tan^2 \theta$

                             $= (1 + \tan^2 \theta) - \tan^2 \theta$

                             $= 1$

41. $\dfrac{\sin \theta}{\cos \theta} + \dfrac{\cos \theta}{\sin \theta} = \dfrac{\sin^2 \theta + \cos^2 \theta}{\sin \theta \cos \theta}$

               $= \dfrac{1}{\sin \theta \cos \theta}$

               $= \dfrac{1}{\sin \theta} \cdot \dfrac{1}{\cos \theta}$

               $= \csc \theta \sec \theta$

43. (a) $\tan 23.5° \approx 0.4348$

  (b) $\cot 66.5° = \dfrac{1}{\tan 66.5°} \approx 0.4348$

45. (a) $\cos 16°18' = \cos\left(16 + \dfrac{18}{60}\right)° \approx 0.9598$

(b) $\sin 73°56' = \sin\left(73 + \dfrac{56}{60}\right)° \approx 0.9609$

47. Make sure that your calculator is in radian mode.

(a) $\cot \dfrac{\pi}{16} = \dfrac{1}{\tan \dfrac{\pi}{16}} \approx 5.0273$

(b) $\tan \dfrac{\pi}{16} \approx 0.1989$

49. Make sure that your calculator is in radian mode.

(a) $\csc 1 = \dfrac{1}{\sin 1} \approx 1.1884$

(b) $\tan \dfrac{1}{2} \approx 0.5463$

51. (a) $\sin \theta = \dfrac{1}{2} \implies \theta = 30° = \dfrac{\pi}{6}$

(b) $\csc \theta = 2 \implies \theta = 30° = \dfrac{\pi}{6}$

53. (a) $\sec \theta = 2 \implies \theta = 60° = \dfrac{\pi}{3}$

(b) $\cot \theta = 1 \implies \theta = 45° = \dfrac{\pi}{4}$

55. (a) $\csc \theta = \dfrac{2\sqrt{3}}{3} \implies \theta = 60° = \dfrac{\pi}{3}$

(b) $\sin \theta = \dfrac{\sqrt{2}}{2} \implies \theta = 45° = \dfrac{\pi}{4}$

57.

$\tan 30° = \dfrac{30}{x}$

$\dfrac{1}{\sqrt{3}} = \dfrac{30}{x}$

$x = 30\sqrt{3}$

59.

$\tan 60° = \dfrac{32}{x}$

$\sqrt{3} = \dfrac{32}{x}$

$\sqrt{3}\,x = 32$

$x = \dfrac{32}{\sqrt{3}} = \dfrac{32\sqrt{3}}{3}$

61.

$\tan 82° = \dfrac{x}{45}$

$x = 45 \tan 82°$

Height of the building:
$123 + 45 \tan 82° \approx 443.2$ meters

Distance between friends:

$\cos 82° = \dfrac{45}{y} \implies y = \dfrac{45}{\cos 82°}$

≈ 323.34 meters

63.

$\sin \theta = \dfrac{1500}{3000} = \dfrac{1}{2}$

$\theta = 30° = \dfrac{\pi}{6}$

65. (a) $\sin 23° = \dfrac{150}{x}$

$x = \dfrac{150}{\sin 23°} \approx 383.9$ feet

(b) $\tan 23° = \dfrac{150}{y}$

$y = \dfrac{150}{\tan 23°} \approx 353.4$ feet

(c) Moving down the line: $\dfrac{150/\sin 23°}{6} \approx 63.98$ feet per second

Dropping vertically: $\dfrac{150}{6} \approx 25$ feet per second

67.

$$\sin 30° = \frac{y_1}{56}$$

$$y_1 = (\sin 30°)(56) = \left(\frac{1}{2}\right)(56) = 28$$

$$\cos 30° = \frac{x_1}{56}$$

$$x_1 = \cos 30°(56) = \frac{\sqrt{3}}{2}(56) = 28\sqrt{3}$$

$$(x_1, y_1) = \left(28\sqrt{3}, 28\right)$$

$$\sin 60° = \frac{y_2}{56}$$

$$y_2 = \sin 60°(56) = \left(\frac{\sqrt{3}}{2}\right)(56) = 28\sqrt{3}$$

$$\cos 60° = \frac{x_2}{56}$$

$$x_2 = (\cos 60°)(56) = \left(\frac{1}{2}\right)(56) = 28$$

$$(x_2, y_2) = \left(28, 28\sqrt{3}\right)$$

69. (a)

(b) $\sin 85° = \dfrac{h}{20}$

(c) $h = 20 \sin 85° \approx 19.9$ meters

(d) The side of the triangle labeled h will become shorter.

(e)

Angle, θ	Height (in meters)
80°	19.7
70°	18.8
60°	17.3
50°	15.3
40°	12.9
30°	10.0
20°	6.8
10°	3.5

(f) The height of the balloon decreases.

71. $\sin 60° \csc 60° = 1$

True, $\csc x = \dfrac{1}{\sin x} \implies \sin 60° \csc 60° = \sin 60°\left(\dfrac{1}{\sin 60°}\right) = 1$

73. $\sin 45° + \sin 45° = 1$

False, $\dfrac{\sqrt{2}}{2} + \dfrac{\sqrt{2}}{2} = \sqrt{2} \neq 1$

75. $\dfrac{\sin 60°}{\sin 30°} = \sin 2°$

False, $\dfrac{\sin 60°}{\sin 30°} = \dfrac{\cos 30°}{\sin 30°} = \cot 30° \approx 1.7321$

$\sin 2° \approx 0.0349$

77. This is true because the corresponding sides of similar triangles are proportional.

79. (a)

θ	0.1	0.2	0.3	0.4	0.5
$\sin \theta$	0.0998	0.1987	0.2955	0.3894	0.4794

(b) In the interval $(0, 0.5]$, $\theta > \sin \theta$.

(c) As $\theta \to 0$, $\sin \theta \to 0$, and $\dfrac{\theta}{\sin \theta} \to 1$.

81. $\dfrac{x^2 - 6x}{x^2 + 4x - 12} \cdot \dfrac{x^2 + 12x + 36}{x^2 - 36} = \dfrac{x\cancel{(x-6)}}{\cancel{(x+6)}(x-2)} \cdot \dfrac{(x+6)\cancel{(x+6)}}{\cancel{(x+6)}\cancel{(x-6)}}$

$$= \dfrac{x}{x-2}, x \neq \pm 6$$

83. $\dfrac{3}{x+2} - \dfrac{2}{x-2} + \dfrac{x}{x^2 + 4x + 4} = \dfrac{3(x+2)(x-2) - 2(x+2)^2 + x(x-2)}{(x-2)(x+2)^2}$

$$= \dfrac{3(x^2 - 4) - 2(x^2 + 4x + 4) + x^2 - 2x}{(x-2)(x+2)^2}$$

$$= \dfrac{2x^2 - 10x - 20}{(x-2)(x+2)^2} = \dfrac{2(x^2 - 5x - 10)}{(x-2)(x+2)^2}$$

Section 6.3   Trigonometric Functions of Any Angle

■ Know the Definitions of Trigonometric Functions of Any Angle.

If θ is in standard position, (x, y) a point on the terminal side and $r = \sqrt{x^2 + y^2} \neq 0$, then

$$\sin \theta = \dfrac{y}{r} \qquad\qquad \csc \theta = \dfrac{r}{y}, \ y \neq 0$$

$$\cos \theta = \dfrac{x}{r} \qquad\qquad \sec \theta = \dfrac{r}{x}, \ x \neq 0$$

$$\tan \theta = \dfrac{y}{x}, \ x \neq 0 \qquad\qquad \cot \theta = \dfrac{x}{y}, \ y \neq 0$$

■ You should know the signs of the trigonometric functions in each quadrant.

■ You should know the trigonometric function values of the quadrant angles 0, $\dfrac{\pi}{2}$, π, and $\dfrac{3\pi}{2}$.

■ You should be able to find reference angles.

■ You should be able to evaluate trigonometric functions of any angle. (Use reference angles.)

■ You should know that the period of sine and cosine is 2π.

■ You should know which trigonometric functions are odd and even.

Even: $\cos x$ and $\sec x$

Odd: $\sin x$, $\tan x$, $\cot x$, $\csc x$

Vocabulary Check

1. reference

2. periodic

3. period

4. even; odd

1. (a) $(x, y) = (4, 3)$

$r = \sqrt{16 + 9} = 5$

$\sin \theta = \dfrac{y}{r} = \dfrac{3}{5}$ $\qquad \csc \theta = \dfrac{r}{y} = \dfrac{5}{3}$

$\cos \theta = \dfrac{x}{r} = \dfrac{4}{5}$ $\qquad \sec \theta = \dfrac{r}{x} = \dfrac{5}{4}$

$\tan \theta = \dfrac{y}{x} = \dfrac{3}{4}$ $\qquad \cot \theta = \dfrac{x}{y} = \dfrac{4}{3}$

(b) $(x, y) = (8, -15)$

$r = \sqrt{64 + 225} = 17$

$\sin \theta = \dfrac{y}{r} = -\dfrac{15}{17}$ $\qquad \csc \theta = \dfrac{r}{y} = -\dfrac{17}{15}$

$\cos \theta = \dfrac{x}{r} = \dfrac{8}{17}$ $\qquad \sec \theta = \dfrac{r}{x} = \dfrac{17}{8}$

$\tan \theta = \dfrac{y}{x} = -\dfrac{15}{8}$ $\qquad \cot \theta = \dfrac{x}{y} = -\dfrac{8}{15}$

3. (a) $(x, y) = \left(-\sqrt{3}, -1\right)$

$r = \sqrt{3 + 1} = 2$

$\sin\theta = \dfrac{y}{r} = -\dfrac{1}{2}$ $\qquad$ $\csc\theta = \dfrac{r}{y} = -2$

$\cos\theta = \dfrac{x}{r} = -\dfrac{\sqrt{3}}{2}$ $\qquad$ $\sec\theta = \dfrac{r}{x} = -\dfrac{2\sqrt{3}}{3}$

$\tan\theta = \dfrac{y}{x} = \dfrac{\sqrt{3}}{3}$ $\qquad$ $\cot\theta = \dfrac{x}{y} = \sqrt{3}$

(b) $(x, y) = (-4, 1)$

$r = \sqrt{16 + 1} = \sqrt{17}$

$\sin\theta = \dfrac{y}{r} = \dfrac{\sqrt{17}}{17}$ $\qquad$ $\csc\theta = \dfrac{r}{y} = \sqrt{17}$

$\cos\theta = \dfrac{x}{r} = -\dfrac{4\sqrt{17}}{17}$ $\qquad$ $\sec\theta = \dfrac{r}{x} = -\dfrac{\sqrt{17}}{4}$

$\tan\theta = \dfrac{y}{x} = -\dfrac{1}{4}$ $\qquad$ $\cot\theta = \dfrac{x}{y} = -4$

5. $(x, y) = (7, 24)$

$r = \sqrt{49 + 576} = 25$

$\sin\theta = \dfrac{y}{r} = \dfrac{24}{25}$ $\qquad$ $\csc\theta = \dfrac{r}{y} = \dfrac{25}{24}$

$\cos\theta = \dfrac{x}{r} = \dfrac{7}{25}$ $\qquad$ $\sec\theta = \dfrac{r}{x} = \dfrac{25}{7}$

$\tan\theta = \dfrac{y}{x} = \dfrac{24}{7}$ $\qquad$ $\cot\theta = \dfrac{x}{y} = \dfrac{7}{24}$

7. $(x, y) = (-4, 10)$

$r = \sqrt{16 + 100} = 2\sqrt{29}$

$\sin\theta = \dfrac{y}{r} = \dfrac{5\sqrt{29}}{29}$ $\qquad$ $\csc\theta = \dfrac{r}{y} = \dfrac{\sqrt{29}}{5}$

$\cos\theta = \dfrac{x}{r} = -\dfrac{2\sqrt{29}}{29}$ $\qquad$ $\sec\theta = \dfrac{r}{x} = -\dfrac{\sqrt{29}}{2}$

$\tan\theta = \dfrac{y}{x} = -\dfrac{5}{2}$ $\qquad$ $\cot\theta = \dfrac{x}{y} = -\dfrac{2}{5}$

9. $(x, y) = (-3.5, 6.8)$

$r = \sqrt{(-3.5)^2 + (6.8)^2} = \sqrt{58.49}$

$\sin\theta = \dfrac{y}{r} = \dfrac{6.8}{\sqrt{58.49}} = \dfrac{6.8\sqrt{58.49}}{58.49}$ $\qquad$ $\csc\theta = \dfrac{r}{y} = \dfrac{\sqrt{58.49}}{6.8}$

$\cos\theta = \dfrac{x}{r} = -\dfrac{3.5}{\sqrt{58.49}} = -\dfrac{3.5\sqrt{58.49}}{58.49}$ $\qquad$ $\sec\theta = \dfrac{r}{x} = -\dfrac{\sqrt{58.49}}{3.5}$

$\tan\theta = \dfrac{y}{x} = -\dfrac{6.8}{3.5}$ $\qquad$ $\cot\theta = \dfrac{x}{y} = -\dfrac{3.5}{6.8}$

11. $\sin\theta < 0 \implies \theta$ lies in Quadrant III or in Quadrant IV.

$\cos\theta < 0 \implies \theta$ lies in Quadrant II or in Quadrant III.

$\sin\theta < 0$ and $\cos\theta < 0 \implies \theta$ lies in Quadrant III.

13. $\sin\theta > 0 \implies \theta$ lies in Quadrant I or in Quadrant II.

$\tan\theta < 0 \implies \theta$ lies in Quadrant II or in Quadrant IV.

$\sin\theta > 0$ and $\tan\theta < 0 \implies \theta$ lies in Quadrant II.

15. $\sin\theta = \dfrac{y}{r} = \dfrac{3}{5} \implies x^2 = 25 - 9 = 16$

θ in Quadrant II $\implies x = -4$

$\sin\theta = \dfrac{y}{r} = \dfrac{3}{5}$ $\qquad$ $\csc\theta = \dfrac{r}{y} = \dfrac{5}{3}$

$\cos\theta = \dfrac{x}{r} = -\dfrac{4}{5}$ $\qquad$ $\sec\theta = \dfrac{r}{x} = -\dfrac{5}{4}$

$\tan\theta = \dfrac{y}{x} = -\dfrac{3}{4}$ $\qquad$ $\cot\theta = \dfrac{x}{y} = -\dfrac{4}{3}$

17. $\sin\theta < 0$ and $\tan\theta < 0 \implies \theta$ is in Quadrant IV $\implies$ $y < 0$ and $x > 0$.

$\tan\theta = \dfrac{y}{x} = \dfrac{-15}{8} \implies r = 17$

$\sin\theta = \dfrac{y}{r} = -\dfrac{15}{17}$ $\qquad$ $\csc\theta = \dfrac{r}{y} = -\dfrac{17}{15}$

$\cos\theta = \dfrac{x}{r} = \dfrac{8}{17}$ $\qquad$ $\sec\theta = \dfrac{r}{x} = \dfrac{17}{8}$

$\tan\theta = \dfrac{y}{x} = -\dfrac{15}{8}$ $\qquad$ $\cot\theta = \dfrac{x}{y} = -\dfrac{8}{15}$

19. $\cot \theta = \dfrac{x}{y} = -\dfrac{3}{1} = \dfrac{3}{-1}$

$\cos \theta > 0 \implies \theta$ is in Quadrant IV $\implies x$ is positive;
$x = 3, y = -1, r = \sqrt{10}$

$\sin \theta = \dfrac{y}{r} = -\dfrac{\sqrt{10}}{10}$ $\csc \theta = \dfrac{r}{y} = -\sqrt{10}$

$\cos \theta = \dfrac{x}{r} = \dfrac{3\sqrt{10}}{10}$ $\sec \theta = \dfrac{r}{x} = \dfrac{\sqrt{10}}{3}$

$\tan \theta = \dfrac{y}{x} = -\dfrac{1}{3}$ $\cot \theta = \dfrac{x}{y} = -3$

21. $\sec \theta = \dfrac{r}{x} = \dfrac{2}{-1} \implies y^2 = 4 - 1 = 3$

$\sin \theta > 0 \implies \theta$ is in Quadrant II $\implies y = \sqrt{3}$

$\sin \theta = \dfrac{y}{r} = \dfrac{\sqrt{3}}{2}$ $\csc \theta = \dfrac{r}{y} = \dfrac{2\sqrt{3}}{3}$

$\cos \theta = \dfrac{x}{r} = -\dfrac{1}{2}$ $\sec \theta = \dfrac{r}{x} = -2$

$\tan \theta = \dfrac{y}{x} = -\sqrt{3}$ $\cot \theta = \dfrac{x}{y} = -\dfrac{\sqrt{3}}{3}$

23. $\cot \theta$ is undefined, $\dfrac{\pi}{2} \le \theta \le \dfrac{3\pi}{2} \implies y = 0 \implies \theta = \pi$

$\sin \theta = 0$ $\csc \theta$ is undefined.

$\cos \theta = -1$ $\sec \theta = -1$

$\tan \theta = 0$ $\cot \theta$ is undefined.

25. To find a point on the terminal side of θ, use any point on the line $y = -x$ that lies in Quadrant II. $(-1, 1)$ is one such point.

$x = -1, y = 1, r = \sqrt{2}$

$\sin \theta = \dfrac{1}{\sqrt{2}} = \dfrac{\sqrt{2}}{2}$ $\csc \theta = \sqrt{2}$

$\cos \theta = -\dfrac{1}{\sqrt{2}} = -\dfrac{\sqrt{2}}{2}$ $\sec \theta = -\sqrt{2}$

$\tan \theta = -1$ $\cot \theta = -1$

27. To find a point on the terminal side of θ, use any point on the line $y = 2x$ that lies in Quadrant III. $(-1, -2)$ is one such point.

$x = -1, y = -2, r = \sqrt{5}$

$\sin \theta = -\dfrac{2}{\sqrt{5}} = -\dfrac{2\sqrt{5}}{5}$ $\csc \theta = \dfrac{\sqrt{5}}{-2} = -\dfrac{\sqrt{5}}{2}$

$\cos \theta = -\dfrac{1}{\sqrt{5}} = -\dfrac{\sqrt{5}}{5}$ $\sec \theta = \dfrac{\sqrt{5}}{-1} = -\sqrt{5}$

$\tan \theta = \dfrac{-2}{-1} = 2$ $\cot \theta = \dfrac{-1}{-2} = \dfrac{1}{2}$

29. $(x, y) = (-1, 0), r = 1$

$\sin \pi = \dfrac{y}{r} = \dfrac{0}{1} = 0$

31. $(x, y) = (0, -1), r = 1$

$\sec \dfrac{3\pi}{2} = \dfrac{r}{x} = \dfrac{1}{0} \implies$ undefined

33. $(x, y) = (0, 1), r = 1$

$\sin \dfrac{\pi}{2} = \dfrac{y}{r} = \dfrac{1}{1} = 1$

35. $(x, y) = (-1, 0), r = 1$

$\csc \pi = \dfrac{r}{y} = \dfrac{1}{0} \implies$ undefined

37. $\theta = 203°$

$\theta' = 203° - 180° = 23°$

39. $\theta = -245°$

$360° - 245° = 115°$ (coterminal angle)

$\theta' = 180° - 115° = 65°$

41. $\theta = \dfrac{2\pi}{3}$

$\theta' = \pi - \dfrac{2\pi}{3} = \dfrac{\pi}{3}$

43. $\theta = 3.5$

$\theta' = 3.5 - \pi$

45. $\theta = 225°$, $\theta' = 45°$, Quadrant III

$\sin 225° = -\sin 45° = -\dfrac{\sqrt{2}}{2}$

$\cos 225° = -\cos 45° = -\dfrac{\sqrt{2}}{2}$

$\tan 225° = \tan 45° = 1$

47. $\theta = 750°$, $\theta' = 30°$, Quadrant I

$\sin 750° = \sin 30° = \dfrac{1}{2}$

$\cos 750° = \cos 30° = \dfrac{\sqrt{3}}{2}$

$\tan 750° = \tan 30° = \dfrac{\sqrt{3}}{3}$

49. $\theta = -150°$, $\theta' = 30°$, Quadrant III

$\sin(-150°) = -\sin 30° = -\dfrac{1}{2}$

$\cos(-150°) = -\cos 30° = -\dfrac{\sqrt{3}}{2}$

$\tan(-150°) = \tan 30° = \dfrac{\sqrt{3}}{3}$

51. $\theta = \dfrac{4\pi}{3}$, $\theta' = \dfrac{\pi}{3}$, Quadrant III

$\sin \dfrac{4\pi}{3} = -\sin \dfrac{\pi}{3} = -\dfrac{\sqrt{3}}{2}$

$\cos \dfrac{4\pi}{3} = -\cos \dfrac{\pi}{3} = -\dfrac{1}{2}$

$\tan \dfrac{4\pi}{3} = \tan \dfrac{\pi}{3} = \sqrt{3}$

53. $\theta = -\dfrac{\pi}{6}$, $\theta' = \dfrac{\pi}{6}$, Quadrant IV

$\sin\left(-\dfrac{\pi}{6}\right) = -\sin \dfrac{\pi}{6} = -\dfrac{1}{2}$

$\cos\left(-\dfrac{\pi}{6}\right) = \cos \dfrac{\pi}{6} = \dfrac{\sqrt{3}}{2}$

$\tan\left(-\dfrac{\pi}{6}\right) = -\tan \dfrac{\pi}{6} = -\dfrac{\sqrt{3}}{3}$

55. $\theta = \dfrac{11\pi}{4}$, $\theta' = \dfrac{\pi}{4}$, Quadrant II

$\sin \dfrac{11\pi}{4} = \sin \dfrac{\pi}{4} = \dfrac{\sqrt{2}}{2}$

$\cos \dfrac{11\pi}{4} = -\cos \dfrac{\pi}{4} = -\dfrac{\sqrt{2}}{2}$

$\tan \dfrac{11\pi}{4} = -\tan \dfrac{\pi}{4} = -1$

57. $\theta = -\dfrac{3\pi}{2}$, $\theta' = \dfrac{\pi}{2}$

$\sin\left(-\dfrac{3\pi}{2}\right) = \sin \dfrac{\pi}{2} = 1$

$\cos\left(-\dfrac{3\pi}{2}\right) = \cos \dfrac{\pi}{2} = 0$

$\tan\left(-\dfrac{3\pi}{2}\right) = \tan \dfrac{\pi}{2}$ which is undefined.

59. $\sin \theta = -\dfrac{3}{5}$

$\sin^2 \theta + \cos^2 \theta = 1$

$\cos^2 \theta = 1 - \sin^2 \theta$

$\cos^2 \theta = 1 - \left(-\dfrac{3}{5}\right)^2$

$\cos^2 \theta = 1 - \dfrac{9}{25}$

$\cos^2 \theta = \dfrac{16}{25}$

$\cos \theta > 0$ in Quadrant IV.

$\cos \theta = \dfrac{4}{5}$

61. $\tan \theta = \dfrac{3}{2}$

$\sec^2 \theta = 1 + \tan^2 \theta$

$\sec^2 \theta = 1 + \left(\dfrac{3}{2}\right)^2$

$\sec^2 \theta = 1 + \dfrac{9}{4}$

$\sec^2 \theta = \dfrac{13}{4}$

$\sec \theta < 0$ in Quadrant III.

$\sec \theta = -\dfrac{\sqrt{13}}{2}$

63. $\cos \theta = \dfrac{5}{8}$

$\cos \theta = \dfrac{1}{\sec \theta} \Rightarrow \sec \theta = \dfrac{1}{\cos \theta}$

$\sec \theta = \dfrac{1}{\frac{5}{8}} = \dfrac{8}{5}$

65. $\sin 10° \approx 0.1736$

67. $\cos(-110°) \approx -0.3420$

69. $\tan 304° \approx -1.4826$

71. $\sec 72° = \dfrac{1}{\cos 72°} \approx 3.2361$

73. $\tan 4.5 \approx 4.6373$

75. $\cos\left(\dfrac{\pi}{9}\right) \approx 0.9397$

77. $\sin(-0.65) \approx -0.6052$

79. $\cot\left(-\dfrac{11\pi}{8}\right) = \dfrac{1}{\tan(-11\pi/8)} \approx -0.4142$

81. (a) $\sin\theta = \dfrac{1}{2} \implies$ reference angle is $30°$ or $\dfrac{\pi}{6}$ and θ is in Quadrant I or Quadrant II.

Values in degrees: $30°, 150°$

Values in radian: $\dfrac{\pi}{6}, \dfrac{5\pi}{6}$

(b) $\sin\theta = -\dfrac{1}{2} \implies$ reference angle is $30°$ or $\dfrac{\pi}{6}$ and θ is in Quadrant III or Quadrant IV.

Values in degrees: $210°, 330°$

Values in radians: $\dfrac{7\pi}{6}, \dfrac{11\pi}{6}$

83. (a) $\csc\theta = \dfrac{2\sqrt{3}}{3} \implies$ reference angle is $60°$ or $\dfrac{\pi}{3}$ and θ is in Quadrant I or Quadrant II.

Values in degrees: $60°, 120°$

Values in radians: $\dfrac{\pi}{3}, \dfrac{2\pi}{3}$

(b) $\cot\theta = -1 \implies$ reference angle is $45°$ or $\dfrac{\pi}{4}$ and θ is in Quadrant II or Quadrant IV.

Values in degrees: $135°, 315°$

Values in radians: $\dfrac{3\pi}{4}, \dfrac{7\pi}{4}$

85. (a) $\tan\theta = 1 \implies$ reference angle is $45°$ or $\dfrac{\pi}{4}$ and θ is in Quadrant I or Quadrant III.

Values in degrees: $45°, 225°$

Values in radians: $\dfrac{\pi}{4}, \dfrac{5\pi}{4}$

(b) $\cot\theta = -\sqrt{3} \implies$ reference angle is $30°$ or $\dfrac{\pi}{6}$ and θ is in Quadrant II or Quadrant IV.

Values in degrees: $150°, 330°$

Values in radians: $\dfrac{5\pi}{6}, \dfrac{11\pi}{6}$

87. $\left(\dfrac{\sqrt{2}}{2}, \dfrac{\sqrt{2}}{2}\right)$ corresponds to $t = \dfrac{\pi}{4}$ on the unit circle.

$\sin\dfrac{\pi}{4} = \dfrac{\sqrt{2}}{2}$ since $\sin t = y$

$\cos\dfrac{\pi}{4} = \dfrac{\sqrt{2}}{2}$ since $\cos t = x$

$\tan\dfrac{\pi}{4} = 1$ since $\tan t = \dfrac{y}{x}$

89. $\left(-\dfrac{\sqrt{3}}{2}, \dfrac{1}{2}\right)$ corresponds to $t = \dfrac{5\pi}{6}$ on the unit circle.

$\sin\dfrac{5\pi}{6} = \dfrac{1}{2}$ since $\sin t = y$

$\cos\dfrac{5\pi}{6} = -\dfrac{\sqrt{3}}{2}$ since $\cos t = x$

$\tan\dfrac{5\pi}{6} = -\dfrac{\sqrt{3}}{3}$ since $\tan t = \dfrac{y}{x}$

91. $\left(-\dfrac{1}{2}, -\dfrac{\sqrt{3}}{2}\right)$ corresponds to $t = \dfrac{4\pi}{3}$ on the unit circle.

$\sin\dfrac{4\pi}{3} = -\dfrac{\sqrt{3}}{2}$ since $\sin t = y$.

$\cos\dfrac{4\pi}{3} = -\dfrac{1}{2}$ since $\cos t = x$.

$\tan\dfrac{4\pi}{3} = \sqrt{3}$ since $\tan t = \dfrac{y}{x}$.

93. $(0, -1)$ corresponds to $t = \dfrac{3\pi}{2}$ on the unit circle.

$\sin\dfrac{3\pi}{2} = -1$ since $\sin t = y$

$\cos\dfrac{3\pi}{2} = 0$ since $\cos t = x$

$\tan\dfrac{3\pi}{2}$ is undefined   since $\tan t = \dfrac{y}{x}$

95. (a) $\sin 5 \approx -1$

(b) $\cos 2 \approx -0.4$

97. (a) $\sin t = 0.25$

$t \approx 0.25$ or 2.89

(b) $\cos t = -0.25$

$t \approx 1.82$ or 4.46

99. (a) New York City: $N \approx 22.1 \sin(0.52t - 2.22) + 55.01$

Fairbanks: $F \approx 36.6 \sin(0.50t - 1.83) + 25.61$

(c) The periods are about the same for both models, approximately 12 months.

(b)

Month	New York City	Fairbanks
February	35°	−1°
March	41°	14°
May	63°	48°
June	72°	59°
August	76°	56°
September	69°	42°
November	47°	7°

101. $y(t) = 2 \cos 6t$

(a) $y(0) = 2 \cos 0 = 2$ centimeters

(b) $y\left(\frac{1}{4}\right) = 2 \cos\left(\frac{3}{2}\right) \approx 0.14$ centimeter

(c) $y\left(\frac{1}{2}\right) = 2 \cos 3 \approx -1.98$ centimeters

103. $I = 5e^{-2t} \sin t$

$I(0.7) = 5e^{-1.4} \sin 0.7 \approx 0.79$

105. False. In each of the four quadrants, the sign of the secant function and the cosine function will be the same since they are reciprocals of each other.

107. $h(t) = f(t)g(t)$

$h(-t) = f(-t)g(-t)$

$\qquad = -f(t)g(t)$

$\qquad = -h(t)$

Therefore, $h(t)$ is odd.

109. (a)

θ	0°	20°	40°	60°	80°
$\sin \theta$	0	0.342	0.643	0.866	0.985
$\sin(180° - \theta)$	0	0.342	0.643	0.866	0.985

(b) Conjecture: $\sin \theta = \sin(180° - \theta)$

111. If θ is obtuse, then $90° < \theta < 180°$. The reference angle is $\theta' = 180° - \theta$ and we have the following:

$\sin \theta = \sin \theta' \qquad \csc \theta = \csc \theta'$

$\cos \theta = -\cos \theta' \qquad \sec \theta = -\sec \theta'$

$\tan \theta = -\tan \theta' \qquad \cot \theta = -\cot \theta'$

113. $f(x) = \dfrac{2}{4 - x}$

Vertical asymptote: $x = 4$

Horizontal asymptote: $y = 0$

x	0	1	2	3	5	6
y	$\frac{1}{2}$	$\frac{2}{3}$	1	2	−2	−1

115. $h(x) = \dfrac{2x^2 - 12x}{x - 6} = \dfrac{2x(x - 6)}{x - 6} = 2x, x \neq 6$

No asymptotes

x	-2	0	2	4	8
y	-4	0	4	8	16

The graph has a hole at $(2, 12)$.

117. $y = 2^{x-1}$

x	-1	0	1	2	3
y	$\frac{1}{4}$	$\frac{1}{2}$	1	2	4

Domain: all real numbers x

Range: $y > 0$

y-intercept: $\left(0, \frac{1}{2}\right)$

Horizontal asymptote: $y = 0$

119. $y = 3^{-x/2}$

x	-4	-2	0	2	4
y	9	3	1	$\frac{1}{3}$	$\frac{1}{9}$

Domain: all real numbers x

Range: $y > 0$

y-intercept: $(0, 1)$

Horizontal asymptote: $y = 0$

121. $y = \ln(x - 1)$

x	1.1	1.5	2	3	4
y	-2.30	-0.69	0	0.69	1.10

Domain: $x - 1 > 0 \implies x > 1$

Range: all real numbers

x-intercept: $(2, 0)$

Vertical asymptote: $x = 1$

123. $y = \log_{10}(x + 2)$

x	-1.5	-1	0	1	2
y	≈ -0.301	0	≈ 0.301	≈ 0.477	≈ 0.602

Domain: $x + 2 > 0 \implies x > -2$

Range: all real numbers

x-intercept: $(-1, 0)$

y-intercept: $(0, 0.301)$

Section 6.4   Graphs of Sine and Cosine Functions

- ■ You should be able to graph $y = a \sin(bx - c)$ and $y = a \cos(bx - c)$. (Assume $b > 0$.)

- ■ Amplitude: $|a|$

- ■ Period: $\dfrac{2\pi}{b}$

- ■ Shift: Solve $bx - c = 0$ and $bx - c = 2\pi$.

- ■ Key increments: $\dfrac{1}{4}$ (period)

Vocabulary Check

1. cycle

2. amplitude

3. $\dfrac{2\pi}{b}$

4. phase shift

5. vertical shift

---

1. $y = 3 \sin 2x$

Period: $\dfrac{2\pi}{2} = \pi$

Amplitude: $|3| = 3$

3. $y = \dfrac{5}{2} \cos \dfrac{x}{2}$

Period: $\dfrac{2\pi}{\frac{1}{2}} = 4\pi$

Amplitude: $\left|\dfrac{5}{2}\right| = \dfrac{5}{2}$

5. $y = \dfrac{1}{2} \sin \dfrac{\pi x}{3}$

Period: $\dfrac{2\pi}{\frac{\pi}{3}} = 6$

Amplitude: $\left|\dfrac{1}{2}\right| = \dfrac{1}{2}$

7. $y = -2 \sin x$

Period: $\dfrac{2\pi}{1} = 2\pi$

Amplitude: $|-2| = 2$

9. $y = 3 \sin 10x$

Period: $\dfrac{2\pi}{10} = \dfrac{\pi}{5}$

Amplitude: $|3| = 3$

11. $y = \dfrac{1}{2} \cos \dfrac{2x}{3}$

Period: $\dfrac{2\pi}{2/3} = 3\pi$

Amplitude: $\left|\dfrac{1}{2}\right| = \dfrac{1}{2}$

13. $y = \dfrac{1}{4} \sin 2\pi x$

Period: $\dfrac{2\pi}{2\pi} = 1$

Amplitude: $\left|\dfrac{1}{4}\right| = \dfrac{1}{4}$

15. $f(x) = \sin x$

$g(x) = \sin(x - \pi)$

The graph of g is a horizontal shift to the right π units of the graph of f (a phase shift).

17. $f(x) = \cos 2x$

$g(x) = -\cos 2x$

The graph of g is a reflection in the x-axis of the graph of f.

19. $f(x) = \cos x$

$g(x) = \cos 2x$

The period of f is twice that of g.

21. $f(x) = \sin 2x$

$f(x) = 3 + \sin 2x$

The graph of g is a vertical shift three units upward of the graph of f.

23. The graph of g has twice the amplitude as the graph of f. The period is the same.

25. The graph of g is a horizontal shift π units to the right of the graph of f.

27. $f(x) = -2 \sin x$

Period: $\dfrac{2\pi}{b} = \dfrac{2\pi}{1} = 2\pi$

Amplitude: 2

Symmetry: origin

Key points:	Intercept	Minimum	Intercept	Maximum	Intercept
	$(0, 0)$	$\left(\dfrac{\pi}{2}, -2\right)$	$(\pi, 0)$	$\left(\dfrac{3\pi}{2}, 0\right)$	$(2\pi, 0)$

Since $g(x) = 4 \sin x = (-2)f(x)$, generate key points for the graph of $g(x)$ by multiplying the y-coordinate of each key point of $f(x)$ by -2.

29. $f(x) = \cos x$

Period: $\dfrac{2\pi}{b} = \dfrac{2\pi}{1} = 2\pi$

Amplitude: 1

Symmetry: y-axis

Key points:	Maximum	Intercept	Minimum	Intercept	Maximum
	$(0, 1)$	$\left(\dfrac{\pi}{2}, 0\right)$	$(\pi, -1)$	$\left(\dfrac{3\pi}{2}, 0\right)$	$(2\pi, 1)$

Since $g(x) = 1 + \cos(x) = f(x) + 1$, the graph of $g(x)$ is the graph of $f(x)$, but translated upward by one unit. Generate key points for the graph of $g(x)$ by adding 1 to the y-coordinate of each key point of $f(x)$.

31. $f(x) = -\dfrac{1}{2} \sin \dfrac{x}{2}$

Period: $\dfrac{2\pi}{b} = \dfrac{2\pi}{1/2} = 4\pi$

Amplitude: $\dfrac{1}{2}$

Symmetry: origin

Key points:	Intercept	Minimum	Intercept	Maximum	Intercept
	$(0, 0)$	$\left(\pi, -\dfrac{1}{2}\right)$	$(2\pi, 0)$	$\left(3\pi, \dfrac{1}{2}\right)$	$(4\pi, 0)$

Since $g(x) = 3 - \dfrac{1}{2} \sin \dfrac{x}{2} = 3 - f(x)$, the graph of $g(x)$ is the graph of $f(x)$, but translated upward by three units.

Generate key points for the graph of $g(x)$ by adding 3 to the y-coordinate of each key point of $f(x)$.

33. $f(x) = 2 \cos x$

Period: $\dfrac{2\pi}{b} = \dfrac{2\pi}{1} = 2\pi$

Amplitude: 2

Symmetry: y-axis

Key points:	Maximum	Intercept	Minimum	Intercept	Maximum
	$(0, 2)$	$\left(\dfrac{\pi}{2}, 0\right)$	$(\pi, -2)$	$\left(\dfrac{3\pi}{2}, 0\right)$	$(2\pi, 2)$

Since $g(x) = 2 \cos(x + \pi) = f(x + \pi)$, the graph of $g(x)$ is the graph of $f(x)$, but with a phase shift (horizontal translation) of $-\pi$. Generate key points for the graph of $g(x)$ by shifting each key point of $f(x)$ π units to the left.

35. $y = 3 \sin x$

Period: 2π

Amplitude: 3

Key points:

$(0, 0), \left(\dfrac{\pi}{2}, 3\right), (\pi, 0),$

$\left(\dfrac{3\pi}{2}, -3\right), (2\pi, 0)$

37. $y = \dfrac{1}{3} \cos x$

Period: 2π

Amplitude: $\dfrac{1}{3}$

Key points:

$\left(0, \dfrac{1}{3}\right), \left(\dfrac{\pi}{2}, 0\right), \left(\pi, -\dfrac{1}{3}\right),$

$\left(\dfrac{3\pi}{2}, 0\right), \left(2\pi, \dfrac{1}{3}\right)$

39. $y = \cos \dfrac{x}{2}$

Period: 4π

Amplitude: 1

Key points:

$(0, 1), (\pi, 0), (2\pi, -1),$

$(3\pi, 0), (4\pi, 1)$

41. $y = \cos 2\pi x$

Period: $\dfrac{2\pi}{2\pi} = 1$

Amplitude: 1

Key points:

$(0, 1), \left(\dfrac{1}{4}, 0\right), \left(\dfrac{1}{2}, -1\right), \left(\dfrac{3}{4}, 0\right)$

43. $y = -\sin \dfrac{2\pi x}{3}$; $a = -1, b = \dfrac{2\pi}{3}, c = 0$

Period: $\dfrac{2\pi}{2\pi/3} = 3$

Amplitude: 1

Key points: $(0, 0), \left(\dfrac{3}{4}, -1\right), \left(\dfrac{3}{2}, 0\right), \left(\dfrac{9}{4}, 1\right), (3, 0)$

45. $y = \sin\left(x - \dfrac{\pi}{4}\right)$; $a = 1, b = 1, c = \dfrac{\pi}{4}$

Period: 2π

Amplitude: 1

Shift: Set $x - \dfrac{\pi}{4} = 0$ and $x - \dfrac{\pi}{4} = 2\pi$

$x = \dfrac{\pi}{4}$ $x = \dfrac{9\pi}{4}$

Key points: $\left(\dfrac{\pi}{4}, 0\right), \left(\dfrac{3\pi}{4}, 1\right), \left(\dfrac{5\pi}{4}, 0\right), \left(\dfrac{7\pi}{4}, -1\right), \left(\dfrac{9\pi}{4}, 0\right)$

47. $y = 3 \cos(x + \pi)$

Period: 2π

Amplitude: 3

Shift: Set $x + \pi = 0$ and $x + \pi = 2\pi$

$x = -\pi$ $x = \pi$

Key points: $(-\pi, 3), \left(-\dfrac{\pi}{2}, 0\right), (0, -3), \left(\dfrac{\pi}{2}, 0\right), (\pi, 3)$

49. $y = 2 - \sin\dfrac{2\pi x}{3}$

Period: 3

Amplitude: 1

Key points: $(0, 2), \left(\dfrac{3}{4}, 1\right), \left(\dfrac{3}{2}, 2\right), \left(\dfrac{9}{4}, 3\right), (3, 2)$

51. $y = 2 + \dfrac{1}{10}\cos 60\pi x$

Period: $\dfrac{2\pi}{60\pi} = \dfrac{1}{30}$

Amplitude: $\dfrac{1}{10}$

Vertical shift two units upward

Key points:

$(0, 2.1), \left(\dfrac{1}{120}, 2\right), \left(\dfrac{1}{60}, 1.9\right), \left(\dfrac{1}{40}, 2\right), \left(\dfrac{1}{30}, 2.1\right)$

53. $y = 3\cos(x + \pi) - 3$

Period: 2π

Amplitude: 3

Shift: Set $x + \pi = 0$ and $x + \pi = 2\pi$

$\qquad\qquad x = -\pi \qquad\qquad\quad x = \pi$

Key points: $(-\pi, 0), \left(-\dfrac{\pi}{2}, -3\right), (0, -6), \left(\dfrac{\pi}{2}, -3\right), (\pi, 0)$

55. $y = \dfrac{2}{3}\cos\left(\dfrac{x}{2} - \dfrac{\pi}{4}\right); \ a = \dfrac{2}{3}, \ b = \dfrac{1}{2}, \ c = \dfrac{\pi}{4}$

Period: 4π

Amplitude: $\dfrac{2}{3}$

Shift: $\dfrac{x}{2} - \dfrac{\pi}{4} = 0$ and $\dfrac{x}{2} - \dfrac{\pi}{4} = 2\pi$

$\qquad\quad x = \dfrac{\pi}{2} \qquad\qquad x = \dfrac{9\pi}{2}$

Key points: $\left(\dfrac{\pi}{2}, \dfrac{2}{3}\right), \left(\dfrac{3\pi}{2}, 0\right), \left(\dfrac{5\pi}{2}, \dfrac{-2}{3}\right), \left(\dfrac{7\pi}{2}, 0\right), \left(\dfrac{9\pi}{2}, \dfrac{2}{3}\right)$

57. $y = -2\sin(4x + \pi)$

59. $y = \cos\left(2\pi x - \dfrac{\pi}{2}\right) + 1$

61. $y = -0.1 \sin\left(\dfrac{\pi x}{10} + \pi\right)$

63. $f(x) = a \cos x + d$

Amplitude: $\dfrac{1}{2}[3 - (-1)] = 2 \implies a = 2$

Vertical shift one unit upward of
$g(x) = 2 \cos x \implies d = 1$. Thus, $f(x) = 2 \cos x + 1$.

65. $f(x) = a \cos x + d$

Amplitude: $\dfrac{1}{2}[8 - 0] = 4$

Since $f(x)$ is the graph of $g(x) = 4 \cos x$ reflected in the
x-axis and shifted vertically four units upward, we have
$a = -4$ and $d = 4$. Thus, $f(x) = -4 \cos x + 4$.

67. $y = a \sin(bx - c)$

Amplitude: $|a| = |3|$ Since the graph is reflected in the
x-axis, we have $a = -3$.

Period: $\dfrac{2\pi}{b} = \pi \implies b = 2$

Phase shift: $c = 0$

Thus, $y = -3 \sin 2x$.

69. $y = a \sin(bx - c)$

Amplitude: $a = 2$

Period: $2\pi \implies b = 1$

Phase shift: $bx - c = 0$ when $x = -\dfrac{\pi}{4}$

$$(1)\left(\dfrac{-\pi}{4}\right) - c = 0 \implies c = -\dfrac{\pi}{4}$$

Thus, $y = 2 \sin\left(x + \dfrac{\pi}{4}\right)$.

71. $y_1 = \sin x$

$y_2 = -\dfrac{1}{2}$

In the interval $[-2\pi, 2\pi]$,

$\sin x = -\dfrac{1}{2}$ when $x = -\dfrac{5\pi}{6}, -\dfrac{\pi}{6}, \dfrac{7\pi}{6}, \dfrac{11\pi}{6}$.

73. $y = 0.85 \sin \dfrac{\pi t}{3}$

(a) Time for one cycle $= \dfrac{2\pi}{\pi/3} = 6$ sec

(b) Cycles per min $= \dfrac{60}{6} = 10$ cycles per min

(c) Amplitude: 0.85; Period: 6

Key points: $(0, 0), \left(\dfrac{3}{2}, 0.85\right), (3, 0), \left(\dfrac{9}{2}, -0.85\right), (6, 0)$

75. $y = 0.001 \sin 880\pi t$

(a) Period: $\dfrac{2\pi}{880\pi} = \dfrac{1}{440}$ seconds

(b) $f = \dfrac{1}{p} = 440$ cycles per second

77. (a) $a = \frac{1}{2}[\text{high} - \text{low}] = \frac{1}{2}[83.5 - 29.6] = 26.95$

$p = 2[\text{high time} - \text{low time}] = 2[7 - 1] = 12$

$b = \frac{2\pi}{p} = \frac{2\pi}{12} = \frac{\pi}{6}$

$\frac{c}{b} = 7 \implies c = 7\left(\frac{\pi}{6}\right) \approx 3.67$

$d = \frac{1}{2}[\text{high} + \text{low}] = \frac{1}{2}[83.5 + 29.6] = 56.55$

$C(t) = 56.55 + 26.95 \cos\left(\frac{\pi t}{6} - 3.67\right)$

(b)

The model is a good fit.

(c)

The model is a good fit.

(d) Tallahassee average maximum: 77.90°

Chicago average maximum: 56.55°

The constant term, d, gives the average maximum temperature.

(e) The period for both models is $\frac{2\pi}{\pi/6} = 12$ months.

This is as we expected since one full period is one year.

(f) Chicago has the greater variability in temperature throughout the year. The amplitude, a, determines this variability since it is $\frac{1}{2}[\text{high temp} - \text{low temp}]$.

79. $C = 30.3 + 21.6 \sin\left(\frac{2\pi t}{365} + 10.9\right)$

(a) Period $= \frac{2\pi}{\dfrac{2\pi}{365}} = 365$

Yes, this is what is expected because there are 365 days in a year.

(b) The average daily fuel consumption is given by the amount of the vertical shift (from 0) which is given by the constant 30.3.

(c)

The consumption exceeds 40 gallons per day when $124 < x < 252$.

81. False. The graph of $\sin(x + 2\pi)$ is the graph of $\sin(x)$ translated to the *left* by one period, and the graphs are indeed identical.

83. True.

Since $\cos x = \sin\left(x + \frac{\pi}{2}\right)$, $y = -\cos x = -\sin\left(x + \frac{\pi}{2}\right)$, and so is a reflection in the x-axis of $y = \sin\left(x + \frac{\pi}{2}\right)$.

85.

Since the graphs are the same, the conjecture is that

$\sin(x) = \cos\left(x - \frac{\pi}{2}\right)$.

87. (a)

The graphs are nearly the same for $-\dfrac{\pi}{2} < x < \dfrac{\pi}{2}$.

(b)

The graphs are nearly the same for $-\dfrac{\pi}{2} < x < \dfrac{\pi}{2}$.

(c) $\sin x \approx x - \dfrac{x^3}{3!} + \dfrac{x^5}{5!} - \dfrac{x^7}{7!}$

$\cos x \approx 1 - \dfrac{x^2}{2!} + \dfrac{x^4}{4!} - \dfrac{x^6}{6!}$

The graphs now agree over a wider range, $-\dfrac{3\pi}{4} < x < \dfrac{3\pi}{4}$.

89. $\log_{10} \sqrt{x-2} = \log_{10}(x-2)^{1/2} = \dfrac{1}{2}\log_{10}(x-2)$

91. $\ln \dfrac{t^3}{t-1} = \ln t^3 - \ln(t-1) = 3\ln t - \ln(t-1)$

93. $\dfrac{1}{2}(\log_{10} x + \log_{10} y) = \dfrac{1}{2}\log_{10}(xy)$
$= \log_{10}\sqrt{xy}$

95. $\ln 3x - 4\ln y = \ln 3x - \ln y^4$
$= \ln\left(\dfrac{3x}{y^4}\right)$

97. Answers will vary.

Section 6.5   Graphs of Other Trigonometric Functions

■ You should be able to graph

$y = a\tan(bx - c)$ $y = a\cot(bx - c)$

$y = a\sec(bx - c)$ $y = a\csc(bx - c)$

■ When graphing $y = a\sec(bx - c)$ or $y = a\csc(bx - c)$ you should first graph $y = a\cos(bx - c)$ or $y = a\sin(bx - c)$ because

  (a) The x-intercepts of sine and cosine are the vertical asymptotes of cosecant and secant.

  (b) The maximums of sine and cosine are the local minimums of cosecant and secant.

  (c) The minimums of sine and cosine are the local maximums of cosecant and secant.

■ You should be able to graph using a damping factor.

Vocabulary Check

1. vertical

3. damping

5. $x \neq n\pi$

7. 2π

2. reciprocal

4. π

6. $(-\infty, -1] \cup [1, \infty)$

1. $y = \sec 2x$

Period: $\dfrac{2\pi}{2} = \pi$

Matches graph (e).

3. $y = \dfrac{1}{2}\cot \pi x$

Period: $\dfrac{\pi}{\pi} = 1$

Matches graph (a).

5. $y = \dfrac{1}{2}\sec \dfrac{\pi x}{2}$

Period: $\dfrac{2\pi}{b} = \dfrac{2\pi}{\frac{\pi}{2}} = 4$

Asymptotes: $x = -1, x = 1$

Matches graph (f).

7. $y = \dfrac{1}{3}\tan x$

Period: π

Two consecutive asymptotes:

$x = -\dfrac{\pi}{2}$ and $x = \dfrac{\pi}{2}$

x	$-\dfrac{\pi}{4}$	0	$\dfrac{\pi}{4}$
y	$-\dfrac{1}{3}$	0	$\dfrac{1}{3}$

9. $y = \tan 3x$

Period: $\dfrac{\pi}{3}$

Two consecutive asymptotes:

$3x = -\dfrac{\pi}{2} \Rightarrow x = -\dfrac{\pi}{6}$

$3x - \dfrac{\pi}{2} \Rightarrow x = \dfrac{\pi}{6}$

x	$-\dfrac{\pi}{12}$	0	$\dfrac{\pi}{12}$
y	-1	0	1

11. $y = -\dfrac{1}{2}\sec x$

Period: 2π

Two consecutive asymptotes:

$x = -\dfrac{\pi}{2}, x = \dfrac{\pi}{2}$

x	$-\dfrac{\pi}{3}$	0	$\dfrac{\pi}{3}$
y	-1	$-\dfrac{1}{2}$	-1

13. $y = \csc \pi x$

Period: $\dfrac{2\pi}{\pi} = 2$

Two consecutive asymptotes:

$x = 0, x = 1$

x	$\dfrac{1}{6}$	$\dfrac{1}{2}$	$\dfrac{5}{6}$
y	2	1	2

15. $y = \sec \pi x - 1$

Period: $\dfrac{2\pi}{\pi} = 2$

Two consecutive asymptotes:

$x = -\dfrac{1}{2}, x = \dfrac{1}{2}$

x	$-\dfrac{1}{3}$	0	$\dfrac{1}{3}$
y	1	0	1

17. $y = \csc \dfrac{x}{2}$

Period: $\dfrac{2\pi}{1/2} = 4\pi$

Two consecutive asymptotes:

$x = 0, x = 2\pi$

x	$\dfrac{\pi}{3}$	π	$\dfrac{5\pi}{3}$
y	2	1	2

19. $y = \cot \dfrac{x}{2}$

Period: $\dfrac{\pi}{1/2} = 2\pi$

Two consecutive
asymptotes:

$\dfrac{x}{2} = 0 \implies x = 0$

$\dfrac{x}{2} = \pi \implies x = 2\pi$

x	$\dfrac{\pi}{2}$	π	$\dfrac{3\pi}{2}$
y	1	0	-1

21. $y = \dfrac{1}{2} \sec 2x$

Period: $\dfrac{2\pi}{2} = \pi$

x	$-\dfrac{\pi}{6}$	0	$\dfrac{\pi}{6}$
y	1	$\dfrac{1}{2}$	1

23. $y = \tan \dfrac{\pi x}{4}$

Period: $\dfrac{\pi}{\pi/4} = 4$

Two consecutive asymptotes:

$\dfrac{\pi x}{4} = -\dfrac{\pi}{2} \implies x = -2$

$\dfrac{\pi x}{4} = \dfrac{\pi}{2} \implies x = 2$

x	-1	0	1
y	-1	0	1

25. $y = \csc(\pi - x)$

Period: 2π

Two consecutive
asymptotes:
$x = 0, x = \pi$

x	$\dfrac{\pi}{6}$	$\dfrac{\pi}{2}$	$\dfrac{5\pi}{6}$
y	2	1	2

27. $y = 2 \sec(x + \pi)$

Period: 2π

Two consecutive
asymptotes:

$x = -\dfrac{\pi}{2}, x = \dfrac{\pi}{2}$

x	$-\dfrac{\pi}{3}$	0	$\dfrac{\pi}{3}$
y	-4	-2	-4

29. $y = \dfrac{1}{4} \csc\left(x + \dfrac{\pi}{4}\right)$

Period: 2π

Two consecutive
asymptotes:

$x = -\dfrac{\pi}{4}, x = \dfrac{3\pi}{4}$

x	$-\dfrac{\pi}{12}$	$\dfrac{\pi}{4}$	$\dfrac{7\pi}{12}$
y	$\dfrac{1}{2}$	$\dfrac{1}{4}$	$\dfrac{1}{2}$

31. $y = \tan \dfrac{x}{3}$

33. $y = -2 \sec 4x = \dfrac{-2}{\cos 4x}$

35. $y = \tan\left(x - \dfrac{\pi}{4}\right)$

37. $y = -\csc(4x - \pi)$

$y = \dfrac{-1}{\sin(4x - \pi)}$

39. $y = 0.1 \tan\left(\dfrac{\pi x}{4} + \dfrac{\pi}{4}\right)$

41. $\tan x = 1$

$x = -\dfrac{7\pi}{4}, -\dfrac{3\pi}{4}, \dfrac{\pi}{4}, \dfrac{5\pi}{4}$

43. $\cot x = -\dfrac{\sqrt{3}}{3}$

$x = -\dfrac{4\pi}{3}, -\dfrac{\pi}{3}, \dfrac{2\pi}{3}, \dfrac{5\pi}{3}$

45. $\sec x = -2$

$x = \pm\dfrac{2\pi}{3}, \pm\dfrac{4\pi}{3}$

47. $\csc x = \sqrt{2}$

$x = -\dfrac{7\pi}{4}, -\dfrac{5\pi}{4}, \dfrac{\pi}{4}, \dfrac{3\pi}{4}$

49. $f(x) = \sec x = \dfrac{1}{\cos x}$

$f(-x) = \sec(-x)$

$\quad = \dfrac{1}{\cos(-x)}$

$\quad = \dfrac{1}{\cos x}$

$\quad = f(x)$

Thus, $f(x) = \sec x$ is an even function and the graph has y-axis symmetry.

51. $f(x) = 2\sin x$

$g(x) = \dfrac{1}{2}\csc x$

(a)

(b) $f > g$ on the interval, $\dfrac{\pi}{6} < x < \dfrac{5\pi}{6}$

(c) As $x \to \pi$, $f(x) = 2\sin x \to 0$ and

$g(x) = \dfrac{1}{2}\csc x \to \pm\infty$ since $g(x)$ is the reciprocal of $f(x)$.

53. $y_1 = \sin x \csc x$ and $y_2 = 1$

$\sin x \csc x = \sin x\left(\dfrac{1}{\sin x}\right) = 1, \sin x \neq 0$

The expressions are equivalent except when $\sin x = 0$ and y_1 is undefined.

55. $y_1 = \dfrac{\cos x}{\sin x}$ and $y_2 = \cot x = \dfrac{1}{\tan x}$

$\cot x = \dfrac{\cos x}{\sin x}$

The expressions are equivalent.

57. $f(x) = |x \cos x|$

As $x \to 0, f(x) \to 0$ and $f(x) > 0$.

Matches graph (d).

59. $g(x) = |x| \sin x$

As $x \to 0, g(x) \to 0$ and $g(x)$ is odd.

Matches graph (b).

61. $f(x) = \sin x + \cos\left(x + \dfrac{\pi}{2}\right), g(x) = 0$

$f(x) = g(x)$, The graph is the line $y = 0$.

63. $f(x) = \sin^2 x, g(x) = \dfrac{1}{2}(1 - \cos 2x)$

$f(x) = g(x)$

65. $g(x) = e^{-x^2/2} \sin x$

$-e^{-x^2/2} \le g(x) \le e^{-x^2/2}$

The damping factor is $y = e^{-x^2/2}$.

As $x \to \infty, g(x) \to 0$.

67. $f(x) = 2^{-x/4} \cos \pi x$

$-2^{-x/4} \le f(x) \le 2^{-x/4}$

Damping factor: $y = 2^{-x/4}$.

As $x \to \infty, f(x) \to 0$.

69. $y = \dfrac{6}{x} + \cos x, \ x > 0$

As $x \to 0, y \to \infty$.

71. $g(x) = \dfrac{\sin x}{x}$

As $x \to 0, \ g(x) \to 1$.

73. $f(x) = \sin \dfrac{1}{x}$

As $x \to 0, f(x)$ oscillates between -1 and 1.

75. $\tan x = \dfrac{7}{d}$

$d = \dfrac{7}{\tan x} = 7 \cot x$

77. $C = 5000 + 2000 \sin \dfrac{\pi t}{12}, R = 25{,}000 + 15{,}000 \cos \dfrac{\pi t}{12}$

(a)

(b) As the predator population increases, the number of prey decreases. When the number of prey is small, the number of predators decreases.

(c) The period for both C and R is:

$$p = \dfrac{2\pi}{\pi/12} = 24 \text{ months}$$

When the prey population is highest, the predator population is increasing most rapidly.
When the prey population is lowest, the predator population is decreasing most rapidly.
When the predator population is lowest, the prey population is increasing most rapidly.
When the predator population is highest, the prey population is decreasing most rapidly.

In addition, weather, food sources for the prey, hunting, all affect the populations of both the predator and the prey.

79. $H(t) = 54.33 - 20.38 \cos \dfrac{\pi t}{6} - 15.69 \sin \dfrac{\pi t}{6}$

$L(t) = 39.36 - 15.70 \cos \dfrac{\pi t}{6} - 14.16 \sin \dfrac{\pi t}{6}$

(a) Period of $\cos \dfrac{\pi t}{6}$: $\dfrac{2\pi}{\pi/6} = 12$

Period of $\sin \dfrac{\pi t}{6}$: $\dfrac{2\pi}{\pi/6} = 12$

Period of $H(t)$: 12 months

Period of $L(t)$: 12 months

(b) From the graph, it appears that the greatest difference between high and low temperatures occurs in summer. The smallest difference occurs in winter.

(c) The highest high and low temperatures appear to occur around the middle of July, roughly one month after the time when the sun is northernmost in the sky.

81. True. Since

$$y = \csc x = \dfrac{1}{\sin x},$$

for a given value of x, the y-coordinate of $\csc x$ is the reciprocal of the y-coordinate of $\sin x$.

83. As $x \to \dfrac{\pi}{2}$ from the left, $f(x) = \tan x \to \infty$.

As $x \to \dfrac{\pi}{2}$ from the right, $f(x) = \tan x \to -\infty$.

85. $f(x) = x - \cos x$

(a)

The zero between 0 and 1 occurs at $x \approx 0.7391$.

(b) $x_n = \cos(x_{n-1})$

$x_0 = 1$

$x_1 = \cos 1 \approx 0.5403$

$x_2 = \cos 0.5403 \approx 0.8576$

$x_3 = \cos 0.8576 \approx 0.6543$

$x_4 = \cos 0.6543 \approx 0.7935$

$x_5 = \cos 0.7935 \approx 0.7014$

$x_6 = \cos 0.7014 \approx 0.7640$

$x_7 = \cos 0.7640 \approx 0.7221$

$x_8 = \cos 0.7221 \approx 0.7504$

$x_9 = \cos 0.7504 \approx 0.7314$

$\vdots$

This sequence appears to be approaching the zero of f: $x \approx 0.7391$.

87. $y_1 = \sec x$

$y_2 = 1 + \dfrac{x^2}{2!} + \dfrac{5x^4}{4!}$

The graph appears to coincide on the interval $-1.1 \le x \le 1.1$.

89. $e^{2x} = 54$

$2x = \ln 54$

$x = \dfrac{\ln 54}{2} \approx 1.994$

91. $\dfrac{300}{1 + e^{-x}} = 100$

$\dfrac{300}{100} = 1 + e^{-x}$

$3 = 1 + e^{-x}$

$2 = e^{-x}$

$\ln 2 = -x$

$x = -\ln 2 \approx -0.693$

93. $\ln(3x - 2) = 73$

$3x - 2 = e^{73}$

$3x = 2 + e^{73}$

$x = \dfrac{2 + e^{73}}{3}$

$\approx 1.684 \times 10^{31}$

95. $\ln(x^2 + 1) = 3.2$

$x^2 + 1 = e^{3.2}$

$x^2 = e^{3.2} - 1$

$x = \pm\sqrt{e^{3.2} - 1} \approx \pm 4.851$

97. $\log_8 x + \log_8(x - 1) = \frac{1}{3}$

$\log_8[x(x - 1)] = \frac{1}{3}$

$x(x - 1) = 8^{1/3}$

$x^2 - x = 2$

$x^2 - x - 2 = 0$

$(x - 2)(x + 1) = 0$

$x = 2, -1$

$x = -1$ is extraneous (not in the domain of $\log_8 x$) so only $x = 2$ is a solution.

Section 6.6 Inverse Trigonometric Functions

■ You should know the definitions, domains, and ranges of $y = \arcsin x$, $y = \arccos x$, and $y = \arctan x$.

Function	Domain	Range
$y = \arcsin x \Rightarrow x = \sin y$	$-1 \leq x \leq 1$	$-\dfrac{\pi}{2} \leq y \leq \dfrac{\pi}{2}$
$y = \arccos x \Rightarrow x = \cos y$	$-1 \leq x \leq 1$	$0 \leq y \leq \pi$
$y = \arctan x \Rightarrow x = \tan y$	$-\infty < x < \infty$	$-\dfrac{\pi}{2} < x < \dfrac{\pi}{2}$

■ You should know the inverse properties of the inverse trigonometric functions.

$\sin(\arcsin x) = x$ and $\arcsin(\sin y) = y$, $-\dfrac{\pi}{2} \leq y \leq \dfrac{\pi}{2}$

$\cos(\arccos x) = x$ and $\arccos(\cos y) = y$, $0 \leq y \leq \pi$

$\tan(\arctan x) = x$ and $\arctan(\tan y) = y$, $-\dfrac{\pi}{2} < y < \dfrac{\pi}{2}$

■ You should be able to use the triangle technique to convert trigonometric functions of inverse trigonometric functions into algebraic expressions.

Vocabulary Check

Function	Alternative Notation	Domain	Range
1. $y = \arcsin x$	$y = \sin^{-1} x$	$-1 \leq x \leq 1$	$-\dfrac{\pi}{2} \leq y \leq \dfrac{\pi}{2}$
2. $y = \arccos x$	$y = \cos^{-1} x$	$-1 \leq x \leq 1$	$0 \leq y \leq \pi$
3. $y = \arctan x$	$y = \tan^{-1} x$	$-\infty < x < \infty$	$-\dfrac{\pi}{2} < y < \dfrac{\pi}{2}$

1. $y = \arcsin \dfrac{1}{2} \Rightarrow \sin y = \dfrac{1}{2}$ for $-\dfrac{\pi}{2} \leq y \leq \dfrac{\pi}{2} \Rightarrow y = \dfrac{\pi}{6}$

3. $y = \arccos \dfrac{1}{2} \Rightarrow \cos y = \dfrac{1}{2}$ for $0 \leq y \leq \pi \Rightarrow y = \dfrac{\pi}{3}$

5. $y = \arctan \dfrac{\sqrt{3}}{3} \Rightarrow \tan y = \dfrac{\sqrt{3}}{3}$ for $-\dfrac{\pi}{2} < y < \dfrac{\pi}{2} \Rightarrow y = \dfrac{\pi}{6}$

7. $y = \arccos\left(-\dfrac{\sqrt{3}}{2}\right) \Rightarrow \cos y = -\dfrac{\sqrt{3}}{2}$ for $0 \leq y \leq \pi \Rightarrow y = \dfrac{5\pi}{6}$

9. $y = \arctan(-\sqrt{3}) \Rightarrow \tan y = -\sqrt{3}$ for $-\dfrac{\pi}{2} < y < \dfrac{\pi}{2} \Rightarrow y = -\dfrac{\pi}{3}$

11. $y = \arccos\left(-\dfrac{1}{2}\right) \Rightarrow \cos y = -\dfrac{1}{2}$ for $0 \leq y \leq \pi \Rightarrow y = \dfrac{2\pi}{3}$

13. $y = \arcsin \dfrac{\sqrt{3}}{2} \Rightarrow \sin y = \dfrac{\sqrt{3}}{2}$ for

$-\dfrac{\pi}{2} \le y \le \dfrac{\pi}{2} \Rightarrow y = \dfrac{\pi}{3}$

15. $y = \arctan 0 \Rightarrow \tan y = 0$ for $-\dfrac{\pi}{2} < y < \dfrac{\pi}{2} \Rightarrow y = 0$

17. $f(x) = \sin x$

$g(x) = \arcsin x$

$y = x$

19. $\arccos 0.28 = \cos^{-1} 0.28 \approx 1.29$

21. $\arcsin(-0.75) = \sin^{-1}(-0.75) \approx -0.85$

23. $\arctan(-3) = \tan^{-1}(-3) \approx -1.25$

25. $\arcsin 0.31 = \sin^{-1} 0.31 \approx 0.32$

27. $\arccos(-0.41) = \cos^{-1}(-0.41) \approx 1.99$

29. $\arctan 0.92 = \tan^{-1} 0.92 \approx 0.74$

31. $\arcsin\left(\dfrac{3}{4}\right) = \sin^{-1}(0.75) \approx 0.85$

33. $\arctan\left(\dfrac{7}{2}\right) = \tan^{-1}(3.5) \approx 1.29$

35. This is the graph of $y = \arctan x$. The coordinates are

$\left(-\sqrt{3}, -\dfrac{\pi}{3}\right), \left(-\dfrac{\sqrt{3}}{3}, -\dfrac{\pi}{6}\right)$, and $\left(1, \dfrac{\pi}{4}\right)$.

37. $\tan \theta = \dfrac{x}{4}$

$\theta = \arctan \dfrac{x}{4}$

39. $\sin \theta = \dfrac{x + 2}{5}$

$\theta = \arcsin\left(\dfrac{x + 2}{5}\right)$

41. $\cos \theta = \dfrac{x + 3}{2x}$

$\theta = \arccos\left(\dfrac{x + 3}{2x}\right)$

43. $\sin(\arcsin 0.3) = 0.3$

45. $\cos[\arccos(-0.1)] = -0.1$

47. $\arcsin(\sin 3\pi) = \arcsin(0) = 0$

Note: 3π is not in the range of the arcsine function.

49. Let $y = \arctan \dfrac{3}{4}$.

$\tan y = \dfrac{3}{4}, \ 0 < y < \dfrac{\pi}{2}$

and $\sin y = \dfrac{3}{5}$.

51. Let $y = \arctan 2$.

$\tan y = 2 = \dfrac{2}{1}, \ 0 < y < \dfrac{\pi}{2}$

and $\cos y = \dfrac{1}{\sqrt{5}} = \dfrac{\sqrt{5}}{5}$.

53. Let $y = \arcsin \dfrac{5}{13}$.

$\sin y = \dfrac{5}{13}, \ 0 < y < \dfrac{\pi}{2}$

and $\cos y = \dfrac{12}{13}$.

55. Let $y = \arctan\left(-\dfrac{3}{5}\right)$.

$\tan y = -\dfrac{3}{5}, \ -\dfrac{\pi}{2} < y < 0$

and $\sec y = \dfrac{\sqrt{34}}{5}$.

57. Let $y = \arccos\left(-\dfrac{2}{3}\right)$.

$\cos y = -\dfrac{2}{3}, \dfrac{\pi}{2} < y < \pi$

and $\sin y = \dfrac{\sqrt{5}}{3}$.

59. Let $y = \arctan x$.

$\tan y = x = \dfrac{x}{1}$

and $\cot y = \dfrac{1}{x}$.

61. Let $y = \arcsin(2x)$.

$\sin y = 2x = \dfrac{2x}{1}$

and $\cos y = \sqrt{1 - 4x^2}$.

63. Let $y = \arccos x$.

$\cos y = x = \dfrac{x}{1}$

and $\sin y = \sqrt{1 - x^2}$.

65. Let $y = \arccos\left(\dfrac{x}{3}\right)$.

$\cos y = \dfrac{x}{3}$

and $\tan y = \dfrac{\sqrt{9 - x^2}}{x}$.

67. Let $y = \arctan \dfrac{x}{\sqrt{2}}$.

$\tan y = \dfrac{x}{\sqrt{2}}$

and $\csc y = \dfrac{\sqrt{x^2 + 2}}{x}$.

69. $f(x) = \sin(\arctan 2x)$, $g(x) = \dfrac{2x}{\sqrt{1 + 4x^2}}$

They are equal. Let $y = \arctan 2x$.

$\tan y = 2x = \dfrac{2x}{1}$

and $\sin y = \dfrac{2x}{\sqrt{1 + 4x^2}}$.

$g(x) = \dfrac{2x}{\sqrt{1 + 4x^2}} = f(x)$

The graph has horizontal asymptotes at $y = \pm 1$.

71. Let $y = \arctan \dfrac{9}{x}$.

$\tan y = \dfrac{9}{x}$ and $\sin y = \dfrac{9}{\sqrt{x^2 + 81}}, x > 0; \dfrac{-9}{\sqrt{x^2 + 81}}, x < 0$.

Thus,

$\arcsin y = \dfrac{9}{\sqrt{x^2 + 81}}, x > 0$;

$\arcsin y = \dfrac{-9}{\sqrt{x^2 + 81}}, x < 0$.

73. Let $y = \arccos \dfrac{3}{\sqrt{x^2 - 2x + 10}}$. Then,

$\cos y = \dfrac{3}{\sqrt{x^2 - 2x + 10}} = \dfrac{3}{\sqrt{(x - 1)^2 + 9}}$

and $\sin y = \dfrac{|x - 1|}{\sqrt{(x - 1)^2 + 9}}$.

Thus, $y = \arcsin \dfrac{|x - 1|}{\sqrt{x^2 - 2x + 10}}$.

75. $y = 2 \arccos x$

Domain: $-1 \le x \le 1$

Range: $0 \le y \le 2\pi$

This is the graph of $f(x) = \arccos x$ with a factor of 2.

77. $f(x) = \arcsin(x - 1)$

Domain: $0 \le x \le 2$

Range: $-\dfrac{\pi}{2} \le y \le \dfrac{\pi}{2}$

This is the graph of $g(x) = \arcsin(x)$ shifted one unit to the right.

79. $f(x) = \arctan 2x$

Domain: all real numbers

Range: $-\dfrac{\pi}{2} < y < \dfrac{\pi}{2}$

This is the graph of $g(x) = \arctan(x)$ with a horizontal stretch of a factor of 2.

81. $h(v) = \tan(\arccos v) = \dfrac{\sqrt{1 - v^2}}{v}$

Domain: $-1 \le v \le 1, v \ne 0$

Range: all real numbers

83. $f(x) = 2\arccos(2x)$

85. $f(x) = \arctan(2x - 3)$

87. $f(x) = \pi - \arcsin\left(\tfrac{2}{3}\right) \approx 2.412$

89. $f(t) = 3\cos 2t + 3\sin 2t = \sqrt{3^2 + 3^2}\,\sin\left(2t + \arctan\dfrac{3}{3}\right)$

$$= 3\sqrt{2}\,\sin(2t + \arctan 1)$$

$$= 3\sqrt{2}\,\sin\left(2t + \dfrac{\pi}{4}\right)$$

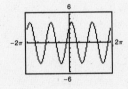

The graphs are the same and implies that the identity is true.

91. (a) $\sin\theta = \dfrac{5}{s}$

$\theta = \arcsin\dfrac{5}{s}$

(b) $s = 40$: $\theta = \arcsin\dfrac{5}{40} \approx 0.13$

$s = 20$: $\theta = \arcsin\dfrac{5}{20} \approx 0.25$

93. $\beta = \arctan\dfrac{3x}{x^2 + 4}$

(a)

(b) β is maximum when $x = 2$ feet.

(c) The graph has a horizontal asymptote at $\beta = 0$. As x increases, β decreases.

95.

(a) $\tan \theta = \dfrac{20}{41}$

$\theta = \arctan\left(\dfrac{20}{41}\right) \approx 26.0°$

(b) $\tan 26° = \dfrac{h}{50}$

$h = 50 \tan 26° \approx 24.39$ feet

99. False.

$\dfrac{5\pi}{4}$ is not in the range of the arctangent function.

$\arctan 1 = \dfrac{\pi}{4}$

97. (a) $\tan \theta = \dfrac{x}{20}$

$\theta = \arctan \dfrac{x}{20}$

(b) $x = 5$: $\theta = \arctan \dfrac{5}{20} \approx 14.0°$

$x = 12$: $\theta = \arctan \dfrac{12}{20} \approx 31.0°$

101. $y = \operatorname{arccot} x$ if and only if $\cot y = x$.

Domain: $-\infty < x < \infty$

Range: $0 < x < \pi$

103. $y = \operatorname{arccsc} x$ if and only if $\csc y = x$.

Domain: $(-\infty, -1] \cup [1, \infty)$

Range: $\left[-\dfrac{\pi}{2}, 0\right) \cup \left(0, \dfrac{\pi}{2}\right]$

105. Area $= \arctan b - \arctan a$

(a) $a = 0, b = 1$

Area $= \arctan 1 - \arctan 0 = \dfrac{\pi}{4} - 0 = \dfrac{\pi}{4}$

(b) $a = -1, b = 1$

Area $= \arctan 1 - \arctan(-1)$

$= \dfrac{\pi}{4} - \left(-\dfrac{\pi}{4}\right) = \dfrac{\pi}{2}$

(c) $a = 0, b = 3$

Area $= \arctan 3 - \arctan 0 \approx 1.25 - 0 = 1.25$

(d) $a = -1, b = 3$

Area $= \arctan 3 - \arctan(-1)$

$\approx 1.25 - \left(-\dfrac{\pi}{4}\right) \approx 2.03$

107. $f(x) = \sin(x), f^{-1}(x) = \arcsin(x)$

(a) $f \cdot f^{-1} = \sin(\arcsin x)$ $f^{-1} \cdot f = \arcsin(\sin x)$

(b) The graphs coincide with the graph of $y = x$ only for certain values of x.

$f \cdot f^{-1} = x$ over its entire domain, $-1 \leq x \leq 1$.

$f^{-1} \cdot f = x$ over the region $-\dfrac{\pi}{2} \leq x \leq \dfrac{\pi}{2}$, corresponding to the region where $\sin x$ is one-to-one and thus has an inverse.

109. $(8.2)^{3.4} \approx 1279.284$

111. $(1.1)^{50} \approx 117.391$

113.
$$\sin \theta = \frac{3}{4} = \frac{\text{opp}}{\text{hyp}}$$
$$(\text{adj})^2 + (3)^2 = (4)^2$$
$$(\text{adj})^2 + 9 = 16$$
$$(\text{adj})^2 = 7$$
$$\text{adj} = \sqrt{7}$$
$$\cos \theta = \frac{\sqrt{7}}{4}$$
$$\tan \theta = \frac{3}{\sqrt{7}} = \frac{3\sqrt{7}}{7}$$
$$\cot \theta = \frac{\sqrt{7}}{3}$$
$$\sec \theta = \frac{4}{\sqrt{7}} = \frac{4\sqrt{7}}{7}$$
$$\csc \theta = \frac{4}{3}$$

115.
$$\cos \theta = \frac{5}{6} = \frac{\text{adj}}{\text{hyp}}$$
$$(\text{opp})^2 + (5)^2 = (6)^2$$
$$(\text{opp})^2 + 25 = 36$$
$$(\text{opp})^2 = 11$$
$$\text{opp} = \sqrt{11}$$
$$\sin \theta = \frac{\sqrt{11}}{6}$$
$$\tan \theta = \frac{\sqrt{11}}{5}$$
$$\cot \theta = \frac{5}{\sqrt{11}} = \frac{5\sqrt{11}}{11}$$
$$\sec \theta = \frac{6}{5}$$
$$\csc \theta = \frac{6}{\sqrt{11}} = \frac{6\sqrt{11}}{11}$$

117. Let x = the number of people presently in the group: Each person's share is now $\frac{250{,}000}{x}$. If two more join the group, each person's share would then be $\frac{250{,}000}{x+2}$.

$$\begin{array}{l}\text{Share per person with}\\ \text{two more people}\end{array} = \begin{array}{l}\text{Original share}\\ \text{per person}\end{array} - 6250$$

$$\frac{250{,}000}{x+2} = \frac{250{,}000}{x} - 6250$$
$$250{,}000x = 250{,}000(x+2) - 6250x(x+2)$$
$$250{,}000x = 250{,}000x + 500{,}000 - 6250x^2 - 12500x$$
$$6250x^2 + 12500x - 500{,}000 = 0$$
$$6250(x^2 + 2x - 80) = 0$$
$$6250(x + 10)(x - 8) = 0$$
$$x = -10 \quad \text{or} \quad x = 8$$

$x = -10$ is not possible.

There were 8 people in the original group.

119. (a) $A = 15{,}000\left(1 + \frac{0.035}{4}\right)^{(4)(10)} \approx \$21{,}253.63$

(b) $A = 15{,}000\left(1 + \frac{0.035}{12}\right)^{(12)(10)} \approx \$21{,}275.17$

(c) $A = 15{,}000\left(1 + \frac{0.035}{365}\right)^{(365)(10)} \approx \$21{,}285.66$

(d) $A = 15{,}000e^{(0.035)(10)} \approx \$21{,}286.01$

Section 6.7   Applications and Models

- You should be able to solve right triangles.
- You should be able to solve right triangle applications.
- You should be able to solve applications of simple harmonic motion.

Vocabulary Check

1. elevation; depression

2. bearing

3. harmonic motion

---

1. Given: $A = 20°$, $b = 10$

$$\tan A = \frac{a}{b} \implies a = b \tan A = 10 \tan 20° \approx 3.64$$

$$\cos A = \frac{b}{c} \implies c = \frac{b}{\cos A} = \frac{10}{\cos 20°} \approx 10.64$$

$$B = 90° - 20° = 70°$$

3. Given: $B = 71°$, $b = 24$

$$\tan B = \frac{b}{a} \implies a = \frac{b}{\tan B} = \frac{24}{\tan 71°} \approx 8.26$$

$$\sin B = \frac{b}{c} \implies c = \frac{b}{\sin B} = \frac{24}{\sin 71°} \approx 25.38$$

$$A = 90° - 71° = 19°$$

5. Given: $a = 6$, $b = 10$

$$c^2 = a^2 + b^2 \implies c = \sqrt{36 + 100}$$
$$= 2\sqrt{34} \approx 11.66$$

$$\tan A = \frac{a}{b} = \frac{6}{10} \implies A = \arctan \frac{3}{5} \approx 30.96°$$

$$B = 90° - 30.96° = 59.04°$$

7. Given: $b = 16$, $c = 52$

$$a = \sqrt{52^2 - 16^2}$$
$$= \sqrt{2448} = 12\sqrt{17} \approx 49.48$$

$$\cos A = \frac{16}{52}$$

$$A = \arccos \frac{16}{52} \approx 72.08°$$

$$B = 90° - 72.08° \approx 17.92°$$

9. Given: $A = 12°15'$, $c = 430.5$

$$B = 90° - 12°15' = 77°45'$$

$$\sin 12°15' = \frac{a}{430.5}$$

$$a = 430.5 \sin 12°15' \approx 91.34$$

$$\cos 12°15' = \frac{b}{430.5}$$

$$b = 430.5 \cos 12°15' \approx 420.70$$

11. $\tan \theta = \dfrac{h}{(1/2)b} \implies h = \dfrac{1}{2}b \tan \theta$

$$h = \frac{1}{2}(4) \tan 52° \approx 2.56 \text{ inches}$$

13. $\tan \theta = \dfrac{h}{(1/2)b} \Rightarrow h = \dfrac{1}{2}b \tan \theta$

$h = \dfrac{1}{2}(46) \tan 41° \approx 19.99$ inches

15. $\tan 25° = \dfrac{50}{x}$

$x = \dfrac{50}{\tan 25°}$

≈ 107.2 feet

17. $\sin 80° = \dfrac{h}{20}$

$20 \sin 80° = h$

$h \approx 19.7$ feet

19. (a)

(b) Let the height of the church $= x$ and the height of the church and steeple $= y$. Then,

$\tan 35° = \dfrac{x}{50}$ and $\tan 47° 40' = \dfrac{y}{50}$

$x = 50 \tan 35°$ and $y = 50 \tan 47° 40'$

$h = y - x = 50(\tan 47° 40' - \tan 35°)$.

(c) $h \approx 19.9$ feet

21. $\sin 34° = \dfrac{x}{4000}$

$x = 4000 \sin 34°$

≈ 2236.8 feet

23. (a)

(b) $\tan \theta = \dfrac{12\frac{1}{2}}{17\frac{1}{3}}$

(c) $\theta = \arctan \dfrac{12\frac{1}{2}}{17\frac{1}{3}} \approx 35.8°$

The angle of elevation of the sum is 35.8°.

25. $1200 \text{ feet} + 150 \text{ feet} - 400 \text{ feet} = 950 \text{ feet}$

$5 \text{ miles} = 5 \text{ miles}\left(\dfrac{5280 \text{ feet}}{1 \text{ mile}}\right) = 26{,}400 \text{ feet}$

$\tan \theta = \dfrac{950}{26{,}400}$

$\theta = \arctan\left(\dfrac{950}{26{,}400}\right) \approx 2.06°$

Not drawn to scale

27. $\sin 10.5° = \dfrac{x}{4}$

$x = 4 \sin 10.5°$

≈ 0.73 mile

29. The plane has traveled $1.5(600) = 900$ miles.

$$\sin 38° = \frac{a}{900} \implies a \approx 554 \text{ miles north}$$

$$\cos 38° = \frac{b}{900} \implies b \approx 709 \text{ miles east}$$

31.

(a) $\cos 29° = \dfrac{a}{120} \implies a \approx 104.95$ nautical miles south

$\sin 29° = \dfrac{b}{120} \implies b \approx 58.18$ nautical miles west

(b) $\tan \theta = \dfrac{20 + b}{a} \approx \dfrac{78.18}{104.95} \implies \theta \approx 36.7°$

Bearing: S 36.7° W

Distance: $d \approx \sqrt{104.95^2 + 78.18^2}$

≈ 130.9 nautical miles from port

33. $\theta = 32°, \ \phi = 68°$

(a) $\alpha = 90° - 32° = 58°$

Bearing from A to C: N 58° E

35. $\tan \theta = \dfrac{45}{30} \implies \theta \approx 56.3°$

Bearing: N 56.3° W

(b) $\beta = \theta = 32°$

$\gamma = 90° - \phi = 22°$

$C = \beta + \gamma = 54°$

$\tan C = \dfrac{d}{50} \implies \tan 54°$

$= \dfrac{d}{50} \implies d \approx 68.82 \text{ meters}$

37. $\tan 6.5° = \dfrac{350}{d} \implies d \approx 3071.91 \text{ ft}$

$\tan 4° = \dfrac{350}{D} \implies D \approx 5005.23 \text{ ft}$

Distance between ships: $D - d \approx 1933.3 \text{ ft}$

39. $\tan 57° = \dfrac{a}{x} \implies x = a \cot 57°$

$\tan 16° = \dfrac{a}{x + \frac{55}{6}}$

$\tan 16° = \dfrac{a}{a \cot 57° + \frac{55}{6}}$

$\cot 16° = \dfrac{a \cot 57° + \frac{55}{6}}{a}$

$a \cot 16° - a \cot 57° = \dfrac{55}{6} \implies a \approx 3.23 \text{ miles}$

$\approx 17,054 \text{ ft}$

41. L_1: $3x - 2y = 5 \implies y = \frac{3}{2}x - \frac{5}{2} \implies m_1 = \frac{3}{2}$

L_2: $x + y = 1 \implies y = -x + 1 \implies m_2 = -1$

$\tan \alpha = \left| \dfrac{-1 - \frac{3}{2}}{1 + (-1)\left(\frac{3}{2}\right)} \right| = \left| \dfrac{-\frac{5}{2}}{-\frac{1}{2}} \right| = 5$

$\alpha = \arctan 5 \approx 78.7°$

43. The diagonal of the base has a length of $\sqrt{a^2 + a^2} = \sqrt{2}a$. Now, we have

$\tan \theta = \dfrac{a}{\sqrt{2}a} = \dfrac{1}{\sqrt{2}}$

$\theta = \arctan \dfrac{1}{\sqrt{2}}$

$\theta \approx 35.3°$.

45. $\sin 36° = \dfrac{d}{25} \implies d \approx 14.69$

Length of side: $2d \approx 29.4$ inches

47.

$\cos 30° = \dfrac{b}{r}$

$b = r \cos 30°$

$b = \dfrac{\sqrt{3}r}{2}$

$y = 2b = 2\left(\dfrac{\sqrt{3}r}{2}\right) = \sqrt{3}r$

49.

$\tan 35° = \dfrac{b}{10}$

$b = 10 \tan 35° \approx 7$

$\cos 35° = \dfrac{10}{a}$

$a = \dfrac{10}{\cos 35°} \approx 12.2$

51. $d = 0$ when $t = 0$, $a = 4$, Period $= 2$

Use $d = a \sin \omega t$ since $d = 0$ when $t = 0$.

$\dfrac{2\pi}{\omega} = 2 \implies \omega = \pi$

Thus, $d = 4 \sin(\pi t)$.

53. $d = 3$ when $t = 0$; $a = 3$, Period $= 1.5$

Use $d = a \cos \omega t$ since $d = 3$ when $t = 0$.

$\dfrac{2\pi}{\omega} = 1.5 \implies \omega = \dfrac{4\pi}{3}$

Thus, $d = 3 \cos\left(\dfrac{4\pi}{3}t\right) = 3 \cos\left(\dfrac{4\pi t}{3}\right)$.

55. $d = 4 \cos 8\pi t$

(a) Maximum displacement = amplitude = 4

(b) Frequency $= \dfrac{\omega}{2\pi} = \dfrac{8\pi}{2\pi}$

$\qquad\qquad = 4$ cycles per unit of time

(c) $d = 4 \cos 40\pi = 4$

(d) $8\pi t = \dfrac{\pi}{2} \implies t = \dfrac{1}{16}$

57. $d = \dfrac{1}{16} \sin 120\pi t$

(a) Maximum displacement = amplitude $= \dfrac{1}{16}$

(b) Frequency $= \dfrac{\omega}{2\pi} = \dfrac{120\pi}{2\pi}$

$\qquad\qquad = 60$ cycles per unit of time

(c) $d = \dfrac{1}{16} \sin 600\pi = 0$

(d) $120\pi t = \pi \implies t = \dfrac{1}{120}$

59. $d = a \sin \omega t$

Frequency $= \dfrac{\omega}{2\pi}$

$264 = \dfrac{\omega}{2\pi}$

$\omega = 2\pi(264) = 528\pi$

61. $y = \dfrac{1}{4} \cos 16t, \ t > 0$

(a)

(b) Period: $\dfrac{2\pi}{16} = \dfrac{\pi}{8}$

(c) $\dfrac{1}{4} \cos 16t = 0$ when $16t = \dfrac{\pi}{2} \implies t = \dfrac{\pi}{32}$

63. (a) and (b)

Base 1	Base 2	Altitude	Area
8	$8 + 16 \cos 10°$	$8 \sin 10°$	22.1
8	$8 + 16 \cos 20°$	$8 \sin 20°$	42.5
8	$8 + 16 \cos 30°$	$8 \sin 30°$	59.7
8	$8 + 16 \cos 40°$	$8 \sin 40°$	72.7
8	$8 + 16 \cos 50°$	$8 \sin 50°$	80.5
8	$8 + 16 \cos 60°$	$8 \sin 60°$	83.1
8	$8 + 16 \cos 70°$	$8 \sin 70°$	80.7

The maximum occurs when $\theta = 60°$ and is approximately 83.1 square feet.

(c) $A(\theta) = \left[8 + (8 + 16 \cos \theta)\right]\left[\dfrac{8 \sin \theta}{2}\right]$

$= (16 + 16 \cos \theta)(4 \sin \theta)$

$= 64(1 + \cos \theta)(\sin \theta)$

(d)

The maximum of 83.1 square feet occurs when $\theta = \dfrac{\pi}{3} = 60°$.

65. False. Since the tower is not exactly vertical, a right triangle with sides 191 ft and d is not formed.

67. No. N 24° E means 24° east of north.

69. $m = 4$, passes through $(-1, 2)$

$y - 2 = 4(x - (-1))$

$y - 2 = 4x + 4$

$y = 4x + 6$

71. Passes through $(-2, 6)$ and $(3, 2)$

$m = \dfrac{2 - 6}{3 - (-2)} = -\dfrac{4}{5}$

$y - 6 = -\dfrac{4}{5}[x - (-2)]$

$y - 6 = -\dfrac{4}{5}x - \dfrac{8}{5}$

$y = -\dfrac{4}{5}x + \dfrac{22}{5}$

Review Exercises for Chapter 6

1. $\theta \approx 40°$

3. (a) $\theta = 70°$

    (b) The angle lies in Quadrant I.

    (c) Coterminal angles: $70° + 360° = 430°$
$$70° - 360° = -290°$$

5. (a) $\theta = -110°$

    (b) The angle lies in Quadrant III.

    (c) Coterminal angles: $-110° + 360° = 250°$
$$-110° - 360° = -470°$$

7. (a) $\theta = \dfrac{11\pi}{4}$

    (b) The angle lies in Quadrant II.

    (c) Coterminal angles: $\dfrac{11\pi}{4} - 2\pi = \dfrac{3\pi}{4}$
$$\dfrac{3\pi}{4} - 2\pi = -\dfrac{5\pi}{4}$$

9. (a) $\theta = -\dfrac{4\pi}{3}$

    (b) The angle lies in Quadrant II.

    (c) Coterminal angles: $-\dfrac{4\pi}{3} + 2\pi = \dfrac{2\pi}{3}$
$$-\dfrac{4\pi}{3} - 2\pi = -\dfrac{10\pi}{3}$$

11. $480° = 480° \cdot \dfrac{\pi\,\text{rad}}{180°} = \dfrac{8\pi}{3}\,\text{rad} \approx 8.378\,\text{rad}$

13. $-16.5° = -16.5° \cdot \dfrac{\pi\,\text{rad}}{180°} \approx -0.288\,\text{rad}$

15. $-33°45' = -33.75° = -33.75° \cdot \dfrac{\pi\,\text{rad}}{180°} = -\dfrac{3\pi}{16}\,\text{rad} \approx -0.589\,\text{rad}$

17. $84°15' = 84.25° = 84.25° \cdot \dfrac{\pi\,\text{rad}}{180°} \approx 1.470\,\text{rad}$

19. $\dfrac{5\pi\,\text{rad}}{7} = \dfrac{5\pi\,\text{rad}}{7} \cdot \dfrac{180°}{\pi\,\text{rad}} \approx 128.571°$

21. $-\dfrac{3\pi}{5}\,\text{rad} = -\dfrac{3\pi\,\text{rad}}{5} \cdot \dfrac{180°}{\pi\,\text{rad}} = -108°$

23. $-3.5\,\text{rad} = -3.5\,\text{rad} \cdot \dfrac{180°}{\pi\,\text{rad}} \approx -200.535°$

25. $1.75 \text{ rad} = \dfrac{1.75}{1} \text{ rad} \cdot \dfrac{180°}{\pi \text{ rad}} \approx 100.268°$

27. $138° = \dfrac{138\pi}{180} = \dfrac{23\pi}{30}$ radians

$s = r\theta = 20\left(\dfrac{23\pi}{30}\right) \approx 48.17$ inches

29. (a) Angular speed $= \dfrac{\left(33\frac{1}{3}\right)(2\pi) \text{ radians}}{1 \text{ minute}}$

$= 66\frac{2}{3}\pi$ radians per minute

(b) Linear speed $= \dfrac{6\left(66\frac{2}{3}\pi\right) \text{ inches}}{1 \text{ minute}}$

$= 400\pi$ inches per minute

31. $120° = \dfrac{120\pi}{180} = \dfrac{2\pi}{3}$ radians

$A = \dfrac{1}{2}r^2\theta = \dfrac{1}{2}(18)^2\left(\dfrac{2\pi}{3}\right) \approx 339.29$ square inches

33. opp $= 4$, adj $= 5$, hyp $= \sqrt{4^2 + 5^2} = \sqrt{41}$

$\sin\theta = \dfrac{\text{opp}}{\text{hyp}} = \dfrac{4}{\sqrt{41}} = \dfrac{4\sqrt{41}}{41}$

$\cos\theta = \dfrac{\text{adj}}{\text{hyp}} = \dfrac{5}{\sqrt{41}} = \dfrac{5\sqrt{41}}{41}$

$\tan\theta = \dfrac{\text{opp}}{\text{adj}} = \dfrac{4}{5}$

$\csc\theta = \dfrac{\text{hyp}}{\text{opp}} = \dfrac{\sqrt{41}}{4}$

$\sec\theta = \dfrac{\text{hyp}}{\text{adj}} = \dfrac{\sqrt{41}}{5}$

$\cot\theta = \dfrac{\text{adj}}{\text{opp}} = \dfrac{5}{4}$

35. adj $= 4$, hyp $= 8$, opp $= \sqrt{8^2 - 4^2} = \sqrt{48} = 4\sqrt{3}$

$\sin\theta = \dfrac{\text{opp}}{\text{hyp}} = \dfrac{4\sqrt{3}}{8} = \dfrac{\sqrt{3}}{2}$

$\cos\theta = \dfrac{\text{adj}}{\text{hyp}} = \dfrac{4}{8} = \dfrac{1}{2}$

$\tan\theta = \dfrac{\text{opp}}{\text{adj}} = \dfrac{4\sqrt{3}}{4} = \sqrt{3}$

$\csc\theta = \dfrac{\text{hyp}}{\text{opp}} = \dfrac{8}{4\sqrt{3}} = \dfrac{2\sqrt{3}}{3}$

$\sec\theta = \dfrac{\text{hyp}}{\text{adj}} = \dfrac{8}{4} = 2$

$\cot\theta = \dfrac{\text{adj}}{\text{opp}} = \dfrac{4}{4\sqrt{3}} = \dfrac{\sqrt{3}}{3}$

37. $\sin\theta = \dfrac{1}{3}$

(a) $\csc\theta = \dfrac{1}{\sin\theta} = 3$

(b) $\sin^2\theta + \cos^2\theta = 1$

$\left(\dfrac{1}{3}\right)^2 + \cos^2\theta = 1$

$\cos^2\theta = 1 - \dfrac{1}{9}$

$\cos^2\theta = \dfrac{8}{9}$

$\cos\theta = \sqrt{\dfrac{8}{9}}$

$\cos\theta = \dfrac{2\sqrt{2}}{3}$

(c) $\sec\theta = \dfrac{1}{\cos\theta} = \dfrac{3}{2\sqrt{2}} = \dfrac{3\sqrt{2}}{4}$

(d) $\tan\theta = \dfrac{\sin\theta}{\cos\theta} = \dfrac{\frac{1}{3}}{\frac{2\sqrt{2}}{3}} = \dfrac{1}{2\sqrt{2}} = \dfrac{\sqrt{2}}{4}$

39. $\csc\theta = 4$

(a) $\sin\theta = \dfrac{1}{\csc\theta} = \dfrac{1}{4}$

(b) $\sin^2\theta + \cos^2\theta = 1$

$\left(\dfrac{1}{4}\right)^2 + \cos^2\theta = 1$

$\cos^2\theta = 1 - \dfrac{1}{16}$

$\cos^2\theta = \dfrac{15}{16}$

$\cos\theta = \sqrt{\dfrac{15}{16}}$

$\cos\theta = \dfrac{\sqrt{15}}{4}$

(c) $\sec\theta = \dfrac{1}{\cos\theta} = \dfrac{4}{\sqrt{15}} = \dfrac{4\sqrt{15}}{15}$

(d) $\tan\theta = \dfrac{\sin\theta}{\cos\theta} = \dfrac{\frac{1}{4}}{\frac{\sqrt{15}}{4}} = \dfrac{1}{\sqrt{15}} = \dfrac{\sqrt{15}}{15}$

41. $\tan 33° \approx 0.6494$

43. $\sin 34.2° \approx 0.5621$

45. $\cot 15°14' \approx \cot 15.2333° = \dfrac{1}{\tan 15.2333°} \approx 3.6722$

47. $\cos\left(\dfrac{\pi}{18}\right) \approx 0.9848$

49. $\sin 1°10' = \dfrac{x}{3.5}$

$x = 3.5 \sin 1°10'(1000)$

≈ 71.3 meters

51. $x = 12, y = 16, r = \sqrt{144 + 256} = \sqrt{400} = 20$

$\sin \theta = \dfrac{y}{r} = \dfrac{4}{5}$　　　$\csc \theta = \dfrac{r}{y} = \dfrac{5}{4}$

$\cos \theta = \dfrac{x}{r} = \dfrac{3}{5}$　　　$\sec \theta = \dfrac{r}{x} = \dfrac{5}{3}$

$\tan \theta = \dfrac{y}{x} = \dfrac{4}{3}$　　　$\cot \theta = \dfrac{x}{y} = \dfrac{3}{4}$

53. $x = \dfrac{2}{3}, y = \dfrac{5}{2}$

$r = \sqrt{\left(\dfrac{2}{3}\right)^2 + \left(\dfrac{5}{2}\right)^2} = \dfrac{\sqrt{241}}{6}$

$\sin \theta = \dfrac{y}{r} = \dfrac{5/2}{\sqrt{241}/6} = \dfrac{15}{\sqrt{241}} = \dfrac{15\sqrt{241}}{241}$

$\cos \theta = \dfrac{x}{r} = \dfrac{2/3}{\sqrt{241}/6} = \dfrac{4}{\sqrt{241}} = \dfrac{4\sqrt{241}}{241}$

$\tan \theta = \dfrac{y}{x} = \dfrac{5/2}{2/3} = \dfrac{15}{4}$

$\csc \theta = \dfrac{r}{y} = \dfrac{\sqrt{241}/6}{5/2} = \dfrac{2\sqrt{241}}{30} = \dfrac{\sqrt{241}}{15}$

$\sec \theta = \dfrac{r}{x} = \dfrac{\sqrt{241}/6}{2/3} = \dfrac{\sqrt{241}}{4}$

$\cot \theta = \dfrac{x}{y} = \dfrac{2/3}{5/2} = \dfrac{4}{15}$

55. $x = -0.5 = -\dfrac{1}{2}, y = 4.5 = \dfrac{9}{2}$

$r = \sqrt{\left(-\dfrac{1}{2}\right)^2 + \left(\dfrac{9}{2}\right)^2} = \dfrac{\sqrt{82}}{2}$

$\sin \theta = \dfrac{y}{r} = \dfrac{9/2}{\sqrt{82}/2} = \dfrac{9}{\sqrt{82}} = \dfrac{9\sqrt{82}}{82}$

$\cos \theta = \dfrac{x}{r} = \dfrac{-1/2}{\sqrt{82}/2} = -\dfrac{1}{\sqrt{82}} = -\dfrac{\sqrt{82}}{82}$

$\tan \theta = \dfrac{y}{x} = \dfrac{9/2}{-1/2} = -9$

$\csc \theta = \dfrac{r}{y} = \dfrac{\sqrt{82}/2}{9/2} = \dfrac{\sqrt{82}}{9}$

$\sec \theta = \dfrac{r}{x} = \dfrac{\sqrt{82}/2}{-1/2} = -\sqrt{82}$

$\cot \theta = \dfrac{x}{y} = \dfrac{-1/2}{9/2} = -\dfrac{1}{9}$

57. $(x, 4x), x > 0$

$x = x, y = 4x$

$r = \sqrt{x^2 + (4x)^2} = \sqrt{17}x$

$\sin \theta = \dfrac{y}{r} = \dfrac{4x}{\sqrt{17}x} = \dfrac{4\sqrt{17}}{17}$

$\cos \theta = \dfrac{x}{r} = \dfrac{x}{\sqrt{17}x} = \dfrac{\sqrt{17}}{17}$

$\tan \theta = \dfrac{y}{x} = \dfrac{4x}{x} = 4$

$\csc \theta = \dfrac{r}{y} = \dfrac{\sqrt{17}x}{4x} = \dfrac{\sqrt{17}}{4}$

$\sec \theta = \dfrac{r}{x} = \dfrac{\sqrt{17}x}{x} = \sqrt{17}$

$\cot \theta = \dfrac{x}{y} = \dfrac{x}{4x} = \dfrac{1}{4}$

59. $\sec \theta = \dfrac{6}{5}, \tan \theta < 0 \implies \theta$ is in Quadrant IV.

$r = 6, x = 5, y = -\sqrt{36 - 25} = -\sqrt{11}$

$\sin \theta = \dfrac{y}{r} = -\dfrac{\sqrt{11}}{6}$　　　$\csc \theta = \dfrac{r}{y} = -\dfrac{6\sqrt{11}}{11}$

$\cos \theta = \dfrac{x}{r} = \dfrac{5}{6}$　　　$\sec \theta = \dfrac{r}{x} = \dfrac{6}{5}$

$\tan \theta = \dfrac{y}{x} = -\dfrac{\sqrt{11}}{5}$　　　$\cot \theta = \dfrac{x}{y} = -\dfrac{5\sqrt{11}}{11}$

61. $\tan \theta = \dfrac{5}{4}$, $\cos \theta < 0 \Rightarrow \theta$ is in Quadrant III.

$x = -4$, $y = -5$

$r = \sqrt{(-4)^2 + (-5)^2} = \sqrt{41}$

$\sin \theta = \dfrac{y}{r} = \dfrac{-5}{\sqrt{41}} = -\dfrac{5\sqrt{41}}{41}$

$\cos \theta = \dfrac{x}{r} = \dfrac{-4}{\sqrt{41}} = -\dfrac{4\sqrt{41}}{41}$

$\tan \theta = \dfrac{5}{4}$

$\csc \theta = \dfrac{r}{y} = \dfrac{\sqrt{41}}{-5} = -\dfrac{\sqrt{41}}{5}$

$\sec \theta = \dfrac{r}{x} = \dfrac{\sqrt{41}}{-4} = -\dfrac{\sqrt{41}}{4}$

$\cot \theta = \dfrac{x}{y} = \dfrac{-4}{-5} = \dfrac{4}{5}$

63. $\tan \theta = \dfrac{y}{x} = -\dfrac{12}{5} \Rightarrow r = 13$

$\sin \theta > 0 \Rightarrow \theta$ is in Quadrant II $\Rightarrow y = 12$, $x = -5$

$\sin \theta = \dfrac{y}{r} = \dfrac{12}{13}$ $\csc \theta = \dfrac{r}{y} = \dfrac{13}{12}$

$\cos \theta = \dfrac{x}{r} = -\dfrac{5}{13}$ $\sec \theta = \dfrac{r}{x} = -\dfrac{13}{5}$

$\tan \theta = \dfrac{y}{x} = -\dfrac{12}{5}$ $\cot \theta = \dfrac{x}{y} = -\dfrac{5}{12}$

65. $\theta = 264°$

$\theta' = 264° - 180° = 84°$

67. $\theta = -\dfrac{6\pi}{5}$

$-\dfrac{6\pi}{5} + 2\pi = \dfrac{4\pi}{5}$

$\theta' = \pi - \dfrac{4\pi}{5} = \dfrac{\pi}{5}$

69. $\sin \dfrac{\pi}{3} = \dfrac{\sqrt{3}}{2}$

$\cos \dfrac{\pi}{3} = \dfrac{1}{2}$

$\tan \dfrac{\pi}{3} = \sqrt{3}$

71. $\sin \dfrac{5\pi}{6} = \sin \dfrac{\pi}{6} = \dfrac{1}{2}$

$\cos \dfrac{5\pi}{6} = -\cos \dfrac{\pi}{6} = -\dfrac{\sqrt{3}}{2}$

$\tan \dfrac{5\pi}{6} = -\tan \dfrac{\pi}{6} = -\dfrac{\sqrt{3}}{3}$

73. $\sin\left(-\dfrac{7\pi}{3}\right) = -\sin \dfrac{\pi}{3} = -\dfrac{\sqrt{3}}{2}$

$\cos\left(-\dfrac{7\pi}{3}\right) = \cos \dfrac{\pi}{3} = \dfrac{1}{2}$

$\tan\left(-\dfrac{7\pi}{3}\right) = -\tan \dfrac{\pi}{3} = -\sqrt{3}$

75. $\sin 495° = \sin 45° = \dfrac{\sqrt{2}}{2}$

$\cos 495° = -\cos 45° = -\dfrac{\sqrt{2}}{2}$

$\tan 495° = -\tan 45° = -1$

77. $\sin(-150°) = -\sin 30° = -\dfrac{1}{2}$

$\cos(-150°) = -\cos 30° = -\dfrac{\sqrt{3}}{2}$

$\tan(-150°) = \tan 30° = \dfrac{\sqrt{3}}{3}$

79. $\sin 4° \approx 0.0698$

81. $\sec 2.8 = \dfrac{1}{\cos 2.8} \approx -1.0613$

83. $\sin\left(-\dfrac{17\pi}{15}\right) \approx 0.4067$

85. $t = \dfrac{2\pi}{3}$, $(x, y) = \left(-\dfrac{1}{2}, \dfrac{\sqrt{3}}{2}\right)$

$\sin \dfrac{2\pi}{3} = y = \dfrac{\sqrt{3}}{2}$

$\cos \dfrac{2\pi}{3} = x = -\dfrac{1}{2}$

$\tan \dfrac{2\pi}{3} = \dfrac{y}{x} = \dfrac{\sqrt{3}/2}{-1/2} = -\sqrt{3}$

87. $t = \dfrac{7\pi}{6}$, $(x, y) = \left(-\dfrac{\sqrt{3}}{2}, -\dfrac{1}{2}\right)$

$\sin \dfrac{7\pi}{6} = y = -\dfrac{1}{2}$

$\cos \dfrac{7\pi}{6} = x = -\dfrac{\sqrt{3}}{2}$

$\tan \dfrac{7\pi}{6} = \dfrac{y}{x} = \dfrac{-1/2}{-\sqrt{3}/2} = \dfrac{1}{\sqrt{3}} = \dfrac{\sqrt{3}}{3}$

89. $y = \sin x$

Amplitude: 1

Period: 2π

91. $y = 3 \cos 2\pi x$

Amplitude: 3

Period: $\dfrac{2\pi}{2\pi} = 1$

93. $f(x) = 5 \sin \dfrac{2x}{5}$

Amplitude: 5

Period: $\dfrac{2\pi}{\frac{2}{5}} = 5\pi$

95. $y = 2 + \sin x$

Amplitude: 1

Period: 2π

Shift the graph of $y = \sin x$
two units upward.

97. $g(t) = \frac{5}{2} \sin(t - \pi)$

Amplitude: $\frac{5}{2}$

Period: 2π

99. $y = a \sin bx$

   (a) $a = 2$, $\dfrac{2\pi}{b} = \dfrac{1}{264} \Rightarrow b = 528\pi$

      $y = 2 \sin(528\pi x)$

   (b) $f = \dfrac{1}{\frac{1}{264}} = 264$ cycles per second

101. $f(x) = \tan x$

103. $f(x) = \cot x$

105. $f(x) = \sec x$

Graph $y = \cos x$ first.

107. $f(x) = \csc x$

Graph $y = \sin x$ first.

109. $f(x) = x \cos x$

Graph $y_1 = x \cos(x)$

$y_2 = x$

$y_3 = -x$

As $x \to \infty, f(x) \to \infty.$

111. $\arcsin\left(-\dfrac{1}{2}\right) = -\arcsin\dfrac{1}{2} = -\dfrac{\pi}{6}$

113. $\arcsin 0.4 \approx 0.41$ radian

115. $\sin^{-1}(-0.44) \approx -0.46$ radian

117. $\arccos\dfrac{\sqrt{3}}{2} = \dfrac{\pi}{6}$

119. $\cos^{-1}(-1) = \pi$

121. $\arccos 0.324 \approx 1.24$ radians

123. $\tan^{-1}(-1.5) \approx -0.98$ radian

125. $f(x) = 2 \arcsin x = 2 \sin^{-1}(x)$

127. $f(x) = \arctan\left(\dfrac{x}{2}\right) = \tan^{-1}\left(\dfrac{x}{2}\right)$

129. $\cos\left(\arctan\dfrac{3}{4}\right) = \dfrac{4}{5}.$ Use a right triangle.

Let $\theta = \arctan\dfrac{3}{4}.$ Then

$\tan\theta = \dfrac{3}{4}$ and $\cos\theta = \dfrac{4}{5}.$

131. $\sec\left(\arctan\dfrac{12}{5}\right) = \dfrac{13}{5}$

Use a right triangle. Let

$\theta = \arctan\dfrac{12}{5}.$ Then

$\tan\theta = \dfrac{12}{5}$ and $\sec\theta = \dfrac{13}{5}.$

133. Let $y = \arccos\left(\dfrac{x}{2}\right).$

Then $\cos y = \dfrac{x}{2}$ and

$\tan y = \tan\left(\arccos\left(\dfrac{x}{2}\right)\right)$

$= \dfrac{\sqrt{4 - x^2}}{x}.$

135. $\tan\theta = \dfrac{70}{30}$

$\theta = \arctan\left(\dfrac{70}{30}\right) \approx 66.8°$

137.

$\tan 21° = \dfrac{h}{25}$

$h = 25 \tan 21° \approx 9.6$ feet

139. $\cos 45° = \dfrac{a}{60} \implies$

$a \approx 42.43$ nautical miles north

$\sin 45° = \dfrac{b}{60} \implies$

$b \approx 42.43$ nautical miles east

141. False. The sine or cosine functions are often useful for modeling simple harmonic motion.

143. False. For each θ there corresponds exactly one value of y.

145. $y = 3 \sin x$

Amplitude: 3

Period: 2π

Matches graph (d).

147. $y = 2 \sin \pi x$

Amplitude: 2

Period: 2

Matches graph (b).

149. $f(\theta) = \sec \theta$ is undefined at the zeros of $g(\theta) = \cos \theta$

since $\sec \theta = \dfrac{1}{\cos \theta}$.

151. The ranges for the other four trigonometric functions are not bounded. For $y = \tan x$ and $y = \cot x$, the range is $(-\infty, \infty)$. For $y = \sec x$ and $y = \csc x$, the range is $(-\infty, -1] \cup [1, \infty)$.

153. (a) $\tan \theta = \dfrac{x}{12}$

$x = 12 \tan \theta$

Area = Area of triangle − Area of sector

$= \left(\dfrac{1}{2}bh\right) - \left(\dfrac{1}{2}r^2\theta\right)$

$= \dfrac{1}{2}(12)(12 \tan \theta) - \dfrac{1}{2}(12^2)(\theta)$

$= 72 \tan \theta - 72\theta$

$= 72(\tan \theta - \theta)$

(b)

As $\theta \Rightarrow \dfrac{\pi}{2}, A \Rightarrow \infty$.

The area increases without bound as θ approaches $\dfrac{\pi}{2}$.

155. Answers will vary.

Problem Solving for Chapter 6

1. (a) $8{:}57 - 6{:}45 = 2$ hours 12 minutes $= 132$ minutes

$\dfrac{132}{48} = \dfrac{11}{4}$ revolutions

$\theta = \left(\dfrac{11}{4}\right)(2\pi) = \dfrac{11\pi}{2}$ radians or $990°$

(b) $s = r\theta = 47.25(5.5\pi) \approx 816.42$ feet

3. (a) $\sin 39° = \dfrac{3000}{d}$

$d = \dfrac{3000}{\sin 39°} \approx 4767$ feet

(b) $\tan 39° = \dfrac{3000}{x}$

$x = \dfrac{3000}{\tan 39°} \approx 3705$ feet

(c) $\tan 63° = \dfrac{w + 3705}{3000}$

$3000 \tan 63° = w + 3705$

$w = 3000 \tan 63° - 3705 \approx 2183$ feet

5. (a) $h(x) = \cos^2 x$

    h is even.

(b) $h(x) = \sin^2 x$

    h is even.

9. Physical (23 days): $P = \sin \dfrac{2\pi t}{23}, t \geq 0$

Emotional (28 days): $E = \sin \dfrac{2\pi t}{28}, t \geq 0$

Intellectual (33 days): $I = \sin \dfrac{2\pi t}{33}, t \geq 0$

(a)

(b) Number of days since birth until September 1, 2006:

$$t = \underbrace{365 \times 20}_{20 \text{ years}} + \underbrace{5}_{\text{leap years}} + \underbrace{11}_{\substack{\text{remaining} \\ \text{July days}}} + \underbrace{31}_{\text{August days}} + \underbrace{1}_{\substack{\text{day in} \\ \text{September}}}$$

$t = 7348$

All three drop early in the month, then peak toward the middle of the month, and drop again toward the latter part of the month.

(c) For September 22, 2006, use $t = 7369$.

$P \approx 0.631$

$E \approx 0.901$

$I \approx 0.945$

7. If we alter the model so that $h = 1$ when $t = 0$, we can use either a sine or a cosine model.

$$a = \frac{1}{2}[\text{max} - \text{min}] = \frac{1}{2}[101 - 1] = 50$$

$$d = \frac{1}{2}[\text{max} + \text{min}] = \frac{1}{2}[101 + 1] = 51$$

$$b = 8\pi$$

For the cosine model we have: $h = 51 - 50 \cos(8\pi t)$

For the sine model we have: $h = 51 - 50 \sin\left(8\pi t + \dfrac{\pi}{2}\right)$

Notice that we needed the horizontal shift so that the sine value was one when $t = 0$.

Another model would be: $h = 51 + 50 \sin\left(8\pi t + \dfrac{3\pi}{2}\right)$

Here we wanted the sine value to be 1 when $t = 0$.

11. (a) Both graphs have a period of 2 and intersect when
$x = 5.35$. They should also intersect when
$x = 5.35 - 2 = 3.35$ and $x = 5.35 + 2 = 7.35$.

(b) The graphs intersect when $x = 5.35 - 3(2) = -0.65$.

(c) Since $13.35 = 5.35 + 4(2)$ and $-4.65 = 5.35 - 5(2)$
the graphs will intersect again at these values. Therefore
$f(13.35) = g(-4.65)$.

13.

(a) $\dfrac{\sin \theta_1}{\sin \theta_2} = 1.333$

$\sin \theta_2 = \dfrac{\sin \theta_1}{1.333} = \dfrac{\sin 60°}{1.333} \approx 0.6497$

$\theta_2 \approx 40.52°$

(b) $\tan \theta_2 = \dfrac{x}{2} \implies x = 2 \tan 40.52° \approx 1.71$ feet

$\tan \theta_1 = \dfrac{y}{2} \implies y = 2 \tan 60° \approx 3.46$ feet

(c) $d = y - x = 3.46 - 1.71 = 1.75$ feet

(d) As you more closer to the rock, θ_1 decreases, which
causes y to decrease, which in turn causes d to
decrease.

Chapter 6   Practice Test

1. Express 350° in radian measure.

2. Express $(5\pi)/9$ in degree measure.

3. Convert $135°14'12''$ to decimal form.

4. Convert $-22.569°$ to D°M′S″ form.

5. If $\cos\theta = \frac{2}{3}$, use the trigonometric identities to find $\tan\theta$.

6. Find θ given $\sin\theta = 0.9063$.

7. Solve for x in the figure below.

8. Find the magnitude of the reference angle for $\theta = (6\pi)/5$.

9. Evaluate $\csc 3.92$.

10. Find $\sec\theta$ given that θ lies in Quadrant III and $\tan\theta = 6$.

11. Graph $y = 3\sin\dfrac{x}{2}$.

12. Graph $y = -2\cos(x - \pi)$.

13. Graph $y = \tan 2x$.

14. Graph $y = -\csc\left(x + \dfrac{\pi}{4}\right)$.

15. Graph $y = 2x + \sin x$, using a graphing calculator.

16. Graph $y = 3x\cos x$, using a graphing calculator.

17. Evaluate $\arcsin 1$.

18. Evaluate $\arctan(-3)$.

19. Evaluate $\sin\left(\arccos\dfrac{4}{\sqrt{35}}\right)$.

20. Write an algebraic expression for $\cos\left(\arcsin\dfrac{x}{4}\right)$.

For Exercises 21–23, solve the right triangle.

21. $A = 40°$, $c = 12$

22. $B = 6.84°$, $a = 21.3$

23. $a = 5$, $b = 9$

24. A 20-foot ladder leans against the side of a barn. Find the height of the top of the ladder if the angle of elevation of the ladder is 67°.

25. An observer in a lighthouse 250 feet above sea level spots a ship off the shore. If the angle of depression to the ship is 5°, how far out is the ship?

CHAPTER 7
Analytic Trigonometry

CHAPTER 7
Analytic Trigonometry

Section 7.1   Using Fundamental Identities

- You should know the fundamental trigonometric identities.

  (a) Reciprocal Identities

  $$\sin u = \frac{1}{\csc u} \qquad\qquad \csc u = \frac{1}{\sin u}$$

  $$\cos u = \frac{1}{\sec u} \qquad\qquad \sec u = \frac{1}{\cos u}$$

  $$\tan u = \frac{1}{\cot u} = \frac{\sin u}{\cos u} \qquad\qquad \cot u = \frac{1}{\tan u} = \frac{\cos u}{\sin u}$$

  (b) Pythagorean Identities

  $$\sin^2 u + \cos^2 u = 1$$
  $$1 + \tan^2 u = \sec^2 u$$
  $$1 + \cot^2 u = \csc^2 u$$

  (c) Cofunction Identities

  $$\sin\left(\frac{\pi}{2} - u\right) = \cos u \qquad\qquad \cos\left(\frac{\pi}{2} - u\right) = \sin u$$

  $$\tan\left(\frac{\pi}{2} - u\right) = \cot u \qquad\qquad \cot\left(\frac{\pi}{2} - u\right) = \tan u$$

  $$\sec\left(\frac{\pi}{2} - u\right) = \csc u \qquad\qquad \csc\left(\frac{\pi}{2} - u\right) = \sec u$$

  (d) Even/Odd Identities

  $$\sin(-x) = -\sin x \qquad\qquad \csc(-x) = -\csc x$$
  $$\cos(-x) = \cos x \qquad\qquad \sec(-x) = \sec x$$
  $$\tan(-x) = -\tan x \qquad\qquad \cot(-x) = -\cot x$$

- You should be able to use these fundamental identities to find function values.
- You should be able to convert trigonometric expressions to equivalent forms by using the fundamental identities.

Vocabulary Check

1. $\tan u$ **2.** $\cos u$ **3.** $\cot u$

4. $\csc u$ **5.** $\cot^2 u$ **6.** $\sec^2 u$

7. $\cos u$ **8.** $\csc u$ **9.** $\cos u$

10. $-\tan u$

---

1. $\sin x = \dfrac{\sqrt{3}}{2}$, $\cos x = -\dfrac{1}{2} \implies x$ is in Quadrant II.

$$\tan x = \frac{\sin x}{\cos x} = \frac{\sqrt{3}/2}{-1/2} = -\sqrt{3}$$

$$\cot x = \frac{1}{\tan x} = -\frac{1}{\sqrt{3}} = -\frac{\sqrt{3}}{3}$$

$$\sec x = \frac{1}{\cos x} = \frac{1}{-1/2} = -2$$

$$\csc x = \frac{1}{\sin x} = \frac{1}{\sqrt{3}/2} = \frac{2}{\sqrt{3}} = \frac{2\sqrt{3}}{3}$$

3. $\sec \theta = \sqrt{2}$, $\sin \theta = -\dfrac{\sqrt{2}}{2} \implies \theta$ is in Quadrant IV.

$$\cos \theta = \frac{1}{\sec \theta} = \frac{1}{\sqrt{2}} = \frac{\sqrt{2}}{2}$$

$$\tan \theta = \frac{\sin \theta}{\cos \theta} = \frac{-\sqrt{2}/2}{\sqrt{2}/2} = -1$$

$$\cot \theta = \frac{1}{\tan \theta} = -1$$

$$\csc \theta = \frac{1}{\sin \theta} = -\sqrt{2}$$

5. $\tan x = \dfrac{5}{12}$, $\sec x = -\dfrac{13}{12} \implies x$ is in

Quadrant III.

$$\cos x = \frac{1}{\sec x} = -\frac{12}{13}$$

$$\sin x = -\sqrt{1 - \cos^2 x} = -\sqrt{1 - \frac{144}{169}} = -\frac{5}{13}$$

$$\cot x = \frac{1}{\tan x} = \frac{12}{5}$$

$$\csc x = \frac{1}{\sin x} = -\frac{13}{5}$$

7. $\sec \phi = \dfrac{3}{2}$, $\csc \phi = -\dfrac{3\sqrt{5}}{5} \implies \phi$ is in Quadrant IV.

$$\sin \phi = \frac{1}{\csc \phi} = \frac{1}{-3\sqrt{5}/5} = -\frac{\sqrt{5}}{3}$$

$$\cos \phi = \frac{1}{\sec \phi} = \frac{1}{3/2} = \frac{2}{3}$$

$$\tan \phi = \frac{\sin \phi}{\cos \phi} = \frac{-\sqrt{5}/3}{2/3} = -\frac{\sqrt{5}}{2}$$

$$\cot \phi = \frac{1}{\tan \phi} = \frac{1}{-\sqrt{5}/2} = -\frac{2}{\sqrt{5}} = -\frac{2\sqrt{5}}{5}$$

9. $\sin(-x) = -\dfrac{1}{3} \implies \sin x = \dfrac{1}{3}$, $\tan x = -\dfrac{\sqrt{2}}{4} \implies x$ is

in Quadrant II.

$$\cos x = -\sqrt{1 - \sin^2 x} = -\sqrt{1 - \frac{1}{9}} = -\frac{2\sqrt{2}}{3}$$

$$\cot x = \frac{1}{\tan x} = \frac{1}{-\sqrt{2}/4} = -2\sqrt{2}$$

$$\sec x = \frac{1}{\cos x} = \frac{1}{-2\sqrt{2}/3} = -\frac{3\sqrt{2}}{4}$$

$$\csc x = \frac{1}{\sin x} = \frac{1}{1/3} = 3$$

11. $\tan \theta = 2$, $\sin \theta < 0 \implies \theta$ is in Quadrant III.

$$\sec \theta = -\sqrt{\tan^2 \theta + 1} = -\sqrt{4 + 1} = -\sqrt{5}$$

$$\cos \theta = \frac{1}{\sec \theta} = -\frac{1}{\sqrt{5}} = -\frac{\sqrt{5}}{5}$$

$$\sin \theta = -\sqrt{1 - \cos^2 \theta}$$

$$= -\sqrt{1 - \frac{1}{5}} = -\frac{2}{\sqrt{5}} = -\frac{2\sqrt{5}}{5}$$

$$\csc \theta = \frac{1}{\sin \theta} = -\frac{\sqrt{5}}{2}$$

$$\cot \theta = \frac{1}{\tan \theta} = \frac{1}{2}$$

13. $\sin \theta = -1$, $\cot \theta = 0 \implies \theta = \dfrac{3\pi}{2}$

$$\cos \theta = \sqrt{1 - \sin^2 \theta} = 0$$

$\sec \theta$ is undefined.

$\tan \theta$ is undefined.

$$\csc \theta = -1$$

15. $\sec x \cos x = \sec x \cdot \dfrac{1}{\sec x} = 1$

The expression is matched with (d).

17. $\cot^2 x - \csc^2 x = \cot^2 x - (1 + \cot^2 x) = -1$

The expression is matched with (b).

19. $\dfrac{\sin(-x)}{\cos(-x)} = \dfrac{-\sin x}{\cos x} = -\tan x$

The expression is matched with (e).

21. $\sin x \sec x = \sin x \cdot \dfrac{1}{\cos x} = \tan x$

The expression is matched with (b).

23. $\sec^4 x - \tan^4 x = (\sec^2 x + \tan^2 x)(\sec^2 x - \tan^2 x)$

$\qquad\qquad\qquad = (\sec^2 x + \tan^2 x)(1) = \sec^2 x + \tan^2 x$

The expression is matched with (f).

25. $\dfrac{\sec^2 x - 1}{\sin^2 x} = \dfrac{\tan^2 x}{\sin^2 x} = \dfrac{\sin^2 x}{\cos^2 x} \cdot \dfrac{1}{\sin^2 x} = \sec^2 x$

The expression is matched with (e).

27. $\cot \theta \sec \theta = \dfrac{\cos \theta}{\sin \theta} \cdot \dfrac{1}{\cos \theta} = \dfrac{1}{\sin \theta} = \csc \theta$

29. $\sin \phi \, (\csc \phi - \sin \phi) = (\sin \phi)\dfrac{1}{\sin \phi} - \sin^2 \phi$

$\qquad\qquad\qquad\qquad = 1 - \sin^2 \phi = \cos^2 \phi$

31. $\dfrac{\cot x}{\csc x} = \dfrac{\cos x/\sin x}{1/\sin x}$

$\qquad = \dfrac{\cos x}{\sin x} \cdot \dfrac{\sin x}{1} = \cos x$

33. $\dfrac{1 - \sin^2 x}{\csc^2 x - 1} = \dfrac{\cos^2 x}{\cot^2 x} = \cos^2 x \tan^2 x = (\cos^2 x)\dfrac{\sin^2 x}{\cos^2 x}$

$\qquad\qquad\qquad\qquad = \sin^2 x$

35. $\sec \alpha \dfrac{\sin \alpha}{\tan \alpha} = \dfrac{1}{\cos \alpha}(\sin \alpha) \cot \alpha$

$\qquad\qquad\qquad = \dfrac{1}{\cos \alpha}(\sin \alpha)\left(\dfrac{\cos \alpha}{\sin \alpha}\right) = 1$

37. $\cos\left(\dfrac{\pi}{2} - x\right) \sec x = (\sin x)(\sec x) = (\sin x)\left(\dfrac{1}{\cos x}\right) = \dfrac{\sin x}{\cos x} = \tan x$

39. $\dfrac{\cos^2 y}{1 - \sin y} = \dfrac{1 - \sin^2 y}{1 - \sin y} = \dfrac{(1 + \sin y)(1 - \sin y)}{1 - \sin y} = 1 + \sin y$

41. $\sin \beta \tan \beta + \cos \beta = (\sin \beta)\dfrac{\sin \beta}{\cos \beta} + \cos \beta$

$\qquad\qquad\qquad = \dfrac{\sin^2 \beta}{\cos \beta} + \dfrac{\cos^2 \beta}{\cos \beta}$

$\qquad\qquad\qquad = \dfrac{\sin^2 \beta + \cos^2 \beta}{\cos \beta}$

$\qquad\qquad\qquad = \dfrac{1}{\cos \beta}$

$\qquad\qquad\qquad = \sec \beta$

43. $\cot u \sin u + \tan u \cos u = \dfrac{\cos u}{\sin u}(\sin u) + \dfrac{\sin u}{\cos u}(\cos u)$

$\qquad\qquad\qquad\qquad = \cos u + \sin u$

45. $\tan^2 x - \tan^2 x \sin^2 x = \tan^2 x(1 - \sin^2 x)$

$\qquad\qquad\qquad = \tan^2 x \cos^2 x$

$\qquad\qquad\qquad = \dfrac{\sin^2 x}{\cos^2 x} \cdot \cos^2 x$

$\qquad\qquad\qquad = \sin^2 x$

47. $\sin^2 x \sec^2 x - \sin^2 x = \sin^2 x(\sec^2 x - 1)$

$\qquad\qquad\qquad = \sin^2 x \tan^2 x$

49. $\dfrac{\sec^2 x - 1}{\sec x - 1} = \dfrac{(\sec x + 1)(\sec x - 1)}{\sec x - 1} = \sec x + 1$

51. $\tan^4 x + 2\tan^2 x + 1 = (\tan^2 x + 1)^2$

$\qquad\qquad\qquad = (\sec^2 x)^2$

$\qquad\qquad\qquad = \sec^4 x$

53. $\sin^4 x - \cos^4 x = (\sin^2 x + \cos^2 x)(\sin^2 x - \cos^2 x)$

$$= (1)(\sin^2 x - \cos^2 x)$$

$$= \sin^2 x - \cos^2 x$$

55. $\csc^3 x - \csc^2 x - \csc x + 1 = \csc^2 x(\csc x - 1) - 1(\csc x - 1)$

$$= (\csc^2 x - 1)(\csc x - 1)$$

$$= \cot^2 x(\csc x - 1)$$

57. $(\sin x + \cos x)^2 = \sin^2 x + 2 \sin x \cos x + \cos^2 x$

$$= (\sin^2 x + \cos^2 x) + 2 \sin x \cos x$$

$$= 1 + 2 \sin x \cos x$$

59. $(2 \csc x + 2)(2 \csc x - 2) = 4 \csc^2 x - 4$

$$= 4(\csc^2 x - 1)$$

$$= 4 \cot^2 x$$

61. $\dfrac{1}{1 + \cos x} + \dfrac{1}{1 - \cos x} = \dfrac{1 - \cos x + 1 + \cos x}{(1 + \cos x)(1 - \cos x)}$

$$= \dfrac{2}{1 - \cos^2 x}$$

$$= \dfrac{2}{\sin^2 x}$$

$$= 2 \csc^2 x$$

63. $\dfrac{\cos x}{1 + \sin x} + \dfrac{1 + \sin x}{\cos x} = \dfrac{\cos^2 x + (1 + \sin x)^2}{\cos x(1 + \sin x)} = \dfrac{\cos^2 x + 1 + 2 \sin x + \sin^2 x}{\cos x(1 + \sin x)}$

$$= \dfrac{2 + 2 \sin x}{\cos x(1 + \sin x)}$$

$$= \dfrac{2(1 + \sin x)}{\cos x(1 + \sin x)}$$

$$= \dfrac{2}{\cos x}$$

$$= 2 \sec x$$

65. $\dfrac{\sin^2 y}{1 - \cos y} = \dfrac{1 - \cos^2 y}{1 - \cos y}$

$$= \dfrac{(1 + \cos y)(1 - \cos y)}{1 - \cos y}$$

$$= 1 + \cos y$$

67. $\dfrac{3}{\sec x - \tan x} \cdot \dfrac{\sec x + \tan x}{\sec x + \tan x} = \dfrac{3(\sec x + \tan x)}{\sec^2 x - \tan^2 x}$

$$= \dfrac{3(\sec x + \tan x)}{1}$$

$$= 3(\sec x + \tan x)$$

69. $y_1 = \cos\left(\dfrac{\pi}{2} - x\right),\ y_2 = \sin x$

x	0.2	0.4	0.6	0.8	1.0	1.2	1.4
y_1	0.1987	0.3894	0.5646	0.7174	0.8415	0.9320	0.9854
y_2	0.1987	0.3894	0.5646	0.7174	0.8415	0.9320	0.9854

Conclusion: $y_1 = y_2$

71. $y_1 = \dfrac{\cos x}{1 - \sin x}$, $y_2 = \dfrac{1 + \sin x}{\cos x}$

x	0.2	0.4	0.6	0.8	1.0	1.2	1.4
y_1	1.2230	1.5085	1.8958	2.4650	3.4082	5.3319	11.6814
y_2	1.2230	1.5085	1.8958	2.4650	3.4082	5.3319	11.6814

Conclusion: $y_1 = y_2$

73. $y_1 = \cos x \cot x + \sin x = \csc x$

$$\cos x \cot x + \sin x = \cos x\left(\frac{\cos x}{\sin x}\right) + \sin x$$

$$= \frac{\cos^2 x}{\sin x} + \frac{\sin^2 x}{\sin x}$$

$$= \frac{\cos^2 x + \sin^2 x}{\sin x} = \frac{1}{\sin x} = \csc x$$

75. $y_1 = \dfrac{1}{\sin x}\left(\dfrac{1}{\cos x} - \cos x\right) = \tan x$

$$\frac{1}{\sin x}\left(\frac{1}{\cos x} - \cos x\right) = \frac{1}{\sin x \cos x} - \frac{\cos x}{\sin x}$$

$$= \frac{1 - \cos^2 x}{\sin x \cos x} = \frac{\sin^2 x}{\sin x \cos x} = \frac{\sin x}{\cos x} = \tan x$$

77. Let $x = 3 \cos \theta$, then

$$\sqrt{9 - x^2} = \sqrt{9 - (3 \cos \theta)^2} = \sqrt{9 - 9\cos^2 \theta} = \sqrt{9(1 - \cos^2 \theta)}$$

$$= \sqrt{9 \sin^2 \theta} = 3 \sin \theta.$$

79. Let $x = 3 \sec \theta$, then

$$\sqrt{x^2 - 9} = \sqrt{(3 \sec \theta)^2 - 9}$$

$$= \sqrt{9 \sec^2 \theta - 9}$$

$$= \sqrt{9\,(\sec^2 \theta - 1)}$$

$$= \sqrt{9 \tan^2 \theta}$$

$$= 3 \tan \theta.$$

81. Let $x = 5 \tan \theta$, then

$$\sqrt{x^2 + 25} = \sqrt{(5 \tan \theta)^2 + 25}$$

$$= \sqrt{25 \tan^2 \theta + 25}$$

$$= \sqrt{25(\tan^2 \theta + 1)}$$

$$= \sqrt{25 \sec^2 \theta}$$

$$= 5 \sec \theta.$$

83. Let $x = 3 \sin \theta$, then $\sqrt{9 - x^2} = 3$ becomes

$$\sqrt{9 - (3 \sin \theta)^2} = 3$$

$$\sqrt{9 - 9 \sin^2 \theta} = 3$$

$$\sqrt{9(1 - \sin^2 \theta)} = 3$$

$$\sqrt{9 \cos^2 \theta} = 3$$

$$3 \cos \theta = 3$$

$$\cos \theta = 1$$

$$\sin \theta = \sqrt{1 - \cos^2 \theta} = \sqrt{1 - (1)^2} = 0.$$

85. Let $x = 2 \cos \theta$, then $\sqrt{16 - 4x^2} = 2\sqrt{2}$ becomes

$$\sqrt{16 - 4(2 \cos \theta)^2} = 2\sqrt{2}$$

$$\sqrt{16 - 16 \cos^2 \theta} = 2\sqrt{2}$$

$$\sqrt{16(1 - \cos^2 \theta)} = 2\sqrt{2}$$

$$\sqrt{16 \sin^2 \theta} = 2\sqrt{2}$$

$$4 \sin \theta = 2\sqrt{2}$$

$$\sin \theta = \frac{\sqrt{2}}{2}$$

$$\cos \theta = \sqrt{1 - \sin^2 \theta}$$

$$= \sqrt{1 - \frac{1}{2}}$$

$$= \sqrt{\frac{1}{2}}$$

$$= \frac{\sqrt{2}}{2}.$$

87. $\sin \theta = \sqrt{1 - \cos^2 \theta}$

Let $y_1 = \sin x$ and $y_2 = \sqrt{1 - \cos^2 x}$, $0 \le x \le 2\pi$.

$y_1 = y_2$ for $0 \le x \le \pi$, so we have

$\sin \theta = \sqrt{1 - \cos^2 \theta}$ for $0 \le \theta \le \pi$.

89. $\sec \theta = \sqrt{1 + \tan^2 \theta}$

Let $y_1 = \frac{1}{\cos x}$ and $y_2 = \sqrt{1 + \tan^2 x}$, $0 \le x \le 2\pi$.

$y_1 = y_2$ for $0 \le x < \frac{\pi}{2}$ and $\frac{3\pi}{2} < x \le 2\pi$, so we have

$\sec \theta = \sqrt{1 + \tan^2 \theta}$ for $0 \le \theta < \frac{\pi}{2}$ and $\frac{3\pi}{2} < \theta < 2\pi$.

91. $\ln|\cos x| - \ln|\sin x| = \ln \left| \frac{\cos x}{\sin x} \right| = \ln|\cot x|$

93. $\ln|\cot t| + \ln(1 + \tan^2 t) = \ln\left[|\cot t|(1 + \tan^2 t)\right]$

$$= \ln |\cot t \sec^2 t|$$

$$= \ln \left| \frac{\cos t}{\sin t} \cdot \frac{1}{\cos^2 t} \right|$$

$$= \ln \left| \frac{1}{\sin t \cos t} \right| = \ln|\csc t \sec t|$$

95. (a) $\csc^2 132° - \cot^2 132° \approx 1.8107 - 0.8107 = 1$

(b) $\csc^2 \frac{2\pi}{7} - \cot^2 \frac{2\pi}{7} \approx 1.6360 - 0.6360 = 1$

97. $\cos\left(\frac{\pi}{2} - \theta\right) = \sin \theta$

(a) $\theta = 80°$

$$\cos(90° - 80°) = \sin 80°$$

$$0.9848 = 0.9848$$

(b) $\theta = 0.8$

$$\cos\left(\frac{\pi}{2} - 0.8\right) = \sin 0.8$$

$$0.7174 = 0.7174$$

99. $\mu W \cos \theta = W \sin \theta$

$$\mu = \frac{W \sin \theta}{W \cos \theta} = \tan \theta$$

101. True. For example, $\sin(-x) = -\sin x$ means that the graph of $\sin x$ is symmetric about the origin.

103. As $x \to \dfrac{\pi^-}{2}$, $\sin x \to 1$

and $\csc x \to 1$.

105. As $x \to \dfrac{\pi^-}{2}$, $\tan x \to \infty$

and $\cot x \to 0$.

107. $\cos \theta = \sqrt{1 - \sin^2 \theta}$ *is not* an identity.

$\cos^2 \theta + \sin^2 \theta = 1 \implies \cos \theta = \pm\sqrt{1 - \sin^2 \theta}$

109. $\dfrac{\sin k\theta}{\cos k\theta} = \tan \theta$ *is not* an identity.

$\dfrac{\sin k\theta}{\cos k\theta} = \tan k\theta$

111. $\sin \theta \csc \theta = 1$ *is* an identity.

$\sin \theta \cdot \dfrac{1}{\sin \theta} = 1$, provided $\sin \theta \neq 0$.

113. Let (x, y) be any point on the terminal side of θ.

Then, $r = \sqrt{x^2 + y^2}$ and

$$\sin^2 \theta + \cos^2 \theta = \left(\frac{y}{r}\right)^2 + \left(\frac{x}{r}\right)^2$$

$$= \frac{y^2 + x^2}{r^2}$$

$$= \frac{r^2}{r^2}$$

$$= 1.$$

115. $\left(\sqrt{x} + 5\right)\left(\sqrt{x} - 5\right) = \left(\sqrt{x}\right)^2 - (5)^2 = x - 25$

117. $\dfrac{1}{x + 5} + \dfrac{x}{x - 8} = \dfrac{(x - 8) + x(x + 5)}{(x + 5)(x - 8)}$

$$= \frac{x^2 + 6x - 8}{(x + 5)(x - 8)}$$

119. $\dfrac{2x}{x^2 - 4} - \dfrac{7}{x + 4} = \dfrac{2x(x + 4) - 7(x^2 - 4)}{(x^2 - 4)(x + 4)}$

$$= \frac{2x^2 + 8x - 7x^2 + 28}{(x^2 - 4)(x + 4)}$$

$$= \frac{-5x^2 + 8x + 28}{(x^2 - 4)(x + 4)}$$

121. $f(x) = \dfrac{1}{2}\sin(\pi x)$

Amplitude: $\dfrac{1}{2}$

Period: $\dfrac{2\pi}{\pi} = 2$

Key points:

$(0, 0), \left(\dfrac{1}{2}, \dfrac{1}{2}\right), (1, 0), \left(\dfrac{3}{2}, -\dfrac{1}{2}\right), (2, 0)$

123. $f(x) = \dfrac{1}{2}\sec\left(x + \dfrac{\pi}{4}\right)$

Sketch the graph of $y = \dfrac{1}{2}\cos\left(x + \dfrac{\pi}{4}\right)$ first.

Amplitude: $\dfrac{1}{2}$

Period: 2π

One cycle: $x + \dfrac{\pi}{4} = 0 \Rightarrow x = -\dfrac{\pi}{4}$

$\qquad\qquad x + \dfrac{\pi}{4} = 2\pi \Rightarrow x = \dfrac{7\pi}{4}$

The x-intercepts of $y = \dfrac{1}{2}\cos\left(x + \dfrac{\pi}{4}\right)$ correspond to the vertical asymptotes of $f(x) = \dfrac{1}{2}\sec\left(x + \dfrac{\pi}{4}\right)$.

$x = \dfrac{\pi}{4}, x = \dfrac{5\pi}{4}, \ldots$

Section 7.2   Verifying Trigonometric Identities

- ■ You should know the difference between an expression, a conditional equation, and an identity.
- ■ You should be able to solve trigonometric identities, using the following techniques.
  - (a) Work with *one* side at a time. Do not "cross" the equal sign.
  - (b) Use algebraic techniques such as combining fractions, factoring expressions, rationalizing denominators, and squaring binomials.
  - (c) Use the fundamental identities.
  - (d) Convert all the terms into sines and cosines.

Vocabulary Check

1. identity

2. conditional equation

3. $\tan u$

4. $\cot u$

5. $\cos^2 u$

6. $\sin u$

7. $-\csc u$

8. $\sec u$

---

1. $\sin t \csc t = \sin t\left(\dfrac{1}{\sin t}\right) = 1$

3. $(1 + \sin\alpha)(1 - \sin\alpha) = 1 - \sin^2\alpha = \cos^2\alpha$

5. $\cos^2\beta - \sin^2\beta = (1 - \sin^2\beta) - \sin^2\beta$

$\qquad\qquad\quad = 1 - 2\sin^2\beta$

7. $\sin^2\alpha - \sin^4\alpha = \sin^2\alpha(1 - \sin^2\alpha)$

$\qquad\qquad\quad = (1 - \cos^2\alpha)(\cos^2\alpha)$

$\qquad\qquad\quad = \cos^2\alpha - \cos^4\alpha$

9. $\dfrac{\csc^2 \theta}{\cot \theta} = \csc^2 \theta \left(\dfrac{1}{\cot \theta} \right) = \csc^2 \theta \tan \theta$

$\qquad = \left(\dfrac{1}{\sin^2 \theta} \right) \left(\dfrac{\sin \theta}{\cos \theta} \right) = \left(\dfrac{1}{\sin \theta} \right) \left(\dfrac{1}{\cos \theta} \right)$

$\qquad = \csc \theta \sec \theta$

11. $\dfrac{\cot^2 t}{\csc t} = \dfrac{\cos^2 t}{\sin^2 t} \cdot \sin t$

$\qquad = \dfrac{\cos^2 t}{\sin t}$

$\qquad = \dfrac{1 - \sin^2 t}{\sin t} = \dfrac{1}{\sin t} - \dfrac{\sin^2 t}{\sin t}$

$\qquad = \csc t - \sin t$

13. $\sin^{1/2} x \cos x - \sin^{5/2} x \cos x = \sin^{1/2} x \cos x (1 - \sin^2 x) = \sin^{1/2} x \cos x \cdot \cos^2 x = \cos^3 x \sqrt{\sin x}$

15. $\dfrac{1}{\sec x \tan x} = \cos x \cot x = \cos x \cdot \dfrac{\cos x}{\sin x}$

$\qquad = \dfrac{\cos^2 x}{\sin x}$

$\qquad = \dfrac{1 - \sin^2 x}{\sin x}$

$\qquad = \dfrac{1}{\sin x} - \sin x$

$\qquad = \csc x - \sin x$

17. $\csc x - \sin x = \dfrac{1}{\sin x} - \dfrac{\sin^2 x}{\sin x}$

$\qquad = \dfrac{1 - \sin^2 x}{\sin x}$

$\qquad = \dfrac{\cos^2 x}{\sin x}$

$\qquad = \dfrac{\cos x}{1} \cdot \dfrac{\cos x}{\sin x}$

$\qquad = \cos x \cot x$

19. $\dfrac{1}{\tan x} + \dfrac{1}{\cot x} = \dfrac{\cot x + \tan x}{\tan x \cot x}$

$\qquad = \dfrac{\cot x + \tan x}{1}$

$\qquad = \tan x + \cot x$

21. $\dfrac{\cos \theta \cot \theta}{1 - \sin \theta} - 1 = \dfrac{\cos \theta \cot \theta - (1 - \sin \theta)}{1 - \sin \theta}$

$\qquad = \dfrac{\cos \theta \left(\dfrac{\cos \theta}{\sin \theta} \right) - 1 + \sin \theta}{1 - \sin \theta} \cdot \dfrac{\sin \theta}{\sin \theta}$

$\qquad = \dfrac{\cos^2 \theta - \sin \theta + \sin^2 \theta}{\sin \theta (1 - \sin \theta)}$

$\qquad = \dfrac{1 - \sin \theta}{\sin \theta (1 - \sin \theta)}$

$\qquad = \dfrac{1}{\sin \theta}$

$\qquad = \csc \theta$

23. $\dfrac{1}{\sin x + 1} + \dfrac{1}{\csc x + 1} = \dfrac{\csc x + 1 + \sin x + 1}{(\sin x + 1)(\csc x + 1)}$

$\qquad = \dfrac{\sin x + \csc x + 2}{\sin x \csc x + \sin x + \csc x + 1}$

$\qquad = \dfrac{\sin x + \csc x + 2}{1 + \sin x + \csc x + 1}$

$\qquad = \dfrac{\sin x + \csc x + 2}{\sin x + \csc x + 2}$

$\qquad = 1$

25. $\tan \left(\dfrac{\pi}{2} - \theta \right) \tan \theta = \cot \theta \tan \theta = \left(\dfrac{1}{\tan \theta} \right) \tan \theta = 1$

27. $\dfrac{\csc(-x)}{\sec(-x)} = \dfrac{1/\sin(-x)}{1/\cos(-x)}$

$\qquad = \dfrac{\cos(-x)}{\sin(-x)}$

$\qquad = \dfrac{\cos x}{-\sin x}$

$\qquad = -\cot x$

29. $\dfrac{\tan x \cot x}{\cos x} = \dfrac{1}{\cos x} = \sec x$

31. $\dfrac{\tan x + \cot y}{\tan x \cot y} = \dfrac{\dfrac{1}{\cot x} + \dfrac{1}{\tan y}}{\dfrac{1}{\cot x} \cdot \dfrac{1}{\tan y}} \cdot \dfrac{\cot x \tan y}{\cot x \tan y}$

$\qquad\qquad = \tan y + \cot x$

33. $\sqrt{\dfrac{1 + \sin \theta}{1 - \sin \theta}} = \sqrt{\dfrac{1 + \sin \theta}{1 - \sin \theta} \cdot \dfrac{1 + \sin \theta}{1 + \sin \theta}}$

$\qquad = \sqrt{\dfrac{(1 + \sin \theta)^2}{1 - \sin^2 \theta}}$

$\qquad = \sqrt{\dfrac{(1 + \sin \theta)^2}{\cos^2 \theta}}$

$\qquad = \dfrac{1 + \sin \theta}{|\cos \theta|}$

35. $\cos^2 \beta + \cos^2\left(\dfrac{\pi}{2} - \beta\right) = \cos^2 \beta + \sin^2 \beta = 1$

37. $\sin t \csc\left(\dfrac{\pi}{2} - t\right) = \sin t \sec t = \sin t\left(\dfrac{1}{\cos t}\right)$

$\qquad\qquad = \dfrac{\sin t}{\cos t} = \tan t$

39. (a)

(b)

Identity

Let $y_1 = \dfrac{2}{(\cos x)^2} - \dfrac{2(\sin x)^2}{(\cos x)^2} - (\sin x)^2 - (\cos x)^2$ and $y_2 = 1$.

Identity

(c) $2\sec^2 x - 2\sec^2 x \sin^2 x - \sin^2 x - \cos^2 x = 2\sec^2 x(1 - \sin^2 x) - (\sin^2 x + \cos^2 x)$

$\qquad\qquad = 2\sec^2 x(\cos^2 x) - 1$

$\qquad\qquad = 2 \cdot \dfrac{1}{\cos^2 x} \cdot \cos^2 x - 1$

$\qquad\qquad = 2 - 1$

$\qquad\qquad = 1$

41. (a)

(b)

Not an identity

Let $y_1 = 2 + (\cos x)^2 - 3(\cos x)^4$ and

$\quad y_2 = (\sin x)^2(3 + 2(\cos x)^2)$.

Not an identity

(c) $2 + \cos^2 x - 3\cos^4 x = (1 - \cos^2 x)(2 + 3\cos^2 x)$

$\qquad\qquad = \sin^2 x(2 + 3\cos^2 x)$

$\qquad\qquad \neq \sin^2 x(3 + 2\cos^2 x)$

43. (a)

Let $y_1 = \dfrac{1}{(\sin x)^4} - \dfrac{2}{(\sin x)^2} + 1$ and $y_2 = \dfrac{1}{(\tan x)^4}$.

Identity

(b)

Identity

(c) $\csc^4 x - 2\csc^2 x + 1 = (\csc^2 x - 1)^2 = (\cot^2 x)^2 = \cot^4 x$

45. (a)

Let $y_1 = \dfrac{\cos x}{(1 - \sin x)}$ and $y_2 = \dfrac{(1 - \sin x)}{\cos x}$.

Not an identity

(b)

Not an identity

(c) $\dfrac{\cos x}{1 - \sin x} = \dfrac{\cos x}{1 - \sin x} \cdot \dfrac{1 + \sin x}{1 + \sin x}$

$\qquad = \dfrac{\cos x(1 + \sin x)}{1 - \sin^2 x}$

$\qquad = \dfrac{\cos x(1 + \sin x)}{\cos^2 x}$

$\qquad = \dfrac{1 + \sin x}{\cos x}$

47. $\tan^3 x \sec^2 x - \tan^3 x = \tan^3 x(\sec^2 x - 1)$

$\qquad\qquad\qquad\qquad\quad = \tan^3 x \tan^2 x$

$\qquad\qquad\qquad\qquad\quad = \tan^5 x$

49. $(\sin^2 x - \sin^4 x)\cos x = \sin^2 x(1 - \sin^2 x)\cos x$

$\qquad\qquad\qquad\qquad\quad = \sin^2 x \cos^2 x \cos x$

$\qquad\qquad\qquad\qquad\quad = \sin^2 x \cos^3 x$

51. $\sin^2 25° + \sin^2 65° = \sin^2 25° + \cos^2(90° - 65°)$

$\qquad\qquad\qquad\qquad = \sin^2 25° + \cos^2 25°$

$\qquad\qquad\qquad\qquad = 1$

53. $\cos^2 20° + \cos^2 52° + \cos^2 38° + \cos^2 70° = \cos^2 20° + \cos^2 52° + \sin^2(90° - 38°) + \sin^2(90° - 70°)$

$\qquad\qquad\qquad\qquad\qquad\qquad\qquad\quad = \cos^2 20° + \cos^2 52° + \sin^2 52° + \sin^2 20°$

$\qquad\qquad\qquad\qquad\qquad\qquad\qquad\quad = (\cos^2 20° + \sin^2 20°) + (\cos^2 52° + \sin^2 52°)$

$\qquad\qquad\qquad\qquad\qquad\qquad\qquad\quad = 1 + 1$

$\qquad\qquad\qquad\qquad\qquad\qquad\qquad\quad = 2$

21.
$$\cos^3 x = \cos x$$
$$\cos^3 x - \cos x = 0$$
$$\cos x(\cos^2 x - 1) = 0$$
$$\cos x = 0 \qquad \text{or} \qquad \cos^2 x - 1 = 0$$
$$x = \frac{\pi}{2}, \frac{3\pi}{2} \qquad\qquad \cos x = \pm 1$$
$$x = 0, \pi$$

23.
$$3\tan^3 x - \tan x = 0$$
$$\tan x(3\tan^2 x - 1) = 0$$
$$\tan x = 0 \qquad \text{or} \qquad 3\tan^2 x - 1 = 0$$
$$x = 0, \pi \qquad\qquad \tan x = \pm\frac{\sqrt{3}}{3}$$
$$x = \frac{\pi}{6}, \frac{5\pi}{6}, \frac{7\pi}{6}, \frac{11\pi}{6}$$

25.
$$\sec^2 x - \sec x - 2 = 0$$
$$(\sec x - 2)(\sec x + 1) = 0$$
$$\sec x - 2 = 0 \qquad \text{or} \quad \sec x + 1 = 0$$
$$\sec x = 2 \qquad\qquad \sec x = -1$$
$$x = \frac{\pi}{3}, \frac{5\pi}{3} \qquad\qquad x = \pi$$

27. $2\sin x + \csc x = 0$
$$2\sin x + \frac{1}{\sin x} = 0$$
$$2\sin^2 x + 1 = 0$$
$$\sin^2 x = -\frac{1}{2} \implies \text{No solution}$$

29.
$$2\cos^2 x + \cos x - 1 = 0$$
$$(2\cos x - 1)(\cos x + 1) = 0$$
$$2\cos x - 1 = 0 \quad \text{or} \quad \cos x + 1 = 0$$
$$\cos x = \frac{1}{2} \qquad\qquad \cos x = -1$$
$$x = \frac{\pi}{3}, \frac{5\pi}{3} \qquad\qquad x = \pi$$

31.
$$2\sec^2 x + \tan^2 x - 3 = 0$$
$$2(\tan^2 x + 1) + \tan^2 x - 3 = 0$$
$$3\tan^2 x - 1 = 0$$
$$\tan x = \pm\frac{\sqrt{3}}{3}$$
$$x = \frac{\pi}{6}, \frac{5\pi}{6}, \frac{7\pi}{6}, \frac{11\pi}{6}$$

33.
$$\csc x + \cot x = 1$$
$$(\csc x + \cot x)^2 = 1^2$$
$$\csc^2 x + 2\csc x \cot x + \cot^2 x = 1$$
$$\cot^2 x + 1 + 2\csc x \cot x + \cot^2 x = 1$$
$$2\cot^2 x + 2\csc x \cot x = 0$$
$$2\cot x(\cot x + \csc x) = 0$$
$$2\cot x = 0 \qquad \text{or} \quad \cot x + \csc x = 0$$
$$x = \frac{\pi}{2}, \frac{3\pi}{2} \qquad\qquad \frac{\cos x}{\sin x} = -\frac{1}{\sin x}$$
$$\cos x = -1$$
$$x = \pi$$

35. $\cos 2x = \dfrac{1}{2}$
$$2x = \frac{\pi}{3} + 2n\pi \quad \text{or} \quad 2x = \frac{5\pi}{3} + 2n\pi$$
$$x = \frac{\pi}{6} + n\pi \qquad\qquad x = \frac{5\pi}{6} + n\pi$$

By checking in the original equation, we find that $x = \pi$ and $x = 3\pi/2$ are extraneous. The only solution to the equation in the interval $[0, 2\pi)$ is $x = \pi/2$.

37. $\tan 3x = 1$

$$3x = \frac{\pi}{4} + 2n\pi \quad \text{or} \quad 3x = \frac{5\pi}{4} + 2n\pi$$

$$x = \frac{\pi}{12} + \frac{2n\pi}{3} \qquad\qquad x = \frac{5\pi}{12} + \frac{2n\pi}{3}$$

These can be combined as $x = \frac{\pi}{12} + \frac{n\pi}{3}$.

39. $\cos\left(\frac{x}{2}\right) = \frac{\sqrt{2}}{2}$

$$\frac{x}{2} = \frac{\pi}{4} + 2n\pi \quad \text{or} \quad \frac{x}{2} = \frac{7\pi}{4} + 2n\pi$$

$$x = \frac{\pi}{2} + 4n\pi \qquad\qquad x = \frac{7\pi}{2} + 4n\pi$$

41. $y = \sin\frac{\pi x}{2} + 1$

From the graph in the textbook we see that the curve has x-intercepts at $x = -1$ and at $x = 3$.

In general, we have: $\sin\left(\frac{\pi x}{2}\right) = -1$

$$\frac{\pi x}{2} = \frac{3\pi}{2} + 2n\pi$$

$$x = 3 + 4n$$

43. $y = \tan^2\left(\frac{\pi x}{6}\right) - 3$

From the graph in the textbook we see that the curve has x-intercepts at $x = \pm 2$.

In general, we have: $\tan^2\left(\frac{\pi x}{6}\right) = 3$

$$\tan\left(\frac{\pi x}{6}\right) = \pm\sqrt{3}$$

$$\frac{\pi x}{6} = \pm\frac{\pi}{3} + n\pi$$

$$x = \pm 2 + 6n$$

45. Graph $y_1 = 2\sin x + \cos x$.

The x-intercepts occur at $x \approx 2.678$ and $x \approx 5.820$.

47. Graph $y_1 = \dfrac{1 + \sin x}{\cos x} + \dfrac{\cos x}{1 + \sin x} - 4$.

The x-intercepts occur at

$$x = \frac{\pi}{3} \approx 1.047 \text{ and } x = \frac{5\pi}{3} \approx 5.236.$$

49. $x \tan x - 1 = 0$

Graph $y_1 = x \tan x - 1$.

The x-intercepts occur at $x \approx 0.860$ and $x \approx 3.426$.

51. $\sec^2 x + 0.5 \tan x - 1 = 0$

Graph $y_1 = \dfrac{1}{(\cos x)^2} + 0.5 \tan x - 1$.

The x-intercepts occur at

$x = 0$, $x \approx 2.678$,

$x = \pi \approx 3.142$, and

$x \approx 5.820$.

53. Graph $y_1 = 2\tan^2 x + 7\tan x - 15$.

The x-intercepts occur at
$x \approx 0.983$, $x \approx 1.768$,
$x \approx 4.124$ and $x \approx 4.910$.

55. $12 \sin^2 x - 13 \sin x + 3 = 0$

$$\sin x = \frac{-(-13) \pm \sqrt{(-13)^2 - 4(12)(3)}}{2(12)}$$

$$= \frac{13 \pm 5}{24}$$

$\sin x = \frac{1}{3}$ or $\sin x = \frac{3}{4}$

$x \approx 0.3398, \ 2.8018$ $x \approx 0.8481, \ 2.2935$

Graph $y_1 = 12 \sin^2 x - 13 \sin x + 3$.

The *x*-intercepts occur at
$x \approx 0.3398$, $x \approx 0.8481$, $x \approx 2.2935$, and $x \approx 2.8018$.

57. $\tan^2 x + 3 \tan x + 1 = 0$

$$\tan x = \frac{-3 \pm \sqrt{3^2 - 4(1)(1)}}{2(1)} = \frac{-3 \pm \sqrt{5}}{2}$$

$\tan x = \dfrac{-3 - \sqrt{5}}{2}$ or $\tan x = \dfrac{-3 + \sqrt{5}}{2}$

$x \approx 1.9357, \ 5.0773$ $x \approx 2.7767, \ 5.9183$

Graph $y_1 = \tan^2 x + 3 \tan x + 1$.

The *x*-intercepts occur at $x \approx 1.9357$, $x \approx 2.7767$,
$x \approx 5.0773$, and $x \approx 5.9183$.

59. $\tan^2 x - 6 \tan x + 5 = 0$

$(\tan x - 1)(\tan x - 5) = 0$

$\tan x - 1 = 0$ or $\tan x - 5 = 0$

$\tan x = 1$ $\tan x = 5$

$x = \dfrac{\pi}{4}, \dfrac{5\pi}{4}$ $x = \arctan 5, \ \arctan 5 + \pi$

61. $2 \cos^2 x - 5 \cos x + 2 = 0$

$(2 \cos x - 1)(\cos x - 2) = 0$

$2 \cos x - 1 = 0$ or $\cos x - 2 = 0$

$\cos x = \dfrac{1}{2}$ $\cos x = 2$

$x = \dfrac{\pi}{3}, \dfrac{5\pi}{3}$ No solution

63. (a) $f(x) = \sin x + \cos x$

Maximum: $\left(\dfrac{\pi}{4}, \ \sqrt{2} \right)$

Minimum: $\left(\dfrac{5\pi}{4}, \ -\sqrt{2} \right)$

(b) $\cos x - \sin x = 0$

$\cos x = \sin x$

$1 = \dfrac{\sin x}{\cos x}$

$\tan x = 1$

$x = \dfrac{\pi}{4}, \dfrac{5\pi}{4}$

$f \left(\dfrac{\pi}{4} \right) = \sin \dfrac{\pi}{4} + \cos \dfrac{\pi}{4} = \dfrac{\sqrt{2}}{2} + \dfrac{\sqrt{2}}{2} = \sqrt{2}$

$f \left(\dfrac{5\pi}{4} \right) = \sin \dfrac{5\pi}{4} + \cos \dfrac{5\pi}{4} = -\sin \dfrac{\pi}{4} + \left(-\cos \dfrac{\pi}{4} \right) = -\dfrac{\sqrt{2}}{2} - \dfrac{\sqrt{2}}{2} = -\sqrt{2}$

Therefore, the maximum point in the interval $[0, 2\pi)$ is $\left(\pi/4, \ \sqrt{2} \right)$ and the minimum point is $\left(5\pi/4, \ -\sqrt{2} \right)$.

65. $f(x) = \tan \dfrac{\pi x}{4}$

Since $\tan \pi/4 = 1$, $x = 1$ is the smallest nonnegative fixed point.

67. $f(x) = \cos \dfrac{1}{x}$

(a) The domain of $f(x)$ is all real numbers x except $x = 0$.

(b) The graph has y-axis symmetry and a horizontal asymptote at $y = 1$.

(c) As $x \rightarrow 0$, $f(x)$ oscillates between -1 and 1.

(d) There are infinitely many solutions in the interval $[-1, 1]$. They occur at $x = \dfrac{2}{(2n + 1)\pi}$ where n is any integer.

(e) The greatest solution appears to occur at $x \approx 0.6366$.

69.
$$y = \frac{1}{12}(\cos 8t - 3 \sin 8t)$$

$$\frac{1}{12}(\cos 8t - 3 \sin 8t) = 0$$

$$\cos 8t = 3 \sin 8t$$

$$\frac{1}{3} = \tan 8t$$

$$8t \approx 0.32175 + n\pi$$

$$t \approx 0.04 + \frac{n\pi}{8}$$

In the interval $0 \le t \le 1$, $t \approx 0.04$, 0.43, and 0.83.

71. $S = 74.50 + 43.75 \sin \dfrac{\pi t}{6}$

t	1	2	3	4	5	6	7	8	9	10	11	12
S	96.4	112.4	118.3	112.4	96.4	74.5	52.6	36.6	30.8	36.6	52.6	74.5

Sales exceed 100,000 units during February, March, and April.

73.
$$\text{Range} = 300 \text{ feet}$$

$$v_0 = 100 \text{ feet per second}$$

$$r = \tfrac{1}{32}v_0^2 \sin 2\theta$$

$$\tfrac{1}{32}(100)^2 \sin 2\theta = 300$$

$$\sin 2\theta = 0.96$$

$$2\theta = \arcsin(0.96) \approx 73.74°$$

$$\theta \approx 36.9°$$

or

$$2\theta = 180° - \arcsin(0.96) \approx 106.26°$$

$$\theta \approx 53.1°$$

75. $h(t) = 53 + 50 \sin\left(\dfrac{\pi}{16}t - \dfrac{\pi}{2}\right)$

(a) $h(t) = 53$ when $50 \sin\left(\dfrac{\pi}{16}t - \dfrac{\pi}{2}\right) = 0$.

$\dfrac{\pi}{16}t - \dfrac{\pi}{2} = 0 \qquad$ or $\qquad \dfrac{\pi}{16}t - \dfrac{\pi}{2} = \pi$

$\qquad \dfrac{\pi}{16}t = \dfrac{\pi}{2} \qquad\qquad\qquad \dfrac{\pi}{16}t = \dfrac{3\pi}{2}$

$\qquad\qquad t = 8 \qquad\qquad\qquad\qquad t = 24$

The Ferris wheel will be 53 feet above ground at 8 seconds and at 24 seconds.

(b) The person will be at the top of the Ferris wheel when

$\sin\left(\dfrac{\pi}{16}t - \dfrac{\pi}{2}\right) = 1$.

$\dfrac{\pi}{16}t - \dfrac{\pi}{2} = \dfrac{\pi}{2}$

$\qquad \dfrac{\pi}{16}t = \pi$

$\qquad\qquad t = 16$

The first time this occurs is after 16 seconds. The period of this function is $\dfrac{2\pi}{\pi/16} = 32$. During 160 seconds, 5 cycles will take place and the person will be at the top of the ride 5 times, spaced 32 seconds apart. The times are: 16 seconds, 48 seconds, 80 seconds, 112 seconds, and 144 seconds.

77. $A = 2x \cos x$, $0 < x < \dfrac{\pi}{2}$

(a)

The maximum area of $A \approx 1.12$ occurs when $x \approx 0.86$.

(b) $A \geq 1$ for $0.6 < x < 1.1$

79. True. The period of $2 \sin 4t - 1$ is $\dfrac{\pi}{2}$ and the period of $2 \sin t - 1$ is 2π.

In the interval $[0, 2\pi)$ the first equation has four cycles whereas the second equation has only one cycle, thus the first equation has four times the x-intercepts (solutions) as the second equation.

81. $y_1 = 2 \sin x$

$y_2 = 3x + 1$

From the graph we see that there is only one point of intersection.

83.

$C = 90° - 66° = 24°$

$\cos 66° = \dfrac{22.3}{a}$

$a \cos 66° = 22.3$

$\qquad a = \dfrac{22.3}{\cos 66°} \approx 54.8$

$\tan 66° = \dfrac{b}{22.3}$

$\qquad b = 22.3 \tan 66° \approx 50.1$

85. $\theta = 390°$, $\theta' = 390° - 360° = 30°$, θ is in Quadrant I.

$$\sin 390° = \sin 30° = \frac{1}{2}$$

$$\cos 390° = \cos 30° = \frac{\sqrt{3}}{2}$$

$$\tan 390° = \tan 30° = \frac{1}{\sqrt{3}} = \frac{\sqrt{3}}{3}$$

87. $\theta = -1845°$, $\theta' = 45°$, θ is in Quadrant IV.

$$\sin(-1845°) = -\sin 45° = -\frac{\sqrt{2}}{2}$$

$$\cos(-1845°) = \cos 45° = \frac{\sqrt{2}}{2}$$

$$\tan(-1845°) = -\tan 45° = -1$$

89. $\tan \theta = \frac{250 \text{ feet}}{2 \text{ miles}} \times \frac{1 \text{ mile}}{5280 \text{ feet}} \approx 0.02367$

$$\theta \approx 1.36°$$

250 ft   2 mi

Not drawn to scale

91. Answers will vary.

Section 7.4   Sum and Difference Formulas

- You should know the sum and difference formulas.

  $$\sin(u \pm v) = \sin u \cos v \pm \cos u \sin v$$

  $$\cos(u \pm v) = \cos u \cos v \mp \sin u \sin v$$

  $$\tan(u \pm v) = \frac{\tan u \pm \tan v}{1 \mp \tan u \tan v}$$

- You should be able to use these formulas to find the values of the trigonometric functions of angles whose sums or differences are special angles.

- You should be able to use these formulas to solve trigonometric equations.

Vocabulary Check

1. $\sin u \cos v - \cos u \sin v$

2. $\cos u \cos v - \sin u \sin v$

3. $\frac{\tan u + \tan v}{1 - \tan u \tan v}$

4. $\sin u \cos v + \cos u \sin v$

5. $\cos u \cos v + \sin u \sin v$

6. $\frac{\tan u - \tan v}{1 + \tan u \tan v}$

1. (a) $\cos(120° + 45°) = \cos 120° \cos 45° - \sin 120° \sin 45°$

$$= \left(-\frac{1}{2}\right)\left(\frac{\sqrt{2}}{2}\right) - \left(\frac{\sqrt{3}}{2}\right)\left(\frac{\sqrt{2}}{2}\right)$$

$$= \frac{-\sqrt{2} - \sqrt{6}}{4}$$

(b) $\cos 120° + \cos 45° = -\frac{1}{2} + \frac{\sqrt{2}}{2} = \frac{-1 + \sqrt{2}}{2}$

3. (a) $\cos\left(\frac{\pi}{4} + \frac{\pi}{3}\right) = \cos \frac{\pi}{4} \cos \frac{\pi}{3} - \sin \frac{\pi}{4} \sin \frac{\pi}{3}$

$$= \frac{\sqrt{2}}{2} \cdot \frac{1}{2} - \frac{\sqrt{2}}{2} \cdot \frac{\sqrt{3}}{2}$$

$$= \frac{\sqrt{2} - \sqrt{6}}{4}$$

(b) $\cos \frac{\pi}{4} + \cos \frac{\pi}{3} = \frac{\sqrt{2}}{2} + \frac{1}{2} = \frac{\sqrt{2} + 1}{2}$

5. (a) $\sin\left(\frac{7\pi}{6} - \frac{\pi}{3}\right) = \sin \frac{5\pi}{6} = \sin \frac{\pi}{6} = \frac{1}{2}$

(b) $\sin \frac{7\pi}{6} - \sin \frac{\pi}{3} = -\frac{1}{2} - \frac{\sqrt{3}}{2} = \frac{-1 - \sqrt{3}}{2}$

7. $\sin 105° = \sin(60° + 45°)$

$\qquad = \sin 60° \cos 45° + \cos 60° \sin 45°$

$\qquad = \dfrac{\sqrt{3}}{2} \cdot \dfrac{\sqrt{2}}{2} + \dfrac{1}{2} \cdot \dfrac{\sqrt{2}}{2}$

$\qquad = \dfrac{\sqrt{2}}{4}\left(\sqrt{3} + 1\right)$

$\cos 105° = \cos(60° + 45°)$

$\qquad = \cos 60° \cos 45° - \sin 60° \sin 45°$

$\qquad = \dfrac{1}{2} \cdot \dfrac{\sqrt{2}}{2} - \dfrac{\sqrt{3}}{2} \cdot \dfrac{\sqrt{2}}{2}$

$\qquad = \dfrac{\sqrt{2}}{4}\left(1 - \sqrt{3}\right)$

$\tan 105° = \tan(60° + 45°)$

$\qquad = \dfrac{\tan 60° + \tan 45°}{1 - \tan 60° \tan 45°}$

$\qquad = \dfrac{\sqrt{3} + 1}{1 - \sqrt{3}} = \dfrac{\sqrt{3} + 1}{1 - \sqrt{3}} \cdot \dfrac{1 + \sqrt{3}}{1 + \sqrt{3}}$

$\qquad = \dfrac{4 + 2\sqrt{3}}{-2} = -2 - \sqrt{3}$

9. $\sin 195° = \sin(225° - 30°)$

$\qquad = \sin 225° \cos 30° - \cos 225° \sin 30°$

$\qquad = -\sin 45° \cos 30° + \cos 45° \sin 30°$

$\qquad = -\dfrac{\sqrt{2}}{2} \cdot \dfrac{\sqrt{3}}{2} + \dfrac{\sqrt{2}}{2} \cdot \dfrac{1}{2}$

$\qquad = \dfrac{\sqrt{2}}{4}\left(1 - \sqrt{3}\right)$

$\cos 195° = \cos(225° - 30°)$

$\qquad = \cos 225° \cos 30° + \sin 225° \sin 30°$

$\qquad = -\cos 45° \cos 30° - \sin 45° \sin 30°$

$\qquad = -\dfrac{\sqrt{2}}{2} \cdot \dfrac{\sqrt{3}}{2} - \dfrac{\sqrt{2}}{2} \cdot \dfrac{1}{2}$

$\qquad = -\dfrac{\sqrt{2}}{4}\left(\sqrt{3} + 1\right)$

$\tan 195° = \tan(225° - 30°)$

$\qquad = \dfrac{\tan 225° - \tan 30°}{1 + \tan 225° \tan 30°}$

$\qquad = \dfrac{\tan 45° - \tan 30°}{1 + \tan 45° \tan 30°}$

$\qquad = \dfrac{1 - \left(\dfrac{\sqrt{3}}{3}\right)}{1 + \left(\dfrac{\sqrt{3}}{3}\right)} = \dfrac{3 - \sqrt{3}}{3 + \sqrt{3}} \cdot \dfrac{3 - \sqrt{3}}{3 - \sqrt{3}}$

$\qquad = \dfrac{12 - 6\sqrt{3}}{6} = 2 - \sqrt{3}$

11. $\sin \dfrac{11\pi}{12} = \sin\left(\dfrac{3\pi}{4} + \dfrac{\pi}{6}\right)$

$\qquad = \sin \dfrac{3\pi}{4} \cos \dfrac{\pi}{6} + \cos \dfrac{3\pi}{4} \sin \dfrac{\pi}{6}$

$\qquad = \dfrac{\sqrt{2}}{2} \cdot \dfrac{\sqrt{3}}{2} + \left(-\dfrac{\sqrt{2}}{2}\right)\dfrac{1}{2}$

$\qquad = \dfrac{\sqrt{2}}{4}\left(\sqrt{3} - 1\right)$

$\cos \dfrac{11\pi}{12} = \cos\left(\dfrac{3\pi}{4} + \dfrac{\pi}{6}\right)$

$\qquad = \cos \dfrac{3\pi}{4} \cos \dfrac{\pi}{6} - \sin \dfrac{3\pi}{4} \sin \dfrac{\pi}{6}$

$\qquad = -\dfrac{\sqrt{2}}{2} \cdot \dfrac{\sqrt{3}}{2} - \dfrac{\sqrt{2}}{2} \cdot \dfrac{1}{2} = -\dfrac{\sqrt{2}}{4}\left(\sqrt{3} + 1\right)$

$\tan 195° = \tan(225° - 30°)$

$\qquad = \dfrac{\tan 225° - \tan 30°}{1 + \tan 225° \tan 30°}$

$\qquad = \dfrac{\tan 45° - \tan 30°}{1 + \tan 45° \tan 30°}$

$\qquad = \dfrac{1 - \left(\dfrac{\sqrt{3}}{3}\right)}{1 + \left(\dfrac{\sqrt{3}}{3}\right)} = \dfrac{3 - \sqrt{3}}{3 + \sqrt{3}} \cdot \dfrac{3 - \sqrt{3}}{3 - \sqrt{3}}$

$\qquad = \dfrac{12 - 6\sqrt{3}}{6} = 2 - \sqrt{3}$

13. $\sin \dfrac{17\pi}{12} = \sin\left(\dfrac{9\pi}{4} - \dfrac{5\pi}{6}\right)$

$\qquad = \sin\dfrac{9\pi}{4}\cos\dfrac{5\pi}{6} - \cos\dfrac{9\pi}{4}\sin\dfrac{5\pi}{6}$

$\qquad = \dfrac{\sqrt{2}}{2}\left(-\dfrac{\sqrt{3}}{2}\right) - \left(\dfrac{\sqrt{2}}{2}\right)\left(\dfrac{1}{2}\right)$

$\qquad = -\dfrac{\sqrt{2}}{4}\left(\sqrt{3} + 1\right)$

$\cos \dfrac{17\pi}{12} = \cos\left(\dfrac{9\pi}{4} - \dfrac{5\pi}{6}\right)$

$\qquad = \cos\dfrac{9\pi}{4}\cos\dfrac{5\pi}{6} + \sin\dfrac{9\pi}{4}\sin\dfrac{5\pi}{6}$

$\qquad = \dfrac{\sqrt{2}}{2}\left(-\dfrac{\sqrt{3}}{2}\right) + \dfrac{\sqrt{2}}{2}\left(\dfrac{1}{2}\right)$

$\qquad = \dfrac{\sqrt{2}}{4}\left(1 - \sqrt{3}\right)$

$\tan \dfrac{17\pi}{12} = \tan\left(\dfrac{9\pi}{4} - \dfrac{5\pi}{6}\right)$

$\qquad = \dfrac{\tan(9\pi/4) - \tan(5\pi/6)}{1 + \tan(9\pi/4)\tan(5\pi/6)}$

$\qquad = \dfrac{1 - \left(-\sqrt{3}/3\right)}{1 + \left(-\sqrt{3}/3\right)}$

$\qquad = \dfrac{3 + \sqrt{3}}{3 - \sqrt{3}} \cdot \dfrac{3 + \sqrt{3}}{3 + \sqrt{3}}$

$\qquad = \dfrac{12 + 6\sqrt{3}}{6} = 2 + \sqrt{3}$

15. $285 = 225 + 60$

$\sin 285° = \sin(225° + 60°)$

$\qquad = \sin 225°\cos 60° + \cos 225°\sin 60°$

$\qquad = -\dfrac{\sqrt{2}}{2}\left(\dfrac{1}{2}\right) - \dfrac{\sqrt{2}}{2}\left(\dfrac{\sqrt{3}}{2}\right) = -\dfrac{\sqrt{2}}{4}\left(\sqrt{3} + 1\right)$

$\cos 285° = \cos(225° + 60°)$

$\qquad = \cos 225°\cos 60° - \sin 225°\sin 60°$

$\qquad = -\dfrac{\sqrt{2}}{2}\left(\dfrac{1}{2}\right) - \left(-\dfrac{\sqrt{2}}{2}\right)\left(\dfrac{\sqrt{3}}{2}\right) = \dfrac{\sqrt{2}}{4}\left(\sqrt{3} - 1\right)$

$\tan 285° = \tan(225° + 60°)$

$\qquad = \dfrac{\tan 225° + \tan 60°}{1 - \tan 225°\tan 60°} = \dfrac{1 + \sqrt{3}}{1 - \sqrt{3}} \cdot \dfrac{1 + \sqrt{3}}{1 + \sqrt{3}}$

$\qquad = \dfrac{4 + 2\sqrt{3}}{-2} = -2 - \sqrt{3} = -\left(2 + \sqrt{3}\right)$

17. $-165° = -(120° + 45°)$

$\sin(-165°) = \sin[-(120° + 45°)]$

$\qquad = -\sin(120° + 45°)$

$\qquad = -[\sin 120°\cos 45° + \cos 120°\sin 45°]$

$\qquad = -\left[\dfrac{\sqrt{3}}{2} \cdot \dfrac{\sqrt{2}}{2} - \dfrac{1}{2} \cdot \dfrac{\sqrt{2}}{2}\right]$

$\qquad = -\dfrac{\sqrt{2}}{4}\left(\sqrt{3} - 1\right)$

$\cos(-165°) = \cos[-(120° + 45°)]$

$\qquad = \cos(120° + 45°)$

$\qquad = \cos 120°\cos 45° - \sin 120°\sin 45°$

$\qquad = -\dfrac{1}{2} \cdot \dfrac{\sqrt{2}}{2} - \dfrac{\sqrt{3}}{2} \cdot \dfrac{\sqrt{2}}{2}$

$\qquad = -\dfrac{\sqrt{2}}{4}\left(1 + \sqrt{3}\right)$

$\tan(-165°) = \tan[-(120° + 45°)]$

$\qquad = -\tan(120° + \tan 45°)$

$\qquad = -\dfrac{\tan 120° + \tan 45°}{1 - \tan 120°\tan 45°}$

$\qquad = -\dfrac{-\sqrt{3} + 1}{1 - \left(-\sqrt{3}\right)(1)}$

$\qquad = -\dfrac{1 - \sqrt{3}}{1 + \sqrt{3}} \cdot \dfrac{1 - \sqrt{3}}{1 - \sqrt{3}}$

$\qquad = -\dfrac{4 - 2\sqrt{3}}{-2}$

$\qquad = 2 - \sqrt{3}$

19. $\dfrac{13\pi}{12} = \dfrac{3\pi}{4} + \dfrac{\pi}{3}$

$\sin \dfrac{13\pi}{12} = \sin\left(\dfrac{3\pi}{4} + \dfrac{\pi}{3}\right)$

$\qquad = \sin \dfrac{3\pi}{4} \cos \dfrac{\pi}{3} + \cos \dfrac{3\pi}{4} \sin \dfrac{\pi}{3}$

$\qquad = \dfrac{\sqrt{2}}{2} \cdot \dfrac{1}{2} + \left(-\dfrac{\sqrt{2}}{2}\right)\left(\dfrac{\sqrt{3}}{2}\right)$

$\qquad = \dfrac{\sqrt{2}}{4}\left(1 - \sqrt{3}\right)$

$\cos \dfrac{13\pi}{12} = \cos\left(\dfrac{3\pi}{4} + \dfrac{\pi}{3}\right)$

$\qquad = \cos \dfrac{3\pi}{4} \cos \dfrac{\pi}{3} - \sin \dfrac{3\pi}{4} \sin \dfrac{\pi}{3}$

$\qquad = -\dfrac{\sqrt{2}}{2} \cdot \dfrac{1}{2} - \dfrac{\sqrt{2}}{2} \cdot \dfrac{\sqrt{3}}{2} = -\dfrac{\sqrt{2}}{4}\left(1 + \sqrt{3}\right)$

$\tan \dfrac{13\pi}{12} = \tan\left(\dfrac{3\pi}{4} + \dfrac{\pi}{3}\right)$

$\qquad = \dfrac{\tan\left(\dfrac{3\pi}{4}\right) + \tan\left(\dfrac{\pi}{3}\right)}{1 - \tan\left(\dfrac{3\pi}{4}\right)\tan\left(\dfrac{\pi}{3}\right)}$

$\qquad = \dfrac{-1 + \sqrt{3}}{1 - (-1)\left(\sqrt{3}\right)}$

$\qquad = -\dfrac{1 - \sqrt{3}}{1 + \sqrt{3}} \cdot \dfrac{1 - \sqrt{3}}{1 - \sqrt{3}}$

$\qquad = -\dfrac{4 - 2\sqrt{3}}{-2}$

$\qquad = 2 - \sqrt{3}$

21. $-\dfrac{13\pi}{12} = -\left(\dfrac{3\pi}{4} + \dfrac{\pi}{3}\right)$

$\sin\left[-\left(\dfrac{3\pi}{4} + \dfrac{\pi}{3}\right)\right] = -\sin\left(\dfrac{3\pi}{4} + \dfrac{\pi}{3}\right)$

$\qquad = -\left[\sin \dfrac{3\pi}{4} \cos \dfrac{\pi}{3} + \cos \dfrac{3\pi}{4} \sin \dfrac{\pi}{3}\right]$

$\qquad = -\left[\dfrac{\sqrt{2}}{2}\left(\dfrac{1}{2}\right) + \left(-\dfrac{\sqrt{2}}{2}\right)\left(\dfrac{\sqrt{3}}{2}\right)\right]$

$\qquad = -\dfrac{\sqrt{2}}{4}\left(1 - \sqrt{3}\right) = \dfrac{\sqrt{2}}{4}\left(\sqrt{3} - 1\right)$

$\cos\left[-\left(\dfrac{3\pi}{4} + \dfrac{\pi}{3}\right)\right] = \cos\left(\dfrac{3\pi}{4} + \dfrac{\pi}{3}\right)$

$\qquad = \cos \dfrac{3\pi}{4} \cos \dfrac{\pi}{3} - \sin \dfrac{3\pi}{4} \sin \dfrac{\pi}{3}$

$\qquad = -\dfrac{\sqrt{2}}{2}\left(\dfrac{1}{2}\right) - \dfrac{\sqrt{2}}{2}\left(\dfrac{\sqrt{3}}{2}\right) = -\dfrac{\sqrt{2}}{4}\left(\sqrt{3} + 1\right)$

$\tan\left[-\left(\dfrac{3\pi}{4} + \dfrac{\pi}{3}\right)\right] = -\tan\left(\dfrac{3\pi}{4} + \dfrac{\pi}{3}\right)$

$\qquad = -\dfrac{\tan \dfrac{3\pi}{4} + \tan \dfrac{\pi}{3}}{1 - \tan \dfrac{3\pi}{4} \tan \dfrac{\pi}{3}} = -\dfrac{-1 + \sqrt{3}}{1 - (-\sqrt{3})}$

$\qquad = \dfrac{1 - \sqrt{3}}{1 + \sqrt{3}} \cdot \dfrac{1 - \sqrt{3}}{1 - \sqrt{3}} = \dfrac{4 - 2\sqrt{3}}{-2} = -2 + \sqrt{3}$

23. $\cos 25° \cos 15° - \sin 25° \sin 15° = \cos(25° + 15°) = \cos 40°$

25. $\dfrac{\tan 325° - \tan 86°}{1 + \tan 325° \tan 86°} = \tan(325° - 86°) = \tan 239°$

27. $\sin 3 \cos 1.2 - \cos 3 \sin 1.2 = \sin(3 - 1.2) = \sin 1.8$

29. $\dfrac{\tan 2x + \tan x}{1 - \tan 2x \tan x} = \tan(2x + x) = \tan 3x$

31. $\sin 330° \cos 30° - \cos 330° \sin 30° = \sin(330° - 30°)$

$$= \sin 300°$$

$$= -\dfrac{\sqrt{3}}{2}$$

33. $\sin \dfrac{\pi}{12} \cos \dfrac{\pi}{4} + \cos \dfrac{\pi}{12} \sin \dfrac{\pi}{4} = \sin\left(\dfrac{\pi}{12} + \dfrac{\pi}{4}\right)$

$$= \sin \dfrac{\pi}{3}$$

$$= \dfrac{\sqrt{3}}{2}$$

35. $\dfrac{\tan 25° + \tan 110°}{1 - \tan 25° \tan 110°} = \tan(25° + 110°)$

$$= \tan 135°$$

$$= -1$$

For Exercises 37–43, we have:

$\sin u = \dfrac{5}{13}$, u in Quadrant II $\Rightarrow$ $\cos u = -\dfrac{12}{13}$, $\tan u = -\dfrac{5}{12}$

$\cos v = -\dfrac{3}{5}$, v in Quadrant II $\Rightarrow$ $\sin v = \dfrac{4}{5}$, $\tan v = -\dfrac{4}{3}$,

Figures for Exercises 37–43

37. $\sin(u + v) = \sin u \cos v + \cos u \sin v$

$$= \left(\dfrac{5}{13}\right)\left(-\dfrac{3}{5}\right) + \left(-\dfrac{12}{13}\right)\left(\dfrac{4}{5}\right)$$

$$= -\dfrac{63}{65}$$

39. $\cos(u + v) = \cos u \cos v - \sin u \sin v$

$$= \left(-\dfrac{12}{13}\right)\left(-\dfrac{3}{5}\right) - \left(\dfrac{5}{13}\right)\left(\dfrac{4}{5}\right)$$

$$= \dfrac{16}{65}$$

41. $\tan(u + v) = \dfrac{\tan u + \tan v}{1 - \tan u \tan v} = \dfrac{-\dfrac{5}{12} + \left(-\dfrac{4}{3}\right)}{1 - \left(-\dfrac{5}{12}\right)\left(-\dfrac{4}{3}\right)} = \dfrac{-\dfrac{21}{12}}{1 - \dfrac{5}{9}}$

$$= \left(-\dfrac{7}{4}\right)\left(\dfrac{9}{4}\right) = -\dfrac{63}{16}$$

43. $\sec(v - u) = \dfrac{1}{\cos(v - u)} = \dfrac{1}{\cos v \cos u + \sin v \sin u}$

$$= \dfrac{1}{\left(-\dfrac{3}{5}\right)\left(-\dfrac{12}{13}\right) + \left(\dfrac{4}{5}\right)\left(\dfrac{5}{13}\right)} = \dfrac{1}{\left(\dfrac{36}{65}\right) + \left(\dfrac{20}{65}\right)} = \dfrac{1}{\dfrac{56}{65}}$$

$$= \dfrac{65}{56}$$

For Exercises 45–49, we have:

$\sin u = -\frac{7}{25}$, u in Quadrant III $\Longrightarrow \cos u = -\frac{24}{25}$, $\tan u = \frac{7}{24}$

$\cos v = -\frac{4}{5}$, v in Quadrant III $\Longrightarrow \sin v = -\frac{3}{5}$, $\tan v = \frac{3}{4}$

Figures for Exercises 45–49

45. $\cos(u + v) = \cos u \cos v - \sin u \sin v$

$= \left(-\frac{24}{25}\right)\left(-\frac{4}{5}\right) - \left(-\frac{7}{25}\right)\left(-\frac{3}{5}\right)$

$= \frac{3}{5}$

47. $\tan(u - v) = \dfrac{\tan u - \tan v}{1 + \tan u \tan v}$

$= \dfrac{\frac{7}{24} - \frac{3}{4}}{1 + \left(\frac{7}{24}\right)\left(\frac{3}{4}\right)} = \dfrac{-\frac{11}{24}}{\frac{39}{32}} = -\dfrac{44}{117}$

49. $\sec(u + v) = \dfrac{1}{\cos(u + v)} = \dfrac{1}{\frac{3}{5}} = \dfrac{5}{3}$

Use Exercise 45 for $\cos(u + v)$.

51. $\sin(\arcsin x + \arccos x) = \sin(\arcsin x)\cos(\arccos x) + \sin(\arccos x)\cos(\arcsin x)$

$= x \cdot x + \sqrt{1 - x^2} \cdot \sqrt{1 - x^2}$

$= x^2 + 1 - x^2$

$= 1$

$\theta = \arcsin x$

$\theta = \arccos x$

53. $\cos(\arccos x + \arcsin x) = \cos(\arccos x)\cos(\arcsin x) - \sin(\arccos x)\sin(\arcsin x)$

$= x \cdot \sqrt{1 - x^2} - \sqrt{1 - x^2} \cdot x$

$= 0$

(Use the triangles in Exercise 51.)

55. $\sin(3\pi - x) = \sin 3\pi \cos x - \sin x \cos 3\pi$

$= (0)(\cos x) - (-1)(\sin x)$

$= \sin x$

57. $\sin\left(\dfrac{\pi}{6} + x\right) = \sin \dfrac{\pi}{6} \cos x + \cos \dfrac{\pi}{6} \sin x$

$= \dfrac{1}{2}\left(\cos x + \sqrt{3}\sin x\right)$

59. $\cos(\pi - \theta) + \sin\left(\frac{\pi}{2} + \theta\right) = \cos\pi\cos\theta + \sin\pi\sin\theta + \sin\frac{\pi}{2}\cos\theta + \cos\frac{\pi}{2}\sin\theta$

$$= (-1)(\cos\theta) + (0)(\sin\theta) + (1)(\cos\theta) + (\sin\theta)(0)$$

$$= -\cos\theta + \cos\theta$$

$$= 0$$

61. $\cos(x + y)\cos(x - y) = (\cos x\cos y - \sin x\sin y)(\cos x\cos y + \sin x\sin y)$

$$= \cos^2 x\cos^2 y - \sin^2 x\sin^2 y$$

$$= \cos^2 x(1 - \sin^2 y) - \sin^2 x\sin^2 y$$

$$= \cos^2 x - \cos^2 x\sin^2 y - \sin^2 x\sin^2 y$$

$$= \cos^2 x - \sin^2 y(\cos^2 x + \sin^2 x)$$

$$= \cos^2 x - \sin^2 y$$

63. $\sin(x + y) + \sin(x - y) = \sin x\cos y + \cos x\sin y + \sin x\cos y - \cos x\sin y$

$$= 2\sin x\cos y$$

65. $\cos\left(\frac{3\pi}{2} - x\right) = \cos\frac{3\pi}{2}\cos x + \sin\frac{3\pi}{2}\sin x$

$$= (0)(\cos x) + (-1)(\sin x)$$

$$= -\sin x$$

67. $\sin\left(\frac{3\pi}{2} + \theta\right) = \sin\frac{3\pi}{2}\cos\theta + \cos\frac{3\pi}{2}\sin\theta$

$$= (-1)(\cos\theta) + (0)(\sin\theta)$$

$$= -\cos\theta$$

69.
$$\sin\left(x + \frac{\pi}{3}\right) + \sin\left(x - \frac{\pi}{3}\right) = 1$$

$$\sin x\cos\frac{\pi}{3} + \cos x\sin\frac{\pi}{3} + \sin x\cos\frac{\pi}{3} - \cos x\sin\frac{\pi}{3} = 1$$

$$2\sin x(0.5) = 1$$

$$\sin x = 1$$

$$x = \frac{\pi}{2}$$

71.
$$\cos\left(x + \frac{\pi}{4}\right) - \cos\left(x - \frac{\pi}{4}\right) = 1$$

$$\cos x\cos\frac{\pi}{4} - \sin x\sin\frac{\pi}{4} - \left(\cos x\cos\frac{\pi}{4} + \sin x\sin\frac{\pi}{4}\right) = 1$$

$$-2\sin x\left(\frac{\sqrt{2}}{2}\right) = 1$$

$$-\sqrt{2}\sin x = 1$$

$$\sin x = -\frac{1}{\sqrt{2}}$$

$$\sin x = -\frac{\sqrt{2}}{2}$$

$$x = \frac{5\pi}{4}, \frac{7\pi}{4}$$

73. Analytically: $\cos\left(x + \dfrac{\pi}{4}\right) + \cos\left(x - \dfrac{\pi}{4}\right) = 1$

$$\cos x \cos\frac{\pi}{4} - \sin x \sin\frac{\pi}{4} + \cos x \cos\frac{\pi}{4} + \sin x \sin\frac{\pi}{4} = 1$$

$$2\cos x\left(\frac{\sqrt{2}}{2}\right) = 1$$

$$\sqrt{2}\cos x = 1$$

$$\cos x = \frac{1}{\sqrt{2}}$$

$$\cos x = \frac{\sqrt{2}}{2}$$

$$x = \frac{\pi}{4}, \frac{7\pi}{4}$$

Graphically: Graph $y_1 = \cos\left(x + \dfrac{\pi}{4}\right) + \cos\left(x - \dfrac{\pi}{4}\right)$ and $y_2 = 1$.

The points of intersection occur at $x = \dfrac{\pi}{4}$ and $x = \dfrac{7\pi}{4}$.

75. $y = \dfrac{1}{3}\sin 2t + \dfrac{1}{4}\cos 2t$

(a) $a = \dfrac{1}{3},\ b = \dfrac{1}{4},\ B = 2$

$\qquad C = \arctan\dfrac{b}{a} = \arctan\dfrac{3}{4} \approx 0.6435$

$\qquad y \approx \sqrt{\left(\dfrac{1}{3}\right)^2 + \left(\dfrac{1}{4}\right)^2}\,\sin(2t + 0.6435)$

$\qquad = \dfrac{5}{12}\sin(2t + 0.6435)$

(b) Amplitude: $\dfrac{5}{12}$ feet

(c) Frequency: $\dfrac{1}{\text{period}} = \dfrac{B}{2\pi} = \dfrac{2}{2\pi} = \dfrac{1}{\pi}$ cycle per second

77. False.

$\sin(u \pm v) = \sin u \cos v \pm \cos u \sin v$

79. False. $\cos\left(x - \dfrac{\pi}{2}\right) = \cos x \cos\dfrac{\pi}{2} + \sin x \sin\dfrac{\pi}{2}$

$$= (\cos x)(0) + (\sin x)(1)$$

$$= \sin x$$

81. $\cos(n\pi + \theta) = \cos n\pi \cos\theta - \sin n\pi \sin\theta$

$$= (-1)^n(\cos\theta) - (0)(\sin\theta)$$

$$= (-1)^n(\cos\theta), \text{ where } n \text{ is an integer.}$$

83. $C = \arctan\dfrac{b}{a} \implies \sin C = \dfrac{b}{\sqrt{a^2 + b^2}},\ \cos C = \dfrac{a}{\sqrt{a^2 + b^2}}$

$\sqrt{a^2 + b^2}\,\sin(B\theta + C) = \sqrt{a^2 + b^2}\left(\sin B\theta \cdot \dfrac{a}{\sqrt{a^2 + b^2}} + \dfrac{b}{\sqrt{a^2 + b^2}} \cdot \cos B\theta\right) = a\sin B\theta + b\cos B\theta$

85. $\sin\theta + \cos\theta$

$a = 1,\ b = 1,\ B = 1$

(a) $C = \arctan\dfrac{b}{a} = \arctan 1 = \dfrac{\pi}{4}$

$\sin\theta + \cos\theta = \sqrt{a^2 + b^2}\,\sin(B\theta + C)$

$\qquad\qquad = \sqrt{2}\,\sin\!\left(\theta + \dfrac{\pi}{4}\right)$

(b) $C = \arctan\dfrac{a}{b} = \arctan 1 = \dfrac{\pi}{4}$

$\sin\theta + \cos\theta = \sqrt{a^2 + b^2}\,\cos(B\theta - C)$

$\qquad\qquad = \sqrt{2}\,\cos\!\left(\theta - \dfrac{\pi}{4}\right)$

87. $12\sin 3\theta + 5\cos 3\theta$

$a = 12,\ b = 5,\ B = 3$

(a) $C = \arctan\dfrac{b}{a} = \arctan\dfrac{5}{12} \approx 0.3948$

$12\sin 3\theta + 5\cos 3\theta = \sqrt{a^2 + b^2}\,\sin(B\theta + C)$

$\qquad\qquad\qquad \approx 13\sin(3\theta + 0.3948)$

(b) $C = \arctan\dfrac{a}{b} = \arctan\dfrac{12}{5} \approx 1.1760$

$12\sin 3\theta + 5\cos 3\theta = \sqrt{a^2 + b^2}\,\cos(B\theta - C)$

$\qquad\qquad\qquad \approx 13\cos(3\theta - 1.1760)$

89. $C = \arctan\dfrac{b}{a} = \dfrac{\pi}{2} \implies a = 0$

$\sqrt{a^2 + b^2} = 2 \implies b = 2$

$B = 1$

$2\sin\!\left(\theta + \dfrac{\pi}{2}\right) = (0)(\sin\theta) + (2)(\cos\theta) = 2\cos\theta$

91. $\dfrac{\cos(x + h) - \cos x}{h} = \dfrac{\cos x\cos h - \sin x\sin h - \cos x}{h}$

$\qquad\qquad = \dfrac{\cos x\cos h - \cos x - \sin x\sin h}{h}$

$\qquad\qquad = \dfrac{\cos x(\cos h - 1) - \sin x\sin h}{h}$

$\qquad\qquad = \dfrac{\cos x(\cos h - 1)}{h} - \dfrac{\sin x\sin h}{h}$

93.

$m_1 = \tan\alpha$ and $m_2 = \tan\beta$

$\beta + \delta = 90° \implies \delta = 90° - \beta$

$\alpha + \theta + \delta = 90° \implies \alpha + \theta + (90° - \beta) = 90° \implies \theta = \beta - \alpha$

Therefore, $\theta = \arctan m_2 - \arctan m_1$.

For $y = x$ and $y = \sqrt{3}x$ we have $m_1 = 1$ and $m_2 = \sqrt{3}$.

$\theta = \arctan\sqrt{3} - \arctan 1$

$\quad = 60° - 45°$

$\quad = 15°$

95.

Conjecture: $\sin^2\!\left(\theta + \dfrac{\pi}{4}\right) + \sin^2\!\left(\theta - \dfrac{\pi}{4}\right) = 1$

$\sin^2\!\left(\theta + \dfrac{\pi}{4}\right) + \sin^2\!\left(\theta - \dfrac{\pi}{4}\right) = \left[\sin\theta\cos\dfrac{\pi}{4} + \cos\theta\sin\dfrac{\pi}{4}\right]^2 + \left[\sin\theta\cos\dfrac{\pi}{4} - \cos\theta\sin\dfrac{\pi}{4}\right]^2$

$\qquad\qquad = \left[\dfrac{\sin\theta}{\sqrt{2}} + \dfrac{\cos\theta}{\sqrt{2}}\right]^2 + \left[\dfrac{\sin\theta}{\sqrt{2}} - \dfrac{\cos\theta}{\sqrt{2}}\right]^2$

$\qquad\qquad = \dfrac{\sin^2\theta}{2} + \sin\theta\cos\theta + \dfrac{\cos^2\theta}{2} + \dfrac{\sin^2\theta}{2} - \sin\theta\cos\theta + \dfrac{\cos^2\theta}{2}$

$\qquad\qquad = \sin^2\theta + \cos^2\theta$

$\qquad\qquad = 1$

97.

$$f(x) = 5(x - 3)$$

$$y = 5(x - 3)$$

$$\frac{y}{5} = x - 3$$

$$\frac{y}{5} + 3 = x$$

$$\frac{x}{5} + 3 = y$$

$$f^{-1}(x) = \frac{x + 15}{5}$$

$$f(f^{-1}(x)) = f\left(\frac{x + 15}{5}\right) = 5\left[\frac{x + 15}{5} - 3\right]$$

$$= 5\left(\frac{x + 15}{5}\right) - 5(3)$$

$$= x + 15 - 15$$

$$= x$$

$$f^{-1}(f(x)) = f^{-1}(5(x - 3)) = \frac{5(x - 3) + 15}{5}$$

$$= \frac{5x - 15 + 15}{5}$$

$$= \frac{5x}{5}$$

$$= x$$

99. $f(x) = x^2 - 8$

f is not one-to-one so f^{-1} does not exist.

101. $\log_3 3^{4x-3} = 4x - 3$

103. $e^{\ln(6x-3)} = 6x - 3$

Section 7.5   Multiple-Angle and Product-to-Sum Formulas

- You should know the following double-angle formulas.

  (a) $\sin 2u = 2 \sin u \cos u$

  (b) $\cos 2u = \cos^2 u - \sin^2 u$

     $= 2 \cos^2 u - 1$

     $= 1 - 2 \sin^2 u$

  (c) $\tan 2u = \dfrac{2 \tan u}{1 - \tan^2 u}$

- You should be able to reduce the power of a trigonometric function.

  (a) $\sin^2 u = \dfrac{1 - \cos 2u}{2}$

  (b) $\cos^2 u = \dfrac{1 + \cos 2u}{2}$

  (c) $\tan^2 u = \dfrac{1 - \cos 2u}{1 + \cos 2u}$

- You should be able to use the half-angle formulas. The signs of $\sin \dfrac{u}{2}$ and $\cos \dfrac{u}{2}$ depend on the quadrant in which $\dfrac{u}{2}$ lies.

  (a) $\sin \dfrac{u}{2} = \pm \sqrt{\dfrac{1 - \cos u}{2}}$

  (b) $\cos \dfrac{u}{2} = \pm \sqrt{\dfrac{1 + \cos u}{2}}$

  (c) $\tan \dfrac{u}{2} = \dfrac{1 - \cos u}{\sin u} = \dfrac{\sin u}{1 + \cos u}$

—CONTINUED—

—CONTINUED—

■ You should be able to use the product-sum formulas.

(a) $\sin u \sin v = \frac{1}{2}[\cos(u - v) - \cos(u + v)]$

(b) $\cos u \cos v = \frac{1}{2}[\cos(u - v) + \cos(u + v)]$

(c) $\sin u \cos v = \frac{1}{2}[\sin(u + v) + \sin(u - v)]$

(d) $\cos u \sin v = \frac{1}{2}[\sin(u + v) - \sin(u - v)]$

■ You should be able to use the sum-product formulas.

(a) $\sin x + \sin y = 2 \sin\left(\dfrac{x + y}{2}\right) \cos\left(\dfrac{x - y}{2}\right)$

(b) $\sin x - \sin y = 2 \cos\left(\dfrac{x + y}{2}\right) \sin\left(\dfrac{x - y}{2}\right)$

(c) $\cos x + \cos y = 2 \cos\left(\dfrac{x + y}{2}\right) \cos\left(\dfrac{x - y}{2}\right)$

(d) $\cos x - \cos y = -2 \sin\left(\dfrac{x + y}{2}\right) \sin\left(\dfrac{x - y}{2}\right)$

Vocabulary Check

1. $2 \sin u \cos u$

2. $\cos^2 u$

3. $\cos^2 u - \sin^2 u = 2 \cos^2 u - 1 = 1 - 2 \sin^2 u$

4. $\tan^2 u$

5. $\pm\sqrt{\dfrac{1 - \cos u}{2}}$

6. $\dfrac{1 - \cos u}{\sin u} = \dfrac{\sin u}{1 + \cos u}$

7. $\dfrac{1}{2}[\cos(u - v) + \cos(u + v)]$

8. $\dfrac{1}{2}[\sin(u + v) + \sin(u - v)]$

9. $2 \sin\left(\dfrac{u + v}{2}\right) \cos\left(\dfrac{u - v}{2}\right)$

10. $-2 \sin\left(\dfrac{u + v}{2}\right) \sin\left(\dfrac{u - v}{2}\right)$

Figure for Exercises 1–7

$\sin \theta = \dfrac{\sqrt{17}}{17}$

$\cos \theta = \dfrac{4\sqrt{17}}{17}$

$\tan \theta = \dfrac{1}{4}$

1. $\sin \theta = \dfrac{\sqrt{17}}{17}$

3. $\cos 2\theta = 2 \cos^2 \theta - 1$

$= 2\left(\dfrac{4\sqrt{17}}{17}\right)^2 - 1$

$= \dfrac{32}{17} - 1$

$= \dfrac{15}{17}$

5. $\tan 2\theta = \dfrac{2 \tan \theta}{1 - \tan^2 \theta}$

$= \dfrac{2\left(\dfrac{1}{4}\right)}{1 - \left(\dfrac{1}{4}\right)^2}$

$= \dfrac{\dfrac{1}{2}}{1 - \dfrac{1}{16}}$

$= \dfrac{1}{2} \cdot \dfrac{16}{15}$

$= \dfrac{8}{15}$

7. $\csc 2\theta = \dfrac{1}{\sin 2\theta} = \dfrac{1}{2\sin\theta\cos\theta} = \dfrac{1}{2\left(\dfrac{\sqrt{17}}{17}\right)\left(\dfrac{4\sqrt{17}}{17}\right)} = \dfrac{17}{8}$

9.
$$\sin 2x - \sin x = 0$$
$$2\sin x\cos x - \sin x = 0$$
$$\sin x(2\cos x - 1) = 0$$
$$\sin x = 0 \quad \text{or} \quad 2\cos x - 1 = 0$$
$$x = 0, \pi \qquad\qquad \cos x = \frac{1}{2}$$
$$x = \frac{\pi}{3}, \frac{5\pi}{3}$$
$$x = 0, \frac{\pi}{3}, \pi, \frac{5\pi}{3}$$

11. $4\sin x\cos x = 1$
$$2\sin 2x = 1$$
$$\sin 2x = \frac{1}{2}$$
$$2x = \frac{\pi}{6} + 2n\pi \quad \text{or} \quad 2x = \frac{5\pi}{6} + 2n\pi$$
$$x = \frac{\pi}{12} + n\pi \qquad\qquad x = \frac{5\pi}{12} + n\pi$$
$$x = \frac{\pi}{12}, \frac{13\pi}{12} \qquad\qquad x = \frac{5\pi}{12}, \frac{17\pi}{12}$$

13.
$$\cos 2x - \cos x = 0$$
$$\cos 2x = \cos x$$
$$\cos^2 x - \sin^2 x = \cos x$$
$$\cos^2 x - (1 - \cos^2 x) - \cos x = 0$$
$$2\cos^2 x - \cos x - 1 = 0$$
$$(2\cos x + 1)(\cos x - 1) = 0$$
$$2\cos x + 1 = 0 \qquad \text{or} \quad \cos x - 1 = 0$$
$$\cos x = -\frac{1}{2} \qquad\qquad \cos x = 1$$
$$x = \frac{2\pi}{3}, \frac{4\pi}{3} \qquad\qquad x = 0$$

15.
$$\tan 2x - \cot x = 0$$
$$\frac{2\tan x}{1 - \tan^2 x} = \cot x$$
$$2\tan x = \cot x(1 - \tan^2 x)$$
$$2\tan x = \cot x - \cot x\tan^2 x$$
$$2\tan x = \cot x - \tan x$$
$$3\tan x = \cot x$$
$$3\tan x - \cot x = 0$$
$$3\tan x - \frac{1}{\tan x} = 0$$
$$\frac{3\tan^2 x - 1}{\tan x} = 0$$
$$\frac{1}{\tan x}(3\tan^2 x - 1) = 0$$
$$\cot x(3\tan^2 x - 1) = 0$$

$$\cot x = 0 \qquad \text{or} \quad 3\tan^2 x - 1 = 0$$
$$x = \frac{\pi}{2}, \frac{3\pi}{2} \qquad\qquad \tan^2 x = \frac{1}{3}$$
$$\tan x = \pm\frac{\sqrt{3}}{3}$$
$$x = \frac{\pi}{6}, \frac{5\pi}{6}, \frac{7\pi}{6}, \frac{11\pi}{6}$$

$$x = \frac{\pi}{6}, \frac{\pi}{2}, \frac{5\pi}{6}, \frac{7\pi}{6}, \frac{3\pi}{2}, \frac{11\pi}{6}$$

17.
$$\sin 4x = -2 \sin 2x$$
$$\sin 4x + 2 \sin 2x = 0$$
$$2 \sin 2x \cos 2x + 2 \sin 2x = 0$$
$$2 \sin 2x(\cos 2x + 1) = 0$$

$2 \sin 2x = 0$ or $\cos 2x + 1 = 0$

$\sin 2x = 0$ $\cos 2x = -1$

$2x = n\pi$ $2x = \pi + 2n\pi$

$x = \dfrac{n}{2}\pi$ $x = \dfrac{\pi}{2} + n\pi$

$x = 0, \dfrac{\pi}{2}, \pi, \dfrac{3\pi}{2}$ $x = \dfrac{\pi}{2}, \dfrac{3\pi}{2}$

19. $6 \sin x \cos x = 3(2 \sin x \cos x)$
$$= 3 \sin 2x$$

21. $4 - 8 \sin^2 x = 4(1 - 2 \sin^2 x)$
$$= 4 \cos 2x$$

23. $\sin u = -\dfrac{4}{5}, \ \pi < u < \dfrac{3\pi}{2} \implies \cos u = -\dfrac{3}{5}$

$\sin 2u = 2 \sin u \cos u = 2\left(-\dfrac{4}{5}\right)\left(-\dfrac{3}{5}\right) = \dfrac{24}{25}$

$\cos 2u = \cos^2 u - \sin^2 u = \dfrac{9}{25} - \dfrac{16}{25} = -\dfrac{7}{25}$

$\tan 2u = \dfrac{2 \tan u}{1 - \tan^2 u} = \dfrac{2\left(\frac{4}{3}\right)}{1 - \frac{16}{9}} = \dfrac{8}{3}\left(-\dfrac{9}{7}\right) = -\dfrac{24}{7}$

25. $\tan u = \dfrac{3}{4}, \ 0 < u < \dfrac{\pi}{2} \implies \sin u = \dfrac{3}{5}$ and $\cos u = \dfrac{4}{5}$

$\sin 2u = 2 \sin u \cos u = 2\left(\dfrac{3}{5}\right)\left(\dfrac{4}{5}\right) = \dfrac{24}{25}$

$\cos 2u = \cos^2 u - \sin^2 u = \dfrac{16}{25} - \dfrac{9}{25} = \dfrac{7}{25}$

$\tan 2u = \dfrac{2 \tan u}{1 - \tan^2 u} = \dfrac{2\left(\frac{3}{4}\right)}{1 - \frac{9}{16}} = \dfrac{3}{2}\left(\dfrac{16}{7}\right) = \dfrac{24}{7}$

27. $\sec u = -\dfrac{5}{2}, \ \dfrac{\pi}{2} < u < \pi \implies \sin u = \dfrac{\sqrt{21}}{5}$ and $\cos u = -\dfrac{2}{5}$

$\sin 2u = 2 \sin u \cos u = 2\left(\dfrac{\sqrt{21}}{5}\right)\left(-\dfrac{2}{5}\right) = -\dfrac{4\sqrt{21}}{25}$

$\cos 2u = \cos^2 u - \sin^2 u = \left(-\dfrac{2}{5}\right)^2 - \left(\dfrac{\sqrt{21}}{5}\right)^2 = -\dfrac{17}{25}$

$\tan 2u = \dfrac{2 \tan u}{1 - \tan^2 u} = \dfrac{2\left(-\dfrac{\sqrt{21}}{2}\right)}{1 - \left(-\dfrac{\sqrt{21}}{2}\right)^2}$

$= \dfrac{-\sqrt{21}}{1 - \dfrac{21}{4}} = \dfrac{4\sqrt{21}}{-17}$

29. $\cos^4 x = (\cos^2 x)(\cos^2 x) = \left(\dfrac{1 + \cos 2x}{2}\right)\left(\dfrac{1 + \cos 2x}{2}\right) = \dfrac{1 + 2\cos 2x + \cos^2 2x}{4}$

$$= \dfrac{1 + 2\cos 2x + \dfrac{1 + \cos 4x}{2}}{4}$$

$$= \dfrac{2 + 4\cos 2x + 1 + \cos 4x}{8}$$

$$= \dfrac{3 + 4\cos 2x + \cos 4x}{8}$$

$$= \dfrac{1}{8}(3 + 4\cos 2x + \cos 4x)$$

31. $(\sin^2 x)(\cos^2 x) = \left(\dfrac{1 - \cos 2x}{2}\right)\left(\dfrac{1 + \cos 2x}{2}\right)$

$$= \dfrac{1 - \cos^2 2x}{4}$$

$$= \dfrac{1}{4}\left(1 - \dfrac{1 + \cos 4x}{2}\right)$$

$$= \dfrac{1}{8}(2 - 1 - \cos 4x)$$

$$= \dfrac{1}{8}(1 - \cos 4x)$$

33. $\sin^2 x \cos^4 x = \sin^2 x \cos^2 x \cos^2 x = \left(\dfrac{1 - \cos 2x}{2}\right)\left(\dfrac{1 + \cos 2x}{2}\right)\left(\dfrac{1 + \cos 2x}{2}\right)$

$$= \dfrac{1}{8}(1 - \cos 2x)(1 + \cos 2x)(1 + \cos 2x)$$

$$= \dfrac{1}{8}(1 - \cos^2 2x)(1 + \cos 2x)$$

$$= \dfrac{1}{8}(1 + \cos 2x - \cos^2 2x - \cos^3 2x)$$

$$= \dfrac{1}{8}\left[1 + \cos 2x - \left(\dfrac{1 + \cos 4x}{2}\right) - \cos 2x\left(\dfrac{1 + \cos 4x}{2}\right)\right]$$

$$= \dfrac{1}{16}[2 + 2\cos 2x - 1 - \cos 4x - \cos 2x - \cos 2x \cos 4x]$$

$$= \dfrac{1}{16}(1 + \cos 2x - \cos 4x - \cos 2x \cos 4x)$$

Figure for Exercises 35–39

$\sin \theta = \frac{8}{17}$

$\cos \theta = \frac{15}{17}$

35. $\cos\dfrac{\theta}{2} = \sqrt{\dfrac{1+\cos\theta}{2}} = \sqrt{\dfrac{1+\frac{15}{17}}{2}} = \sqrt{\dfrac{\frac{32}{17}}{2}} = \sqrt{\dfrac{16}{17}} = \dfrac{4\sqrt{17}}{17}$

37. $\tan\dfrac{\theta}{2} = \dfrac{\sin\theta}{1+\cos\theta} = \dfrac{\frac{8}{17}}{1+\frac{15}{17}} = \dfrac{8}{17}\cdot\dfrac{17}{32} = \dfrac{1}{4}$

39. $\csc\dfrac{\theta}{2} = \dfrac{1}{\sin\frac{\theta}{2}} = \dfrac{1}{\sqrt{\dfrac{(1-\cos\theta)}{2}}} = \dfrac{1}{\sqrt{\dfrac{1-\frac{15}{17}}{2}}} = \dfrac{1}{\sqrt{\dfrac{1}{17}}} = \sqrt{17}$

41. $\sin 75° = \sin\left(\dfrac{1}{2}\cdot 150°\right) = \sqrt{\dfrac{1-\cos 150°}{2}} = \sqrt{\dfrac{1+\frac{\sqrt{3}}{2}}{2}}$

$\qquad = \dfrac{1}{2}\sqrt{2+\sqrt{3}}$

$\quad\cos 75° = \cos\left(\dfrac{1}{2}\cdot 150°\right) = \sqrt{\dfrac{1+\cos 150°}{2}} = \sqrt{\dfrac{1-\frac{\sqrt{3}}{2}}{2}}$

$\qquad = \dfrac{1}{2}\sqrt{2-\sqrt{3}}$

$\quad\tan 75° = \tan\left(\dfrac{1}{2}\cdot 150°\right) = \dfrac{\sin 150°}{1+\cos 150°} = \dfrac{\frac{1}{2}}{1-\frac{\sqrt{3}}{2}}$

$\qquad = \dfrac{1}{2-\sqrt{3}}\cdot\dfrac{2+\sqrt{3}}{2+\sqrt{3}} = \dfrac{2+\sqrt{3}}{4-3} = 2+\sqrt{3}$

43. $\sin 112° 30' = \sin\left(\dfrac{1}{2}\cdot 225°\right) = \sqrt{\dfrac{1-\cos 225°}{2}} = \sqrt{\dfrac{1+\frac{\sqrt{2}}{2}}{2}} = \dfrac{1}{2}\sqrt{2+\sqrt{2}}$

$\quad\cos 112° 30' = \cos\left(\dfrac{1}{2}\cdot 225°\right) = -\sqrt{\dfrac{1+\cos 225°}{2}} = -\sqrt{\dfrac{1-\frac{\sqrt{2}}{2}}{2}} = -\dfrac{1}{2}\sqrt{2-\sqrt{2}}$

$\quad\tan 112° 30' = \tan\left(\dfrac{1}{2}\cdot 225°\right) = \dfrac{\sin 225°}{1+\cos 225°} = \dfrac{-\frac{\sqrt{2}}{2}}{1-\frac{\sqrt{2}}{2}} = -1-\sqrt{2}$

45. $\sin\dfrac{\pi}{8} = \sin\left[\dfrac{1}{2}\left(\dfrac{\pi}{4}\right)\right] = \sqrt{\dfrac{1-\cos\frac{\pi}{4}}{2}} = \dfrac{1}{2}\sqrt{2-\sqrt{2}}$

$\quad\cos\dfrac{\pi}{8} = \cos\left[\dfrac{1}{2}\left(\dfrac{\pi}{4}\right)\right] = \sqrt{\dfrac{1+\cos\frac{\pi}{4}}{2}} = \dfrac{1}{2}\sqrt{2+\sqrt{2}}$

$\quad\tan\dfrac{\pi}{8} = \tan\left[\dfrac{1}{2}\left(\dfrac{\pi}{4}\right)\right] = \dfrac{\sin\frac{\pi}{4}}{1+\cos\frac{\pi}{4}} = \dfrac{\frac{\sqrt{2}}{2}}{1+\frac{\sqrt{2}}{2}} = \sqrt{2}-1$

47. $\sin\dfrac{3\pi}{8} = \sin\left(\dfrac{1}{2} \cdot \dfrac{3\pi}{4}\right) = \sqrt{\dfrac{1 - \cos\dfrac{3\pi}{4}}{2}} = \sqrt{\dfrac{1 + \dfrac{\sqrt{2}}{2}}{2}} = \dfrac{1}{2}\sqrt{2 + \sqrt{2}}$

$\cos\dfrac{3\pi}{8} = \cos\left(\dfrac{1}{2} \cdot \dfrac{3\pi}{4}\right) = \sqrt{\dfrac{1 + \cos\dfrac{3\pi}{4}}{2}} = \sqrt{\dfrac{1 - \dfrac{\sqrt{2}}{2}}{2}} = \dfrac{1}{2}\sqrt{2 - \sqrt{2}}$

$\tan\dfrac{3\pi}{8} = \tan\left(\dfrac{1}{2} \cdot \dfrac{3\pi}{4}\right) = \dfrac{\sin\dfrac{3\pi}{4}}{1 + \cos\dfrac{3\pi}{4}} = \dfrac{\dfrac{\sqrt{2}}{2}}{1 - \dfrac{\sqrt{2}}{2}} = \dfrac{\dfrac{\sqrt{2}}{2}}{\dfrac{(2 - \sqrt{2})}{2}} = \dfrac{\sqrt{2}}{2 - \sqrt{2}} = \sqrt{2} + 1$

49. $\sin u = \dfrac{5}{13}, \ \dfrac{\pi}{2} < u < \pi \implies \cos u = -\dfrac{12}{13}$

$\sin\left(\dfrac{u}{2}\right) = \sqrt{\dfrac{1 - \cos u}{2}} = \sqrt{\dfrac{1 + \dfrac{12}{13}}{2}} = \dfrac{5\sqrt{26}}{26}$

$\cos\left(\dfrac{u}{2}\right) = \sqrt{\dfrac{1 + \cos u}{2}} = \sqrt{\dfrac{1 - \dfrac{12}{13}}{2}} = \dfrac{\sqrt{26}}{26}$

$\tan\left(\dfrac{u}{2}\right) = \dfrac{\sin u}{1 + \cos u} = \dfrac{\dfrac{5}{13}}{1 - \dfrac{12}{13}} = 5$

51. $\tan u = -\dfrac{5}{8}, \ \dfrac{3\pi}{2} < u < 2\pi \implies \sin u = -\dfrac{5}{\sqrt{89}}$ and $\cos u = \dfrac{8}{\sqrt{89}}$

$\sin\left(\dfrac{u}{2}\right) = \sqrt{\dfrac{1 - \cos u}{2}} = \sqrt{\dfrac{1 - \dfrac{8}{\sqrt{89}}}{2}} \sqrt{\dfrac{\sqrt{89} - 8}{2\sqrt{89}}} = \sqrt{\dfrac{89 - 8\sqrt{89}}{178}}$

$\cos\left(\dfrac{u}{2}\right) = -\sqrt{\dfrac{1 + \cos u}{2}} = -\sqrt{\dfrac{1 + \dfrac{8}{\sqrt{89}}}{2}} = -\sqrt{\dfrac{\sqrt{89} + 8}{2\sqrt{89}}} = -\sqrt{\dfrac{89 + 8\sqrt{89}}{178}}$

$\tan\left(\dfrac{u}{2}\right) = \dfrac{1 - \cos u}{\sin u} = \dfrac{1 - \dfrac{8}{\sqrt{89}}}{-\dfrac{5}{\sqrt{89}}} = \dfrac{8 - \sqrt{89}}{5}$

53. $\csc u = -\dfrac{5}{3}, \ \pi < u < \dfrac{3\pi}{2} \implies \sin u = -\dfrac{3}{5}$ and $\cos u = -\dfrac{4}{5}$

$\sin\left(\dfrac{u}{2}\right) = \sqrt{\dfrac{1 - \cos u}{2}} = \sqrt{\dfrac{1 + \dfrac{4}{5}}{2}} = \dfrac{3\sqrt{10}}{10}$

$\cos\left(\dfrac{u}{2}\right) = -\sqrt{\dfrac{1 + \cos u}{2}} = -\sqrt{\dfrac{1 - \dfrac{4}{5}}{2}} = -\dfrac{\sqrt{10}}{10}$

$\tan\left(\dfrac{u}{2}\right) = \dfrac{1 - \cos u}{\sin u} = \dfrac{1 + \dfrac{4}{5}}{-\dfrac{3}{5}} = -3$

55. $\sqrt{\dfrac{1 - \cos 6x}{2}} = |\sin 3x|$

57. $-\sqrt{\dfrac{1 - \cos 8x}{1 + \cos 8x}} = -\dfrac{\sqrt{\dfrac{1 - \cos 8x}{2}}}{\sqrt{\dfrac{1 + \cos 8x}{2}}}$

$= -\left|\dfrac{\sin 4x}{\cos 4x}\right| = -|\tan 4x|$

59. $\sin \dfrac{x}{2} + \cos x = 0$

$\pm \sqrt{\dfrac{1 - \cos x}{2}} = -\cos x$

$\dfrac{1 - \cos x}{2} = \cos^2 x$

$0 = 2\cos^2 x + \cos x - 1$

$ = (2\cos x - 1)(\cos x + 1)$

$\cos x = \dfrac{1}{2} \quad \text{or} \quad \cos x = -1$

$x = \dfrac{\pi}{3}, \dfrac{5\pi}{3} \qquad x = \pi$

By checking these values in the original equation, we see that $x = \pi/3$ and $x = 5\pi/3$ are extraneous, and $x = \pi$ is the only solution.

61. $\cos \dfrac{x}{2} - \sin x = 0$

$\pm \sqrt{\dfrac{1 + \cos x}{2}} = \sin x$

$\dfrac{1 + \cos x}{2} = \sin^2 x$

$1 + \cos x = 2\sin^2 x$

$1 + \cos x = 2 - 2\cos^2 x$

$2\cos^2 x + \cos x - 1 = 0$

$(2\cos x - 1)(\cos x + 1) = 0$

$2\cos x - 1 = 0 \quad \text{or} \quad \cos x + 1 = 0$

$\cos x = \dfrac{1}{2} \qquad\qquad \cos x = -1$

$x = \dfrac{\pi}{3}, \dfrac{5\pi}{3} \qquad\qquad x = \pi$

$x = \dfrac{\pi}{3}, \ \pi, \ \dfrac{5\pi}{3}$

$\pi/3$, π, and $5\pi/3$ are all solutions to the equation.

63. $6 \sin \dfrac{\pi}{4} \cos \dfrac{\pi}{4} = 6 \cdot \dfrac{1}{2}\left[\sin\left(\dfrac{\pi}{4} + \dfrac{\pi}{4}\right) + \sin\left(\dfrac{\pi}{4} - \dfrac{\pi}{4}\right) \right] = 3\left(\sin \dfrac{\pi}{2} + \sin 0 \right)$

65. $10 \cos 75° \cos 15° = 10\left(\dfrac{1}{2}\right)\left[\cos(75° - 15°) + \cos(75° + 15°)\right] = 5[\cos 60° + \cos 90°]$

67. $\cos 4\theta \sin 6\theta = \dfrac{1}{2}[\sin(4\theta + 6\theta) - \sin(4\theta - 6\theta)] = \dfrac{1}{2}[\sin 10\theta - \sin(-2\theta)] = \dfrac{1}{2}(\sin 10\theta + \sin 2\theta)$

69. $5\cos(-5\beta)\cos 3\beta = 5 \cdot \dfrac{1}{2}[\cos(-5\beta - 3\beta) + \cos(-5\beta + 3\beta)] = \dfrac{5}{2}[\cos(-8\beta) + \cos(-2\beta)]$

$ = \dfrac{5}{2}(\cos 8\beta + \cos 2\beta)$

71. $\sin(x + y)\sin(x - y) = \dfrac{1}{2}(\cos 2y - \cos 2x)$

73. $\cos(\theta - \pi)\sin(\theta + \pi) = \dfrac{1}{2}[\sin 2\theta - \sin(-2\pi)]$

$ = \dfrac{1}{2}(\sin 2\theta + \sin 2\pi)$

75. $\sin 5\theta - \sin 3\theta = 2\cos\left(\dfrac{5\theta + 3\theta}{2}\right)\sin\left(\dfrac{5\theta - 3\theta}{2}\right)$

$ = 2\cos 4\theta \sin \theta$

77. $\cos 6x + \cos 2x = 2\cos\left(\dfrac{6x + 2x}{2}\right)\cos\left(\dfrac{6x - 2x}{2}\right)$

$ = 2\cos 4x \cos 2x$

79. $\sin(\alpha + \beta) - \sin(\alpha - \beta) = 2\cos\left(\dfrac{\alpha + \beta + \alpha - \beta}{2}\right)\sin\left(\dfrac{\alpha + \beta - \alpha + \beta}{2}\right) = 2\cos \alpha \sin \beta$

81. $\cos\left(\theta + \dfrac{\pi}{2}\right) - \cos\left(\theta - \dfrac{\pi}{2}\right) = -2\sin\left[\dfrac{\left(\theta + \dfrac{\pi}{2}\right) + \left(\theta - \dfrac{\pi}{2}\right)}{2}\right]\sin\left[\dfrac{\left(\theta + \dfrac{\pi}{2}\right) - \left(\theta - \dfrac{\pi}{2}\right)}{2}\right] = -2\sin\theta\sin\dfrac{\pi}{2}$

83. $\sin 60° + \sin 30° = 2\sin\left(\dfrac{60° + 30°}{2}\right)\cos\left(\dfrac{60° - 30°}{2}\right) = 2\sin 45°\cos 15°$

$\sin 60° + \sin 30° = \dfrac{\sqrt{3}}{2} + \dfrac{1}{2} = \dfrac{\sqrt{3} + 1}{2}$

85. $\cos\dfrac{3\pi}{4} - \cos\dfrac{\pi}{4} = -2\sin\left(\dfrac{\dfrac{3\pi}{4} + \dfrac{\pi}{4}}{2}\right)\sin\left(\dfrac{\dfrac{3\pi}{4} - \dfrac{\pi}{4}}{2}\right) = -2\sin\dfrac{\pi}{2}\sin\dfrac{\pi}{4}$

$\cos\dfrac{3\pi}{4} - \cos\dfrac{\pi}{4} = -\dfrac{\sqrt{2}}{2} - \dfrac{\sqrt{2}}{2} = -\sqrt{2}$

87. $\sin 6x + \sin 2x = 0$

$2\sin\left(\dfrac{6x + 2x}{2}\right)\cos\left(\dfrac{6x - 2x}{2}\right) = 0$

$2(\sin 4x)\cos 2x = 0$

$\sin 4x = 0 \quad \text{or} \quad \cos 2x = 0$

$4x = n\pi \qquad\qquad 2x = \dfrac{\pi}{2} + n\pi$

$x = \dfrac{n\pi}{4} \qquad\qquad x = \dfrac{\pi}{4} + \dfrac{n\pi}{2}$

In the interval $[0, 2\pi)$ we have

$x = 0,\ \dfrac{\pi}{4},\ \dfrac{\pi}{2},\ \dfrac{3\pi}{4},\ \pi,\ \dfrac{5\pi}{4},\ \dfrac{3\pi}{2},\ \dfrac{7\pi}{4}.$

89. $\dfrac{\cos 2x}{\sin 3x - \sin x} - 1 = 0$

$\dfrac{\cos 2x}{\sin 3x - \sin x} = 1$

$\dfrac{\cos 2x}{2\cos 2x\sin x} = 1$

$2\sin x = 1$

$\sin x = \dfrac{1}{2}$

$x = \dfrac{\pi}{6},\ \dfrac{5\pi}{6}$

Figure for Exercises 91–93

91. $\sin^2\alpha = \left(\dfrac{5}{13}\right)^2 = \dfrac{25}{169}$

$\sin^2\alpha = 1 - \cos^2\alpha = 1 - \left(\dfrac{12}{13}\right)^2$

$= 1 - \dfrac{144}{169} = \dfrac{25}{169}$

93. $\sin\alpha\cos\beta = \left(\dfrac{5}{13}\right)\left(\dfrac{4}{5}\right) = \dfrac{4}{13}$

$\sin\alpha\cos\beta = \cos\left(\dfrac{\pi}{2} - \alpha\right)\sin\left(\dfrac{\pi}{2} - \beta\right)$

$= \left(\dfrac{5}{13}\right)\left(\dfrac{4}{5}\right) = \dfrac{4}{13}$

95. $\csc 2\theta = \dfrac{1}{\sin 2\theta}$

$\qquad = \dfrac{1}{2\sin\theta\cos\theta}$

$\qquad = \dfrac{1}{\sin\theta}\cdot\dfrac{1}{2\cos\theta}$

$\qquad = \dfrac{\csc\theta}{2\cos\theta}$

97. $\cos^2 2\alpha - \sin^2 2\alpha = \cos[2(2\alpha)]$

$\qquad\qquad\qquad\qquad = \cos 4\alpha$

99. $(\sin x + \cos x)^2 = \sin^2 x + 2\sin x\cos x + \cos^2 x$

$\qquad\qquad\qquad = (\sin^2 x + \cos^2 x) + 2\sin x\cos x$

$\qquad\qquad\qquad = 1 + \sin 2x$

101. $1 + \cos 10y = 1 + \cos^2 5y - \sin^2 5y$

$\qquad\qquad = 1 + \cos^2 5y - (1 - \cos^2 5y)$

$\qquad\qquad = 2\cos^2 5y$

103. $\sec\dfrac{u}{2} = \dfrac{1}{\cos\dfrac{u}{2}}$

$\qquad = \pm\sqrt{\dfrac{2}{1 + \cos u}}$

$\qquad = \pm\sqrt{\dfrac{2\sin u}{\sin u(1 + \cos u)}}$

$\qquad = \pm\sqrt{\dfrac{2\sin u}{\sin u + \sin u\cos u}}$

$\qquad = \pm\sqrt{\dfrac{\dfrac{2\sin u}{\cos u}}{\dfrac{\sin u}{\cos u} + \dfrac{\sin u\cos u}{\cos u}}}$

$\qquad = \pm\sqrt{\dfrac{2\tan u}{\tan u + \sin u}}$

105. $\dfrac{\sin x \pm \sin y}{\cos x + \cos y} = \dfrac{2\sin\left(\dfrac{x \pm y}{2}\right)\cos\left(\dfrac{x \mp y}{2}\right)}{2\cos\left(\dfrac{x + y}{2}\right)\cos\left(\dfrac{x - y}{2}\right)}$

$\qquad\qquad\qquad = \tan\left(\dfrac{x \pm y}{2}\right)$

107. $\dfrac{\cos 4x + \cos 2x}{\sin 4x + \sin 2x} = \dfrac{2\cos\left(\dfrac{4x + 2x}{2}\right)\cos\left(\dfrac{4x - 2x}{2}\right)}{2\sin\left(\dfrac{4x + 2x}{2}\right)\cos\left(\dfrac{4x - 2x}{2}\right)}$

$\qquad\qquad\qquad = \dfrac{2\cos 3x\cos x}{2\sin 3x\cos x} = \cot 3x$

109. $\sin\left(\dfrac{\pi}{6} + x\right) + \sin\left(\dfrac{\pi}{6} - x\right) = 2\sin\dfrac{\pi}{6}\cos x$

$\qquad\qquad\qquad\qquad = 2\cdot\dfrac{1}{2}\cos x$

$\qquad\qquad\qquad\qquad = \cos x$

111.

Let $y_1 = \cos(3x)$ and
$$y_2 = (\cos x)^3 - 3(\sin x)^2 \cos x.$$

$\cos 3\beta = \cos(2\beta + \beta)$

$\quad = \cos 2\beta \cos\beta - \sin 2\beta \sin \beta$

$\quad = (\cos^2 \beta - \sin^2 \beta) \cos \beta - 2 \sin \beta \cos \beta \sin \beta$

$\quad = \cos^3 \beta - \sin^2 \beta \cos \beta - 2 \sin^2 \beta \cos \beta$

$\quad = \cos^3 \beta - 3 \sin^2 \beta \cos \beta$

113.

Let $y_1 = \dfrac{(\cos 4x - \cos 2x)}{(2 \sin 3x)}$

and $y_2 = -\sin x.$

$$\frac{\cos 4x - \cos 2x}{2 \sin 3x} = \frac{-2 \sin\left(\dfrac{4x + 2x}{2}\right) \sin\left(\dfrac{4x - 2x}{2}\right)}{2 \sin 3x}$$

$$= \frac{-2 \sin 3x \sin x}{2 \sin 3x} = -\sin x$$

115. $\sin^2 x = \dfrac{1 - \cos 2x}{2} = \dfrac{1}{2} - \dfrac{\cos 2x}{2}$

117. $\sin(2 \arcsin x) = 2 \sin(\arcsin x) \cos(\arcsin x)$

$\qquad\qquad\qquad\; = 2x\sqrt{1 - x^2}$

119. $\dfrac{1}{32}(75)^2 \sin 2\theta = 130$

$\qquad \sin 2\theta = \dfrac{130(32)}{75^2}$

$\qquad\quad \theta = \dfrac{1}{2} \sin^{-1}\left(\dfrac{130(32)}{75^2}\right)$

$\qquad\quad \theta \approx 23.85°$

121. $\sin \dfrac{\theta}{2} = \dfrac{1}{M}$

(a) $\sin \dfrac{\theta}{2} = 1$

$\quad \dfrac{\theta}{2} = \arcsin 1$

$\quad \dfrac{\theta}{2} = \dfrac{\pi}{2}$

$\quad \theta = \pi$

(c) $\dfrac{S}{760} = 1$

$\quad S = 760$ miles per hour

$\quad \dfrac{S}{760} = 4.5$

$\quad S = 3420$ miles per hour

(b) $\sin \dfrac{\theta}{2} = \dfrac{1}{4.5}$

$\quad \dfrac{\theta}{2} = \arcsin\left(\dfrac{1}{4.5}\right)$

$\quad \theta = 2 \arcsin\left(\dfrac{1}{4.5}\right)$

$\quad \theta \approx 0.4482$

(d) $\sin \dfrac{\theta}{2} = \dfrac{1}{M}$

$\quad \dfrac{\theta}{2} = \arcsin\left(\dfrac{1}{M}\right)$

$\quad \theta = 2 \arcsin\left(\dfrac{1}{M}\right)$

123. False. For $u < 0$,

$$\sin 2u = -\sin(-2u)$$

$$= -2\sin(-u)\cos(-u)$$

$$= -2(-\sin u)\cos u$$

$$= 2\sin u \cos u.$$

125. (a) $y = 4\sin\dfrac{x}{2} + \cos x$

Maximum: $(\pi, 3)$

(b) $\qquad 2\cos\dfrac{x}{2} - \sin x = 0$

$$2\left(\pm\sqrt{\dfrac{1 + \cos x}{2}}\right) = \sin x$$

$$4\left(\dfrac{1 + \cos x}{2}\right) = \sin^2 x$$

$$2(1 + \cos x) = 1 - \cos^2 x$$

$$\cos^2 x + 2\cos x + 1 = 0$$

$$(\cos x + 1)^2 = 0$$

$$\cos x = -1$$

$$x = \pi$$

127. $f(x) = \sin^4 x + \cos^4 x$

(a) $\sin^4 x + \cos^4 x = (\sin^2 x)^2 + (\cos^2 x)^2$

$$= \left(\dfrac{1 - \cos 2x}{2}\right)^2 + \left(\dfrac{1 + \cos 2x}{2}\right)^2$$

$$= \dfrac{1}{4}[(1 - \cos 2x)^2 + (1 + \cos 2x)^2]$$

$$= \dfrac{1}{4}(1 - 2\cos 2x + \cos^2 2x + 1 + 2\cos 2x + \cos^2 2x)$$

$$= \dfrac{1}{4}(2 + 2\cos^2 2x)$$

$$= \dfrac{1}{4}\left[2 + 2\left(\dfrac{1 + \cos 2(2x)}{2}\right)\right]$$

$$= \dfrac{1}{4}(3 + \cos 4x)$$

(b) $\sin^4 x + \cos^4 x = (\sin^2 x)^2 + \cos^4 x$

$$= (1 - \cos^2 x)^2 + \cos^4 x$$

$$= 1 - 2\cos^2 x + \cos^4 x + \cos^4 x$$

$$= 2\cos^4 x - 2\cos^2 x + 1$$

(c) $\sin^4 x + \cos^4 x = \sin^4 x + 2\sin^2 x \cos^2 x + \cos^4 x - 2\sin^2 x \cos^2 x$

$$= (\sin^2 x + \cos^2 x)^2 - 2\sin^2 x \cos^2 x$$

$$= 1 - 2\sin^2 x \cos^2 x$$

(d) $1 - 2\sin^2 x \cos^2 x = 1 - (2\sin x \cos x)(\sin x \cos x)$

$$= 1 - (\sin 2x)\left(\dfrac{1}{2}\sin 2x\right)$$

$$= 1 - \dfrac{1}{2}\sin^2 2x$$

(e) No, it does not mean that one of you is wrong. There is often more than one way to rewrite a trigonometric expression.

129. (a)

(b) $d = \sqrt{(-1 - 5)^2 + (4 - 2)^2} = \sqrt{(-6)^2 + (2)^2}$

$= \sqrt{40} = 2\sqrt{10}$

(c) Midpoint: $\left(\dfrac{5 + (-1)}{2}, \dfrac{2 + 4}{2}\right) = (2, 3)$

131. (a)

(b) $d = \sqrt{\left(\dfrac{4}{3} - 0\right)^2 + \left(\dfrac{5}{2} - \dfrac{1}{2}\right)^2} = \sqrt{\dfrac{16}{9} + 4}$

$= \sqrt{\dfrac{52}{9}} = \dfrac{2\sqrt{13}}{3}$

(c) Midpoint: $\left(\dfrac{0 + \frac{4}{3}}{2}, \dfrac{\frac{1}{2} + \frac{5}{2}}{2}\right) = \left(\dfrac{2}{3}, \dfrac{3}{2}\right)$

133. (a) Complement: $90° - 55° = 35°$

Supplement: $180° - 55° = 125°$

(b) Complement: Not possible. $162° > 90°$

Supplement: $180° - 162° = 18°$

135. (a) Complement: $\dfrac{\pi}{2} - \dfrac{\pi}{18} = \dfrac{4\pi}{9}$

Supplement: $\pi - \dfrac{\pi}{18} = \dfrac{17\pi}{18}$

(b) Complement: $\dfrac{\pi}{2} - \dfrac{9\pi}{20} = \dfrac{\pi}{20}$

Supplement: $\pi - \dfrac{9\pi}{20} = \dfrac{11\pi}{20}$

137. Let x = profit for September,
then $x + 0.16x$ = profit for October.

$x + (x + 0.16x) = 507,600$

$2.16x = 507,600$

$x = 235,000$

$x + 0.16x = 272,600$

Profit for September: \$235,000

Profit for October: \$272,600

139.

$d^2 = 90^2 + 90^2$

$= 16,200$

$d = \sqrt{16,200}$

$= 90\sqrt{2}$

≈ 127 feet

Review Exercises for Chapter 7

1. $\dfrac{1}{\cos x} = \sec x$

3. $\dfrac{1}{\sec x} = \cos x$

5. $\dfrac{\cos x}{\sin x} = \cot x$

7. $\sin x = \dfrac{3}{5},\ \cos x = \dfrac{4}{5}$

$\tan x = \dfrac{\sin x}{\cos x} = \dfrac{\frac{3}{5}}{\frac{4}{5}} = \dfrac{3}{4}$

$\cot x = \dfrac{1}{\tan x} = \dfrac{4}{3}$

$\sec x = \dfrac{1}{\cos x} = \dfrac{5}{4}$

$\csc x = \dfrac{1}{\sin x} = \dfrac{5}{3}$

9. $\sin\left(\dfrac{\pi}{2} - x\right) = \dfrac{\sqrt{2}}{2} \Rightarrow \cos x = \dfrac{1}{\sqrt{2}} = \dfrac{\sqrt{2}}{2}$

$\sin x = -\dfrac{\sqrt{2}}{2}$

$\tan x = \dfrac{\sin x}{\cos x} = \dfrac{-\dfrac{1}{\sqrt{2}}}{\dfrac{1}{\sqrt{2}}} = -1$

$\cot x = \dfrac{1}{\tan x} = -1$

$\sec x = \dfrac{1}{\cos x} = \sqrt{2}$

$\csc x = \dfrac{1}{\sin x} = -\sqrt{2}$

11. $\dfrac{1}{\cot^2 x + 1} = \dfrac{1}{\csc^2 x} = \sin^2 x$

13. $\tan^2 x(\csc^2 x - 1) = \tan^2 x(\cot^2 x) = \tan^2 x\left(\dfrac{1}{\tan^2 x}\right) = 1$

15. $\dfrac{\sin\left(\dfrac{\pi}{2} - \theta\right)}{\sin \theta} = \dfrac{\cos \theta}{\sin \theta} = \cot \theta$

17. $\cos^2 x + \cos^2 x \cot^2 x = \cos^2 x(1 + \cot^2 x) = \cos^2 x(\csc^2 x)$

$= \cos^2 x\left(\dfrac{1}{\sin^2 x}\right) = \dfrac{\cos^2 x}{\sin^2 x} = \cot^2 x$

19. $(\tan x + 1)^2 \cos x = (\tan^2 x + 2\tan x + 1)\cos x$

$= (\sec^2 x + 2\tan x)\cos x$

$= \sec^2 x \cos x + 2\left(\dfrac{\sin x}{\cos x}\right)\cos x = \sec x + 2\sin x$

21. $\dfrac{1}{\csc \theta + 1} - \dfrac{1}{\csc \theta - 1} = \dfrac{(\csc \theta - 1) - (\csc \theta + 1)}{(\csc \theta + 1)(\csc \theta - 1)}$

$= \dfrac{-2}{\csc^2 \theta - 1}$

$= \dfrac{-2}{\cot^2 \theta}$

$= -2\tan^2 \theta$

23. $\csc^2 x - \csc x \cot x = \dfrac{1}{\sin^2 x} - \left(\dfrac{1}{\sin x}\right)\left(\dfrac{\cos x}{\sin x}\right)$

$= \dfrac{1 - \cos x}{\sin^2 x}$

25. $\cos x(\tan^2 x + 1) = \cos x \sec^2 x$

$= \dfrac{1}{\sec x}\sec^2 x$

$= \sec x$

27. $\cos\left(x + \dfrac{\pi}{2}\right) = \cos x \cos \dfrac{\pi}{2} - \sin x \sin \dfrac{\pi}{2}$

$= (\cos x)(0) - (\sin x)(1)$

$= -\sin x$

29. $\dfrac{1}{\tan \theta \csc \theta} = \dfrac{1}{\dfrac{\sin \theta}{\cos \theta} \cdot \dfrac{1}{\sin \theta}} = \cos \theta$

31. $\sin^5 x \cos^2 x = \sin^4 x \cos^2 x \sin x$

$= (1 - \cos^2 x)^2 \cos^2 x \sin x$

$= (1 - 2\cos^2 x + \cos^4 x) \cos^2 x \sin x$

$= (\cos^2 x - 2\cos^4 x + \cos^6 x) \sin x$

33. $\sin x = \sqrt{3} - \sin x$

$$\sin x = \frac{\sqrt{3}}{2}$$

$$x = \frac{\pi}{3} + 2\pi n, \frac{2\pi}{3} + 2\pi n$$

35. $3\sqrt{3} \tan u = 3$

$$\tan u = \frac{1}{\sqrt{3}}$$

$$u = \frac{\pi}{6} + n\pi$$

37. $3 \csc^2 x = 4$

$$\csc^2 x = \frac{4}{3}$$

$$\sin x = \pm\frac{\sqrt{3}}{2}$$

$$x = \frac{\pi}{3} + 2\pi n, \frac{2\pi}{3} + 2\pi n, \frac{4\pi}{3} + 2\pi n, \frac{5\pi}{3} + 2\pi n$$

These can be combined as:

$$x = \frac{\pi}{3} + n\pi \quad \text{or} \quad x = \frac{2\pi}{3} + n\pi$$

39. $\qquad 2\cos^2 x - \cos x = 1$

$$2\cos^2 x - \cos x - 1 = 0$$

$$(2\cos x + 1)(\cos x - 1) = 0$$

$$2\cos x + 1 = 0 \qquad\qquad \cos x - 1 = 0$$

$$\cos x = -\frac{1}{2} \qquad\qquad \cos x = 1$$

$$x = \frac{2\pi}{3}, \frac{4\pi}{3} \qquad\qquad x = 0$$

41. $\qquad \cos^2 x + \sin x = 1$

$$1 - \sin^2 x + \sin x - 1 = 0$$

$$-\sin x(\sin x - 1) = 0$$

$$\sin x = 0 \qquad \sin x - 1 = 0$$

$$x = 0, \pi \qquad \sin x = 1$$

$$x = \frac{\pi}{2}$$

43. $2\sin 2x - \sqrt{2} = 0$

$$\sin 2x = \frac{\sqrt{2}}{2}$$

$$2x = \frac{\pi}{4} + 2\pi n, \frac{3\pi}{4} + 2\pi n$$

$$x = \frac{\pi}{8} + \pi n, \frac{3\pi}{8} + \pi n$$

$$x = \frac{\pi}{8}, \frac{3\pi}{8}, \frac{9\pi}{8}, \frac{11\pi}{8}$$

45. $\cos 4x(\cos x - 1) = 0$

$$\cos 4x = 0 \qquad\qquad\qquad \cos x - 1 = 0$$

$$4x = \frac{\pi}{2} + 2\pi n, \frac{3\pi}{2} + 2\pi n \qquad\qquad \cos x = 1$$

$$x = \frac{\pi}{8} + \frac{\pi}{2}n, \frac{3\pi}{8} + \frac{\pi}{2}n \qquad\qquad x = 0$$

$$x = 0, \frac{\pi}{8}, \frac{3\pi}{8}, \frac{5\pi}{8}, \frac{7\pi}{8}, \frac{9\pi}{8}, \frac{11\pi}{8}, \frac{13\pi}{8}, \frac{15\pi}{8}$$

47. $\sin^2 x - 2\sin x = 0$

$$\sin x(\sin x - 2) = 0$$

$$\sin x = 0 \qquad \sin x - 2 = 0$$

$$x = 0, \pi \qquad \text{No solution}$$

49. $\qquad \tan^2 \theta + \tan \theta - 12 = 0$

$$(\tan \theta + 4)(\tan \theta - 3) = 0$$

$$\tan \theta + 4 = 0 \qquad\qquad \tan \theta - 3 = 0$$

$$\theta = \arctan(-4) + n\pi \qquad\qquad \theta = \arctan 3 + n\pi$$

$$\theta = \arctan(-4) + \pi, \arctan(-4) + 2\pi, \arctan 3, \arctan 3 + \pi$$

51. $\sin 285° = \sin(315° - 30°)$

$\qquad = \sin 315° \cos 30° - \cos 315° \sin 30°$

$\qquad = \left(-\dfrac{\sqrt{2}}{2}\right)\left(\dfrac{\sqrt{3}}{2}\right) - \left(\dfrac{\sqrt{2}}{2}\right)\left(\dfrac{1}{2}\right)$

$\qquad = -\dfrac{\sqrt{2}}{4}\left(\sqrt{3} + 1\right)$

$\quad \cos 285° = \cos(315° - 30°)$

$\qquad = \cos 315° \cos 30° + \sin 315° \sin 30°$

$\qquad = \left(\dfrac{\sqrt{2}}{2}\right)\left(\dfrac{\sqrt{3}}{2}\right) + \left(-\dfrac{\sqrt{2}}{2}\right)\left(\dfrac{1}{2}\right)$

$\qquad = \dfrac{\sqrt{2}}{4}\left(\sqrt{3} - 1\right)$

$\quad \tan 285° = \tan(315° - 30°) = \dfrac{\tan 315° - \tan 30°}{1 + \tan 315° \tan 30°}$

$\qquad = \dfrac{(-1) - \left(\dfrac{\sqrt{3}}{3}\right)}{1 + (-1)\left(\dfrac{\sqrt{3}}{3}\right)} = -2 - \sqrt{3}$

53. $\sin \dfrac{25\pi}{12} = \sin\left(\dfrac{11\pi}{6} + \dfrac{\pi}{4}\right) = \sin \dfrac{11\pi}{6} \cos \dfrac{\pi}{4} + \cos \dfrac{11\pi}{6} \sin \dfrac{\pi}{4}$

$\qquad = \left(-\dfrac{1}{2}\right)\left(\dfrac{\sqrt{2}}{2}\right) + \left(\dfrac{\sqrt{3}}{2}\right)\left(\dfrac{\sqrt{2}}{2}\right) = \dfrac{\sqrt{2}}{4}\left(\sqrt{3} - 1\right)$

$\quad \cos \dfrac{25\pi}{12} = \cos\left(\dfrac{11\pi}{6} + \dfrac{\pi}{4}\right) = \cos \dfrac{11\pi}{6} \cos \dfrac{\pi}{4} - \sin \dfrac{11\pi}{6} \sin \dfrac{\pi}{4}$

$\qquad = \left(\dfrac{\sqrt{3}}{2}\right)\left(\dfrac{\sqrt{2}}{2}\right) - \left(-\dfrac{1}{2}\right)\left(\dfrac{\sqrt{2}}{2}\right) = \dfrac{\sqrt{2}}{4}\left(\sqrt{3} + 1\right)$

$\quad \tan \dfrac{25\pi}{12} = \tan\left(\dfrac{11\pi}{6} + \dfrac{\pi}{4}\right) = \dfrac{\tan \dfrac{11\pi}{6} + \tan \dfrac{\pi}{4}}{1 - \tan \dfrac{11\pi}{6} \tan \dfrac{\pi}{4}}$

$\qquad = \dfrac{\left(-\dfrac{\sqrt{3}}{3}\right) + 1}{1 - \left(-\dfrac{\sqrt{3}}{3}\right)(1)} = 2 - \sqrt{3}$

55. $\sin 60° \cos 45° - \cos 60° \sin 45° = \sin(60° - 45°) = \sin 15°$

57. $\dfrac{\tan 25° + \tan 10°}{1 - \tan 25° \tan 10°} = \tan(25° + 10°) = \tan 35°$

Figures for Exercises 59–63

59. $\sin(u + v) = \sin u \cos v + \cos u \sin v$

$\qquad = \left(\dfrac{3}{4}\right)\left(-\dfrac{5}{13}\right) + \left(-\dfrac{\sqrt{7}}{4}\right)\left(\dfrac{12}{13}\right)$

$\qquad = -\dfrac{3}{52}\left(5 + 4\sqrt{7}\right)$

61. $\cos(u - v) = \cos u \cos v + \sin u \sin v$

$\qquad = \left(-\dfrac{\sqrt{7}}{4}\right)\left(-\dfrac{5}{13}\right) + \left(\dfrac{3}{4}\right)\left(\dfrac{12}{13}\right)$

$\qquad = \dfrac{1}{52}\left(5\sqrt{7} + 36\right)$

63. $\cos(u + v) = \cos u \cos v - \sin u \sin v$

$\qquad = \left(-\dfrac{\sqrt{7}}{4}\right)\left(-\dfrac{5}{13}\right) - \left(\dfrac{3}{4}\right)\left(\dfrac{12}{13}\right)$

$\qquad = \dfrac{1}{52}\left(5\sqrt{7} - 36\right)$

65. $\cos\left(x + \dfrac{\pi}{2}\right) = \cos x \cos \dfrac{\pi}{2} - \sin x \sin \dfrac{\pi}{2}$

$\qquad = \cos x(0) - \sin x(1)$

$\qquad = -\sin x$

67. $\cot\left(\dfrac{\pi}{2} - x\right) = \tan x$ by the cofunction identity.

69. $\cos 3x = \cos(2x + x)$

$\qquad = \cos 2x \cos x - \sin 2x \sin x$

$\qquad = (\cos^2 x - \sin^2 x)\cos x - (2\sin x \cos x)\sin x$

$\qquad = \cos^3 x - \sin^2 x \cos x - 2\sin^2 x \cos x$

$\qquad = \cos^3 x - 3\sin^2 x \cos x$

$\qquad = \cos^3 x - 3(1 - \cos^2 x)\cos x$

$\qquad = \cos^3 x - 3\cos x + 3\cos^3 x$

$\qquad = 4\cos^3 x - 3\cos x$

71. $\sin\left(x + \dfrac{\pi}{4}\right) - \sin\left(x - \dfrac{\pi}{4}\right) = 1$

$\qquad 2\cos x \sin\dfrac{\pi}{4} = 1$

$\qquad \cos x = \dfrac{\sqrt{2}}{2}$

$\qquad x = \dfrac{\pi}{4}, \dfrac{7\pi}{4}$

73. $\sin\left(x + \dfrac{\pi}{2}\right) - \sin\left(x - \dfrac{\pi}{2}\right) = \sqrt{3}$

$\qquad 2\cos x \sin\dfrac{\pi}{2} = \sqrt{3}$

$\qquad \cos x = \dfrac{\sqrt{3}}{2}$

$\qquad x = \dfrac{\pi}{6}, \dfrac{11\pi}{6}$

75. $\sin u = -\dfrac{4}{5}, \ \pi < u < \dfrac{3\pi}{2}$

$\cos u = -\sqrt{1 - \sin^2 u} = \dfrac{-3}{5}$

$\tan u = \dfrac{\sin u}{\cos u} = \dfrac{4}{3}$

$\sin 2u = 2\sin u \cos u = 2\left(-\dfrac{4}{5}\right)\left(-\dfrac{3}{5}\right) = \dfrac{24}{25}$

$\cos 2u = \cos^2 u - \sin^2 u = \left(-\dfrac{3}{5}\right)^2 - \left(-\dfrac{4}{5}\right)^2 = -\dfrac{7}{25}$

$\tan 2u = \dfrac{2\tan u}{1 - \tan^2 u} = \dfrac{2\left(\dfrac{4}{3}\right)}{1 - \left(\dfrac{4}{3}\right)^2} = -\dfrac{24}{7}$

77. $\sin 4x = 2\sin 2x \cos 2x$

$\qquad = 2[2\sin x \cos x(\cos^2 x - \sin^2 x)]$

$\qquad = 4\sin x \cos x(2\cos^2 x - 1)$

$\qquad = 8\cos^3 x \sin x - 4\cos x \sin x$

79. $\tan^2 2x = \dfrac{\sin^2 2x}{\cos^2 2x} = \dfrac{\dfrac{1 - \cos 4x}{2}}{\dfrac{1 + \cos 4x}{2}} = \dfrac{1 - \cos 4x}{1 + \cos 4x}$

81. $\sin^2 x \tan^2 x = \sin^2 x\left(\dfrac{\sin^2 x}{\cos^2 x}\right) = \dfrac{\sin^4 x}{\cos^2 x}$

$\qquad = \dfrac{\left(\dfrac{1 - \cos 2x}{2}\right)^2}{\dfrac{1 + \cos 2x}{2}} = \dfrac{\dfrac{1 - 2\cos 2x + \cos^2 2x}{4}}{\dfrac{1 + \cos 2x}{2}}$

$\qquad = \dfrac{1 - 2\cos 2x + \dfrac{1 + \cos 4x}{2}}{2(1 + \cos 2x)}$

$\qquad = \dfrac{2 - 4\cos 2x + 1 + \cos 4x}{4(1 + \cos 2x)}$

$\qquad = \dfrac{3 - 4\cos 2x + \cos 4x}{4(1 + \cos 2x)}$

83. $\sin(-75°) = -\sqrt{\dfrac{1 - \cos 150°}{2}} = -\sqrt{\dfrac{1 - \left(-\dfrac{\sqrt{3}}{2}\right)}{2}} = -\dfrac{\sqrt{2 + \sqrt{3}}}{2}$

$\qquad\qquad\quad = -\dfrac{1}{2}\sqrt{2 + \sqrt{3}}$

$\cos(-75°) = \sqrt{\dfrac{1 + \cos 150°}{2}} = \sqrt{\dfrac{1 + \left(-\dfrac{\sqrt{3}}{2}\right)}{2}} = \dfrac{\sqrt{2 - \sqrt{3}}}{2}$

$\qquad\qquad\quad = \dfrac{1}{2}\sqrt{2 - \sqrt{3}}$

$\tan(-75°) = -\left(\dfrac{1 - \cos 150°}{\sin 150°}\right) = -\left(\dfrac{1 - \left(-\dfrac{\sqrt{3}}{2}\right)}{\dfrac{1}{2}}\right) = -\left(2 + \sqrt{3}\right)$

$\qquad\qquad\quad = -2 - \sqrt{3}$

85. $\sin\left(\dfrac{19\pi}{12}\right) = -\sqrt{\dfrac{1 - \cos\dfrac{19\pi}{6}}{2}} = -\sqrt{\dfrac{1 - \left(-\dfrac{\sqrt{3}}{2}\right)}{2}} = -\dfrac{\sqrt{2 + \sqrt{3}}}{2}$

$\qquad\qquad\quad = -\dfrac{1}{2}\sqrt{2 + \sqrt{3}}$

$\cos\left(\dfrac{19\pi}{12}\right) = \sqrt{\dfrac{1 + \cos\dfrac{19\pi}{6}}{2}} = \sqrt{\dfrac{1 + \left(-\dfrac{\sqrt{3}}{2}\right)}{2}} = \dfrac{\sqrt{2 - \sqrt{3}}}{2}$

$\qquad\qquad\quad = \dfrac{1}{2}\sqrt{2 - \sqrt{3}}$

$\tan\left(\dfrac{19\pi}{12}\right) = \dfrac{1 - \cos\dfrac{19\pi}{6}}{\sin\dfrac{19\pi}{6}} = \dfrac{1 - \left(-\dfrac{\sqrt{3}}{2}\right)}{-\dfrac{1}{2}} = -2 - \sqrt{3}$

87. Given $\sin u = \dfrac{3}{5}, 0 < u < \dfrac{\pi}{2} \Rightarrow \cos u = \dfrac{4}{5}$ and $\dfrac{u}{2}$ is in Quadrant I.

$\sin\left(\dfrac{u}{2}\right) = \sqrt{\dfrac{1 - \cos u}{2}} = \sqrt{\dfrac{1 - 4/5}{2}} = \sqrt{\dfrac{1}{10}} = \dfrac{\sqrt{10}}{10}$

$\cos\left(\dfrac{u}{2}\right) = \sqrt{\dfrac{1 + \cos u}{2}} = \sqrt{\dfrac{1 + 4/5}{2}} = \sqrt{\dfrac{9}{10}} = \dfrac{3\sqrt{10}}{10}$

$\tan\left(\dfrac{u}{2}\right) = \dfrac{1 - \cos u}{\sin u} = \dfrac{1 - 4/5}{3/5} = \dfrac{1}{3}$

89. Given $\cos u = -\dfrac{2}{7}, \dfrac{\pi}{2} < u < \pi \Rightarrow \sin u = \dfrac{3\sqrt{5}}{7}$ and $\dfrac{u}{2}$ is in Quadrant I.

$\sin\left(\dfrac{u}{2}\right) = \sqrt{\dfrac{1 - \cos u}{2}} = \sqrt{\dfrac{1 - (-2/7)}{2}} = \sqrt{\dfrac{9}{14}} = \dfrac{3}{\sqrt{14}} = \dfrac{3\sqrt{14}}{14}$

$\cos\left(\dfrac{u}{2}\right) = \sqrt{\dfrac{1 + \cos u}{2}} = \sqrt{\dfrac{1 + (-2/7)}{2}} = \sqrt{\dfrac{5}{14}} = \dfrac{\sqrt{70}}{14}$

$\tan\left(\dfrac{u}{2}\right) = \dfrac{1 - \cos u}{\sin u} = \dfrac{1 - (-2/7)}{3\sqrt{5}/7} = \dfrac{9/7}{3\sqrt{5}/7} = \dfrac{3}{\sqrt{5}} = \dfrac{3\sqrt{5}}{5}$

91. $-\sqrt{\dfrac{1 + \cos 10x}{2}} = -\left|\cos \dfrac{10x}{2}\right| = -|\cos 5x|$

93. $\cos \dfrac{\pi}{6} \sin \dfrac{\pi}{6} = \dfrac{1}{2}\left[\sin \dfrac{\pi}{3} - \sin 0\right] = \dfrac{1}{2}\sin \dfrac{\pi}{3}$

95. $\cos 5\theta \cos 3\theta = \dfrac{1}{2}[\cos 2\theta + \cos 8\theta]$

97. $\sin 4\theta - \sin 2\theta = 2 \cos\left(\dfrac{4\theta + 2\theta}{2}\right) \sin\left(\dfrac{4\theta - 2\theta}{2}\right)$

$$= 2 \cos 3\theta \sin \theta$$

99. $\cos\left(x + \dfrac{\pi}{6}\right) - \cos\left(x - \dfrac{\pi}{6}\right) = -2 \sin x \, \sin \dfrac{\pi}{6}$

101. $r = \dfrac{1}{32}v_0{}^2 \sin 2\theta$

range $= 100$ feet

$v_0 = 80$ feet per second

$r = \dfrac{1}{32}(80)^2 \sin 2\theta = 100$

$\sin 2\theta = 0.5$

$2\theta = 30°$

$\theta = 15°$ or $\dfrac{\pi}{12}$

103. $y = 1.5 \sin 8t - 0.5 \cos 8t$

105. Amplitude $= \dfrac{\sqrt{10}}{2}$ feet

107. False. If $\dfrac{\pi}{2} < \theta < \pi$, then $\dfrac{\pi}{4} < \dfrac{\theta}{2} < \dfrac{\pi}{2}$ and $\dfrac{\theta}{2}$ is in Quadrant I.

$\cos \dfrac{\theta}{2} > 0$

109. True. $4 \sin(-x)\cos(-x) = 4(-\sin x)\cos x$

$$= -4 \sin x \cos x = -2(2 \sin x \cos x)$$

$$= -2 \sin 2x$$

111. Reciprocal Identities: $\sin \theta = \dfrac{1}{\csc \theta}$ $\quad$ $\csc \theta = \dfrac{1}{\sin \theta}$

$\cos \theta = \dfrac{1}{\sec \theta}$ $\quad$ $\sec \theta = \dfrac{1}{\cos \theta}$

$\tan \theta = \dfrac{1}{\cot \theta}$ $\quad$ $\cot \theta = \dfrac{1}{\tan \theta}$

Quotient Identities: $\tan \theta = \dfrac{\sin \theta}{\cos \theta}$ $\quad$ $\cot \theta = \dfrac{\cos \theta}{\sin \theta}$

Pythagorean Identities: $\sin^2 \theta + \cos^2 \theta = 1$

$$1 + \tan^2 \theta = \sec^2 \theta$$

$$1 + \cot^2 \theta = \csc^2 \theta$$

113. $a \sin x - b = 0$

$\sin x = \dfrac{b}{a}$

If $|b| > |a|$, then $\left|\dfrac{b}{a}\right| > 1$ and there is no solution

since $|\sin x| \leq 1$ for all x.

115. The graph of y_1 is a vertical shift of the graph of y_2 one unit upward so $y_1 = y_2 + 1$.

117. $y = \sqrt{x + 3} + 4 \cos x$

Zeros: $x \approx -1.8431,\ 2.1758,$
$3.9903,\ 8.8935,\ 9.8820$

Problem Solving for Chapter 7

1. (a) Since $\sin^2 \theta + \cos^2 \theta = 1$ and $\cos^2 \theta = 1 - \sin^2 \theta$:

$$\cos \theta = \pm \sqrt{1 - \sin^2 \theta}$$

$$\tan \theta = \frac{\sin \theta}{\cos \theta} = \pm \frac{\sin \theta}{\sqrt{1 - \sin^2 \theta}}$$

$$\cot \theta = \frac{1}{\tan \theta} = \pm \frac{\sqrt{1 - \sin^2 \theta}}{\sin \theta}$$

$$\sec \theta = \frac{1}{\cos \theta} = \pm \frac{1}{\sqrt{1 - \sin^2 \theta}}$$

$$\cos \theta = \frac{1}{\sin \theta}$$

We also have the following relationships:

$$\cos \theta = \sin\left(\frac{\pi}{2} - \theta\right)$$

$$\tan \theta = \frac{\sin \theta}{\sin\left(\frac{\pi}{2} - \theta\right)}$$

$$\cot \theta = \frac{\sin\left(\frac{\pi}{2} - \theta\right)}{\sin \theta}$$

$$\sec \theta = \frac{1}{\sin\left(\frac{\pi}{2} - \theta\right)}$$

$$\csc \theta = \frac{1}{\sin \theta}$$

(b) $\sin \theta = \pm \sqrt{1 - \cos^2 \theta}$

$$\tan \theta = \frac{\sin \theta}{\cos \theta} = \pm \frac{\sqrt{1 - \cos^2 \theta}}{\cos \theta}$$

$$\csc \theta = \frac{1}{\sin \theta} = \pm \frac{1}{\sqrt{1 - \cos^2 \theta}}$$

$$\sec \theta = \frac{1}{\cos \theta}$$

$$\cot \theta = \frac{1}{\tan \theta} = \pm \frac{\cos \theta}{\sqrt{1 - \cos^2 \theta}}$$

We also have the following relationships:

$$\sin \theta = \cos\left(\frac{\pi}{2} - \theta\right)$$

$$\tan \theta = \frac{\cos\left(\frac{\pi}{2} - \theta\right)}{\cos \theta}$$

$$\csc \theta = \frac{1}{\cos\left(\frac{\pi}{2} - \theta\right)}$$

$$\sec \theta = \frac{1}{\cos \theta}$$

$$\cot \theta = \frac{\cos \theta}{\cos\left(\frac{\pi}{2} - \theta\right)}$$

3. $\sin\left[\dfrac{(12n + 1)\pi}{6}\right] = \sin\left[\dfrac{1}{6}(12n\pi + \pi)\right]$

$$= \sin\left(2n\pi + \frac{\pi}{6}\right)$$

$$= \sin\frac{\pi}{6} = \frac{1}{2}$$

Thus, $\sin\left[\dfrac{(12n + 1)\pi}{6}\right] = \dfrac{1}{2}$ for all integers n.

5. From the figure, it appears that $u + v = w$. Assume that u, v, and w are all in Quadrant I. From the figure:

$$\tan u = \frac{s}{3s} = \frac{1}{3}$$

$$\tan v = \frac{s}{2s} = \frac{1}{2}$$

$$\tan w = \frac{s}{s} = 1$$

$$\tan(u + v) = \frac{\tan u + \tan v}{1 - \tan u \tan v} = \frac{1/3 + 1/2}{1 - (1/3)(1/2)} = \frac{5/6}{1 - (1/6)} = 1 = \tan w.$$

Thus, $\tan(u + v) = \tan w$. Because u, v, and w are all in Quadrant I, we have

$$\arctan[\tan(u + v)] = \arctan[\tan w]$$

$$u + v = w.$$

7.

The hypotenuse of the larger right triangle is:

$$\sqrt{\sin^2 \theta + (1 + \cos \theta)^2} = \sqrt{\sin^2 \theta + 1 + 2\cos \theta + \cos^2 \theta}$$

$$= \sqrt{2 + 2\cos \theta}$$

$$= \sqrt{2(1 + \cos \theta)}$$

$$\sin\left(\frac{\theta}{2}\right) = \frac{\sin \theta}{\sqrt{2(1 + \cos \theta)}} = \frac{\sin \theta}{\sqrt{2(1 + \cos \theta)}} \cdot \frac{\sqrt{1 - \cos \theta}}{\sqrt{1 - \cos \theta}}$$

$$= \frac{\sin \theta \sqrt{1 - \cos \theta}}{\sqrt{2(1 - \cos^2 \theta)}} = \frac{\sin \theta \sqrt{1 - \cos \theta}}{\sqrt{2} \sin \theta}$$

$$= \sqrt{\frac{1 - \cos \theta}{2}}$$

$$\cos\left(\frac{\theta}{2}\right) = \frac{1 + \cos \theta}{\sqrt{2(1 + \cos \theta)}} = \frac{\sqrt{(1 + \cos \theta)^2}}{\sqrt{2(1 + \cos \theta)}} = \sqrt{\frac{1 + \cos \theta}{2}}$$

$$\tan\left(\frac{\theta}{2}\right) = \frac{\sin \theta}{1 + \cos \theta}$$

9. Seward: $D = 12.2 - 6.4 \cos\left[\frac{\pi(t + 0.2)}{182.6}\right]$

New Orleans: $D = 12.2 - 1.9 \cos\left[\frac{\pi(t + 0.2)}{182.6}\right]$

(a)

(b) The graphs intersect when $t \approx 91$ and when $t \approx 274$. These values correspond to April 1 and October 1, the spring equinox and the fall equinox.

(c) Seward has the greater variation in the number of daylight hours. This is determined by the amplitudes, 6.4 and 1.9.

(d) Period: $\dfrac{2\pi}{\dfrac{\pi}{182.6}} = 365.2$ days

11. (a) Let $y_1 = \sin x$ and $y_2 = 0.5$.

$\sin x \geq 0.5$ on the interval $\left[\dfrac{\pi}{6}, \dfrac{5\pi}{6}\right]$.

(c) Let $y_1 = \tan x$ and $y_2 = \sin x$.

$\tan x < \sin x$ on the intervals $\left(\dfrac{\pi}{2}, \pi\right)$ and $\left(\dfrac{3\pi}{2}, 2\pi\right)$.

(b) Let $y_1 = \cos x$ and $y_2 = -0.5$.

$\cos x \leq -0.5$ on the interval $\left[\dfrac{2\pi}{3}, \dfrac{4\pi}{3}\right]$.

(d) Let $y_1 = \cos x$ and $y_2 = \sin x$.

$\cos x \geq \sin x$ on the intervals $\left[0, \dfrac{\pi}{4}\right]$ and $\left[\dfrac{5\pi}{4}, 2\pi\right]$.

13. (a) $\sin(u + v + w) = \sin[(u + v) + w]$

$= \sin(u + v)\cos w + \cos(u + v)\sin w$

$= [\sin u \cos v + \cos u \sin v]\cos w + [\cos u \cos v - \sin u \sin v]\sin w$

$= \sin u \cos v \cos w + \cos u \sin v \cos w + \cos u \cos v \sin w - \sin u \sin v \sin w$

(b) $\tan(u + v + w) = \tan[(u + v) + w]$

$= \dfrac{\tan(u + v) + \tan w}{1 - \tan(u + v)\tan w}$

$= \dfrac{\left[\dfrac{\tan u + \tan v}{1 - \tan u \tan v}\right] + \tan w}{1 - \left[\dfrac{\tan u + \tan v}{1 - \tan u \tan v}\right]\tan w} \cdot \dfrac{(1 - \tan u \tan v)}{(1 - \tan u \tan v)}$

$= \dfrac{\tan u + \tan v + (1 - \tan u \tan v)\tan w}{(1 - \tan u \tan v) - (\tan u + \tan v)\tan w}$

$= \dfrac{\tan u + \tan v + \tan w - \tan u \tan v \tan w}{1 - \tan u \tan v - \tan u \tan w - \tan v \tan w}$

15. $h_1 = 3.75 \sin 733t + 7.5$

$h_2 = 3.75 \sin 733\left(t + \dfrac{4\pi}{3}\right) + 7.5$

(a)

(b) The period for h_1 and h_2 is $\dfrac{2\pi}{733} \approx 0.0086$.

The graphs intersect twice per cycle.

There are $\dfrac{1}{2\pi/733} \approx 116.66$ cycles in the interval $[0, 1]$, so the graphs intersect approximately 233.3 times.

Chapter 7   Practice Test

1. Find the value of the other five trigonometric functions, given $\tan x = \frac{4}{11}$, $\sec x < 0$.

2. Simplify $\dfrac{\sec^2 x + \csc^2 x}{\csc^2 x(1 + \tan^2 x)}$.

3. Rewrite as a single logarithm and simplify $\ln|\tan \theta| - \ln|\cot \theta|$.

4. True or false:
$$\cos\left(\frac{\pi}{2} - x\right) = \frac{1}{\csc x}$$

5. Factor and simplify: $\sin^4 x + (\sin^2 x)\cos^2 x$

6. Multiply and simplify: $(\csc x + 1)(\csc x - 1)$

7. Rationalize the denominator and simplify:
$$\frac{\cos^2 x}{1 - \sin x}$$

8. Verify:
$$\frac{1 + \cos \theta}{\sin \theta} + \frac{\sin \theta}{1 + \cos \theta} = 2\csc \theta$$

9. Verify:
$$\tan^4 x + 2\tan^2 x + 1 = \sec^4 x$$

10. Use the sum or difference formulas to determine:

    (a) $\sin 105°$ (b) $\tan 15°$

11. Simplify: $(\sin 42°)\cos 38° - (\cos 42°)\sin 38°$

12. Verify $\tan\left(\theta + \dfrac{\pi}{4}\right) = \dfrac{1 + \tan \theta}{1 - \tan \theta}$.

13. Write $\sin(\arcsin x - \arccos x)$ as an algebraic expression in x.

14. Use the double-angle formulas to determine:

    (a) $\cos 120°$ (b) $\tan 300°$

15. Use the half-angle formulas to determine:

    (a) $\sin 22.5°$ (d) $\tan \dfrac{\pi}{12}$

16. Given $\sin = 4/5$, θ lies in Quadrant II, find $\cos(\theta/2)$.

17. Use the power-reducing identities to write $(\sin^2 x)\cos^2 x$ in terms of the first power of cosine.

18. Rewrite as a sum: $6(\sin 5\theta)\cos 2\theta$.

19. Rewrite as a product:
$$\sin(x + \pi) + \sin(x - \pi).$$

20. Verify $\dfrac{\sin 9x + \sin 5x}{\cos 9x - \cos 5x} = -\cot 2x$.

21. Verify:
$$(\cos u)\sin v = \tfrac{1}{2}[\sin(u + v) - \sin(u - v)].$$

22. Find all solutions in the interval $[0, 2\pi)$:
$$4\sin^2 x = 1$$

23. Find all solutions in the interval $[0, 2\pi)$:
$$\tan^2 \theta + \left(\sqrt{3} - 1\right)\tan\theta - \sqrt{3} = 0$$

24. Find all solutions in the interval $[0, 2\pi)$:
$$\sin 2x = \cos x$$

25. Use the quadratic formula to find all solutions in the interval $[0, 2\pi)$:
$$\tan^2 x - 6\tan x + 4 = 0$$

C H A P T E R   8
Additional Topics in Trigonometry

CHAPTER 8
Additional Topics in Trigonometry

Section 8.1 Law of Sines

- If ABC is any oblique triangle with sides a, b, and c, then

$$\frac{a}{\sin A} = \frac{b}{\sin B} = \frac{c}{\sin C}.$$

- You should be able to use the Law of Sines to solve an oblique triangle for the remaining three parts, given:

  (a) Two angles and any side (AAS or ASA)

  (b) Two sides and an angle opposite one of them (SSA)

  1. If A is acute and $h = b \sin A$:

     (a) $a < h$, no triangle is possible.

     (b) $a = h$ or $a > b$, one triangle is possible.

     (c) $h < a < b$, two triangles are possible.

  2. If A is obtuse and $h = b \sin A$:

     (a) $a \leq b$, no triangle is possible.

     (b) $a > b$, one triangle is possible.

- The area of any triangle equals one-half the product of the lengths of two sides and the sine of their included angle.

$$A = \tfrac{1}{2}ab \sin C = \tfrac{1}{2}ac \sin B = \tfrac{1}{2}bc \sin A$$

Vocabulary Check

1. oblique

2. $\dfrac{b}{\sin B}$

3. $\dfrac{1}{2} ac \sin B$

1.

Given: $A = 30°$, $B = 45°$, $a = 20$

$C = 180° - A - B = 105°$

$b = \dfrac{a}{\sin A}(\sin B) = \dfrac{20 \sin 45°}{\sin 30°} = 20\sqrt{2} \approx 28.28$

$c = \dfrac{a}{\sin A}(\sin C) = \dfrac{20 \sin 105°}{\sin 30°} \approx 38.64$

3.

Given: $A = 25°$, $B = 35°$, $a = 3.5$

$C = 180° - A - B = 120°$

$b = \dfrac{a}{\sin A}(\sin B) = \dfrac{3.5}{\sin 25°}(\sin 35°) \approx 4.75$

$c = \dfrac{a}{\sin A}(\sin C) = \dfrac{3.5}{\sin 25°}(\sin 120°) \approx 7.17$

5. Given: $A = 36°$, $a = 8$, $b = 5$

$$\sin B = \frac{b \sin A}{a} = \frac{5 \sin 36°}{8} \approx 0.36737 \implies B \approx 21.55°$$

$$C = 180° - A - B \approx 180° - 36° - 21.55 = 122.45°$$

$$c = \frac{a}{\sin A}(\sin C) = \frac{8}{\sin 36°}(\sin 122.45°) \approx 11.49$$

7. Given: $A = 102.4°$, $C = 16.7°$, $a = 21.6$

$$B = 180° - A - C = 60.9°$$

$$b = \frac{a}{\sin A}(\sin B) = \frac{21.6}{\sin 102.4°}(\sin 60.9°) \approx 19.32$$

$$c = \frac{a}{\sin A}(\sin C) = \frac{21.6}{\sin 102.4°}(\sin 16.7°) \approx 6.36$$

9. Given: $A = 83° 20'$, $C = 54.6°$, $c = 18.1$

$$B = 180° - A - C = 180° - 83° 20' - 54° 36' = 42° 4'$$

$$a = \frac{c}{\sin C}(\sin A) = \frac{18.1}{\sin 54.6°}(\sin 83° 20') \approx 22.05$$

$$b = \frac{c}{\sin C}(\sin B) = \frac{18.1}{\sin 54.6°}(\sin 42° 4') \approx 14.88$$

11. Given: $B = 15° 30'$, $a = 4.5$, $b = 6.8$

$$\sin A = \frac{a \sin B}{b} = \frac{4.5 \sin 15° 30'}{6.8} \approx 0.17685 \implies A \approx 10° 11'$$

$$C = 180° - A - B \approx 180° - 10° 11' - 15° 30' = 154° 19'$$

$$c = \frac{b}{\sin B}(\sin C) = \frac{6.8}{\sin 15° 30'}(\sin 154° 19') \approx 11.03$$

13. Given: $C = 145°$, $b = 4$, $c = 14$

$$\sin B = \frac{b \sin C}{c} = \frac{4 \sin 145°}{14} \approx 0.16388 \implies B \approx 9.43°$$

$$A = 180° - B - C \approx 180° - 9.43° - 145° = 25.57°$$

$$a = \frac{c}{\sin C}(\sin A) \approx \frac{14}{\sin 145°}(\sin 25.57°) \approx 10.53$$

15. Given: $A = 110° 15'$, $a = 48$, $b = 16$

$$\sin B = \frac{b \sin A}{a} = \frac{16 \sin 110° 15'}{48} \approx 0.31273 \implies B \approx 18° 13'$$

$$C = 180° - A - B \approx 180° - 110° 15' - 18° 13' = 51° 32'$$

$$c = \frac{a}{\sin A}(\sin C) = \frac{48}{\sin 110° 15'}(\sin 51° 32') \approx 40.06$$

17. Given: $A = 55°$, $B = 42°$, $c = \frac{3}{4}$

$$C = 180° - A - B = 83°$$

$$a = \frac{c}{\sin C}(\sin A) = \frac{0.75}{\sin 83°}(\sin 55°) \approx 0.62$$

$$b = \frac{c}{\sin C}(\sin B) = \frac{0.75}{\sin 83°}(\sin 42°) \approx 0.51$$

19. Given: $A = 110°$, $a = 125$, $b = 100$

$$\sin B = \frac{b \sin A}{a} = \frac{100 \sin 110°}{125} \approx 0.75175 \implies B \approx 48.74°$$

$$C = 180° - A - B \approx 21.26°$$

$$c = \frac{a \sin C}{\sin A} \approx \frac{125 \sin 21.26°}{\sin 110°} \approx 48.23$$

21. Given: $a = 18, b = 20, A = 76°$

$h = 20 \sin 76° \approx 19.41$

Since $a < h$, no triangle is formed.

23. Given: $A = 58°, a = 11.4, c = 12.8$

$\sin B = \dfrac{b \sin A}{a} = \dfrac{12.8 \sin 58°}{11.4} \approx 0.9522 \implies B \approx 72.21°$ or $B \approx 107.79°$

Case 1

$B \approx 72.21°$

$C = 180° - A - B \approx 49.79°$

$c = \dfrac{a}{\sin A}(\sin C) \approx \dfrac{11.4 \sin 49.79°}{\sin 58°} \approx 10.27$

Case 2

$B \approx 107.79°$

$C = 180° - A - B \approx 14.21°$

$c = \dfrac{a}{\sin A}(\sin C) \approx \dfrac{11.4 \sin 14.21°}{\sin 58°} \approx 3.30$

25. Given: $A = 36°, \ a = 5$

(a) One solution if $b \le 5$ or $b = \dfrac{5}{\sin 36°}$

(b) Two solutions if $5 < b < \dfrac{5}{\sin 36°}$

(c) No solution if $b > \dfrac{5}{\sin 36°}$

27. Given: $A = 10°, a = 10.8$

(a) One solution if $b \le 10.8$ or $b = \dfrac{10.8}{\sin 10°}$

(b) Two solutions if $10.8 < b < \dfrac{10.8}{\sin 10°}$

(c) No solution if $b > \dfrac{10.8}{\sin 10°}$

29. Area $= \frac{1}{2} ab \sin C = \frac{1}{2}(4)(6) \sin 120° \approx 10.4$

31. Area $= \frac{1}{2} bc \sin A = \frac{1}{2}(57)(85) \sin 43° \ 45' \approx 1675.2$

33. Area $= \dfrac{1}{2} ac \sin B = \dfrac{1}{2}(105)(64)\sin(72°30') \approx 3204.5$

35. $C = 180° - 23° - 94° = 63°$

$h = \dfrac{35}{\sin 63°}(\sin 23°) \approx 15.3$ meters

37. $\dfrac{\sin(42° - \theta)}{10} = \dfrac{\sin 48°}{17}$

$\sin(42° - \theta) \approx 0.43714$

$42° - \theta \approx 25.9°$

$\theta \approx 16.1°$

39. Given: $c = 100$

$A = 74° - 28° = 46°$,

$B = 180° - 41° - 74° = 65°$,

$C = 180° - 46° - 65° = 69°$

$a = \dfrac{c}{\sin C}(\sin A) = \dfrac{100}{\sin 69°}(\sin 46°) \approx 77$ meters

41. (a)

Not drawn to scale

(b) $\dfrac{x}{\sin 17.5°} = \dfrac{9000}{\sin 1.3°}$

$x \approx 119{,}289.1261$ feet ≈ 22.6 miles

(c) $\dfrac{y}{\sin 71.2°} = \dfrac{x}{\sin 90°}$

$y = x \sin 71.2° \approx 119{,}289.1261 \sin 71.2°$

$\approx 112{,}924.963$ feet ≈ 21.4 miles

(d) $z = x \sin 18.8° \approx 119{,}289.1261 \sin 18.8°$

$\approx 38{,}443$ feet ≈ 7.3 miles

43.

In 15 minutes the boat has traveled

$$(10 \text{ mph})\left(\tfrac{1}{4} \text{ hr}\right) = \tfrac{10}{4} \text{ miles.}$$

$$\theta = 180° - 20° - (90° + 63°)$$

$$\theta = 7°$$

$$\frac{10/4}{\sin 7°} = \frac{y}{\sin 20°}$$

$$y \approx 7.0161$$

$$\sin 27° = \frac{d}{7.0161}$$

$$d \approx 3.2 \text{ miles}$$

45. True. If one angle of a triangle is obtuse, then there is less than 90° left for the other two angles, so it cannot contain a right angle. It must be oblique.

47. (a) $\dfrac{\sin \alpha}{9} = \dfrac{\sin \beta}{18}$

$\sin \alpha = 0.5 \sin \beta$

$\alpha = \arcsin(0.5 \sin \beta)$

(b)

Domain: $0 < \beta < \pi$

Range: $0 < \alpha \le \dfrac{\pi}{6}$

(c) $\gamma = \pi - \alpha - \beta = \pi - \beta - \arcsin(0.5 \sin \beta)$

$\dfrac{c}{\sin \gamma} = \dfrac{18}{\sin \beta}$

$c = \dfrac{18 \sin \gamma}{\sin \beta} = \dfrac{18 \sin[\pi - \beta - \arcsin(0.5 \sin \beta)]}{\sin \beta}$

(d)

Domain: $0 < \beta < \pi$

Range: $9 < c < 27$

(e)

β	0.4	0.8	1.2	1.6	2.0	2.4	2.8
α	0.1960	0.3669	0.4848	0.5234	0.4720	0.3445	0.1683
c	25.95	23.07	19.19	15.33	12.29	10.31	9.27

As $\beta \to 0$, $c \to 27$

As $\beta \to \pi$, $c \to 9$

49. $\sin x \cot x = \sin x \dfrac{\cos x}{\sin x} = \cos x$

51. $1 - \sin^2\left(\dfrac{\pi}{2} - x\right) = 1 - \cos^2 x = \sin^2 x$

53. $6 \sin 8\theta \cos 3\theta = (6)\left(\tfrac{1}{2}\right)[\sin(8\theta + 3\theta) + \sin(8\theta - 3\theta)] = 3(\sin 11\theta + \sin 5\theta)$

Section 8.2   Law of Cosines

- If ABC is any oblique triangle with sides a, b, and c, the following equations are valid.

  (a) $a^2 = b^2 + c^2 - 2bc \cos A$ or $\cos A = \dfrac{b^2 + c^2 - a^2}{2bc}$

  (b) $b^2 = a^2 + c^2 - 2ac \cos B$ or $\cos B = \dfrac{a^2 + c^2 - b^2}{2ac}$

  (c) $c^2 = a^2 + b^2 - 2ab \cos C$ or $\cos C = \dfrac{a^2 + b^2 - c^2}{2ab}$

- You should be able to use the Law of Cosines to solve an oblique triangle for the remaining three parts, given:

  (a) Three sides (SSS)

  (b) Two sides and their included angle (SAS)

- Given any triangle with sides of length a, b, and c, the area of the triangle is

  $$\text{Area} = \sqrt{s(s - a)(s - b)(s - c)}, \text{ where } s = \frac{a + b + c}{2}. \qquad \text{(Heron's Formula)}$$

Vocabulary Check

1. Cosines

2. $b^2 = a^2 + c^2 - 2ac \cos B$

3. Heron's Area

---

1. Given: $a = 7, b = 10, c = 15$

$\cos C = \dfrac{a^2 + b^2 - c^2}{2ab} = \dfrac{49 + 100 - 225}{2(7)(10)} \approx -0.5429 \implies C \approx 122.88°$

$\sin B = \dfrac{b \sin C}{c} = \dfrac{10 \sin 122.88°}{15} \approx 0.5599 \implies B \approx 34.05°$

$A \approx 180° - 34.05° - 122.88° \approx 23.07°$

3. Given: $A = 30°$, $b = 15$, $c = 30$

$a^2 = b^2 + c^2 - 2bc \cos A$

$\quad = 225 + 900 - 2(15)(30) \cos 30° \approx 345.5771$

$a \approx 18.59$

$\cos C = \dfrac{a^2 + b^2 - c^2}{2ab} \approx \dfrac{(18.59)^2 + 15^2 - 30^2}{2(18.59)(15)} \approx -0.5907 \implies C \approx 126.21°$

$B \approx 180° - 30° - 126.21° = 13.79°$

5. $a = 11, b = 14, c = 20$

$\cos C = \dfrac{a^2 + b^2 - c^2}{2ab} = \dfrac{121 + 196 - 400}{2(11)(14)} \approx -0.2695 \implies C \approx 105.63°$

$\sin B = \dfrac{b \sin C}{c} = \dfrac{14 \sin 105.63°}{20} \approx 0.6741 \implies B \approx 42.38°$

$A \approx 180° - 42.38° - 105.63° \approx 31.99°$

7. Given: $a = 75.4$, $b = 52$, $c = 52$

$$\cos A = \frac{b^2 + c^2 - a^2}{2bc} = \frac{52^2 + 52^2 - 75.4^2}{2(52)(52)} = -0.05125 \implies A \approx 92.94°$$

$$\sin B = \frac{b \sin A}{a} \approx \frac{52(0.9987)}{75.4} \approx 0.68876 \implies B \approx 43.53°$$

$$C = B \approx 43.53°$$

9. Given: $A = 135°$, $b = 4$, $c = 9$

$$a^2 = b^2 + c^2 - 2bc \cos A = 16 + 81 - 2(4)(9)\cos 135° \approx 147.9117 \implies a \approx 12.16$$

$$\sin B = \frac{b \sin A}{a} = \frac{4 \sin 135°}{12.16} \approx 0.2326 \implies B \approx 13.45°$$

$$C \approx 180° - 135° - 13.45° \approx 31.55°$$

11. Given: $B = 10° \, 35'$, $a = 40$, $c = 30$

$$b^2 = a^2 + c^2 - 2ac \cos B = 1600 + 900 - 2(40)(30)\cos 10° \, 35' \approx 140.8268 \implies b \approx 11.87$$

$$\sin C = \frac{c \sin B}{b} = \frac{30 \sin 10° \, 35'}{11.87} \approx 0.4642 \implies C \approx 27.66° \approx 27° \, 40'$$

$$A \approx 180° - 10° \, 35' - 27° \, 40' = 141° \, 45'$$

13. Given: $B = 125° \, 40'$, $a = 32$, $c = 32$

$$b^2 = a^2 + c^2 - 2ac \cos B \approx 32^2 + 32^2 - 2(32)(32) \cos 125° \, 40' \approx 3242.1888 \implies b \approx 56.94$$

$$A = C \implies 2A = 180° - 125° \, 40' = 54° \, 20' \implies A = C = 27° \, 10'$$

15. $C = 43°$, $a = \dfrac{4}{9}$, $b = \dfrac{7}{9}$

$$c^2 = a^2 + b^2 - 2ab \cos C = \left(\frac{4}{9}\right)^2 + \left(\frac{7}{9}\right)^2 - 2\left(\frac{4}{9}\right)\left(\frac{7}{9}\right)\cos 43° \approx 0.2968 \implies c \approx 0.54$$

$$\sin A = \frac{a \sin C}{c} = \frac{(4/9) \sin 43°}{0.5448} \approx 0.5564 \implies A \approx 33.80°$$

$$B \approx 180° - 43° - 33.8° \approx 103.20°$$

17. $d^2 = 5^2 + 8^2 - 2(5)(8)\cos 45° \approx 32.4315 \implies d \approx 5.69$

$2\phi = 360° - 2(45°) = 270° \implies \phi = 135°$

$c^2 = 5^2 + 8^2 - 2(5)(8)\cos 135° \approx 145.5685 \implies c \approx 12.07$

19. $\cos \phi = \dfrac{10^2 + 14^2 - 20^2}{2(10)(14)}$

$\phi \approx 111.8°$

$2\theta \approx 360° - 2(111.8°)$

$\theta = 68.2°$

$d^2 = 10^2 + 14^2 - 2(10)(14) \cos 68.2°$

$d \approx 13.86$

21. $\cos \alpha = \dfrac{(12.5)^2 + (15)^2 - 10^2}{2(12.5)(15)} = 0.75 \implies \alpha \approx 41.41°$

$\cos \beta = \dfrac{10^2 + 15^2 - (12.5)^2}{2(10)(15)} = 0.5625 \implies \beta \approx 55.77°$

$z = 180° - \alpha - \beta = 82.82°$

$u = 180° - z = 97.18°$

$b^2 = 12.5^2 + 10^2 - 2(12.5)(10)\cos 97.18° \approx 287.4967 \implies b \approx 16.96$

$\cos \delta = \dfrac{12.5^2 + 16.96^2 - 10^2}{2(12.5)(16.96)} \approx 0.8111 \implies \delta \approx 35.80°$

$\theta = \alpha + \delta = 41.41° + 35.80° \approx 77.2°$

$2\phi = 360° - 2\theta \implies \phi = \dfrac{360° - 2(77.21°)}{2} = 102.8°$

23. $a = 5,\ b = 7,\ c = 10 \implies s = \dfrac{a + b + c}{2} = 11$

Area $= \sqrt{s(s - a)(s - b)(s - c)} = \sqrt{11(6)(4)(1)} \approx 16.25$

25. $a = 2.5, b = 10.2, c = 9 \implies s = \dfrac{a + b + c}{2} = 10.85$

Area $= \sqrt{s(s - a)(s - b)(s - c)} = \sqrt{10.85(8.35)(0.65)(1.85)} \approx 10.4$

27. $a = 12.32, b = 8.46, c = 15.05 \implies s = \dfrac{a + b + c}{2} = 17.915$

Area $= \sqrt{s(s - a)(s - b)(s - c)} = \sqrt{17.915(5.595)(9.455)(2.865)} \approx 52.11$

29. $\cos B = \dfrac{1700^2 + 3700^2 - 3000^2}{2(1700)(3700)} \implies B \approx 52.9°$

Bearing: $90° - 52.9° = $ N 37.1° E

$\cos C = \dfrac{1700^2 + 3000^2 - 3700^2}{2(1700)(3000)} \implies C \approx 100.2°$

Bearing: $90° - 26.9° = $ S 63.1° E

31. $b^2 = 220^2 + 250^2 - 2(220)(250)\cos 105° \implies b \approx 373.3$ meters

33.

The largest angle is across from the largest side.

$\cos C = \dfrac{650^2 + 575^2 - 725^2}{2(650)(575)}$

$C \approx 72.3°$

35. $C = 180° - 53° - 67° = 60°$

$c^2 = a^2 + b^2 - 2ab \cos C$

$= 36^2 + 48^2 - 2(36)(48)(0.5)$

$= 1872$

$c \approx 43.3$ mi

37. (a) $\cos \theta = \dfrac{273^2 + 178^2 - 235^2}{2(273)(178)}$

$\theta \approx 58.4°$

Bearing: N 58.4° W

(b) $\cos \phi = \dfrac{235^2 + 178^2 - 273^2}{2(235)(178)}$

$\phi \approx 81.5°$

Bearing: S 81.5° W

39. $d^2 = 60.5^2 + 90^2 - 2(60.5)(90) \cos 45° \approx 4059.8572 \implies d \approx 63.7$ ft

41. $a^2 = 35^2 + 20^2 - 2(35)(20)\cos 42° \implies a \approx 24.2$ miles

43. $\overline{RS} = \sqrt{8^2 + 10^2} = \sqrt{164} = 2\sqrt{41} \approx 12.8$ ft

$\overline{PQ} = \frac{1}{2}\sqrt{16^2 + 10^2} = \frac{1}{2}\sqrt{356} = \sqrt{89} \approx 9.4$ ft

$\tan P = \dfrac{10}{16} = \dfrac{\overline{QS}}{\overline{PS}} = \dfrac{\overline{QS}}{8} \implies \overline{QS} = 5$

45. $d^2 = 10^2 + 7^2 - 2(10)(7) \cos \theta$

$\theta = \arccos\left[\dfrac{10^2 + 7^2 - d^2}{2(10)(7)}\right]$

$s = \dfrac{360° - \theta}{360°}(2\pi r) = \dfrac{(360° - \theta)\pi}{45°}$

d (inches)	9	10	12	13	14	15	16
θ (degrees)	60.9°	69.5°	88.0°	98.2°	109.6°	122.9°	139.8°
s (inches)	20.88	20.28	18.99	18.28	17.48	16.55	15.37

47. $a = 200$

$b = 500$

$c = 600 \implies s = \dfrac{200 + 500 + 600}{2} = 650$

Area $= \sqrt{650(450)(150)(50)} \approx 46{,}837.5$ square feet

49. $s = \dfrac{510 + 840 + 1120}{2} = 1235$

Area $= \sqrt{1235(1235 - 510)(1235 - 840)(1235 - 1120)}$

$\approx 201{,}674$ square yards

Cost $\approx \left(\dfrac{201{,}674}{4840}\right)(2000) \approx \$83{,}336.36$

51. False. The average of the three sides of a triangle is

$\dfrac{a + b + c}{3}$, not $\dfrac{a + b + c}{2} = s$.

53. False. If $a = 10$, $b = 16$, and $c = 5$, then by the Law of Cosines, we would have:

$$\cos A = \frac{16^2 + 5^2 - 10^2}{2(16)(5)} = 1.13125 > 1$$

This is not possible. In general, if the sum of any two sides is less than the third side, then they cannot form a triangle. Here $10 + 5$ is less than 16.

55. $a = 25$, $b = 55$, $c = 72$

(a) area of triangle: $s = \frac{1}{2}(25 + 55 + 72) = 76$

area $= \sqrt{76(51)(21)(4)} \approx 570.60$

(b) area of circumscribed circle:

$$\cos C = \frac{25^2 + 55^2 - 72^2}{2(25)(55)} \approx -0.5578 \implies C \approx 123.9°$$

$$R = \frac{1}{2}\left(\frac{c}{\sin C}\right) \approx 43.37 \quad \text{(see \#54)}$$

area $= \pi R^2 \approx 5909.2$

(c) area of inscribed circle:

$$r = \sqrt{\frac{(s - a)(s - b)(s - c)}{s}} = \sqrt{\frac{(51)(21)(4)}{76}} \approx 7.51 \quad \text{(see \#54)}$$

area $= \pi r^2 \approx 177.09$

57. $\dfrac{1}{2}bc(1 + \cos A) = \dfrac{1}{2}bc\left[1 + \dfrac{b^2 + c^2 - a^2}{2bc}\right]$

$\qquad = \dfrac{1}{2}bc\left[\dfrac{2bc + b^2 + c^2 - a^2}{2bc}\right]$

$\qquad = \dfrac{1}{4}[(b + c)^2 - a^2]$

$\qquad = \dfrac{1}{4}[(b + c) + a][(b + c) - a]$

$\qquad = \dfrac{b + c + a}{2} \cdot \dfrac{b + c - a}{2}$

$\qquad = \dfrac{a + b + c}{2} \cdot \dfrac{-a + b + c}{2}$

59. $\arcsin(-1) = -\dfrac{\pi}{2}$

61. $\arctan \sqrt{3} = \dfrac{\pi}{3}$

63. $\arcsin\left(-\dfrac{\sqrt{3}}{2}\right) = -\dfrac{\pi}{3}$

65. Let $\theta = \arcsin 2x$, then

$\sin \theta = 2x = \dfrac{2x}{1}$ and

$\sec \theta = \dfrac{1}{\sqrt{1 - 4x^2}}$.

67. Let $\theta = \arctan(x - 2)$, then

$\tan \theta = x - 2 = \dfrac{x - 2}{1}$ and

$\cot \theta = \dfrac{1}{x - 2}$.

69. $5 = \sqrt{25 - x^2}, x = 5 \sin \theta$

$5 = \sqrt{25 - (5 \sin \theta)^2}$

$5 = \sqrt{25(1 - \sin^2 \theta)}$

$5 = 5 \cos \theta$

$\cos \theta = 1$

$\sec \theta = \frac{1}{\cos \theta} = 1$

$\csc \theta$ is undefined.

71. $-\sqrt{3} = \sqrt{x^2 - 9}, x = 3 \sec \theta$

$-\sqrt{3} = \sqrt{(3 \sec \theta)^2 - 9}$

$-\sqrt{3} = \sqrt{9(\sec^2 \theta - 1)}$

$-\sqrt{3} = 3 \tan \theta$

$\tan \theta = -\frac{\sqrt{3}}{3}$

$\sec \theta = \sqrt{1 + \tan^2 \theta} = \sqrt{1 + \left(-\frac{\sqrt{3}}{3}\right)^2} = \frac{2\sqrt{3}}{3}$

$\cot \theta = \frac{1}{\tan \theta} = -\sqrt{3}$

$\csc \theta = -\sqrt{1 + \cot^2 \theta} = -\sqrt{1 + (-\sqrt{3})^2} = -2$

73. $\cos \frac{5\pi}{6} - \cos \frac{\pi}{3} = -2 \sin\left(\frac{\frac{5\pi}{6} + \frac{\pi}{3}}{2}\right) \sin\left(\frac{\frac{5\pi}{6} - \frac{\pi}{3}}{2}\right)$

$= -2 \sin \frac{7\pi}{12} \sin \frac{\pi}{4}$

Section 8.3   Vectors in the Plane

- A vector **v** is the collection of all directed line segments that are equivalent to a given directed line segment $\overrightarrow{PQ}$.

- You should be able to *geometrically* perform the operations of vector addition and scalar multiplication.

- The component form of the vector with initial point $P = (p_1, p_2)$ and terminal point $Q = (q_1, q_2)$ is

  $\overrightarrow{PQ} = \langle q_1 - p_1, q_2 - p_2 \rangle = \langle v_1, v_2 \rangle = \mathbf{v}.$

- The magnitude of $\mathbf{v} = \langle v_1, v_2 \rangle$ is given by $\|\mathbf{v}\| = \sqrt{v_1^2 + v_2^2}$.

- If $\|\mathbf{v}\| = 1$, **v** is a unit vector.

- You should be able to perform the operations of scalar multiplication and vector addition in component form.

  (a) $\mathbf{u} + \mathbf{v} = \langle u_1 + v_1, u_2 + v_2 \rangle$ (b) $k\mathbf{u} = \langle ku_1, ku_2 \rangle$

- You should know the following properties of vector addition and scalar multiplication.

  (a) $\mathbf{u} + \mathbf{v} = \mathbf{v} + \mathbf{u}$

  (b) $(\mathbf{u} + \mathbf{v}) + \mathbf{w} = \mathbf{u} + (\mathbf{v} + \mathbf{w})$

  (c) $\mathbf{u} + \mathbf{0} = \mathbf{u}$

  (d) $\mathbf{u} + (-\mathbf{u}) = \mathbf{0}$

  (e) $c(d\mathbf{u}) = (cd)\mathbf{u}$

  (f) $(c + d)\mathbf{u} = c\mathbf{u} + d\mathbf{u}$

  (g) $c(\mathbf{u} + \mathbf{v}) = c\mathbf{u} + c\mathbf{v}$

  (h) $1(\mathbf{u}) = \mathbf{u}, 0\mathbf{u} = \mathbf{0}$

  (i) $\|c\mathbf{v}\| = |c|\, \|\mathbf{v}\|$

- A unit vector in the direction of **v** is $\mathbf{u} = \frac{\mathbf{v}}{\|\mathbf{v}\|}$.

- The standard unit vectors are $\mathbf{i} = \langle 1, 0 \rangle$ and $\mathbf{j} = \langle 0, 1 \rangle$. $\mathbf{v} = \langle v_1, v_2 \rangle$ can be written as $\mathbf{v} = v_1\mathbf{i} + v_2\mathbf{j}$.

- A vector **v** with magnitude $\|\mathbf{v}\|$ and direction θ can be written as $\mathbf{v} = a\mathbf{i} + b\mathbf{j} = \|\mathbf{v}\|(\cos \theta)\mathbf{i} + \|\mathbf{v}\|(\sin \theta)\mathbf{j}$, where $\tan \theta = b/a$.

Vocabulary Check

1. directed line segment

2. initial; terminal

3. magnitude

4. vector

5. standard position

6. unit vector

7. multiplication; addition

8. resultant

9. linear combination; horizontal; vertical

---

1. $\mathbf{v} = \langle 4 - 0, 1 - 0 \rangle = \langle 4, 1 \rangle$

   $\mathbf{u} = \mathbf{v}$

3. Initial point: $(0, 0)$

   Terminal point: $(3, 2)$

   $\mathbf{v} = \langle 3 - 0, 2 - 0 \rangle = \langle 3, 2 \rangle$

   $\|\mathbf{v}\| = \sqrt{3^2 + 2^2} = \sqrt{13}$

5. Initial point: $(2, 2)$

   Terminal point: $(-1, 4)$

   $\mathbf{v} = \langle -1 - 2, 4 - 2 \rangle = \langle -3, 2 \rangle$

   $\|\mathbf{v}\| = \sqrt{(-3)^2 + 2^2} = \sqrt{13}$

7. Initial point: $(3, -2)$

   Terminal point: $(3, 3)$

   $\mathbf{v} = \langle 3 - 3, 3 - (-2) \rangle = \langle 0, 5 \rangle$

   $\|\mathbf{v}\| = \sqrt{0^2 + 5^2} = \sqrt{25} = 5$

9. Initial point: $(-1, 5)$

   Terminal point: $(15, 12)$

   $\mathbf{v} = \langle 15 - (-1), 12 - 5 \rangle = \langle 16, 7 \rangle$

   $\|\mathbf{v}\| = \sqrt{16^2 + 7^2} = \sqrt{305}$

11. Initial point: $(-3, -5)$

    Terminal point: $(5, 1)$

    $\mathbf{v} = \langle 5 - (-3), 1 - (-5) \rangle = \langle 8, 6 \rangle$

    $\|\mathbf{v}\| = \sqrt{8^2 + 6^2} = \sqrt{100} = 10$

13. Initial point: $(1, 3)$

    Terminal point: $(-8, -9)$

    $\mathbf{v} = \langle -8 - 1, -9 - 3 \rangle = \langle -9, -12 \rangle$

    $\|\mathbf{v}\| = \sqrt{(-9)^2 + (-12)^2} = \sqrt{225} = 15$

15.

17.

19. $\mathbf{u} + 2\mathbf{v}$

21. $\mathbf{u} = \langle 2, 1 \rangle, \mathbf{v} = \langle 1, 3 \rangle$

    (a) $\mathbf{u} + \mathbf{v} = \langle 3, 4 \rangle$

    (b) $\mathbf{u} - \mathbf{v} = \langle 1, -2 \rangle$

    (c) $2\mathbf{u} - 3\mathbf{v} = \langle 4, 2 \rangle - \langle 3, 9 \rangle = \langle 1, -7 \rangle$

23. $\mathbf{u} = \langle -5, 3 \rangle$, $\mathbf{v} = \langle 0, 0 \rangle$

    (a) $\mathbf{u} + \mathbf{v} = \langle -5, 3 \rangle = \mathbf{u}$
    (b) $\mathbf{u} - \mathbf{v} = \langle -5, 3 \rangle = \mathbf{u}$
    (c) $2\mathbf{u} - 3\mathbf{v} = 2\mathbf{u} = \langle -10, 6 \rangle$

25. $\mathbf{u} = \mathbf{i} + \mathbf{j}$, $\mathbf{v} = 2\mathbf{i} - 3\mathbf{j}$

    (a) $\mathbf{u} + \mathbf{v} = 3\mathbf{i} - 2\mathbf{j}$
    (b) $\mathbf{u} - \mathbf{v} = -\mathbf{i} + 4\mathbf{j}$
    (c) $2\mathbf{u} - 3\mathbf{v} = (2\mathbf{i} + 2\mathbf{j}) - (6\mathbf{i} - 9\mathbf{j})$

$$= -4\mathbf{i} + 11\mathbf{j}$$

27. $\mathbf{u} = 2\mathbf{i}$, $\mathbf{v} = \mathbf{j}$

    (a) $\mathbf{u} + \mathbf{v} = 2\mathbf{i} + \mathbf{j}$
    (b) $\mathbf{u} - \mathbf{v} = 2\mathbf{i} - \mathbf{j}$
    (c) $2\mathbf{u} - 3\mathbf{v} = 4\mathbf{i} - 3\mathbf{j}$

29. $\mathbf{v} = \dfrac{1}{\|\mathbf{u}\|}\mathbf{u} = \dfrac{1}{\sqrt{3^2 + 0^2}}\langle 3, 0 \rangle = \dfrac{1}{3}\langle 3, 0 \rangle = \langle 1, 0 \rangle$

31. $\mathbf{u} = \dfrac{1}{\|\mathbf{v}\|}\mathbf{v} = \dfrac{1}{\sqrt{(-2)^2 + 2^2}}\langle -2, 2 \rangle = \dfrac{1}{2\sqrt{2}}\langle -2, 2 \rangle$

$$= \left\langle -\dfrac{1}{\sqrt{2}}, \dfrac{1}{\sqrt{2}} \right\rangle$$

$$= \left\langle -\dfrac{\sqrt{2}}{2}, \dfrac{\sqrt{2}}{2} \right\rangle$$

33. $\mathbf{u} = \dfrac{1}{\|\mathbf{v}\|}\mathbf{v} = \dfrac{1}{\sqrt{6^2 + (-2)^2}}(6\mathbf{i} - 2\mathbf{j}) = \dfrac{1}{\sqrt{40}}(6\mathbf{i} - 2\mathbf{j})$

$$= \dfrac{1}{2\sqrt{10}}(6\mathbf{i} - 2\mathbf{j}) = \dfrac{3}{\sqrt{10}}\mathbf{i} - \dfrac{1}{\sqrt{10}}\mathbf{j}$$

$$= \dfrac{3\sqrt{10}}{10}\mathbf{i} - \dfrac{\sqrt{10}}{10}\mathbf{j}$$

35. $\mathbf{u} = \dfrac{1}{\|\mathbf{w}\|}\mathbf{w} = \dfrac{1}{4}(4\mathbf{j}) = \mathbf{j}$

37. $\mathbf{u} = \dfrac{1}{\|\mathbf{w}\|}\mathbf{w} = \dfrac{1}{\sqrt{1^2 + (-2)^2}}(\mathbf{i} - 2\mathbf{j}) = \dfrac{1}{\sqrt{5}}(\mathbf{i} - 2\mathbf{j})$

$= \dfrac{1}{\sqrt{5}}\mathbf{i} - \dfrac{2}{\sqrt{5}}\mathbf{j} = \dfrac{\sqrt{5}}{5}\mathbf{i} - \dfrac{2\sqrt{5}}{5}\mathbf{j}$

39. $5\left(\dfrac{1}{\|\mathbf{u}\|}\mathbf{u}\right) = 5\left(\dfrac{1}{\sqrt{3^2 + 3^2}}\langle 3, 3\rangle\right) = \dfrac{5}{3\sqrt{2}}\langle 3, 3\rangle$

$= \left\langle \dfrac{5}{\sqrt{2}}, \dfrac{5}{\sqrt{2}} \right\rangle = \left\langle \dfrac{5\sqrt{2}}{2}, \dfrac{5\sqrt{2}}{2} \right\rangle$

41. $9\left(\dfrac{1}{\|\mathbf{u}\|}\mathbf{u}\right) = 9\left(\dfrac{1}{\sqrt{2^2 + 5^2}}\langle 2, 5\rangle\right) = \dfrac{9}{\sqrt{29}}\langle 2, 5\rangle$

$= \left\langle \dfrac{18}{\sqrt{29}}, \dfrac{45}{\sqrt{29}} \right\rangle = \left\langle \dfrac{18\sqrt{29}}{29}, \dfrac{45\sqrt{29}}{29} \right\rangle$

43. $\mathbf{u} = \langle 4 - (-3), 5 - 1\rangle$

$= \langle 7, 4\rangle$

$= 7\mathbf{i} + 4\mathbf{j}$

45. $\mathbf{u} = \langle 2 - (-1), 3 - (-5)\rangle$

$= \langle 3, 8\rangle$

$= 3\mathbf{i} + 8\mathbf{j}$

47. $\mathbf{v} = \frac{3}{2}\mathbf{u}$

$= \frac{3}{2}(2\mathbf{i} - \mathbf{j})$

$= 3\mathbf{i} - \frac{3}{2}\mathbf{j} = \langle 3, -\frac{3}{2}\rangle$

49. $\mathbf{v} = \mathbf{u} + 2\mathbf{w}$

$= (2\mathbf{i} - \mathbf{j}) + 2(\mathbf{i} + 2\mathbf{j})$

$= 4\mathbf{i} + 3\mathbf{j} = \langle 4, 3\rangle$

51. $\mathbf{v} = \frac{1}{2}(3\mathbf{u} + \mathbf{w})$

$= \frac{1}{2}(6\mathbf{i} - 3\mathbf{j} + \mathbf{i} + 2\mathbf{j})$

$= \frac{7}{2}\mathbf{i} - \frac{1}{2}\mathbf{j} = \langle \frac{7}{2}, -\frac{1}{2}\rangle$

53. $\mathbf{v} = 3(\cos 60°\mathbf{i} + \sin 60°\mathbf{j})$

$\|\mathbf{v}\| = 3, \ \theta = 60°$

55. $\mathbf{v} = 6\mathbf{i} - 6\mathbf{j}$

$\|\mathbf{v}\| = \sqrt{6^2 + (-6)^2} = \sqrt{72}$

$= 6\sqrt{2}$

$\tan \theta = \dfrac{-6}{6} = -1$

Since $\mathbf{v}$ lies in Quadrant IV,
$\theta = 315°$.

57. $\mathbf{v} = \langle 3\cos 0°, 3\sin 0°\rangle$

$= \langle 3, 0\rangle$

59. $\mathbf{v} = \left\langle \dfrac{7}{2}\cos 150°, \dfrac{7}{2}\sin 150° \right\rangle$

$= \left\langle -\dfrac{7\sqrt{3}}{4}, \dfrac{7}{4} \right\rangle$

61. $\mathbf{v} = \langle 3\sqrt{2}\cos 150°, \ 3\sqrt{2}\sin 150°\rangle$

$= \left\langle -\dfrac{3\sqrt{6}}{2}, \dfrac{3\sqrt{2}}{2} \right\rangle$

63. $\mathbf{v} = 2\left(\dfrac{1}{\sqrt{1^2 + 3^2}}\right)(\mathbf{i} + 3\mathbf{j})$

$= \dfrac{2}{\sqrt{10}}(\mathbf{i} + 3\mathbf{j})$

$= \dfrac{\sqrt{10}}{5}\mathbf{i} + \dfrac{3\sqrt{10}}{5}\mathbf{j} = \left\langle \dfrac{\sqrt{10}}{5}, \dfrac{3\sqrt{10}}{5} \right\rangle$

65. $\mathbf{u} = \langle 5 \cos 0°, 5 \sin 0° \rangle = \langle 5, 0 \rangle$

$\mathbf{v} = \langle 5 \cos 90°, 5 \sin 90° \rangle = \langle 0, 5 \rangle$

$\mathbf{u} + \mathbf{v} = \langle 5, 5 \rangle$

67. $\mathbf{u} = \langle 20 \cos 45°, 20 \sin 45° \rangle = \left\langle 10\sqrt{2}, 10\sqrt{2} \right\rangle$

$\mathbf{v} = \langle 50 \cos 180°, 50 \sin 180° \rangle = \langle -50, 0 \rangle$

$\mathbf{u} + \mathbf{v} = \left\langle 10\sqrt{2} - 50, 10\sqrt{2} \right\rangle$

69. $\mathbf{v} = \mathbf{i} + \mathbf{j}$

$\mathbf{w} = 2\mathbf{i} - 2\mathbf{j}$

$\mathbf{u} = \mathbf{v} - \mathbf{w} = -\mathbf{i} + 3\mathbf{j}$

$\|\mathbf{v}\| = \sqrt{2}$

$\|\mathbf{w}\| = 2\sqrt{2}$

$\|\mathbf{v} - \mathbf{w}\| = \sqrt{10}$

$\cos \alpha = \dfrac{\|\mathbf{v}\|^2 + \|\mathbf{w}\|^2 - \|\mathbf{v} - \mathbf{w}\|^2}{2\|\mathbf{v}\|\,\|\mathbf{w}\|} = \dfrac{2 + 8 - 10}{2\sqrt{2} \cdot 2\sqrt{2}} = 0$

$\alpha = 90°$

71. Force One: $\mathbf{u} = 45\mathbf{i}$

Force Two: $\mathbf{v} = 60 \cos \theta \mathbf{i} + 60 \sin \theta \mathbf{j}$

Resultant Force: $\mathbf{u} + \mathbf{v} = (45 + 60 \cos \theta)\mathbf{i} + 60 \sin \theta \mathbf{j}$

$\|\mathbf{u} + \mathbf{v}\| = \sqrt{(45 + 60 \cos \theta)^2 + (60 \sin \theta)^2} = 90$

$2025 + 5400 \cos \theta + 3600 = 8100$

$5400 \cos \theta = 2475$

$\cos \theta = \dfrac{2475}{5400} \approx 0.4583$

$\theta \approx 62.7°$

73. $\mathbf{u} = 300\mathbf{i}$

$\mathbf{v} = (125 \cos 45°)\mathbf{i} + (125 \sin 45°)\mathbf{j} = \dfrac{125}{\sqrt{2}}\mathbf{i} + \dfrac{125}{\sqrt{2}}\mathbf{j}$

$\mathbf{R} = \mathbf{u} + \mathbf{v} = \left(300 + \dfrac{125}{\sqrt{2}}\right)\mathbf{i} + \dfrac{125}{\sqrt{2}}\mathbf{j}$

$\|\mathbf{R}\| = \sqrt{\left(300 + \dfrac{125}{\sqrt{2}}\right)^2 + \left(\dfrac{125}{\sqrt{2}}\right)^2} \approx 398.32$ newtons

$\tan \theta = \dfrac{\dfrac{125}{\sqrt{2}}}{300 + \left(\dfrac{125}{\sqrt{2}}\right)} \Rightarrow \theta \approx 12.8°$

75. $\mathbf{u} = (75 \cos 30°)\mathbf{i} + (75 \sin 30°)\mathbf{j} \approx 64.95\mathbf{i} + 37.5\mathbf{j}$

$\mathbf{v} = (100 \cos 45°)\mathbf{i} + (100 \sin 45°)\mathbf{j} \approx 70.71\mathbf{i} + 70.71\mathbf{j}$

$\mathbf{w} = (125 \cos 120°)\mathbf{i} + (125 \sin 120°)\mathbf{j} \approx -62.5\mathbf{i} + 108.3\mathbf{j}$

$\mathbf{u} + \mathbf{v} + \mathbf{w} \approx 73.16\mathbf{i} + 216.5\mathbf{j}$

$\|\mathbf{u} + \mathbf{v} + \mathbf{w}\| \approx 228.5$ pounds

$\tan \theta \approx \dfrac{216.5}{73.16} \approx 2.9593$

$\theta \approx 71.3°$

77. Horizontal component of velocity: $70 \cos 35° \approx 57.34$ feet per second

Vertical component of velocity: $70 \sin 35° \approx 40.15$ feet per second

79. Cable $\overrightarrow{AC}$: $\mathbf{u} = \|\mathbf{u}\|(\cos 50°\mathbf{i} - \sin 50°\mathbf{j})$

Cable $\overrightarrow{BC}$: $\mathbf{v} = \|\mathbf{v}\|(-\cos 30°\mathbf{i} - \sin 30°\mathbf{j})$

Resultant: $\mathbf{u} + \mathbf{v} = -2000\mathbf{j}$

$\|\mathbf{u}\| \cos 50° - \|\mathbf{v}\| \cos 30° = 0$

$-\|\mathbf{u}\| \sin 50° - \|\mathbf{v}\|\sin 30° = -2000$

Solving this system of equations yields:

$T_{AC} = \|\mathbf{u}\| \approx 1758.8$ pounds

$T_{BC} = \|\mathbf{v}\| \approx 1305.4$ pounds

81. Towline 1: $\mathbf{u} = \|\mathbf{u}\|(\cos 18°\mathbf{i} + \sin 18°\mathbf{j})$

Towline 2: $\mathbf{v} = \|\mathbf{u}\|(\cos 18°\mathbf{i} - \sin 18°\mathbf{j})$

Resultant: $\mathbf{u} + \mathbf{v} = 6000\mathbf{i}$

$\|\mathbf{u}\| \cos 18° + \|\mathbf{u}\| \cos 18° = 6000$

$\|\mathbf{u}\| \approx 3154.4$

Therefore, the tension on each towline is $\|\mathbf{u}\| \approx 3154.4$ pounds.

83. Airspeed: $\mathbf{u} = (875 \cos 58°)\mathbf{i} - (875 \sin 58°)\mathbf{j}$

Groundspeed: $\mathbf{v} = (800 \cos 50°)\mathbf{i} - (800 \sin 50°)\mathbf{j}$

Wind: $\mathbf{w} = \mathbf{v} - \mathbf{u} = (800 \cos 50° - 875 \cos 58°)\mathbf{i} + (-800 \sin 50° + 875 \sin 58°)\mathbf{j}$

$\approx 50.5507\mathbf{i} + 129.2065\mathbf{j}$

Wind speed: $\|\mathbf{w}\| \approx \sqrt{(50.5507)^2 + (129.2065)^2}$

≈ 138.7 kilometers per hour

Wind direction: $\tan \theta \approx \dfrac{129.2065}{50.5507}$

$\theta \approx 68.6°; \; 90° - \theta = 21.4°$

Bearing: N 21.4° E

85. $W = FD = (100 \cos 50°)(30) = 1928.4$ foot–pounds

87. True. See Example 1.

89. (a) The angle between them is 0°.

(b) The angle between them is 180°.

(c) No. At most it can be equal to the sum when the angle between them is 0°.

91. Let $\mathbf{v} = (\cos \theta)\mathbf{i} + (\sin \theta)\mathbf{j}$.

$\|\mathbf{v}\| = \sqrt{\cos^2 \theta + \sin^2 \theta} = \sqrt{1} = 1$

Therefore, $\mathbf{v}$ is a unit vector for any value of θ.

93. $\mathbf{u} = \langle 5 - 1, 2 - 6 \rangle = \langle 4, -4 \rangle$

$\mathbf{v} = \langle 9 - 4, 4 - 5 \rangle = \langle 5, -1 \rangle$

$\mathbf{u} - \mathbf{v} = \langle -1, -3 \rangle$ or $\mathbf{v} - \mathbf{u} = \langle 1, 3 \rangle$

95. $\sqrt{x^2 - 64} = \sqrt{(8 \sec \theta)^2 - 64}$

$= \sqrt{64(\sec^2 \theta - 1)}$

$= 8\sqrt{\tan^2 \theta}$

$= 8 \tan \theta$ for $0 < \theta < \dfrac{\pi}{2}$

97. $\sqrt{x^2 + 36} = \sqrt{(6\tan\theta)^2 + 36}$

$\qquad\qquad = \sqrt{36(\tan^2\theta + 1)}$

$\qquad\qquad = 6\sqrt{\sec^2\theta}$

$\qquad\qquad = 6\sec\theta \quad \text{for} \;\; 0 < \theta < \dfrac{\pi}{2}$

99. $\cos x(\cos x + 1) = 0$

$\qquad \cos x = 0 \quad \text{or} \quad \cos x + 1 = 0$

$\qquad\qquad x = \dfrac{\pi}{2} + n\pi \qquad \cos x = -1$

$\qquad\qquad\qquad\qquad\qquad\qquad x = \pi + 2n\pi$

101. $3\sec x \sin x - 2\sqrt{3}\sin x = 0$

$\qquad \sin x\left(3\sec x - 2\sqrt{3}\right) = 0$

$\qquad \sin x = 0 \quad \text{or} \quad 3\sec x - 2\sqrt{3} = 0$

$\qquad\quad x = n\pi \qquad\qquad \sec x = \dfrac{2\sqrt{3}}{3}$

$\qquad\qquad\qquad\qquad\qquad \cos x = \dfrac{3}{2\sqrt{3}} = \dfrac{\sqrt{3}}{2}$

$\qquad\qquad\qquad\qquad\qquad\qquad x = \dfrac{\pi}{6} + 2n\pi$

$\qquad\qquad\qquad\qquad\qquad\qquad x = \dfrac{11\pi}{6} + 2n\pi$

Section 8.4   Vectors and Dot Products

- Know the definition of the dot product of $\mathbf{u} = \langle u_1, u_2 \rangle$ and $\mathbf{v} = \langle v_1, v_2 \rangle$.

  $\mathbf{u} \cdot \mathbf{v} = u_1 v_1 + u_2 v_2$

- Know the following properties of the dot product:

  1. $\mathbf{u} \cdot \mathbf{v} = \mathbf{v} \cdot \mathbf{u}$

  2. $\mathbf{0} \cdot \mathbf{v} = 0$

  3. $\mathbf{u} \cdot (\mathbf{v} + \mathbf{w}) = \mathbf{u} \cdot \mathbf{v} + \mathbf{u} \cdot \mathbf{w}$

  4. $\mathbf{v} \cdot \mathbf{v} = \|\mathbf{v}\|^2$

  5. $c(\mathbf{u} \cdot \mathbf{v}) = c\mathbf{u} \cdot \mathbf{v} = \mathbf{u} \cdot c\mathbf{v}$

- If θ is the angle between two nonzero vectors $\mathbf{u}$ and $\mathbf{v}$, then

  $\cos\theta = \dfrac{\mathbf{u} \cdot \mathbf{v}}{\|\mathbf{u}\|\,\|\mathbf{v}\|}.$

- The vectors $\mathbf{u}$ and $\mathbf{v}$ are orthogonal if $\mathbf{u} \cdot \mathbf{v} = 0$.

- Know the definition of vector components.

  $\mathbf{u} = \mathbf{w}_1 + \mathbf{w}_2$ where $\mathbf{w}_1$ and $\mathbf{w}_2$ are orthogonal, and $\mathbf{w}_1$ is parallel to $\mathbf{v}$. $\mathbf{w}_1$ is called the projection of $\mathbf{u}$ onto $\mathbf{v}$

  and is denoted by $\mathbf{w}_1 = \text{proj}_{\mathbf{v}}\mathbf{u} = \left(\dfrac{\mathbf{u} \cdot \mathbf{v}}{\|\mathbf{v}\|^2}\right)\mathbf{v}$. Then we have $\mathbf{w}_2 = \mathbf{u} - \mathbf{w}_1$.

- Know the definition of work.

  1. Projection form: $w = \|\text{proj}_{\overrightarrow{PQ}}\,\mathbf{F}\|\,\|\overrightarrow{PQ}\|$

  2. Dot product form: $w = \mathbf{F} \cdot \overrightarrow{PQ}$

Vocabulary Check

1. dot product

2. $\dfrac{\mathbf{u} \cdot \mathbf{v}}{\|\mathbf{u}\| \, \|\mathbf{v}\|}$

3. orthogonal

4. $\left(\dfrac{\mathbf{u} \cdot \mathbf{v}}{\|\mathbf{v}\|^2}\right)\mathbf{v}$

5. $\|\mathrm{proj}_{\overrightarrow{PQ}}\mathbf{F}\| \, \|\overrightarrow{PQ}\|$; $\mathbf{F} \cdot \overrightarrow{PQ}$

---

1. $\mathbf{u} = \langle 6, 1 \rangle$, $\mathbf{v} = \langle -2, 3 \rangle$

$\mathbf{u} \cdot \mathbf{v} = 6(-2) + 1(3) = -9$

3. $\mathbf{u} = \langle -4, 1 \rangle$, $\mathbf{v} = \langle 2, -3 \rangle$

$\mathbf{u} \cdot \mathbf{v} = -4(2) + 1(-3) = -11$

5. $\mathbf{u} = 4\mathbf{i} - 2\mathbf{j}$, $\mathbf{v} = \mathbf{i} - \mathbf{j}$

$\mathbf{u} \cdot \mathbf{v} = 4(1) + (-2)(-1) = 6$

7. $\mathbf{u} = 3\mathbf{i} + 2\mathbf{j}$, $\mathbf{v} = -2\mathbf{i} - 3\mathbf{j}$

$\mathbf{u} \cdot \mathbf{v} = 3(-2) + 2(-3) = -12$

9. $\mathbf{u} = \langle 2, 2 \rangle$

$\mathbf{u} \cdot \mathbf{u} = 2(2) + 2(2) = 8$

The result is a scalar.

11. $\mathbf{u} = \langle 2, 2 \rangle$, $\mathbf{v} = \langle -3, 4 \rangle$

$(\mathbf{u} \cdot \mathbf{v})\mathbf{v} = [(2)(-3) + 2(4)]\langle -3, 4 \rangle$

$= 2\langle -3, 4 \rangle = \langle -6, 8 \rangle$

The result is a vector.

13. $\mathbf{u} = \langle 2, 2 \rangle$, $\mathbf{v} = \langle -3, 4 \rangle$, $\mathbf{w} = \langle 1, -2 \rangle$

$(3\mathbf{w} \cdot \mathbf{v})\mathbf{u} = [3(1)(-3) + 3(-2)(4)]\langle 2, 2 \rangle$

$= -33\langle 2, 2 \rangle$

$= \langle -66, -66 \rangle$

The result is a vector.

15. $\mathbf{w} = \langle 1, -2 \rangle$

$\|\mathbf{w}\| - 1 = \sqrt{(1)^2 + (-2)^2} - 1 = \sqrt{5} - 1$

The result is a scalar.

17. $\mathbf{u} = \langle 2, 2 \rangle$, $\mathbf{v} = \langle -3, 4 \rangle$, $\mathbf{w} = \langle 1, -2 \rangle$

$(\mathbf{u} \cdot \mathbf{v}) - (\mathbf{u} \cdot \mathbf{w}) = [2(-3) + 2(4)] - [2(1) + 2(-2)]$

$= 2 - (-2)$

$= 4$

The result is a scalar.

19. $\mathbf{u} = \langle -5, 12 \rangle$

$\|\mathbf{u}\| = \sqrt{\mathbf{u} \cdot \mathbf{u}} = \sqrt{(-5)^2 + 12^2} = 13$

21. $\mathbf{u} = 20\mathbf{i} + 25\mathbf{j}$

$\|\mathbf{u}\| = \sqrt{\mathbf{u} \cdot \mathbf{u}} = \sqrt{(20)^2 + (25)^2} = \sqrt{1025} = 5\sqrt{41}$

23. $\mathbf{u} = 6\mathbf{j}$

$\|\mathbf{u}\| = \sqrt{\mathbf{u} \cdot \mathbf{u}} = \sqrt{(0)^2 + (6)^2} = \sqrt{36} = 6$

25. $\mathbf{u} = \langle 1, 0 \rangle$, $\mathbf{v} = \langle 0, -2 \rangle$

$\cos \theta = \dfrac{\mathbf{u} \cdot \mathbf{v}}{\|\mathbf{u}\| \, \|\mathbf{v}\|} = \dfrac{0}{(1)(2)} = 0$

$\theta = 90°$

27. $\mathbf{u} = 3\mathbf{i} + 4\mathbf{j}$, $\mathbf{v} = -2\mathbf{j}$

$\cos \theta = \dfrac{\mathbf{u} \cdot \mathbf{v}}{\|\mathbf{u}\| \, \|\mathbf{v}\|} = -\dfrac{8}{(5)(2)}$

$\theta = \arccos\left(-\dfrac{4}{5}\right)$

$\theta \approx 143.13°$

29. $\mathbf{u} = 2\mathbf{i} - \mathbf{j}$, $\mathbf{v} = 6\mathbf{i} + 4\mathbf{j}$

$\cos \theta = \dfrac{\mathbf{u} \cdot \mathbf{v}}{\|\mathbf{u}\| \, \|\mathbf{v}\|} = \dfrac{8}{\sqrt{5}\sqrt{52}} = 0.4961$

$\theta = 60.26°$

31. $\mathbf{u} = 5\mathbf{i} + 5\mathbf{j}$, $\mathbf{v} = -6\mathbf{i} + 6\mathbf{j}$

$\cos \theta = \dfrac{\mathbf{u} \cdot \mathbf{v}}{\|\mathbf{u}\| \, \|\mathbf{v}\|} = 0$

$\theta = 90°$

33. $\mathbf{u} = \left(\cos \dfrac{\pi}{3}\right)\mathbf{i} + \left(\sin \dfrac{\pi}{3}\right)\mathbf{j} = \dfrac{1}{2}\mathbf{i} + \dfrac{\sqrt{3}}{2}\mathbf{j}$

$\mathbf{v} = \left(\cos \dfrac{3\pi}{4}\right)\mathbf{i} + \left(\sin \dfrac{3\pi}{4}\right)\mathbf{j} = -\dfrac{\sqrt{2}}{2}\mathbf{i} + \dfrac{\sqrt{2}}{2}\mathbf{j}$

$\|\mathbf{u}\| = \|\mathbf{v}\| = 1$

$\cos \theta = \dfrac{\mathbf{u} \cdot \mathbf{v}}{\|\mathbf{u}\|\,\|\mathbf{v}\|} = \mathbf{u} \cdot \mathbf{v}$

$= \left(\dfrac{1}{2}\right)\left(-\dfrac{\sqrt{2}}{2}\right) + \left(\dfrac{\sqrt{3}}{2}\right)\left(\dfrac{\sqrt{2}}{2}\right) = \dfrac{-\sqrt{2} + \sqrt{6}}{4}$

$\theta = \arccos\left(\dfrac{-\sqrt{2} + \sqrt{6}}{4}\right) = 75° = \dfrac{5\pi}{12}$

35. $\mathbf{u} = 3\mathbf{i} + 4\mathbf{j}$

$\mathbf{v} = -7\mathbf{i} + 5\mathbf{j}$

$\cos \theta = \dfrac{\mathbf{u} \cdot \mathbf{v}}{\|\mathbf{u}\|\,\|\mathbf{v}\|}$

$= \dfrac{3(-7) + 4(5)}{3\sqrt{74}}$

$= \dfrac{-1}{5\sqrt{74}} \approx -0.0232$

$\theta \approx 91.3°$

37. $\mathbf{u} = 5\mathbf{i} + 5\mathbf{j}$

$\mathbf{v} = -8\mathbf{i} + 8\mathbf{j}$

$\cos \theta = \dfrac{\mathbf{u} \cdot \mathbf{v}}{\|\mathbf{u}\|\,\|\mathbf{v}\|}$

$= \dfrac{5(-8) + 5(8)}{\sqrt{50}\sqrt{128}}$

$= 0$

$\theta = 90°$

39. $P = (1, 2),\ Q = (3, 4),\ R = (2, 5)$

$\overrightarrow{PQ} = \langle 2, 2\rangle,\ \overrightarrow{PR} = \langle 1, 3\rangle,\ \overrightarrow{QR} = \langle -1, 1\rangle$

$\cos \alpha = \dfrac{\overrightarrow{PQ} \cdot \overrightarrow{PR}}{\|\overrightarrow{PQ}\|\,\|\overrightarrow{PR}\|} = \dfrac{8}{(2\sqrt{2})(\sqrt{10})} \Rightarrow \alpha = \arccos \dfrac{2}{\sqrt{5}} \approx 26.57°$

$\cos \beta = \dfrac{\overrightarrow{PQ} \cdot \overrightarrow{QR}}{\|\overrightarrow{PQ}\|\,\|\overrightarrow{QR}\|} = 0 \Rightarrow \beta = 90°.$ Thus, $\gamma = 180° - 26.57° - 90° = 63.43°.$

41. $P = (-3, 0),\ Q = (2, 2),\ R = (0, 6)$

$\overrightarrow{QP} = \langle -5, -2\rangle,\ \overrightarrow{PR} = \langle 3, 6\rangle,\ \overrightarrow{QR} = \langle -2, 4\rangle,\ \overrightarrow{PQ} = \langle 5, 2\rangle$

$\cos \alpha = \dfrac{\overrightarrow{PQ} \cdot \overrightarrow{PR}}{\|\overrightarrow{PQ}\|\,\|\overrightarrow{PR}\|} = \dfrac{27}{\sqrt{29}\sqrt{45}} \Rightarrow \alpha \approx 41.63°$

$\cos \beta = \dfrac{\overrightarrow{QP} \cdot \overrightarrow{QR}}{\|\overrightarrow{QP}\|\,\|\overrightarrow{PR}\|} = \dfrac{2}{\sqrt{29}\sqrt{20}} \Rightarrow \beta \approx 85.24°$

$\delta = 180° - 41.63° - 85.24° = 53.13°$

43. $\mathbf{u} \cdot \mathbf{v} = \|\mathbf{u}\|\,\|\mathbf{v}\|\cos \theta$

$= (4)(10)\cos \dfrac{2\pi}{3}$

$= 40\left(-\dfrac{1}{2}\right)$

$= -20$

45. $\mathbf{u} \cdot \mathbf{v} = \|\mathbf{u}\|\,\|\mathbf{v}\|\cos \theta$

$= (9)(36)\cos \dfrac{3\pi}{4}$

$= 324\left(-\dfrac{\sqrt{2}}{2}\right)$

$= -162\sqrt{2} \approx -229.1$

47. $\mathbf{u} = \langle -12, 30\rangle,\ \mathbf{v} = \left\langle \dfrac{1}{2}, -\dfrac{5}{4}\right\rangle$

$\mathbf{u} = -24\mathbf{v} \Rightarrow \mathbf{u}$ and $\mathbf{v}$ are parallel.

49. $\mathbf{u} = \frac{1}{4}(3\mathbf{i} - \mathbf{j})$, $\mathbf{v} = 5\mathbf{i} + 6\mathbf{j}$

$\mathbf{u} \neq k\mathbf{v} \implies$ Not parallel

$\mathbf{u} \cdot \mathbf{v} \neq 0 \implies$ Not orthogonal

Neither

51. $\mathbf{u} = 2\mathbf{i} - 2\mathbf{j}$, $\mathbf{v} = -\mathbf{i} - \mathbf{j}$

$\mathbf{u} \cdot \mathbf{v} = 0 \implies$ $\mathbf{u}$ and $\mathbf{v}$ are orthogonal.

53. $\mathbf{u} = \langle 2, 2 \rangle$, $\mathbf{v} = \langle 6, 1 \rangle$

$\mathbf{w}_1 = \text{proj}_{\mathbf{v}}\mathbf{u} = \left(\dfrac{\mathbf{u} \cdot \mathbf{v}}{\|\mathbf{v}\|^2} \right)\mathbf{v} = \dfrac{14}{37}\langle 6, 1 \rangle = \dfrac{1}{37}\langle 84, 14 \rangle$

$\mathbf{w}_2 = \mathbf{u} - \mathbf{w}_1 = \langle 2, 2 \rangle - \dfrac{14}{37}\langle 6, 1 \rangle = \left\langle -\dfrac{10}{37}, \dfrac{60}{37} \right\rangle = \dfrac{10}{37}\langle -1, 6 \rangle = \dfrac{1}{37}\langle -10, 60 \rangle$

$\mathbf{u} = \dfrac{1}{37}\langle 84, 14 \rangle + \dfrac{1}{37}\langle -10, 60 \rangle = \langle 2, 2 \rangle$

55. $\mathbf{u} = \langle 0, 3 \rangle$, $\mathbf{v} = \langle 2, 15 \rangle$

$\mathbf{w}_1 = \text{proj}_{\mathbf{v}}\mathbf{u} = \left(\dfrac{\mathbf{u} \cdot \mathbf{v}}{\|\mathbf{v}\|^2} \right)\mathbf{v} = \dfrac{45}{229}\langle 2, 15 \rangle$

$\mathbf{w}_2 = \mathbf{u} - \mathbf{w}_1 = \langle 0, 3 \rangle - \dfrac{45}{229}\langle 2, 15 \rangle = \left\langle -\dfrac{90}{229}, \dfrac{12}{229} \right\rangle$

$\qquad\qquad = \dfrac{6}{229}\langle -15, 2 \rangle$

$\mathbf{u} = \dfrac{45}{229}\langle 2, 15 \rangle + \dfrac{6}{229}\langle -15, 2 \rangle = \langle 0, 3 \rangle$

57. $\text{proj}_{\mathbf{v}}\mathbf{u} = \mathbf{0}$ since they are perpendicular.

Since $\mathbf{u}$ and $\mathbf{v}$ are orthogonal, $\mathbf{u} \cdot \mathbf{v} = 0$ and $\text{proj}_{\mathbf{v}}\mathbf{u} = \mathbf{0}$.

$\text{proj}_{\mathbf{v}}\mathbf{u} = \dfrac{\mathbf{u} \cdot \mathbf{v}}{\|\mathbf{v}\|^2}\mathbf{v} = 0$, since $\mathbf{u} \cdot \mathbf{v} = \mathbf{0}$.

59. $\mathbf{u} = \langle 3, 5 \rangle$

For $\mathbf{v}$ to be orthogonal to $\mathbf{u}$, $\mathbf{u} \cdot \mathbf{v}$ must equal 0.

Two possibilities: $\langle -5, 3 \rangle$ and $\langle 5, -3 \rangle$

61. $\mathbf{u} = \frac{1}{2}\mathbf{i} - \frac{2}{3}\mathbf{j}$

For $\mathbf{u}$ and $\mathbf{v}$ to be orthogonal, $\mathbf{u} \cdot \mathbf{v}$ must equal 0.

Two possibilities: $\mathbf{v} = \frac{2}{3}\mathbf{i} + \frac{1}{2}\mathbf{j}$ and $\mathbf{v} = -\frac{2}{3}\mathbf{i} - \frac{1}{2}\mathbf{j}$

63. $\mathbf{w} = \| \text{proj}_{\overrightarrow{PQ}} \mathbf{v}\| \|\overrightarrow{PQ}\|$ where $\overrightarrow{PQ} = \langle 4, 7 \rangle$ and $\mathbf{v} = \langle 1, 4 \rangle$.

$\text{proj}_{\overrightarrow{PQ}} \mathbf{v} = \left(\dfrac{\mathbf{v} \cdot \overrightarrow{PQ}}{\|\overrightarrow{PQ}\|^2} \right)\overrightarrow{PQ} = \left(\dfrac{32}{65} \right)\langle 4, 7 \rangle$

$\mathbf{w} = \| \text{proj}_{\overrightarrow{PQ}} \mathbf{v}\| \|\overrightarrow{PQ}\| = \left(\dfrac{32\sqrt{65}}{65} \right)\left(\sqrt{65} \right) = 32$

65. (a) $\mathbf{u} = \langle 1650, 3200 \rangle$, $\mathbf{v} = \langle 15.25, 10.50 \rangle$

$\mathbf{u} \cdot \mathbf{v} = 1650(15.25) + 3200(10.50) = \$58,762.50$

This gives the total revenue that can be earned by selling all of the pans.

(b) Increase prices by 5%: $1.05\mathbf{v}$ The operation is scalar multiplication.

$\mathbf{u} \cdot 1.05\mathbf{v} = 1.05\mathbf{u} \cdot \mathbf{v}$

$\qquad\quad = 1.05[1650(15.25) + 3200(10.50)]$

$\qquad\quad = 1.05(58,762.50)$

$\qquad\quad = 61,700.63$

67. (a) Force due to gravity:

$$\mathbf{F} = -30{,}000\mathbf{j}$$

Unit vector along hill:

$$\mathbf{v} = (\cos d)\mathbf{i} + (\sin d)\mathbf{j}$$

Projection of $\mathbf{F}$ onto $\mathbf{v}$:

$$\mathbf{w}_1 = \text{proj}_\mathbf{v}\,\mathbf{F} = \left(\frac{\mathbf{F}\cdot\mathbf{v}}{\|\mathbf{v}\|^2}\right)\mathbf{v} = (\mathbf{F}\cdot\mathbf{v})\mathbf{v} = -30{,}000\sin d\,\mathbf{v}$$

The magnitude of the force is $30{,}000\sin d$.

(b)

d	0°	1°	2°	3°	4°	5°	6°	7°	8°	9°	10°
Force	0	523.6	1047.0	1570.1	2092.7	2614.7	3135.9	3656.1	4175.2	4693.0	5209.4

(c) Force perpendicular to the hill when $d = 5°$:

$$\text{Force} = \sqrt{(30{,}000)^2 - (2614.7)^2} \approx 29{,}885.8 \text{ pounds}$$

69. $\mathbf{w} = (245)(3) = 735$ newton-meters

71. $\mathbf{w} = (\cos 30°)(45)(20) \approx 779.4$ foot-pounds

73. $\mathbf{w} = (\cos 30°)(250)(100) \approx 21{,}650.64$ foot-pounds

75. False. Work is represented by a scalar.

77. (a) $\mathbf{u}\cdot\mathbf{v} = 0 \Rightarrow \mathbf{u}$ and $\mathbf{v}$ are orthogonal and $\theta = \dfrac{\pi}{2}$.

(b) $\mathbf{u}\cdot\mathbf{v} > 0 \Rightarrow \cos\theta > 0 \Rightarrow 0 \le \theta < \dfrac{\pi}{2}$

(c) $\mathbf{u}\cdot\mathbf{v} < 0 \Rightarrow \cos\theta < 0 \Rightarrow \dfrac{\pi}{2} < \theta \le \pi$

79. In a rhombus, $\|\mathbf{u}\| = \|\mathbf{v}\|$. The diagonals are $\mathbf{u} + \mathbf{v}$ and $\mathbf{u} - \mathbf{v}$.

$$(\mathbf{u}+\mathbf{v})\cdot(\mathbf{u}-\mathbf{v}) = (\mathbf{u}+\mathbf{v})\cdot\mathbf{u} - (\mathbf{u}+\mathbf{v})\cdot\mathbf{v}$$

$$= \mathbf{u}\cdot\mathbf{u} + \mathbf{v}\cdot\mathbf{u} - \mathbf{u}\cdot\mathbf{v} - \mathbf{v}\cdot\mathbf{v}$$

$$= \|\mathbf{u}\|^2 - \|\mathbf{v}\|^2 = 0$$

Therefore, the diagonals are orthogonal.

81. $\sqrt{42}\cdot\sqrt{24} = \sqrt{1008}$

$\qquad\qquad = \sqrt{144\cdot 7}$

$\qquad\qquad = 12\sqrt{7}$

83. $\sqrt{-3}\,\sqrt{-8} = \left(\sqrt{3}i\right)\left(2\sqrt{2}i\right)$

$\qquad\qquad = 2\sqrt{6}i^2$

$\qquad\qquad = -2\sqrt{6}$

85. $\sin 2x - \sqrt{3}\sin x = 0$

$2\sin x\cos x - \sqrt{3}\sin x = 0$

$\sin x(2\cos x - \sqrt{3}) = 0$

$\sin x = 0 \quad\text{or}\quad 2\cos x - \sqrt{3} = 0$

$x = 0, \pi \qquad\qquad \cos x = \dfrac{\sqrt{3}}{2}$

$\qquad\qquad\qquad x = \dfrac{\pi}{6}, \dfrac{11\pi}{6}$

87. $2\tan x = \tan 2x$

$2\tan x = \dfrac{2\tan x}{1 - \tan^2 x}$

$2\tan x(1 - \tan^2 x) = 2\tan x$

$2\tan x(1 - \tan^2 x) - 2\tan x = 0$

$2\tan x[(1 - \tan^2 x) - 1] = 0$

$2\tan x(-\tan^2 x) = 0$

$-2\tan^3 x = 0$

$\tan x = 0$

$x = 0, \pi$

For Exercises 89–91:

$\sin u = -\frac{12}{13}$, u in Quadrant IV $\Rightarrow$ $\cos u = \frac{5}{13}$ $\cos v = \frac{24}{25}$, v in Quadrant IV $\Rightarrow$ $\sin v = -\frac{7}{25}$

89. $\sin(u - v) = \sin u \cos v - \cos u \sin v$

$\qquad = \left(-\frac{12}{13}\right)\left(\frac{24}{25}\right) - \left(\frac{5}{13}\right)\left(-\frac{7}{25}\right)$

$\qquad = -\frac{253}{325}$

91. $\cos(v - u) = \cos v \cos u + \sin v \sin u$

$\qquad = \left(\frac{24}{25}\right)\left(\frac{5}{13}\right) + \left(-\frac{7}{25}\right)\left(-\frac{12}{13}\right)$

$\qquad = \frac{204}{325}$

Section 8.5   Trigonometric Form of a Complex Number

- You should be able to graphically represent complex numbers and know the following facts about them.

- The absolute value of the complex number $z = a + bi$ is $|z| = \sqrt{a^2 + b^2}$.

- The trigonometric form of the complex number $z = a + bi$ is $z = r(\cos \theta + i \sin \theta)$ where

  (a) $a = r \cos \theta$

  (b) $b = r \sin \theta$

  (c) $r = \sqrt{a^2 + b^2}$; r is called the modulus of z.

  (d) $\tan \theta = \dfrac{b}{a}$; θ is called the argument of z.

- Given $z_1 = r_1(\cos \theta_1 + i \sin \theta_1)$ and $z_2 = r_2(\cos \theta_2 + i \sin \theta_2)$:

  (a) $z_1 z_2 = r_1 r_2 [\cos(\theta_1 + \theta_2) + i \sin(\theta_1 + \theta_2)]$

  (b) $\dfrac{z_1}{z_2} = \dfrac{r_1}{r_2}[\cos(\theta_1 - \theta_2) + i \sin(\theta_1 - \theta_2)]$, $z_2 \neq 0$

- You should know DeMoivre's Theorem: If $z = r(\cos \theta + i \sin \theta)$, then for any positive integer n,

  $z^n = r^n (\cos n\theta + i \sin n\theta)$.

- You should know that for any positive integer n, $z = r(\cos \theta + i \sin \theta)$ has n distinct nth roots given by

  $$\sqrt[n]{r}\left[\cos\left(\frac{\theta + 2\pi k}{n}\right) + i \sin\left(\frac{\theta + 2\pi k}{n}\right)\right]$$

  where $k = 0, 1, 2, \ldots, n - 1$.

Vocabulary Check

1. absolute value

2. trigonometric form; modulus; argument

3. DeMoivre's

4. n^{th} root

1. $|-7i| = \sqrt{0^2 + (-7)^2}$

$\qquad = \sqrt{49} = 7$

3. $|-4 + 4i| = \sqrt{(-4)^2 + (4)^2}$

$\qquad = \sqrt{32} = 4\sqrt{2}$

5. $|6 - 7i| = \sqrt{6^2 + (-7)^2}$

$\qquad = \sqrt{85}$

7. $z = 3i$

$r = \sqrt{0^2 + 3^2} = \sqrt{9} = 3$

$\tan \theta = \dfrac{3}{0}$, undefined $\implies \theta = \dfrac{\pi}{2}$

$z = 3\left(\cos \dfrac{\pi}{2} + i \sin \dfrac{\pi}{2}\right)$

9. $z = 3 - i$

$r = \sqrt{(3)^2 + (-1)^2} = \sqrt{10}$

$\tan \theta = -\dfrac{1}{3}$, θ is in Quadrant IV.

$\theta \approx 5.96$ radians

$z \approx \sqrt{10}(\cos 5.96 + i \sin 5.96)$

11. $z = 3 - 3i$

$r = \sqrt{3^2 + (-3)^2} = \sqrt{18} = 3\sqrt{2}$

$\tan \theta = \dfrac{-3}{3} = -1$, θ is in Quadrant IV $\implies \theta = \dfrac{7\pi}{4}$.

$z = 3\sqrt{2}\left(\cos \dfrac{7\pi}{4} + i \sin \dfrac{7\pi}{4}\right)$

13. $z = \sqrt{3} + i$

$r = \sqrt{\left(\sqrt{3}\right)^2 + 1^2} = \sqrt{4} = 2$

$\tan \theta = \dfrac{1}{\sqrt{3}} = \dfrac{\sqrt{3}}{3} \implies \theta = \dfrac{\pi}{6}$

$z = 2\left(\cos \dfrac{\pi}{6} + i \sin \dfrac{\pi}{6}\right)$

15. $z = -2\left(1 + \sqrt{3}i\right)$

$r = \sqrt{(-2)^2 + \left(-2\sqrt{3}\right)^2} = \sqrt{16} = 4$

$\tan \theta = \dfrac{\sqrt{3}}{1} = \sqrt{3}$, θ is in Quadrant III $\implies \theta = \dfrac{4\pi}{3}$.

$z = 4\left(\cos \dfrac{4\pi}{3} + i \sin \dfrac{4\pi}{3}\right)$

17. $z = -5i$

$r = \sqrt{0^2 + (-5)^2} = \sqrt{25} = 5$

$\tan \theta = \dfrac{-5}{0}$, undefined $\implies \theta = \dfrac{3\pi}{2}$

$z = 5\left(\cos \dfrac{3\pi}{2} + i \sin \dfrac{3\pi}{2}\right)$

19. $z = -7 + 4i$

$r = \sqrt{(-7)^2 + (4)^2} = \sqrt{65}$

$\tan \theta = \dfrac{4}{-7}$, θ is in Quadrant II $\implies \theta \approx 2.62$.

$z \approx \sqrt{65}(\cos 2.62 + i \sin 2.62)$

21. $z = 7 + 0i$

$r = \sqrt{(7)^2 + (0)^2} = \sqrt{49} = 7$

$\tan \theta = \dfrac{0}{7} = 0 \implies \theta = 0$

$z = 7(\cos 0 + i \sin 0)$

23. $z = 3 + \sqrt{3}i$

$r = \sqrt{(3)^2 + \left(\sqrt{3}\right)^2} = \sqrt{12}$

$\qquad\qquad = 2\sqrt{3}$

$\tan \theta = \dfrac{\sqrt{3}}{3} \implies \theta = \dfrac{\pi}{6}$

$z = 2\sqrt{3}\left(\cos \dfrac{\pi}{6} + i \sin \dfrac{\pi}{6}\right)$

25. $z = -3 - i$

$r = \sqrt{(-3)^2 + (-1)^2} = \sqrt{10}$

$\tan \theta = \dfrac{-1}{-3} = \dfrac{1}{3}$, θ is in Quadrant III $\implies \theta \approx 3.46$.

$z \approx \sqrt{10}\,(\cos 3.46 + i \sin 3.46)$

27. $z = 5 + 2i$

$r = \sqrt{5^2 + 2^2} = \sqrt{29}$

$\tan \theta = \dfrac{2}{5}$

$\qquad \theta \approx 0.38$

$z \approx \sqrt{29}(\cos 0.38 + i \sin 0.38)$

29. $z = -8 - 5\sqrt{3}i$

$r = \sqrt{(-8)^2 + \left(-5\sqrt{3}\right)^2} = \sqrt{139}$

$\tan \theta = \dfrac{5\sqrt{3}}{8}$

$\qquad \theta \approx 3.97$

$z \approx \sqrt{139}(\cos 3.97 + i \sin 3.97)$

31. $3(\cos 120° + i \sin 120°) = 3\left(-\dfrac{1}{2} + \dfrac{\sqrt{3}}{2}i\right)$

$\qquad\qquad\qquad\qquad\qquad = -\dfrac{3}{2} + \dfrac{3\sqrt{3}}{2}i$

33. $\dfrac{3}{2}(\cos 300° + i \sin 300°) = \dfrac{3}{2}\left[\dfrac{1}{2} + i\left(-\dfrac{\sqrt{3}}{2}\right)\right]$

$$= \dfrac{3}{4} - \dfrac{3\sqrt{3}}{4}i$$

35. $3.75\left(\cos\dfrac{3\pi}{4} + i\sin\dfrac{3\pi}{4}\right) = -\dfrac{15\sqrt{2}}{8} + \dfrac{15\sqrt{2}}{8}i$

37. $8\left(\cos\dfrac{\pi}{2} + i\sin\dfrac{\pi}{2}\right) = 8(0 + i) = 8i$

39. $3[\cos(18°45') + i\sin(18°45')] \approx 2.8408 + 0.9643i$

41. $5\left(\cos\dfrac{\pi}{9} + i\sin\dfrac{\pi}{9}\right) \approx 4.6985 + 1.7101i$

43. $3(\cos 165.5° + i\sin 165.5°) \approx -2.9044 + 0.7511i$

45. $z = \dfrac{\sqrt{2}}{2}(1 + i) = \cos 45° + i\sin 45°$

$z^2 = \cos 90° + i\sin 90° = i$

$z^3 = \cos 135° + i\sin 135° = \dfrac{\sqrt{2}}{2}(-1 + i)$

$z^4 = \cos 180° + i\sin 180° = -1$

The absolute value of each is 1, and consecutive powers of z are each 45° apart.

47. $\left[2\left(\cos\dfrac{\pi}{4} + i\sin\dfrac{\pi}{4}\right)\right]\left[6\left(\cos\dfrac{\pi}{12} + i\sin\dfrac{\pi}{12}\right)\right] = (2)(6)\left[\cos\left(\dfrac{\pi}{4} + \dfrac{\pi}{12}\right) + i\sin\left(\dfrac{\pi}{4} + \dfrac{\pi}{12}\right)\right]$

$$= 12\left(\cos\dfrac{\pi}{3} + i\sin\dfrac{\pi}{3}\right)$$

49. $\left[\dfrac{5}{3}(\cos 140° + i\sin 140°)\right]\left[\dfrac{2}{3}(\cos 60° + i\sin 60°)\right] = \left(\dfrac{5}{3}\right)\left(\dfrac{2}{3}\right)[\cos(140° + 60°) + i\sin(140° + 60°)]$

$$= \dfrac{10}{9}(\cos 200° + i\sin 200°)$$

51. $[0.45(\cos 310° + i\sin 310°)][0.60(\cos 200° + i\sin 200°)] = (0.45)(0.60)[\cos(310° + 200°) + i\sin(310° + 200°)]$

$$= 0.27(\cos 510° + i\sin 510°)$$

$$= 0.27(\cos 150° + i\sin 150°)$$

53. $\dfrac{\cos 50° + i \sin 50°}{\cos 20° + i \sin 20°} = \cos(50° - 20°) + i \sin(50° - 20°) = \cos 30° + i \sin 30°$

55. $\dfrac{\cos\dfrac{5\pi}{3} + i \sin\dfrac{5\pi}{3}}{\cos \pi + i \sin \pi} = \cos\left(\dfrac{5\pi}{3} - \pi\right) + i \sin\left(\dfrac{5\pi}{3} - \pi\right) = \cos\left(\dfrac{2\pi}{3}\right) + i \sin\left(\dfrac{2\pi}{3}\right)$

57. $\dfrac{12(\cos 52° + i \sin 52°)}{3(\cos 110° + i \sin 110°)} = 4[\cos(52° - 110°) + i \sin(52° - 110°)]$

$$= 4[\cos(-58°) + i \sin(-58°)]$$

$$= 4(\cos 302° + i \sin 302°)$$

59. (a) $2 + 2i = 2\sqrt{2}\left(\cos\dfrac{\pi}{4} + i \sin\dfrac{\pi}{4}\right)$

$1 - i = \sqrt{2}\left[\cos\left(-\dfrac{\pi}{4}\right) + i \sin\left(-\dfrac{\pi}{4}\right)\right] = \sqrt{2}\left(\cos\dfrac{7\pi}{4} + i \sin\dfrac{7\pi}{4}\right)$

(b) $(2 + 2i)(1 - i) = \left[2\sqrt{2}\left(\cos\dfrac{\pi}{4} + i \sin\dfrac{\pi}{4}\right)\right]\left[\sqrt{2}\left(\cos\left(\dfrac{7\pi}{4}\right) + i \sin\left(\dfrac{7\pi}{4}\right)\right)\right] = 4(\cos 2\pi + i \sin 2\pi)$

$$= 4(\cos 0 + i \sin 0) = 4$$

(c) $(2 + 2i)(1 - i) = 2 - 2i + 2i - 2i^2 = 2 + 2 = 4$

61. (a) $-2i = 2\left[\cos\left(-\dfrac{\pi}{2}\right) + i \sin\left(-\dfrac{\pi}{2}\right)\right] = 2\left(\cos\dfrac{3\pi}{2} + i \sin\dfrac{3\pi}{2}\right)$

$1 + i = \sqrt{2}\left(\cos\dfrac{\pi}{4} + i \sin\dfrac{\pi}{4}\right)$

(b) $-2i(1 + i) = 2\left[\cos\left(\dfrac{3\pi}{2}\right) + i \sin\left(\dfrac{3\pi}{2}\right)\right]\left[\sqrt{2}\left(\cos\dfrac{\pi}{4} + i \sin\dfrac{\pi}{4}\right)\right]$

$$= 2\sqrt{2}\left[\cos\left(\dfrac{7\pi}{4}\right) + i \sin\left(\dfrac{7\pi}{4}\right)\right]$$

$$= 2\sqrt{2}\left[\dfrac{1}{\sqrt{2}} - \dfrac{1}{\sqrt{2}}i\right] = 2 - 2i$$

(c) $-2i(1 + i) = -2i - 2i^2 = -2i + 2 = 2 - 2i$

63. (a) $3 + 4i \approx 5(\cos 0.93 + i \sin 0.93)$

$1 - \sqrt{3}i = 2\left(\cos\dfrac{5\pi}{3} + i \sin\dfrac{5\pi}{3}\right)$

(c) $\dfrac{3 + 4i}{1 - \sqrt{3}i} = \dfrac{3 + 4i}{1 - \sqrt{3}i} \cdot \dfrac{1 + \sqrt{3}i}{1 + \sqrt{3}i}$

$$= \dfrac{3 + \left(4 + 3\sqrt{3}\right)i + 4\sqrt{3}i^2}{1 + 3}$$

$$= \dfrac{3 - 4\sqrt{3}}{4} + \dfrac{4 + 3\sqrt{3}}{4}i$$

$$\approx -0.982 + 2.299i$$

(b) $\dfrac{3 + 4i}{1 - \sqrt{3}i} \approx \dfrac{5(\cos 0.93 + i \sin 0.93)}{2\left(\cos\dfrac{5\pi}{3} + i \sin\dfrac{5\pi}{3}\right)}$

$$\approx 2.5[\cos(-4.31) + i \sin(-4.31)]$$

$$= \dfrac{5}{2}(\cos 1.97 + i \sin 1.97)$$

$$\approx -0.982 + 2.299i$$

65. (a) $5 = 5(\cos 0 + i \sin 0)$

$2 + 3i \approx \sqrt{13}(\cos 0.98 + i \sin 0.98)$

(b) $\dfrac{5}{2 + 3i} \approx \dfrac{5(\cos 0 + i \sin 0)}{\sqrt{13}(\cos 0.98 + i \sin 0.98)} = \dfrac{5}{\sqrt{13}}[\cos(-0.98) + i \sin(-0.98)] = \dfrac{5}{\sqrt{13}}(\cos 5.30 + i \sin 5.30) \approx 0.769 - 1.154i$

(c) $\dfrac{5}{2 + 3i} = \dfrac{5}{2 + 3i} \cdot \dfrac{2 - 3i}{2 - 3i} = \dfrac{10 - 15i}{13} = \dfrac{10}{13} - \dfrac{15}{13}i \approx 0.769 - 1.154i$

67. Let $z = x + iy$ such that:

$|z| = 2 \implies 2 = \sqrt{x^2 + y^2}$

$\implies 4 = x^2 + y^2$:

circle with radius of 2

69. Let $\theta = \dfrac{\pi}{6}$.

Since $r \geq 0$, we have the portion of the line $\theta = \pi/6$ in Quadrant I.

71. $(1 + i)^5 = \left[\sqrt{2}\left(\cos \dfrac{\pi}{4} + i \sin \dfrac{\pi}{4} \right) \right]^5$

$= \left(\sqrt{2} \right)^5 \left(\cos \dfrac{5\pi}{4} + i \sin \dfrac{5\pi}{4} \right)$

$= 4\sqrt{2}\left(-\dfrac{\sqrt{2}}{2} - \dfrac{\sqrt{2}}{2}i \right)$

$= -4 - 4i$

73. $(-1 + i)^{10} = \left[\sqrt{2}\left(\cos \dfrac{3\pi}{4} + i \sin \dfrac{3\pi}{4} \right) \right]^{10}$

$= \left(\sqrt{2} \right)^{10}\left(\cos \dfrac{30\pi}{4} + i \sin \dfrac{30\pi}{4} \right)$

$= 32\left[\cos\left(\dfrac{3\pi}{2} + 6\pi \right) + i \sin\left(\dfrac{3\pi}{2} + 6\pi \right) \right]$

$= 32\left(\cos \dfrac{3\pi}{2} + i \sin \dfrac{3\pi}{2} \right)$

$= 32[0 + i(-1)] = -32i$

75. $2\left(\sqrt{3} + i \right)^7 = 2\left[2\left(\cos \dfrac{\pi}{6} + i \sin \dfrac{\pi}{6} \right) \right]^7$

$= 2\left[2^7\left(\cos \dfrac{7\pi}{6} + i \sin \dfrac{7\pi}{6} \right) \right]$

$= 256\left(-\dfrac{\sqrt{3}}{2} - \dfrac{1}{2}i \right)$

$= -128\sqrt{3} - 128i$

77. $[5(\cos 20° + i \sin 20°)]^3 = 5^3(\cos 60° + i \sin 60°)$

$= \dfrac{125}{2} + \dfrac{125\sqrt{3}}{2}i$

79. $\left(\cos \dfrac{\pi}{4} + i \sin \dfrac{\pi}{4} \right)^{12} = \cos \dfrac{12\pi}{4} + i \sin \dfrac{12\pi}{4}$

$= \cos 3\pi + i \sin 3\pi$

$= -1$

81. $[5(\cos 3.2 + i \sin 3.2)]^4 = 5^4(\cos 12.8 + i \sin 12.8)$

$\approx 608.02 + 144.69i$

83. $(3 - 2i)^5 \approx [3.6056[\cos(-0.588) + i \sin(-0.588)]]^5$

$\approx (3.6056)^5[\cos(-2.94) + i \sin(-2.94)]$

$\approx -597 - 122i$

85. $[3(\cos 15° + i \sin 15°)]^4 = 81(\cos 60° + i \sin 60°)$

$= \dfrac{81}{2} + \dfrac{81\sqrt{3}}{2}i$

87. $\left[2\left(\cos \dfrac{\pi}{10} + i \sin \dfrac{\pi}{10} \right) \right]^5 = 2^5\left(\cos \dfrac{\pi}{2} + i \sin \dfrac{\pi}{2} \right) = 32i$

89. (a) Square roots of $5(\cos 120° + i \sin 120°)$:

$$\sqrt{5}\left[\cos\left(\frac{120° + 360°k}{2}\right) + i \sin\left(\frac{120° + 360°k}{2}\right)\right], \ k = 0, \ 1$$

$k = 0$: $\sqrt{5}(\cos 60° + i \sin 60°)$

$k = 1$: $\sqrt{5}(\cos 240° + i \sin 240°)$

(c) $\dfrac{\sqrt{5}}{2} + \dfrac{\sqrt{15}}{2}i, \ -\dfrac{\sqrt{5}}{2} - \dfrac{\sqrt{15}}{2}i$

(b)

91. (a) Cube roots of $8\left(\cos\dfrac{2\pi}{3} + i \sin\dfrac{2\pi}{3}\right)$:

$$\sqrt[3]{8}\left[\cos\left(\frac{(2\pi/3) + 2\pi k}{3}\right) + i \sin\left(\frac{(2\pi/3) + 2\pi k}{3}\right)\right], \ k = 0, 1, 2$$

$k = 0$: $2\left(\cos\dfrac{2\pi}{9} + i \sin\dfrac{2\pi}{9}\right)$

$k = 1$: $2\left(\cos\dfrac{8\pi}{9} + i \sin\dfrac{8\pi}{9}\right)$

$k = 2$: $2\left(\cos\dfrac{14\pi}{9} + i \sin\dfrac{14\pi}{9}\right)$

(c) $1.5321 + 1.2856i, \ -1.8794 + 0.6840i, \ 0.3473 - 1.9696i$

(b)

93. (a) Square roots of $-25i = 25\left(\cos\dfrac{3\pi}{2} + i \sin\dfrac{3\pi}{2}\right)$:

$$\sqrt{25}\left[\cos\left(\frac{\frac{3\pi}{2} + 2k\pi}{2}\right) + i \sin\left(\frac{\frac{3\pi}{2} + 2k\pi}{2}\right)\right], \ k = 0, 1$$

$k = 0$: $5\left(\cos\dfrac{3\pi}{4} + i \sin\dfrac{3\pi}{4}\right)$

$k = 1$: $5\left(\cos\dfrac{7\pi}{4} + i \sin\dfrac{7\pi}{4}\right)$

(c) $-\dfrac{5\sqrt{2}}{2} + \dfrac{5\sqrt{2}}{2}i, \ \dfrac{5\sqrt{2}}{2} - \dfrac{5\sqrt{2}}{2}i$

(b)

95. (a) Cube roots of $-\dfrac{125}{2}(1 + \sqrt{3}i) = 125\left(\cos\dfrac{4\pi}{3} + i \sin\dfrac{4\pi}{3}\right)$:

$$\sqrt[3]{125}\left[\cos\left(\frac{\frac{4\pi}{3} + 2k\pi}{3}\right) + i \sin\left(\frac{\frac{4\pi}{3} + 2k\pi}{3}\right)\right], \ k = 0, 1, 2$$

$k = 0$: $5\left(\cos\dfrac{4\pi}{9} + i \sin\dfrac{4\pi}{9}\right)$

$k = 1$: $5\left(\cos\dfrac{10\pi}{9} + i \sin\dfrac{10\pi}{9}\right)$

$k = 2$: $5\left(\cos\dfrac{16\pi}{9} + i \sin\dfrac{16\pi}{9}\right)$

(c) $0.8682 + 4.9240i, \ -4.6985 - 1.7101i, \ 3.8302 - 3.2140i$

(b)

97. (a) Fourth roots of $16 = 16(\cos 0 + i \sin 0)$:

$$\sqrt[4]{16}\left[\cos\frac{0 + 2\pi k}{4} + i \sin\frac{0 + 2\pi k}{4}\right], k = 0, 1, 2, 3$$

$k = 0$: $2(\cos 0 + i \sin 0)$

$k = 1$: $2\left(\cos\frac{\pi}{2} + i \sin\frac{\pi}{2}\right)$

$k = 2$: $2(\cos \pi + i \sin \pi)$

$k = 3$: $2\left(\cos\frac{3\pi}{2} + i \sin\frac{3\pi}{2}\right)$

(c) $2, 2i, -2, -2i$

(b)

99. (a) Fifth roots of $1 = \cos 0 + i \sin 0$:

$$\cos\left(\frac{2k\pi}{5}\right) + i \sin\left(\frac{2k\pi}{5}\right), k = 0, 1, 2, 3, 4$$

$k = 0$: $\cos 0 + i \sin 0$

$k = 1$: $\cos\frac{2\pi}{5} + i \sin\frac{2\pi}{5}$

$k = 2$: $\cos\frac{4\pi}{5} + i \sin\frac{4\pi}{5}$

$k = 3$: $\cos\frac{6\pi}{5} + i \sin\frac{6\pi}{5}$

$k = 4$: $\cos\frac{8\pi}{5} + i \sin\frac{8\pi}{5}$

(c) $1, 0.3090 + 0.9511i, -0.8090 + 0.5878i, -0.8090 - 0.5878i, 0.3090 - 0.9511i$

(b)

101. (a) Cube roots of $-125 = 125(\cos \pi + i \sin \pi)$:

$$\sqrt[3]{125}\left[\cos\left(\frac{\pi + 2\pi k}{3}\right) + i \sin\left(\frac{\pi + 2\pi k}{3}\right)\right], k = 0, 1, 2$$

$k = 0$: $5\left(\cos\frac{\pi}{3} + i \sin\frac{\pi}{3}\right)$

$k = 1$: $5(\cos \pi + i \sin \pi)$

$k = 2$: $5\left(\cos\frac{5\pi}{3} + i \sin\frac{5\pi}{3}\right)$

(c) $\frac{5}{2} + \frac{5\sqrt{3}}{2}i, -5, \frac{5}{2} - \frac{5\sqrt{3}}{2}i$

(b)

103. (a) Fifth roots of $128(-1 + i) = 128\sqrt{2}\left(\cos\dfrac{3\pi}{4} + i\sin\dfrac{3\pi}{4}\right) = 2^{15/2}\left(\cos\dfrac{3\pi}{4} + i\sin\dfrac{3\pi}{4}\right)$

$$2^{3/2}\left[\cos\left(\dfrac{\dfrac{3\pi}{4} + 2\pi k}{5}\right) + i\sin\left(\dfrac{\dfrac{3\pi}{4} + 2\pi k}{5}\right)\right], \ k = 0, 1, 2, 3, 4$$

$k = 0$: $2\sqrt{2}\left(\cos\dfrac{3\pi}{20} + i\sin\dfrac{3\pi}{20}\right)$

$k = 1$: $2\sqrt{2}\left(\cos\dfrac{11\pi}{20} + i\sin\dfrac{11\pi}{20}\right)$

$k = 2$: $2\sqrt{2}\left(\cos\dfrac{19\pi}{20} + i\sin\dfrac{19\pi}{20}\right)$

$k = 3$: $2\sqrt{2}\left(\cos\dfrac{27\pi}{20} + i\sin\dfrac{27\pi}{20}\right)$

$k = 4$: $2\sqrt{2}\left(\cos\dfrac{7\pi}{4} + i\sin\dfrac{7\pi}{4}\right)$

(b)

(c) $2.5201 + 1.2841i, \ -0.4425 + 2.7936i,$
$\quad -2.7936 + 0.4425i, \ -1.2841 - 2.5201i, \ 2 - 2i$

105. $x^4 + i = 0$

$\quad x^4 = -i$

The solutions are the fourth roots of $i = \cos\dfrac{3\pi}{2} + i\sin\dfrac{3\pi}{2}$:

$$\sqrt[4]{1}\left[\cos\left(\dfrac{\dfrac{3\pi}{2} + 2k\pi}{4}\right) + i\sin\left(\dfrac{\dfrac{3\pi}{2} + 2k\pi}{4}\right)\right], \ k = 0, 1, 2, 3$$

$k = 0$: $\cos\dfrac{3\pi}{8} + i\sin\dfrac{3\pi}{8} \approx 0.3827 + 0.9239i$

$k = 1$: $\cos\dfrac{7\pi}{8} + i\sin\dfrac{7\pi}{8} \approx -0.9239 + 0.3827i$

$k = 2$: $\cos\dfrac{11\pi}{8} + i\sin\dfrac{11\pi}{8} \approx -0.3827 - 0.9239i$

$k = 3$: $\cos\dfrac{15\pi}{8} + i\sin\dfrac{15\pi}{8} \approx 0.9239 - 0.3827i$

107. $x^5 + 243 = 0$

$\quad x^5 = -243$

The solutions are the fifth roots of $-243 = 243(\cos\pi + i\sin\pi)$:

$$\sqrt[5]{243}\left[\cos\left(\dfrac{\pi + 2k\pi}{5}\right) + i\sin\left(\dfrac{\pi + 2k\pi}{5}\right)\right], \ k = 0, 1, 2, 3, 4$$

$k = 0$: $3\left(\cos\dfrac{\pi}{5} + i\sin\dfrac{\pi}{5}\right) \approx 2.4271 + 1.7634i$

$k = 1$: $3\left(\cos\dfrac{3\pi}{5} + i\sin\dfrac{3\pi}{5}\right) \approx -0.9271 + 2.8532i$

$k = 2$: $3(\cos\pi + i\sin\pi) = -3$

$k = 3$: $3\left(\cos\dfrac{7\pi}{5} + i\sin\dfrac{7\pi}{5}\right) \approx -0.9271 - 2.8532i$

$k = 4$: $3\left(\cos\dfrac{9\pi}{5} + i\sin\dfrac{9\pi}{5}\right) \approx 2.4271 - 1.7634i$

109. $x^4 + 16i = 0$

$$x^4 = -16i$$

The solutions are the fourth roots of $-16i = 16\left(\cos\dfrac{3\pi}{2} + i\sin\dfrac{3\pi}{2}\right)$:

$$\sqrt[4]{16}\left[\cos\dfrac{\dfrac{3\pi}{2} + 2\pi k}{4} + i\sin\dfrac{\dfrac{3\pi}{2} + 2\pi k}{4}\right], k = 0, 1, 2, 3$$

$k = 0$: $2\left(\cos\dfrac{3\pi}{8} + i\sin\dfrac{3\pi}{8}\right) \approx 0.7654 + 1.8478i$

$k = 1$: $2\left(\cos\dfrac{7\pi}{8} + i\sin\dfrac{7\pi}{8}\right) \approx -1.8478 + 0.7654i$

$k = 2$: $2\left(\cos\dfrac{11\pi}{8} + i\sin\dfrac{11\pi}{8}\right) \approx -0.7654 - 1.8478i$

$k = 3$: $2\left(\cos\dfrac{15\pi}{8} + i\sin\dfrac{15\pi}{8}\right) \approx 1.8478 - 0.7654i$

111. $x^3 - (1 - i) = 0$

$$x^3 = 1 - i = \sqrt{2}\left(\cos\dfrac{7\pi}{4} + i\sin\dfrac{7\pi}{4}\right)$$

The solutions are the cube roots of $1 - i$:

$$\sqrt[3]{\sqrt{2}}\left[\cos\left(\dfrac{\dfrac{7\pi}{4} + 2\pi k}{3}\right) + i\sin\left(\dfrac{\dfrac{7\pi}{4} + 2\pi k}{3}\right)\right], k = 0, 1, 2$$

$k = 0$: $\sqrt[6]{2}\left(\cos\dfrac{7\pi}{12} + i\sin\dfrac{7\pi}{12}\right) \approx -0.2905 + 1.0842i$

$k = 1$: $\sqrt[6]{2}\left(\cos\dfrac{5\pi}{4} + i\sin\dfrac{5\pi}{4}\right) \approx -0.7937 - 0.7937i$

$k = 2$: $\sqrt[6]{2}\left(\cos\dfrac{23\pi}{12} + i\sin\dfrac{23\pi}{12}\right) \approx 1.0842 - 0.2905i$

113. True, by the definition of the absolute value of a complex number.

115. True. $z_1 z_2 = r_1 r_2[\cos(\theta_1 + \theta_2) + i\sin(\theta_1 + \theta_2)]$ and $z_1 z_2 = 0$ if and only if $r_1 = 0$ and/or $r_2 = 0$.

117. $\dfrac{z_1}{z_2} = \dfrac{r_1(\cos\theta_1 + i\sin\theta_1)}{r_2(\cos\theta_2 + i\sin\theta_2)} \cdot \dfrac{\cos\theta_2 - i\sin\theta_2}{\cos\theta_2 - i\sin\theta_2}$

$$= \dfrac{r_1}{r_2(\cos^2\theta_2 + \sin^2\theta_2)}[\cos\theta_1\cos\theta_2 + \sin\theta_1\sin\theta_2 + i(\sin\theta_1\cos\theta_2 - \sin\theta_2\cos\theta_1)]$$

$$= \dfrac{r_1}{r_2}[\cos(\theta_1 - \theta_2) + i\sin(\theta_1 - \theta_2)]$$

119. (a) $z\bar{z} = [r(\cos\theta + i\sin\theta)][r(\cos(-\theta) + i\sin(-\theta))]$

$$= r^2[\cos(\theta - \theta) + i\sin(\theta - \theta)]$$

$$= r^2[\cos 0 + i\sin 0]$$

$$= r^2$$

(b) $\dfrac{z}{\bar{z}} = \dfrac{r(\cos\theta + i\sin\theta)}{r[\cos(-\theta) + i\sin(-\theta)]}$

$$= \dfrac{r}{r}[\cos(\theta - (-\theta)) + i\sin(\theta - (-\theta))]$$

$$= \cos 2\theta + i\sin 2\theta$$

121. $-\frac{1}{2}\left(1 + \sqrt{3}i\right) = -\left(\cos\frac{4\pi}{3} + i\sin\frac{4\pi}{3}\right)$

$\left[-\frac{1}{2}\left(1 + \sqrt{3}i\right)\right]^6 = \left[-\left(\cos\frac{4\pi}{3} + i\sin\frac{4\pi}{3}\right)\right]^6$

$= \cos 8\pi + i\sin 8\pi$

$= 1$

123. (a) $2(\cos 30° + i\sin 30°)$

$2(\cos 150° + i\sin 150°)$

$2(\cos 270° + i\sin 270°)$

(b) These are the cube roots of $8i$.

125. $A = 22°, a = 8$

$B = 90° - A = 68°$

$\tan 22° = \dfrac{8}{b} \implies b = \dfrac{8}{\tan 22°} \approx 19.80$

$\sin 22° = \dfrac{8}{c} \implies c = \dfrac{8}{\sin 22°} \approx 21.36$

127. $A = 30°, b = 112.6$

$B = 90° - A = 60°$

$\tan 30° = \dfrac{a}{112.6} \implies a = 112.6 \tan 30° \approx 65.01$

$\cos 30° = \dfrac{112.6}{c} \implies c = \dfrac{112.6}{\cos 30°} \approx 130.02$

129. $A = 42°15' = 42.25°, c = 11.2$

$B = 90° - A = 47°45'$

$\sin 42.25° = \dfrac{a}{11.2} \implies a = 11.2 \sin 42.25° \approx 7.53$

$\cos 42.25° = \dfrac{b}{11.2} \implies b = 11.2 \cos 42.25° \approx 8.29$

131. $d = 16 \cos\dfrac{\pi}{4}t$

Maximum displacement: $|16| = 16$

$16 \cos\dfrac{\pi}{4}t = 0 \implies \dfrac{\pi}{4}t = \dfrac{\pi}{2} \implies t = 2$

133. $d = \frac{1}{16}\sin\left(\frac{5}{4}\pi t\right)$

Maximum displacement: $\left|\dfrac{1}{16}\right| = \dfrac{1}{16}$

$\frac{1}{16}\sin\left(\frac{5}{4}\pi t\right) = 0$

$\frac{5}{4}\pi t = \pi$

$t = \frac{4}{5}$

135. $6\sin 8\theta \cos 3\theta = (6)\left(\frac{1}{2}\right)[\sin(8\theta + 3\theta) + \sin(8\theta - 3\theta)]$

$= 3(\sin 11\theta + \sin 5\theta)$

Review Exercises for Chapter 8

1. Given: $A = 35°, B = 71°, a = 8$

$C = 180° - 35° - 71° = 74°$

$b = \dfrac{a \sin B}{\sin A} = \dfrac{8 \sin 71°}{\sin 35°} \approx 13.19$

$c = \dfrac{a \sin C}{\sin A} = \dfrac{8 \sin 74°}{\sin 35°} \approx 13.41$

3. Given: $B = 72°, C = 82°, b = 54$

$A = 180° - 72° - 82° = 26°$

$a = \dfrac{b \sin A}{\sin B} = \dfrac{54 \sin 26°}{\sin 72°} \approx 24.89$

$c = \dfrac{b \sin C}{\sin B} = \dfrac{54 \sin 82°}{\sin 72°} \approx 56.23$

5. Given: $A = 16°, B = 98°, c = 8.4$

$C = 180° - 16° - 98° = 66°$

$a = \dfrac{c \sin A}{\sin C} = \dfrac{8.4 \sin 16°}{\sin 66°} \approx 2.53$

$b = \dfrac{c \sin B}{\sin C} = \dfrac{8.4 \sin 98°}{\sin 66°} \approx 9.11$

7. Given: $A = 24°, C = 48°, b = 27.5$

$B = 180° - 24° - 48° = 108°$

$a = \dfrac{b \sin A}{\sin B} = \dfrac{27.5 \sin 24°}{\sin 108°} \approx 11.76$

$c = \dfrac{b \sin C}{\sin B} = \dfrac{27.5 \sin 48°}{\sin 108°} \approx 21.49$

9. Given: $B = 150°$, $b = 30$, $c = 10$

$$\sin C = \frac{c \sin B}{b} = \frac{10 \sin 150°}{30} \approx 0.1667 \implies C \approx 9.59°$$

$$A \approx 180° - 150° - 9.59° = 20.41°$$

$$a = \frac{b \sin A}{\sin B} = \frac{30 \sin 20.41°}{\sin 150°} \approx 20.92$$

11. $A = 75°$, $a = 51.2$, $b = 33.7$

$$\sin B = \frac{b \sin A}{a} = \frac{33.7 \sin 75°}{51.2} \approx 0.6358 \implies B \approx 39.48°$$

$$C \approx 180° - 75° - 39.48° = 65.52°$$

$$c = \frac{a \sin C}{\sin A} = \frac{51.2 \sin 65.52°}{\sin 75°} \approx 48.24$$

13. Area $= \frac{1}{2}bc \sin A = \frac{1}{2}(5)(7)\sin 27° \approx 7.9$

15. Area $= \frac{1}{2}ab \sin C = \frac{1}{2}(16)(5)\sin 123° \approx 33.5$

17. $\tan 17° = \dfrac{h}{x + 50} \implies h = (x + 50) \tan 17°$

$$h = x \tan 17° + 50 \tan 17°$$

$$\tan 31° = \frac{h}{x} \implies h = x \tan 31°$$

$$x \tan 17° + 50 \tan 17° = x \tan 31°$$

$$50 \tan 17° = x(\tan 31° - \tan 17°)$$

$$\frac{50 \tan 17°}{\tan 31° - \tan 17°} = x$$

$$x \approx 51.7959$$

$h = x \tan 31° \approx 51.7959 \tan 31° \approx 31.1$ meters

The height of the building is approximately 31.1 meters.

19. $\dfrac{h}{\sin 17°} = \dfrac{75}{\sin 45°}$

$$h = \frac{75 \sin 17°}{\sin 45°}$$

$$h \approx 31.01 \text{ feet}$$

21. Given: $a = 5$, $b = 8$, $c = 10$

$$\cos C = \frac{a^2 + b^2 - c^2}{2ab} = -0.1375 \implies C \approx 97.90°$$

$$\cos B = \frac{a^2 + c^2 - b^2}{2ac} = 0.61 \implies B \approx 52.41°$$

$$A = 180° - B - C \approx 29.69°$$

23. Given: $a = 2.5$, $b = 5.0$, $c = 4.5$

$$\cos B = \frac{a^2 + c^2 - b^2}{2ac} = 0.0667 \implies B \approx 86.18°$$

$$\cos C = \frac{a^2 + b^2 - c^2}{2ab} = 0.44 \implies C \approx 63.90°$$

$$A = 180° - B - C \approx 29.92°$$

25. Given: $B = 110°$, $a = 4$, $c = 4$

$$b = \sqrt{a^2 + c^2 - 2ac \cos B} \approx 6.55$$

$$A = C = \frac{1}{2}(180° - 110°) = 35°$$

27. Given: $C = 43°$, $a = 22.5$, $b = 31.4$

$$c = \sqrt{a^2 + b^2 - 2ab \cos C} \approx 21.42$$

$$\cos B = \frac{a^2 + c^2 - b^2}{2ac} \approx -0.02169 \implies B \approx 91.24°$$

$$A = 180° - B - C \approx 45.76°$$

29.

$a^2 = 5^2 + 8^2 - 2(5)(8)\cos 28° \approx 18.364$

$a \approx 4.3$ feet

$b^2 = 8^2 + 5^2 - 2(8)(5)\cos 152° \approx 159.636$

$b \approx 12.6$ feet

31. Length of AC $= \sqrt{300^2 + 425^2 - 2(300)(425)\cos 115°}$

≈ 615.1 meters

33. $a = 4,\ b = 5,\ c = 7$

$$s = \frac{a + b + c}{2} = \frac{4 + 5 + 7}{2} = 8$$

$$\text{Area} = \sqrt{s(s - a)(s - b)(s - c)}$$

$$= \sqrt{8(4)(3)(1)} \approx 9.80$$

35. $a = 12.3,\ b = 15.8,\ c = 3.7$

$$s = \frac{a + b + c}{2} = \frac{12.3 + 15.8 + 3.7}{2} = 15.9$$

$$\text{Area} = \sqrt{s(s - a)(s - b)(s - c)}$$

$$= \sqrt{15.9(3.6)(0.1)(12.2)} = 8.36$$

37. $\|\mathbf{u}\| = \sqrt{(4 - (-2))^2 + (6 - 1)^2} = \sqrt{61}$

$\|\mathbf{v}\| = \sqrt{(6 - 0)^2 + (3 - (-2))^2} = \sqrt{61}$

$\mathbf{u}$ is directed along a line with a slope of $\dfrac{6 - 1}{4 - (-2)} = \dfrac{5}{6}$.

$\mathbf{v}$ is directed along a line with a slope of $\dfrac{3 - (-2)}{6 - 0} = \dfrac{5}{6}$.

Since $\mathbf{u}$ and $\mathbf{v}$ have identical magnitudes and directions, $\mathbf{u} = \mathbf{v}$.

39. Initial point: $(-5, 4)$

Terminal point: $(2, -1)$

$\mathbf{v} = \langle 2 - (-5), -1 - 4 \rangle = \langle 7, -5 \rangle$

41. Initial point: $(0, 10)$

Terminal point: $(7, 3)$

$\mathbf{v} = \langle 7 - 0, 3 - 10 \rangle = \langle 7, -7 \rangle$

43. $\|\mathbf{v}\| = 8,\ \theta = 120°$

$\langle 8 \cos 120°, 8 \sin 120° \rangle = \langle -4, 4\sqrt{3} \rangle$

45. $\mathbf{u} = \langle -1, -3 \rangle,\ \mathbf{v} = \langle -3, 6 \rangle$

(a) $\mathbf{u} + \mathbf{v} = \langle -1, -3 \rangle + \langle -3, 6 \rangle = \langle -4, 3 \rangle$

(b) $\mathbf{u} - \mathbf{v} = \langle -1, -3 \rangle - \langle -3, 6 \rangle = \langle 2, -9 \rangle$

(c) $3\mathbf{u} = 3\langle -1, -3 \rangle = \langle -3, -9 \rangle$

(d) $2\mathbf{v} + 5\mathbf{u} = 2\langle -3, 6 \rangle + 5\langle -1, -3 \rangle$

$= \langle -6, 12 \rangle + \langle -5, -15 \rangle = \langle -11, -3 \rangle$

47. $\mathbf{u} = \langle -5, 2 \rangle,\ \mathbf{v} = \langle 4, 4 \rangle$

(a) $\mathbf{u} + \mathbf{v} = \langle -5, 2 \rangle + \langle 4, 4 \rangle = \langle -1, 6 \rangle$

(b) $\mathbf{u} - \mathbf{v} = \langle -5, 2 \rangle - \langle 4, 4 \rangle = \langle -9, -2 \rangle$

(c) $3\mathbf{u} = 3\langle -5, 2 \rangle = \langle -15, 6 \rangle$

(d) $2\mathbf{v} + 5\mathbf{u} = 2\langle 4, 4 \rangle + 5\langle -5, 2 \rangle$

$= \langle 8, 8 \rangle + \langle -25, 10 \rangle = \langle -17, 18 \rangle$

49. $\mathbf{u} = 2\mathbf{i} - \mathbf{j},\ \mathbf{v} = 5\mathbf{i} + 3\mathbf{j}$

(a) $\mathbf{u} + \mathbf{v} = (2\mathbf{i} - \mathbf{j}) + (5\mathbf{i} + 3\mathbf{j}) = 7\mathbf{i} + 2\mathbf{j}$

(b) $\mathbf{u} - \mathbf{v} = (2\mathbf{i} - \mathbf{j}) - (5\mathbf{i} + 3\mathbf{j}) = -3\mathbf{i} - 4\mathbf{j}$

(c) $3\mathbf{u} = 3(2\mathbf{i} - \mathbf{j}) = 6\mathbf{i} - 3\mathbf{j}$

(d) $2\mathbf{v} + 5\mathbf{u} = 2(5\mathbf{i} + 3\mathbf{j}) + 5(2\mathbf{i} - \mathbf{j})$

$= (10\mathbf{i} + 6\mathbf{j}) + (10\mathbf{i} - 5\mathbf{j}) = 20\mathbf{i} + \mathbf{j}$

51. $\mathbf{u} = 4\mathbf{i},\ \mathbf{v} = -\mathbf{i} + 6\mathbf{j}$

(a) $\mathbf{u} + \mathbf{v} = 4\mathbf{i} + (-\mathbf{i} + 6\mathbf{j}) = 3\mathbf{i} + 6\mathbf{j}$

(b) $\mathbf{u} - \mathbf{v} = 4\mathbf{i} - (-\mathbf{i} + 6\mathbf{j}) = 5\mathbf{i} - 6\mathbf{j}$

(c) $3\mathbf{u} = 3(4\mathbf{i}) = 12\mathbf{i}$

(d) $2\mathbf{v} + 5\mathbf{u} = 2(-\mathbf{i} + 6\mathbf{j}) + 5(4\mathbf{i})$

$= (-2\mathbf{i} + 12\mathbf{j}) + 20\mathbf{i} = 18\mathbf{i} + 12\mathbf{j}$

53. $\mathbf{u} = 6\mathbf{i} - 5\mathbf{j},\ \mathbf{v} = 10\mathbf{i} + 3\mathbf{j}$

$2\mathbf{u} + \mathbf{v} = 2(6\mathbf{i} - 5\mathbf{j}) + (10\mathbf{i} + 3\mathbf{j})$

$= 22\mathbf{i} - 7\mathbf{j}$

$= \langle 22, -7 \rangle$

55. $\mathbf{v} = 10\mathbf{i} + 3\mathbf{j}$

$3\mathbf{v} = 3(10\mathbf{i} + 3\mathbf{j})$

$= 30\mathbf{i} + 9\mathbf{j}$

$= \langle 30, 9 \rangle$

57. $\mathbf{u} = \langle -3, 4 \rangle = -3\mathbf{i} + 4\mathbf{j}$

59. Initial point: $(3, 4)$

Terminal point: $(9, 8)$

$\mathbf{u} = (9 - 3)\mathbf{i} + (8 - 4)\mathbf{j} = 6\mathbf{i} + 4\mathbf{j}$

61. $\mathbf{v} = -10\mathbf{i} + 10\mathbf{j}$

$\|\mathbf{v}\| = \sqrt{(-10)^2 + (10)^2} = \sqrt{200} = 10\sqrt{2}$

$\tan \theta = \dfrac{10}{-10} = -1 \implies \theta = 135°$ since

$\mathbf{v}$ is in Quadrant II.

$\mathbf{v} = 10\sqrt{2}(\mathbf{i} \cos 135° + \mathbf{j} \sin 135°)$

63. $\mathbf{v} = 7(\cos 60° \, \mathbf{i} + \sin 60° \, \mathbf{j})$

$\|\mathbf{v}\| = 7$

$\theta = 60°$

65. $\mathbf{v} = 5\mathbf{i} + 4\mathbf{j}$

$\|\mathbf{v}\| = \sqrt{5^2 + 4^2} = \sqrt{41}$

$\tan \theta = \dfrac{4}{5} \implies \theta \approx 38.7°$

67. $\mathbf{v} = -3\mathbf{i} - 3\mathbf{j}$

$\|\mathbf{v}\| = \sqrt{(-3)^2 + (-3)^2} = 3\sqrt{2}$

$\tan \theta = \dfrac{-3}{-3} = 1 \implies \theta = 225°$

69. Magnitude of resultant:

$c = \sqrt{85^2 + 50^2 - 2(85)(50) \cos 165°}$

≈ 133.92 pounds

Let θ be the angle between the resultant and the 85-pound force.

$\cos \theta \approx \dfrac{(133.92)^2 + 85^2 - 50^2}{2(133.92)(85)}$

≈ 0.9953

$\implies \theta \approx 5.6°$

71. Airspeed: $\mathbf{u} = 430(\cos 45° \mathbf{i} - \sin 45° \mathbf{j}) = 215\sqrt{2}(\mathbf{i} - \mathbf{j})$

Wind: $\mathbf{w} = 35(\cos 60° + \sin 60° \, \mathbf{j}) = \dfrac{35}{2}(\mathbf{i} + \sqrt{3}\mathbf{j})$

Groundspeed: $\mathbf{u} + \mathbf{w} = \left(215\sqrt{2} + \dfrac{35}{2}\right)\mathbf{i} + \left(\dfrac{35\sqrt{3}}{2} - 215\sqrt{2}\right)$

$\|\mathbf{u} + \mathbf{w}\| = \sqrt{\left(215\sqrt{2} + \dfrac{35}{2}\right)^2 + \left(\dfrac{35\sqrt{3}}{2} - 215\sqrt{2}\right)^2}$

≈ 422.30 miles per hour

Bearing: $\tan \theta' = \dfrac{17.5\sqrt{3} - 215\sqrt{2}}{215\sqrt{2} + 17.5}$

$\theta' \approx -40.4°$

$\theta = 90° + |\theta'| = 130.4°$

73. $\mathbf{u} = \langle 6, 7 \rangle, \mathbf{v} = \langle -3, 9 \rangle$

$\mathbf{u} \cdot \mathbf{v} = 6(-3) + 7(9) = 45$

75. $\mathbf{u} = 3\mathbf{i} + 7\mathbf{j}, \mathbf{v} = 11\mathbf{i} - 5\mathbf{j}$

$\mathbf{u} \cdot \mathbf{v} = 3(11) + 7(-5) = -2$

77. $\mathbf{u} = \langle -3, 4 \rangle$

$2\mathbf{u} = \langle -6, 8 \rangle$

$2\mathbf{u} \cdot \mathbf{u} = (-6)(-3) + 8(4) = 50$

The result is a scalar.

79. $\mathbf{u} = \langle -3, 4 \rangle$, $\mathbf{v} = \langle 2, 1 \rangle$

$\mathbf{u} \cdot \mathbf{v} = (-3)(2) + 4(1) = -2$

$\mathbf{u}(\mathbf{u} \cdot \mathbf{v}) = \mathbf{u}(-2) = -2\mathbf{u} = \langle 6, -8 \rangle$

The result is a vector.

81. $\mathbf{u} = \cos \dfrac{7\pi}{4}\mathbf{i} + \sin \dfrac{7\pi}{4}\mathbf{j} = \left\langle \dfrac{1}{\sqrt{2}}, -\dfrac{1}{\sqrt{2}} \right\rangle$

$\mathbf{v} = \cos \dfrac{5\pi}{6}\mathbf{i} + \sin \dfrac{5\pi}{6}\mathbf{j} = \left\langle -\dfrac{\sqrt{3}}{2}, \dfrac{1}{2} \right\rangle$

$\cos \theta = \dfrac{\mathbf{u} \cdot \mathbf{v}}{\|\mathbf{u}\| \, \|\mathbf{v}\|} = \dfrac{-\sqrt{3} - 1}{2\sqrt{2}} \implies \theta = \dfrac{11\pi}{12}$

83. $\mathbf{u} = \langle 2\sqrt{2}, -4 \rangle$, $\mathbf{v} = \langle -\sqrt{2}, 1 \rangle$

$\cos \theta = \dfrac{\mathbf{u} \cdot \mathbf{v}}{\|\mathbf{u}\| \, \|\mathbf{v}\|} = \dfrac{-8}{\left(\sqrt{24}\right)\left(\sqrt{3}\right)} \implies \theta \approx 160.5°$

85. $\mathbf{u} = \langle -3, 8 \rangle$

$\mathbf{v} = \langle 8, 3 \rangle$

$\mathbf{u} \cdot \mathbf{v} = -3(8) + 8(3) = 0$

$\mathbf{u}$ and $\mathbf{v}$ are orthogonal.

87. $\mathbf{u} = -\mathbf{i}$

$\mathbf{v} = \mathbf{i} + 2\mathbf{j}$

$\mathbf{u} \cdot \mathbf{v} \neq 0 \implies$ Not orthogonal

$\mathbf{v} \neq k\mathbf{u} \implies$ Not parallel

Neither

89. $\mathbf{u} = \langle -4, 3 \rangle$, $\mathbf{v} = \langle -8, -2 \rangle$

$\mathbf{w}_1 = \text{proj}_{\mathbf{v}}\mathbf{u} = \left(\dfrac{\mathbf{u} \cdot \mathbf{v}}{\|\mathbf{v}\|^2} \right)\mathbf{v} = \left(\dfrac{26}{68} \right)\langle -8, -2 \rangle = -\dfrac{13}{17}\langle 4, 1 \rangle$

$\mathbf{w}_2 = \mathbf{u} - \mathbf{w}_1 = \langle -4, 3 \rangle - \left(-\dfrac{13}{17} \right)\langle 4, 1 \rangle = \dfrac{16}{17}\langle -1, 4 \rangle$

$\mathbf{u} = \mathbf{w}_1 + \mathbf{w}_2 = -\dfrac{13}{17}\langle 4, 1 \rangle + \dfrac{16}{17}\langle -1, 4 \rangle$

91. $\mathbf{u} = \langle 2, 7 \rangle$, $\mathbf{v} = \langle 1, -1 \rangle$

$\mathbf{w}_1 = \text{proj}_{\mathbf{v}}\mathbf{u} = \left(\dfrac{\mathbf{u} \cdot \mathbf{v}}{\|\mathbf{v}\|^2} \right)\mathbf{v} = -\dfrac{5}{2}\langle 1, -1 \rangle = \dfrac{5}{2}\langle -1, 1 \rangle$

$\mathbf{w}_2 = \mathbf{u} - \mathbf{w}_1 = \langle 2, 7 \rangle - \left(\dfrac{5}{2} \right)\langle -1, 1 \rangle = \dfrac{9}{2}\langle 1, 1 \rangle$

$\mathbf{u} = \mathbf{w}_1 + \mathbf{w}_2 = \dfrac{5}{2}\langle -1, 1 \rangle + \dfrac{9}{2}\langle 1, 1 \rangle$

93. $P = (5, 3)$, $Q = (8, 9) \implies \overrightarrow{PQ} = \langle 3, 6 \rangle$

$W = \mathbf{v} \cdot \overrightarrow{PQ} = \langle 2, 7 \rangle \cdot \langle 3, 6 \rangle = 48$

95. $w = (18{,}000)\left(\dfrac{48}{12} \right) = 72{,}000$ foot-pounds

97. $|7i| = \sqrt{0^2 + 7^2} = 7$

99. $|5 + 3i| = \sqrt{5^2 + 3^2}$

$= \sqrt{34}$

101. $5 - 5i$

$r = \sqrt{5^2 + (-5)^2} = \sqrt{50} = 5\sqrt{2}$

$\tan \theta = \dfrac{-5}{5} = -1 \implies \theta = \dfrac{7\pi}{4}$ since the

complex number is in Quadrant IV.

$5 - 5i = 5\sqrt{2}\left(\cos \dfrac{7\pi}{4} + i \sin \dfrac{7\pi}{4} \right)$

103. $-3\sqrt{3} + 3i$

$r = \sqrt{\left(-3\sqrt{3}\right)^2 + 3^2} = \sqrt{36} = 6$

$\tan \theta = \dfrac{3}{-3\sqrt{3}} = -\dfrac{1}{\sqrt{3}} \implies \theta = \dfrac{5\pi}{6}$

since the complex number is in Quadrant II.

$-3\sqrt{3} + 3i = 6\left(\cos \dfrac{5\pi}{6} + i \sin \dfrac{5\pi}{6} \right)$

105. (a) $z_1 = 2\sqrt{3} - 2i = 4\left(\cos\frac{11\pi}{6} + i\sin\frac{11\pi}{6}\right)$

$z_2 = -10i = 10\left(\cos\frac{3\pi}{2} + i\sin\frac{3\pi}{2}\right)$

(b) $z_1z_2 = \left[4\left(\cos\frac{11\pi}{6} + i\sin\frac{11\pi}{6}\right)\right]\left[10\left(\cos\frac{3\pi}{2} + i\sin\frac{3\pi}{2}\right)\right]$

$= 40\left(\cos\frac{10\pi}{3} + i\sin\frac{10\pi}{3}\right)$

$\dfrac{z_1}{z_2} = \dfrac{4\left(\cos\dfrac{11\pi}{6} + i\sin\dfrac{11\pi}{6}\right)}{10\left(\cos\dfrac{3\pi}{2} + i\sin\dfrac{3\pi}{2}\right)} = \dfrac{2}{5}\left(\cos\dfrac{\pi}{3} + i\sin\dfrac{\pi}{3}\right)$

107. $\left[5\left(\cos\frac{\pi}{12} + i\sin\frac{\pi}{12}\right)\right]^4 = 5^4\left(\cos\frac{4\pi}{12} + i\sin\frac{4\pi}{12}\right)$

$= 625\left(\cos\frac{\pi}{3} + i\sin\frac{\pi}{3}\right)$

$= 625\left(\frac{1}{2} + \frac{\sqrt{3}}{2}i\right)$

$= \frac{625}{2} + \frac{625\sqrt{3}}{2}i$

109. $(2 + 3i)^6 \approx \left[\sqrt{13}(\cos 56.3° + i\sin 56.3°)\right]^6$

$= 13^3(\cos 337.9° + i\sin 337.9°)$

$\approx 13^3(0.9263 - 0.3769i)$

$\approx 2035 - 828i$

111. Sixth roots of $-729i = 729\left(\cos\frac{3\pi}{2} + i\sin\frac{3\pi}{2}\right)$:

(a) and (c)

$\sqrt[6]{729}\left[\cos\left(\dfrac{\dfrac{3\pi}{2} + 2k\pi}{6}\right) + i\sin\left(\dfrac{\dfrac{3\pi}{2} + 2k\pi}{6}\right)\right], k = 0, 1, 2, 3, 4, 5$

$k = 0: 3\left(\cos\frac{\pi}{4} + i\sin\frac{\pi}{4}\right) = \frac{3\sqrt{2}}{2} + \frac{3\sqrt{2}}{2}i$

$k = 1: 3\left(\cos\frac{7\pi}{12} + i\sin\frac{7\pi}{12}\right) \approx -0.776 + 2.898i$

$k = 2: 3\left(\cos\frac{11\pi}{12} + i\sin\frac{11\pi}{12}\right) \approx -2.898 + 0.776i$

$k = 3: 3\left(\cos\frac{5\pi}{4} + i\sin\frac{5\pi}{4}\right) = -\frac{3\sqrt{2}}{2} - \frac{3\sqrt{2}}{2}i$

$k = 4: 3\left(\cos\frac{19\pi}{12} + i\sin\frac{19\pi}{12}\right) \approx 0.776 - 2.898i$

$k = 5: 3\left(\cos\frac{23\pi}{12} + i\sin\frac{23\pi}{12}\right) \approx 2.898 - 0.776i$

(b)

113. Cube roots of $8 = 8(\cos 0 + i \sin 0)$, $k = 0, 1, 2$

(a) and (c)

$$\sqrt[3]{8}\left[\cos\left(\frac{0 + 2\pi k}{3}\right) + i \sin\left(\frac{0 + 2\pi k}{3}\right)\right]$$

$k = 0: 2(\cos 0 + i \sin 0) = 2$

$k = 1: 2\left(\cos\frac{2\pi}{3} + i \sin\frac{2\pi}{3}\right) = -1 + \sqrt{3}i$

$k = 2: 2\left(\cos\frac{4\pi}{3} + i \sin\frac{4\pi}{3}\right) = -1 - \sqrt{3}i$

(b)

115. $x^4 + 81 = 0$

$\quad x^4 = -81 \qquad$ Solve by finding the fourth roots of -81.

$\quad -81 = 81(\cos \pi + i \sin \pi)$

$$\sqrt[4]{-81} = \sqrt[4]{81}\left[\cos\left(\frac{\pi + 2\pi k}{4}\right) + i \sin\left(\frac{\pi + 2\pi k}{4}\right)\right], \ k = 0, 1, 2, 3$$

$k = 0: 3\left(\cos\frac{\pi}{4} + i \sin\frac{\pi}{4}\right) = \frac{3\sqrt{2}}{2} + \frac{3\sqrt{2}}{2}i$

$k = 1: 3\left(\cos\frac{3\pi}{4} + i \sin\frac{3\pi}{4}\right) = -\frac{3\sqrt{2}}{2} + \frac{3\sqrt{2}}{2}i$

$k = 2: 3\left(\cos\frac{5\pi}{4} + i \sin\frac{5\pi}{4}\right) = -\frac{3\sqrt{2}}{2} - \frac{3\sqrt{2}}{2}i$

$k = 3: 3\left(\cos\frac{7\pi}{4} + i \sin\frac{7\pi}{4}\right) = \frac{3\sqrt{2}}{2} - \frac{3\sqrt{2}}{2}i$

117. $x^3 + 8i = 0$

$\quad x^3 = -8i \qquad$ Solve by finding the cube roots of $-8i$.

$\quad -8i = 8\left(\cos\frac{3\pi}{2} + i \sin\frac{3\pi}{2}\right)$

$$\sqrt[3]{-8i} = \sqrt[3]{8}\left[\cos\left(\frac{\frac{3\pi}{2} + 2\pi k}{3}\right) + i \sin\left(\frac{\frac{3\pi}{2} + 2\pi k}{3}\right)\right], \ k = 0, 1, 2$$

$k = 0: 2\left(\cos\frac{\pi}{2} + i \sin\frac{\pi}{2}\right) = 2i$

$k = 1: 2\left(\cos\frac{7\pi}{6} + i \sin\frac{7\pi}{6}\right) = -\sqrt{3} - i$

$k = 2: 2\left(\cos\frac{11\pi}{6} + i \sin\frac{11\pi}{6}\right) = \sqrt{3} - i$

119. True. $\sin 90°$ is defined in the Law of Sines.

121. True, by the definition of a unit vector.

$$\mathbf{u} = \frac{\mathbf{v}}{\|\mathbf{v}\|} \ \text{ so } \ \mathbf{v} = \|\mathbf{v}\|\mathbf{u}$$

123. False. $x = \sqrt{3} + i$ is a solution to $x^3 - 8i = 0$, not $x^2 - 8i = 0$.

$\quad$ Also, $(\sqrt{3} + i)^2 - 8i = 2 + (2\sqrt{3} - 8)i \neq 0$.

125. $a^2 = b^2 + c^2 - 2bc \cos A$

$\quad b^2 = a^2 + c^2 - 2ac \cos B$

$\quad c^2 = a^2 + b^2 - 2ab \cos C$

127. *A* and *C* appear to have the same magnitude and direction.

129. If $k > 0$, the direction of $k\mathbf{u}$ is the same, and the magnitude is $k\|\mathbf{u}\|$.

If $k < 0$, the direction of $k\mathbf{u}$ is the opposite direction of $\mathbf{u}$, and the magnitude is $|k|\,\|\mathbf{u}\|$.

131. (a) The trigonometric form of the three roots shown is:

$4(\cos 60° + i \sin 60°)$

$4(\cos 180° + i \sin 180°)$

$4(\cos 300° + i \sin 300°)$

(b) Since there are three evenly spaced roots on the circle of radius 4, they are cube roots of a complex number of modulus $4^3 = 64$.

Cubing them yields -64.

$[4(\cos 60° + i \sin 60°)]^3 = -64$

$[4(\cos 180° + i \sin 180°)]^3 = -64$

$[4(\cos 300° + i \sin 300°)]^3 = -64$

133. $z_1 = 2(\cos \theta + i \sin \theta)$

$z_2 = 2(\cos(\pi - \theta) + i \sin(\pi - \theta))$

$z_1 z_2 = (2)(2)[\cos(\theta + (\pi - \theta)) + i \sin(\theta + (\pi - \theta))]$

$\quad = 4(\cos \pi + i \sin \pi)$

$\quad = -4$

$\dfrac{z_1}{z_2} = \dfrac{2(\cos \theta + i \sin \theta)}{2(\cos(\pi - \theta) + i \sin(\pi - \theta))}$

$\quad = 1[\cos(\theta - (\pi - \theta)) + i \sin(\theta - (\pi - \theta))]$

$\quad = \cos(2\theta - \pi) + i \sin(2\theta - \pi)$

$\quad = \cos 2\theta \cos \pi + \sin 2\theta \sin \pi + i(\sin 2\theta \cos \pi - \cos 2\theta \sin \pi)$

$\quad = -\cos 2\theta - i \sin 2\theta$

Problem Solving for Chapter 8

1. $\overrightarrow{PQ}^2 = 4.7^2 + 6^2 - 2(4.7)(6)\cos 25°$

$\overrightarrow{PQ} \approx 2.6409$ feet

$\dfrac{\sin \alpha}{4.7} = \dfrac{\sin 25°}{2.6409} \Rightarrow \alpha \approx 48.78°$

$\theta + \beta = 180° - 25° - 48.78° = 106.22°$

$(\theta + \beta) + \theta = 180° \Rightarrow \theta = 180° - 106.22° = 73.78°$

$\beta = 106.22° - 73.78° = 32.44°$

$\gamma = 180° - \alpha - \beta = 180° - 48.78° - 32.44° = 98.78°$

$\phi = 180° - \gamma = 180° - 98.78° = 81.22°$

$\dfrac{\overrightarrow{PT}}{\sin 25°} = \dfrac{4.7}{\sin 81.22°}$

$\overrightarrow{PT} \approx 2.01$ feet

3. (a)

A ———75 mi——— B
30° 135° 15°
x y 75°
60° Lost party

(c) $z^2 = (27.45)^2 + (20)^2 - 2(27.45)(20)\cos 20°$

$z \approx 11.03$ miles

$\dfrac{\sin \theta}{27.45} = \dfrac{\sin 20°}{11.03}$

A — 10°
20 mi   Rescue
80° 20°   party
60°   z
27.452 mi

Lost
party

$\sin \theta \approx 0.8511$

$\theta = 180° - \sin^{-1}(0.8511)$

$\theta \approx 121.7°$

To find the bearing, we have $\theta - 10° - 90° \approx 21.7°$.

Bearing: S 21.7° E

(b) $\dfrac{x}{\sin 15°} = \dfrac{75}{\sin 135°}$ and $\dfrac{y}{\sin 30°} = \dfrac{75}{\sin 135°}$

$x \approx 27.45$ miles $y \approx 53.03$ miles

5. If $\mathbf{u} \neq 0$, $\mathbf{v} \neq 0$, and $\mathbf{u} + \mathbf{v} \neq 0$, then $\left\| \dfrac{\mathbf{u}}{\|\mathbf{u}\|} \right\| = \left\| \dfrac{\mathbf{v}}{\|\mathbf{v}\|} \right\| = \left\| \dfrac{\mathbf{u} + \mathbf{v}}{\|\mathbf{u} + \mathbf{v}\|} \right\| = 1$ since all of these are magnitudes of **unit** vectors.

(a) $\mathbf{u} = \langle 1, -1 \rangle$, $\mathbf{v} = \langle -1, 2 \rangle$, $\mathbf{u} + \mathbf{v} = \langle 0, 1 \rangle$

$\|\mathbf{u}\| = \sqrt{2}$, $\|\mathbf{v}\| = \sqrt{5}$, $\|\mathbf{u} + \mathbf{v}\| = 1$

(b) $\mathbf{u} = \langle 0, 1 \rangle$, $\mathbf{v} = \langle 3, -3 \rangle$, $\mathbf{u} + \mathbf{v} = \langle 3, -2 \rangle$

$\|\mathbf{u}\| = 1$, $\|\mathbf{v}\| = \sqrt{18} = 3\sqrt{2}$, $\|\mathbf{u} + \mathbf{v}\| = \sqrt{13}$

(c) $\mathbf{u} = \left\langle 1, \dfrac{1}{2} \right\rangle$, $\mathbf{v} = \langle 2, 3 \rangle$, $\mathbf{u} + \mathbf{v} = \left\langle 3, \dfrac{7}{2} \right\rangle$

$\|\mathbf{u}\| = \dfrac{\sqrt{5}}{2}$, $\|\mathbf{v}\| = \sqrt{13}$, $\|\mathbf{u} + \mathbf{v}\| = \sqrt{9 + \dfrac{49}{4}} = \dfrac{\sqrt{85}}{2}$

(d) $\mathbf{u} = \langle 2, -4 \rangle$, $\mathbf{v} = \langle 5, 5 \rangle$, $\mathbf{u} + \mathbf{v} = \langle 7, 1 \rangle$

$\|\mathbf{u}\| = \sqrt{20} = 2\sqrt{5}$, $\|\mathbf{v}\| = \sqrt{50} = 5\sqrt{2}$, $\|\mathbf{u} + \mathbf{v}\| = \sqrt{50} = 5\sqrt{2}$

7. Initial point: $(0, 0)$

Terminal point: $\left(\dfrac{\mathbf{u}_1 + \mathbf{v}_1}{2}, \dfrac{\mathbf{u}_2 + \mathbf{v}_2}{2} \right)$

$\mathbf{w} = \left\langle \dfrac{\mathbf{u}_1 + \mathbf{v}_1}{2}, \dfrac{\mathbf{u}_2 + \mathbf{v}_2}{2} \right\rangle = \dfrac{1}{2}(\mathbf{u} + \mathbf{v})$

Initial point: $(\mathbf{u}_1, \mathbf{u}_2)$

Terminal point: $\dfrac{1}{2}(\mathbf{u}_1 + \mathbf{v}_1, \mathbf{u}_2 + \mathbf{v}_2)$

$\mathbf{w} = \left\langle \dfrac{\mathbf{u}_1 + \mathbf{v}_1}{2} - \mathbf{u}_1, \dfrac{\mathbf{u}_2 + \mathbf{v}_2}{2} - \mathbf{u}_2 \right\rangle = \left\langle \dfrac{\mathbf{v}_1 - \mathbf{u}_1}{2}, \dfrac{\mathbf{v}_2 - \mathbf{u}_2}{2} \right\rangle = \dfrac{1}{2}(\mathbf{v} - \mathbf{u})$

9. $W = (\cos \theta)\|F\| \|\overrightarrow{PQ}\|$ and $\|F_1\| = \|F_2\|$

(a)

If $\theta_1 = -\theta_2$ then the work is the same since $\cos(-\theta) = \cos \theta$.

(b)

If $\theta_1 = 60°$ then $W_1 = \dfrac{1}{2}\|F_1\| \|\overrightarrow{PQ}\|$

If $\theta_2 = 30°$ then $W_2 = \dfrac{\sqrt{3}}{2}\|F_2\| \|\overrightarrow{PQ}\|$

$W_2 = \sqrt{3} \, W_1$

The amount of work done by F_2 is $\sqrt{3}$ times as great as the amount of work done by F_1.

Chapter 8   Practice Test

For Exercises 1 and 2, use the Law of Sines to find the remaining sides and angles of the triangle.

1. $A = 40°$, $B = 12°$, $b = 100$

2. $C = 150°$, $a = 5$, $c = 20$

3. Find the area of the triangle: $a = 3$, $b = 6$, $C = 130°$.

4. Determine the number of solutions to the triangle: $a = 10$, $b = 35$, $A = 22.5°$.

For Exercises 5 and 6, use the Law of Cosines to find the remaining sides and angles of the triangle.

5. $a = 49$, $b = 53$, $c = 38$

6. $C = 29°$, $a = 100$, $b = 300$

7. Use Heron's Formula to find the area of the triangle: $a = 4.1$, $b = 6.8$, $c = 5.5$.

8. A ship travels 40 miles due east, then adjusts its course 12° southward. After traveling 70 miles in that direction, how far is the ship from its point of departure?

9. $\mathbf{w} = 4\mathbf{u} - 7\mathbf{v}$ where $\mathbf{u} = 3\mathbf{i} + \mathbf{j}$ and $\mathbf{v} = -\mathbf{i} + 2\mathbf{j}$.   Find $\mathbf{w}$.

10. Find a unit vector in the direction of $\mathbf{v} = 5\mathbf{i} - 3\mathbf{j}$.

11. Find the dot product and the angle between $\mathbf{u} = 6\mathbf{i} + 5\mathbf{j}$ and $\mathbf{v} = 2\mathbf{i} - 3\mathbf{j}$.

12. $\mathbf{v}$ is a vector of magnitude 4 making an angle of 30° with the positive x-axis. Find $\mathbf{v}$ in component form.

13. Find the projection of $\mathbf{u}$ onto $\mathbf{v}$ given $\mathbf{u} = \langle 3, -1 \rangle$ and $\mathbf{v} = \langle -2, 4 \rangle$.

14. Give the trigonometric form of $z = 5 - 5i$.

15. Give the standard form of $z = 6(\cos 225° + i \sin 225°)$.

16. Multiply $[7(\cos 23° + i \sin 23°)][4(\cos 7° + i \sin 7°)]$.

17. Divide $\dfrac{9\left(\cos \dfrac{5\pi}{4} + i \sin \dfrac{5\pi}{4}\right)}{3(\cos \pi + i \sin \pi)}$.

18. Find $(2 + 2i)^8$.

19. Find the cube roots of $8\left(\cos \dfrac{\pi}{3} + i \sin \dfrac{\pi}{3}\right)$.

20. Find all the solutions to $x^4 + i = 0$.

CHAPTER 9
Systems of Equations and Inequalities

CHAPTER 9
Systems of Equations and Inequalities

Section 9.1   Linear and Nonlinear Systems of Equations

- You should be able to solve systems of equations by the method of substitution.
    1. Solve one of the equations for one of the variables.
    2. Substitute this expression into the other equation and solve.
    3. Back-substitute into the first equation to find the value of the other variable.
    4. Check your answer in each of the original equations.
- You should be able to find solutions graphically. (See Example 5 in textbook.)

Vocabulary Check

1. system of equations
2. solution
3. solving
4. substitution
5. point of intersection
6. break-even

---

1. $\begin{cases} 4x - y = 1 \\ 6x + y = -6 \end{cases}$

    (a) $4(0) - (-3) \neq 1$

    $(0, -3)$ *is not* a solution.

    (b) $4(-1) - (-4) \neq 1$

    $(-1, -4)$ *is not* a solution.

    (c) $4\left(-\frac{3}{2}\right) - (-2) \neq 1$

    $\left(-\frac{3}{2}, -2\right)$ *is not* a solution.

    (d) $4\left(-\frac{1}{2}\right) - (-3) = 1$

    $6\left(-\frac{1}{2}\right) + (-3) = -6$

    $\left(-\frac{1}{2}, -3\right)$ *is* a solution.

3. $\begin{cases} y = -2e^x \\ 3x - y = 2 \end{cases}$

    (a) $0 \neq -2e^{-2}$

    $(-2, 0)$ *is not* a solution.

    (b) $\qquad -2 = -2e^0$

    $3(0) - (-2) = 2$

    $(0, -2)$ *is* a solution.

    (c) $-3 \neq -2e^0$

    $(0, -3)$ *is not* a solution.

    (d) $2 \neq -2e^{-1}$

    $(-1, 2)$ *is not* a solution.

5. $\begin{cases} 2x + y = 6 & \text{Equation 1} \\ -x + y = 0 & \text{Equation 2} \end{cases}$

    Solve for y in Equation 1: $y = 6 - 2x$

    Substitute for y in Equation 2: $-x + (6 - 2x) = 0$

    Solve for x: $-3x + 6 = 0 \implies x = 2$

    Back-substitute $x = 2$: $y = 6 - 2(2) = 2$

    Solution: $(2, 2)$

7. $\begin{cases} x - y = -4 & \text{Equation 1} \\ x^2 - y = -2 & \text{Equation 2} \end{cases}$

Solve for y in Equation 1: $y = x + 4$

Substitute for y in Equation 2: $x^2 - (x + 4) = -2$

Solve for x: $x^2 - x - 2 = 0 \implies (x + 1)(x - 2) = 0 \implies x = -1, 2$

Back-substitute $x = -1$: $y = -1 + 4 = 3$

Back-substitute $x = 2$: $y = 2 + 4 = 6$

Solutions: $(-1, 3), (2, 6)$

9. $\begin{cases} -2x + y = -5 & \text{Equation 1} \\ x^2 + y^2 = 25 & \text{Equation 2} \end{cases}$

Solve for y in Equation 1: $y = 2x - 5$

Substitute for y in Equation 2: $x^2 + (2x - 5)^2 = 25$

Solve for x:

$$5x^2 - 20x = 0 \implies 5x(x - 4) = 0 \implies x = 0, 4$$

Back-substitute $x = 0$: $y = 2(0) - 5 = -5$

Back-substitute $x = 4$: $y = 2(4) - 5 = 3$

Solutions: $(0, -5), (4, 3)$

11. $\begin{cases} x^2 + y = 0 & \text{Equation 1} \\ x^2 - 4x - y = 0 & \text{Equation 2} \end{cases}$

Solve for y in Equation 1: $y = -x^2$

Substitute for y in Equation 2: $x^2 - 4x - (-x^2) = 0$

Solve for x: $2x^2 - 4x = 0 \implies 2x(x - 2) = 0 \implies x = 0, 2$

Back-substitute $x = 0$: $y = -0^2 = 0$

Back-substitute $x = 2$: $y = -2^2 = -4$

Solutions: $(0, 0), (2, -4)$

13. $\begin{cases} y = x^3 - 3x^2 + 1 & \text{Equation 1} \\ y = x^2 - 3x + 1 & \text{Equation 2} \end{cases}$

Substitute for y in Equation 2:

$$x^3 - 3x^2 + 1 = x^2 - 3x + 1$$

$$x^3 - 4x^2 + 3x = 0$$

$$x(x - 1)(x - 3) = 0 \implies x = 0, 1, 3$$

Back-substitute $x = 0$: $y = 0^3 - 3(0)^2 + 1 = 1$

Back-substitute $x = 1$: $y = 1^3 - 3(1)^2 + 1 = -1$

Back-substitute $x = 3$: $y = 3^3 - 3(3)^2 + 1 = 1$

Solutions: $(0, 1), (1, -1), (3, 1)$

15. $\begin{cases} x - y = 0 & \text{Equation 1} \\ 5x - 3y = 10 & \text{Equation 2} \end{cases}$

Solve for y in Equation 1: $y = x$

Substitute for y in Equation 2: $5x - 3x = 10$

Solve for x: $2x = 10 \implies x = 5$

Back-substitute in Equation 1: $y = x = 5$

Solution: $(5, 5)$

17. $\begin{cases} 2x - y + 2 = 0 & \text{Equation 1} \\ 4x + y - 5 = 0 & \text{Equation 2} \end{cases}$

Solve for y in Equation 1: $y = 2x + 2$

Substitute for y in Equation 2: $4x + (2x + 2) - 5 = 0$

Solve for x: $6x - 3 = 0 \implies x = \frac{1}{2}$

Back-substitute $x = \frac{1}{2}$: $y = 2x + 2 = 2\left(\frac{1}{2}\right) + 2 = 3$

Solution: $\left(\frac{1}{2}, 3\right)$

19. $\begin{cases} 1.5x + 0.8y = 2.3 & \text{Equation 1} \\ 0.3x - 0.2y = 0.1 & \text{Equation 2} \end{cases}$

Multiply the equations by 10.

$\quad 15x + 8y = 23 \qquad$ Revised Equation 1

$\quad\ \ 3x - 2y = 1 \qquad$ Revised Equation 2

Solve for y in revised Equation 2: $y = \frac{3}{2}x - \frac{1}{2}$

Substitute for y in revised Equation 1:
$15x + 8\left(\frac{3}{2}x - \frac{1}{2}\right) = 23$

Solve for x:
$15x + 12x - 4 = 23 \implies 27x = 27 \implies x = 1$

Back-substitute $x = 1$: $y = \frac{3}{2}(1) - \frac{1}{2} = 1$

Solution: $(1, 1)$

21. $\begin{cases} \frac{1}{5}x + \frac{1}{2}y = 8 & \text{Equation 1} \\ x + y = 20 & \text{Equation 2} \end{cases}$

Solve for x in Equation 2: $x = 20 - y$

Substitute for x in Equation 1: $\frac{1}{5}(20 - y) + \frac{1}{2}y = 8$

Solve for y: $4 + \frac{3}{10}y = 8 \implies y = \frac{40}{3}$

Back-substitute $y = \frac{40}{3}$: $x = 20 - y = 20 - \frac{40}{3} = \frac{20}{3}$

Solution: $\left(\frac{20}{3}, \frac{40}{3}\right)$

23. $\begin{cases} 6x + 5y = -3 & \text{Equation 1} \\ -x - \frac{5}{6}y = -7 & \text{Equation 2} \end{cases}$

Solve for x in Equation 2: $x = 7 - \frac{5}{6}y$

Substitute for x in Equation 1: $6\left(7 - \frac{5}{6}y\right) + 5y = -3$

Solve for y: $42 - 5y + 5y = -3 \implies 42 = -3$ (False)

No solution

25. $\begin{cases} x^2 - y = 0 & \text{Equation 1} \\ 2x + y = 0 & \text{Equation 2} \end{cases}$

Solve for y in Equation 2: $y = -2x$

Substitute for y in Equation 1: $x^2 - (-2x) = 0$

Solve for x:
$x^2 + 2x = 0 \implies x(x + 2) = 0 \implies x = 0, -2$

Back-substitute $x = 0$: $y = -2(0) = 0$

Back-substitute $x = -2$: $y = -2(-2) = 4$

Solutions: $(0, 0), (-2, 4)$

27. $\begin{cases} x - y = -1 & \text{Equation 1} \\ x^2 - y = -4 & \text{Equation 2} \end{cases}$

Solve for y in Equation 1: $y = x + 1$

Substitute for y in Equation 2: $x^2 - (x + 1) = -4$

Solve for x: $x^2 - x - 1 = -4 \implies x^2 - x + 3 = 0$

The Quadratic Formula yields no real solutions.

29. $\begin{cases} -x + 2y = 2 \implies y = \dfrac{x + 2}{2} \\ 3x + y = 15 \implies y = -3x + 15 \end{cases}$

Point of intersection: $(4, 3)$

31. $\begin{cases} x - 3y = -2 \implies y = \frac{1}{3}(x + 2) \\ \end{cases}$

Point of intersection: $\left(\frac{5}{2}, \frac{3}{2}\right)$

33. $\begin{cases} x + y = 4 \implies y = -x + 4 \\ x^2 + y^2 - 4x = 0 \implies (x - 2)^2 + y^2 = 4 \end{cases}$

Points of intersection: $(2, 2), (4, 0)$

35. $\begin{cases} x - y + 3 = 0 \implies y = x + 3 \\ y = x^2 - 4x + 7 \implies y = (x - 2)^2 + 3 \end{cases}$

Points of intersection: $(1, 4), (4, 7)$

37. $\begin{cases} 7x + 8y = 24 \implies y = -\dfrac{7}{8}x + 3 \\ x - 8y = 8 \implies y = \dfrac{1}{8}x - 1 \end{cases}$

Point of intersection: $\left(4, -\dfrac{1}{2}\right)$

39. $\begin{cases} 3x - 2y = 0 \implies y = \dfrac{3}{2}x \\ x^2 - y^2 = 4 \implies \dfrac{x^2}{4} - \dfrac{y^2}{4} = 1 \end{cases}$

No points of intersection $\implies$ No solution

41. $\begin{cases} x^2 + y^2 = 25 \\ 3x^2 - 16y = 0 \implies y = \dfrac{3}{16}x^2 \end{cases}$

Points of intersection: $(-4, 3)$ and $(4, 3)$

Algebraically we have:

$$x^2 = 25 - y^2$$
$$\tfrac{16}{3}y = 25 - y^2$$
$$16y = 75 - 3y^2$$
$$3y^2 + 16y - 75 = 0$$
$$(3y + 25)(y - 3) = 0$$
$$y = -\tfrac{25}{3} \implies x^2 = -\tfrac{400}{9}, \quad \text{No real solution}$$
$$y = 3 \implies x^2 = 16$$

Solutions: $(\pm 4, 3)$

43. $\begin{cases} y = e^x \\ x - y + 1 = 0 \implies y = x + 1 \end{cases}$

Point of intersection: $(0, 1)$

45. $\begin{cases} x + 2y = 8 \implies y = -\dfrac{1}{2}x + 4 \\ y = \log_2 x \implies y = \dfrac{\ln x}{\ln 2} \end{cases}$

Point of intersection: $(4, 2)$

47. $\begin{cases} x^2 + y^2 = 169 \implies y_1 = \sqrt{169 - x^2} \text{ and } y_2 = -\sqrt{169 - x^2} \\ x^2 - 8y = 104 \implies y_3 = \dfrac{1}{8}x^2 - 13 \end{cases}$

Points of intersection: $(0, -13), (\pm 12, 5)$

49. $\begin{cases} y = 2x & \text{Equation 1} \\ y = x^2 + 1 & \text{Equation 2} \end{cases}$

Substitute for y in Equation 2: $2x = x^2 + 1$

Solve for x: $x^2 - 2x + 1 = (x - 1)^2 = 0 \implies x = 1$

Back-substitute $x = 1$ in Equation 1: $y = 2x = 2$

Solution: $(1, 2)$

51. $\begin{cases} 3x - 7y + 6 = 0 & \text{Equation 1} \\ x^2 - y^2 = 4 & \text{Equation 2} \end{cases}$

Solve for y in Equation 1: $y = \dfrac{3x + 6}{7}$

Substitute for y in Equation 2: $x^2 - \left(\dfrac{3x + 6}{7}\right)^2 = 4$

Solve for x: $x^2 - \left(\dfrac{9x^2 + 36x + 36}{49}\right) = 4$

$$49x^2 - (9x^2 + 36x + 36) = 196$$

$$40x^2 - 36x - 232 = 0$$

$$4(10x - 29)(x + 2) = 0 \implies x = \frac{29}{10}, -2$$

Back-substitute $x = \dfrac{29}{10}$: $y = \dfrac{3x + 6}{7} = \dfrac{3(29/10) + 6}{7} = \dfrac{21}{10}$

Back-substitute $x = -2$: $y = \dfrac{3x + 6}{7} = \dfrac{3(-2) + 6}{7} = 0$

Solutions: $\left(\dfrac{29}{10}, \dfrac{21}{10}\right)$, $(-2, 0)$

53. $\begin{cases} x - 2y = 4 & \text{Equation 1} \\ x^2 - y = 0 & \text{Equation 2} \end{cases}$

Solve for y in Equation 2: $y = x^2$

Substitute for y in Equation 1: $x - 2x^2 = 4$

Solve for x: $0 = 2x^2 - x + 4 \implies x = \dfrac{1 \pm \sqrt{1 - 4(2)(4)}}{2(2)}$

$$\implies x = \frac{1 \pm \sqrt{-31}}{4}$$

The discriminant in the Quadratic Formula is negative.

No real solution

55. $\begin{cases} y - e^{-x} = 1 \implies y = e^{-x} + 1 \\ y - \ln x = 3 \implies y = \ln x + 3 \end{cases}$

Point of intersection: approximately $(0.287, 1.751)$

57. $\begin{cases} y = x^4 - 2x^2 + 1 & \text{Equation 1} \\ y = 1 - x^2 & \text{Equation 2} \end{cases}$

Substitute for y in Equation 1: $1 - x^2 = x^4 - 2x^2 + 1$

Solve for x: $x^4 - x^2 = 0 \implies x^2(x^2 - 1) = 0$

$$\implies x = 0, \pm 1$$

Back-substitute $x = 0$: $1 - x^2 = 1 - 0^2 = 1$

Back-substitute $x = 1$: $1 - x^2 = 1 - 1^2 = 0$

Back-substitute $x = -1$: $1 - x^2 = 1 - (-1)^2 = 0$

Solutions: $(0, 1)$, $(\pm 1, 0)$

59. $\begin{cases} xy - 1 = 0 & \text{Equation 1} \\ 2x - 4y + 7 = 0 & \text{Equation 2} \end{cases}$

Solve for y in Equation 1: $y = \dfrac{1}{x}$

Substitute for y in Equation 2: $2x - 4\left(\dfrac{1}{x}\right) + 7 = 0$

Solve for x:

$$2x^2 - 4 + 7x = 0 \implies (2x - 1)(x + 4) = 0$$

$$\implies x = \frac{1}{2}, -4$$

Back-substitute $x = \dfrac{1}{2}$: $y = \dfrac{1}{1/2} = 2$

Back-substitute $x = -4$: $y = \dfrac{1}{-4} = -\dfrac{1}{4}$

Solutions: $\left(\dfrac{1}{2}, 2\right), \left(-4, -\dfrac{1}{4}\right)$

61. $C = 8650x + 250{,}000, \ R = 9950x$

$R = C$

$9950x = 8650x + 250{,}000$

$1300x = 250{,}000$

$x \approx 192$ units

63. $C = 35.45x + 16{,}000, \ R = 55.95x$

(a) $R = C$

$55.95x = 35.45x + 16{,}000$

$20.50x = 16{,}000$

$x \approx 781$ units

(b) $P = R - C$

$60{,}000 = 55.95x - (35.45x + 16{,}000)$

$60{,}000 = 20.50x - 16{,}000$

$76{,}000 = 20.50x$

$x \approx 3708$ units

65. $\begin{cases} R = 360 - 24x & \text{Equation 1} \\ R = 24 + 18x & \text{Equation 2} \end{cases}$

(a) Substitute for R in Equation 2: $360 - 24x = 24 + 18x$

Solve for x: $336 = 42x \implies x = 8$ weeks

(b)

Weeks	1	2	3	4	5	6	7	8	9	10
$R = 360 - 24x$	336	312	288	264	240	216	192	168	144	120
$R = 24 + 18x$	42	60	78	96	114	132	150	168	186	204

The rentals are equal when $x = 8$ weeks.

67. $0.06x = 0.03x + 350$

$0.03x = 350$

$x \approx \$11{,}666.67$

To make the straight commission offer the better offer, you would have to sell more than $11,666.67 per week.

69. (a) $\begin{cases} x + y = 25{,}000 \\ 0.06x + 0.085y = 2000 \end{cases}$

(b) $y_1 = 25{,}000 - x$

$y_2 = \dfrac{2000 - 0.06x}{0.085}$

As the amount at 6% increases, the amount at 8.5% decreases. The amount of interest is fixed at $2000.

(c) The point of intersection occurs when $x = 5000$, so the most that can be invested at 6% and still earn $2000 per year in interest is $5000.

71.

t	8	9	10	11	12	13
Solar	70	69	66	65	64	63
Wind	31	46	57	68	105	108

(a) Solar: $C \approx 0.1429t^2 - 4.46t + 96.8$

Wind: $C \approx 16.371t - 102.7$

(b)

(c) Point of intersection: $(10.3, 66.01)$

During the year 2000, the consumption of solar energy will equal the consumption of wind energy.

(d) $0.1429t^2 - 4.46t + 96.8 = 16.371t - 102.7$

$0.1429t^2 - 20.831t + 199.5 = 0$

By the Quadratic Formula we obtain $t \approx 10.3$ and $t \approx 135.47$.

(e) The results are the same for $t \approx 10.3$. The other "solution", $t \approx 135.47$, is too large to consider as a reasonable answer.

(f) Answers will vary.

73.
$$2l + 2w = 30 \implies l + w = 15$$
$$l = w + 3 \implies (w + 3) + w = 15$$
$$2w = 12$$
$$w = 6$$

$l = w + 3 = 9$

Dimensions: 6×9 meters

75.
$$2l + 2w = 42 \implies l + w = 21$$
$$w = \tfrac{3}{4}l \implies l + \tfrac{3}{4}l = 21$$
$$\tfrac{7}{4}l = 21$$
$$l = 12$$

$w = \tfrac{3}{4}l = 9$

Dimensions: 9×12 inches

77.
$$2l + 2w = 40 \implies l + w = 20 \implies w = 20 - l$$
$$lw = 96 \implies l(20 - l) = 96$$
$$20l - l^2 = 96$$
$$0 = l^2 - 20l + 96$$
$$0 = (l - 8)(l - 12)$$
$$l = 8 \text{ or } l = 12$$

If $l = 8$, then $w = 12$.

If $l = 12$, then $w = 8$.

Since the length is supposed to be greater than the width, we have $l = 12$ kilometers and $w = 8$ kilometers.
Dimensions: 8×12 kilometers

79. False. To solve a system of equations by substitution, you can solve for either variable in one of the two equations and then back-substitute.

81. To solve a system of equations by substitution, use the following steps.

1. Solve one of the equations for one variable in terms of the other.

2. Substitute this expression into the other equation to obtain an equation in one variable.

3. Solve this equation.

4. Back-substitute the value(s) found in Step 3 into the expression found in Step 1 to find the value(s) of the other variable.

5. Check your solution(s) in each of the original equations.

83. $y = x^2$

    (a) Line with two points of intersection

        $y = 2x$

        $(0, 0)$ and $(2, 4)$

    (b) Line with one point of intersection

        $y = 0$

        $(0, 0)$

    (c) Line with no points of intersection

        $y = x - 2$

85. $(-2, 7), (5, 5)$

$$m = \frac{5 - 7}{5 - (-2)} = -\frac{2}{7}$$

$$y - 7 = -\frac{2}{7}(x - (-2))$$

$$7y - 49 = -2x - 4$$

$$2x + 7y - 45 = 0$$

87. $(6, 3), (10, 3)$

$$m = \frac{3 - 3}{10 - 6} = 0 \implies \text{The line is horizontal.}$$

$$y = 3$$

$$y - 3 = 0$$

89. $\left(\frac{3}{5}, 0\right), (4, 6)$

$$m = \frac{6 - 0}{4 - (3/5)} = \frac{6}{17/5} = \frac{30}{17}$$

$$y - 6 = \frac{30}{17}(x - 4)$$

$$17y - 102 = 30x - 120$$

$$0 = 30x - 17y - 18$$

$$30x - 17y - 18 = 0$$

91. $f(x) = \dfrac{5}{x - 6}$

Domain: All real numbers except $x = 6$

Horizontal asymptote: $y = 0$

Vertical asymptote: $x = 6$

93. $f(x) = \dfrac{x^2 + 2}{x^2 - 16}$

Domain: All real numbers except $x = \pm 4$.

Horizontal asymptote: $y = 1$

Vertical asymptotes: $x = \pm 4$

Section 9.2   Two-Variable Linear Systems

> ■ You should be able to solve a linear system by the method of elimination.
> 1. Obtain coefficients for either x or y that differ only in sign. This is done by multiplying all the terms of one or both equations by appropriate constants.
> 2. Add the equations to eliminate one of the variables and then solve for the remaining variable.
> 3. Use back-substitution into either original equation and solve for the other variable.
> 4. Check your answer.
>
> ■ You should know that for a system of two linear equations, one of the following is true.
> 1. There are infinitely many solutions; the lines are identical. The system is consistent. The slopes are equal.
> 2. There is no solution; the lines are parallel. The system is inconsistent. The slopes are equal.
> 3. There is one solution; the lines intersect at one point. The system is consistent. The slopes are not equal.

Vocabulary Check

1. elimination

2. equivalent

3. consistent; inconsistent

4. equilibrium price

1. $\begin{cases} 2x + y = 5 & \text{Equation 1} \\ x - y = 1 & \text{Equation 2} \end{cases}$

Add to eliminate y: $3x = 6 \implies x = 2$

Substitute $x = 2$ in Equation 2: $2 - y = 1 \implies y = 1$

Solution: $(2, 1)$

3. $\begin{cases} x + y = 0 & \text{Equation 1} \\ 3x + 2y = 1 & \text{Equation 2} \end{cases}$

Multiply Equation 1 by -2: $-2x - 2y = 0$

Add this to Equation 2 to eliminate y: $x = 1$

Substitute $x = 1$ in Equation 1: $1 + y = 0 \implies y = -1$

Solution: $(1, -1)$

5. $\begin{cases} x - y = 2 & \text{Equation 1} \\ -2x + 2y = 5 & \text{Equation 2} \end{cases}$

Multiply Equation 1 by 2: $2x - 2y = 4$

Add this to Equation 2: $0 = 9$

There are no solutions.

7. $\begin{cases} 3x - 2y = 5 & \text{Equation 1} \\ -6x + 4y = -10 & \text{Equation 2} \end{cases}$

Multiply Equation 1 by 2 and add to Equation 2: $0 = 0$

The equations are dependent. There are infinitely many solutions.

Let $x = a$, then $y = \dfrac{3a - 5}{2} = \dfrac{3}{2}a - \dfrac{5}{2}$.

Solution: $\left(a, \dfrac{3}{2}a - \dfrac{5}{2}\right)$ where

a is any real number.

9. $\begin{cases} 9x + 3y = 1 & \text{Equation 1} \\ 3x - 6y = 5 & \text{Equation 2} \end{cases}$

Multiply Equation 2 by (-3): $\quad 9x + 3y = 1$

$\qquad\qquad\qquad\qquad -9x + 18y = -15$

Add to eliminate x: $21y = -14 \implies y = -\frac{2}{3}$

Substitute $y = -\frac{2}{3}$ in Equation 1: $9x + 3\left(-\frac{2}{3}\right) = 1$

$\qquad\qquad\qquad\qquad\qquad\qquad x = \frac{1}{3}$

Solution: $\left(\frac{1}{3}, -\frac{2}{3}\right)$

11. $\begin{cases} x + 2y = 4 & \text{Equation 1} \\ x - 2y = 1 & \text{Equation 2} \end{cases}$

Add to eliminate y:

$\quad 2x = 5$

$\quad\ x = \frac{5}{2}$

Substitute $x = \frac{5}{2}$ in Equation 1:

$\quad \frac{5}{2} + 2y = 4 \implies y = \frac{3}{4}$

Solution: $\left(\frac{5}{2}, \frac{3}{4}\right)$

13. $\begin{cases} 2x + 3y = 18 & \text{Equation 1} \\ 5x - y = 11 & \text{Equation 2} \end{cases}$

Multiply Equation 2 by 3: $15x - 3y = 33$

Add this to Equation 1 to eliminate y:

$$17x = 51 \implies x = 3$$

Substitute $x = 3$ in Equation 1:

$$6 + 3y = 18 \implies y = 4$$

Solution: $(3, 4)$

15. $\begin{cases} 3x + 2y = 10 & \text{Equation 1} \\ 2x + 5y = 3 & \text{Equation 2} \end{cases}$

Multiply Equation 1 by 2 and
Equation 2 by (-3):

$$\begin{cases} 6x + 4y = 20 \\ -6x - 15y = -9 \end{cases}$$

Add to eliminate x: $-11y = 11 \implies y = -1$

Substitute $y = -1$ in Equation 1:

$$3x - 2 = 10 \implies x = 4$$

Solution: $(4, -1)$

17. $\begin{cases} 5u + 6v = 24 & \text{Equation 1} \\ 3u + 5v = 18 & \text{Equation 2} \end{cases}$

Multiply Equation 1 by 5 and Equation 2 by -6:

$$\begin{cases} 25u + 30v = 120 \\ -18u - 30v = -108 \end{cases}$$

Add to eliminate v: $7u = 12 \implies u = \frac{12}{7}$

Substitute $u = \frac{12}{7}$ in Equation 1:

$$5\left(\frac{12}{7}\right) + 6v = 24 \implies 6v = \frac{108}{7} \implies v = \frac{18}{7}$$

Solution: $\left(\frac{12}{7}, \frac{18}{7}\right)$

19. $\begin{cases} \frac{9}{5}x + \frac{6}{5}y = 4 & \text{Equation 1} \\ 9x + 6y = 3 & \text{Equation 2} \end{cases}$

Multiply Equation 1 by 10 and Equation 2 by -2:

$$\begin{cases} 18x + 12y = 40 \\ -18x - 12y = -6 \end{cases}$$

Add to eliminate x and y: $0 = 34$

Inconsistent

No solution

21. $\begin{cases} \frac{x}{4} + \frac{y}{6} = 1 & \text{Equation 1} \\ x - y = 3 & \text{Equation 2} \end{cases}$

Multiply Equation 1 by 6: $\frac{3}{2}x + y = 6$

Add this to Equation 2 to eliminate y:

$$\frac{5}{2}x = 9 \implies x = \frac{18}{5}$$

Substitute $x = \frac{18}{5}$ in Equation 2:

$$\frac{18}{5} - y = 3$$

$$y = \frac{3}{5}$$

Solution: $\left(\frac{18}{5}, \frac{3}{5}\right)$

23. $\begin{cases} -5x + 6y = -3 & \text{Equation 1} \\ 20x - 24y = 12 & \text{Equation 2} \end{cases}$

Multiply Equation 1 by 4:

$$\begin{cases} -20x + 24y = -12 \\ 20x - 24y = 12 \end{cases}$$

Add to eliminate x and y: $0 = 0$

The equations are dependent. There are infinitely many solutions.

Let $x = a$, then

$$-5a + 6y = -3 \implies y = \frac{5a - 3}{6} = \frac{5}{6}a - \frac{1}{2}.$$

Solution: $\left(a, \frac{5}{6}a - \frac{1}{2}\right)$ where a is any real number

25. $\begin{cases} 0.05x - 0.03y = 0.21 & \text{Equation 1} \\ 0.07x + 0.02y = 0.16 & \text{Equation 2} \end{cases}$

Multiply Equation 1 by 200 and Equation 2 by 300:

$\begin{cases} 10x - 6y = 42 \\ 21x + 6y = 48 \end{cases}$

Add to eliminate y: $31x = 90$

$$x = \frac{90}{31}$$

Substitute $x = \frac{90}{31}$ in Equation 2:

$$0.07\left(\frac{90}{31}\right) + 0.02y = 0.16$$

$$y = -\frac{67}{31}$$

Solution: $\left(\frac{90}{31}, -\frac{67}{31}\right)$

27. $\begin{cases} 4b + 3m = 3 & \text{Equation 1} \\ 3b + 11m = 13 & \text{Equation 2} \end{cases}$

Multiply Equation 1 by 3 and Equation 2 by (-4):

$\begin{cases} 12b + 9m = 9 \\ -12b - 44m = -52 \end{cases}$

Add to eliminate b: $-35m = -43$

$$m = \frac{43}{35}$$

Substitute $m = \frac{43}{35}$ in Equation 1:

$$4b + 3\left(\frac{43}{35}\right) = 3 \implies b = -\frac{6}{35}$$

Solution: $\left(-\frac{6}{35}, \frac{43}{35}\right)$

29. $\begin{cases} \dfrac{x+3}{4} + \dfrac{y-1}{3} = 1 & \text{Equation 1} \\ 2x - y = 12 & \text{Equation 2} \end{cases}$

Multiply Equation 1 by 12 and Equation 2 by 4:

$\begin{cases} 3x + 4y = 7 \\ 8x - 4y = 48 \end{cases}$

Add to eliminate y: $11x = 55 \implies x = 5$

Substitute $x = 5$ into Equation 2:

$$2(5) - y = 12 \implies y = -2$$

Solution: $(5, -2)$

31. $\begin{cases} 2x - 5y = 0 \\ x - y = 3 \end{cases}$

Multiply Equation 2 by -5:

$\begin{cases} 2x - 5y = 0 \\ -5x + 5y = -15 \end{cases}$

Add to eliminate y: $-3x = -15 \implies x = 5$

Matches graph (b).

Number of solutions: One

Consistent

33. $\begin{cases} 2x - 5y = 0 \\ 2x - 3y = -4 \end{cases}$

Multiply Equation 1 by -1:

$\begin{cases} -2x + 5y = 0 \\ 2x - 3y = -4 \end{cases}$

Add to eliminate x: $2y = -4 \implies y = -2$

Matches graph (c).

Number of solutions: One

Consistent

35. $\begin{cases} 3x - 5y = 7 & \text{Equation 1} \\ 2x + y = 9 & \text{Equation 2} \end{cases}$

Multiply Equation 2 by 5:

$$10x + 5y = 45$$

Add this to Equation 1:

$$13x = 52 \implies x = 4$$

Back-substitute $x = 4$ into Equation 2:

$$2(4) + y = 9 \implies y = 1$$

Solution: $(4, 1)$

37. $\begin{cases} y = 2x - 5 & \text{Equation 1} \\ y = 5x - 11 & \text{Equation 2} \end{cases}$

Since both equations are solved for y, set them equal to one another and solve for x.

$$2x - 5 = 5x - 11$$

$$6 = 3x$$

$$2 = x$$

Back-substitute $x = 2$ into Equation 1:

$$y = 2(2) - 5 = -1$$

Solution: $(2, -1)$

39. $\begin{cases} x - 5y = 21 & \text{Equation 1} \\ 6x + 5y = 21 & \text{Equation 2} \end{cases}$

Add the equations: $7x = 42 \implies x = 6$

Back-substitute $x = 6$ into Equation 1:

$$6 - 5y = 21 \implies -5y = 15 \implies y = -3$$

Solution: $(6, -3)$

41. $\begin{cases} -2x + 8y = 19 & \text{Equation 1} \\ y = x - 3 & \text{Equation 2} \end{cases}$

Substitute the expression for y from Equation 2
into Equation 1.

$$-2x + 8(x - 3) = 19 \implies -2x + 8x - 24 = 19$$
$$6x = 43$$
$$x = \tfrac{43}{6}$$

Back-substitute $x = \tfrac{43}{6}$ into Equation 2: $y = \tfrac{43}{6} - 3 \implies y = \tfrac{25}{6}$

Solution: $\left(\tfrac{43}{6}, \tfrac{25}{6}\right)$

43. Let $r_1 =$ the air speed of the plane
and $r_2 =$ the wind air speed.

$$\begin{array}{llll} 3.6(r_1 - r_2) = 1800 & \text{Equation 1} & \implies & r_1 - r_2 = 500 \\ 3(r_1 + r_2) = 1800 & \text{Equation 2} & \implies & \underline{r_1 + r_2 = 600} \end{array}$$

$$\begin{aligned} 2r_1 &= 1100 \qquad \text{Add the equations.}\\ r_1 &= 550 \\ 550 + r_2 &= 600 \\ r_2 &= 50 \end{aligned}$$

The air speed of the plane is 550 mph and the speed of the wind is 50 mph.

45.
$$50 - 0.5x = 0.125x$$
$$50 = 0.625x$$
$$x = 80 \text{ units}$$
$$p = \$10$$

Solution: $(80, 10)$

47.
$$140 - 0.00002x = 80 + 0.00001x$$
$$60 = 0.00003x$$
$$x = 2{,}000{,}000 \text{ units}$$
$$p = \$100.00$$

Solution: $(2{,}000{,}000, 100)$

49. Let $x =$ number of calories in a cheeseburger

$y =$ number of calories in a small order of french fries

$$\begin{cases} 2x + y = 850 & \text{Equation 1} \\ 3x + 2y = 1390 & \text{Equation 2} \end{cases}$$

Multiply Equation 1 by -2:

$$\begin{cases} -4x - 2y = -1700 \\ 3x + 2y = 1390 \end{cases}$$

$$\begin{aligned} -x &= -310 \qquad \text{Add the equations.} \\ x &= 310 \\ y &= 230 \end{aligned}$$

Solution: The cheeseburger contains 310 calories and the
fries contain 230 calories.

51. Let $x =$ the number of liters at 20%

Let $y =$ the number of liters at 50%.

(a) $\begin{cases} x + y = 10 \\ 0.2x + 0.5y = 0.3(10) \end{cases}$

$$\begin{array}{ll} -2 \cdot \text{Equation 1:} & -2x - 2y = -20 \\ 10 \cdot \text{Equation 2:} & \underline{2x + 5y = 30} \end{array}$$

$$\begin{aligned} 3y &= 10 \\ y &= \tfrac{10}{3} \\ x + \tfrac{10}{3} &= 10 \\ x &= \tfrac{20}{3} \end{aligned}$$

(b)

As x increases, y decreases.

(c) In order to obtain the specified concentration of
the final mixture, $6\tfrac{2}{3}$ liters of the 20% solution and
$3\tfrac{1}{3}$ liters of the 50% solution are required.

53. Let x = amount invested at 7.5%

y = amount invested at 9%.

$$\begin{cases} x + y = 12{,}000 & \text{Equation 1} \\ 0.075x + 0.09y = 990 & \text{Equation 2} \end{cases}$$

Multiply Equation 1 by 9 and Equation 2 by -100.

$$\begin{cases} 9x + 9y = 108{,}000 \\ -7.5x - 9y = -99{,}000 \end{cases}$$

$$1.5x = 9{,}000 \qquad \text{Add the equations.}$$

$$x = \$6000$$

$$y = \$6000$$

The most that can be invested at 7.5% is \$6000.

57.
$$\begin{cases} 5b + 10a = 20.2 \implies -10b - 20a = -40.4 \\ 10b + 30a = 50.1 \implies \underline{10b + 30a = 50.1} \end{cases}$$

$$10a = 9.7$$

$$a = 0.97$$

$$b = 2.10$$

Least squares regression line: $y = 0.97x + 2.10$

61. $(0, 4), (1, 3), (1, 1), (2, 0)$

$$n = 4, \ \sum_{i=1}^{4} x_i = 4, \ \sum_{i=1}^{4} y_i = 8, \ \sum_{i=1}^{4} x_i^2 = 6, \ \sum_{i=1}^{4} x_i y_i = 4$$

$$\begin{cases} 4b + 4a = 8 \implies 4b + 4a = 8 \\ 4b + 6a = 4 \implies \underline{-4b - 6a = -4} \end{cases}$$

$$-2a = 4$$

$$a = -2$$

$$b = 4$$

Least squares regression line: $y = -2x + 4$

63. $(5, 66.65), (6, 70.93), (7, 75.31), (8, 78.62), (9, 81.33), (10, 85.89), (11, 88.27)$

(a) $n = 7, \ \sum_{i=1}^{7} x_i = 56, \ \sum_{i=1}^{7} x_i^2 = 476, \ \sum_{i=1}^{7} y_i = 547, \ \sum_{i=1}^{7} x_i y_i = 4476.8$

$$\begin{cases} 7b + 56a = 547 \\ 56b + 476a = 4476.8 \end{cases}$$

Multiply Equation 1 by -8.

$$\begin{cases} -56b - 448a = -4376 \\ 56b + 476a = 4476.8 \end{cases}$$

$$28a = 100.8$$

$$a = 3.6$$

$$b \approx 49.343$$

Least squares regression line: $y \approx 3.6t + 49.343$

(b) $y \approx 3.6t + 49.343$, This agrees with part (a).

55. Let x = number of student tickets

y = number of adult tickets.

$$\begin{cases} x + y = 1435 & \text{Equation 1} \\ 1.50x + 5.00y = 3552.50 & \text{Equation 2} \end{cases}$$

Multiply Equation 1 by -1.50.

$$\begin{cases} -1.50x - 1.50y = -2152.50 \\ 1.50x + 5.00y = 3552.50 \end{cases}$$

$$3.50y = 1400.00 \qquad \text{Add the equations.}$$

$$y = 400$$

$$x = 1035$$

Solution: 1035 student tickets and 400 adult tickets were sold.

59.
$$\begin{cases} 7b + 21a = 35.1 \implies -21b - 63a = -105.3 \\ 21b + 91a = 114.2 \implies \underline{21b + 91a = 114.2} \end{cases}$$

$$28a = 8.9$$

$$a = \frac{89}{280}$$

$$b = \frac{1137}{280}$$

Least squares regression line: $y = \frac{1}{280}(89x + 1137)$

$$y \approx 0.32x + 4.1$$

—CONTINUED—

63. —CONTINUED—

(c)

t	Actual room rate	Model approximation
5	$66.65	$67.34
6	$70.93	$70.94
7	$75.31	$74.54
8	$78.62	$78.14
9	$81.33	$81.74
10	$85.89	$85.34
11	$88.27	$88.94

The model is a good fit to the data.

(d) When $t = 12$: $y \approx \$92.54$

This is a little off from the actual rate.

(e) $3.6t + 49.343 = 100$

$$3.6t = 50.657$$

$$t \approx 14.1$$

According to the model, room rates will average $100.00 during the year 2004.

65. False. Two lines that coincide have infinitely many points of intersection.

67. No, it is not possible for a consistent system of linear equations to have exactly two solutions. Either the lines will intersect once or they will coincide and then the system would have infinite solutions.

69. $\begin{cases} 100y - x = 200 & \text{Equation 1} \\ 99y - x = -198 & \text{Equation 2} \end{cases}$

Subtract Equation 2 from Equation 1 to eliminate x:

$100y - x = 200$

$\underline{-99y + x = 198}$

$\qquad y = 398$

Substitute $y = 398$ into Equation 1:

$100(398) - x = 200 \Rightarrow x = 39,600$

Solution: $(39,600, 398)$

The lines are not parallel. The scale on the axes must be changed to see the point of intersection.

71. $\begin{cases} 4x - 8y = -3 & \text{Equation 1} \\ 2x + ky = 16 & \text{Equation 2} \end{cases}$

Multiply Equation 2 by -2: $-4x - 2ky = -32$

Add this to Equation 1: $4x - 8y = -3$

$\qquad\qquad \underline{-4x - 2ky = -32}$

$\qquad\qquad -8y - 2ky = -35$

The system is inconsistent if $-8y - 2ky = 0$. This occurs when $k = -4$.

73. $-11 - 6x \geq 33$

$\qquad -6x \geq 44$

$\qquad x \leq -\frac{22}{3}$

75. $8x - 15 \leq -4(2x - 1)$

$8x - 15 \leq -8x + 4$

$16x \leq 19$

$x \leq \frac{19}{16}$

77. $|x - 8| < 10$

$-10 < x - 8 < 10$

$-2 < x < 18$

79. $2x^2 + 3x - 35 < 0$

$(2x - 7)(x + 5) < 0$

Critical numbers: $x = -5, \frac{7}{2}$

Test intervals: $(-\infty, -5), \left(-5, \frac{7}{2}\right), \left(\frac{7}{2}, \infty\right)$

Test: Is $(2x - 7)(x + 5) < 0$?

Solution: $-5 < x < \frac{7}{2}$

81. $\ln x + \ln 6 = \ln(6x)$

83. $\log_9 12 - \log_9 x = \log_9\left(\dfrac{12}{x}\right)$

85. $\begin{cases} 2x - y = 4 \implies y = 2x - 4 \\ -4x + 2y = -12 \end{cases}$

$-4x + 2(2x - 4) = -12$

$-4x + 4x - 8 = -12$

$-8 = -12$

There are no solutions.

87. Answers will vary.

Section 9.3   Multivariable Linear Systems

- You should know the operations that lead to equivalent systems of equations:
  (a) Interchange any two equations.
  (b) Multiply all terms of an equation by a nonzero constant.
  (c) Replace an equation by the sum of itself and a constant multiple of any other equation in the system.
- You should be able to use the method of Gaussian elimination with back-substitution.

Vocabulary Check

1. row-echelon

2. ordered triple

3. Gaussian

4. row operation

5. nonsquare

6. position

1. $\begin{cases} 3x - y + z = 1 \\ 2x \quad\;\; - 3z = -14 \\ \quad\;\; 5y + 2z = 8 \end{cases}$

(a) $3(2) - (0) + (-3) \neq 1$

$(2, 0, -3)$ *is not* a solution.

(b) $3(-2) - (0) + 8 \neq 1$

$(-2, 0, 8)$ *is not* a solution.

(c) $3(0) - (-1) + 3 \neq 1$

$(0, -1, 3)$ *is not* a solution.

(d) $3(-1) - \;\;(0) + \;\;4 = \;\;1$
$2(-1) \qquad\;\; - 3(4) = -14$
$\qquad\;\; 5(0) + 2(4) = \;\;8$

$(-1, 0, 4)$ *is a* solution.

3. $\begin{cases} 4x + y - z = 0 \\ -8x - 6y + z = -\frac{7}{4} \\ 3x - y \quad\;\; = -\frac{9}{4} \end{cases}$

(a) $4\left(\frac{1}{2}\right) + \left(-\frac{3}{4}\right) - \left(-\frac{7}{4}\right) \neq 0$

$\left(\frac{1}{2}, -\frac{3}{4}, -\frac{7}{4}\right)$ *is not* a solution.

(b) $4\left(-\frac{3}{2}\right) + \left(\frac{5}{4}\right) - \left(-\frac{5}{4}\right) \neq 0$

$\left(-\frac{3}{2}, \frac{5}{4}, -\frac{5}{4}\right)$ *is not* a solution.

(c) $4\left(-\frac{1}{2}\right) + \;\;\left(\frac{3}{4}\right) - \left(-\frac{5}{4}\right) = \;\;0$
$-8\left(-\frac{1}{2}\right) - 6\left(\frac{3}{4}\right) + \left(-\frac{5}{4}\right) = -\frac{7}{4}$
$3\left(-\frac{1}{2}\right) - \;\;\left(\frac{3}{4}\right) \qquad\;\; = -\frac{9}{4}$

$\left(-\frac{1}{2}, \frac{3}{4}, -\frac{5}{4}\right)$ *is a* solution.

(d) $4\left(-\frac{1}{2}\right) + \left(\frac{1}{6}\right) - \left(-\frac{3}{4}\right) \neq 0$

$\left(-\frac{1}{2}, \frac{1}{6}, -\frac{3}{4}\right)$ *is not* a solution.

5. $\begin{cases} 2x - y + 5z = 24 & \text{Equation 1} \\ y + 2z = 6 & \text{Equation 2} \\ z = 4 & \text{Equation 3} \end{cases}$

Back-substitute $z = 4$ into Equation 2:

$y + 2(4) = 6$

$y = -2$

Back-substitute $y = -2$ and $z = 4$ into Equation 1:

$2x - (-2) + 5(4) = 24$

$2x + 22 = 24$

$x = 1$

Solution: $(1, -2, 4)$

7. $\begin{cases} 2x + y - 3z = 10 & \text{Equation 1} \\ y + z = 12 & \text{Equation 2} \\ z = 2 & \text{Equation 3} \end{cases}$

Substitute $z = 2$ into Equation 2: $y + (2) = 12 \implies y = 10$

Substitute $y = 10$ and $z = 2$ into Equation 1:

$2x + (10) - 3(2) = 10$

$2x + 4 = 10$

$2x = 6$

$x = 3$

Solution: $(3, 10, 2)$

9. $\begin{cases} 4x - 2y + z = 8 & \text{Equation 1} \\ -y + z = 4 & \text{Equation 2} \\ z = 2 & \text{Equation 3} \end{cases}$

Substitute $z = 2$ into Equation 2:

$-y + (2) = 4 \implies y = -2$

Substitute $y = -2$ and $z = 2$ into Equation 1:

$4x - 2(-2) + (2) = 8$

$4x + 6 = 8$

$4x = 2$

$x = \frac{1}{2}$

Solution: $\left(\frac{1}{2}, -2, 2\right)$

11. $\begin{cases} x - 2y + 3z = 5 & \text{Equation 1} \\ -x + 3y - 5z = 4 & \text{Equation 2} \\ 2x \quad\;\; - 3z = 0 & \text{Equation 3} \end{cases}$

Add Equation 1 to Equation 2:

$\begin{cases} x - 2y + 3z = 5 \\ y - 2z = 9 \\ 2x \quad\;\; - 3z = 0 \end{cases}$

This is the first step in putting the system in row-echelon form.

13. $\begin{cases} x + y + z = 6 & \text{Equation 1} \\ 2x - y + z = 3 & \text{Equation 2} \\ 3x \quad - z = 0 & \text{Equation 3} \end{cases}$

$\begin{cases} x + y + z = 6 \\ \quad -3y - z = -9 & -2\text{Eq.1} + \text{Eq.2} \\ \quad -3y - 4z = -18 & -3\text{Eq.1} + \text{Eq.3} \end{cases}$

$\begin{cases} x + y + z = 6 \\ \quad -3y - z = -9 \\ \quad\quad -3z = -9 & -\text{Eq.2} + \text{Eq.3} \end{cases}$

$\begin{cases} x + y + z = 6 \\ \quad -3y - z = -9 \\ \quad\quad z = 3 & -\frac{1}{3}\text{Eq.3} \end{cases}$

$-3y - 3 = -9 \implies y = 2$

$x + 2 + 3 = 6 \implies x = 1$

Solution: $(1, 2, 3)$

15. $\begin{cases} 2x \quad + 2z = 2 \\ 5x + 3y \quad = 4 \\ \quad 3y - 4z = 4 \end{cases}$

$\begin{cases} x \quad + z = 1 & \frac{1}{2}\text{Eq.1} \\ 5x + 3y \quad = 4 \\ \quad 3y - 4z = 4 \end{cases}$

$\begin{cases} x \quad + z = 1 \\ \quad 3y - 5z = -1 & -5\text{Eq.1} + \text{Eq.2} \\ \quad 3y - 4z = 4 \end{cases}$

$\begin{cases} x \quad + z = 1 \\ \quad 3y - 5z = -1 \\ \quad\quad z = 5 & -\text{Eq.2} + \text{Eq.3} \end{cases}$

$3y - 5(5) = -1 \implies y = 8$

$x + 5 = 1 \implies x = -4$

Solution: $(-4, 8, 5)$

17. $\begin{cases} 3x + 3y \quad = 9 \\ 2x \quad - 3z = 10 \\ \quad 6y + 4z = -12 \end{cases}$ Interchange equations.

$\begin{cases} x + y \quad = 3 & \frac{1}{3}\text{Eq.1} \\ 2x \quad - 3z = 10 \\ \quad 6y + 4z = -12 \end{cases}$

$\begin{cases} x + y \quad = 3 \\ \quad -2y - 3z = 4 & -2\text{Eq.1} + \text{Eq.2} \\ \quad 6y + 4z = -12 \end{cases}$

$\begin{cases} x + y \quad = 3 \\ \quad -2y - 3z = 4 \\ \quad\quad -5z = 0 & 3\text{Eq.2} + \text{Eq.3} \end{cases}$

$\begin{cases} x + y \quad = 3 \\ \quad -2y - 3z = 4 \\ \quad\quad z = 0 & -\frac{1}{5}\text{Eq.3} \end{cases}$

$-2y - 3(0) = 4 \implies y = -2$

$x - 2 = 3 \implies x = 5$

Solution: $(5, -2, 0)$

19. $\begin{cases} x - 2y + 2z = -9 \\ 2x + y - z = 7 \\ 3x - y + z = 5 \end{cases}$ Interchange equations.

$\begin{cases} x - 2y + 2z = -9 \\ \quad 5y - 5z = 25 & -2\text{Eq.1} + \text{Eq.2} \\ \quad 5y - 5z = 32 & -3\text{Eq.1} + \text{Eq.3} \end{cases}$

$\begin{cases} x - 2y + 2z = -9 \\ \quad 5y - 5z = 25 \\ \quad\quad 0 = 7 & -\text{Eq.2} + \text{Eq.3} \end{cases}$

Inconsistent, no solution

21. $\begin{cases} 3x - 5y + 5z = 1 \\ 5x - 2y + 3z = 0 \\ 7x - y + 3z = 0 \end{cases}$

$\begin{cases} 6x - 10y + 10z = 2 & 2\text{Eq.1} \\ 5x - 2y + 3z = 0 \\ 7x - y + 3z = 0 \end{cases}$

$\begin{cases} x - 8y + 7z = 2 & -\text{Eq.2} + \text{Eq.1} \\ 5x - 2y + 3z = 0 \\ 7x - y + 3z = 0 \end{cases}$

$\begin{cases} x - 8y + 7z = 2 \\ \quad 38y - 32z = -10 & -5\text{Eq.1} + \text{Eq.2} \\ \quad 55y - 46z = -14 & -7\text{Eq.1} + \text{Eq.3} \end{cases}$

$\begin{cases} x - 8y + 7z = 2 \\ \quad 2090y - 1760z = -550 & 55\text{Eq.2} \\ \quad -2090y + 1748z = 532 & -38\text{Eq.3} \end{cases}$

$\begin{cases} x - 8y + 7z = 2 \\ \quad 2090y - 1760z = -550 \\ \quad\quad -12z = -18 & \text{Eq.2} + \text{Eq.3} \end{cases}$

$-12z = -18 \implies z = \frac{3}{2}$

$38y - 32\left(\frac{3}{2}\right) = -10 \implies y = 1$

$x - 8(1) + 7\left(\frac{3}{2}\right) = 2 \implies x = -\frac{1}{2}$

Solution: $\left(-\frac{1}{2}, 1, \frac{3}{2}\right)$

23. $\begin{cases} x + 2y - 7z = -4 \\ 2x + y + z = 13 \\ 3x + 9y - 36z = -33 \end{cases}$

$\begin{cases} x + 2y - 7z = -4 \\ \quad -3y + 15z = 21 \\ \quad 3y - 15z = -21 \end{cases} \quad \begin{matrix} -2\text{Eq.1} + \text{Eq.2} \\ -3\text{Eq.1} + \text{Eq.3} \end{matrix}$

$\begin{cases} x + 2y - 7z = -4 \\ \quad -3y + 15z = 21 \\ \quad\quad\quad 0 = 0 \end{cases} \quad \text{Eq.2} + \text{Eq.3}$

$\begin{cases} x + 2y - 7z = -4 \\ \quad y - 5z = -7 \end{cases} \quad -\tfrac{1}{3}\text{Eq.2}$

$\begin{cases} x \quad + 3z = 10 \\ \quad y - 5z = -7 \end{cases} \quad -2\text{Eq.2} + \text{Eq.1}$

Let $z = a$, then:

$y = 5a - 7$

$x = -3a + 10$

Solution: $(-3a + 10, 5a - 7, a)$

27. $\begin{cases} x - 2y + 5z = 2 \\ 4x \quad - z = 0 \end{cases}$

Let $z = a$, then $x = \tfrac{1}{4}a$.

$\tfrac{1}{4}a - 2y + 5a = 2$

$a - 8y + 20a = 8$

$-8y = -21a + 8$

$y = \tfrac{21}{8}a - 1$

Answer: $\left(\tfrac{1}{4}a, \tfrac{21}{8}a - 1, a\right)$

To avoid fractions, we could go back and let $z = 8a$, then $4x - 8a = 0 \implies x = 2a$.

$2a - 2y + 5(8a) = 2$

$-2y + 42a = 2$

$y = 21a - 1$

Solution: $(2a, 21a - 1, 8a)$

25. $\begin{cases} 3x - 3y + 6z = 6 \\ x + 2y - z = 5 \\ 5x - 8y + 13z = 7 \end{cases}$

$\begin{cases} x - y + 2z = 2 \\ x + 2y - z = 5 \\ 5x - 8y + 13z = 7 \end{cases} \quad \tfrac{1}{3}\text{Eq.1}$

$\begin{cases} x - y + 2z = 2 \\ \quad 3y - 3z = 3 \\ \quad -3y + 3z = -3 \end{cases} \quad \begin{matrix} -\text{Eq.1} + \text{Eq.2} \\ -5\text{Eq.1} + \text{Eq.3} \end{matrix}$

$\begin{cases} x - y + 2z = 2 \\ \quad y - z = 1 \\ \quad\quad\quad 0 = 0 \end{cases} \quad \begin{matrix} \tfrac{1}{3}\text{Eq.2} \\ \text{Eq.2} + \text{Eq.3} \end{matrix}$

$\begin{cases} x \quad + z = 3 \\ \quad y - z = 1 \end{cases} \quad \text{Eq.2} + \text{Eq.1}$

Let $z = a$, then:

$y = a + 1$

$x = -a + 3$

Solution: $(-a + 3, a + 1, a)$

29. $\begin{cases} 2x - 3y + z = -2 \\ -4x + 9y = 7 \end{cases}$

$\begin{cases} 2x - 3y + z = -2 \\ \quad 3y + 2z = 3 \end{cases} \quad 2\text{Eq.1} + \text{Eq.2}$

$\begin{cases} 2x \quad + 3z = 1 \\ \quad 3y + 2z = 3 \end{cases} \quad \text{Eq.2} + \text{Eq.1}$

Let $z = a$, then:

$y = -\tfrac{2}{3}a + 1$

$x = -\tfrac{3}{2}a + \tfrac{1}{2}$

Solution: $\left(-\tfrac{3}{2}a + \tfrac{1}{2}, -\tfrac{2}{3}a + 1, a\right)$

31. $\begin{cases} x \qquad\qquad + 3w = 4 \\ \quad 2y - \ z - \ w = 0 \\ \quad 3y \qquad - 2w = 1 \\ 2x - \ y + 4z \qquad = 5 \end{cases}$

$\begin{cases} x \qquad\qquad + 3w = 4 \\ \quad 2y - \ z - \ w = 0 \\ \quad 3y \qquad - 2w = 1 \\ \quad -y + 4z - 6w = -3 \qquad -2\text{Eq.1} + \text{Eq.4} \end{cases}$

$\begin{cases} x \qquad\qquad + 3w = 4 \\ \quad y - 4z + 6w = 3 \qquad -\text{Eq.4 and interchange} \\ \quad 2y - \ z - \ w = 0 \qquad\ \text{the equations.} \\ \quad 3y \qquad - 2w = 1 \end{cases}$

$\begin{cases} x \qquad\qquad + 3w = 4 \\ \quad y - 4z + 6w = 3 \\ \qquad 7z - 13w = -6 \qquad -\text{Eq.2} + \text{Eq.3} \\ \qquad 12z - 20w = -8 \qquad -3\text{Eq.2} + \text{Eq.4} \end{cases}$

$\begin{cases} x \qquad\qquad + 3w = 4 \\ \quad y - 4z + 6w = 3 \\ \qquad z - 3w = -2 \qquad -\frac{1}{2}\text{Eq.4} + \text{Eq.3} \\ \qquad 12z - 20w = -8 \end{cases}$

$\begin{cases} x \qquad\qquad + 3w = 4 \\ \quad y - 4z + 6w = 3 \\ \qquad z - 3w = -2 \\ \qquad\qquad 16w = 16 \qquad -12\text{Eq.3} + \text{Eq.4} \end{cases}$

$16w = 16 \implies w = 1$

$z - 3(1) = -2 \implies z = 1$

$y - 4(1) + 6(1) = 3 \implies y = 1$

$x + 3(1) = 4 \implies x = 1$

Solution: $(1, 1, 1, 1)$

33. $\begin{cases} x \qquad + 4z = 1 \\ x + y + 10z = 10 \\ 2x - y + 2z = -5 \end{cases}$

$\begin{cases} x \qquad + 4z = 1 \\ \quad y + 6z = 9 \qquad -\text{Eq.1} + \text{Eq.2} \\ \quad -y - 6z = -7 \qquad -2\text{Eq.1} + \text{Eq.3} \end{cases}$

$\begin{cases} x \qquad + 4z = 1 \\ \quad y + 6z = 9 \\ \qquad\quad 0 = 2 \qquad \text{Eq.2} + \text{Eq.3} \end{cases}$

No solution, inconsistent

35. $\begin{cases} 2x + 3y \qquad = 0 \\ 4x + 3y - \ z = 0 \\ 8x + 3y + 3z = 0 \end{cases}$

$\begin{cases} 2x + 3y \qquad = 0 \\ \quad -3y - \ z = 0 \qquad -2\text{Eq.1} + \text{Eq.2} \\ \quad -9y + 3z = 0 \qquad -4\text{Eq.1} + \text{Eq.3} \end{cases}$

$\begin{cases} 2x + 3y \qquad = 0 \\ \quad -3y - \ z = 0 \\ \qquad\quad 6z = 0 \qquad -3\text{Eq.2} + \text{Eq.3} \end{cases}$

$6z = 0 \implies z = 0$

$-3y - 0 = 0 \implies y = 0$

$2x + 3(0) = 0 \implies x = 0$

Solution: $(0, 0, 0)$

37. $\begin{cases} 12x + \ 5y + \ z = 0 \\ 23x + \ 4y - \ z = 0 \end{cases}$

$\begin{cases} 24x + 10y + 2z = 0 \qquad 2\text{Eq.1} \\ 23x + \ 4y - \ z = 0 \end{cases}$

$\begin{cases} x + \ 6y + 3z = 0 \qquad -\text{Eq.2} + \text{Eq.1} \\ 23x + \ 4y - \ z = 0 \end{cases}$

$\begin{cases} x + \ 6y + 3z = 0 \\ \quad -134y - 70z = 0 \qquad -23\text{Eq.1} + \text{Eq.2} \end{cases}$

$\begin{cases} x + \ 6y + 3z = 0 \\ \quad -67y - 35z = 0 \qquad \frac{1}{2}\text{Eq.2} \end{cases}$

To avoid fractions, let $z = 67a$, then:

$-67y - 35(67a) = 0$

$y = -35a$

$x + 6(-35a) + 3(67a) = 0$

$x = 9a$

Solution: $(9a, -35a, 67a)$

39. $s = \frac{1}{2}at^2 + v_0 t + s_0$

$(1, 128), (2, 80), (3, 0)$

$128 = \frac{1}{2}a + v_0 + s_0 \implies a + 2v_0 + 2s_0 = 256$

$80 = 2a + 2v_0 + s_0 \implies 2a + 2v_0 + s_0 = 80$

$0 = \frac{9}{2}a + 3v_0 + s_0 \implies 9a + 6v_0 + 2s_0 = 0$

Solving this system yields $a = -32, v_0 = 0, s_0 = 144$.

Thus, $s = \frac{1}{2}(-32)t^2 + (0)t + 144 = -16t^2 + 144$.

41. $s = \frac{1}{2}at^2 + v_0t + s_0$

$(1, 452), (2, 372), (3, 260)$

$452 = \frac{1}{2}a + \ v_0 + s_0 \implies \ a + 2v_0 + 2s_0 = 904$

$372 = 2a + 2v_0 + s_0 \implies 2a + 2v_0 + \ s_0 = 372$

$260 = \frac{9}{2}a + 3v_0 + s_0 \implies 9a + 6v_0 + 2s_0 = 520$

Solving this system yields $a = -32$, $v_0 = -32$, $s_0 = 500$.

Thus, $s = \frac{1}{2}(-32)t^2 + (-32)t + 500$

$\qquad = -16t^2 - 32t + 500.$

43. $y = ax^2 + bx + c$ passing through $(0, 0), (2, -2), (4, 0)$

$(0, 0):\ 0 = c$

$(2, -2):\ -2 = \ 4a + 2b + c \implies \ -1 = 2a + b$

$(4, 0):\ 0 = 16a + 4b + c \implies \ 0 = 4a + b$

Solution: $a = \frac{1}{2}, b = -2, c = 0$

The equation of the parabola is $y = \frac{1}{2}x^2 - 2x$.

45. $y = ax^2 + bx + c$ passing through $(2, 0), (3, -1), (4, 0)$

$(2, 0):\ 0 = 4a + 2b + c$

$(3, -1):\ -1 = 9a + 3b + c$

$(4, 0):\ 0 = 16a + 4b + c$

$\begin{cases} 0 = \ 4a + 2b + c \\ -1 = \ 5a + \ b \qquad\qquad -\text{Eq.1} + \text{Eq.2} \\ 0 = 12a + 2b \qquad\quad\ -\text{Eq.1} + \text{Eq.3} \end{cases}$

$\begin{cases} 0 = \ 4a + 2b + c \\ -1 = \ 5a + \ b \\ 2 = \ 2a \qquad\qquad\qquad -2\text{Eq.2} + \text{Eq.3} \end{cases}$

Solution: $a = 1, b = -6, c = 8$

The equation of the parabola is $y = x^2 - 6x + 8$.

47. $x^2 + y^2 + Dx + Ey + F = 0$ passing through $(0, 0), (2, 2), (4, 0)$

$(0, 0):\ F = 0$

$(2, 2):\ 8 + 2D + 2E + F = 0 \implies D + E = -4$

$(4, 0):\ 16 + 4D + F = 0 \implies D = -4$ and $E = 0$

The equation of the circle is $x^2 + y^2 - 4x = 0$.

To graph, let $y_1 = \sqrt{4x - x^2}$ and $y_2 = -\sqrt{4x - x^2}$.

49. $x^2 + y^2 + Dx + Ey + F = 0$ passing through $(-3, -1), (2, 4), (-6, 8)$

$(-3, -1):\ 10 - 3D - E + F = 0 \implies 10 = 3D + E - F$

$(2, 4):\ 20 + 2D + 4E + F = 0 \implies 20 = -2D - 4E - F$

$(-6, 8):\ 100 - 6D + 8E + F = 0 \implies 100 = 6D - 8E - F$

Solution: $D = 6, E = -8, F = 0$

The equation of the circle is $x^2 + y^2 + 6x - 8y = 0$. To graph, complete the squares first, then solve for y.

$(x^2 + 6x + 9) + (y^2 - 8y + 16) = 0 + 9 + 16$

$\qquad\qquad (x + 3)^2 + (y - 4)^2 = 25$

$\qquad\qquad\qquad\quad (y - 4)^2 = 25 - (x + 3)^2$

$\qquad\qquad\qquad\qquad y - 4 = \pm\sqrt{25 - (x + 3)^2}$

$\qquad\qquad\qquad\qquad\quad y = 4 \pm \sqrt{25 - (x + 3)^2}$

Let $y_1 = 4 + \sqrt{25 - (x + 3)^2}$ and $y_2 = 4 - \sqrt{25 - (x + 3)^2}$.

51. Let x = number of touchdowns.

Let y = number of extra-point kicks.

Let z = number of field goals.

$$\begin{cases} x + y + z = 13 \\ 6x + y + 3z = 45 \\ x - y \quad\quad = 0 \\ x \quad\quad - 6z = 0 \end{cases}$$

$$\begin{cases} x + y + z = 13 \\ \quad -5y - 3z = -33 \quad -6\text{Eq.1} + \text{Eq.2} \\ \quad -2y - z = -13 \quad -\text{Eq.1} + \text{Eq.3} \\ \quad -y - 7z = -13 \quad -\text{Eq.1} + \text{Eq.4} \end{cases}$$

$$\begin{cases} x + y + z = 13 \\ \quad -y - 7z = -13 \quad \text{Interchange Eq.2 and Eq.4.} \\ \quad -2y - z = -13 \\ \quad -5y - 3z = -33 \end{cases}$$

$$\begin{cases} x + y + z = 13 \\ \quad y + 7z = 13 \quad -\text{Eq.2} \\ \quad -2y - z = -13 \\ \quad -5y - 3z = -33 \end{cases}$$

$$\begin{cases} x + y + z = 13 \\ \quad y + 7z = 13 \\ \quad\quad 13z = 13 \quad 2\text{Eq.2} + \text{Eq.3} \\ \quad\quad 32z = 32 \quad 5\text{Eq.2} + \text{Eq.4} \end{cases}$$

$z = 1$

$y + 7(1) = 13 \implies y = 6$

$x + 6 + 1 = 13 \implies x = 6$

Thus, 6 touchdowns, 6 extra-point kicks, and 1 field goal were scored.

55. Let C = amount in certificates of deposit.

Let M = amount in municipal bonds.

Let B = amount in blue-chip stocks.

Let G = amount in growth or speculative stocks.

$$\begin{cases} C + M + B + G = 500,000 \\ 0.10C + 0.08M + 0.12B + 0.13G = 0.10(500,000) \\ B + G = \frac{1}{4}(500,000) \end{cases}$$

This system has infinitely many solutions.

Let $G = s$, then $B = 125,000 - s$

$M = 125,000 + \frac{1}{2}s$

$C = 250,000 - \frac{1}{2}s$

One possible solution is to let $s = 50,000$.

Certificates of deposit: $225,000

Municipal bonds: $150,000

Blue-chip stocks: $75,000

Growth or speculative stocks: $50,000

53. Let x = amount at 8%.

Let y = amount at 9%.

Let z = amount at 10%.

$$\begin{cases} x + y + z = 775,000 \\ 0.08x + 0.09y + 0.10z = 67,500 \\ x = 4z \end{cases}$$

$y + 5z = 775,000$

$0.09y + 0.42z = 67,500$

$z = 75,000$

$y = 775,000 - 5z = 400,000$

$x = 4z = 300,000$

$300,000 was borrowed at 8%.

$400,000 was borrowed at 9%.

$75,000 was borrowed at 10%.

57. Let x = pounds of brand X.

Let y = pounds of brand Y.

Let z = pounds of brand Z.

Fertilizer A: $\quad \frac{1}{3}y + \frac{2}{9}z = 5$

Fertilizer B: $\frac{1}{2}x + \frac{2}{3}y + \frac{5}{9}z = 13$

Fertilizer C: $\frac{1}{2}x \quad + \frac{2}{9}z = 4$

$$\begin{cases} \frac{1}{2}x + \frac{2}{3}y + \frac{5}{9}z = 13 \quad \text{Interchange Eq.1 and Eq.2.} \\ \frac{1}{3}y + \frac{2}{9}z = 5 \\ \frac{1}{2}x \quad + \frac{2}{9}z = 4 \end{cases}$$

$$\begin{cases} \frac{1}{2}x + \frac{2}{3}y + \frac{5}{9}z = 13 \\ \frac{1}{3}y + \frac{2}{9}z = 5 \\ -\frac{2}{3}y - \frac{1}{3}z = -9 \quad -\text{Eq.1} + \text{Eq.3} \end{cases}$$

$$\begin{cases} \frac{1}{2}x + \frac{2}{3}y + \frac{5}{9}z = 13 \\ \frac{1}{3}y + \frac{2}{9}z = 5 \\ \frac{1}{9}z = 1 \quad 2\text{Eq.2} + \text{Eq.3} \end{cases}$$

$z = 9$

$\frac{1}{3}y + \frac{2}{9}(9) = 5 \implies y = 9$

$\frac{1}{2}x + \frac{2}{3}(9) + \frac{5}{9}(9) = 13 \implies x = 4$

4 pounds of brand X, 9 pounds of brand Y, and 9 pounds of brand Z are needed to obtain the desired mixture.

59. Let $x =$ pounds of Vanilla coffee.

Let $y =$ pounds of Hazelnut coffee.

Let $z =$ pounds of French Roast coffee.

$$\begin{cases} x + y + z = 10 \\ 2x + 2.50y + 3z = 26 \\ y - z = 0 \end{cases}$$

$$\begin{cases} x + y + z = 10 \\ 0.5y + z = 6 \qquad -2\text{Eq.1} + \text{Eq.2} \\ y - z = 0 \end{cases}$$

$$\begin{cases} x + y + z = 10 \\ 0.5y + z = 6 \\ -3z = -12 \qquad -2\text{Eq.2} + \text{Eq.3} \end{cases}$$

$z = 4$

$0.5y + 4 = 6 \implies y = 4$

$x + 4 + 4 = 10 \implies x = 2$

2 pounds of Vanilla coffee, 4 pounds of Hazelnut coffee, and 4 pounds of French Roast coffee are needed to obtain the desired mixture.

61. Let $x =$ number of television ads.

Let $y =$ number of radio ads.

Let $z =$ number of local newspaper ads.

$$\begin{cases} x + y + z = 60 \\ 1000x + 200y + 500z = 42,000 \\ x - y - z = 0 \end{cases}$$

$$\begin{cases} x + y + z = 60 \\ -800y - 500z = -18,000 \qquad -1000\text{Eq.1} + \text{Eq.2} \\ -2y - 2z = -60 \qquad -\text{Eq.1} + \text{Eq.3} \end{cases}$$

$$\begin{cases} x + y + z = 60 \\ -2y - 2z = -60 \qquad \text{Interchange} \\ -800y - 500z = -18,000 \qquad \text{Eq.2 and Eq.3.} \end{cases}$$

$$\begin{cases} x + y + z = 60 \\ -2y - 2z = -60 \\ 300z = 6000 \qquad -400\text{Eq.2} + \text{Eq.3} \end{cases}$$

$z = 20$

$-2y - 2(20) = -60 \implies y = 10$

$x + 10 + 20 = 60 \implies x = 30$

30 television ads, 10 radio ads, and 20 newspaper ads can be run each month.

63. (a) To use 2 liters of the 50% solution:

Let $x =$ amount of 10% solution.

Let $y =$ amount of 20% solution.

$x + y = 8 \implies y = 8 - x$

$x(0.10) + y(0.20) + 2(0.50) = 10(0.25)$

$0.10x + 0.20(8 - x) + 1 = 2.5$

$0.10x + 1.6 - 0.20x + 1 = 2.5$

$-0.10x = -0.1$

$x = 1$ liter of 10% solution

$y = 7$ liters of 20% solution

Given: 2 liters of 50% solution

(b) To use as little of the 50% solution as possible, the chemist should use no 10% solution.

Let $x =$ amount of 20% solution.

Let $y =$ amount of 50% solution.

$x + y = 10 \implies y = 10 - x$

$x(0.20) + y(0.50) = 10(0.25)$

$x(0.20) + (10 - x)(0.50) = 10(0.25)$

$x(0.20) + 5 - 0.50x = 2.5$

$-0.30x = -2.5$

$x = 8\frac{1}{3}$ liters of 20% solution

$y = 1\frac{2}{3}$ liters of 50% solution

(c) To use as much of the 50% solution as possible, the chemist should use no 20% solution.

Let $x =$ amount of 10% solution.

Let $y =$ amount of 50% solution.

$x + y = 10 \implies y = 10 - x$

$x(0.10) + y(0.50) = 10(0.25)$

$0.10x + 0.50(10 - x) = 2.5$

$0.10x + 5 - 0.50x = 2.5$

$-0.40x = -2.5$

$x = 6\frac{1}{4}$ liters of 10% solution

$y = 3\frac{3}{4}$ liters of 50% solution

65. $\begin{cases} I_1 - I_2 + I_3 = 0 & \text{Equation 1} \\ 3I_1 + 2I_2 = 7 & \text{Equation 2} \\ 2I_2 + 4I_3 = 8 & \text{Equation 3} \end{cases}$

$\begin{cases} I_1 - I_2 + I_3 = 0 \\ 5I_2 - 3I_3 = 7 & (-3)\text{Eq.1} + \text{Eq.2} \\ 2I_2 + 4I_3 = 8 \end{cases}$

$\begin{cases} I_1 - I_2 + I_3 = 0 \\ 10I_2 - 6I_3 = 14 & 2\text{Eq.2} \\ 10I_2 + 20I_3 = 40 & 5\text{Eq.3} \end{cases}$

$\begin{cases} I_1 - I_2 + I_3 = 0 \\ 10I_2 - 6I_3 = 14 \\ 26I_3 = 26 & (-1)\text{Eq.2} + \text{Eq.3} \end{cases}$

$26I_3 = 26 \implies I_3 = 1$

$10I_2 - 6(1) = 14 \implies I_2 = 2$

$I_1 - 2 + 1 = 0 \implies I_1 = 1$

Solution: $I_1 = 1, I_2 = 2, I_3 = 1$

67. $(-4, 5), (-2, 6), (2, 6), (4, 2)$

$n = 4, \sum_{i=1}^{4} x_i = 0, \sum_{i=1}^{4} x_i^2 = 40, \sum_{i=1}^{4} x_i^3 = 0, \sum_{i=1}^{4} x_i^4 = 544, \sum_{i=1}^{4} y_i = 19, \sum_{i=1}^{4} x_i y_i = -12, \sum_{i=1}^{4} x_i^2 y_i = 160$

$\begin{cases} 4c + 40a = 19 \\ 40b = -12 \\ 40c + 544a = 160 \end{cases}$

$\begin{cases} 4c + 40a = 19 \\ 40b = -12 \\ 144a = -30 & -10\text{Eq.1} + \text{Eq.3} \end{cases}$

$144a = -30 \implies a = -\frac{5}{24}$

$40b = -12 \implies b = -\frac{3}{10}$

$4c + 40\left(-\frac{5}{24}\right) = 19 \implies c = \frac{41}{6}$

Least squares regression parabola: $y = -\frac{5}{24}x^2 - \frac{3}{10}x + \frac{41}{6}$

69. $(0, 0), (2, 2), (3, 6), (4, 12)$

$n = 4, \sum_{i=1}^{4} x_i = 9, \sum_{i=1}^{4} x_i^2 = 29, \sum_{i=1}^{4} x_i^3 = 99, \sum_{i=1}^{4} x_i^4 = 353, \sum_{i=1}^{4} y_i = 20, \sum_{i=1}^{4} x_i y_i = 70, \sum_{i=1}^{4} x_i^2 y_i = 254$

$\begin{cases} 4c + 9b + 29a = 20 \\ 9c + 29b + 99a = 70 \\ 29c + 99b + 353a = 254 \end{cases}$

$\begin{cases} 9c + 29b + 99a = 70 & \text{Interchange equations.} \\ 4c + 9b + 29a = 20 \\ 29c + 99b + 353a = 254 \end{cases}$

$\begin{cases} c + 11b + 41a = 30 & -2\text{Eq.2} + \text{Eq.1} \\ -35b - 135a = -100 & -4\text{Eq.1} + \text{Eq.2} \\ -220b - 836a = -616 & -29\text{Eq.1} + \text{Eq.3} \end{cases}$

$\begin{cases} c + 11b + 41a = 30 \\ 1540b + 5940a = 4400 & -44\text{Eq.2} \\ -1540b - 5852a = -4312 & 7\text{Eq.3} \end{cases}$

—CONTINUED—

69. —CONTINUED—

$$\begin{cases} c + \quad 11b + \quad 41a = \quad 30 \\ \qquad\quad 1540b + 5940a = \quad 4400 \\ \qquad\qquad\qquad\quad 88a = \qquad 88 \quad \text{Eq.2} + \text{Eq.3} \end{cases}$$

$$88a = \quad 88 \implies a = \quad 1$$

$$1540b + 5940(1) = 4400 \implies b = -1$$

$$c + 11(-1) + 41(1) = \quad 30 \implies c = \quad 0$$

Least squares regression parabola: $y = x^2 - x$

71. (a) $(100, 75)$, $(120, 68)$, $(140, 55)$

$$n = 3, \sum_{i=1}^{3} x_i = 360, \sum_{i=1}^{3} x_i^2 = 44{,}000, \sum_{i=1}^{3} x_i^3 = 5{,}472{,}000$$

$$\sum_{i=1}^{3} x_i^4 = 691{,}520{,}000, \sum_{i=1}^{3} y_i = 198, \sum_{i=1}^{3} x_i y_i = 23{,}360,$$

$$\sum_{i=1}^{3} x_i^2 y_i = 2{,}807{,}200$$

$$\begin{cases} 3c + \qquad 360b + \qquad 44{,}000a = \qquad 198 \\ 360c + \quad 44{,}000b + \quad 5{,}472{,}000a = \quad 23{,}360 \\ 44{,}000c + 5{,}472{,}000b + 691{,}520{,}000a = 2{,}807{,}200 \end{cases}$$

Solving this system yields $a = -0.0075$, $b = 1.3$
and $c = 20$.

Least squares regression parabola:
$y = -0.0075x^2 + 1.3x + 20$

(b)

(c)

x	Actual Percent y	Model Approximation y
100	75	75
120	68	68
140	55	55

The model is a good fit to the actual data.
The values are the same.

(d) For $x = 170$:

$$y = -0.0075(170)^2 + 1.3(170) + 20$$

$$= 24.25\%$$

(e) For $y = 40$:

$$40 = -0.0075x^2 + 1.3x + 20$$

$$0.0075x^2 - 1.3x + 20 = 0$$

By the Quadratic Formula we have $x \approx 17$ or $x \approx 156$.

Choosing the value that fits with our data, we have
156 females.

73. Let $x = $ number of touchdowns.

Let $y = $ number of extra-point kicks.

Let $z = $ number of two-point conversions.

Let $w = $ number of field goals.

$$\begin{cases} x + y + \quad z + \quad w = 16 \\ 6x + y + 2z + 3w = 32 + 29 \\ x \qquad\qquad - 4w = 0 \quad \implies x = 4w \\ \qquad\quad 2z - \quad w = 0 \quad \implies z = \tfrac{1}{2}w \end{cases}$$

$$\begin{cases} 4w + y + \quad \tfrac{1}{2}w + \quad w = 16 \implies 5.5w + y = 16 \\ 6(4w) + y + 2\left(\tfrac{1}{2}\right)w + 3w = 61 \implies 28w + y = 61 \end{cases}$$

$$28w + y = 61$$
$$\underline{-5.5w - y = -16}$$
$$22.5w \qquad = 45$$
$$w = 2$$
$$y = 5$$
$$x = 4w = 8$$
$$z = \tfrac{1}{2}w = 1$$

Thus, 8 touchdowns, 5 extra-point kicks, 1 two-point
conversion, and 2 field goals were scored.

75. $$\begin{cases} \qquad y + \lambda = 0 \\ x \qquad + \lambda = 0 \\ x + y - 10 = 0 \end{cases} \begin{matrix} \implies x = y = -\lambda \\ \implies 2x - 10 = \quad 0 \\ x = \quad 5 \\ y = \quad 5 \\ \lambda = -5 \end{matrix}$$

77. $\begin{cases} 2x - 2x\lambda = 0 \implies 2x(1 - \lambda) = 0 \implies \lambda = 1 \text{ or } x = 0 \\ -2y + \quad \lambda = 0 \\ \quad y - \quad x^2 = 0 \end{cases}$

If $\lambda = 1$:

$2y = \lambda \implies y = \dfrac{1}{2}$

$x^2 = y \implies x = \pm\sqrt{\dfrac{1}{2}} = \pm\dfrac{\sqrt{2}}{2}$

If $x = 0$:

$x^2 = y \implies y = 0$

$2y = \lambda \implies \lambda = 0$

Solution: $x = \pm\dfrac{\sqrt{2}}{2}$ or $x = 0$

$\qquad\qquad y = \dfrac{1}{2} \qquad\qquad y = 0$

$\qquad\qquad \lambda = 1 \qquad\qquad \lambda = 0$

79. False. Equation 2 does not have a leading coefficient of 1.

81. No, they are not equivalent. There are two arithmetic errors. The constant in the second equation should be -11 and the coefficient of z in the third equation should be 2.

83. There are an infinite number of linear systems that have $(4, -1, 2)$ as their solution. Two such systems are as follows:

$\begin{cases} 3x + y - z = 9 \\ x + 2y - z = 0 \\ -x + y + 3z = 1 \end{cases}$ $\begin{cases} x + y + z = 5 \\ x \quad\quad - 2z = 0 \\ \quad 2y + z = 0 \end{cases}$

85. There are an infinite number of linear systems that have $\left(3, -\dfrac{1}{2}, \dfrac{7}{4}\right)$ as their solution. Two such systems are as follows:

$\begin{cases} x + 2y - 4z = -5 \\ -x - 4y + 8z = 13 \\ x + 6y + 4z = 7 \end{cases}$ $\begin{cases} x + 2y + 4z = 9 \\ \quad y + 2z = 3 \\ x \quad\quad - 4z = -4 \end{cases}$

87. $(0.075)(85) = 6.375$

89. $(0.005)n = 400$

$\qquad n = 80{,}000$

91. $(7 - i) + (4 + 2i) = (7 + 4) + (-i + 2i) = 11 + i$

93. $(4 - i)(5 + 2i) = 20 + 8i - 5i - 2i^2 = 20 + 3i + 2$

$\qquad\qquad\qquad\qquad\qquad\quad = 22 + 3i$

95. $\dfrac{i}{1 + i} + \dfrac{6}{1 - i} = \dfrac{i(1 - i) + 6(1 + i)}{(1 + i)(1 - i)}$

$\qquad\qquad\qquad = \dfrac{i - i^2 + 6 + 6i}{1 - i^2}$

$\qquad\qquad\qquad = \dfrac{7 + 7i}{2}$

$\qquad\qquad\qquad = \dfrac{7}{2} + \dfrac{7}{2}i$

97. $f(x) = x^3 + x^2 - 12x$

(a) $x^3 + x^2 - 12x = 0$

$\qquad x(x^2 + x - 12) = 0$

$\qquad x(x + 4)(x - 3) = 0$

Zeros: $x = -4, 0, 3$

(b)

99. $f(x) = 2x^3 + 5x^2 - 21x - 36$

(a) $2x^3 + 5x^2 - 21x - 36 = 0$

$$
\begin{array}{r|rrrr}
3 & 2 & 5 & -21 & -36 \\
  &   & 6 & 33 & 36 \\
\hline
  & 2 & 11 & 12 & 0
\end{array}
$$

$f(x) = (x - 3)(2x^2 + 11x + 12)$

$ = (x - 3)(x + 4)(2x + 3)$

Zeros: $x = -4, -\frac{3}{2}, 3$

(b)

101. $y = 4^{x-4} - 5$

x	-2	0	2	4	5
y	-4.9998	-4.996	-4.938	-4	-1

Horizontal asymptote: $y = -5$

103. $y = 1.9^{-0.8x} + 3$

x	-2	-1	0	1	2
y	5.793	4.671	4	3.598	3.358

Horizontal asymptote: $y = 3$

105. $\begin{cases} 2x + y = 120 & \text{Equation 1} \\ x + 2y = 120 & \text{Equation 2} \end{cases}$

$$
\begin{array}{rl}
2x + y = & 120 \\
-2x - 4y = & -240 \qquad (-2)\text{Eq.2} \\
\hline
-3y = & -120 \\
y = & 40
\end{array}
$$

$x + 2(40) = 120 \implies x = 40$

Solution: $(40, 40)$

107. Answers will vary.

Section 9.4   Partial Fractions

- ■   You should know how to decompose a rational function $\dfrac{N(x)}{D(x)}$ into partial fractions.

  (a) If the fraction is improper, divide to obtain

  $$\frac{N(x)}{D(x)} = p(x) + \frac{N_1(x)}{D(x)}$$

  where $p(x)$ is a polynomial.

  (b) Factor the denominator completely into linear and irreducible quadratic factors.

  (c) For each factor of the form $(px + q)^m$, the partial fraction decomposition includes the terms

  $$\frac{A_1}{(px + q)} + \frac{A_2}{(px + q)^2} + \cdots + \frac{A_m}{(px + q)^m}.$$

  (d) For each factor of the form $(ax^2 + bx + c)^n$, the partial fraction decomposition includes the terms

  $$\frac{B_1 x + C_1}{ax^2 + bx + c} + \frac{B_2 x + C_2}{(ax^2 + bx + c)^2} + \cdots + \frac{B_n x + C_n}{(ax^2 + bx + c)^n}.$$

- ■   You should know how to determine the values of the constants in the numerators.

  (a) Set $\dfrac{N_1(x)}{D(x)} = $ partial fraction decomposition.

  (b) Multiply both sides by $D(x)$ to obtain the basic equation.

  (c) For distinct linear factors, substitute the zeros of the distinct linear factors into the basic equation.

  (d) For repeated linear factors, use the coefficients found in part (c) to rewrite the basic equation. Then use other values of x to solve for the remaining coefficients.

  (e) For quadratic factors, expand the basic equation, collect like terms, and then equate the coefficients of like terms.

Vocabulary Check

1. partial fraction decomposition

2. improper

3. m; n; irreducible

4. basic equation

---

1. $\dfrac{3x - 1}{x(x - 4)} = \dfrac{A}{x} + \dfrac{B}{x - 4}$

Matches (b).

3. $\dfrac{3x - 1}{x(x^2 + 4)} = \dfrac{A}{x} + \dfrac{Bx + C}{x^2 + 4}$

Matches (d).

5. $\dfrac{7}{x^2 - 14x} = \dfrac{7}{x(x - 14)} = \dfrac{A}{x} + \dfrac{B}{x - 14}$

7. $\dfrac{12}{x^3 - 10x^2} = \dfrac{12}{x^2(x - 10)} = \dfrac{A}{x} + \dfrac{B}{x^2} + \dfrac{C}{x - 10}$

9. $\dfrac{4x^2 + 3}{(x - 5)^3} = \dfrac{A}{x - 5} + \dfrac{B}{(x - 5)^2} + \dfrac{C}{(x - 5)^3}$

11. $\dfrac{2x - 3}{x^3 + 10x} = \dfrac{2x - 3}{x(x^2 + 10)} = \dfrac{A}{x} + \dfrac{Bx + C}{x^2 + 10}$

13. $\dfrac{x - 1}{x(x^2 + 1)^2} = \dfrac{A}{x} + \dfrac{Bx + C}{x^2 + 1} + \dfrac{Dx + E}{(x^2 + 1)^2}$

15. $\dfrac{1}{x^2 - 1} = \dfrac{A}{x + 1} + \dfrac{B}{x - 1}$

$\qquad 1 = A(x - 1) + B(x + 1)$

$\qquad$ Let $x = -1$: $1 = -2A \implies A = -\dfrac{1}{2}$

$\qquad$ Let $x = 1$: $1 = 2B \implies B = \dfrac{1}{2}$

$\qquad \dfrac{1}{x^2 - 1} = \dfrac{1/2}{x - 1} - \dfrac{1/2}{x + 1} = \dfrac{1}{2}\left(\dfrac{1}{x - 1} - \dfrac{1}{x + 1}\right)$

17. $\dfrac{1}{x^2 + x} = \dfrac{A}{x} + \dfrac{B}{x + 1}$

$\qquad 1 = A(x + 1) + Bx$

$\qquad$ Let $x = 0$: $1 = A$

$\qquad$ Let $x = -1$: $1 = -B \implies B = -1$

$\qquad \dfrac{1}{x^2 + x} = \dfrac{1}{x} - \dfrac{1}{x + 1}$

19. $\dfrac{1}{2x^2 + x} = \dfrac{A}{2x + 1} + \dfrac{B}{x}$

$\qquad 1 = Ax + B(2x + 1)$

$\qquad$ Let $x = -\dfrac{1}{2}$: $1 = -\dfrac{1}{2}A \implies A = -2$

$\qquad$ Let $x = 0$: $1 = B$

$\qquad \dfrac{1}{2x^2 + x} = \dfrac{1}{x} - \dfrac{2}{2x + 1}$

21. $\dfrac{3}{x^2 + x - 2} = \dfrac{A}{x - 1} + \dfrac{B}{x + 2}$

$\qquad 3 = A(x + 2) + B(x - 1)$

$\qquad$ Let $x = 1$: $3 = 3A \implies A = 1$

$\qquad$ Let $x = -2$: $3 = -3B \implies B = -1$

$\qquad \dfrac{3}{x^2 + x - 2} = \dfrac{1}{x - 1} - \dfrac{1}{x + 2}$

23. $\dfrac{x^2 + 12x + 12}{x^3 - 4x} = \dfrac{A}{x} + \dfrac{B}{x + 2} + \dfrac{C}{x - 2}$

$x^2 + 12x + 12 = A(x + 2)(x - 2) + Bx(x - 2) + Cx(x + 2)$

Let $x = 0$: $12 = -4A \implies A = -3$

Let $x = -2$: $-8 = 8B \implies B = -1$

Let $x = 2$: $40 = 8C \implies C = 5$

$\dfrac{x^2 + 12x + 12}{x^3 - 4x} = -\dfrac{3}{x} - \dfrac{1}{x + 2} + \dfrac{5}{x - 2}$

25. $\dfrac{4x^2 + 2x - 1}{x^2(x + 1)} = \dfrac{A}{x} + \dfrac{B}{x^2} + \dfrac{C}{x + 1}$

$4x^2 + 2x - 1 = Ax(x + 1) + B(x + 1) + Cx^2$

Let $x = 0$: $-1 = B$

Let $x = -1$: $1 = C$

Let $x = 1$: $5 = 2A + 2B + C$

$\qquad\qquad\quad 5 = 2A - 2 + 1$

$\qquad\qquad\quad 6 = 2A$

$\qquad\qquad\quad 3 = A$

$\dfrac{4x^2 + 2x - 1}{x^2(x + 1)} = \dfrac{3}{x} - \dfrac{1}{x^2} + \dfrac{1}{x + 1}$

27. $\dfrac{3x}{(x - 3)^2} = \dfrac{A}{x - 3} + \dfrac{B}{(x - 3)^2}$

$\qquad 3x = A(x - 3) + B$

$\qquad$ Let $x = 3$: $9 = B$

$\qquad$ Let $x = 0$: $0 = -3A + B$

$\qquad\qquad\qquad\; 0 = -3A + 9$

$\qquad\qquad\qquad\; 3 = A$

$\qquad \dfrac{3x}{(x - 3)^2} = \dfrac{3}{x - 3} + \dfrac{9}{(x - 3)^2}$

29. $\dfrac{x^2-1}{x(x^2+1)} = \dfrac{A}{x} + \dfrac{Bx+C}{x^2+1}$

$\quad x^2 - 1 = A(x^2+1) + (Bx+C)x$

$\qquad = Ax^2 + A + Bx^2 + Cx$

$\qquad = (A+B)x^2 + Cx + A$

Equating coefficients of like terms gives
$1 = A + B, 0 = C,$ and $-1 = A.$

Therefore, $A = -1, B = 2,$ and $C = 0.$

$\dfrac{x^2-1}{x(x^2+1)} = -\dfrac{1}{x} + \dfrac{2x}{x^2+1}$

31. $\dfrac{x}{x^3 - x^2 - 2x + 2} = \dfrac{x}{(x-1)(x^2-2)} = \dfrac{A}{x-1} + \dfrac{Bx+C}{x^2-2}$

$\quad x = A(x^2-2) + (Bx+C)(x-1)$

$\qquad = Ax^2 - 2A + Bx^2 - Bx + Cx - C$

$\qquad = (A+B)x^2 + (C-B)x - (2A+C)$

Equating coefficients of like terms gives
$0 = A + B, 1 = C - B,$ and $0 = 2A + C.$

Therefore, $A = -1, B = 1,$ and $C = 2.$

$\dfrac{x}{x^3 - x^2 - 2x + 2} = \dfrac{-1}{x-1} + \dfrac{x+2}{x^2-2}$

33. $\dfrac{x^2}{x^4 - 2x^2 - 8} = \dfrac{x^2}{(x^2-4)(x^2+2)} = \dfrac{x^2}{(x+2)(x-2)(x^2+2)}$

$\qquad = \dfrac{A}{x+2} + \dfrac{B}{x-2} + \dfrac{Cx+D}{x^2+2}$

$\quad x^2 = A(x-2)(x^2+2) + B(x+2)(x^2+2) + (Cx+D)(x+2)(x-2)$

$\qquad = A(x^3 - 2x^2 + 2x - 4) + B(x^3 + 2x^2 + 2x + 4) + (Cx+D)(x^2-4)$

$\qquad = Ax^3 - 2Ax^2 + 2Ax - 4A + Bx^3 + 2Bx^2 + 2Bx + 4B + Cx^3 + Dx^2 - 4Cx - 4D$

$\qquad = (A+B+C)x^3 + (-2A+2B+D)x^2 + (2A+2B-4C)x + (-4A+4B-4D)$

Equating coefficients of like terms gives

$0 = A+B+C, 1 = -2A+2B+D, 0 = 2A+2B-4C,$ and $0 = -4A+4B-4D.$

Using the first and third equation, we have $A + B + C = 0$ and $A + B - 2C = 0$;
by subtraction, $C = 0.$ Using the second and fourth equation, we have $-2A + 2B + D = 1$
and $-2A + 2B - 2D = 0$; by subtraction, $3D = 1,$ so $D = \frac{1}{3}.$ Substituting 0 for C and $\frac{1}{3}$ for
D in the first and second equations, we have

$A + B = 0$ and $-2A + 2B = \frac{2}{3},$ so $A = -\frac{1}{6}$ and $B = \frac{1}{6}.$

$\dfrac{x^2}{x^4 - 2x^2 - 8} = \dfrac{-\frac{1}{6}}{x+2} + \dfrac{\frac{1}{6}}{x-2} + \dfrac{\frac{1}{3}}{x^2+2}$

$\qquad = \dfrac{1}{3(x^2+2)} - \dfrac{1}{6(x+2)} + \dfrac{1}{6(x-2)}$

$\qquad = \dfrac{1}{6}\left(\dfrac{2}{x^2+2} - \dfrac{1}{x+2} + \dfrac{1}{x-2}\right)$

35. $\dfrac{x}{16x^4 - 1} = \dfrac{x}{(4x^2-1)(4x^2+1)} = \dfrac{x}{(2x+1)(2x-1)(4x^2+1)}$

$\qquad = \dfrac{A}{2x+1} + \dfrac{B}{2x-1} + \dfrac{Cx+D}{4x^2+1}$

$\quad x = A(2x-1)(4x^2+1) + B(2x+1)(4x^2+1) + (Cx+D)(2x+1)(2x-1)$

$\qquad = A(8x^3 - 4x^2 + 2x - 1) + B(8x^3 + 4x^2 + 2x + 1) + (Cx+D)(4x^2-1)$

$\qquad = 8Ax^3 - 4Ax^2 + 2Ax - A + 8Bx^3 + 4Bx^2 + 2Bx + B + 4Cx^3 + 4Dx^2 - Cx - D$

$\qquad = (8A+8B+4C)x^3 + (-4A+4B+4D)x^2 + (2A+2B-C)x + (-A+B-D)$

—CONTINUED—

35. **—CONTINUED—**

Equating coefficients of like terms gives $0 = 8A + 8B + 4C$, $0 = -4A + 4B + 4D$, $1 = 2A + 2B - C$, and $0 = -A + B - D$.

Using the first and third equations, we have $2A + 2B + C = 0$ and $2A + 2B - C = 1$; by subtraction, $2C = -1$, so $C = -\frac{1}{2}$.

Using the second and fourth equations, we have $-A + B + D = 0$ and $-A + B - D = 0$; by subtraction $2D = 0$, so $D = 0$.

Substituting $-\frac{1}{2}$ for C and 0 for D in the first and second equations, we have $8A + 8B = 2$ and $-4A + 4B = 0$, so $A = \frac{1}{8}$ and $B = \frac{1}{8}$.

$$\frac{x}{16x^4 - 1} = \frac{\frac{1}{8}}{2x + 1} + \frac{\frac{1}{8}}{2x - 1} + \frac{\left(-\frac{1}{2}\right)x}{4x^2 + 1}$$

$$= \frac{1}{8(2x + 1)} + \frac{1}{8(2x - 1)} - \frac{x}{2(4x^2 + 1)}$$

$$= \frac{1}{8}\left(\frac{1}{2x + 1} + \frac{1}{2x - 1} - \frac{4x}{4x^2 + 1}\right)$$

37. $$\frac{x^2 + 5}{(x + 1)(x^2 - 2x + 3)} = \frac{A}{x + 1} + \frac{Bx + C}{x^2 - 2x + 3}$$

$$x^2 + 5 = A(x^2 - 2x + 3) + (Bx + C)(x + 1)$$

$$= Ax^2 - 2Ax + 3A + Bx^2 + Bx + Cx + C$$

$$= (A + B)x^2 + (-2A + B + C)x + (3A + C)$$

Equating coefficients of like terms gives $1 = A + B$, $0 = -2A + B + C$, and $5 = 3A + C$.

Subtracting both sides of the second equation from the first gives $1 = 3A - C$; combining this with the third equation gives $A = 1$ and $C = 2$. Since $A + B = 1$, we also have $B = 0$.

$$\frac{x^2 + 5}{(x + 1)(x^2 - 2x + 3)} = \frac{1}{x + 1} + \frac{2}{x^2 - 2x + 3}$$

39. $$\frac{x^2 - x}{x^2 + x + 1} = 1 + \frac{-2x - 1}{x^2 + x + 1} = 1 - \frac{2x + 1}{x^2 + x + 1}$$

41. $$\frac{2x^3 - x^2 + x + 5}{x^2 + 3x + 2} = 2x - 7 + \frac{18x + 19}{(x + 1)(x + 2)}$$

$$\frac{18x + 19}{(x + 1)(x + 2)} = \frac{A}{x + 1} + \frac{B}{x + 2}$$

$$18x + 19 = A(x + 2) + B(x + 1)$$

Let $x = -1$: $1 = A$

Let $x = -2$: $-17 = -B \implies B = 17$

$$\frac{2x^3 - x^2 + x + 5}{x^2 + 3x + 2} = 2x - 7 + \frac{1}{x + 1} + \frac{17}{x + 2}$$

43.
$$\frac{x^4}{(x-1)^3} = \frac{x^4}{x^3 - 3x^2 + 3x - 1} = x + 3 + \frac{6x^2 - 8x + 3}{(x-1)^3}$$

$$\frac{6x^2 - 8x + 3}{(x-1)^3} = \frac{A}{x-1} + \frac{B}{(x-1)^2} + \frac{C}{(x-1)^3}$$

$$6x^2 - 8x + 3 = A(x-1)^2 + B(x-1) + C$$

Let $x = 1$: $1 = C$

Let $x = 0$: $3 = A - B + 1$ $\left.\begin{array}{l} \end{array}\right\}$ $A - B = 2$

Let $x = 2$: $11 = A + B + 1$ $\left.\begin{array}{l} \end{array}\right\}$ $A + B = 10$

So, $A = 6$ and $B = 4$.

$$\frac{x^4}{(x-1)^3} = x + 3 + \frac{6}{x-1} + \frac{4}{(x-1)^2} + \frac{1}{(x-1)^3}$$

45. $\dfrac{5-x}{2x^2 + x - 1} = \dfrac{A}{2x-1} + \dfrac{B}{x+1}$

$$-x + 5 = A(x+1) + B(2x-1)$$

Let $x = \dfrac{1}{2}$: $\dfrac{9}{2} = \dfrac{3}{2}A \implies A = 3$

Let $x = -1$: $6 = -3B \implies B = -2$

$$\frac{5-x}{2x^2 + x - 1} = \frac{3}{2x-1} - \frac{2}{x+1}$$

47. $\dfrac{x-1}{x^3 + x^2} = \dfrac{A}{x} + \dfrac{B}{x^2} + \dfrac{C}{x+1}$

$$x - 1 = Ax(x+1) + B(x+1) + Cx^2$$

Let $x = -1$: $-2 = C$

Let $x = 0$: $-1 = B$

Let $x = 1$: $0 = 2A + 2B + C$

$$0 = 2A - 2 - 2$$

$$2 = A$$

$$\frac{x-1}{x^3 + x^2} = \frac{2}{x} - \frac{1}{x^2} - \frac{2}{x+1}$$

49. $\dfrac{x^2 + x + 2}{(x^2 + 2)^2} = \dfrac{Ax + B}{x^2 + 2} + \dfrac{Cx + D}{(x^2 + 2)^2}$

$$x^2 + x + 2 = (Ax + B)(x^2 + 2) + Cx + D$$

$$x^2 + x + 2 = Ax^3 + Bx^2 + (2A + C)x + (2B + D)$$

Equating coefficients of like powers:

$$0 = A$$

$$1 = B$$

$$1 = 2A + C \implies C = 1$$

$$2 = 2B + D \implies D = 0$$

$$\frac{x^2 + x + 2}{(x^2 + 2)^2} = \frac{1}{x^2 + 2} + \frac{x}{(x^2 + 2)^2}$$

51. $\dfrac{2x^3 - 4x^2 - 15x + 5}{x^2 - 2x - 8} = 2x + \dfrac{x+5}{(x+2)(x-4)}$

$$\frac{x+5}{(x+2)(x-4)} = \frac{A}{x+2} + \frac{B}{x-4}$$

$$x + 5 = A(x-4) + B(x+2)$$

Let $x = -2$: $3 = -6A \implies A = -\dfrac{1}{2}$

Let $x = 4$: $9 = 6B \implies B = \dfrac{3}{2}$

$$\frac{2x^3 - 4x^2 - 15x + 5}{x^2 - 2x - 8} = 2x + \frac{1}{2}\left(\frac{3}{x-4} - \frac{1}{x+2}\right)$$

53. (a) $\dfrac{x-12}{x(x-4)} = \dfrac{A}{x} + \dfrac{B}{x-4}$

$\qquad x - 12 = A(x-4) + Bx$

Let $x = 0$: $-12 = -4A \implies A = 3$

Let $x = 4$: $-8 = 4B \implies B = -2$

$\dfrac{x-12}{x(x-4)} = \dfrac{3}{x} - \dfrac{2}{x-4}$

(b) $y = \dfrac{x-12}{x(x-4)}$ $\qquad\qquad y = \dfrac{3}{x}$ $\qquad\qquad y = -\dfrac{2}{x-4}$

Vertical asymptotes: $x = 0$ and $x = 4$ $\qquad$ Vertical asymptote: $x = 0$ $\qquad$ Vertical asymptote: $x = 4$

(c) The combination of the vertical asymptotes of the terms of the decomposition are the same as the vertical asymptotes of the rational function.

55. (a) $\dfrac{2(4x-3)}{x^2-9} = \dfrac{A}{x-3} + \dfrac{B}{x+3}$

$\qquad 2(4x-3) = A(x+3) + B(x-3)$

Let $x = 3$: $18 = 6A \implies A = 3$

Let $x = -3$: $-30 = -6B \implies B = 5$

$\dfrac{2(4x-3)}{x^2-9} = \dfrac{3}{x-3} + \dfrac{5}{x+3}$

(b) $y = \dfrac{2(4x-3)}{x^2-9}$ $\qquad\qquad y = \dfrac{3}{x-3}$ $\qquad\qquad y = \dfrac{5}{x+3}$

Vertical asymptotes: $x = \pm 3$ $\qquad$ Vertical asymptote: $x = 3$ $\qquad$ Vertical asymptote: $x = -3$

(c) The combination of the vertical asymptotes of the terms of the decomposition are the same as the vertical asymptotes of the rational function.

57. (a) $\dfrac{2000(4 - 3x)}{(11 - 7x)(7 - 4x)} = \dfrac{A}{11 - 7x} + \dfrac{B}{7 - 4x}$, $0 < x \leq 1$

$2000(4 - 3x) = A(7 - 4x) + B(11 - 7x)$

Let $x = \dfrac{11}{7}$: $-\dfrac{10,000}{7} = \dfrac{5}{7}A \implies A = -2000$

Let $x = \dfrac{7}{4}$: $-2500 = -\dfrac{5}{4}B \implies B = 2000$

$\dfrac{2000(4 - 3x)}{(11 - 7x)(7 - 4x)} = \dfrac{-2000}{11 - 7x} + \dfrac{2000}{7 - 4x}$

$= \dfrac{2000}{7 - 4x} - \dfrac{2000}{11 - 7x}$, $0 < x \leq 1$

(b). $Y_{\max} = \left| \dfrac{2000}{7 - 4x} \right|$

$Y_{\min} = \left| \dfrac{2000}{11 - 7x} \right|$

(c)

59. False. The partial fraction decomposition is $\dfrac{A}{x + 10} + \dfrac{B}{x - 10} + \dfrac{C}{(x - 10)^2}$.

61. $\dfrac{1}{a^2 - x^2} = \dfrac{A}{a + x} + \dfrac{B}{a - x}$, a is a constant.

$1 = A(a - x) + B(a + x)$

Let $x = -a$: $1 = 2aA \implies A = \dfrac{1}{2a}$

Let $x = a$: $1 = 2aB \implies B = \dfrac{1}{2a}$

$\dfrac{1}{a^2 - x^2} = \dfrac{1}{2a}\left(\dfrac{1}{a + x} + \dfrac{1}{a - x} \right)$

63. $\dfrac{1}{y(a - y)} = \dfrac{A}{y} + \dfrac{B}{a - y}$

$1 = A(a - y) + By$

Let $y = 0$: $1 = aA \implies A = \dfrac{1}{a}$

Let $y = a$: $1 = aB \implies B = \dfrac{1}{a}$

$\dfrac{1}{y(a - y)} = \dfrac{1}{a}\left(\dfrac{1}{y} + \dfrac{1}{a - y} \right)$

65. $f(x) = x^2 - 9x + 18 = (x - 6)(x - 3)$

Intercepts: $(0, 18)$, $(3, 0)$, $(6, 0)$

Graph rises to the left and rises to the right.

67. $f(x) = -x^2(x - 3)$

Intercepts: $(0, 0)$, $(3, 0)$

Graph rises to the left and falls to the right.

69. $f(x) = \dfrac{x^2 + x - 6}{x + 5}$

x-intercepts: $(-3, 0)$, $(2, 0)$

y-intercept: $\left(0, -\dfrac{6}{5}\right)$

Vertical asymptote: $x = -5$

Slant asymptote: $y = x - 4$

No horizontal asymptote.

Section 9.5   Systems of Inequalities

■ You should be able to sketch the graph of an inequality in two variables.

    (a) Replace the inequality with an equal sign and graph the equation. Use a dashed line for < or >, a solid line for ≤ or ≥.

    (b) Test a point in each region formed by the graph. If the point satisfies the inequality, shade the whole region.

■ You should be able to sketch systems of inequalities.

Vocabulary Check

1. solution   **2.** graph

3. linear   **4.** solution

5. consumer surplus

---

1. $y < 2 - x^2$

Using a dashed line, graph $y = 2 - x^2$ and shade inside the parabola.

3. $x \geq 2$

Using a solid line, graph the vertical line $x = 2$ and shade to the right of this line.

5. $y \geq -1$

Using a solid line, graph the horizontal line $y = -1$ and shade above this line.

7. $y < 2 - x$

Using a dashed line, graph $y = 2 - x$, and then shade below the line. (Use $(0, 0)$ as a test point.)

9. $2y - x \geq 4$

Using a solid line, graph $2y - x = 4$, and then shade above the line. (Use $(0, 0)$ as a test point.)

11. $(x + 1)^2 + (y - 2)^2 < 9$

Using a dashed line, sketch the circle $(x + 1)^2 + (y - 2)^2 = 9$.

Center: $(-1, 2)$

Radius: 3

Test point: $(0, 0)$

Shade the inside of the circle.

13. $y \leq \dfrac{1}{1 + x^2}$

Using a solid line, graph

$y = \dfrac{1}{1 + x^2}$, and then shade below the curve. (Use $(0, 0)$ as a test point.)

15. $y < \ln x$

17. $y < 3^{-x-4}$

19. $y \geq \frac{2}{3}x - 1$

21. $y \leq -3.8x + 1.1$

23. $x^2 + 5y - 10 \leq 0$

$y \leq 2 - \dfrac{x^2}{5}$

25. $\dfrac{5}{2}y - 3x^2 - 6 \geq 0$

$y \geq \dfrac{2}{5}(3x^2 + 6)$

27. The line through $(-4, 0)$ and $(0, 2)$ is $y = \frac{1}{2}x + 2$. For the shaded region below the line, we have $y \leq \frac{1}{2}x + 2$.

29. The line through $(0, 2)$ and $(3, 0)$ is $y = -\frac{2}{3}x + 2$. For the shaded region above the line, we have $y \geq -\frac{2}{3}x + 2$.

31. $\begin{cases} x \geq -4 \\ y > -3 \\ y \leq -8x - 3 \end{cases}$

(a) $0 \leq -8(0) - 3$, False

   $(0, 0)$ *is not* a solution.

(b) $-3 > -3$, False

   $(-1, -3)$ *is not* a solution.

(c) $-4 \geq -4$, True

   $0 > -3$, True

   $0 \leq -8(-4) - 3$, True

   $(-4, 0)$ *is* a solution.

(d) $-3 \geq -4$, True

   $11 > -3$, True

   $11 < -8(-3) - 3$, True

   $(-3, 11)$ *is* a solution.

33. $\begin{cases} 3x + y > 1 \\ -y - \frac{1}{2}x^2 \le -4 \\ -15x + 4y > 0 \end{cases}$

(a) $3(0) + (10) > 1$, True

$-10 - \frac{1}{2}(0)^2 \le -4$, True

$-15(0) + 4(10) > 0$, True

$(0, 10)$ *is* a solution.

(b) $3(0) + (-1) > 1$, False $\Rightarrow$ $(0, -1)$ *is not* a solution.

(c) $3(2) + (9) > 1$, True

$-9 - \frac{1}{2}(2)^2 \le -4$, True

$-15(2) + 4(9) > 0$, True

$(2, 9)$ *is* a solution.

(d) $3(-1) + 6 > 1$, True

$-6 - \frac{1}{2}(-1)^2 \le -4$, True

$-15(-1) + 4(6) > 0$, True

$(-1, 6)$ *is* a solution.

35. $\begin{cases} x + y \le 1 \\ -x + y \le 1 \\ y \ge 0 \end{cases}$

First, find the points of intersection of each pair of equations.

Vertex A	Vertex B	Vertex C
$x + y = 1$	$x + y = 1$	$-x + y = 1$
$-x + y = 1$	$y = 0$	$y = 0$
$(0, 1)$	$(1, 0)$	$(-1, 0)$

37. $\begin{cases} x^2 + y \le 5 \\ x \ge -1 \\ y \ge 0 \end{cases}$

First, find the points of intersection of each pair of equations.

Vertex A	Vertex B	Vertex C
$x^2 + y = 5$	$x^2 + y = 5$	$x = -1$
$x = -1$	$y = 0$	$y = 0$
$(-1, 4)$	$(\pm\sqrt{5}, 0)$	$(-1, 0)$

39. $\begin{cases} 2x + y > 2 \\ 6x + 3y < 2 \end{cases}$

The graphs of $2x + y = 2$ and $6x + 3y = 2$ are parallel lines. The first inequality has the region above the line shaded. The second inequality has the region below the line shaded. There are no points that satisfy both inequalities.

No solution

41. $\begin{cases} -3x + 2y < 6 \\ x - 4y > -2 \\ 2x + y < 3 \end{cases}$

First, find the points of intersection of each pair of equations.

Vertex A	Point B	Vertex C
$-3x + 2y = 6$	$-3x + 2y = 6$	$x - 4y = -2$
$x - 4y = -2$	$2x + y = 3$	$2x + y = 3$
$(-2, 0)$	$(0, 3)$	$\left(\frac{10}{9}, \frac{7}{9}\right)$

Note that B is not a vertex of the solution region.

43. $\begin{cases} x > y^2 \\ x < y + 2 \end{cases}$

Points of intersection:

$$y^2 = y + 2$$
$$y^2 - y - 2 = 0$$
$$(y + 1)(y - 2) = 0$$
$$y = -1, 2$$

$(1, -1), (4, 2)$

45. $\begin{cases} x^2 + y^2 \leq 9 \\ x^2 + y^2 \geq 1 \end{cases}$

There are no points of intersection. The region common to both inequalities is the region between the circles.

47. $3x + 4 \geq y^2$

$\quad x - y < 0$

Points of intersection:

$x - y = 0 \Longrightarrow y = x$

$3y + 4 = y^2$

$0 = y^2 - 3y - 4$

$0 = (y - 4)(y + 1)$

$y = 4 \text{ or } y = -1$

$x = 4 \qquad x = -1$

$(4, 4) \text{ and } (-1, -1)$

49. $\begin{cases} y \leq \sqrt{3x} + 1 \\ y \geq \quad x^2 + 1 \end{cases}$

51. $\begin{cases} y < x^3 - 2x + 1 \\ y > -2x \\ x \leq 1 \end{cases}$

53. $\begin{cases} x^2 y \geq 1 \Longrightarrow y \geq \dfrac{1}{x^2} \\ 0 < x \leq 4 \\ \quad y \leq 4 \end{cases}$

55. $\begin{cases} y \leq 4 - x \\ x \geq 0 \\ y \geq 0 \end{cases}$

57. Line through points $(0, 4)$ and $(4, 0)$: $y = 4 - x$

Line through points $(0, 2)$ and $(8, 0)$: $y = 2 - \frac{1}{4}x$

$\begin{cases} y \geq 4 - x \\ y \geq 2 - \frac{1}{4}x \\ x \geq 0 \\ y \geq 0 \end{cases}$

59. $\begin{cases} x^2 + y^2 \leq 16 \\ x \geq \quad 0 \\ y \geq \quad 0 \end{cases}$

61. Rectangular region with vertices at $(2, 1), (5, 1), (5, 7),$ and $(2, 7)$

$\begin{cases} x \geq 2 \\ x \leq 5 \\ y \geq 1 \\ y \leq 7 \end{cases}$

This system may be written as:

$\begin{cases} 2 \leq x \leq 5 \\ 1 \leq y \leq 7 \end{cases}$

63. Triangle with vertices at $(0, 0), (5, 0), (2, 3)$

$(0, 0), (5, 0)$ Line: $y = 0$

$(0, 0), (2, 3)$ Line: $y = \frac{3}{2}x$

$(2, 3), (5, 0)$ Line: $y = -x + 5$

$\begin{cases} y \leq \quad \frac{3}{2}x \\ y \leq -x + 5 \\ y \geq \quad 0 \end{cases}$

65. (a)   Demand = Supply

$$50 - 0.5x = 0.125x$$

$$50 = 0.625x$$

$$80 = x$$

$$10 = p$$

Point of equilibrium: (80, 10)

(b) The consumer surplus is the area of the triangular region defined by

$$\begin{cases} p \le 50 - 0.5x \\ p \ge 10 \\ x \ge 0. \end{cases}$$

Consumer surplus $= \frac{1}{2}$(base)(height)

$$= \frac{1}{2}(80)(40)$$

$$= \$1600$$

The producer surplus is the area of the triangular region defined by

$$\begin{cases} p \ge 0.125x \\ p \le 10 \\ x \ge 0. \end{cases}$$

Producer surplus $= \frac{1}{2}$(base)(height)

$$= \frac{1}{2}(80)(10)$$

$$= \$400$$

69. x = number of tables

y = number of chairs

$$\begin{cases} x + \frac{3}{2}y \le 12 & \text{Assembly center} \\ \frac{4}{3}x + \frac{3}{2}y \le 15 & \text{Finishing center} \\ \quad x \ge 0 \\ \quad y \ge 0 \end{cases}$$

67. (a)   Demand = Supply

$$140 - 0.00002x = 80 + 0.00001x$$

$$60 = 0.00003x$$

$$2,000,000 = x$$

$$100 = p$$

Point of equilibrium: (2,000,000, 100)

(b) The consumer surplus is the area of the triangular region defined by

$$\begin{cases} p \le 140 - 0.00002x \\ p \ge 100 \\ x \ge 0. \end{cases}$$

Consumer surplus $= \frac{1}{2}$(base)(height)

$$= \frac{1}{2}(2,000,000)(40)$$

$$= \$40,000,000 \text{ or } \$40 \text{ million}$$

The producer surplus is the area of the triangular region defined by

$$\begin{cases} p \ge 80 + 0.00001x \\ p \le 100 \\ x \ge 0. \end{cases}$$

Producer surplus $= \frac{1}{2}$(base)(height)

$$= \frac{1}{2}(2,000,000)(20)$$

$$= \$20,000,000 \text{ or } \$20 \text{ million}$$

71. x = amount in smaller account

y = amount in larger account

Account constraints:

$$\begin{cases} x + y \le 20,000 \\ \quad y \ge 2x \\ \quad x \ge 5,000 \\ \quad y \ge 5,000 \end{cases}$$

73. x = number of packages of gravel

y = number of bags of stone

$$\begin{cases} 55x + 70y \le 7500 \quad \text{Weight} \\ x \ge 50 \\ y \ge 40 \end{cases}$$

75. (a) x = number of ounces of food X

y = number of ounces of food Y

$$\begin{cases} 20x + 10y \ge 300 \quad \text{(calcium)} \\ 15x + 10y \ge 150 \quad \text{(iron)} \\ 10x + 20y \ge 200 \quad \text{(vitamin B)} \\ x \ge 0 \\ y \ge 0 \end{cases}$$

(b)

(c) Answers will vary. Some possible solutions which would satisfy the minimum daily requirements for calcium, iron, and vitamin B:

$(0, 30) \implies 30$ ounces of food Y

$(20, 0) \implies 20$ ounces of food X

$\left(13\frac{1}{3}, 3\frac{1}{3}\right) \implies 13\frac{1}{3}$ ounces of food X and $3\frac{1}{3}$ ounces of food Y

77. (a) $(9, 125.8), (10, 145.6), (11, 164.1),$ $(12, 182.7), (13, 203.1)$

Linear model:
$y = 19.17t - 46.61$

(b)
$$\begin{cases} y \le 19.17t - 46.61 \\ t \ge 8.5 \\ t \le 13.5 \\ y \ge 0 \end{cases}$$

(c) Area of a trapezoid: $A = \dfrac{h}{2}(a + b)$

$h = 13.5 - 8.5 = 5$

$a = 19.17(8.5) - 46.61 = 116.335$

$b = 19.17(13.5) - 46.61 = 212.185$

$A = \dfrac{5}{2}(116.335 + 212.185)$

$= \$821.3$ billion .

79. True. The figure is a rectangle with length of 9 units and width of 11 units.

81. The graph is a half-line on the real number line; on the rectangular coordinate system, the graph is a half-plane.

83. x = radius of smaller circle

y = radius of larger circle

(a) Constraints on circles:

$\pi y^2 - \pi x^2 \ge 10$

$y > x$

$x > 0$

(b)

(c) The line is an asymptote to the boundary. The larger the circles, the closer the radii can be and the constraint still be satisfied.

85. $\begin{cases} x^2 + y^2 \le 16 \implies \text{region inside the circle} \\ x + y \ge 4 \implies \text{region above the line} \end{cases}$

Matches graph (d).

87. $\begin{cases} x^2 + y^2 \ge 16 \implies \text{region outside the circle} \\ x + y \ge 4 \implies \text{region above the line} \end{cases}$

Matches graph (c).

89. $(-2, 6), (4, -4)$

$$m = \frac{-4 - 6}{4 - (-2)} = \frac{-10}{6} = -\frac{5}{3}$$

$$y - (-4) = -\frac{5}{3}(x - 4)$$

$$3y + 12 = -5x + 20$$

$$5x + 3y - 8 = 0$$

91. $\left(\frac{3}{4}, -2\right), \left(-\frac{7}{2}, 5\right)$

$$m = \frac{5 - (-2)}{-\frac{7}{2} - \frac{3}{4}} = \frac{7}{-\frac{17}{4}} = -\frac{28}{17}$$

$$y - (-2) = -\frac{28}{17}\left(x - \frac{3}{4}\right)$$

$$17y + 34 = -28x + 21$$

$$28x + 17y + 13 = 0$$

93. $(3.4, -5.2), (-2.6, 0.8)$

$$m = \frac{0.8 - (-5.2)}{-2.6 - 3.4} = \frac{6}{-6} = -1$$

$$y - 0.8 = -1(x - (-2.6))$$

$$y - 0.8 = -x - 2.6$$

$$x + y + 1.8 = 0$$

95. (a) $(8, 39.43), (9, 41.24), (10, 45.27), (11, 47.37), (12, 48.40), (13, 49.91)$

Linear model: $y \approx 2.17t + 22.5$

Quadratic model: $y \approx -0.241t^2 + 7.23t - 3.4$

Exponential model: $y \approx 27(1.05^t)$

(c) The quadratic model is the best fit for the actual data.

(d) For 2008, use $t = 18$: $y \approx -0.241(18)^2 + 7.23(18) - 3.4 \approx \$48.66$

(b)

Section 9.6   Linear Programming

■ To solve a linear programming problem:

1. Sketch the solution set for the system of constraints.

2. Find the vertices of the region.

3. Test the objective function at each of the vertices.

Vocabulary Check

1. optimization

3. objective

5. vertex

2. linear programming

4. constraints; feasible solutions

1. $z = 4x + 3y$

At $(0, 5)$: $z = 4(0) + 3(5) = 15$

At $(0, 0)$: $z = 4(0) + 3(0) = 0$

At $(5, 0)$: $z = 4(5) + 3(0) = 20$

The minimum value is 0 at $(0, 0)$.

The maximum value is 20 at $(5, 0)$.

3. $z = 3x + 8y$

At $(0, 5)$: $z = 3(0) + 8(5) = 40$

At $(0, 0)$: $z = 3(0) + 8(0) = 0$

At $(5, 0)$: $z = 3(5) + 8(0) = 15$

The minimum value is 0 at $(0, 0)$.

The maximum value is 40 at $(0, 5)$.

5. $z = 3x + 2y$

At $(0, 5)$: $z = 3(0) + 2(5) = 10$

At $(4, 0)$: $z = 3(4) + 2(0) = 12$

At $(3, 4)$: $z = 3(3) + 2(4) = 17$

At $(0, 0)$: $z = 3(0) + 2(0) = 0$

The minimum value is 0 at $(0, 0)$.

The maximum value is 17 at $(3, 4)$.

7. $z = 5x + 0.5y$

At $(0, 5)$: $z = 5(0) + \frac{5}{2} = \frac{5}{2}$

At $(4, 0)$: $z = 5(4) + \frac{0}{2} = 20$

At $(3, 4)$: $z = 5(3) + \frac{4}{2} = 17$

At $(0, 0)$: $z = 5(0) + \frac{0}{2} = 0$

The minimum value is 0 at $(0, 0)$.

The maximum value is 20 at $(4, 0)$.

9. $z = 10x + 7y$

At $(0, 45)$: $z = 10(0) + 7(45) = 315$

At $(30, 45)$: $z = 10(30) + 7(45) = 615$

At $(60, 20)$: $z = 10(60) + 7(20) = 740$

At $(60, 0)$: $z = 10(60) + 7(0) = 600$

At $(0, 0)$: $z = 10(0) + 7(0) = 0$

The minimum value is 0 at $(0, 0)$.

The maximum value is 740 at $(60, 20)$.

11. $z = 25x + 30y$

At $(0, 45)$: $z = 25(0) + 30(45) = 1350$

At $(30, 45)$: $z = 25(30) + 30(45) = 2100$

At $(60, 20)$: $z = 25(60) + 30(20) = 2100$

At $(60, 0)$: $z = 25(60) + 30(0) = 1500$

At $(0, 0)$: $z = 25(0) + 30(0) = 0$

The minimum value is 0 at $(0, 0)$.

The maximum value is 2100 at any point along the line segment connecting $(30, 45)$ and $(60, 20)$.

13. $z = 6x + 10y$

At $(0, 2)$: $z = 6(0) + 10(2) = 20$

At $(5, 0)$: $z = 6(5) + 10(0) = 30$

At $(0, 0)$: $z = 6(0) + 10(0) = 0$

The minimum value is 0 at $(0, 0)$.

The maximum value is 30 at $(5, 0)$.

15. $z = 9x + 24y$

At $(0, 2)$: $z = 9(0) + 24(2) = 48$

At $(5, 0)$: $z = 9(5) + 24(0) = 45$

At $(0, 0)$: $z = 9(0) + 24(0) = 0$

The minimum value is 0 at $(0, 0)$.

The maximum value is 48 at $(0, 2)$.

17. $z = 4x + 5y$

At $(10, 0)$: $z = 4(10) + 5(0) = 40$

At $(5, 3)$: $z = 4(5) + 5(3) = 35$

At $(0, 8)$: $z = 4(0) + 5(8) = 40$

The minimum value is 35 at $(5, 3)$.

The region is unbounded. There is no maximum.

19. $z = 2x + 7y$

At $(10, 0)$: $z = 2(10) + 7(0) = 20$

At $(5, 3)$: $z = 2(5) + 7(3) = 31$

At $(0, 8)$: $z = 2(0) + 7(8) = 56$

The minimum value is 20 at $(10, 0)$.

The region is unbounded. There is no maximum.

21. $z = 4x + y$

At $(36, 0)$: $z = 4(36) + 0 = 144$

At $(40, 0)$: $z = 4(40) + 0 = 160$

At $(24, 8)$: $z = 4(24) + 8 = 104$

The minimum value is 104 at $(24, 8)$.

The maximum value is 160 at $(40, 0)$.

23. $z = x + 4y$

At $(36, 0)$: $z = 36 + 4(0) = 36$

At $(40, 0)$: $z = 40 + 4(0) = 40$

At $(24, 8)$: $z = 24 + 4(8) = 56$

The minimum value is 36 at $(36, 0)$.

The maximum value is 56 at $(24, 8)$.

Figure for Exercises 25 and 27

25. $z = 2x + y$

At $(0, 10)$: $z = 2(0) + (10) = 10$

At $(3, 6)$: $z = 2(3) + (6) = 12$

At $(5, 0)$: $z = 2(5) + (0) = 10$

At $(0, 0)$: $z = 2(0) + (0) = 0$

The maximum value is 12 at $(3, 6)$.

27. $z = x + y$

At $(0, 10)$: $z = (0) + (10) = 10$

At $(3, 6)$: $z = (3) + (6) = 9$

At $(5, 0)$: $z = (5) + (0) = 5$

At $(0, 0)$: $z = (0) + (0) = 0$

The maximum value is 10 at $(0, 10)$.

Figure for Exercises 29 and 31

29. $z = x + 5y$

At $(0, 5)$: $z = 0 + 5(5) = 25$

At $\left(\frac{22}{3}, \frac{19}{6}\right)$: $z = \frac{22}{3} + 5\left(\frac{19}{6}\right) = \frac{139}{6}$

At $\left(\frac{21}{2}, 0\right)$: $z = \frac{21}{2} + 5(0) = \frac{21}{2}$

At $(0, 0)$: $z = 0 + 5(0) = 0$

The maximum value is 25 at $(0, 5)$.

31. $z = 4x + 5y$

At $(0, 5)$: $z = 4(0) + 5(5) = 25$

At $\left(\frac{22}{3}, \frac{19}{6}\right)$: $z = 4\left(\frac{22}{3}\right) + 5\left(\frac{19}{6}\right) = \frac{271}{6}$

At $\left(\frac{21}{2}, 0\right)$: $z = 4\left(\frac{21}{2}\right) + 5(0) = 42$

At $(0, 0)$: $z = 4(0) + 5(0) = 0$

The maximum value is $\frac{271}{6}$ at $\left(\frac{22}{3}, \frac{19}{6}\right)$.

33. Objective function: $z = 2.5x + y$

Constraints: $x \geq 0, y \geq 0, 3x + 5y \leq 15, 5x + 2y \leq 10$

At $(0, 0)$: $z = 0$

At $(2, 0)$: $z = 5$

At $\left(\frac{20}{19}, \frac{45}{19}\right)$: $z = \frac{95}{19} = 5$

At $(0, 3)$: $z = 3$

The maximum value of 5 occurs at any point on the line segment connecting $(2, 0)$ and $\left(\frac{20}{19}, \frac{45}{19}\right)$.

35. Objective function: $z = -x + 2y$

Constraints: $x \geq 0, y \geq 0, x \leq 10, x + y \leq 7$

At $(0, 0)$: $z = -0 + 2(0) = 0$

At $(0, 7)$: $z = -0 + 2(7) = 14$

At $(7, 0)$: $z = -7 + 2(0) = -7$

The constraint $x \leq 10$ is extraneous.

The maximum value of 14 occurs at $(0, 7)$.

37. Objective function: $z = 3x + 4y$

Constraints: $x \geq 0, y \geq 0, x + y \leq 1, 2x + y \leq 4$

At $(0, 0)$: $z = 3(0) + 4(0) = 0$

At $(0, 1)$: $z = 3(0) + 4(1) = 4$

At $(1, 0)$: $z = 3(1) + 4(0) = 3$

The constraint $2x + y \leq 4$ is extraneous.

The maximum value of 4 occurs at $(0, 1)$.

39. x = number of Model A

y = number of Model B

Constraints: $2x + 2.5y \le 4000$

$\qquad\qquad\quad 4x + y \le 4800$

$\qquad\qquad\quad x + 0.75y \le 1500$

$\qquad\qquad\qquad\qquad x \ge 0$

$\qquad\qquad\qquad\qquad y \ge 0$

Objective function: $P = 45x + 50y$

Vertices:

$(0, 0)$, $(0, 1600)$, $(750, 1000)$, $(1050, 600)$, $(1200, 0)$

At $(0, 0)$: $P = 45(0) + 50(0) = 0$

At $(0, 1600)$: $P = 45(0) + 50(1600) = 80,000$

At $(750, 1000)$: $P = 45(750) + 50(1000) = 83,750$

At $(1050, 600)$: $P = 45(1050) + 50(600) = 77,250$

At $(1200, 0)$: $P = 45(1200) + 50(0) = 54,000$

The optimal profit of $83,750 occurs when 750 units of Model A and 1000 units of Model B are produced.

43. x = number of bags of Brand X

y = number of bags of Brand Y

Constraints:

$2x + y \ge 12$

$2x + 9y \ge 36$

$2x + 3y \ge 24$

$\qquad x \ge 0$

$\qquad y \ge 0$

Objective function:

$C = 25x + 20y$

Vertices: $(0, 12)$, $(3, 6)$, $(9, 2)$, $(18, 0)$

At $(0, 12)$: $C = 25(0) + 20(12) = 240$

At $(3, 6)$: $C = 25(3) + 20(6) = 195$

At $(9, 2)$: $C = 25(9) + 20(2) = 265$

At $(18, 0)$: $C = 25(18) + 20(0) = 450$

To optimize cost, use three bags of Brand X and six bags of Brand Y for an optimal cost of $195.

41. x = number of $250 models

y = number of $300 models

Constraints: $\quad 250x + 300y \le 65,000$

$\qquad\qquad\qquad x + y \le 250$

$\qquad\qquad\qquad\qquad x \ge 0$

$\qquad\qquad\qquad\qquad y \ge 0$

Objective function: $P = 25x + 40y$

Vertices: $(0, 0)$, $(250, 0)$, $(200, 50)$, $\left(0, 216\frac{2}{3}\right)$

At $(0, 0)$: $P = 25(0) + 40(0) = 0$

At $(250, 0)$: $P = 25(250) + 40(0) = 6250$

At $(200, 50)$: $P = 25(200) + 40(50) = 7000$

At $\left(0, 216\frac{2}{3}\right)$: $P = 25(0) + 40\left(216\frac{2}{3}\right) \approx 8666.67$

An optimal profit of $8640 occurs when 0 units of the $250 model and 216 units of the $300 model are stocked in inventory. $\left(\text{Note: A merchant cannot sell } \frac{2}{3} \text{ of a unit.}\right)$

45. x = number of audits

y = number of tax returns

Constraints:

$75x + 12.5y \le 900$

$10x + 2.5y \le 155$

$\qquad x \ge 0$

$\qquad y \ge 0$

Objective function:

$R = 2500x + 350y$

Vertices: $(0, 0)$, $(12, 0)$, $(5, 42)$, $(0, 62)$

At $(0, 0)$: $R = 2500(0) + 350(0) = 0$

At $(12, 0)$: $R = 2500(12) + 350(0) = 30,000$

At $(5, 42)$: $R = 2500(5) + 350(42) = 27,200$

At $(0, 62)$: $R = 2500(0) + 350(62) = 21,700$

The revenue will be optimal if 12 audits and 0 tax returns are done each week. The optimal revenue is $30,000.

47. $x =$ amount of Type A

$y =$ amount of Type B

Constraints: $x + y \leq \quad 250{,}000$

$\qquad\qquad x \quad\;\; \geq \frac{1}{4}(250{,}000)$

$\qquad\qquad\quad\; y \geq \frac{1}{4}(250{,}000)$

Objective Function: $P = 0.08x + 0.10y$

Vertices: $(62{,}500, 62{,}500)$, $(62{,}500, 187{,}500)$, $(187{,}500, 62{,}500)$

At $(62{,}500, 62{,}500)$: $\quad P = 0.08(62{,}500) + 0.10(62{,}500) = \$11{,}250$

At $(62{,}500, 187{,}500)$: $\quad P = 0.08(62{,}500) + 0.10(187{,}500) = \$23{,}750$

At $(187{,}500, 62{,}500)$: $\quad P = 0.08(187{,}500) + 0.10(62{,}500) = \$21{,}250$

To obtain an optimal return the investor should allocate \$62,500 to Type A and \$187,500 to Type B. The optimal return is \$23,750.

49. True. The objective function has a maximum value at any point on the line segment connecting the two vertices. Both of these points are on the line $y = -x + 11$ and lie between $(4, 7)$ and $(8, 3)$.

51. Constraints: $x \geq 0, y \geq 0, x + 3y \leq 15, 4x + y \leq 16$

Vertex	Value of $z = 3x + ty$
$(0, 0)$	$z = 0$
$(0, 5)$	$z = 5t$
$(3, 4)$	$z = 9 + 4t$
$(4, 0)$	$z = 12$

(a) For the maximum value to be at $(0, 5)$, $z = 5t$ must be greater than or equal to $z = 9 + 4t$ and $z = 12$.

$\quad 5t \geq 9 + 4t$ and $5t \geq 12$

$\qquad t \geq 9 \qquad\qquad\; t \geq \frac{12}{5}$

Thus, $t \geq 9$.

(b) For the maximum value to be at $(3, 4)$, $z = 9 + 4t$ must be greater than or equal to $z = 5t$ and $z = 12$.

$\quad 9 + 4t \geq 5t$ and $\;\; 9 + 4t \geq 12$

$\qquad 9 \geq t \qquad\qquad 4t \geq 3$

$\qquad\qquad\qquad\qquad\qquad t \geq \frac{3}{4}$

Thus, $\frac{3}{4} \leq t \leq 9$.

53. There are an infinite number of objective functions that would have a maximum at $(0, 4)$. One such objective function is $z = x + 5y$.

55. There are an infinite number of objective functions that would have a maximum at $(5, 0)$. One such objective function is $z = 4x + y$.

57. $\dfrac{\frac{9}{x}}{\left(\frac{6}{x} + 2\right)} = \dfrac{\frac{9}{x}}{\frac{6 + 2x}{x}} = \dfrac{9}{x} \cdot \dfrac{x}{2(3 + x)} = \dfrac{9}{2(3 + x)} = \dfrac{9}{2(x + 3)}, \quad x \neq 0$

59. $\dfrac{\left(\dfrac{4}{x^2 - 9} + \dfrac{2}{x - 2}\right)}{\left(\dfrac{1}{x + 3} + \dfrac{1}{x - 3}\right)} = \dfrac{\dfrac{4(x - 2) + 2(x^2 - 9)}{(x - 2)(x^2 - 9)}}{\dfrac{(x - 3) + (x + 3)}{x^2 - 9}}$

$$= \frac{2x^2 + 4x - 26}{(x - 2)(x^2 - 9)} \cdot \frac{x^2 - 9}{2x}$$

$$= \frac{2(x^2 + 2x - 13)}{(x - 2)(2x)}$$

$$= \frac{x^2 + 2x - 13}{x(x - 2)}, \quad x \neq \pm 3$$

61. $e^{2x} + 2e^x - 15 = 0$

$(e^x + 5)(e^x - 3) = 0$

$e^x = -5$ or $e^x = 3$

No real solution. $x = \ln 3$
            $x \approx 1.099$

63. $8(62 - e^{x/4}) = 192$

$62 - e^{x/4} = 24$

$-e^{x/4} = -38$

$e^{x/4} = 38$

$\dfrac{x}{4} = \ln 38$

$x = 4 \ln 38$

$x \approx 14.550$

65. $7 \ln 3x = 12$

$\ln 3x = \dfrac{12}{7}$

$3x = e^{12/7}$

$x = \dfrac{e^{12/7}}{3}$

$x \approx 1.851$

67. $\begin{cases} -x - 2y + 3z = -23 \\ 2x + 6y - z = 17 \\ 5y + z = 8 \end{cases}$

$\begin{cases} -x - 2y + 3z = -23 \\ 2y + 5z = -29 \quad\quad 2\text{Eq.1} + \text{Eq.2} \\ 5y + z = 8 \end{cases}$

$\begin{cases} -x - 2y + 3z = -23 \\ 2y + 5z = -29 \\ -\frac{23}{2}z = \frac{161}{2} \quad\quad -\frac{5}{2}\text{Eq.2} + \text{Eq.3} \end{cases}$

$-\frac{23}{2}z = \frac{161}{2} \implies z = -7$

$2y + 5(-7) = -29 \implies y = 3$

$-x - 2(3) + 3(-7) = -23 \implies -x - 27 = -23$

$\implies x = -4$

Solution: $(-4, 3, -7)$

Review Exercises for Chapter 9

1. $\begin{cases} x + y = 2 \\ x - y = 0 \implies x = y \end{cases}$

$x + x = 2$

$2x = 2$

$x = 1$

$y = 1$

Solution: $(1, 1)$

3. $\begin{cases} 0.5x + y = 0.75 \implies y = 0.75 - 0.5x \\ 1.25x - 4.5y = -2.5 \end{cases}$

$1.25x - 4.5(0.75 - 0.5x) = -2.5$

$1.25x - 3.375 + 2.25x = -2.5$

$3.50x = 0.875$

$x = 0.25$

$y = 0.625$

Solution: $(0.25, 0.625)$

5. $\begin{cases} x^2 - y^2 = 9 \\ x - y = 1 \implies x = y + 1 \end{cases}$

$(y + 1)^2 - y^2 = 9$

$2y + 1 = 9$

$y = 4$

$x = 5$

Solution: $(5, 4)$

7. $\begin{cases} y = 2x^2 \\ y = x^4 - 2x^2 \implies 2x^2 = x^4 - 2x^2 \end{cases}$

$0 = x^4 - 4x^2$

$0 = x^2(x^2 - 4)$

$0 = x^2(x + 2)(x - 2)$

$x = 0, x = -2, x = 2$

$y = 0, y = 8, y = 8$

Solutions: $(0, 0), (-2, 8), (2, 8)$

9. $\begin{cases} 2x - y = 10 \\ x + 5y = -6 \end{cases}$

Point of intersection: $(4, -2)$

11. $\begin{cases} y = 2x^2 - 4x + 1 \\ y = x^2 - 4x + 3 \end{cases}$

Point of intersection: $(1.41, -0.66), (-1.41, 10.66)$

13. $\begin{cases} y = -2e^{-x} \\ 2e^x + y = 0 \implies y = -2e^x \end{cases}$

Point of intersection: $(0, -2)$

15. Let $x =$ number of kits.

$C = 12x + 50,000$

$R = 25x$

Break-even: $R = C$

$25x = 12x + 50,000$

$13x = 50,000$

$x \approx 3846.15$

You would need to sell 3847 kits to cover your costs.

17. $2l + 2w = 480$

$l = 1.50w$

$2(1.50w) + 2w = 480$

$5w = 480$

$w = 96$

$l = 144$

The dimensions are 96×144 meters.

19. $\begin{cases} 2x - y = 2 \implies 16x - 8y = 16 \\ 6x + 8y = 39 \implies \underline{6x + 8y = 39} \\ 22x = 55 \\ x = \frac{55}{22} = \frac{5}{2} \end{cases}$

Back-substitute $x = \frac{5}{2}$ into Equation 1.

$2\left(\frac{5}{2}\right) - y = 2$

$y = 3$

Solution: $\left(\frac{5}{2}, 3\right)$

21. $\begin{cases} 0.2x + 0.3y = 0.14 \implies 20x + 30y = 14 \implies \quad 20x + 30y = 14 \\ 0.4x + 0.5y = 0.20 \implies 4x + 5y = 2 \implies \underline{-20x - 25y = -10} \end{cases}$

$$5y = 4$$

$$y = \tfrac{4}{5}$$

Back-substitute $y = \tfrac{4}{5}$ into Equation 2.

$$4x + 5\left(\tfrac{4}{5}\right) = 2$$

$$4x = -2$$

$$x = -\tfrac{1}{2}$$

Solution: $\left(-\tfrac{1}{2}, \tfrac{4}{5}\right) = (-0.5, 0.8)$

23. $\begin{cases} 3x - 2y = 0 \implies 3x - 2y = 0 \\ 3x + 2(y + 5) = 10 \implies \underline{3x + 2y = 0} \end{cases}$

$$6x = 0$$

$$x = 0$$

Back-substitute $x = 0$ into Equation 1.

$$3(0) - 2y = 0$$

$$2y = 0$$

$$y = 0$$

Solution: $(0, 0)$

25. $\begin{cases} 1.25x - 2y = 3.5 \implies \quad 5x - 8y = 14 \\ 5x - 8y = 14 \implies \underline{-5x + 8y = -14} \end{cases}$

$$0 = 0$$

There are infinitely many solutions.

Let $y = a$, then $5x - 8a = 14 \implies x = \tfrac{8}{5}a + \tfrac{14}{5}$.

Solution: $\left(\tfrac{8}{5}a + \tfrac{14}{5}, a\right)$ where a is any real number.

27. $\begin{cases} x + 5y = 4 \implies \quad x + 5y = 4 \\ x - 3y = 6 \implies \underline{-x + 3y = -6} \end{cases}$

$$8y = -2 \implies y = -\tfrac{1}{4}$$

Matches graph (d). The system has one solution and is consistent.

29. $\begin{cases} 3x - y = 7 \implies \quad 6x - 2y = 14 \\ -6x + 2y = 8 \implies \underline{-6x + 2y = 8} \end{cases}$

$$0 \neq 22$$

Matches graph (b). The system has no solution and is inconsistent.

31. $37 - 0.0002x = 22 + 0.00001x$

$$15 = 0.00021x$$

$$x = \frac{500,000}{7}, p = \frac{159}{7}$$

Point of equilibrium: $\left(\dfrac{500,000}{7}, \dfrac{159}{7}\right)$

33. $\begin{cases} x - 4y + 3z = 3 \\ -y + z = -1 \\ z = -5 \end{cases}$

$$-y + (-5) = -1 \implies y = -4$$

$$x - 4(-4) + 3(-5) = 3 \implies x = 2$$

Solution: $(2, -4, -5)$

35. $\begin{cases} x + 2y + 6z = 4 \\ -3x + 2y - z = -4 \\ 4x + 2z = 16 \end{cases}$

$\begin{cases} x + 2y + 6z = 4 \\ 8y + 17z = 8 \quad\quad 3\text{Eq.1} + \text{Eq.2} \\ -8y - 22z = 0 \quad\quad -4\text{Eq.1} + \text{Eq.3} \end{cases}$

$\begin{cases} x + 2y + 6z = 4 \\ 8y + 17z = 8 \\ -5z = 8 \quad\quad \text{Eq.2} + \text{Eq.3} \end{cases}$

$\begin{cases} x + 2y + 6z = 4 \\ 8y + 17z = 8 \\ z = -\frac{8}{5} \quad\quad -\frac{1}{5}\text{Eq.3} \end{cases}$

$8y + 17\left(-\frac{8}{5}\right) = 8 \Rightarrow y = \frac{22}{5}$

$x + 2\left(\frac{22}{5}\right) + 6\left(-\frac{8}{5}\right) = 4 \Rightarrow x = \frac{24}{5}$

Solution: $\left(\frac{24}{5}, \frac{22}{5}, -\frac{8}{5}\right)$

39. $\begin{cases} 5x - 12y + 7z = 16 \Rightarrow \\ 3x - 7y + 4z = 9 \Rightarrow \end{cases} \begin{cases} 15x - 36y + 21z = 48 \\ -15x + 35y - 20z = -45 \\ \hline -y + z = 3 \end{cases}$

Let $z = a$. Then $y = a - 3$ and

$5x - 12(a - 3) + 7a = 16 \Rightarrow x = a - 4.$

Solution: $(a - 4,\ a - 3,\ a)$ where a is any real number.

37. $\begin{cases} x - 2y + z = -6 \\ 2x - 3y = -7 \\ -x + 3y - 3z = 11 \end{cases}$

$\begin{cases} x - 2y + z = -6 \\ y - 2z = 5 \quad\quad -2\text{Eq.1} + \text{Eq.2} \\ y - 2z = 5 \quad\quad \text{Eq.1} + \text{Eq.3} \end{cases}$

$\begin{cases} x - 2y + z = -6 \\ y - 2z = 5 \\ 0 = 0 \quad\quad -\text{Eq.2} + \text{Eq.3} \end{cases}$

Let $z = a$, then:

$y = 2a + 5$

$x - 2(2a + 5) + a = -6$

$x - 3a - 10 = -6$

$x = 3a + 4$

Solution: $(3a + 4, 2a + 5, a)$ where a is any real number.

41. $y = ax^2 + bx + c$ through $(0, -5)$, $(1, -2)$, and $(2, 5)$.

$(0, -5): -5 = \quad\quad + c \Rightarrow \quad\quad c = -5$

$(1, -2): -2 = a + b + c \Rightarrow \begin{cases} a + b = 3 \\ 2a + b = 5 \end{cases}$

$(2, 5): 5 = 4a + 2b + c \Rightarrow$

$\begin{cases} 2a + b = 5 \\ -a - b = -3 \end{cases}$

$a = 2$

$b = 1$

The equation of the parabola is $y = 2x^2 + x - 5$.

43. $x^2 + y^2 + Dx + Ey + F = 0$ through $(-1, -2)$, $(5, -2)$ and $(2, 1)$.

$(-1, -2): 5 - D - 2E + F = 0 \Rightarrow \begin{cases} D + 2E - F = 5 \\ 5D - 2E + F = -29 \\ 2D + E + F = -5 \end{cases}$

$(5, -2): 29 + 5D - 2E + F = 0 \Rightarrow$

$(2, 1): 5 + 2D + E + F = 0 \Rightarrow$

From the first two equations we have

$6D = -24$

$D = -4.$

Substituting $D = -4$ into the second and third equations yields:

$-20 - 2E + F = -29 \Rightarrow \begin{cases} -2E + F = -9 \\ -E - F = -3 \end{cases}$

$-8 + E + F = -5 \Rightarrow$

$\begin{aligned} -3E &= -12 \\ E &= 4 \\ F &= -1 \end{aligned}$

—CONTINUED—

43. —CONTINUED—

The equation of the circle is $x^2 + y^2 - 4x + 4y - 1 = 0$.

To verify the result using a graphing utility, solve the equation for y.

$$(x^2 - 4x + 4) + (y^2 + 4y + 4) = 1 + 4 + 4$$

$$(x - 2)^2 + (y + 2)^2 = 9$$

$$(y + 2)^2 = 9 - (x - 2)^2$$

$$y = -2 \pm \sqrt{9 - (x - 2)^2}$$

Let $y_1 = -2 + \sqrt{9 - (x - 2)^2}$ and $y_2 = -2 - \sqrt{9 - (x - 2)^2}$.

45. $(3, 101.7), (4, 108.4), (5, 121.1)$

(a) $n = 3, \sum_{i=1}^{3} x_i = 12, \sum_{i=1}^{3} x_i^2 = 50, \sum_{i=1}^{3} x_i^3 = 216, \sum_{i=1}^{3} x_i^4 = 962, \sum_{i=1}^{3} y_i = 331.2, \sum_{i=1}^{3} x_i y_i = 1344.2, \sum_{i=1}^{3} x_i^2 y_i = 5677.2$

$$\begin{cases} 3c + 12b + 50a = 331.2 \\ 12c + 50b + 216a = 1344.2 \\ 50c + 216b + 962a = 5677.2 \end{cases}$$

Solving this system yields $c = 117.6, b = -14.3, a = 3$.

Quadratic model: $y = 3x^2 - 14.3x + 117.6$

(b)

The model is a good fit to the data. The actual points lie on the parabola.

(c) For 2008, use $x = 8$:

$y = 3(8)^2 - 14.3(8) + 117.6$

$= 195.2$ million online shoppers

This answer seems reasonable.

47. Let x = amount invested at 7%

y = amount invested at 9%

z = amount invested at 11%.

$y = x - 3000$ and

$z = x - 5000 \Rightarrow y + z = 2x - 8000$

$$\begin{cases} x + y + z = 40{,}000 \\ 0.07x + 0.09y + 0.11z = 3500 \\ y + z = 2x - 8000 \end{cases}$$

$x + (2x - 8000) = 40{,}000 \Rightarrow x = 16{,}000$

$y = 16{,}000 - 3000 \Rightarrow y = 13{,}000$

$z = 16{,}000 - 5000 \Rightarrow z = 11{,}000$

Thus, \$16,000 was invested at 7%, \$13,000 at 9% and \$11,000 at 11%.

49. $\dfrac{3}{x^2 + 20x} = \dfrac{3}{x(x + 20)} = \dfrac{A}{x} + \dfrac{B}{x + 20}$

51. $\dfrac{3x - 4}{x^3 - 5x^2} = \dfrac{3x - 4}{x^2(x - 5)} = \dfrac{A}{x} + \dfrac{B}{x^2} + \dfrac{C}{x - 5}$

53. $\dfrac{4 - x}{x^2 + 6x + 8} = \dfrac{A}{x + 2} + \dfrac{B}{x + 4}$

$4 - x = A(x + 4) + B(x + 2)$

Let $x = -2$: $6 = 2A \Rightarrow A = 3$

Let $x = -4$: $8 = -2B \Rightarrow B = -4$

$\dfrac{4 - x}{x^2 + 6x + 8} = \dfrac{3}{x + 2} - \dfrac{4}{x + 4}$

55. $\dfrac{x^2}{x^2 + 2x - 15} = 1 - \dfrac{2x - 15}{x^2 + 2x - 15}$

$\dfrac{-2x + 15}{(x + 5)(x - 3)} = \dfrac{A}{x + 5} + \dfrac{B}{x - 3}$

$\qquad -2x + 15 = A(x - 3) + B(x + 5)$

Let $x = -5$: $25 = -8A \implies A = -\dfrac{25}{8}$

Let $x = 3$: $9 = 8B \implies B = \dfrac{9}{8}$

$\dfrac{x^2}{x^2 + 2x - 15} = 1 - \dfrac{25}{8(x + 5)} + \dfrac{9}{8(x - 3)}$

57. $\dfrac{x^2 + 2x}{x^3 - x^2 + x - 1} = \dfrac{x^2 + 2x}{(x - 1)(x^2 + 1)} = \dfrac{A}{x - 1} + \dfrac{Bx + C}{x^2 + 1}$

$x^2 + 2x = A(x^2 + 1) + (Bx + C)(x - 1)$

$\qquad\quad = Ax^2 + A + Bx^2 - Bx + Cx - C$

$\qquad\quad = (A + B)x^2 + (-B + C)x + (A - C)$

Equating coefficients of like terms gives $1 = A + B$, $2 = -B + C$, and $0 = A - C$. Adding both sides of all three equations gives $3 = 2A$. Therefore, $A = \frac{3}{2}$, $B = -\frac{1}{2}$, and $C = \frac{3}{2}$.

$\dfrac{x^2 + 2x}{x^3 - x^2 + x - 1} = \dfrac{\frac{3}{2}}{x - 1} + \dfrac{-\frac{1}{2}x + \frac{3}{2}}{x^2 + 1}$

$\qquad\qquad\qquad = \dfrac{1}{2}\left(\dfrac{3}{x - 1} - \dfrac{x - 3}{x^2 + 1}\right)$

59. $\dfrac{3x^3 + 4x}{(x^2 + 1)^2} = \dfrac{Ax + B}{x^2 + 1} + \dfrac{Cx + D}{(x^2 + 1)^2}$

$3x^3 + 4x = (Ax + B)(x^2 + 1) + Cx + D = Ax^3 + Bx^2 + (A + C)x + (B + D)$

Equating coefficients of like powers:

$3 = A$

$0 = B$

$4 = 3 + C \implies C = 1$

$0 = B + D \implies D = 0$

$\dfrac{3x^3 + 4x}{(x^2 + 1)^2} = \dfrac{3x}{x^2 + 1} + \dfrac{x}{(x^2 + 1)^2}$

61. $y \le 5 - \frac{1}{2}x$

63. $y - 4x^2 > -1$

65. $\begin{cases} x + 2y \le 160 \\ 3x + \ y \le 180 \\ \qquad x \ge 0 \\ \qquad y \ge 0 \end{cases}$

Vertex A	Vertex B	Vertex C	Vertex D
$x + 2y = 160$	$x + 2y = 160$	$3x + y = 180$	$x = 0$
$3x + y = 180$	$x = 0$	$y = 0$	$y = 0$
$(40, 60)$	$(0, 80)$	$(60, 0)$	$(0, 0)$

67. $\begin{cases} 3x + 2y \geq 24 \\ x + 2y \geq 12 \\ 2 \leq x \leq 15 \\ y \leq 15 \end{cases}$

Vertex A	Vertex B	Vertex C	Vertex D	Vertex E
$3x + 2y = 24$	$3x + 2y = 24$	$x = 2$	$x = 15$	$x + 2y = 12$
$x + 2y = 12$	$x = 2$	$y = 15$	$y = 15$	$x = 15$
$(6, 3)$	$(2, 9)$	$(2, 15)$	$(15, 15)$	$\left(15, -\frac{3}{2}\right)$

69. $\begin{cases} y < x + 1 \\ y > x^2 - 1 \end{cases}$

Vertices:

$x + 1 = x^2 - 1$

$0 = x^2 - x - 2 = (x + 1)(x - 2)$

$x = -1$ or $x = 2$

$y = 0 \qquad y = 3$

$(-1, 0) \qquad (2, 3)$

71. $\begin{cases} 2x - 3y \geq 0 \\ 2x - y \leq 8 \\ y \geq 0 \end{cases}$

Vertex A	Vertex B	Vertex C
$2x - 3y = 0$	$2x - 3y = 0$	$2x - y = 8$
$2x - y = 8$	$y = 0$	$y = 0$
$(6, 4)$	$(0, 0)$	$(4, 0)$

73. x = number of units of Product I

y = number of units of Product II

$\begin{cases} 20x + 30y \leq 24,000 \\ 12x + 8y \leq 12,400 \\ x \geq 0 \\ y \geq 0 \end{cases}$

75. (a)

$160 - 0.0001x = 70 + 0.0002x$

$90 = 0.0003x$

$x = 300,000$ units

$p = \$130$

Point of equilibrium: $(300,000, 130)$

(b) Consumer surplus: $\frac{1}{2}(300,000)(30) = \$4,500,000$

Producer surplus: $\frac{1}{2}(300,000)(60) = \$9,000,000$

77. Objective function: $z = 3x + 4y$

Constraints: $\begin{cases} x \geq 0 \\ y \geq 0 \\ 2x + 5y \leq 50 \\ 4x + y \leq 28 \end{cases}$

At $(0, 0)$: $z = 0$

At $(0, 10)$: $z = 40$

At $(5, 8)$: $z = 47$

At $(7, 0)$: $z = 21$

The minimum value is 0 at $(0, 0)$.

The maximum value is 47 at $(5, 8)$.

79. Objective function: $z = 1.75x + 2.25y$

Constraints: $\begin{cases} x \geq 0 \\ y \geq 0 \\ 2x + y \geq 25 \\ 3x + 2y \geq 45 \end{cases}$

At $(0, 25)$: $z = 56.25$

At $(5, 15)$: $z = 42.5$

At $(15, 0)$: $z = 26.25$

The minimum value is 26.25 at $(15, 0)$.

Since the region in unbounded, there is no maximum value.

81. Objective function: $z = 5x + 11y$

Constraints: $\begin{cases} x \geq 0 \\ y \geq 0 \\ x + 3y \leq 12 \\ 3x + 2y \leq 15 \end{cases}$

At $(0, 0)$: $z = 0$

At $(5, 0)$: $z = 25$

At $(3, 3)$: $z = 48$

At $(0, 4)$: $z = 44$

The minimum value is 0 at $(0, 0)$.

The maximum value is 48 at $(3, 3)$.

83. Let x = number of haircuts

y = number of permanents.

Objective function: Optimize $R = 25x + 70y$ subject to the following constraints:

$\begin{cases} x \geq 0 \\ y \geq 0 \\ \left(\frac{20}{60}\right)x + \left(\frac{70}{60}\right)y \leq 24 \implies 2x + 7y \leq 144 \end{cases}$

At $(0, 0)$: $R = 0$

At $(72, 0)$: $R = 1800$

At $\left(0, \frac{144}{7}\right)$: $R = 1440$

The revenue is optimal if the student does 72 haircuts and no permanents. The maximum revenue is $1800.

85. Let x = the number of bags of Brand X, and

y = the number of bags of Brand Y.

Objective function: Optimize $C = 15x + 30y$.

Constraints: $\begin{cases} 8x + 2y \geq 16 \\ x + y \geq 5 \\ 2x + 7y \geq 20 \\ x \geq 0 \\ y \geq 0 \end{cases}$

At $(0, 8)$: $C = 15(0) + 30(8) = 240$

At $(1, 4)$: $C = 15(1) + 30(4) = 135$

At $(3, 2)$: $C = 15(3) + 30(2) = 105$

At $(10, 0)$: $C = 15(10) + 30(0) = 150$

To optimize cost, use three bags of Brand X and two bags of Brand Y. The minimum cost is $105.

87. False. The system $y \leq 5$, $y \geq -2$, $y \leq \frac{7}{2}x - 9$, and $y \leq -\frac{7}{2}x + 26$ represents the region covered by an isosceles trapezoid.

89. There are an infinite number of linear systems with the solution $(-6, 8)$. One possible solution is:
$$\begin{cases} x + y = 2 \\ x - y = -14 \end{cases}$$

91. There are infinite linear systems with the solution $\left(\frac{4}{3}, 3\right)$. One possible solution is:
$$\begin{cases} 3x + y = 7 \\ -6x + 3y = 1 \end{cases}$$

93. There are an infinite number of linear systems with the solution $(4, -1, 3)$. One possible system is as follows:
$$\begin{cases} x + y + z = 6 \\ x + y - z = 0 \\ x - y - z = 2 \end{cases}$$

95. There are an infinite number of linear systems with the solution $\left(5, \frac{3}{2}, 2\right)$. One possible solution is:
$$\begin{cases} 2x + 2y - 3z = 7 \\ x - 2y + z = 4 \\ -x + 4y - z = -1 \end{cases}$$

97. A system of linear equations is inconsistent if it has no solution.

99. If the solution to a system of equations is at fractional or irrational values, then the substitution method may yield an exact answer. The graphical method works well when the solution is at integer values, otherwise we can usually only approximate the solution.

Problem Solving for Chapter 9

1. The longest side of the triangle is a diameter of the circle and has a length of 20.

The lines $y = \frac{1}{2}x + 5$ and $y = -2x + 20$ intersect at the point $(6, 8)$.

The distance between $(-10, 0)$ and $(6, 8)$ is:
$$d_1 = \sqrt{(6 - (-10))^2 + (8 - 0)^2} = \sqrt{320} = 8\sqrt{5}$$

The distance between $(6, 8)$ and $(10, 0)$ is:
$$d_2 = \sqrt{(10 - 6)^2 + (0 - 8)^2} = \sqrt{80} = 4\sqrt{5}$$

Since $\left(\sqrt{320}\right)^2 + \left(\sqrt{80}\right)^2 = (20)^2$
$$400 = 400$$

the sides of the triangle satisfy the Pythagorean Theorem. Thus, the triangle is a right triangle.

3. The system will have exactly one solution when the slopes of the line are *not* equal.
$$\begin{cases} ax + by = e \implies y = -\frac{a}{b}x + \frac{e}{b} \\ cx + dy = f \implies y = -\frac{c}{d}x + \frac{f}{d} \end{cases}$$

$$-\frac{a}{b} \neq -\frac{c}{d}$$

$$\frac{a}{b} \neq \frac{c}{d}$$

$$ad \neq bc$$

5. There are a finite number of solutions.

(a) If both equations are linear, then the maximum number of solutions to a finite system is *one*.

(b) If one equation is linear and the other is quadratic, then the maximum number of solutions is *two*.

(c) If both equations are quadratic, then the maximum number of solutions to a finite system is *four*.

7. The point where the two sections meet is at a depth of 10.1 feet. The distance between $(0, -10.1)$ and $(252.5, 0)$ is:

$$d = \sqrt{(252.5 - 0)^2 + (0 - (-10.1))^2} = \sqrt{63858.26}$$

$$d \approx 252.7$$

Each section is approximately 252.7 feet long.

9. Let $x =$ cost of the cable, per foot.

Let $y =$ cost of a connector.

$$\begin{cases} 6x + 2y = 15.50 \implies & 6x + 2y = 15.50 \\ 3x + 2y = 10.25 \implies & -3x - 2y = -10.25 \end{cases}$$

$$\begin{aligned} 3x &= 5.25 \\ x &= 1.75 \\ y &= 2.50 \end{aligned}$$

For a four-foot cable with a connector on each end the cost should be $4(1.75) + 2(2.50) = \$12.00$

11. Let $X = \dfrac{1}{x}$, $Y = \dfrac{1}{y}$, and $Z = \dfrac{1}{z}$.

(a)
$$\begin{cases} \dfrac{12}{x} - \dfrac{12}{y} = 7 \implies 12X - 12Y = 7 \implies & 12X - 12Y = 7 \\ \dfrac{3}{x} + \dfrac{4}{y} = 0 \implies 3X + 4Y = 0 \implies & \underline{9X + 12Y = 0} \end{cases}$$

$$21X = 7$$

$$X = \frac{1}{3}$$

$$Y = -\frac{1}{4}$$

Thus, $\dfrac{1}{x} = \dfrac{1}{3} \implies x = 3$ and $\dfrac{1}{y} = -\dfrac{1}{4} \implies y = -4$.

Solution: $(3, -4)$

—CONTINUED—

11. —CONTINUED—

(b)
$$\begin{cases} \dfrac{2}{x} + \dfrac{1}{y} - \dfrac{3}{z} = 4 \implies 2X + Y - 3Z = 4 & \text{Eq.1} \\[2mm] \dfrac{4}{x} \phantom{+ \dfrac{1}{y}} + \dfrac{2}{z} = 10 \implies 4X + 2Z = 10 & \text{Eq.2} \\[2mm] -\dfrac{2}{x} + \dfrac{3}{y} - \dfrac{13}{z} = -8 \implies -2X + 3Y - 13Z = -8 & \text{Eq.3} \end{cases}$$

$$\begin{cases} 2X + Y - 3Z = 4 \\ -2Y + 8Z = 2 & -2\text{Eq.1} + \text{Eq.2} \\ 4Y - 16Z = -4 & \text{Eq.1} + \text{Eq.3} \end{cases}$$

$$\begin{cases} 2X + Y - 3Z = 4 \\ -2Y + 8Z = 2 \\ 0 = 0 & 2\text{Eq.2} + \text{Eq.3} \end{cases}$$

The system has infinite solutions.

Let $Z = a$, then $Y = 4a - 1$ and $X = \dfrac{-a + 5}{2}$.

Then $\dfrac{1}{z} = a \implies z = \dfrac{1}{a}, \dfrac{1}{y} = 4a - 1 \implies y = \dfrac{1}{4a - 1}$

$x = \dfrac{-a + 5}{2} \implies x = \dfrac{2}{-a + 5}$.

Solution: $\left(\dfrac{2}{-a + 5}, \dfrac{1}{4a - 1}, \dfrac{1}{a} \right), a \neq 5, \dfrac{1}{4}, 0$

13. Solution: $(1, -1, 2)$

$$\begin{cases} 4x - 2y + 5z = 16 & \text{Equation 1} \\ x + y = 0 & \text{Equation 2} \\ -x - 3y + 2z = 6 & \text{Equation 3} \end{cases}$$

(a) $\begin{cases} 4x - 2y + 5z = 16 \\ x + y = 0 \end{cases}$

$\begin{cases} x + y = 0 \\ 4x - 2y + 5z = 16 \end{cases}$ Interchange the equations.

$\begin{cases} x + y = 0 \\ -6y + 5z = 16 & -4\text{Eq.1} + \text{Eq.2} \end{cases}$

Let $z = a$, then $y = \dfrac{5a - 16}{6}$ and $x = \dfrac{-5a + 16}{6}$.

Solution: $\left(\dfrac{-5a + 16}{6}, \dfrac{5a - 16}{6}, a \right)$

When $a = 2$ we have the original solution.

(c) $\begin{cases} x + y = 0 \\ -x - 3y + 2z = 6 \end{cases}$

$\begin{cases} x + y = 0 \\ -2y + 2z = 6 & \text{Eq.1} + \text{Eq.2} \end{cases}$

Let $z = c$, then $y = c - 3$ and $x = -c + 3$

Solution: $(-c + 3, c - 3, c)$

When $c = 2$ we have the original solution.

(b) $\begin{cases} 4x - 2y + 5z = 16 \\ -x - 3y + 2z = 6 \end{cases}$

$\begin{cases} -x - 3y + 2z = 6 \\ 4x - 2y + 5z = 16 \end{cases}$ Interchange the equations.

$\begin{cases} -x - 3y + 2z = 6 \\ -14y + 13z = 40 & 4\text{Eq.1} + \text{Eq.2} \end{cases}$

Let $z = b$, then $y = \dfrac{13b - 40}{14}$ and $x = \dfrac{-11b + 36}{14}$

Solution: $\left(\dfrac{-11b + 36}{14}, \dfrac{13b - 40}{14}, b \right)$

When $b = 2$ we have the original solution.

(d) Each of these systems has infinite solutions.

15. t = amount of terrestrial vegetation in kilograms

a = amount of aquatic vegetation in kilograms

$$a + t \le 32$$

$$0.15a \ge 1.9$$

$$193a + 4(193)t \ge 11,000$$

17. (a) x = HDL cholesterol (good)

y = LDL cholesterol (bad)

$$\begin{cases} 0 < \ y < 130 \\ x \ge 35 \\ x + \ y \le 200 \end{cases}$$

(b)

(c) $y = 120$ *is* in the region since $0 < y < 130$.

$x = 90$ *is* in the region since $35 < x < 200$.

$x + y = 210$ *is not* in the region since $x + y < 200$.

(d) If the LDL reading is 150 and the HDL reading is 40, then $x \ge 35$ and $x + y \le 200$ but $y \not< 130$.

(e) $\dfrac{x + y}{x} < 4$

$x + y < 4x$

$y < 3x$

The point $(50, 120)$ is in the region and $120 < 3(50)$.

Chapter 9   Practice Test

For Exercises 1–3, solve the given system by the method of substitution.

1. $\begin{cases} x + y = 1 \\ 3x - y = 15 \end{cases}$

2. $\begin{cases} x - 3y = -3 \\ x^2 + 6y = 5 \end{cases}$

3. $\begin{cases} x + y + z = 6 \\ 2x - y + 3z = 0 \\ 5x + 2y - z = -3 \end{cases}$

4. Find the two numbers whose sum is 110 and product is 2800.

5. Find the dimensions of a rectangle if its perimeter is 170 feet and its area is 1500 square feet.

For Exercises 6–8, solve the linear system by elimination.

6. $\begin{cases} 2x + 15y = 4 \\ x - 3y = 23 \end{cases}$

7. $\begin{cases} x + y = 2 \\ 38x - 19y = 7 \end{cases}$

8. $\begin{cases} 0.4x + 0.5y = 0.112 \\ 0.3x - 0.7y = -0.131 \end{cases}$

9. Herbert invests \$17,000 in two funds that pay 11% and 13% simple interest, respectively. If he receives \$2080 in yearly interest, how much is invested in each fund?

10. Find the least squares regression line for the points $(4, 3)$, $(1, 1)$, $(-1, -2)$, and $(-2, -1)$.

For Exercises 11–12, solve the system of equations.

11. $\begin{cases} x + y = -2 \\ 2x - y + z = 11 \\ 4y - 3z = -20 \end{cases}$

12. $\begin{cases} 3x + 2y - z = 5 \\ 6x - y + 5z = 2 \end{cases}$

13. Find the equation of the parabola $y = ax^2 + bx + c$ passing through the points $(0, -1)$, $(1, 4)$ and $(2, 13)$.

For Exercises 14–15, write the partial fraction decomposition of the rational functions.

14. $\dfrac{10x - 17}{x^2 - 7x - 8}$

15. $\dfrac{x^2 + 4}{x^4 + x^2}$

16. Graph $x^2 + y^2 \geq 9$.

17. Graph the solution of the system.

$$\begin{cases} x + y \leq 6 \\ \quad x \geq 2 \\ \quad y \geq 0 \end{cases}$$

18. Derive a set of inequalities to describe the triangle with vertices $(0, 0)$, $(0, 7)$, and $(2, 3)$.

19. Find the maximum value of the objective function, $z = 30x + 26y$, subject to the following constraints.

$$\begin{cases} \quad\quad x \geq 0 \\ \quad\quad y \geq 0 \\ 2x + 3y \leq 21 \\ 5x + 3y \leq 30 \end{cases}$$

20. Graph the system of inequalities.

$$\begin{cases} \quad\quad x^2 + y^2 \leq 4 \\ (x - 2)^2 + y^2 \geq 4 \end{cases}$$

C H A P T E R   1 0
Matrices and Determinants

CHAPTER 10
Matrices and Determinants

Section 10.1   Matrices and Systems of Equations

■ You should be able to use elementary row operations to produce a row-echelon form (or reduced row-echelon form) of a matrix.

    1. Interchange two rows.

    2. Multiply a row by a nonzero constant.

    3. Add a multiple of one row to another row.

■ You should be able to use either Gaussian elimination with back-substitution or Gauss- Jordan elimination to solve a system of linear equations.

Vocabulary Check

1. matrix   **2.** square   **3.** main diagonal

4. row; column   **5.** augmented   **6.** coefficient

7. row-equivalent   **8.** reduced row-echelon form   **9.** Gauss-Jordan elimination

---

1. Since the matrix has one row and two columns, its order is 1×2.

3. Since the matrix has three rows and one column, its order is 3×1.

5. Since the matrix has two rows and two columns, its order is 2×2.

7.
$$\begin{cases} 4x - 3y = -5 \\ -x + 3y = 12 \end{cases}$$
$$\begin{bmatrix} 4 & -3 & \vdots & -5 \\ -1 & 3 & \vdots & 12 \end{bmatrix}$$

9.
$$\begin{cases} x + 10y - 2z = 2 \\ 5x - 3y + 4z = 0 \\ 2x + y = 6 \end{cases}$$
$$\begin{bmatrix} 1 & 10 & -2 & \vdots & 2 \\ 5 & -3 & 4 & \vdots & 0 \\ 2 & 1 & 0 & \vdots & 6 \end{bmatrix}$$

11.
$$\begin{cases} 7x - 5y + z = 13 \\ 19x - 8z = 10 \end{cases}$$
$$\begin{bmatrix} 7 & -5 & 1 & \vdots & 13 \\ 19 & 0 & -8 & \vdots & 10 \end{bmatrix}$$

13.
$$\begin{bmatrix} 1 & 2 & \vdots & 7 \\ 2 & -3 & \vdots & 4 \end{bmatrix}$$
$$\begin{cases} x + 2y = 7 \\ 2x - 3y = 4 \end{cases}$$

15.
$$\begin{bmatrix} 2 & 0 & 5 & \vdots & -12 \\ 0 & 1 & -2 & \vdots & 7 \\ 6 & 3 & 0 & \vdots & 2 \end{bmatrix}$$
$$\begin{cases} 2x + 5z = -12 \\ y - 2z = 7 \\ 6x + 3y = 2 \end{cases}$$

17.
$$\begin{bmatrix} 9 & 12 & 3 & 0 & \vdots & 0 \\ -2 & 18 & 5 & 2 & \vdots & 10 \\ 1 & 7 & -8 & 0 & \vdots & -4 \\ 3 & 0 & 2 & 0 & \vdots & -10 \end{bmatrix}$$
$$\begin{cases} 9x + 12y + 3z = 0 \\ -2x + 18y + 5z + 2w = 10 \\ x + 7y - 8z = -4 \\ 3x + 2z = -10 \end{cases}$$

19.
$$\begin{bmatrix} 1 & 4 & 3 \\ 2 & 10 & 5 \end{bmatrix}$$
$$-2R_1 + R_2 \rightarrow \begin{bmatrix} 1 & 4 & 3 \\ 0 & \boxed{2} & -1 \end{bmatrix}$$

21.
$$\begin{bmatrix} 1 & 1 & 4 & -1 \\ 3 & 8 & 10 & 3 \\ -2 & 1 & 12 & 6 \end{bmatrix}$$

$$\begin{matrix} \\ -3R_1 + R_2 \rightarrow \\ 2R_1 + R_3 \rightarrow \end{matrix} \begin{bmatrix} 1 & 1 & 4 & -1 \\ 0 & 5 & \boxed{-2} & \boxed{6} \\ 0 & 3 & \boxed{20} & \boxed{4} \end{bmatrix}$$

$$\begin{matrix} \\ \tfrac{1}{5}R_2 \rightarrow \\ {} \end{matrix} \begin{bmatrix} 1 & 1 & 4 & -1 \\ 0 & 1 & -\tfrac{2}{5} & \tfrac{6}{5} \\ 0 & 3 & \boxed{20} & \boxed{4} \end{bmatrix}$$

23. $\begin{bmatrix} -2 & 5 & 1 \\ 3 & -1 & -8 \end{bmatrix} \rightarrow \begin{bmatrix} 13 & 0 & -39 \\ 3 & -1 & -8 \end{bmatrix}$

Add 5 times Row 2 to Row 1.

25. $\begin{bmatrix} 0 & -1 & -5 & 5 \\ -1 & 3 & -7 & 6 \\ 4 & -5 & 1 & 3 \end{bmatrix} \rightarrow \begin{bmatrix} -1 & 3 & -7 & 6 \\ 0 & -1 & -5 & 5 \\ 0 & 7 & -27 & 27 \end{bmatrix}$

Interchange Row 1 and Row 2. Then add 4 times the new Row 1 to Row 3.

27. $\begin{bmatrix} 1 & 2 & 3 \\ 2 & -1 & -4 \\ 3 & 1 & -1 \end{bmatrix}$

(a) $\begin{bmatrix} 1 & 2 & 3 \\ 0 & -5 & -10 \\ 3 & 1 & -1 \end{bmatrix}$

(b) $\begin{bmatrix} 1 & 2 & 3 \\ 0 & -5 & -10 \\ 0 & -5 & -10 \end{bmatrix}$

(c) $\begin{bmatrix} 1 & 2 & 3 \\ 0 & -5 & -10 \\ 0 & 0 & 0 \end{bmatrix}$

(d) $\begin{bmatrix} 1 & 2 & 3 \\ 0 & 1 & 2 \\ 0 & 0 & 0 \end{bmatrix}$

(e) $\begin{bmatrix} 1 & 0 & -1 \\ 0 & 1 & 2 \\ 0 & 0 & 0 \end{bmatrix}$ This matrix is in reduced row-echelon form.

29. $\begin{bmatrix} 1 & 0 & 0 & 0 \\ 0 & 1 & 1 & 5 \\ 0 & 0 & 0 & 0 \end{bmatrix}$

This matrix is in reduced row-echelon form.

31. $\begin{bmatrix} 2 & 0 & 4 & 0 \\ 0 & -1 & 3 & 6 \\ 0 & 0 & 1 & 5 \end{bmatrix}$

The first nonzero entries in Rows 1 and 2 are not 1. The matrix is not in row-echelon form.

33.
$$\begin{bmatrix} 1 & 1 & 0 & 5 \\ -2 & -1 & 2 & -10 \\ 3 & 6 & 7 & 14 \end{bmatrix}$$

$$\begin{matrix} 2R_1 + R_2 \rightarrow \\ -3R_1 + R_3 \rightarrow \end{matrix} \begin{bmatrix} 1 & 1 & 0 & 5 \\ 0 & 1 & 2 & 0 \\ 0 & 3 & 7 & -1 \end{bmatrix}$$

$$\begin{matrix} \\ -3R_2 + R_3 \rightarrow \end{matrix} \begin{bmatrix} 1 & 1 & 0 & 5 \\ 0 & 1 & 2 & 0 \\ 0 & 0 & 1 & -1 \end{bmatrix}$$

35.
$$\begin{bmatrix} 1 & -1 & -1 & 1 \\ 5 & -4 & 1 & 8 \\ -6 & 8 & 18 & 0 \end{bmatrix}$$

$$\begin{matrix} -5R_1 + R_2 \rightarrow \\ 6R_1 + R_3 \rightarrow \end{matrix} \begin{bmatrix} 1 & -1 & -1 & 1 \\ 0 & 1 & 6 & 3 \\ 0 & 2 & 12 & 6 \end{bmatrix}$$

$$\begin{matrix} \\ -2R_2 + R_3 \rightarrow \end{matrix} \begin{bmatrix} 1 & -1 & -1 & 1 \\ 0 & 1 & 6 & 3 \\ 0 & 0 & 0 & 0 \end{bmatrix}$$

37. Use the reduced row-echelon form feature of a graphing utility.

$$\begin{bmatrix} 3 & 3 & 3 \\ -1 & 0 & -4 \\ 2 & 4 & -2 \end{bmatrix} \Rightarrow \begin{bmatrix} 1 & 0 & 0 \\ 0 & 1 & 0 \\ 0 & 0 & 1 \end{bmatrix}$$

39. Use the reduced row-echelon form feature of a graphing utility.

$$\begin{bmatrix} 1 & 2 & 3 & -5 \\ 1 & 2 & 4 & -9 \\ -2 & -4 & -4 & 3 \\ 4 & 8 & 11 & -14 \end{bmatrix} \Rightarrow \begin{bmatrix} 1 & 2 & 0 & 0 \\ 0 & 0 & 1 & 0 \\ 0 & 0 & 0 & 1 \\ 0 & 0 & 0 & 0 \end{bmatrix}$$

41. Use the reduced row-echelon form feature of a graphing utility.

$$\begin{bmatrix} -3 & 5 & 1 & 12 \\ 1 & -1 & 1 & 4 \end{bmatrix} \Rightarrow \begin{bmatrix} 1 & 0 & 3 & 16 \\ 0 & 1 & 2 & 12 \end{bmatrix}$$

43. $\begin{cases} x - 2y = 4 \\ y = -3 \end{cases}$

$x - 2(-3) = 4$

$x = -2$

Solution: $(-2, -3)$

45. $\begin{cases} x - y + 2z = 4 \\ y - z = 2 \\ z = -2 \end{cases}$

$y - (-2) = 2$

$y = 0$

$x - 0 + 2(-2) = 4$

$x = 8$

Solution: $(8, 0, -2)$

47. $\begin{bmatrix} 1 & 0 & \vdots & 3 \\ 0 & 1 & \vdots & -4 \end{bmatrix}$

$x = 3$

$y = -4$

Solution: $(3, -4)$

49. $\begin{bmatrix} 1 & 0 & 0 & \vdots & -4 \\ 0 & 1 & 0 & \vdots & -10 \\ 0 & 0 & 1 & \vdots & 4 \end{bmatrix}$

$x = -4$

$y = -10$

$z = 4$

Solution: $(-4, -10, 4)$

51. $\begin{cases} x + 2y = 7 \\ 2x + y = 8 \end{cases}$

$$\begin{bmatrix} 1 & 2 & \vdots & 7 \\ 2 & 1 & \vdots & 8 \end{bmatrix}$$

$-2R_1 + R_2 \rightarrow \begin{bmatrix} 1 & 2 & \vdots & 7 \\ 0 & -3 & \vdots & -6 \end{bmatrix}$

$-\frac{1}{3}R_2 \rightarrow \begin{bmatrix} 1 & 2 & \vdots & 7 \\ 0 & 1 & \vdots & 2 \end{bmatrix}$

$\begin{cases} x + 2y = 7 \\ y = 2 \end{cases}$

$y = 2$

$x + 2(2) = 7 \implies x = 3$

Solution: $(3, 2)$

53. $\begin{cases} 3x - 2y = -27 \\ x + 3y = 13 \end{cases}$

$$\begin{bmatrix} 3 & -2 & \vdots & -27 \\ 1 & 3 & \vdots & 13 \end{bmatrix}$$

$\begin{matrix} R_1 \\ R_2 \end{matrix} \begin{bmatrix} 1 & 3 & \vdots & 13 \\ 3 & -2 & \vdots & -27 \end{bmatrix}$

$-3R_1 + R_2 \rightarrow \begin{bmatrix} 1 & 3 & \vdots & 13 \\ 0 & -11 & \vdots & -66 \end{bmatrix}$

$-\frac{1}{11}R_2 \rightarrow \begin{bmatrix} 1 & 3 & \vdots & 13 \\ 0 & 1 & \vdots & 6 \end{bmatrix}$

$\begin{cases} x + 3y = 13 \\ y = 6 \end{cases}$

$y = 6$

$x + 3(6) = 13 \implies x = -5$

Solution: $(-5, 6)$

55. $\begin{cases} -2x + 6y = -22 \\ x + 2y = -9 \end{cases}$

$$\begin{bmatrix} -2 & 6 & \vdots & -22 \\ 1 & 2 & \vdots & -9 \end{bmatrix}$$

$\begin{matrix} R_1 \\ R_2 \end{matrix} \begin{bmatrix} 1 & 2 & \vdots & -9 \\ -2 & 6 & \vdots & -22 \end{bmatrix}$

$2R_1 + R_2 \rightarrow \begin{bmatrix} 1 & 2 & \vdots & -9 \\ 0 & 10 & \vdots & -40 \end{bmatrix}$

$\frac{1}{10}R_2 \rightarrow \begin{bmatrix} 1 & 2 & \vdots & -9 \\ 0 & 1 & \vdots & -4 \end{bmatrix}$

$\begin{cases} x + 2y = -9 \\ y = -4 \end{cases}$

$y = -4$

$x + 2(-4) = -9 \implies x = -1$

Solution: $(-1, -4)$

57. $\begin{cases} -x + 2y = 1.5 \\ 2x - 4y = 3.0 \end{cases}$

$$\begin{bmatrix} -1 & 2 & \vdots & 1.5 \\ 2 & -4 & \vdots & 3.0 \end{bmatrix}$$

$2R_1 + R_2 \rightarrow \begin{bmatrix} -1 & 2 & \vdots & 1.5 \\ 0 & 0 & \vdots & 6.0 \end{bmatrix}$

The system is inconsistent and there is no solution.

59. $\begin{cases} x \quad\ - 3z = -2 \\ 3x + y - 2z = 5 \\ 2x + 2y + z = 4 \end{cases}$

$$\begin{bmatrix} 1 & 0 & -3 & \vdots & -2 \\ 3 & 1 & -2 & \vdots & 5 \\ 2 & 2 & 1 & \vdots & 4 \end{bmatrix}$$

$\begin{matrix} \\ -3R_1 + R_2 \to \\ -2R_1 + R_3 \to \end{matrix} \begin{bmatrix} 1 & 0 & -3 & \vdots & -2 \\ 0 & 1 & 7 & \vdots & 11 \\ 0 & 2 & 7 & \vdots & 8 \end{bmatrix}$

$\begin{matrix} \\ \\ -2R_2 + R_3 \to \end{matrix} \begin{bmatrix} 1 & 0 & -3 & \vdots & -2 \\ 0 & 1 & 7 & \vdots & 11 \\ 0 & 0 & -7 & \vdots & -14 \end{bmatrix}$

$\begin{matrix} \\ \\ -\frac{1}{7}R_3 \to \end{matrix} \begin{bmatrix} 1 & 0 & -3 & \vdots & -2 \\ 0 & 1 & 7 & \vdots & 11 \\ 0 & 0 & 1 & \vdots & 2 \end{bmatrix}$

$\begin{cases} x \ - 3z = -2 \\ y + 7z = 11 \\ \qquad z = 2 \end{cases}$

$z = 2$

$y + 7(2) = 11 \implies y = -3$

$x - 3(2) = -2 \implies x = 4$

Solution: $(4, -3, 2)$

61. $\begin{cases} -x + y - z = -14 \\ 2x - y + z = 21 \\ 3x + 2y + z = 19 \end{cases}$

$$\begin{bmatrix} -1 & 1 & -1 & \vdots & -14 \\ 2 & -1 & 1 & \vdots & 21 \\ 3 & 2 & 1 & \vdots & 19 \end{bmatrix}$$

$-R_1 \to \begin{bmatrix} 1 & -1 & 1 & \vdots & 14 \\ 2 & -1 & 1 & \vdots & 21 \\ 3 & 2 & 1 & \vdots & 19 \end{bmatrix}$

$\begin{matrix} \\ -2R_1 + R_2 \to \\ -3R_1 + R_3 \to \end{matrix} \begin{bmatrix} 1 & -1 & 1 & \vdots & 14 \\ 0 & 1 & -1 & \vdots & -7 \\ 0 & 5 & -2 & \vdots & -23 \end{bmatrix}$

$\begin{matrix} \\ \\ -5R_2 + R_3 \to \end{matrix} \begin{bmatrix} 1 & -1 & 1 & \vdots & 14 \\ 0 & 1 & -1 & \vdots & -7 \\ 0 & 0 & 3 & \vdots & 12 \end{bmatrix}$

$\begin{matrix} \\ \\ \frac{1}{3}R_3 \to \end{matrix} \begin{bmatrix} 1 & -1 & 1 & \vdots & 14 \\ 0 & 1 & -1 & \vdots & -7 \\ 0 & 0 & 1 & \vdots & 4 \end{bmatrix}$

$\begin{cases} x - y + z = 14 \\ y - z = -7 \\ \qquad z = 4 \end{cases}$

$z = 4$

$y - 4 = -7 \implies y = -3$

$x - (-3) + 4 = 14 \implies x = 7$

Solution: $(7, -3, 4)$

63. $\begin{cases} x + 2y - 3z = -28 \\ 4y + 2z = 0 \\ -x + y - z = -5 \end{cases}$

$$\begin{bmatrix} 1 & 2 & -3 & \vdots & -28 \\ 0 & 4 & 2 & \vdots & 0 \\ -1 & 1 & -1 & \vdots & -5 \end{bmatrix}$$

$\begin{matrix} \\ \frac{1}{4}R_2 \to \\ R_1 + R_3 \to \end{matrix} \begin{bmatrix} 1 & 2 & -3 & \vdots & -28 \\ 0 & 1 & \frac{1}{2} & \vdots & 0 \\ 0 & 3 & -4 & \vdots & -33 \end{bmatrix}$

$\begin{matrix} \\ \\ -3R_2 + R_3 \to \end{matrix} \begin{bmatrix} 1 & 2 & -3 & \vdots & -28 \\ 0 & 1 & \frac{1}{2} & \vdots & 0 \\ 0 & 0 & -\frac{11}{2} & \vdots & -33 \end{bmatrix}$

$\begin{matrix} \\ \\ -\frac{2}{11}R_3 \to \end{matrix} \begin{bmatrix} 1 & 2 & -3 & \vdots & -28 \\ 0 & 1 & \frac{1}{2} & \vdots & 0 \\ 0 & 0 & 1 & \vdots & 6 \end{bmatrix}$

$\begin{cases} x + 2y - 3z = -28 \\ y + \frac{1}{2}z = 0 \\ \qquad z = 6 \end{cases}$

$z = 6$

$y + \frac{1}{2}(6) = 0 \implies y = -3$

$x + 2(-3) - 3(6) = -28 \implies x = -4$

Solution: $(-4, -3, 6)$

65. $\begin{cases} x + y - 5z = 3 \\ x \quad\ - 2z = 1 \\ 2x - y - z = 0 \end{cases}$

$$\begin{bmatrix} 1 & 1 & -5 & \vdots & 3 \\ 1 & 0 & -2 & \vdots & 1 \\ 2 & -1 & -1 & \vdots & 0 \end{bmatrix}$$

$\begin{matrix} \\ -R_1 + R_2 \to \\ -2R_1 + R_3 \to \end{matrix} \begin{bmatrix} 1 & 1 & -5 & \vdots & 3 \\ 0 & -1 & 3 & \vdots & -2 \\ 0 & -3 & 9 & \vdots & -6 \end{bmatrix}$

$\begin{matrix} \\ -R_2 \to \\ \\ \end{matrix} \begin{bmatrix} 1 & 1 & -5 & \vdots & 3 \\ 0 & 1 & -3 & \vdots & 2 \\ 0 & -3 & 9 & \vdots & -6 \end{bmatrix}$

$\begin{matrix} -R_2 + R_1 \to \\ \\ 3R_2 + R_3 \to \end{matrix} \begin{bmatrix} 1 & 0 & -2 & \vdots & 1 \\ 0 & 1 & -3 & \vdots & 2 \\ 0 & 0 & 0 & \vdots & 0 \end{bmatrix}$

$\begin{cases} x \ - 2z = 1 \\ y - 3z = 2 \end{cases}$

Let $z = a$.

$y - 3a = 2 \implies y = 3a + 2$

$x - 2a = 1 \implies x = 2a + 1$

Solution: $(2a + 1, 3a + 2, a)$ where a is any real number.

67. $\begin{cases} x + 2y + z + 2w = 8 \\ 3x + 7y + 6z + 9w = 26 \end{cases}$

$$\begin{bmatrix} 1 & 2 & 1 & 2 & \vdots & 8 \\ 3 & 7 & 6 & 9 & \vdots & 26 \end{bmatrix}$$

$-3R_1 + R_2 \rightarrow \begin{bmatrix} 1 & 2 & 1 & 2 & \vdots & 8 \\ 0 & 1 & 3 & 3 & \vdots & 2 \end{bmatrix}$

$-2R_2 + R_1 \rightarrow \begin{bmatrix} 1 & 0 & -5 & -4 & \vdots & 4 \\ 0 & 1 & 3 & 3 & \vdots & 2 \end{bmatrix}$

$\begin{cases} x - 5z - 4w = 4 \\ y + 3z + 3w = 2 \end{cases}$

Let $w = a$ and $z = b$.

$y + 3b + 3a = 2 \implies y = 2 - 3b - 3a$

$x - 5b - 4a = 4 \implies x = 4 + 5b + 4a$

Solution: $(4 + 5b + 4a, 2 - 3b - 3a, b, a)$
where a and b are real numbers

69. $\begin{cases} -x + y = -22 \\ 3x + 4y = 4 \\ 4x - 8y = 32 \end{cases}$

$$\begin{bmatrix} -1 & 1 & \vdots & -22 \\ 3 & 4 & \vdots & 4 \\ 4 & -8 & \vdots & 32 \end{bmatrix}$$

$-R_1 \rightarrow \begin{bmatrix} 1 & -1 & \vdots & 22 \\ 3 & 4 & \vdots & 4 \\ 4 & -8 & \vdots & 32 \end{bmatrix}$

$\begin{matrix} -3R_1 + R_2 \rightarrow \\ -4R_1 + R_3 \rightarrow \end{matrix} \begin{bmatrix} 1 & -1 & \vdots & 22 \\ 0 & 7 & \vdots & -62 \\ 0 & -4 & \vdots & -56 \end{bmatrix}$

$\begin{matrix} \frac{1}{7}R_2 \rightarrow \\ -\frac{1}{4}R_3 \rightarrow \end{matrix} \begin{bmatrix} 1 & -1 & \vdots & 22 \\ 0 & 1 & \vdots & -\frac{62}{7} \\ 0 & 1 & \vdots & 14 \end{bmatrix}$

$-R_2 + R_3 \rightarrow \begin{bmatrix} 1 & -1 & \vdots & 22 \\ 0 & 1 & \vdots & -\frac{62}{7} \\ 0 & 0 & \vdots & \frac{160}{7} \end{bmatrix}$

The system is inconsistent and there is no solution.

71. Use the reduced row-echelon form feature of a graphing utility.

$\begin{cases} 3x + 3y + 12z = 6 \\ x + y + 4z = 2 \\ 2x + 5y + 20z = 10 \\ -x + 2y + 8z = 4 \end{cases}$ $\begin{bmatrix} 3 & 3 & 12 & \vdots & 6 \\ 1 & 1 & 4 & \vdots & 2 \\ 2 & 5 & 20 & \vdots & 10 \\ -1 & 2 & 8 & \vdots & 4 \end{bmatrix} \implies \begin{bmatrix} 1 & 0 & 0 & \vdots & 0 \\ 0 & 1 & 4 & \vdots & 2 \\ 0 & 0 & 0 & \vdots & 0 \\ 0 & 0 & 0 & \vdots & 0 \end{bmatrix} \implies \begin{cases} x = 0 \\ y + 4z = 2 \end{cases}$

Let $z = a$.

$y = 2 - 4a$

$x = 0$

Solution: $(0, 2 - 4a, a)$ where a is any real number

73. Use the reduced row-echelon form feature of a graphing utility.

$\begin{cases} 2x + y - z + 2w = -6 \\ 3x + 4y + w = 1 \\ x + 5y + 2z + 6w = -3 \\ 5x + 2y - z - w = 3 \end{cases}$ $\begin{bmatrix} 2 & 1 & -1 & 2 & \vdots & -6 \\ 3 & 4 & 0 & 1 & \vdots & 1 \\ 1 & 5 & 2 & 6 & \vdots & -3 \\ 5 & 2 & -1 & -1 & \vdots & 3 \end{bmatrix} \implies \begin{bmatrix} 1 & 0 & 0 & 0 & \vdots & 1 \\ 0 & 1 & 0 & 0 & \vdots & 0 \\ 0 & 0 & 1 & 0 & \vdots & 4 \\ 0 & 0 & 0 & 1 & \vdots & -2 \end{bmatrix}$

$x = 1$

$y = 0$

$z = 4$

$w = -2$

Solution: $(1, 0, 4, -2)$

75. Use the reduced row-echelon form feature of a graphing utility.

$$\begin{cases} x + y + z + w = 0 \\ 2x + 3y + z - 2w = 0 \\ 3x + 5y + z = 0 \end{cases} \quad \begin{bmatrix} 1 & 1 & 1 & 1 & \vdots & 0 \\ 2 & 3 & 1 & -2 & \vdots & 0 \\ 3 & 5 & 1 & 0 & \vdots & 0 \end{bmatrix} \Rightarrow \begin{bmatrix} 1 & 0 & 2 & 0 & \vdots & 0 \\ 0 & 1 & -1 & 0 & \vdots & 0 \\ 0 & 0 & 0 & 1 & \vdots & 0 \end{bmatrix}$$

$$\begin{cases} x + 2z = 0 \\ y - z = 0 \\ w = 0 \end{cases}$$

Let $z = a$. Then $x = -2a$ and $y = a$.

Solution: $(-2a, a, a, 0)$ where a is a real number

77. (a) $\begin{cases} x - 2y + z = -6 \\ y - 5z = 16 \\ z = -3 \end{cases}$

$$y - 5(-3) = 16$$

$$y = 1$$

$$x - 2(1) + (-3) = -6$$

$$x = -1$$

Solution: $(-1, 1, -3)$

(b) $\begin{cases} x + y - 2z = 6 \\ y + 3z = -8 \\ z = -3 \end{cases}$

$$y + 3(-3) = -8$$

$$y = 1$$

$$x + (1) - 2(-3) = 6$$

$$x = -1$$

Solution: $(-1, 1, -3)$

Both systems yield the same solution, namely $(-1, 1, -3)$.

79. (a) $\begin{cases} x - 4y + 5z = 27 \\ y - 7z = -54 \\ z = 8 \end{cases}$

$$y - 7(8) = -54$$

$$y = 2$$

$$x - 4(2) + 5(8) = 27$$

$$x = -5$$

Solution: $(-5, 2, 8)$

(b) $\begin{cases} x - 6y + z = 15 \\ y + 5z = 42 \\ z = 8 \end{cases}$

$$y + 5(8) = 42$$

$$y = 2$$

$$x - 6(2) + (8) = 15$$

$$x = 19$$

Solution: $(19, 2, 8)$

The systems do *not* yield the same solution.

81. $\begin{cases} x + 3y + z = 3 \\ x + 5y + 5z = 1 \\ 2x + 6y + 3z = 8 \end{cases}$

$$\begin{bmatrix} 1 & 3 & 1 & \vdots & 3 \\ 1 & 5 & 5 & \vdots & 1 \\ 2 & 6 & 3 & \vdots & 8 \end{bmatrix}$$

$$\begin{matrix} \\ -R_1 + R_2 \rightarrow \\ -2R_1 + R_3 \rightarrow \end{matrix} \begin{bmatrix} 1 & 3 & 1 & \vdots & 3 \\ 0 & 2 & 4 & \vdots & -2 \\ 0 & 0 & 1 & \vdots & 2 \end{bmatrix}$$

$$\tfrac{1}{2}R_2 \rightarrow \begin{bmatrix} 1 & 3 & 1 & \vdots & 3 \\ 0 & 1 & 2 & \vdots & -1 \\ 0 & 0 & 1 & \vdots & 2 \end{bmatrix}$$ This is a matrix in row-echelon form.

$$\begin{bmatrix} 1 & 3 & \frac{3}{2} & \vdots & 4 \\ 0 & 1 & \frac{7}{4} & \vdots & -\frac{3}{2} \\ 0 & 0 & 1 & \vdots & 2 \end{bmatrix}$$ The row-echelon form feature of a graphing utility yields this form.

There are infinitely many matrices in row-echelon form that correspond to the original system of equations. All such matrices will yield the same solution, namely $(16, -5, 2)$.

83. $\dfrac{4x^2}{(x+1)^2(x-1)} = \dfrac{A}{x-1} + \dfrac{B}{x+1} + \dfrac{C}{(x+1)^2}$

$$4x^2 = A(x+1)^2 + B(x+1)(x-1) + C(x-1)$$

$$4x^2 = A(x^2 + 2x + 1) + B(x^2 - 1) + C(x - 1)$$

$$4x^2 = (A + B)x^2 + (2A + C)x + (A - B - C)$$

System of equations: $\begin{aligned} A + B &= 4 \\ 2A + C &= 0 \\ A - B - C &= 0 \end{aligned}$

$$\begin{bmatrix} 1 & 1 & 0 & \vdots & 4 \\ 2 & 0 & 1 & \vdots & 0 \\ 1 & -1 & -1 & \vdots & 0 \end{bmatrix} \overset{\text{rref}}{\rightarrow} \begin{bmatrix} 1 & 0 & 0 & \vdots & 1 \\ 0 & 1 & 0 & \vdots & 3 \\ 0 & 0 & 1 & \vdots & -2 \end{bmatrix}$$

Thus, $A = 1, B = 3, C = -2$.

So, $\dfrac{4x^2}{(x+1)^2(x-1)} = \dfrac{1}{x-1} + \dfrac{3}{x+1} - \dfrac{2}{(x+1)^2}$.

85. x = amount at 7%

y = amount at 8%,

z = amount at 10%

$z = 4x \implies -4x + z = 0$

$$\begin{cases} x + y + z = 1{,}500{,}000 \\ 0.07x + 0.08y + 0.10z = 130{,}500 \\ -4x + z = 0 \end{cases}$$

$$\begin{bmatrix} 1 & 1 & 1 & 1{,}500{,}000 \\ 0.07 & 0.08 & 0.10 & 130{,}500 \\ -4 & 0 & 1 & 0 \end{bmatrix}$$

$$\begin{array}{c} -0.07R_1 + R_2 \to \\ 4R_1 + R_3 \to \end{array} \begin{bmatrix} 1 & 1 & 1 & \vdots & 1{,}500{,}000 \\ 0 & 0.01 & 0.03 & \vdots & 25{,}500 \\ 0 & 4 & 5 & \vdots & 6{,}000{,}000 \end{bmatrix}$$

$$100R_2 \to \begin{bmatrix} 1 & 1 & 1 & \vdots & 1{,}500{,}000 \\ 0 & 1 & 3 & \vdots & 2{,}550{,}000 \\ 0 & 4 & 5 & \vdots & 6{,}000{,}000 \end{bmatrix}$$

$$-4R_2 + R_3 \to \begin{bmatrix} 1 & 1 & 1 & \vdots & 1{,}500{,}000 \\ 0 & 1 & 3 & \vdots & 2{,}550{,}000 \\ 0 & 0 & -7 & \vdots & -4{,}200{,}000 \end{bmatrix}$$

$$-\tfrac{1}{7}R_3 \to \begin{bmatrix} 1 & 1 & 1 & \vdots & 1{,}500{,}000 \\ 0 & 1 & 3 & \vdots & 2{,}550{,}000 \\ 0 & 0 & 1 & \vdots & 600{,}000 \end{bmatrix}$$

$$\begin{cases} x + y + z = 1{,}500{,}000 \\ y + 3z = 2{,}550{,}000 \\ z = 600{,}000 \end{cases}$$

$y + 3(600{,}000) = 2{,}550{,}000 \implies y = 750{,}000$

$x + 750{,}000 + 600{,}000 = 1{,}500{,}000 \implies x = 150{,}000$

Solution: $150,000 at 7%, $750,000 at 8%, and $600,000 at 10%

87. $y = ax^2 + bx + c$

$$\begin{cases} a + b + c = 8 \\ 4a + 2b + c = 13 \\ 9a + 3b + c = 20 \end{cases}$$

$$\begin{bmatrix} 1 & 1 & 1 & \vdots & 8 \\ 4 & 2 & 1 & \vdots & 13 \\ 9 & 3 & 1 & \vdots & 20 \end{bmatrix}$$

$$\begin{array}{c} -4R_1 + R_2 \to \\ -9R_1 + R_3 \to \end{array} \begin{bmatrix} 1 & 1 & 1 & \vdots & 8 \\ 0 & -2 & -3 & \vdots & -19 \\ 0 & -6 & -8 & \vdots & -52 \end{bmatrix}$$

$$\begin{array}{c} -\tfrac{1}{2}R_2 \to \\ -3R_2 + R_3 \to \end{array} \begin{bmatrix} 1 & 1 & 1 & \vdots & 8 \\ 0 & 1 & \tfrac{3}{2} & \vdots & \tfrac{19}{2} \\ 0 & 0 & 1 & \vdots & 5 \end{bmatrix}$$

$$\begin{cases} a + b + c = 8 \\ b + \tfrac{3}{2}c = \tfrac{19}{2} \\ c = 5 \end{cases}$$

$c = 5$

$b + \tfrac{3}{2}(5) = \tfrac{19}{2} \implies b = 2$

$a + 2 + 5 = 8 \implies a = 1$

Equation of parabola: $y = x^2 + 2x + 5$

89. (a) $(0, 5.0)$, $(15, 9.6)$, $(30, 12.4)$

$y = ax^2 + bx + c$

$$\begin{cases} c = 5 \\ 225a + 15b + c = 9.6 \implies 225a + 15b = 4.6 \\ 900a + 30b + c = 12.4 \implies 900a + 30b = 7.4 \end{cases}$$

$$\begin{bmatrix} 225 & 15 & \vdots & 4.6 \\ 900 & 30 & \vdots & 7.4 \end{bmatrix}$$

$$-4R_1 + R_2 \to \begin{bmatrix} 225 & 15 & \vdots & 4.6 \\ 0 & -30 & \vdots & -11 \end{bmatrix}$$

$$\begin{array}{c} \tfrac{1}{225}R_1 \to \\ \left(-\tfrac{1}{30}\right)R_2 \to \end{array} \begin{bmatrix} 1 & \tfrac{1}{15} & \vdots & \tfrac{23}{1125} \\ 0 & 1 & \vdots & \tfrac{11}{30} \end{bmatrix}$$

$$\begin{cases} a + \tfrac{1}{15}b = \tfrac{23}{1125} \\ b = \tfrac{11}{30} \end{cases}$$

$a + \tfrac{1}{15}\left(\tfrac{11}{30}\right) = \tfrac{23}{1125} \implies a = -\tfrac{1}{250} = -0.004$

Equation of parabola: $y = -0.004x^2 + 0.367x + 5$.

(b)

(c) The maximum height is approximately 13 feet and the ball strikes the ground at approximately 104 feet.

(d) The maximum occurs at the vertex.

$x = -\dfrac{b}{2a} = \dfrac{-0.367}{2(-0.004)} = 45.875$

$y = -0.004(45.875)^2 + 0.367(45.875) + 5$

$= 13.418$ feet

The ball strikes the ground when $y = 0$.

$-0.004x^2 + 0.367x + 5 = 0$

By the Quadratic Formula and using the positive value for x we have $x \approx 103.793$ feet.

(e) The values found in part (d) are more accurate, but still very close to the estimates found in part (c).

91. (a) $x_1 + x_3 = 600$

$x_1 = x_2 + x_4 \implies x_1 - x_2 - x_4 = 0$

$x_2 + x_5 = 500$

$x_3 + x_6 = 600$

$x_4 + x_7 = x_6 \implies x_4 - x_6 + x_7 = 0$

$x_5 + x_7 = 500$

$$\left[\begin{array}{ccccccc:c} 1 & 0 & 1 & 0 & 0 & 0 & 0 & 600 \\ 1 & -1 & 0 & -1 & 0 & 0 & 0 & 0 \\ 0 & 1 & 0 & 0 & 1 & 0 & 0 & 500 \\ 0 & 0 & 1 & 0 & 0 & 1 & 0 & 600 \\ 0 & 0 & 0 & 1 & 0 & -1 & 1 & 0 \\ 0 & 0 & 0 & 0 & 1 & 0 & 1 & 500 \end{array}\right]$$

$$\begin{array}{c} \\ -R_1 + R_2 \to \\ R_2 + R_3 \to \\ R_3 + R_4 \to \\ R_4 + R_5 \to \\ -R_5 + R_6 \to \end{array} \left[\begin{array}{ccccccc:c} 1 & 0 & 1 & 0 & 0 & 0 & 0 & 600 \\ 0 & -1 & -1 & -1 & 0 & 0 & 0 & -600 \\ 0 & 0 & -1 & -1 & 1 & 0 & 0 & -100 \\ 0 & 0 & 0 & -1 & 1 & 1 & 0 & 500 \\ 0 & 0 & 0 & 0 & 1 & 0 & 1 & 500 \\ 0 & 0 & 0 & 0 & 0 & 0 & 0 & 0 \end{array}\right]$$

$$\begin{array}{c} \\ -R_3 + R_2 \to \\ -R_4 + R_3 \to \\ -R_4 \to \\ \\ \end{array} \left[\begin{array}{ccccccc:c} 1 & 0 & 1 & 0 & 0 & 0 & 0 & 600 \\ 0 & -1 & 0 & 0 & -1 & 0 & 0 & -500 \\ 0 & 0 & -1 & 0 & 0 & -1 & 0 & -600 \\ 0 & 0 & 0 & 1 & -1 & -1 & 0 & -500 \\ 0 & 0 & 0 & 0 & 1 & 0 & 1 & 500 \\ 0 & 0 & 0 & 0 & 0 & 0 & 0 & 0 \end{array}\right]$$

$$\begin{array}{c} \\ -R_2 \to \\ -R_3 \to \\ \\ \\ \end{array} \left[\begin{array}{ccccccc:c} 1 & 0 & 1 & 0 & 0 & 0 & 0 & 600 \\ 0 & 1 & 0 & 0 & 1 & 0 & 0 & 500 \\ 0 & 0 & 1 & 0 & 0 & 1 & 0 & 600 \\ 0 & 0 & 0 & 1 & -1 & -1 & 0 & -500 \\ 0 & 0 & 0 & 0 & 1 & 0 & 1 & 500 \\ 0 & 0 & 0 & 0 & 0 & 0 & 0 & 0 \end{array}\right]$$

$$\begin{cases} x_1 + x_3 = 600 \\ x_2 + x_5 = 500 \\ x_3 + x_6 = 600 \\ x_4 - x_5 - x_6 = -500 \\ x_5 + x_7 = 500 \end{cases}$$

Let $x_7 = t$ and $x_6 = s$, then $x_5 = 500 - t$,

$x_4 = -500 + s + (500 - t) = s - t$,

$x_3 = 600 - s$, $x_2 = 500 - (500 - t) = t$,

$x_1 = 600 - (600 - s) = s$.

Solution: $(s, t, 600 - s, s - t, 500 - t, s, t)$

(b) $s = 0, t = 0$: $x_1 = 0, x_2 = 0, x_3 = 600, x_4 = 0, x_5 = 500, x_6 = 0, x_7 = 0$

(c) $s = 0, t = -500$: $x_1 = 0, x_2 = -500, x_3 = 600, x_4 = 500, x_5 = 1000, x_6 = 0, x_7 = -500$

93. False. It is a 2×4 matrix.

95. False. Gaussian elimination reduces a matrix until a row-echelon form is obtained and Gauss-Jordan elimination reduces a matrix until a reduced row-echelon form is obtained.

97. (a) In the row-echelon form of an augmented matrix that corresponds to an inconsistent system of linear equations, there exists a row consisting of all zeros except for the entry in the last column.

(b) In the row-echelon form of an augmented matrix that corresponds to a system with an infinite number of solutions, there are fewer rows with nonzero entries than there are variables and no row has the first non-zero value in the last column.

99. They are the same.

101. $f(x) = \dfrac{2x^2 - 4x}{3x - x^2} = \dfrac{2x - 4}{3 - x}, x \neq 0$

x	-2	-1	0	1	2	3	4	5
$f(x)$	-1.6	-1.5	undef.	-1	0	undef.	-4	-3

Vertical asymptote: $x = 3$

Horizontal asymptote: $y = -2$

Intercept: $(2, 0)$

103. $f(x) = 2^{x-1}$

x	-1	0	1	2	3
$f(x)$	$\frac{1}{4}$	$\frac{1}{2}$	1	2	4

Horizontal asymptote: $y = 0$

Intercept: $\left(0, \frac{1}{2}\right)$

105. $h(x) = \ln(x - 1)$

x	1.5	2	3	4	5
$h(x)$	-0.693	0	0.693	1.099	1.386

Vertical asymptote: $x = 1$

Intercept: $(2, 0)$

Section 10.2   Operations with Matrices

- $A = B$ if and only if they have the same order and $a_{ij} = b_{ij}$.
- You should be able to perform the operations of matrix addition, scalar multiplication, and matrix multiplication.
- Some properties of matrix addition and scalar multiplication are:

  (a) $A + B = B + A$

  (b) $A + (B + C) = (A + B) + C$

  (c) $(cd)A = c(dA)$

  (d) $1A = A$

  (e) $c(A + B) = cA + cB$

  (f) $(c + d)A = cA + dA$

- You should remember that $AB \neq BA$ in general.
- Some properties of matrix multiplication are:

  (a) $A(BC) = (AB)C$

  (b) $A(B + C) = AB + AC$

  (c) $(A + B)C = AC + BC$

  (d) $c(AB) = (cA)B = A(cB)$

- You should know that I_n, the identity matrix of order n, is an $n \times n$ matrix consisting of ones on its main diagonal and zeros elsewhere. If A is an $n \times n$ matrix, then $AI_n = I_nA = A$.

Vocabulary Check

1. equal

2. scalars

3. zero; O

4. identity

5. (a) (iii)   (b) (iv)   (c) (i)   (d) (v)   (e) (ii)

6. (a) (ii)   (b) (iv)   (c) (i)   (d) (iii)

---

1. $x = -4, \ y = 22$

3. $2x + 1 = 5, \ 3x = 6, \ 3y - 5 = 4$

$x = 2, \ y = 3$

5. (a) $A + B = \begin{bmatrix} 1 & -1 \\ 2 & -1 \end{bmatrix} + \begin{bmatrix} 2 & -1 \\ -1 & 8 \end{bmatrix} = \begin{bmatrix} 1+2 & -1-1 \\ 2-1 & -1+8 \end{bmatrix} = \begin{bmatrix} 3 & -2 \\ 1 & 7 \end{bmatrix}$

    (b) $A - B = \begin{bmatrix} 1 & -1 \\ 2 & -1 \end{bmatrix} - \begin{bmatrix} 2 & -1 \\ -1 & 8 \end{bmatrix} = \begin{bmatrix} 1-2 & -1+1 \\ 2+1 & -1-8 \end{bmatrix} = \begin{bmatrix} -1 & 0 \\ 3 & -9 \end{bmatrix}$

    (c) $3A = 3\begin{bmatrix} 1 & -1 \\ 2 & -1 \end{bmatrix} = \begin{bmatrix} 3(1) & 3(-1) \\ 3(2) & 3(-1) \end{bmatrix} = \begin{bmatrix} 3 & -3 \\ 6 & -3 \end{bmatrix}$

    (d) $3A - 2B = \begin{bmatrix} 3 & -3 \\ 6 & -3 \end{bmatrix} - 2\begin{bmatrix} 2 & -1 \\ -1 & 8 \end{bmatrix} = \begin{bmatrix} 3 & -3 \\ 6 & -3 \end{bmatrix} + \begin{bmatrix} -4 & 2 \\ 2 & -16 \end{bmatrix} = \begin{bmatrix} -1 & -1 \\ 8 & -19 \end{bmatrix}$

7. $A = \begin{bmatrix} 6 & -1 \\ 2 & 4 \\ -3 & 5 \end{bmatrix}, \ B = \begin{bmatrix} 1 & 4 \\ -1 & 5 \\ 1 & 10 \end{bmatrix}$

    (a) $A + B = \begin{bmatrix} 7 & 3 \\ 1 & 9 \\ -2 & 15 \end{bmatrix}$ (b) $A - B = \begin{bmatrix} 5 & -5 \\ 3 & -1 \\ -4 & -5 \end{bmatrix}$ (c) $3A = \begin{bmatrix} 18 & -3 \\ 6 & 12 \\ -9 & 15 \end{bmatrix}$

    (d) $3A - 2B = \begin{bmatrix} 18 & -3 \\ 6 & 12 \\ -9 & 15 \end{bmatrix} - \begin{bmatrix} 2 & 8 \\ -2 & 10 \\ 2 & 20 \end{bmatrix} = \begin{bmatrix} 16 & -11 \\ 8 & 2 \\ -11 & -5 \end{bmatrix}$

9. $A = \begin{bmatrix} 2 & 2 & -1 & 0 & 1 \\ 1 & 1 & -2 & 0 & -1 \end{bmatrix}, \ B = \begin{bmatrix} 1 & 1 & -1 & 1 & 0 \\ -3 & 4 & 9 & -6 & -7 \end{bmatrix}$

    (a) $A + B = \begin{bmatrix} 3 & 3 & -2 & 1 & 1 \\ -2 & 5 & 7 & -6 & -8 \end{bmatrix}$ (b) $A - B = \begin{bmatrix} 1 & 1 & 0 & -1 & 1 \\ 4 & -3 & -11 & 6 & 6 \end{bmatrix}$

    (c) $3A = \begin{bmatrix} 6 & 6 & -3 & 0 & 3 \\ 3 & 3 & -6 & 0 & -3 \end{bmatrix}$

    (d) $3A - 2B = \begin{bmatrix} 6 & 6 & -3 & 0 & 3 \\ 3 & 3 & -6 & 0 & -3 \end{bmatrix} - \begin{bmatrix} 2 & 2 & -2 & 2 & 0 \\ -6 & 8 & 18 & -12 & -14 \end{bmatrix} = \begin{bmatrix} 4 & 4 & -1 & -2 & 3 \\ 9 & -5 & -24 & 12 & 11 \end{bmatrix}$

11. $A = \begin{bmatrix} 6 & 0 & 3 \\ -1 & -4 & 0 \end{bmatrix}, \ B = \begin{bmatrix} 8 & -1 \\ 4 & -3 \end{bmatrix}$

    (a) $A + B$ is not possible. A and B do not have the same order.

    (b) $A - B$ is not possible. A and B do not have the same order.

    (c) $3A = \begin{bmatrix} 18 & 0 & 9 \\ -3 & -12 & 0 \end{bmatrix}$

    (d) $3A - 2B$ is not possible. A and B do not have the same order.

13. $\begin{bmatrix} -5 & 0 \\ 3 & -6 \end{bmatrix} + \begin{bmatrix} 7 & 1 \\ -2 & -1 \end{bmatrix} + \begin{bmatrix} -10 & -8 \\ 14 & 6 \end{bmatrix} = \begin{bmatrix} -5+7+(-10) & 0+1+(-8) \\ 3+(-2)+14 & -6+(-1)+6 \end{bmatrix} = \begin{bmatrix} -8 & -7 \\ 15 & -1 \end{bmatrix}$

15. $4\left(\begin{bmatrix} -4 & 0 & 1 \\ 0 & 2 & 3 \end{bmatrix} - \begin{bmatrix} 2 & 1 & -2 \\ 3 & -6 & 0 \end{bmatrix}\right) = 4\begin{bmatrix} -6 & -1 & 3 \\ -3 & 8 & 3 \end{bmatrix} = \begin{bmatrix} -24 & -4 & 12 \\ -12 & 32 & 12 \end{bmatrix}$

17. $-3\left(\begin{bmatrix} 0 & -3 \\ 7 & 2 \end{bmatrix} + \begin{bmatrix} -6 & 3 \\ 8 & 1 \end{bmatrix}\right) - 2\begin{bmatrix} 4 & -4 \\ 7 & -9 \end{bmatrix} = -3\begin{bmatrix} -6 & 0 \\ 15 & 3 \end{bmatrix} - \begin{bmatrix} 8 & -8 \\ 14 & -18 \end{bmatrix} = \begin{bmatrix} 18 & 0 \\ -45 & -9 \end{bmatrix} - \begin{bmatrix} 8 & -8 \\ 14 & -18 \end{bmatrix} = \begin{bmatrix} 10 & 8 \\ -59 & 9 \end{bmatrix}$

19. $\frac{3}{7}\begin{bmatrix} 2 & 5 \\ -1 & -4 \end{bmatrix} + 6\begin{bmatrix} -3 & 0 \\ 2 & 2 \end{bmatrix} \approx \begin{bmatrix} -17.143 & 2.143 \\ 11.571 & 10.286 \end{bmatrix}$

21. $-\begin{bmatrix} 3.211 & 6.829 \\ -1.004 & 4.914 \\ 0.055 & -3.889 \end{bmatrix} - \begin{bmatrix} -1.630 & -3.090 \\ 5.256 & 8.335 \\ -9.768 & 4.251 \end{bmatrix} = \begin{bmatrix} -1.581 & -3.739 \\ -4.252 & -13.249 \\ 9.713 & -0.362 \end{bmatrix}$

23. $X = 3\begin{bmatrix} -2 & -1 \\ 1 & 0 \\ 3 & -4 \end{bmatrix} - 2\begin{bmatrix} 0 & 3 \\ 2 & 0 \\ -4 & -1 \end{bmatrix} = \begin{bmatrix} -6 & -3 \\ 3 & 0 \\ 9 & -12 \end{bmatrix} - \begin{bmatrix} 0 & 6 \\ 4 & 0 \\ -8 & -2 \end{bmatrix} = \begin{bmatrix} -6 & -9 \\ -1 & 0 \\ 17 & -10 \end{bmatrix}$

25. $X = -\frac{3}{2}A + \frac{1}{2}B = -\frac{3}{2}\begin{bmatrix} -2 & -1 \\ 1 & 0 \\ 3 & -4 \end{bmatrix} + \frac{1}{2}\begin{bmatrix} 0 & 3 \\ 2 & 0 \\ -4 & -1 \end{bmatrix} = \begin{bmatrix} 3 & \frac{3}{2} \\ -\frac{3}{2} & 0 \\ -\frac{9}{2} & 6 \end{bmatrix} + \begin{bmatrix} 0 & \frac{3}{2} \\ 1 & 0 \\ -2 & -\frac{1}{2} \end{bmatrix} = \begin{bmatrix} 3 & 3 \\ -\frac{1}{2} & 0 \\ -\frac{13}{2} & \frac{11}{2} \end{bmatrix}$

27. A is 3×2 and B is 3×3. AB is not possible.

29. A is 3×3, B is $3 \times 2 \implies AB$ is 3×2.

$\begin{bmatrix} 0 & -1 & 0 \\ 4 & 0 & 2 \\ 8 & -1 & 7 \end{bmatrix}\begin{bmatrix} 2 & 1 \\ -3 & 4 \\ 1 & 6 \end{bmatrix} = \begin{bmatrix} (0)(2) + (-1)(-3) + (0)(1) & (0)(1) + (-1)(4) + (0)(6) \\ (4)(2) + (0)(-3) + (2)(1) & (4)(1) + (0)(4) + (2)(6) \\ (8)(2) + (-1)(-3) + (7)(1) & (8)(1) + (-1)(4) + (7)(6) \end{bmatrix} = \begin{bmatrix} 3 & -4 \\ 10 & 16 \\ 26 & 46 \end{bmatrix}$

31. A is 3×3, B is $3 \times 3 \implies AB$ is 3×3.

$\begin{bmatrix} 1 & 0 & 0 \\ 0 & 4 & 0 \\ 0 & 0 & -2 \end{bmatrix}\begin{bmatrix} 3 & 0 & 0 \\ 0 & -1 & 0 \\ 0 & 0 & 5 \end{bmatrix} = \begin{bmatrix} (1)(3) + (0)(0) + (0)(0) & (1)(0) + (0)(-1) + (0)(0) & (1)(0) + (0)(0) + (0)(5) \\ (0)(3) + (4)(0) + (0)(0) & (0)(0) + (4)(-1) + (0)(0) & (0)(0) + (4)(0) + (0)(5) \\ (0)(3) + (0)(0) + (-2)(0) & (0)(0) + (0)(-1) + (-2)(0) & (0)(0) + (0)(0) + (-2)(5) \end{bmatrix}$

$= \begin{bmatrix} 3 & 0 & 0 \\ 0 & -4 & 0 \\ 0 & 0 & -10 \end{bmatrix}$

33. A is 3×3, B is $3 \times 3 \implies AB$ is 3×3.

$\begin{bmatrix} 0 & 0 & 5 \\ 0 & 0 & -3 \\ 0 & 0 & 4 \end{bmatrix}\begin{bmatrix} 6 & -11 & 4 \\ 8 & 16 & 4 \\ 0 & 0 & 0 \end{bmatrix} = \begin{bmatrix} (0)(6) + (0)(8) + (5)(0) & (0)(-11) + (0)(16) + (5)(0) & (0)(4) + (0)(4) + (5)(0) \\ (0)(6) + (0)(8) + (-3)(0) & (0)(-11) + (0)(16) + (-3)(0) & (0)(4) + (0)(4) + (-3)(0) \\ (0)(6) + (0)(8) + (4)(0) & (0)(-11) + (0)(16) + (4)(0) & (0)(4) + (0)(4) + (4)(0) \end{bmatrix}$

$= \begin{bmatrix} 0 & 0 & 0 \\ 0 & 0 & 0 \\ 0 & 0 & 0 \end{bmatrix}$

35. $\begin{bmatrix} 5 & 6 & -3 \\ -2 & 5 & 1 \\ 10 & -5 & 5 \end{bmatrix}\begin{bmatrix} 1 & -1 & 2 \\ 8 & 1 & 4 \\ 4 & -2 & 9 \end{bmatrix} = \begin{bmatrix} 41 & 7 & 7 \\ 42 & 5 & 25 \\ -10 & -25 & 45 \end{bmatrix}$

37. $\begin{bmatrix} -3 & 8 & -6 & 8 \\ -12 & 15 & 9 & 6 \\ 5 & -1 & 1 & 5 \end{bmatrix} \begin{bmatrix} 3 & 1 & 6 \\ 24 & 15 & 14 \\ 16 & 10 & 21 \\ 8 & -4 & 10 \end{bmatrix} = \begin{bmatrix} 151 & 25 & 48 \\ 516 & 279 & 387 \\ 47 & -20 & 87 \end{bmatrix}$

39. A is 2×4 and B is $2 \times 4 \implies AB$ is not possible.

41. (a) $AB = \begin{bmatrix} 1 & 2 \\ 4 & 2 \end{bmatrix} \begin{bmatrix} 2 & -1 \\ -1 & 8 \end{bmatrix} = \begin{bmatrix} (1)(2) + (2)(-1) & (1)(-1) + (2)(8) \\ (4)(2) + (2)(-1) & (4)(-1) + (2)(8) \end{bmatrix} = \begin{bmatrix} 0 & 15 \\ 6 & 12 \end{bmatrix}$

(b) $BA = \begin{bmatrix} 2 & -1 \\ -1 & 8 \end{bmatrix} \begin{bmatrix} 1 & 2 \\ 4 & 2 \end{bmatrix} = \begin{bmatrix} (2)(1) + (-1)(4) & (2)(2) + (-1)(2) \\ (-1)(1) + (8)(4) & (-1)(2) + (8)(2) \end{bmatrix} = \begin{bmatrix} -2 & 2 \\ 31 & 14 \end{bmatrix}$

(c) $A^2 = \begin{bmatrix} 1 & 2 \\ 4 & 2 \end{bmatrix} \begin{bmatrix} 1 & 2 \\ 4 & 2 \end{bmatrix} = \begin{bmatrix} (1)(1) + (2)(4) & (1)(2) + (2)(2) \\ (4)(1) + (2)(4) & (4)(2) + (2)(2) \end{bmatrix} = \begin{bmatrix} 9 & 6 \\ 12 & 12 \end{bmatrix}$

43. (a) $AB = \begin{bmatrix} 3 & -1 \\ 1 & 3 \end{bmatrix} \begin{bmatrix} 1 & -3 \\ 3 & 1 \end{bmatrix} = \begin{bmatrix} (3)(1) + (-1)(3) & (3)(-3) + (-1)(1) \\ (1)(1) + (3)(3) & (1)(-3) + (3)(1) \end{bmatrix} = \begin{bmatrix} 0 & -10 \\ 10 & 0 \end{bmatrix}$

(b) $BA = \begin{bmatrix} 1 & -3 \\ 3 & 1 \end{bmatrix} \begin{bmatrix} 3 & -1 \\ 1 & 3 \end{bmatrix} = \begin{bmatrix} (1)(3) + (-3)(1) & (1)(-1) + (-3)(3) \\ (3)(3) + (1)(1) & (3)(-1) + (1)(3) \end{bmatrix} = \begin{bmatrix} 0 & -10 \\ 10 & 0 \end{bmatrix}$

(c) $A^2 = \begin{bmatrix} 3 & -1 \\ 1 & 3 \end{bmatrix} \begin{bmatrix} 3 & -1 \\ 1 & 3 \end{bmatrix} = \begin{bmatrix} (3)(3) + (-1)(1) & (3)(-1) + (-1)(3) \\ (1)(3) + (3)(1) & (1)(-1) + (3)(3) \end{bmatrix} = \begin{bmatrix} 8 & -6 \\ 6 & 8 \end{bmatrix}$

45. (a) $AB = \begin{bmatrix} 7 \\ 8 \\ -1 \end{bmatrix} \begin{bmatrix} 1 & 1 & 2 \end{bmatrix} = \begin{bmatrix} 7(1) & 7(1) & 7(2) \\ 8(1) & 8(1) & 8(2) \\ -1(1) & -1(1) & -1(2) \end{bmatrix} = \begin{bmatrix} 7 & 7 & 14 \\ 8 & 8 & 16 \\ -1 & -1 & -2 \end{bmatrix}$

(b) $BA = \begin{bmatrix} 1 & 1 & 2 \end{bmatrix} \begin{bmatrix} 7 \\ 8 \\ -1 \end{bmatrix} = [(1)(7) + (1)(8) + (2)(-1)] = [13]$

(c) A^2 is not possible.

47. $\begin{bmatrix} 3 & 1 \\ 0 & -2 \end{bmatrix} \begin{bmatrix} 1 & 0 \\ -2 & 2 \end{bmatrix} \begin{bmatrix} 1 & 0 \\ 2 & 4 \end{bmatrix} = \begin{bmatrix} 1 & 2 \\ 4 & -4 \end{bmatrix} \begin{bmatrix} 1 & 0 \\ 2 & 4 \end{bmatrix} = \begin{bmatrix} 5 & 8 \\ -4 & -16 \end{bmatrix}$

49. $\begin{bmatrix} 0 & 2 & -2 \\ 4 & 1 & 2 \end{bmatrix} \left(\begin{bmatrix} 4 & 0 \\ 0 & -1 \\ -1 & 2 \end{bmatrix} + \begin{bmatrix} -2 & 3 \\ -3 & 5 \\ 0 & -3 \end{bmatrix} \right) = \begin{bmatrix} 0 & 2 & -2 \\ 4 & 1 & 2 \end{bmatrix} \begin{bmatrix} 2 & 3 \\ -3 & 4 \\ -1 & -1 \end{bmatrix} = \begin{bmatrix} -4 & 10 \\ 3 & 14 \end{bmatrix}$

51. (a) $\begin{bmatrix} -1 & 1 \\ -2 & 1 \end{bmatrix} \begin{bmatrix} x_1 \\ x_2 \end{bmatrix} = \begin{bmatrix} 4 \\ 0 \end{bmatrix}$

(b) $\begin{bmatrix} -1 & 1 & \vdots & 4 \\ -2 & 1 & \vdots & 0 \end{bmatrix}$

$-R_2 + R_1 \rightarrow \begin{bmatrix} 1 & 0 & \vdots & 4 \\ -2 & 1 & \vdots & 0 \end{bmatrix}$

$2R_1 + R_2 \rightarrow \begin{bmatrix} 1 & 0 & \vdots & 4 \\ 0 & 1 & \vdots & 8 \end{bmatrix}$

$X = \begin{bmatrix} 4 \\ 8 \end{bmatrix}$

53. (a) $\begin{bmatrix} -2 & -3 \\ 6 & 1 \end{bmatrix} \begin{bmatrix} x_1 \\ x_2 \end{bmatrix} = \begin{bmatrix} -4 \\ -36 \end{bmatrix}$

(b) $\begin{bmatrix} -2 & -3 & \vdots & -4 \\ 6 & 1 & \vdots & -36 \end{bmatrix}$

$3R_1 + R_2 \rightarrow \begin{bmatrix} -2 & -3 & \vdots & -4 \\ 0 & -8 & \vdots & -48 \end{bmatrix}$

$\begin{matrix} -\frac{1}{2}R_1 \rightarrow \\ -\frac{1}{8}R_2 \rightarrow \end{matrix} \begin{bmatrix} 1 & \frac{3}{2} & \vdots & 2 \\ 0 & 1 & \vdots & 6 \end{bmatrix}$

$-\frac{3}{2}R_2 + R_1 \rightarrow \begin{bmatrix} 1 & 0 & \vdots & -7 \\ 0 & 1 & \vdots & 6 \end{bmatrix}$

$X = \begin{bmatrix} -7 \\ 6 \end{bmatrix}$

55. (a) $A = \begin{bmatrix} 1 & -2 & 3 \\ -1 & 3 & -1 \\ 2 & -5 & 5 \end{bmatrix} \begin{bmatrix} x_1 \\ x_2 \\ x_3 \end{bmatrix} = \begin{bmatrix} 9 \\ -6 \\ 17 \end{bmatrix}$

(b) $\begin{bmatrix} 1 & -2 & 3 & \vdots & 9 \\ -1 & 3 & -1 & \vdots & -6 \\ 2 & -5 & 5 & \vdots & 17 \end{bmatrix}$

$\begin{matrix} R_1 + R_2 \rightarrow \\ -2R_2 + R_3 \rightarrow \end{matrix} \begin{bmatrix} 1 & -2 & 3 & \vdots & 9 \\ 0 & 1 & 2 & \vdots & 3 \\ 0 & -1 & -1 & \vdots & -1 \end{bmatrix}$

$\begin{matrix} 2R_2 + R_1 \rightarrow \\ \\ R_2 + R_3 \rightarrow \end{matrix} \begin{bmatrix} 1 & 0 & 7 & \vdots & 15 \\ 0 & 1 & 2 & \vdots & 3 \\ 0 & 0 & 1 & \vdots & 2 \end{bmatrix}$

$\begin{matrix} -7R_3 + R_1 \rightarrow \\ -2R_3 + R_2 \rightarrow \end{matrix} \begin{bmatrix} 1 & 0 & 0 & \vdots & 1 \\ 0 & 1 & 0 & \vdots & -1 \\ 0 & 0 & 1 & \vdots & 2 \end{bmatrix}$

$X = \begin{bmatrix} 1 \\ -1 \\ 2 \end{bmatrix}$

57. (a) $\begin{bmatrix} 1 & -5 & 2 \\ -3 & 1 & -1 \\ 0 & -2 & 5 \end{bmatrix} \begin{bmatrix} x_1 \\ x_2 \\ x_3 \end{bmatrix} = \begin{bmatrix} -20 \\ 8 \\ -16 \end{bmatrix}$

(b) $\begin{bmatrix} 1 & -5 & 2 & \vdots & -20 \\ -3 & 1 & -1 & \vdots & 8 \\ 0 & -2 & 5 & \vdots & -16 \end{bmatrix}$

$\begin{matrix} 3R_1 + R_2 \rightarrow \end{matrix} \begin{bmatrix} 1 & -5 & 2 & \vdots & -20 \\ 0 & -14 & 5 & \vdots & -52 \\ 0 & -2 & 5 & \vdots & -16 \end{bmatrix}$

$\begin{matrix} -R_3 + R_2 \rightarrow \end{matrix} \begin{bmatrix} 1 & -5 & 2 & \vdots & -20 \\ 0 & -12 & 0 & \vdots & -36 \\ 0 & -2 & 5 & \vdots & -16 \end{bmatrix}$

$\begin{matrix} -\frac{1}{12}R_2 \rightarrow \end{matrix} \begin{bmatrix} 1 & -5 & 2 & \vdots & -20 \\ 0 & 1 & 0 & \vdots & 3 \\ 0 & -2 & 5 & \vdots & -16 \end{bmatrix}$

$\begin{matrix} 5R_2 + R_1 \rightarrow \\ \\ 2R_2 + R_3 \rightarrow \end{matrix} \begin{bmatrix} 1 & 0 & 2 & \vdots & -5 \\ 0 & 1 & 0 & \vdots & 3 \\ 0 & 0 & 5 & \vdots & -10 \end{bmatrix}$

$\begin{matrix} \\ \\ \frac{1}{5}R_3 \rightarrow \end{matrix} \begin{bmatrix} 1 & 0 & 2 & \vdots & -5 \\ 0 & 1 & 0 & \vdots & 3 \\ 0 & 0 & 1 & \vdots & -2 \end{bmatrix}$

$\begin{matrix} -2R_3 + R_1 \rightarrow \end{matrix} \begin{bmatrix} 1 & 0 & 0 & \vdots & -1 \\ 0 & 1 & 0 & \vdots & 3 \\ 0 & 0 & 1 & \vdots & -2 \end{bmatrix}$

$X = \begin{bmatrix} -1 \\ 3 \\ -2 \end{bmatrix}$

59. $1.2 \begin{bmatrix} 70 & 50 & 25 \\ 35 & 100 & 70 \end{bmatrix} = \begin{bmatrix} 84 & 60 & 30 \\ 42 & 120 & 84 \end{bmatrix}$

61. (a)

	Farmer's Market	Fruit Stand	Fruit Farm	
$A = \begin{bmatrix} 125 \\ 100 \end{bmatrix}$	125	100	75	Apples
	100	175	125	Peaches

$A = \begin{bmatrix} 125 & 100 & 75 \\ 100 & 175 & 125 \end{bmatrix} \begin{matrix} \text{Apples} \\ \text{Peaches} \end{matrix}$

Each entry represents the number of bushels of each type of crop that are shipped to each outlet.

(b) $B = \begin{bmatrix} 3.50 & 6.00 \end{bmatrix}$

Each entry represents the profit per bushel for each type of crop.

(c) $BA = \begin{bmatrix} 3.50 & 6.00 \end{bmatrix} \begin{bmatrix} 125 & 100 & 75 \\ 100 & 175 & 125 \end{bmatrix}$

$= \begin{bmatrix} \$1037.50 & \$1400.00 & \$1012.50 \end{bmatrix}$

The entries in the matrix represent the profits for both crops at each of the three outlets.

63. $ST = \begin{bmatrix} 3 & 2 & 2 & 3 & 0 \\ 0 & 2 & 3 & 4 & 3 \\ 4 & 2 & 1 & 3 & 2 \end{bmatrix} \begin{bmatrix} 840 & 1100 \\ 1200 & 1350 \\ 1450 & 1650 \\ 2650 & 3000 \\ 3050 & 3200 \end{bmatrix} = \begin{bmatrix} \$15,770 & \$18,300 \\ \$26,500 & \$29,250 \\ \$21,260 & \$24,150 \end{bmatrix}$

The entries represent the wholesale and retail inventory values of the inventories at the three outlets.

65. $P^3 = P^2P = \begin{bmatrix} 0.40 & 0.15 & 0.15 \\ 0.28 & 0.53 & 0.17 \\ 0.32 & 0.32 & 0.68 \end{bmatrix} \begin{bmatrix} 0.6 & 0.1 & 0.1 \\ 0.2 & 0.7 & 0.1 \\ 0.2 & 0.2 & 0.8 \end{bmatrix} = \begin{bmatrix} 0.300 & 0.175 & 0.175 \\ 0.308 & 0.433 & 0.217 \\ 0.392 & 0.392 & 0.608 \end{bmatrix}$

$P^4 = P^3P = \begin{bmatrix} 0.300 & 0.175 & 0.175 \\ 0.308 & 0.433 & 0.217 \\ 0.392 & 0.392 & 0.608 \end{bmatrix} \begin{bmatrix} 0.6 & 0.1 & 0.1 \\ 0.2 & 0.7 & 0.1 \\ 0.2 & 0.2 & 0.8 \end{bmatrix} = \begin{bmatrix} 0.250 & 0.188 & 0.188 \\ 0.315 & 0.377 & 0.248 \\ 0.435 & 0.435 & 0.565 \end{bmatrix}$

$P^5 = P^4P = \begin{bmatrix} 0.250 & 0.188 & 0.188 \\ 0.315 & 0.377 & 0.248 \\ 0.435 & 0.435 & 0.565 \end{bmatrix} \begin{bmatrix} 0.6 & 0.1 & 0.1 \\ 0.2 & 0.7 & 0.1 \\ 0.2 & 0.2 & 0.8 \end{bmatrix} = \begin{bmatrix} 0.225 & 0.194 & 0.194 \\ 0.314 & 0.345 & 0.267 \\ 0.461 & 0.461 & 0.539 \end{bmatrix}$

$P^6 = \begin{bmatrix} 0.213 & 0.197 & 0.197 \\ 0.311 & 0.326 & 0.280 \\ 0.477 & 0.477 & 0.523 \end{bmatrix}$

$P^7 = \begin{bmatrix} 0.206 & 0.198 & 0.198 \\ 0.308 & 0.316 & 0.288 \\ 0.486 & 0.486 & 0.514 \end{bmatrix}$

$P^8 = \begin{bmatrix} 0.203 & 0.199 & 0.199 \\ 0.305 & 0.309 & 0.292 \\ 0.492 & 0.492 & 0.508 \end{bmatrix}$

As P is raised to higher and higher powers, the resulting matrices appear to be approaching the matrix

$\begin{bmatrix} 0.2 & 0.2 & 0.2 \\ 0.3 & 0.3 & 0.3 \\ 0.5 & 0.5 & 0.5 \end{bmatrix}$.

67. (a) $AB = \begin{bmatrix} 40 & 64 & 52 \\ 60 & 82 & 76 \\ 76 & 96 & 84 \end{bmatrix} \begin{bmatrix} 2.65 & 0.65 \\ 2.85 & 0.70 \\ 3.05 & 0.85 \end{bmatrix} = \begin{array}{cc} \text{Sales} & \text{Profit} \\ \begin{bmatrix} 447 & 115 \\ 624.50 & 161 \\ 731.20 & 188 \end{bmatrix} & \begin{array}{l} \text{Friday} \\ \text{Saturday} \\ \text{Sunday} \end{array} \end{array}$

The entries in Column 1 represent the total sales of the three kinds of milk for Friday, Saturday, and Sunday. The entries in Column 2 represent each days' total profit.

(b) Total profit for the weekend: $115 + 161 + 188 = \$464$

69. (a) $B = \begin{array}{ccc} \text{Bicycled} & \text{Jogged} & \text{Walked} \\ [2 & 0.5 & 3] \end{array}$ 20-minute time periods

(b) $BA = \begin{bmatrix} 2 & 0.5 & 3 \end{bmatrix} \begin{array}{cc} & \begin{array}{cc} \text{120-pound} & \text{150-pound} \\ \text{person} & \text{person} \end{array} \\ \begin{bmatrix} 109 & 136 \\ 127 & 159 \\ 64 & 79 \end{bmatrix} \end{array} = \begin{bmatrix} 473.5 & 588.5 \end{bmatrix}$ Calories burned

The first entry represents the total calories burned by the 120-pound person and the second entry represents the total calories burned by the 150-pound person.

71. True. The sum of two matrices of different orders is undefined.

For 73–79, A is of order 2×3, B is of order 2×3, C is of order 3×2 and D is of order 2×2.

73. $A + 2C$ is not possible. A and C are not of the same order.

75. AB is not possible. The number of columns of A does not equal the number of rows of B.

77. $BC - D$ is possible. The resulting order is 2×2.

79. $D(A - 3B)$ is possible. The resulting order is 2×3.

81. $AC = \begin{bmatrix} 0 & 1 \\ 0 & 1 \end{bmatrix}\begin{bmatrix} 2 & 3 \\ 2 & 3 \end{bmatrix} = \begin{bmatrix} 2 & 3 \\ 2 & 3 \end{bmatrix}$

$BC = \begin{bmatrix} 1 & 0 \\ 1 & 0 \end{bmatrix}\begin{bmatrix} 2 & 3 \\ 2 & 3 \end{bmatrix} = \begin{bmatrix} 2 & 3 \\ 2 & 3 \end{bmatrix}$

Thus, $AC = BC$ even though $A \neq B$.

83. The product of two diagonal matrices of the same order is a diagonal matrix whose entries are the products of the corresponding diagonal entries of A and B.

85. $3x^2 + 20x - 32 = 0$

$(3x - 4)(x + 8) = 0$

$3x - 4 = 0$ or $x + 8 = 0$

$x = \frac{4}{3}$ or $\quad x = -8$

Solutions: $\frac{4}{3}, -8$

87. $4x^3 + 10x^2 - 3x = 0$

$x(4x^2 + 10x - 3) = 0$

$x = 0$ or $4x^2 + 10x - 3 = 0$

$x = \frac{-10 \pm \sqrt{10^2 - 4(4)(-3)}}{2(4)} = \frac{-10 \pm \sqrt{148}}{8}$

$= \frac{-5 \pm \sqrt{37}}{4}$ by the Quadratic Formula

Solutions: $0, \dfrac{-5 \pm \sqrt{37}}{4}$

89. $3x^3 - 12x^2 + 5x - 20 = 0$

$3x^2(x - 4) + 5(x - 4) = 0$

$(x - 4)(3x^2 + 5) = 0$

$x - 4 = 0$ $\quad$ or $\quad 3x^2 + 5 = 0$

$x = 4 \qquad\qquad x^2 = -\frac{5}{3}$

$x = \pm\sqrt{-\frac{5}{3}} = \pm\frac{\sqrt{15}}{3}i$

Solutions: $4, \pm\dfrac{\sqrt{15}}{3}i$

91. $\begin{cases} -x + 4y = -9 \\ 5x - 8y = 39 \end{cases}$ $\qquad$ Eq.1

$\qquad\qquad\qquad\qquad\qquad$ Eq.2

$-5x + 20y = -45 \qquad$ (5)Eq.1

$\underline{5x - 8y = 39}$

$\qquad\quad 12y = -6 \qquad$ Add equations.

$\qquad\qquad y = -\frac{1}{2}$

$-x + 4\left(-\frac{1}{2}\right) = -9 \Rightarrow x = 7$

Solution: $\left(7, -\frac{1}{2}\right)$

93. $\begin{cases} -x + 2y = -5 \\ -3x - y = -8 \end{cases}$ $\qquad$ Equation 1

$\qquad\qquad\qquad\qquad\qquad$ Equation 2

$-x + 2y = -5$

$\underline{-6x - 2y = -16} \qquad$ (2)Eq.2

$-7x = -21 \qquad$ Add equations.

$\qquad x = 3$

$-3 + 2y = -5 \Rightarrow y = -1$

Solution: $(3, -1)$

Section 10.3 The Inverse of a Square Matrix

- You should know that the inverse of an $n \times n$ matrix A is the $n \times n$ matrix A^{-1}, if is exists, such that $AA^{-1} = A^{-1}A = I$, where I is the $n \times n$ identity matrix.

- You should be able to find the inverse, if it exists, of a square matrix.

  (a) Write the $n \times 2n$ matrix that consists of the given matrix A on the left and the $n \times n$ identity matrix I on the right to obtain $[A \;\vdots\; I]$. Note that we separate the matrices A and I by a dotted line. We call this process **adjoining** the matrices A and I.

  (b) If possible, row reduce A to I using elementary row operations on the *entire* matrix $[A \;\vdots\; I]$. The result will be the matrix $[I \;\vdots\; A^{-1}]$. If this is not possible, then A is not invertible.

  (c) Check your work by multiplying to see that $AA^{-1} = I = A^{-1}A$.

- The inverse of $A = \begin{bmatrix} a & b \\ c & d \end{bmatrix}$ is $A^{-1} = \dfrac{1}{ad - bc}\begin{bmatrix} d & -b \\ -c & a \end{bmatrix}$ if $ad - cb \neq 0$.

- You should be able to use inverse matrices to solve systems of linear equations if the coefficient matrix is square and invertible.

Vocabulary Check

1. square   **2.** inverse   **3.** nonsingular; singular   **4.** $A^{-1}B$

---

1. $AB = \begin{bmatrix} 2 & 1 \\ 5 & 3 \end{bmatrix}\begin{bmatrix} 3 & -1 \\ -5 & 2 \end{bmatrix} = \begin{bmatrix} 6-5 & -2+2 \\ 15-15 & -5+6 \end{bmatrix} = \begin{bmatrix} 1 & 0 \\ 0 & 1 \end{bmatrix}$

$BA = \begin{bmatrix} 3 & -1 \\ -5 & 2 \end{bmatrix}\begin{bmatrix} 2 & 1 \\ 5 & 3 \end{bmatrix} = \begin{bmatrix} 6-5 & 3-3 \\ -10+10 & -5+6 \end{bmatrix} = \begin{bmatrix} 1 & 0 \\ 0 & 1 \end{bmatrix}$

3. $AB = \begin{bmatrix} 1 & 2 \\ 3 & 4 \end{bmatrix}\begin{bmatrix} -2 & 1 \\ \frac{3}{2} & -\frac{1}{2} \end{bmatrix} = \begin{bmatrix} -2+3 & 1-1 \\ -6+6 & 3-2 \end{bmatrix} = \begin{bmatrix} 1 & 0 \\ 0 & 1 \end{bmatrix}$

$BA = \begin{bmatrix} -2 & 1 \\ \frac{3}{2} & -\frac{1}{2} \end{bmatrix}\begin{bmatrix} 1 & 2 \\ 3 & 4 \end{bmatrix} = \begin{bmatrix} -2+3 & -4+4 \\ \frac{3}{2}-\frac{3}{2} & 3-2 \end{bmatrix} = \begin{bmatrix} 1 & 0 \\ 0 & 1 \end{bmatrix}$

5. $AB = \begin{bmatrix} 2 & -17 & 11 \\ -1 & 11 & -7 \\ 0 & 3 & -2 \end{bmatrix}\begin{bmatrix} 1 & 1 & 2 \\ 2 & 4 & -3 \\ 3 & 6 & -5 \end{bmatrix} = \begin{bmatrix} 2-34+33 & 2-68+66 & 4+51-55 \\ -1+22-21 & -1+44-42 & -2-33+35 \\ 6-6 & 12-12 & -9+10 \end{bmatrix} = \begin{bmatrix} 1 & 0 & 0 \\ 0 & 1 & 0 \\ 0 & 0 & 1 \end{bmatrix}$

$BA = \begin{bmatrix} 1 & 1 & 2 \\ 2 & 4 & -3 \\ 3 & 6 & -5 \end{bmatrix}\begin{bmatrix} 2 & -17 & 11 \\ -1 & 11 & -7 \\ 0 & 3 & -2 \end{bmatrix} = \begin{bmatrix} 2-1 & -17+11+6 & 11-7-4 \\ 4-4 & -34+44-9 & 22-28+6 \\ 6-6 & -51+66-15 & 33-42+10 \end{bmatrix} = \begin{bmatrix} 1 & 0 & 0 \\ 0 & 1 & 0 \\ 0 & 0 & 1 \end{bmatrix}$

7. $AB = \begin{bmatrix} 2 & 0 & 1 & 1 \\ 3 & 0 & 0 & 1 \\ -1 & 1 & -2 & 1 \\ 4 & -1 & 1 & 0 \end{bmatrix}\begin{bmatrix} -1 & 2 & -1 & -1 \\ -4 & 9 & -5 & -6 \\ 0 & 1 & -1 & -1 \\ 3 & -5 & 3 & 3 \end{bmatrix}$

$= \begin{bmatrix} -2+3 & 4+1-5 & -2-1+3 & -2-1+3 \\ 0 & 6-5 & 0 & 0 \\ 1-4+3 & -2+9-2-5 & 1-5+2+3 & 1-6+2+3 \\ 0 & 8-9+1 & -4+5-1 & -4+6-1 \end{bmatrix} = \begin{bmatrix} 1 & 0 & 0 & 0 \\ 0 & 1 & 0 & 0 \\ 0 & 0 & 1 & 0 \\ 0 & 0 & 0 & 1 \end{bmatrix}$

—CONTINUED—

7. —CONTINUED—

$$BA = \begin{bmatrix} -1 & 2 & -1 & -1 \\ -4 & 9 & -5 & -6 \\ 0 & 1 & -1 & -1 \\ 3 & -5 & 3 & 3 \end{bmatrix} \begin{bmatrix} 2 & 0 & 1 & 1 \\ 3 & 0 & 0 & 1 \\ -1 & 1 & -2 & 1 \\ 4 & -1 & 1 & 0 \end{bmatrix}$$

$$= \begin{bmatrix} -2+6+1-4 & 0 & -1+2-1 & -1+2-1 \\ -8+27+5-24 & -5+6 & -4+10-6 & -4+9-5 \\ 3+1-4 & 0 & 2-1 & 0 \\ 6-15-3+12 & 0 & 3-6+3 & 3-5+3 \end{bmatrix} = \begin{bmatrix} 1 & 0 & 0 & 0 \\ 0 & 1 & 0 & 0 \\ 0 & 0 & 1 & 0 \\ 0 & 0 & 0 & 1 \end{bmatrix}$$

9. $AB = \frac{1}{3}\begin{bmatrix} -2 & 2 & 3 \\ 1 & -1 & 0 \\ 0 & 1 & 4 \end{bmatrix} \begin{bmatrix} -4 & -5 & 3 \\ -4 & -8 & 3 \\ 1 & 2 & 0 \end{bmatrix} = \frac{1}{3}\begin{bmatrix} 8-8+3 & 10-16+6 & -6+6 \\ -4+4 & -5+8 & 3-3 \\ -4+4 & -8+8 & 3 \end{bmatrix}$

$$= \frac{1}{3}\begin{bmatrix} 3 & 0 & 0 \\ 0 & 3 & 0 \\ 0 & 0 & 3 \end{bmatrix} = \begin{bmatrix} 1 & 0 & 0 \\ 0 & 1 & 0 \\ 0 & 0 & 1 \end{bmatrix}$$

$$BA = \frac{1}{3}\begin{bmatrix} -4 & -5 & 3 \\ -4 & -8 & 3 \\ 1 & 2 & 0 \end{bmatrix} \begin{bmatrix} -2 & 2 & 3 \\ 1 & -1 & 0 \\ 0 & 1 & 4 \end{bmatrix} = \frac{1}{3}\begin{bmatrix} 8-5 & -8+5+3 & -12+12 \\ 8-8 & -8+8+3 & -12+12 \\ -2+2 & 2-2 & 3 \end{bmatrix} = \begin{bmatrix} 1 & 0 & 0 \\ 0 & 1 & 0 \\ 0 & 0 & 1 \end{bmatrix}$$

11. $[A \ \vdots \ I] = \begin{bmatrix} 2 & 0 & \vdots & 1 & 0 \\ 0 & 3 & \vdots & 0 & 1 \end{bmatrix}$

$\begin{array}{l} \frac{1}{2}R_1 \to \\ \frac{1}{3}R_2 \to \end{array} \begin{bmatrix} 1 & 0 & \vdots & \frac{1}{2} & 0 \\ 0 & 1 & \vdots & 0 & \frac{1}{3} \end{bmatrix} = [I \ \vdots \ A^{-1}]$

$A^{-1} = \begin{bmatrix} \frac{1}{2} & 0 \\ 0 & \frac{1}{3} \end{bmatrix}$

13. $[A \ \vdots \ I] = \begin{bmatrix} 1 & -2 & \vdots & 1 & 0 \\ 2 & -3 & \vdots & 0 & 1 \end{bmatrix}$

$-2R_1 + R_2 \to \begin{bmatrix} 1 & -2 & \vdots & 1 & 0 \\ 0 & 1 & \vdots & -2 & 1 \end{bmatrix}$

$2R_2 + R_1 \to \begin{bmatrix} 1 & 0 & \vdots & -3 & 2 \\ 0 & 1 & \vdots & -2 & 1 \end{bmatrix} = [I \ \vdots \ A^{-1}]$

$A^{-1} = \begin{bmatrix} -3 & 2 \\ -2 & 1 \end{bmatrix}$

15. $[A \ \vdots \ I] = \begin{bmatrix} -1 & 1 & \vdots & 1 & 0 \\ -2 & 1 & \vdots & 0 & 1 \end{bmatrix}$

$-R_2 + R_1 \to \begin{bmatrix} 1 & 0 & \vdots & 1 & -1 \\ -2 & 1 & \vdots & 0 & 1 \end{bmatrix}$

$2R_1 + R_2 \to \begin{bmatrix} 1 & 0 & \vdots & 1 & -1 \\ 0 & 1 & \vdots & 2 & -1 \end{bmatrix} = [I \ \vdots \ A^{-1}]$

$A^{-1} = \begin{bmatrix} 1 & -1 \\ 2 & -1 \end{bmatrix}$

17. $[A \ \vdots \ I] = \begin{bmatrix} 2 & 4 & \vdots & 1 & 0 \\ 4 & 8 & \vdots & 0 & 1 \end{bmatrix}$

$-2R_1 + R_2 \to \begin{bmatrix} 2 & 4 & \vdots & 1 & 0 \\ 0 & 0 & \vdots & -2 & 1 \end{bmatrix}$

The two zeros in the second row imply that the inverse does not exist.

19. $A = \begin{bmatrix} 2 & 7 & 1 \\ -3 & -9 & 2 \end{bmatrix}$

21. $[A \ \vdots \ I] = \begin{bmatrix} 1 & 1 & 1 & \vdots & 1 & 0 & 0 \\ 3 & 5 & 4 & \vdots & 0 & 1 & 0 \\ 3 & 6 & 5 & \vdots & 0 & 0 & 1 \end{bmatrix}$

$\begin{matrix} \\ -3R_1 + R_2 \rightarrow \\ -3R_1 + R_3 \rightarrow \end{matrix} \begin{bmatrix} 1 & 1 & 1 & \vdots & 1 & 0 & 0 \\ 0 & 2 & 1 & \vdots & -3 & 1 & 0 \\ 0 & 3 & 2 & \vdots & -3 & 0 & 1 \end{bmatrix}$

$\begin{matrix} \\ \frac{1}{2}R_2 \rightarrow \\ \\ \end{matrix} \begin{bmatrix} 1 & 1 & 1 & \vdots & 1 & 0 & 0 \\ 0 & 1 & \frac{1}{2} & \vdots & -\frac{3}{2} & \frac{1}{2} & 0 \\ 0 & 3 & 2 & \vdots & -3 & 0 & 1 \end{bmatrix}$

$\begin{matrix} -R_2 + R_1 \rightarrow \\ \\ -3R_2 + R_3 \rightarrow \end{matrix} \begin{bmatrix} 1 & 0 & \frac{1}{2} & \vdots & \frac{5}{2} & -\frac{1}{2} & 0 \\ 0 & 1 & \frac{1}{2} & \vdots & -\frac{3}{2} & \frac{1}{2} & 0 \\ 0 & 0 & \frac{1}{2} & \vdots & \frac{3}{2} & -\frac{3}{2} & 1 \end{bmatrix}$

$\begin{matrix} -R_3 + R_1 \rightarrow \\ -R_3 + R_2 \rightarrow \\ \\ \end{matrix} \begin{bmatrix} 1 & 0 & 0 & \vdots & 1 & 1 & -1 \\ 0 & 1 & 0 & \vdots & -3 & 2 & -1 \\ 0 & 0 & \frac{1}{2} & \vdots & \frac{3}{2} & -\frac{3}{2} & 1 \end{bmatrix}$

$\begin{matrix} \\ \\ 2R_3 \rightarrow \end{matrix} \begin{bmatrix} 1 & 0 & 0 & \vdots & 1 & 1 & -1 \\ 0 & 1 & 0 & \vdots & -3 & 2 & -1 \\ 0 & 0 & 1 & \vdots & 3 & -3 & 2 \end{bmatrix} = [I \ \vdots \ A^{-1}]$

$A^{-1} = \begin{bmatrix} 1 & 1 & -1 \\ -3 & 2 & -1 \\ 3 & -3 & 2 \end{bmatrix}$

23. $[A \ \vdots \ I] = \begin{bmatrix} 1 & 0 & 0 & \vdots & 1 & 0 & 0 \\ 3 & 4 & 0 & \vdots & 0 & 1 & 0 \\ 2 & 5 & 5 & \vdots & 0 & 0 & 1 \end{bmatrix}$

$\begin{matrix} \\ -3R_1 + R_2 \rightarrow \\ -2R_1 + R_3 \rightarrow \end{matrix} \begin{bmatrix} 1 & 0 & 0 & \vdots & 1 & 0 & 0 \\ 0 & 4 & 0 & \vdots & -3 & 1 & 0 \\ 0 & 5 & 5 & \vdots & -2 & 0 & 1 \end{bmatrix}$

$\begin{matrix} \\ \\ -\frac{5}{4}R_2 + R_3 \rightarrow \end{matrix} \begin{bmatrix} 1 & 0 & 0 & \vdots & 1 & 0 & 0 \\ 0 & 4 & 0 & \vdots & -3 & 1 & 0 \\ 0 & 0 & 5 & \vdots & \frac{7}{4} & -\frac{5}{4} & 1 \end{bmatrix}$

$\begin{matrix} \\ \frac{1}{4}R_2 \rightarrow \\ \frac{1}{5}R_3 \rightarrow \end{matrix} \begin{bmatrix} 1 & 0 & 0 & \vdots & 1 & 0 & 0 \\ 0 & 1 & 0 & \vdots & -\frac{3}{4} & \frac{1}{4} & 0 \\ 0 & 0 & 1 & \vdots & \frac{7}{20} & -\frac{1}{4} & \frac{1}{5} \end{bmatrix} = [I \ \vdots \ A^{-1}]$

$A^{-1} = \begin{bmatrix} 1 & 0 & 0 \\ -\frac{3}{4} & \frac{1}{4} & 0 \\ \frac{7}{20} & -\frac{1}{4} & \frac{1}{5} \end{bmatrix}$

25. $[A \ \vdots \ I] = \begin{bmatrix} -8 & 0 & 0 & 0 & \vdots & 1 & 0 & 0 & 0 \\ 0 & 1 & 0 & 0 & \vdots & 0 & 1 & 0 & 0 \\ 0 & 0 & 4 & 0 & \vdots & 0 & 0 & 1 & 0 \\ 0 & 0 & 0 & -5 & \vdots & 0 & 0 & 0 & 1 \end{bmatrix}$

$\begin{matrix} -\frac{1}{8}R_1 \rightarrow \\ \\ \frac{1}{4}R_3 \rightarrow \\ -\frac{1}{5}R_4 \rightarrow \end{matrix} \begin{bmatrix} 1 & 0 & 0 & 0 & \vdots & -\frac{1}{8} & 0 & 0 & 0 \\ 0 & 1 & 0 & 0 & \vdots & 0 & 1 & 0 & 0 \\ 0 & 0 & 1 & 0 & \vdots & 0 & 0 & \frac{1}{4} & 0 \\ 0 & 0 & 0 & 1 & \vdots & 0 & 0 & 0 & -\frac{1}{5} \end{bmatrix} = [I \ \vdots \ A^{-1}]$

$A^{-1} = \begin{bmatrix} -\frac{1}{8} & 0 & 0 & 0 \\ 0 & 1 & 0 & 0 \\ 0 & 0 & \frac{1}{4} & 0 \\ 0 & 0 & 0 & -\frac{1}{5} \end{bmatrix}$

27. $A = \begin{bmatrix} 1 & 2 & -1 \\ 3 & 7 & -10 \\ -5 & -7 & -15 \end{bmatrix}$

$A^{-1} = \begin{bmatrix} -175 & 37 & -13 \\ 95 & -20 & 7 \\ 14 & -3 & 1 \end{bmatrix}$

29. $A = \begin{bmatrix} 1 & 1 & 2 \\ 3 & 1 & 0 \\ -2 & 0 & 3 \end{bmatrix}$

$A^{-1} = \frac{1}{2} \begin{bmatrix} -3 & 3 & 2 \\ 9 & -7 & -6 \\ -2 & 2 & 2 \end{bmatrix} = \begin{bmatrix} -1.5 & 1.5 & 1 \\ 4.5 & -3.5 & -3 \\ -1 & 1 & 1 \end{bmatrix}$

31. $A = \begin{bmatrix} -\frac{1}{2} & \frac{3}{4} & \frac{1}{4} \\ 1 & 0 & -\frac{3}{2} \\ 0 & -1 & \frac{1}{2} \end{bmatrix}$

$A^{-1} = \begin{bmatrix} -12 & -5 & -9 \\ -4 & -2 & -4 \\ -8 & -4 & -6 \end{bmatrix}$

33. $A = \begin{bmatrix} 0.1 & 0.2 & 0.3 \\ -0.3 & 0.2 & 0.2 \\ 0.5 & 0.4 & 0.4 \end{bmatrix}$

$A^{-1} = \frac{5}{11} \begin{bmatrix} 0 & -4 & 2 \\ -22 & 11 & 11 \\ 22 & -6 & -8 \end{bmatrix} = \begin{bmatrix} 0 & -1.\overline{81} & 0.\overline{90} \\ -10 & 5 & 5 \\ 10 & -2.\overline{72} & -3.\overline{63} \end{bmatrix}$

35. $A = \begin{bmatrix} 1 & 0 & 3 & 0 \\ 0 & 2 & 0 & 4 \\ 1 & 0 & 3 & 0 \\ 0 & 2 & 0 & 4 \end{bmatrix}$

A^{-1} does not exist.

37. $A = \begin{bmatrix} -1 & 0 & 1 & 0 \\ 0 & 2 & 0 & -1 \\ 2 & 0 & -1 & 0 \\ 0 & -1 & 0 & 1 \end{bmatrix}$

$A^{-1} = \begin{bmatrix} 1 & 0 & 1 & 0 \\ 0 & 1 & 0 & 1 \\ 2 & 0 & 1 & 0 \\ 0 & 1 & 0 & 2 \end{bmatrix}$

39. $A = \begin{bmatrix} a & b \\ c & d \end{bmatrix}, A^{-1} = \frac{1}{ad - bc} \begin{bmatrix} d & -b \\ -c & a \end{bmatrix}$

$A = \begin{bmatrix} 5 & -2 \\ 2 & 3 \end{bmatrix}$

$ad - bc = (5)(3) - (-2)(2) = 19$

$A^{-1} = \frac{1}{19} \begin{bmatrix} 3 & 2 \\ -2 & 5 \end{bmatrix} = \begin{bmatrix} \frac{3}{19} & \frac{2}{19} \\ -\frac{2}{19} & \frac{5}{19} \end{bmatrix}$

41. $A = \begin{bmatrix} -4 & -6 \\ 2 & 3 \end{bmatrix}$

$ad - bc = (-4)(3) - (-2)(-6) = 0$

Since $ad - bc = 0$, A^{-1} does not exist.

43. $A = \begin{bmatrix} \frac{7}{2} & -\frac{3}{4} \\ \frac{1}{5} & \frac{4}{5} \end{bmatrix}$

$ad - bc = \left(\frac{7}{2}\right)\left(\frac{4}{5}\right) - \left(-\frac{3}{4}\right)\left(\frac{1}{5}\right) = \frac{28}{10} + \frac{3}{20} = \frac{59}{20}$

$A^{-1} = \frac{1}{59/20} \begin{bmatrix} \frac{4}{5} & \frac{3}{4} \\ -\frac{1}{5} & \frac{7}{2} \end{bmatrix} = \frac{20}{59} \begin{bmatrix} \frac{4}{5} & \frac{3}{4} \\ -\frac{1}{5} & \frac{7}{2} \end{bmatrix} = \begin{bmatrix} \frac{16}{59} & \frac{15}{59} \\ -\frac{4}{59} & \frac{70}{59} \end{bmatrix}$

45. $\begin{bmatrix} x \\ y \end{bmatrix} = \begin{bmatrix} -3 & 2 \\ -2 & 1 \end{bmatrix} \begin{bmatrix} 5 \\ 10 \end{bmatrix} = \begin{bmatrix} 5 \\ 0 \end{bmatrix}$

Solution: $(5, 0)$

47. $\begin{bmatrix} x \\ y \end{bmatrix} = \begin{bmatrix} -3 & 2 \\ -2 & 1 \end{bmatrix} \begin{bmatrix} 4 \\ 2 \end{bmatrix} = \begin{bmatrix} -8 \\ -6 \end{bmatrix}$

Solution: $(-8, -6)$

49. $\begin{bmatrix} x \\ y \\ z \end{bmatrix} = \begin{bmatrix} 1 & 1 & -1 \\ -3 & 2 & -1 \\ 3 & -3 & 2 \end{bmatrix} \begin{bmatrix} 0 \\ 5 \\ 2 \end{bmatrix} = \begin{bmatrix} 3 \\ 8 \\ -11 \end{bmatrix}$

Solution: $(3, 8, -11)$

51. $\begin{bmatrix} x_1 \\ x_2 \\ x_3 \\ x_4 \end{bmatrix} = \begin{bmatrix} -24 & 7 & 1 & -2 \\ -10 & 3 & 0 & -1 \\ -29 & 7 & 3 & -2 \\ 12 & -3 & -1 & 1 \end{bmatrix} \begin{bmatrix} 0 \\ 1 \\ -1 \\ 2 \end{bmatrix} = \begin{bmatrix} 2 \\ 1 \\ 0 \\ 0 \end{bmatrix}$

Solution: $(2, 1, 0, 0)$

53. $A = \begin{bmatrix} 3 & 4 \\ 5 & 3 \end{bmatrix}$

$A^{-1} = \dfrac{1}{9 - 20} \begin{bmatrix} 3 & -4 \\ -5 & 3 \end{bmatrix}$

$\begin{bmatrix} x \\ y \end{bmatrix} = -\dfrac{1}{11} \begin{bmatrix} 3 & -4 \\ -5 & 3 \end{bmatrix} \begin{bmatrix} -2 \\ 4 \end{bmatrix} = -\dfrac{1}{11} \begin{bmatrix} -22 \\ 22 \end{bmatrix} = \begin{bmatrix} 2 \\ -2 \end{bmatrix}$

Solution: $(2, -2)$

55. $A = \begin{bmatrix} -0.4 & 0.8 \\ 2 & -4 \end{bmatrix}$

$A^{-1} = \dfrac{1}{1.6 - 1.6} \begin{bmatrix} -4 & -0.8 \\ -2 & -0.4 \end{bmatrix}$

A^{-1} does not exist.

This implies that there is no unique solution; that is, either the system is inconsistent *or* there are infinitely many solutions.

Find the reduced row-echelon form of the matrix corresponding to the system.

$$\begin{bmatrix} -0.4 & 0.8 & \vdots & 1.6 \\ 2 & -4 & \vdots & 5 \end{bmatrix}$$

$-2.5R_1 \rightarrow \begin{bmatrix} 1 & -2 & \vdots & -4 \\ 2 & -4 & \vdots & 5 \end{bmatrix}$

$-2R_1 + R_2 \rightarrow \begin{bmatrix} 1 & -2 & \vdots & -4 \\ 0 & 0 & \vdots & 13 \end{bmatrix}$

The given system is inconsistent and there is no solution.

57. $A = \begin{bmatrix} -\frac{1}{4} & \frac{3}{8} \\ \frac{3}{2} & \frac{3}{4} \end{bmatrix}$

$A^{-1} = \dfrac{1}{-\frac{3}{16} - \frac{9}{16}} \begin{bmatrix} \frac{3}{4} & -\frac{3}{8} \\ -\frac{3}{2} & -\frac{1}{4} \end{bmatrix} = -\dfrac{4}{3} \begin{bmatrix} \frac{3}{4} & -\frac{3}{8} \\ -\frac{3}{2} & -\frac{1}{4} \end{bmatrix} = \begin{bmatrix} -1 & \frac{1}{2} \\ 2 & \frac{1}{3} \end{bmatrix}$

$\begin{bmatrix} x \\ y \end{bmatrix} = \begin{bmatrix} -1 & \frac{1}{2} \\ 2 & \frac{1}{3} \end{bmatrix} \begin{bmatrix} -2 \\ -12 \end{bmatrix} = \begin{bmatrix} -4 \\ -8 \end{bmatrix}$

Solution: $(-4, -8)$

59. $A = \begin{bmatrix} 4 & -1 & 1 \\ 2 & 2 & 3 \\ 5 & -2 & 6 \end{bmatrix}$

Find A^{-1}.

$[A : I] = \begin{bmatrix} 4 & -1 & 1 & \vdots & 1 & 0 & 0 \\ 2 & 2 & 3 & \vdots & 0 & 1 & 0 \\ 5 & -2 & 6 & \vdots & 0 & 0 & 1 \end{bmatrix}$

$\begin{matrix} R_1 \\ \\ R_3 \end{matrix} \begin{bmatrix} 5 & -2 & 6 & \vdots & 0 & 0 & 1 \\ 2 & 2 & 3 & \vdots & 0 & 1 & 0 \\ 4 & -1 & 1 & \vdots & 1 & 0 & 0 \end{bmatrix}$

$-R_3 + R_1 \rightarrow \begin{bmatrix} 1 & -1 & 5 & \vdots & -1 & 0 & 1 \\ 2 & 2 & 3 & \vdots & 0 & 1 & 0 \\ 4 & -1 & 1 & \vdots & 1 & 0 & 0 \end{bmatrix}$

$\begin{matrix} -2R_1 + R_2 \rightarrow \\ -4R_1 + R_3 \rightarrow \end{matrix} \begin{bmatrix} 1 & -1 & 5 & \vdots & -1 & 0 & 1 \\ 0 & 4 & -7 & \vdots & 2 & 1 & -2 \\ 0 & 3 & -19 & \vdots & 5 & 0 & -4 \end{bmatrix}$

$-R_3 + R_2 \rightarrow \begin{bmatrix} 1 & -1 & 5 & \vdots & -1 & 0 & 1 \\ 0 & 1 & 12 & \vdots & -3 & 1 & 2 \\ 0 & 3 & -19 & \vdots & 5 & 0 & -4 \end{bmatrix}$

—CONTINUED—

59. —CONTINUED—

$$R_2 + R_1 \rightarrow \begin{bmatrix} 1 & 0 & 17 & \vdots & -4 & 1 & 3 \\ 0 & 1 & 12 & \vdots & -3 & 1 & 2 \\ -3R_2 + R_3 \rightarrow & 0 & 0 & -55 & \vdots & 14 & -3 & -10 \end{bmatrix}$$

$$-\tfrac{1}{55}R_3 \rightarrow \begin{bmatrix} 1 & 0 & 17 & \vdots & -4 & 1 & 3 \\ 0 & 1 & 12 & \vdots & -3 & 1 & 2 \\ 0 & 0 & 1 & \vdots & -\tfrac{14}{55} & \tfrac{3}{55} & \tfrac{2}{11} \end{bmatrix}$$

$$\begin{aligned} -17R_3 + R_1 \rightarrow \\ -12R_3 + R_2 \rightarrow \end{aligned} \begin{bmatrix} 1 & 0 & 0 & \vdots & \tfrac{18}{55} & \tfrac{4}{55} & -\tfrac{1}{11} \\ 0 & 1 & 0 & \vdots & \tfrac{3}{55} & \tfrac{19}{55} & -\tfrac{2}{11} \\ 0 & 0 & 1 & \vdots & -\tfrac{14}{55} & \tfrac{3}{55} & \tfrac{2}{11} \end{bmatrix} = [I \;\vdots\; A^{-1}]$$

$$A^{-1} = \tfrac{1}{55}\begin{bmatrix} 18 & 4 & -5 \\ 3 & 19 & -10 \\ -14 & 3 & 10 \end{bmatrix}$$

$$\begin{bmatrix} x \\ y \\ z \end{bmatrix} = \tfrac{1}{55}\begin{bmatrix} 18 & 4 & -5 \\ 3 & 19 & -10 \\ -14 & 3 & 10 \end{bmatrix}\begin{bmatrix} -5 \\ 10 \\ 1 \end{bmatrix} = \tfrac{1}{55}\begin{bmatrix} -55 \\ 165 \\ 110 \end{bmatrix} = \begin{bmatrix} -1 \\ 3 \\ 2 \end{bmatrix}$$

Solution: $(-1, 3, 2)$

61. $A = \begin{bmatrix} 5 & -3 & 2 \\ 2 & 2 & -3 \\ 1 & -7 & 8 \end{bmatrix}$

A^{-1} does not exist. This implies that there is no unique solution; that is, either the system is inconsistent *or* the system has infinitely many solutions. Use a graphing utility to find the reduced row-echelon form of the matrix corresponding to the system.

$$\begin{bmatrix} 5 & -3 & 2 & \vdots & 2 \\ 2 & 2 & -3 & \vdots & 3 \\ 1 & -7 & 8 & \vdots & -4 \end{bmatrix}$$

$$\begin{bmatrix} 1 & 0 & -\tfrac{5}{16} & \vdots & \tfrac{13}{16} \\ 0 & 1 & -\tfrac{19}{16} & \vdots & \tfrac{11}{16} \\ 0 & 0 & 0 & \vdots & 0 \end{bmatrix}$$

$$\begin{cases} x - \tfrac{5}{16}z = \tfrac{13}{16} \\ y - \tfrac{19}{16}z = \tfrac{11}{16} \end{cases}$$

Let $z = a$. Then $x = \tfrac{5}{16}a + \tfrac{13}{16}$ and $y = \tfrac{19}{16}a + \tfrac{11}{16}$.

Solution: $\left(\tfrac{5}{16}a + \tfrac{13}{16}, \tfrac{19}{16}a + \tfrac{11}{16}, a\right)$ where a is a real number

63. $A = \begin{bmatrix} 3 & -2 & 1 \\ -4 & 1 & -3 \\ 1 & -5 & 1 \end{bmatrix}$

$$A^{-1} = \begin{bmatrix} 0.56 & 0.12 & -0.2 \\ -0.04 & -0.08 & -0.2 \\ -0.76 & -0.52 & 0.2 \end{bmatrix}$$

$$\begin{bmatrix} x \\ y \\ z \end{bmatrix} = \begin{bmatrix} 0.56 & 0.12 & -0.2 \\ -0.04 & -0.08 & -0.2 \\ -0.76 & -0.52 & 0.2 \end{bmatrix}\begin{bmatrix} -29 \\ 37 \\ -24 \end{bmatrix} = \begin{bmatrix} -7 \\ 3 \\ -2 \end{bmatrix}$$

Solution: $(-7, 3, -2)$

65. $A = \begin{bmatrix} 7 & -3 & 0 & 2 \\ -2 & 1 & 0 & -1 \\ 4 & 0 & 1 & -2 \\ -1 & 1 & 0 & -1 \end{bmatrix}$

$$A^{-1} = \begin{bmatrix} 0 & -1 & 0 & 1 \\ -1 & -5 & 0 & 3 \\ -2 & -4 & 1 & -2 \\ -1 & -4 & 0 & 1 \end{bmatrix}$$

$$\begin{bmatrix} x \\ y \\ z \\ w \end{bmatrix} = \begin{bmatrix} 0 & -1 & 0 & 1 \\ -1 & -5 & 0 & 3 \\ -2 & -4 & 1 & -2 \\ -1 & -4 & 0 & 1 \end{bmatrix}\begin{bmatrix} 41 \\ -13 \\ 12 \\ -8 \end{bmatrix} = \begin{bmatrix} 5 \\ 0 \\ -2 \\ 3 \end{bmatrix}$$

Solution: $(5, 0, -2, 3)$

67. $A = \begin{bmatrix} 1 & 1 & 1 \\ 0.065 & 0.07 & 0.09 \\ 0 & 2 & -1 \end{bmatrix}$

$[A \vdots I] = \begin{bmatrix} 1 & 1 & 1 & \vdots & 1 & 0 & 0 \\ 0.065 & 0.07 & 0.09 & \vdots & 0 & 1 & 0 \\ 0 & 2 & -1 & \vdots & 0 & 0 & 1 \end{bmatrix}$

$200R_2 \rightarrow \begin{bmatrix} 1 & 1 & 1 & \vdots & 1 & 0 & 0 \\ 13 & 14 & 18 & \vdots & 0 & 200 & 0 \\ 0 & 2 & -1 & \vdots & 0 & 0 & 1 \end{bmatrix}$

$-13R_1 + R_2 \rightarrow \begin{bmatrix} 1 & 1 & 1 & \vdots & 1 & 0 & 0 \\ 0 & 1 & 5 & \vdots & -13 & 200 & 0 \\ 0 & 2 & -1 & \vdots & 0 & 0 & 1 \end{bmatrix}$

$\begin{matrix} -R_2 + R_1 \rightarrow \\ \\ -2R_2 + R_3 \rightarrow \end{matrix} \begin{bmatrix} 1 & 0 & -4 & \vdots & 14 & -200 & 0 \\ 0 & 1 & 5 & \vdots & -13 & 200 & 0 \\ 0 & 0 & -11 & \vdots & 26 & -400 & 1 \end{bmatrix}$

$-\frac{1}{11}R_3 \rightarrow \begin{bmatrix} 1 & 0 & -4 & \vdots & 14 & -200 & 0 \\ 0 & 1 & 5 & \vdots & -13 & 200 & 0 \\ 0 & 0 & 1 & \vdots & -\frac{26}{11} & \frac{400}{11} & -\frac{1}{11} \end{bmatrix}$

$\begin{matrix} 4R_3 + R_1 \rightarrow \\ -5R_3 + R_2 \rightarrow \\ \\ \end{matrix} \begin{bmatrix} 1 & 0 & 0 & \vdots & \frac{50}{11} & -\frac{600}{11} & -\frac{4}{11} \\ 0 & 1 & 0 & \vdots & -\frac{13}{11} & \frac{200}{11} & \frac{5}{11} \\ 0 & 0 & 1 & \vdots & -\frac{26}{11} & \frac{400}{11} & -\frac{1}{11} \end{bmatrix} = [I \vdots A^{-1}]$

$X = A^{-1}B = \frac{1}{11}\begin{bmatrix} 50 & -600 & -4 \\ -13 & 200 & 5 \\ -26 & 400 & -1 \end{bmatrix}\begin{bmatrix} 10,000 \\ 705 \\ 0 \end{bmatrix} = \begin{bmatrix} 7000 \\ 1000 \\ 2000 \end{bmatrix}$

Solution: $7000 in AAA-rated bonds, $1000 in A-rated bonds, $2000 in B-rated bonds

69. Use the inverse matrix A^{-1} from Exercise 67.

$X = A^{-1}B = \frac{1}{11}\begin{bmatrix} 50 & -600 & -4 \\ -13 & 200 & 5 \\ -26 & 400 & -1 \end{bmatrix}\begin{bmatrix} 12,000 \\ 835 \\ 0 \end{bmatrix} = \begin{bmatrix} 9000 \\ 1000 \\ 2000 \end{bmatrix}$

Solution: $9000 in AAA-rated bonds, $1000 in A-rated bonds, $2000 in B-rated bonds

71. (a)
$$A = \begin{bmatrix} 2 & 0 & 4 \\ 0 & 1 & 4 \\ 1 & 1 & -1 \end{bmatrix}$$

$$[A \ \vdots \ I] = \begin{bmatrix} 2 & 0 & 4 & \vdots & 1 & 0 & 0 \\ 0 & 1 & 4 & \vdots & 0 & 1 & 0 \\ 1 & 1 & -1 & \vdots & 0 & 0 & 1 \end{bmatrix}$$

$$\begin{matrix} R_1 \\ \\ R_3 \end{matrix} \begin{bmatrix} 1 & 1 & -1 & \vdots & 0 & 0 & 1 \\ 0 & 1 & 4 & \vdots & 0 & 1 & 0 \\ 2 & 0 & 4 & \vdots & 1 & 0 & 0 \end{bmatrix}$$

$$-2R_1 + R_3 \rightarrow \begin{bmatrix} 1 & 1 & -1 & \vdots & 0 & 0 & 1 \\ 0 & 1 & 4 & \vdots & 0 & 1 & 0 \\ 0 & -2 & 6 & \vdots & 1 & 0 & -2 \end{bmatrix}$$

$$\begin{matrix} -R_2 + R_1 \rightarrow \\ \\ 2R_2 + R_3 \rightarrow \end{matrix} \begin{bmatrix} 1 & 0 & -5 & \vdots & 0 & -1 & 1 \\ 0 & 1 & 4 & \vdots & 0 & 1 & 0 \\ 0 & 0 & 14 & \vdots & 1 & 2 & -2 \end{bmatrix}$$

$$\tfrac{1}{14}R_3 \rightarrow \begin{bmatrix} 1 & 0 & -5 & \vdots & 0 & -1 & 1 \\ 0 & 1 & 4 & \vdots & 0 & 1 & 0 \\ 0 & 0 & 1 & \vdots & \frac{1}{14} & \frac{1}{7} & -\frac{1}{7} \end{bmatrix}$$

$$\begin{matrix} 5R_3 + R_1 \rightarrow \\ -4R_3 + R_2 \rightarrow \\ \\ \end{matrix} \begin{bmatrix} 1 & 0 & 0 & \vdots & \frac{5}{14} & -\frac{2}{7} & \frac{2}{7} \\ 0 & 1 & 0 & \vdots & -\frac{2}{7} & \frac{3}{7} & \frac{4}{7} \\ 0 & 0 & 1 & \vdots & \frac{1}{14} & \frac{1}{7} & -\frac{1}{7} \end{bmatrix} = [I \ \vdots \ A^{-1}]$$

$$A^{-1} = \frac{1}{14} \begin{bmatrix} 5 & -4 & 4 \\ -4 & 6 & 8 \\ 1 & 2 & -2 \end{bmatrix}$$

$$\begin{bmatrix} I_1 \\ I_2 \\ I_3 \end{bmatrix} = \frac{1}{14} \begin{bmatrix} 5 & -4 & 4 \\ -4 & 6 & 8 \\ 1 & 2 & -2 \end{bmatrix} \begin{bmatrix} 14 \\ 28 \\ 0 \end{bmatrix} = \begin{bmatrix} -3 \\ 8 \\ 5 \end{bmatrix}$$

Solution: $I_1 = -3$ amperes, $I_2 = 8$ amperes, $I_3 = 5$ amperes

(b)
$$\begin{bmatrix} I_1 \\ I_2 \\ I_3 \end{bmatrix} = \frac{1}{14} \begin{bmatrix} 5 & -4 & 4 \\ -4 & 6 & 8 \\ 1 & 2 & -2 \end{bmatrix} \begin{bmatrix} 24 \\ 23 \\ 0 \end{bmatrix} = \begin{bmatrix} 2 \\ 3 \\ 5 \end{bmatrix}$$

Solution:

$I_1 = 2$ amperes, $I_2 = 3$ amperes, $I_3 = 5$ amperes

73. True. If B is the inverse of A, then $AB = I = BA$.

75. $AA^{-1} = \begin{bmatrix} a & b \\ c & d \end{bmatrix} \left(\dfrac{1}{ad - bc} \right) \begin{bmatrix} d & -b \\ -c & a \end{bmatrix} = \dfrac{1}{ad - bc} \begin{bmatrix} a & b \\ c & d \end{bmatrix} \begin{bmatrix} d & -b \\ -c & a \end{bmatrix}$

$\qquad = \dfrac{1}{ad - bc} \begin{bmatrix} ad - bc & 0 \\ 0 & ad - bc \end{bmatrix} = \begin{bmatrix} 1 & 0 \\ 0 & 1 \end{bmatrix}$

$A^{-1}A = \dfrac{1}{ad - bc} \begin{bmatrix} d & -b \\ -c & a \end{bmatrix} \begin{bmatrix} a & b \\ c & d \end{bmatrix} = \dfrac{1}{ad - bc} \begin{bmatrix} ad - bc & 0 \\ 0 & ad - bc \end{bmatrix} = \begin{bmatrix} 1 & 0 \\ 0 & 1 \end{bmatrix}$

77. $|x + 7| \geq 2$

$\qquad x + 7 \leq -2 \ \text{ or } \ x + 7 \geq 2$

$\qquad \ x \leq -9 \ \text{ or } \qquad x \geq -5$

79. $3^{x/2} = 315$

$\qquad \ln 3^{x/2} = \ln 315$

$\qquad \dfrac{x}{2} \ln 3 = \ln 315$

$\qquad x = \dfrac{2 \ln 315}{\ln 3} \approx 10.472$

81. $\log_2 x - 2 = 4.5$

$\qquad \log_2 x = 6.5$

$\qquad x = 2^{6.5} \approx 90.510$

83. Answers will vary.

Section 10.4   The Determinant of a Square Matrix

- You should be able to determine the determinant of a matrix of order 2×2 by using the difference of the products of the diagonals.
- You should be able to use expansion by cofactors to find the determinant of a matrix of order 3×3 or greater.
- The determinant of a triangular matrix equals the product of the entries on the main diagonal.

Vocabulary Check

1. determinant   **2.** minor   **3.** cofactor   **4.** expanding by cofactors

---

1. 5

3. $\begin{vmatrix} 2 & 1 \\ 3 & 4 \end{vmatrix} = 2(4) - 1(3) = 8 - 3 = 5$

5. $\begin{vmatrix} 5 & 2 \\ -6 & 3 \end{vmatrix} = 5(3) - 2(-6) = 15 + 12 = 27$

7. $\begin{vmatrix} -7 & 0 \\ 3 & 0 \end{vmatrix} = -7(0) - 0(3) = 0$

9. $\begin{vmatrix} 2 & 6 \\ 0 & 3 \end{vmatrix} = 2(3) - 6(0) = 6$

11. $\begin{vmatrix} -3 & -2 \\ -6 & -1 \end{vmatrix} = (-3)(-1) - (-2)(-6) = 3 - 12 = -9$

13. $\begin{vmatrix} 9 & 0 \\ 7 & 8 \end{vmatrix} = 9(8) - 0(7) = 72 - 0 = 72$

15. $\begin{vmatrix} -\frac{1}{2} & \frac{1}{3} \\ -6 & \frac{1}{3} \end{vmatrix} = -\frac{1}{2}\left(\frac{1}{3}\right) - \frac{1}{3}(-6) = -\frac{1}{6} + 2 = \frac{11}{6}$

17. $\begin{vmatrix} 0.3 & 0.2 & 0.2 \\ 0.2 & 0.2 & 0.2 \\ -0.4 & 0.4 & 0.3 \end{vmatrix} = -0.002$

19. $\begin{vmatrix} 0.9 & 0.7 & 0 \\ -0.1 & 0.3 & 1.3 \\ -2.2 & 4.2 & 6.1 \end{vmatrix} = -4.842$

21. $\begin{vmatrix} 1 & 4 & -2 \\ 3 & 6 & -6 \\ -2 & 1 & 4 \end{vmatrix} = 0$

23. $\begin{bmatrix} 3 & 4 \\ 2 & -5 \end{bmatrix}$

    (a) $M_{11} = -5$ (b) $C_{11} = M_{11} = -5$

        $M_{12} = 2$ $C_{12} = -M_{12} = -2$

        $M_{21} = 4$ $C_{21} = -M_{21} = -4$

        $M_{22} = 3$ $C_{22} = M_{22} = 3$

25. $\begin{bmatrix} 3 & 1 \\ -2 & -4 \end{bmatrix}$

    (a) $M_{11} = -4$ (b) $C_{11} = M_{11} = -4$

        $M_{12} = -2$ $C_{12} = -M_{12} = 2$

        $M_{21} = 1$ $C_{21} = -M_{21} = -1$

        $M_{22} = 3$ $C_{22} = M_{22} = 3$

27. $\begin{bmatrix} 4 & 0 & 2 \\ -3 & 2 & 1 \\ 1 & -1 & 1 \end{bmatrix}$

(a) $M_{11} = \begin{vmatrix} 2 & 1 \\ -1 & 1 \end{vmatrix} = 2 - (-1) = 3$

$M_{12} = \begin{vmatrix} -3 & 1 \\ 1 & 1 \end{vmatrix} = -3 - 1 = -4$

$M_{13} = \begin{vmatrix} -3 & 2 \\ 1 & -1 \end{vmatrix} = 3 - 2 = 1$

$M_{21} = \begin{vmatrix} 0 & 2 \\ -1 & 1 \end{vmatrix} = 0 - (-2) = 2$

$M_{22} = \begin{vmatrix} 4 & 2 \\ 1 & 1 \end{vmatrix} = 4 - 2 = 2$

$M_{23} = \begin{vmatrix} 4 & 0 \\ 1 & -1 \end{vmatrix} = -4 - 0 = -4$

$M_{31} = \begin{vmatrix} 0 & 2 \\ 2 & 1 \end{vmatrix} = 0 - 4 = -4$

$M_{32} = \begin{vmatrix} 4 & 2 \\ -3 & 1 \end{vmatrix} = 4 - (-6) = 10$

$M_{33} = \begin{vmatrix} 4 & 0 \\ -3 & 2 \end{vmatrix} = 8 - 0 = 8$

(b) $C_{11} = (-1)^2 M_{11} = 3$

$C_{12} = (-1)^3 M_{12} = 4$

$C_{13} = (-1)^4 M_{13} = 1$

$C_{21} = (-1)^3 M_{21} = -2$

$C_{22} = (-1)^4 M_{22} = 2$

$C_{23} = (-1)^5 M_{23} = 4$

$C_{31} = (-1)^4 M_{31} = -4$

$C_{32} = (-1)^5 M_{32} = -10$

$C_{33} = (-1)^6 M_{33} = 8$

29. $\begin{bmatrix} 3 & -2 & 8 \\ 3 & 2 & -6 \\ -1 & 3 & 6 \end{bmatrix}$

(a) $M_{11} = \begin{vmatrix} 2 & -6 \\ 3 & 6 \end{vmatrix} = 12 + 18 = 30$

$M_{12} = \begin{vmatrix} 3 & -6 \\ -1 & 6 \end{vmatrix} = 18 - 6 = 12$

$M_{13} = \begin{vmatrix} 3 & 2 \\ -1 & 3 \end{vmatrix} = 9 + 2 = 11$

$M_{21} = \begin{vmatrix} -2 & 8 \\ 3 & 6 \end{vmatrix} = -12 - 24 = -36$

$M_{22} = \begin{vmatrix} 3 & 8 \\ -1 & 6 \end{vmatrix} = 18 + 8 = 26$

$M_{23} = \begin{vmatrix} 3 & -2 \\ -1 & 3 \end{vmatrix} = 9 - 2 = 7$

$M_{31} = \begin{vmatrix} -2 & 8 \\ 2 & -6 \end{vmatrix} = 12 - 16 = -4$

$M_{32} = \begin{vmatrix} 3 & 8 \\ 3 & -6 \end{vmatrix} = -18 - 24 = -42$

$M_{33} = \begin{vmatrix} 3 & -2 \\ 3 & 2 \end{vmatrix} = 6 + 6 = 12$

(b) $C_{11} = (-1)^2 M_{11} = 30$

$C_{12} = (-1)^3 M_{12} = -12$

$C_{13} = (-1)^4 M_{13} = 11$

$C_{21} = (-1)^3 M_{21} = 36$

$C_{22} = (-1)^4 M_{22} = 26$

$C_{23} = (-1)^5 M_{23} = -7$

$C_{31} = (-1)^4 M_{31} = -4$

$C_{32} = (-1)^5 M_{32} = 42$

$C_{33} = (-1)^6 M_{33} = 12$

31. (a) $\begin{vmatrix} -3 & 2 & 1 \\ 4 & 5 & 6 \\ 2 & -3 & 1 \end{vmatrix} = -3 \begin{vmatrix} 5 & 6 \\ -3 & 1 \end{vmatrix} - 2 \begin{vmatrix} 4 & 6 \\ 2 & 1 \end{vmatrix} + \begin{vmatrix} 4 & 5 \\ 2 & -3 \end{vmatrix} = -3(23) - 2(-8) - 22 = -75$

(b) $\begin{vmatrix} -3 & 2 & 1 \\ 4 & 5 & 6 \\ 2 & -3 & 1 \end{vmatrix} = -2 \begin{vmatrix} 4 & 6 \\ 2 & 1 \end{vmatrix} + 5 \begin{vmatrix} -3 & 1 \\ 2 & 1 \end{vmatrix} + 3 \begin{vmatrix} -3 & 1 \\ 4 & 6 \end{vmatrix} = -2(-8) + 5(-5) + 3(-22) = -75$

33. (a) $\begin{vmatrix} 5 & 0 & -3 \\ 0 & 12 & 4 \\ 1 & 6 & 3 \end{vmatrix} = 0 \begin{vmatrix} 0 & -3 \\ 6 & 3 \end{vmatrix} + 12 \begin{vmatrix} 5 & -3 \\ 1 & 3 \end{vmatrix} - 4 \begin{vmatrix} 5 & 0 \\ 1 & 6 \end{vmatrix} = 0(18) + 12(18) - 4(30) = 96$

(b) $\begin{vmatrix} 5 & 0 & -3 \\ 0 & 12 & 4 \\ 1 & 6 & 3 \end{vmatrix} = 0 \begin{vmatrix} 0 & 4 \\ 1 & 3 \end{vmatrix} + 12 \begin{vmatrix} 5 & -3 \\ 1 & 3 \end{vmatrix} - 6 \begin{vmatrix} 5 & -3 \\ 0 & 4 \end{vmatrix} = 0(-4) + 12(18) - 6(20) = 96$

35. (a) $\begin{vmatrix} 6 & 0 & -3 & 5 \\ 4 & 13 & 6 & -8 \\ -1 & 0 & 7 & 4 \\ 8 & 6 & 0 & 2 \end{vmatrix} = -4\begin{vmatrix} 0 & -3 & 5 \\ 0 & 7 & 4 \\ 6 & 0 & 2 \end{vmatrix} + 13\begin{vmatrix} 6 & -3 & 5 \\ -1 & 7 & 4 \\ 8 & 0 & 2 \end{vmatrix} - 6\begin{vmatrix} 6 & 0 & 5 \\ -1 & 0 & 4 \\ 8 & 6 & 2 \end{vmatrix} - 8\begin{vmatrix} 6 & 0 & -3 \\ -1 & 0 & 7 \\ 8 & 6 & 0 \end{vmatrix}$

$= -4(-282) + 13(-298) - 6(-174) - 8(-234) = 170$

(b) $\begin{vmatrix} 6 & 0 & -3 & 5 \\ 4 & 13 & 6 & -8 \\ -1 & 0 & 7 & 4 \\ 8 & 6 & 0 & 2 \end{vmatrix} = 0\begin{vmatrix} 4 & 6 & -8 \\ -1 & 7 & 4 \\ 8 & 0 & 2 \end{vmatrix} + 13\begin{vmatrix} 6 & -3 & 5 \\ -1 & 7 & 4 \\ 8 & 0 & 2 \end{vmatrix} + 0\begin{vmatrix} 6 & -3 & 5 \\ 4 & 6 & -8 \\ 8 & 0 & 2 \end{vmatrix} + 6\begin{vmatrix} 6 & -3 & 5 \\ 4 & 6 & -8 \\ -1 & 7 & 4 \end{vmatrix}$

$= 0 + 13(-298) + 0 + 6(674) = 170$

37. Expand along Column 1.

$\begin{vmatrix} 2 & -1 & 0 \\ 4 & 2 & 1 \\ 4 & 2 & 1 \end{vmatrix} = 2\begin{vmatrix} 2 & 1 \\ 2 & 1 \end{vmatrix} - 4\begin{vmatrix} -1 & 0 \\ 2 & 1 \end{vmatrix} + 4\begin{vmatrix} -1 & 0 \\ 2 & 1 \end{vmatrix} = 2(0) - 4(-1) + 4(-1) = 0$

39. Expand along Row 2.

$\begin{vmatrix} 6 & 3 & -7 \\ 0 & 0 & 0 \\ 4 & -6 & 3 \end{vmatrix} = 0\begin{vmatrix} 3 & -7 \\ -6 & 3 \end{vmatrix} - 0\begin{vmatrix} 6 & -7 \\ 4 & 3 \end{vmatrix} + 0\begin{vmatrix} 6 & 3 \\ 4 & -6 \end{vmatrix} = 0$

41. $\begin{vmatrix} -1 & 2 & -5 \\ 0 & 3 & 4 \\ 0 & 0 & 3 \end{vmatrix} = (-1)(3)(3) = -9$ (Upper triangular)

43. Expand along Column 3.

$\begin{vmatrix} 1 & 4 & -2 \\ 3 & 2 & 0 \\ -1 & 4 & 3 \end{vmatrix} = -2\begin{vmatrix} 3 & 2 \\ -1 & 4 \end{vmatrix} + 3\begin{vmatrix} 1 & 4 \\ 3 & 2 \end{vmatrix}$

$= -2(14) + 3(-10) = -58$

45. $\begin{vmatrix} 2 & 4 & 6 \\ 0 & 3 & 1 \\ 0 & 0 & -5 \end{vmatrix} = (2)(3)(-5) = -30$ (Upper triangular)

47. Expand along Column 3.

$\begin{vmatrix} 2 & 6 & 6 & 2 \\ 2 & 7 & 3 & 6 \\ 1 & 5 & 0 & 1 \\ 3 & 7 & 0 & 7 \end{vmatrix} = 6\begin{vmatrix} 2 & 7 & 6 \\ 1 & 5 & 1 \\ 3 & 7 & 7 \end{vmatrix} - 3\begin{vmatrix} 2 & 6 & 2 \\ 1 & 5 & 1 \\ 3 & 7 & 7 \end{vmatrix} = 6(-20) - 3(16) = -168$

49. Expand along Column 1.

$\begin{vmatrix} 5 & 3 & 0 & 6 \\ 4 & 6 & 4 & 12 \\ 0 & 2 & -3 & 4 \\ 0 & 1 & -2 & 2 \end{vmatrix} = 5\begin{vmatrix} 6 & 4 & 12 \\ 2 & -3 & 4 \\ 1 & -2 & 2 \end{vmatrix} - 4\begin{vmatrix} 3 & 0 & 6 \\ 2 & -3 & 4 \\ 1 & -2 & 2 \end{vmatrix} = 5(0) - 4(0) = 0$

51. Expand along Column 2, then along Column 4.

$$\begin{vmatrix} 3 & 2 & 4 & -1 & 5 \\ -2 & 0 & 1 & 3 & 2 \\ 1 & 0 & 0 & 4 & 0 \\ 6 & 0 & 2 & -1 & 0 \\ 3 & 0 & 5 & 1 & 0 \end{vmatrix} = -2 \begin{vmatrix} -2 & 1 & 3 & 2 \\ 1 & 0 & 4 & 0 \\ 6 & 2 & -1 & 0 \\ 3 & 5 & 1 & 0 \end{vmatrix} = (-2)(-2) \begin{vmatrix} 1 & 0 & 4 \\ 6 & 2 & -1 \\ 3 & 5 & 1 \end{vmatrix} = 4(103) = 412$$

53. $\begin{vmatrix} 3 & 8 & -7 \\ 0 & -5 & 4 \\ 8 & 1 & 6 \end{vmatrix} = -126$

55. $\begin{vmatrix} 7 & 0 & -14 \\ -2 & 5 & 4 \\ -6 & 2 & 12 \end{vmatrix} = 0$

57. $\begin{vmatrix} 1 & -1 & 8 & 4 \\ 2 & 6 & 0 & -4 \\ 2 & 0 & 2 & 6 \\ 0 & 2 & 8 & 0 \end{vmatrix} = -336$

59. $\begin{vmatrix} 3 & -2 & 4 & 3 & 1 \\ -1 & 0 & 2 & 1 & 0 \\ 5 & -1 & 0 & 3 & 2 \\ 4 & 7 & -8 & 0 & 0 \\ 1 & 2 & 3 & 0 & 2 \end{vmatrix} = 410$

61. (a) $\begin{vmatrix} -1 & 0 \\ 0 & 3 \end{vmatrix} = -3$

(b) $\begin{vmatrix} 2 & 0 \\ 0 & -1 \end{vmatrix} = -2$

(c) $\begin{bmatrix} -1 & 0 \\ 0 & 3 \end{bmatrix}\begin{bmatrix} 2 & 0 \\ 0 & -1 \end{bmatrix} = \begin{bmatrix} -2 & 0 \\ 0 & -3 \end{bmatrix}$

(d) $\begin{vmatrix} -2 & 0 \\ 0 & -3 \end{vmatrix} = 6$

63. (a) $\begin{vmatrix} 4 & 0 \\ 3 & -2 \end{vmatrix} = -8$

(b) $\begin{vmatrix} -1 & 1 \\ -2 & 2 \end{vmatrix} = 0$

(c) $\begin{bmatrix} 4 & 0 \\ 3 & -2 \end{bmatrix}\begin{bmatrix} -1 & 1 \\ -2 & 2 \end{bmatrix} = \begin{bmatrix} -4 & 4 \\ 1 & -1 \end{bmatrix}$

(d) $\begin{vmatrix} -4 & 4 \\ 1 & -1 \end{vmatrix} = 0$

65. (a) $\begin{vmatrix} 0 & 1 & 2 \\ -3 & -2 & 1 \\ 0 & 4 & 1 \end{vmatrix} = -21$

(b) $\begin{vmatrix} 3 & -2 & 0 \\ 1 & -1 & 2 \\ 3 & 1 & 1 \end{vmatrix} = -19$

(c) $\begin{bmatrix} 0 & 1 & 2 \\ -3 & -2 & 1 \\ 0 & 4 & 1 \end{bmatrix}\begin{bmatrix} 3 & -2 & 0 \\ 1 & -1 & 2 \\ 3 & 1 & 1 \end{bmatrix} = \begin{bmatrix} 7 & 1 & 4 \\ -8 & 9 & -3 \\ 7 & -3 & 9 \end{bmatrix}$

(d) $\begin{vmatrix} 7 & 1 & 4 \\ -8 & 9 & -3 \\ 7 & -3 & 9 \end{vmatrix} = 399$

67. (a) $\begin{vmatrix} -1 & 2 & 1 \\ 1 & 0 & 1 \\ 0 & 1 & 0 \end{vmatrix} = 2$

(b) $\begin{vmatrix} -1 & 0 & 0 \\ 0 & 2 & 0 \\ 0 & 0 & 3 \end{vmatrix} = -6$

(c) $\begin{bmatrix} -1 & 2 & 1 \\ 1 & 0 & 1 \\ 0 & 1 & 0 \end{bmatrix}\begin{bmatrix} -1 & 0 & 0 \\ 0 & 2 & 0 \\ 0 & 0 & 3 \end{bmatrix} = \begin{bmatrix} 1 & 4 & 3 \\ -1 & 0 & 3 \\ 0 & 2 & 0 \end{bmatrix}$

(d) $\begin{vmatrix} 1 & 4 & 3 \\ -1 & 0 & 3 \\ 0 & 2 & 0 \end{vmatrix} = -12$

69. $\begin{vmatrix} w & x \\ y & z \end{vmatrix} = wz - xy$

$-\begin{vmatrix} y & z \\ w & x \end{vmatrix} = -(xy - wz) = wz - xy$

Thus, $\begin{vmatrix} w & x \\ y & z \end{vmatrix} = -\begin{vmatrix} y & z \\ w & x \end{vmatrix}.$

71. $\begin{vmatrix} w & x \\ y & z \end{vmatrix} = wz - xy$

$\begin{vmatrix} w & x + cw \\ y & z + cy \end{vmatrix} = w(z + cy) - y(x + cw) = wz - xy$

Thus, $\begin{vmatrix} w & x \\ y & z \end{vmatrix} = \begin{vmatrix} w & x + cw \\ y & z + cy \end{vmatrix}.$

73. $\begin{vmatrix} 1 & x & x^2 \\ 1 & y & y^2 \\ 1 & z & z^2 \end{vmatrix} = \begin{vmatrix} y & y^2 \\ z & z^2 \end{vmatrix} - \begin{vmatrix} x & x^2 \\ z & z^2 \end{vmatrix} + \begin{vmatrix} x & x^2 \\ y & y^2 \end{vmatrix}$

$= (yz^2 - y^2z) - (xz^2 - x^2z) + (xy^2 - x^2y)$

$= yz^2 - xz^2 - y^2z + x^2z + xy(y - x)$

$= z^2(y - x) - z(y^2 - x^2) + xy(y - x)$

$= z^2(y - x) - z(y - x)(y + x) + xy(y - x)$

$= (y - x)[z^2 - z(y + x) + xy]$

$= (y - x)[z^2 - zy - zx + xy]$

$= (y - x)[z^2 - zx - zy + xy]$

$= (y - x)[z(z - x) - y(z - x)]$

$= (y - x)(z - x)(z - y)$

75. $\begin{vmatrix} x - 1 & 2 \\ 3 & x - 2 \end{vmatrix} = 0$

$(x - 1)(x - 2) - 6 = 0$

$x^2 - 3x - 4 = 0$

$(x + 1)(x - 4) = 0$

$x = -1 \text{ or } x = 4$

77. $\begin{vmatrix} x + 3 & 2 \\ 1 & x + 2 \end{vmatrix} = 0$

$(x + 3)(x + 2) - 2 = 0$

$x^2 + 5x + 4 = 0$

$(x + 1)(x + 4) = 0$

$x = -1 \text{ or } x = -4$

79. $\begin{vmatrix} 4u & -1 \\ -1 & 2v \end{vmatrix} = 8uv - 1$

81. $\begin{vmatrix} e^{2x} & e^{3x} \\ 2e^{2x} & 3e^{3x} \end{vmatrix} = 3e^{5x} - 2e^{5x} = e^{5x}$

83. $\begin{vmatrix} x & \ln x \\ 1 & \dfrac{1}{x} \end{vmatrix} = 1 - \ln x$

85. True. If an entire row is zero, then each cofactor in the expansion is multiplied by zero.

87. Let $A = \begin{bmatrix} 1 & 3 \\ -2 & 4 \end{bmatrix}$ and $B = \begin{bmatrix} -4 & 0 \\ 3 & 5 \end{bmatrix}$.

$|A| = \begin{vmatrix} 1 & 3 \\ -2 & 4 \end{vmatrix} = 10, \ |B| = \begin{vmatrix} -4 & 0 \\ 3 & 5 \end{vmatrix} = -20, \ |A| + |B| = -10$

$A + B = \begin{bmatrix} -3 & 3 \\ 1 & 9 \end{bmatrix}, \ |A + B| = \begin{vmatrix} -3 & 3 \\ 1 & 9 \end{vmatrix} = -30$

Thus, $|A + B| \neq |A| + |B|$. Your answer may differ, depending on how you choose A and B.

89. A square matrix is a square array of numbers. The determinant of a square matrix is a real number.

91. (a) $\begin{vmatrix} 1 & 3 & 4 \\ -7 & 2 & -5 \\ 6 & 1 & 2 \end{vmatrix} = -115$

$-\begin{vmatrix} 1 & 4 & 3 \\ -7 & -5 & 2 \\ 6 & 2 & 1 \end{vmatrix} = -115$

Column 2 and Column 3 were interchanged.

(b) $\begin{vmatrix} 1 & 3 & 4 \\ -2 & 2 & 0 \\ 1 & 6 & 2 \end{vmatrix} = -40$

$-\begin{vmatrix} 1 & 6 & 2 \\ -2 & 2 & 0 \\ 1 & 3 & 4 \end{vmatrix} = -40$

Row 1 and Row 3 were interchanged.

93. (a) $A = \begin{bmatrix} 1 & 2 \\ 2 & -3 \end{bmatrix}$, $B = \begin{bmatrix} 5 & 10 \\ 2 & -3 \end{bmatrix}$

$$|B| = \begin{vmatrix} 5 & 10 \\ 2 & -3 \end{vmatrix} = -35$$

$$5|A| = 5\begin{vmatrix} 1 & 2 \\ 2 & -3 \end{vmatrix} = -35$$

Row 1 was multiplied by 5.

$$|B| = 5|A|$$

(b) $A = \begin{bmatrix} 1 & 2 & -1 \\ 3 & -3 & 2 \\ 7 & 1 & 3 \end{bmatrix}$, $B = \begin{bmatrix} 1 & 8 & -3 \\ 3 & -12 & 6 \\ 7 & 4 & 9 \end{bmatrix}$

$$|B| = \begin{bmatrix} 1 & 8 & -3 \\ 3 & -12 & 6 \\ 7 & 4 & 9 \end{bmatrix} = -300$$

$$12|A| = 12\begin{vmatrix} 1 & 2 & -1 \\ 3 & -3 & 2 \\ 7 & 1 & 3 \end{vmatrix} = -300$$

Column 2 was multiplied by 4 and Column 3 was multiplied by 3.

$$|B| = (4)(3)|A| = 12|A|$$

95. $f(x) = x^3 - 2x$

Since f is a polynomial, the domain is all real numbers x.

97. $\qquad h(x) = \sqrt{16 - x^2}$

$$16 - x^2 \geq 0$$

$$(4 + x)(4 - x) \geq 0$$

Critical numbers: $x = \pm 4$

Test intervals: $(-\infty, -4), (-4, 4), (4, \infty)$

Test: Is $16 - x^2 \geq 0$?

Solution: $[-4, 4]$

Domain of h: $-4 \leq x \leq 4$

99. $g(t) = \ln(t - 1)$

$$t - 1 > 0$$

$$t > 1$$

Domain: all real numbers $t > 1$

101. $\begin{cases} x + y \leq 8 \\ \qquad x \geq -3 \\ 2x - y < 5 \end{cases}$

103. $[A \vdots I] = \begin{bmatrix} -4 & 1 & \vdots & 1 & 0 \\ 8 & -1 & \vdots & 0 & 1 \end{bmatrix}$

$2R_1 + R_2 \rightarrow \begin{bmatrix} -4 & 1 & \vdots & 1 & 0 \\ 0 & 1 & \vdots & 2 & 1 \end{bmatrix}$

$-R_2 + R_1 \rightarrow \begin{bmatrix} -4 & 0 & \vdots & -1 & -1 \\ 0 & 1 & \vdots & 2 & 1 \end{bmatrix}$

$-\frac{1}{4}R_1 \rightarrow \begin{bmatrix} 1 & 0 & \vdots & \frac{1}{4} & \frac{1}{4} \\ 0 & 1 & \vdots & 2 & 1 \end{bmatrix} = [I \vdots A^{-1}]$

$A^{-1} = \begin{bmatrix} \frac{1}{4} & \frac{1}{4} \\ 2 & 1 \end{bmatrix}$

105. $[A \vdots I] = \begin{bmatrix} -7 & 2 & 9 & \vdots & 1 & 0 & 0 \\ 2 & -4 & -6 & \vdots & 0 & 1 & 0 \\ 3 & 5 & 2 & \vdots & 0 & 0 & 1 \end{bmatrix}$

$4R_2 + R_1 \rightarrow \begin{bmatrix} 1 & -14 & -15 & \vdots & 1 & 4 & 0 \\ 2 & -4 & -6 & \vdots & 0 & 1 & 0 \\ 3 & 5 & 2 & \vdots & 0 & 0 & 1 \end{bmatrix}$

$\begin{matrix} -2R_1 + R_2 \rightarrow \\ -3R_1 + R_3 \rightarrow \end{matrix} \begin{bmatrix} 1 & -14 & -15 & \vdots & 1 & 4 & 0 \\ 0 & 24 & 24 & \vdots & -2 & -7 & 0 \\ 0 & 47 & 47 & \vdots & -3 & -12 & 1 \end{bmatrix}$

$-\frac{47}{24}R_2 + R_3 \rightarrow \begin{bmatrix} 1 & -14 & -15 & \vdots & 1 & 4 & 0 \\ 0 & 24 & 24 & \vdots & -2 & -7 & 0 \\ 0 & 0 & 0 & \vdots & \frac{11}{12} & \frac{41}{24} & 1 \end{bmatrix}$

The zeros in Row 3 imply that the inverse does not exist.

Section 10.5   Applications of Matrices and Determinants

- You should be able to use Cramer's Rule to solve a system of linear equations.

- Now you should be able to solve a system of linear equations by graphing, substitution, elimination, elementary row operations on an augmented matrix, using the inverse matrix, or Cramer's Rule.

- You should be able to find the area of a triangle with vertices (x_1, y_1), (x_2, y_2), and (x_3, y_3).

$$\text{Area} = \pm\frac{1}{2} \begin{vmatrix} x_1 & y_1 & 1 \\ x_2 & y_2 & 1 \\ x_3 & y_3 & 1 \end{vmatrix}$$

 The $\pm$ symbol indicates that the appropriate sign should be chosen so that the area is positive.

- You should be able to test to see if three points, (x_1, y_1), (x_2, y_2), and (x_3, y_3), are collinear.

$$\begin{vmatrix} x_1 & y_1 & 1 \\ x_2 & y_2 & 1 \\ x_3 & y_3 & 1 \end{vmatrix} = 0, \text{ if and only if they are collinear.}$$

- You should be able to find the equation of the line through (x_1, y_1) and (x_2, y_2) by evaluating.

$$\begin{vmatrix} x & y & 1 \\ x_1 & y_1 & 1 \\ x_2 & y_2 & 1 \end{vmatrix} = 0$$

- You should be able to encode and decode messages by using an invertible $n \times n$ matrix.

Vocabulary Check

1. Cramer's Rule   **2.** colinear   **3.** $A = \pm\frac{1}{2}\begin{vmatrix} x_1 & y_1 & 1 \\ x_2 & y_2 & 1 \\ x_3 & y_3 & 1 \end{vmatrix}$ **4.** cryptogram   **5.** uncoded; coded

---

1. $\begin{cases} 3x + 4y = -2 \\ 5x + 3y = 4 \end{cases}$

$$x = \dfrac{\begin{vmatrix} -2 & 4 \\ 4 & 3 \end{vmatrix}}{\begin{vmatrix} 3 & 4 \\ 5 & 3 \end{vmatrix}} = \dfrac{-22}{-11} = 2$$

$$y = \dfrac{\begin{vmatrix} 3 & -2 \\ 5 & 4 \end{vmatrix}}{\begin{vmatrix} 3 & 4 \\ 5 & 3 \end{vmatrix}} = \dfrac{22}{-11} = -2$$

Solution: $(2, -2)$

3. $\begin{cases} 3x + 2y = -2 \\ 6x + 4y = 4 \end{cases}$

Since $\begin{vmatrix} 3 & 2 \\ 6 & 4 \end{vmatrix} = 0$, Cramer's Rule does not apply.

The system is inconsistent in this case and has no solution.

5. $\begin{cases} -0.4x + 0.8y = 1.6 \\ 0.2x + 0.3y = 2.2 \end{cases}$

$$x = \frac{\begin{vmatrix} 1.6 & 0.8 \\ 2.2 & 0.3 \end{vmatrix}}{\begin{vmatrix} -0.4 & 0.8 \\ 0.2 & 0.3 \end{vmatrix}} = \frac{-1.28}{-0.28} = \frac{32}{7}$$

$$y = \frac{\begin{vmatrix} -0.4 & 1.6 \\ 0.2 & 2.2 \end{vmatrix}}{\begin{vmatrix} -0.4 & 0.8 \\ 0.2 & 0.3 \end{vmatrix}} = \frac{-1.20}{-0.28} = \frac{30}{7}$$

Solution: $\left(\dfrac{32}{7}, \dfrac{30}{7} \right)$

7. $\begin{cases} 4x - y + z = -5 \\ 2x + 2y + 3z = 10, \\ 5x - 2y + 6z = 1 \end{cases}$ $\quad D = \begin{vmatrix} 4 & -1 & 1 \\ 2 & 2 & 3 \\ 5 & -2 & 6 \end{vmatrix} = 55$

$$x = \frac{\begin{vmatrix} -5 & -1 & 1 \\ 10 & 2 & 3 \\ 1 & -2 & 6 \end{vmatrix}}{55} = \frac{-55}{55} = -1$$

$$y = \frac{\begin{vmatrix} 4 & -5 & 1 \\ 2 & 10 & 3 \\ 5 & 1 & 6 \end{vmatrix}}{55} = \frac{165}{55} = 3$$

$$z = \frac{\begin{vmatrix} 4 & -1 & -5 \\ 2 & 2 & 10 \\ 5 & -2 & 1 \end{vmatrix}}{55} = \frac{110}{55} = 2$$

Solution: $(-1, 3, 2)$

9. $\begin{cases} x + 2y + 3z = -3 \\ -2x + y - z = 6, \\ 3x - 3y + 2z = -11 \end{cases}$ $\quad D = \begin{vmatrix} 1 & 2 & 3 \\ -2 & 1 & -1 \\ 3 & -3 & 2 \end{vmatrix} = 10$

$$x = \frac{\begin{vmatrix} -3 & 2 & 3 \\ 6 & 1 & -1 \\ -11 & -3 & 2 \end{vmatrix}}{10} = \frac{-20}{10} = -2$$

$$y = \frac{\begin{vmatrix} 1 & -3 & 3 \\ -2 & 6 & -1 \\ 3 & -11 & 2 \end{vmatrix}}{10} = \frac{10}{10} = 1$$

$$z = \frac{\begin{vmatrix} 1 & 2 & -3 \\ -2 & 1 & 6 \\ 3 & -3 & -11 \end{vmatrix}}{10} = \frac{-10}{10} = -1$$

Solution: $(-2, 1, -1)$

11. $\begin{cases} 3x + 3y + 5z = 1 \\ 3x + 5y + 9z = 2, \\ 5x + 9y + 17z = 4 \end{cases}$ $\quad D = \begin{vmatrix} 3 & 3 & 5 \\ 3 & 5 & 9 \\ 5 & 9 & 17 \end{vmatrix} = 4$

$$x = \frac{\begin{vmatrix} 1 & 3 & 5 \\ 2 & 5 & 9 \\ 4 & 9 & 17 \end{vmatrix}}{4} = 0$$

$$y = \frac{\begin{vmatrix} 3 & 1 & 5 \\ 3 & 2 & 9 \\ 5 & 4 & 17 \end{vmatrix}}{4} = -\frac{1}{2}$$

$$z = \frac{\begin{vmatrix} 3 & 3 & 1 \\ 3 & 5 & 2 \\ 5 & 9 & 4 \end{vmatrix}}{4} = \frac{1}{2}$$

Solution: $\left(0, -\dfrac{1}{2}, \dfrac{1}{2} \right)$

13. $\begin{cases} 2x + y + 2z = 6 \\ -x + 2y - 3z = 0 \\ 3x + 2y - z = 6 \end{cases}$ $\quad D = \begin{vmatrix} 2 & 1 & 2 \\ -1 & 2 & -3 \\ 3 & 2 & -1 \end{vmatrix} = -18$

$$x = \frac{\begin{vmatrix} 6 & 1 & 2 \\ 0 & 2 & -3 \\ 6 & 2 & -1 \end{vmatrix}}{-18} = 1, \quad y = \frac{\begin{vmatrix} 2 & 6 & 2 \\ -1 & 0 & -3 \\ 3 & 6 & -1 \end{vmatrix}}{-18} = 2, \quad z = \frac{\begin{vmatrix} 2 & 1 & 6 \\ -1 & 2 & 0 \\ 3 & 2 & 6 \end{vmatrix}}{-18} = 1$$

Solution: $(1, 2, 1)$

15. Vertices: $(0, 0), (3, 1), (1, 5)$

$$\text{Area} = \frac{1}{2} \begin{vmatrix} 0 & 0 & 1 \\ 3 & 1 & 1 \\ 1 & 5 & 1 \end{vmatrix} = \frac{1}{2} \begin{vmatrix} 3 & 1 \\ 1 & 5 \end{vmatrix} = 7 \text{ square units}$$

17. Vertices: $(-2, -3), (2, -3), (0, 4)$

$$\text{Area} = \frac{1}{2}\begin{vmatrix} -2 & -3 & 1 \\ 2 & -3 & 1 \\ 0 & 4 & 1 \end{vmatrix} = \frac{1}{2}\left(-2\begin{vmatrix} -3 & 1 \\ 4 & 1 \end{vmatrix} - 2\begin{vmatrix} -3 & 1 \\ 4 & 1 \end{vmatrix}\right) = \frac{1}{2}(14 + 14) = 14 \text{ square units}$$

19. Vertices: $\left(0, \frac{1}{2}\right), \left(\frac{5}{2}, 0\right), (4, 3)$

$$\text{Area} = \frac{1}{2}\begin{vmatrix} 0 & \frac{1}{2} & 1 \\ \frac{5}{2} & 0 & 1 \\ 4 & 3 & 1 \end{vmatrix} = \frac{1}{2}\left(-\frac{1}{2}\begin{vmatrix} \frac{5}{2} & 1 \\ 4 & 1 \end{vmatrix} + 1\begin{vmatrix} \frac{5}{2} & 0 \\ 4 & 3 \end{vmatrix}\right) = \frac{1}{2}\left(\frac{3}{4} + \frac{15}{2}\right) = \frac{33}{8} \text{ square units}$$

21. Vertices: $(-2, 4), (2, 3), (-1, 5)$

$$\text{Area} = \frac{1}{2}\begin{vmatrix} -2 & 4 & 1 \\ 2 & 3 & 1 \\ -1 & 5 & 1 \end{vmatrix} = \frac{1}{2}\left[\begin{vmatrix} 2 & 3 \\ -1 & 5 \end{vmatrix} - \begin{vmatrix} -2 & 4 \\ -1 & 5 \end{vmatrix} + \begin{vmatrix} -2 & 4 \\ 2 & 3 \end{vmatrix}\right] = \frac{1}{2}(13 + 6 - 14) = \frac{5}{2} \text{ square units}$$

23. Vertices: $(-3, 5), (2, 6), (3, -5)$

$$\text{Area} = -\frac{1}{2}\begin{vmatrix} -3 & 5 & 1 \\ 2 & 6 & 1 \\ 3 & -5 & 1 \end{vmatrix} = -\frac{1}{2}\left[\begin{vmatrix} 2 & 6 \\ 3 & -5 \end{vmatrix} - \begin{vmatrix} -3 & 5 \\ 3 & -5 \end{vmatrix} + \begin{vmatrix} -3 & 5 \\ 2 & 6 \end{vmatrix}\right] = -\frac{1}{2}(-28 + 0 - 28) = 28 \text{ square units}$$

25. $4 = \pm\frac{1}{2}\begin{vmatrix} -5 & 1 & 1 \\ 0 & 2 & 1 \\ -2 & y & 1 \end{vmatrix}$

$\pm 8 = -5\begin{vmatrix} 2 & 1 \\ y & 1 \end{vmatrix} - 2\begin{vmatrix} 1 & 1 \\ 2 & 1 \end{vmatrix}$

$\pm 8 = -5(2 - y) - 2(-1)$

$\pm 8 = 5y - 8$

$y = \dfrac{8 \pm 8}{5}$

$y = \dfrac{16}{5}$ or $y = 0$

27. $6 = \pm\frac{1}{2}\begin{vmatrix} -2 & -3 & 1 \\ 1 & -1 & 1 \\ -8 & y & 1 \end{vmatrix}$

$\pm 12 = \begin{vmatrix} 1 & -1 \\ -8 & y \end{vmatrix} - \begin{vmatrix} -2 & -3 \\ -8 & y \end{vmatrix} + \begin{vmatrix} -2 & -3 \\ 1 & -1 \end{vmatrix}$

$\pm 12 = (y - 8) - (-2y - 24) + 5$

$\pm 12 = 3y + 21$

$y = \dfrac{-21 \pm 12}{3} = -7 \pm 4$

$y = -3$ or $y = -11$

29. Vertices: $(0, 25), (10, 0), (28, 5)$

$$\text{Area} = \frac{1}{2}\begin{vmatrix} 0 & 25 & 1 \\ 10 & 0 & 1 \\ 28 & 5 & 1 \end{vmatrix} = 250 \text{ square miles}$$

31. Points: $(3, -1), (0, -3), (12, 5)$

$$\begin{vmatrix} 3 & -1 & 1 \\ 0 & -3 & 1 \\ 12 & 5 & 1 \end{vmatrix} = 3\begin{vmatrix} -3 & 1 \\ 5 & 1 \end{vmatrix} + 12\begin{vmatrix} -1 & 1 \\ -3 & 1 \end{vmatrix} = 3(-8) + 12(2) = 0$$

The points are collinear.

33. Points: $\left(2, -\frac{1}{2}\right)$, $(-4, 4)$, $(6, -3)$

$$\begin{vmatrix} 2 & -\frac{1}{2} & 1 \\ -4 & 4 & 1 \\ 6 & -3 & 1 \end{vmatrix} = \begin{vmatrix} -4 & 4 \\ 6 & -3 \end{vmatrix} - \begin{vmatrix} 2 & -\frac{1}{2} \\ 6 & -3 \end{vmatrix} + \begin{vmatrix} 2 & -\frac{1}{2} \\ -4 & 4 \end{vmatrix} = -12 + 3 + 6 = -3 \neq 0$$

The points are not collinear.

35. Points: $(0, 2)$, $(1, 2.4)$, $(-1, 1.6)$

$$\begin{vmatrix} 0 & 2 & 1 \\ 1 & 2.4 & 1 \\ -1 & 1.6 & 1 \end{vmatrix} = -2\begin{vmatrix} 1 & 1 \\ -1 & 1 \end{vmatrix} + \begin{vmatrix} 1 & 2.4 \\ -1 & 1.6 \end{vmatrix} = -2(2) + 4 = 0$$

The points are collinear.

37.
$$\begin{vmatrix} 2 & -5 & 1 \\ 4 & y & 1 \\ 5 & -2 & 1 \end{vmatrix} = 0$$

$$2\begin{vmatrix} y & 1 \\ -2 & 1 \end{vmatrix} + 5\begin{vmatrix} 4 & 1 \\ 5 & 1 \end{vmatrix} + \begin{vmatrix} 4 & y \\ 5 & -2 \end{vmatrix} = 0$$

$$2(y + 2) + 5(-1) + (-8 - 5y) = 0$$

$$-3y - 9 = 0$$

$$y = -3$$

39. Points: $(0, 0)$, $(5, 3)$

Equation: $\begin{vmatrix} x & y & 1 \\ 0 & 0 & 1 \\ 5 & 3 & 1 \end{vmatrix} = -\begin{vmatrix} x & y \\ 5 & 3 \end{vmatrix} = 5y - 3x = 0 \implies 3x - 5y = 0$

41. Points: $(-4, 3)$, $(2, 1)$

Equation: $\begin{vmatrix} x & y & 1 \\ -4 & 3 & 1 \\ 2 & 1 & 1 \end{vmatrix} = x\begin{vmatrix} 3 & 1 \\ 1 & 1 \end{vmatrix} - y\begin{vmatrix} -4 & 1 \\ 2 & 1 \end{vmatrix} + \begin{vmatrix} -4 & 3 \\ 2 & 1 \end{vmatrix} = 2x + 6y - 10 = 0 \implies x + 3y - 5 = 0$

43. Points: $\left(-\frac{1}{2}, 3\right)$, $\left(\frac{5}{2}, 1\right)$

Equation: $\begin{vmatrix} x & y & 1 \\ -\frac{1}{2} & 3 & 1 \\ \frac{5}{2} & 1 & 1 \end{vmatrix} = x\begin{vmatrix} 3 & 1 \\ 1 & 1 \end{vmatrix} - y\begin{vmatrix} -\frac{1}{2} & 1 \\ \frac{5}{2} & 1 \end{vmatrix} + \begin{vmatrix} -\frac{1}{2} & 3 \\ \frac{5}{2} & 1 \end{vmatrix} = 2x + 3y - 8 = 0$

45. The uncoded row matrices are the rows of the 7×3 matrix on the left.

$$\begin{matrix} \text{T} & \text{R} & \text{O} \\ \text{U} & \text{B} & \text{L} \\ \text{E} & & \text{I} \\ \text{N} & & \text{R} \\ \text{I} & \text{V} & \text{E} \\ \text{R} & & \text{C} \\ \text{I} & \text{T} & \text{Y} \end{matrix} \begin{bmatrix} 20 & 18 & 15 \\ 21 & 2 & 12 \\ 5 & 0 & 9 \\ 14 & 0 & 18 \\ 9 & 22 & 5 \\ 18 & 0 & 3 \\ 9 & 20 & 25 \end{bmatrix} \begin{bmatrix} 1 & -1 & 0 \\ 1 & 0 & -1 \\ -6 & 2 & 3 \end{bmatrix} = \begin{bmatrix} -52 & 10 & 27 \\ -49 & 3 & 34 \\ -49 & 13 & 27 \\ -94 & 22 & 54 \\ 1 & 1 & -7 \\ 0 & -12 & 9 \\ -121 & 41 & 55 \end{bmatrix}$$

Solution: -52 10 27 -49 3 34 -49 13 27 -94 22 54 1 1 -7 0 -12 9 -121 41 55

In Exercises 47 and 49, use the matrix $A = \begin{bmatrix} 1 & 2 & 2 \\ 3 & 7 & 9 \\ -1 & -4 & -7 \end{bmatrix}$.

47. C A L L __ A T __ N O O N
$[3 \quad 1 \quad 12]\,[12 \quad 0 \quad 1]\,[20 \quad 0 \quad 14]\,[15 \quad 15 \quad 14]$

$[3 \quad 1 \quad 12]A = [-6 \ -35 \ -69]$

$[12 \quad 0 \quad 1]A = [11 \quad 20 \quad 17]$

$[20 \quad 0 \quad 14]A = [6 \ -16 \ -58]$

$[15 \quad 15 \quad 14]A = [46 \quad 79 \quad 67]$

Cryptogram: $-6 \ -35 \ -69 \ 11 \ 20 \ 17 \ 6 \ -16 \ -58 \ 46 \ 79 \ 67$

49. H A P P Y __ B I R T H DA Y __
$[8 \quad 1 \quad 16]\,[16 \quad 25 \quad 0]\,[2 \quad 9 \quad 18]\,[20 \quad 8 \quad 4]\,[1 \quad 25 \quad 0]$

$[8 \quad 1 \quad 16]A = [-5 \ -41 \ -87]$

$[16 \quad 25 \quad 0]A = [91 \quad 207 \quad 257]$

$[2 \quad 9 \quad 18]A = [11 \ -5 \ -41]$

$[20 \quad 8 \quad 4]A = [40 \quad 80 \quad 84]$

$[1 \quad 25 \quad 0]A = [76 \quad 177 \quad 227]$

Cryptogram: $-5 \ -41 \ -87 \ 91 \ 207 \ 257 \ 11 \ -5 \ -41 \ 40 \ 80 \ 84 \ 76 \ 177 \ 227$

51. $A^{-1} = \begin{bmatrix} 1 & 2 \\ 3 & 5 \end{bmatrix}^{-1} = \begin{bmatrix} -5 & 2 \\ 3 & -1 \end{bmatrix}$

$\begin{bmatrix} 11 & 21 \\ 64 & 112 \\ 25 & 50 \\ 29 & 53 \\ 23 & 46 \\ 40 & 75 \\ 55 & 92 \end{bmatrix} \begin{bmatrix} -5 & 2 \\ 3 & -1 \end{bmatrix} = \begin{bmatrix} 8 & 1 \\ 16 & 16 \\ 25 & 0 \\ 14 & 5 \\ 23 & 0 \\ 25 & 5 \\ 1 & 18 \end{bmatrix} \begin{matrix} \text{H} & \text{A} \\ \text{P} & \text{P} \\ \text{Y} & \_ \\ \text{N} & \text{E} \\ \text{W} & \_ \\ \text{Y} & \text{E} \\ \text{A} & \text{R} \end{matrix}$ Message: HAPPY NEW YEAR

53. $A^{-1} = \begin{bmatrix} 1 & -1 & 0 \\ 1 & 0 & -1 \\ -6 & 2 & 3 \end{bmatrix}^{-1} = \begin{bmatrix} -2 & -3 & -1 \\ -3 & -3 & -1 \\ -2 & -4 & -1 \end{bmatrix}$

$\begin{bmatrix} 9 & -1 & -9 \\ 38 & -19 & -19 \\ 28 & -9 & -19 \\ -80 & 25 & 41 \\ -64 & 21 & 31 \\ 9 & -5 & -4 \end{bmatrix} \begin{bmatrix} -2 & -3 & -1 \\ -3 & -3 & -1 \\ -2 & -4 & -1 \end{bmatrix} = \begin{bmatrix} 3 & 12 & 1 \\ 19 & 19 & 0 \\ 9 & 19 & 0 \\ 3 & 1 & 14 \\ 3 & 5 & 12 \\ 5 & 4 & 0 \end{bmatrix} \begin{matrix} \text{C} & \text{L} & \text{A} \\ \text{S} & \text{S} & \_ \\ \text{I} & \text{S} & \_ \\ \text{C} & \text{A} & \text{N} \\ \text{C} & \text{E} & \text{L} \\ \text{E} & \text{D} & \_ \end{matrix}$ Message: CLASS IS CANCELED

55. $A^{-1} = \begin{bmatrix} 1 & 2 & 2 \\ 3 & 7 & 9 \\ -1 & -4 & -7 \end{bmatrix}^{-1} = \begin{bmatrix} -13 & 6 & 4 \\ 12 & -5 & -3 \\ -5 & 2 & 1 \end{bmatrix}$

$\begin{bmatrix} 20 & 17 & -15 \\ -12 & -56 & -104 \\ 1 & -25 & -65 \\ 62 & 143 & 181 \end{bmatrix} \begin{bmatrix} -13 & 6 & 4 \\ 12 & -5 & -3 \\ -5 & 2 & 1 \end{bmatrix} = \begin{bmatrix} 19 & 5 & 14 \\ 4 & 0 & 16 \\ 12 & 1 & 14 \\ 5 & 19 & 0 \end{bmatrix} \begin{matrix} \text{S} & \text{E} & \text{N} \\ \text{D} & \_ & \text{P} \\ \text{L} & \text{A} & \text{N} \\ \text{E} & \text{S} & \_ \end{matrix}$ Message: SEND PLANES

57. Let A be the 2×2 matrix needed to decode the message.

$$\begin{bmatrix} -18 & -18 \\ 1 & 16 \end{bmatrix} A = \begin{bmatrix} 0 & 18 \\ 15 & 14 \end{bmatrix} \begin{matrix} \text{R} \\ \text{O} \end{matrix} \begin{matrix} \\ \text{N} \end{matrix}$$

$$A = \begin{bmatrix} -18 & -18 \\ 1 & 16 \end{bmatrix}^{-1} \begin{bmatrix} 0 & 18 \\ 15 & 14 \end{bmatrix} = \begin{bmatrix} -\frac{8}{135} & -\frac{1}{15} \\ \frac{1}{270} & \frac{1}{15} \end{bmatrix} \begin{bmatrix} 0 & 18 \\ 15 & 14 \end{bmatrix} = \begin{bmatrix} -1 & -2 \\ 1 & 1 \end{bmatrix}$$

8	21		13	5	M	E
−15	−10		5	20	E	T
−13	−13		0	13	_	M
5	10		5	0	E	_
5	25	$\begin{bmatrix} -1 & -2 \\ 1 & 1 \end{bmatrix} =$	20	15	T	O
5	19		14	9	N	I
−1	6		7	8	G	H
20	40		20	0	T	_
−18	−18		0	18	_	R
1	16		15	14	O	N

Message: MEET ME TONIGHT RON

59. False. In Cramer's Rule, the **denominator** is the determinant of the coefficient matrix.

61. False. If the determinant of the coefficient matrix is zero, the system has either no solution or infinitely many solutions.

63.
$$\begin{cases} -x - 7y = -22 & \text{Equation 1} \\ 5x + y = -26 & \text{Equation 2} \end{cases}$$

$$\begin{array}{ll} -5x - 35y = -110 & \text{(5)Eq.1} \\ 5x + y = -26 & \\ \hline -34y = -136 & \text{Add equations.} \\ y = 4 & \\ -x - 7(4) = -22 & \\ x = -6 & \end{array}$$

Solution: $(-6, 4)$

65.
$$\begin{cases} -x - 3y + 5z = -14 \\ 4x + 2y - z = -1 \\ 5x - 3y + 2z = -11 \end{cases}$$

$$A^{-1} = \begin{bmatrix} -1 & -3 & 5 \\ 4 & 2 & -1 \\ 5 & -3 & 2 \end{bmatrix}^{-1}$$

$$= \frac{1}{72} \begin{bmatrix} -1 & 9 & 7 \\ 13 & 27 & -19 \\ 22 & 18 & -10 \end{bmatrix}$$

$$\begin{bmatrix} x \\ y \\ z \end{bmatrix} = A^{-1} \begin{bmatrix} -14 \\ -1 \\ -11 \end{bmatrix} = \begin{bmatrix} -1 \\ 0 \\ -3 \end{bmatrix}$$

Solution: $(-1, 0, -3)$

67. Objective function: $z = 6x + 4y$

Constraints:
$$\begin{aligned} x &\geq 0 \\ y &\geq 0 \\ x + 6y &\leq 30 \\ 6x + y &\leq 40 \end{aligned}$$

At $(0, 0)$: $z = 6(0) + 4(0) = 0$

At $(0, 5)$: $z = 6(0) + 4(5) = 20$

At $(6, 4)$: $z = 6(6) + 4(4) = 52$

At $\left(\frac{20}{3}, 0\right)$: $z = 6\left(\frac{20}{3}\right) + 4(0) = 40$

The minimum value of 0 occurs at $(0, 0)$.

The maximum value of 52 occurs at $(6, 4)$.

Review Exercises for Chapter 10

1. $\begin{bmatrix} -4 \\ 0 \\ 5 \end{bmatrix}$

Order: 3×1

3. $[3]$

Order: 1×1

5. $\begin{cases} 3x - 10y = 15 \\ 5x + 4y = 22 \end{cases}$

$\begin{bmatrix} 3 & -10 & \vdots & 15 \\ 5 & 4 & \vdots & 22 \end{bmatrix}$

7. $\begin{bmatrix} 5 & 1 & 7 & \vdots & -9 \\ 4 & 2 & 0 & \vdots & 10 \\ 9 & 4 & 2 & \vdots & 3 \end{bmatrix}$

$\begin{cases} 5x + y + 7z = -9 \\ 4x + 2y = 10 \\ 9x + 4y + 2z = 3 \end{cases}$

9. $\begin{bmatrix} 0 & 1 & 1 \\ 1 & 2 & 3 \\ 2 & 2 & 2 \end{bmatrix}$

$\begin{matrix} R_1 \\ R_2 \end{matrix} \begin{bmatrix} 1 & 2 & 3 \\ 0 & 1 & 1 \\ 2 & 2 & 2 \end{bmatrix}$

$-2R_1 + R_3 \rightarrow \begin{bmatrix} 1 & 2 & 3 \\ 0 & 1 & 1 \\ 0 & -2 & -4 \end{bmatrix}$

$2R_2 + R_3 \rightarrow \begin{bmatrix} 1 & 2 & 3 \\ 0 & 1 & 1 \\ 0 & 0 & -2 \end{bmatrix}$

$-\frac{1}{2}R_3 \rightarrow \begin{bmatrix} 1 & 2 & 3 \\ 0 & 1 & 1 \\ 0 & 0 & 1 \end{bmatrix}$

11. $\begin{bmatrix} 1 & 2 & 3 & \vdots & 9 \\ 0 & 1 & -2 & \vdots & 2 \\ 0 & 0 & 1 & \vdots & 0 \end{bmatrix} \Rightarrow \begin{cases} x + 2y + 3z = 9 \\ y - 2z = 2 \\ z = 0 \end{cases}$

$y - 2(0) = 2 \Rightarrow y = 2$

$x + 2(2) + 3(0) = 9 \Rightarrow x = 5$

Solution: $(5, 2, 0)$

13. $\begin{bmatrix} 1 & -5 & 4 & \vdots & 1 \\ 0 & 1 & 2 & \vdots & 3 \\ 0 & 0 & 1 & \vdots & 4 \end{bmatrix} \Rightarrow \begin{cases} x - 5y + 4z = 1 \\ y + 2z = 3 \\ z = 4 \end{cases}$

$y + 2(4) = 3 \Rightarrow y = -5$

$x - 5(-5) + 4(4) = 1 \Rightarrow x = -40$

Solution: $(-40, -5, 4)$

15. $\begin{bmatrix} 5 & 4 & \vdots & 2 \\ -1 & 1 & \vdots & -22 \end{bmatrix}$

$4R_2 + R_1 \rightarrow \begin{bmatrix} 1 & 8 & \vdots & -86 \\ -1 & 1 & \vdots & -22 \end{bmatrix}$

$R_1 + R_2 \rightarrow \begin{bmatrix} 1 & 8 & \vdots & -86 \\ 0 & 9 & \vdots & -108 \end{bmatrix}$

$\frac{1}{9}R_2 \rightarrow \begin{bmatrix} 1 & 8 & \vdots & -86 \\ 0 & 1 & \vdots & -12 \end{bmatrix}$

$\begin{cases} x + 8y = -86 \\ y = -12 \end{cases}$

$y = -12$

$x + 8(-12) = -86 \Rightarrow x = 10$

Solution: $(10, -12)$

17. $\begin{bmatrix} 0.3 & -0.1 & \vdots & -0.13 \\ 0.2 & -0.3 & \vdots & -0.25 \end{bmatrix}$

$\begin{matrix} 10R_1 \rightarrow \\ 10R_2 \rightarrow \end{matrix} \begin{bmatrix} 3 & -1 & \vdots & -1.3 \\ 2 & -3 & \vdots & -2.5 \end{bmatrix}$

$-R_2 + R_1 \rightarrow \begin{bmatrix} 1 & 2 & \vdots & 1.2 \\ 2 & -3 & \vdots & -2.5 \end{bmatrix}$

$-2R_1 + R_2 \rightarrow \begin{bmatrix} 1 & 2 & \vdots & 1.2 \\ 0 & -7 & \vdots & -4.9 \end{bmatrix}$

$-\frac{1}{7}R_2 \rightarrow \begin{bmatrix} 1 & 2 & \vdots & 1.2 \\ 0 & 1 & \vdots & 0.7 \end{bmatrix}$

$\begin{cases} x + 2y = 1.2 \\ y = 0.7 \end{cases}$

$y = 0.7$

$x + 2(0.7) = 1.2 \Rightarrow x = -0.2$

Solution: $(-0.2, 0.7) = \left(-\frac{1}{5}, \frac{7}{10} \right)$

19.
$$\begin{bmatrix} 2 & 3 & 1 & \vdots & 10 \\ 2 & -3 & -3 & \vdots & 22 \\ 4 & -2 & 3 & \vdots & -2 \end{bmatrix}$$

$$\begin{matrix} \\ -R_1 + R_2 \to \\ -2R_1 + R_3 \to \end{matrix} \begin{bmatrix} 2 & 3 & 1 & \vdots & 10 \\ 0 & -6 & -4 & \vdots & 12 \\ 0 & -8 & 1 & \vdots & -22 \end{bmatrix}$$

$$\begin{matrix} \frac{1}{2}R_1 \to \\ -\frac{1}{6}R_2 \to \\ \end{matrix} \begin{bmatrix} 1 & \frac{3}{2} & \frac{1}{2} & \vdots & 5 \\ 0 & 1 & \frac{2}{3} & \vdots & -2 \\ 0 & -8 & 1 & \vdots & -22 \end{bmatrix}$$

$$\begin{matrix} \\ \\ 8R_2 + R_3 \to \end{matrix} \begin{bmatrix} 1 & \frac{3}{2} & \frac{1}{2} & \vdots & 5 \\ 0 & 1 & \frac{2}{3} & \vdots & -2 \\ 0 & 0 & \frac{19}{3} & \vdots & -38 \end{bmatrix}$$

$$\begin{matrix} \\ \\ \frac{3}{19}R_3 \to \end{matrix} \begin{bmatrix} 1 & \frac{3}{2} & \frac{1}{2} & \vdots & 5 \\ 0 & 1 & \frac{2}{3} & \vdots & -2 \\ 0 & 0 & 1 & \vdots & -6 \end{bmatrix}$$

$$z = -6$$

$$y + \tfrac{2}{3}(-6) = -2 \implies y = 2$$

$$x + \tfrac{3}{2}(2) + \tfrac{1}{2}(-6) = 5 \implies x = 5$$

Solution: $(5, 2, -6)$

21.
$$\begin{bmatrix} 2 & 1 & 2 & \vdots & 4 \\ 2 & 2 & 0 & \vdots & 5 \\ 2 & -1 & 6 & \vdots & 2 \end{bmatrix}$$

$$\begin{matrix} \\ -R_1 + R_2 \to \\ -R_1 + R_3 \to \end{matrix} \begin{bmatrix} 2 & 1 & 2 & \vdots & 4 \\ 0 & 1 & -2 & \vdots & 1 \\ 0 & -2 & 4 & \vdots & -2 \end{bmatrix}$$

$$\begin{matrix} -R_2 + R_1 \to \\ \\ 2R_2 + R_3 \to \end{matrix} \begin{bmatrix} 2 & 0 & 4 & \vdots & 3 \\ 0 & 1 & -2 & \vdots & 1 \\ 0 & 0 & 0 & \vdots & 0 \end{bmatrix}$$

$$\begin{matrix} \frac{1}{2}R_1 \to \\ \\ \end{matrix} \begin{bmatrix} 1 & 0 & 2 & \vdots & \frac{3}{2} \\ 0 & 1 & -2 & \vdots & 1 \\ 0 & 0 & 0 & \vdots & 0 \end{bmatrix}$$

Let $z = a$, then:

$$y - 2a = 1 \implies y = 2a + 1$$

$$x + 2a = \tfrac{3}{2} \implies x = -2a + \tfrac{3}{2}$$

Solution: $\left(-2a + \tfrac{3}{2},\ 2a + 1,\ a\right)$ where a is any real number

23.
$$\begin{bmatrix} 2 & 1 & 1 & 0 & \vdots & 6 \\ 0 & -2 & 3 & -1 & \vdots & 9 \\ 3 & 3 & -2 & -2 & \vdots & -11 \\ 1 & 0 & 1 & 3 & \vdots & 14 \end{bmatrix}$$

$$\begin{matrix} -R_4 + R_1 \to \\ \\ \\ \\ \end{matrix} \begin{bmatrix} 1 & 1 & 0 & -3 & \vdots & -8 \\ 0 & -2 & 3 & -1 & \vdots & 9 \\ 3 & 3 & -2 & -2 & \vdots & -11 \\ 1 & 0 & 1 & 3 & \vdots & 14 \end{bmatrix}$$

$$\begin{matrix} \\ \\ -3R_1 + R_3 \to \\ -R_1 + R_4 \to \end{matrix} \begin{bmatrix} 1 & 1 & 0 & -3 & \vdots & -8 \\ 0 & -2 & 3 & -1 & \vdots & 9 \\ 0 & 0 & -2 & 7 & \vdots & 13 \\ 0 & -1 & 1 & 6 & \vdots & 22 \end{bmatrix}$$

$$\begin{matrix} \\ -3R_4 + R_2 \to \\ \\ \\ \end{matrix} \begin{bmatrix} 1 & 1 & 0 & -3 & \vdots & -8 \\ 0 & 1 & 0 & -19 & \vdots & -57 \\ 0 & 0 & -2 & 7 & \vdots & 13 \\ 0 & -1 & 1 & 6 & \vdots & 22 \end{bmatrix}$$

$$\begin{matrix} \\ \\ \\ R_2 + R_4 \to \end{matrix} \begin{bmatrix} 1 & 1 & 0 & -3 & \vdots & -8 \\ 0 & 1 & 0 & -19 & \vdots & -57 \\ 0 & 0 & -2 & 7 & \vdots & 13 \\ 0 & 0 & 1 & -13 & \vdots & -35 \end{bmatrix}$$

$$\begin{matrix} \\ \\ R_4 \to \\ R_3 \to \end{matrix} \begin{bmatrix} 1 & 1 & 0 & -3 & \vdots & -8 \\ 0 & 1 & 0 & -19 & \vdots & -57 \\ 0 & 0 & 1 & -13 & \vdots & -35 \\ 0 & 0 & -2 & 7 & \vdots & 13 \end{bmatrix}$$

$$\begin{matrix} \\ \\ \\ 2R_3 + R_4 \to \end{matrix} \begin{bmatrix} 1 & 1 & 0 & -3 & \vdots & -8 \\ 0 & 1 & 0 & -19 & \vdots & -57 \\ 0 & 0 & 1 & -13 & \vdots & -35 \\ 0 & 0 & 0 & -19 & \vdots & -57 \end{bmatrix}$$

$$\begin{matrix} \\ \\ \\ \frac{1}{19}R_4 \to \end{matrix} \begin{bmatrix} 1 & 1 & 0 & -3 & \vdots & -8 \\ 0 & 1 & 0 & -19 & \vdots & -57 \\ 0 & 0 & 1 & -13 & \vdots & -35 \\ 0 & 0 & 0 & 1 & \vdots & 3 \end{bmatrix}$$

$$w = 3$$

$$z - 13(3) = -35 \implies z = 4$$

$$y - 19(3) = -57 \implies y = 0$$

$$x + 0 - 3(3) = -8 \implies x = 1$$

Solution: $(1, 0, 4, 3)$

25.
$$\begin{bmatrix} -1 & 1 & 2 & \vdots & 1 \\ 2 & 3 & 1 & \vdots & -2 \\ 5 & 4 & 2 & \vdots & 4 \end{bmatrix}$$

$$-R_1 \rightarrow \begin{bmatrix} 1 & -1 & -2 & \vdots & -1 \\ 2 & 3 & 1 & \vdots & -2 \\ 5 & 4 & 2 & \vdots & 4 \end{bmatrix}$$

$$\begin{matrix} -2R_1 + R_2 \rightarrow \\ -5R_1 + R_3 \rightarrow \end{matrix} \begin{bmatrix} 1 & -1 & -2 & \vdots & -1 \\ 0 & 5 & 5 & \vdots & 0 \\ 0 & 9 & 12 & \vdots & 9 \end{bmatrix}$$

$$\tfrac{1}{5}R_2 \rightarrow \begin{bmatrix} 1 & -1 & -2 & \vdots & -1 \\ 0 & 1 & 1 & \vdots & 0 \\ 0 & 9 & 12 & \vdots & 9 \end{bmatrix}$$

$$\begin{matrix} R_2 + R_1 \rightarrow \\ \\ -9R_2 + R_3 \rightarrow \end{matrix} \begin{bmatrix} 1 & 0 & -1 & \vdots & -1 \\ 0 & 1 & 1 & \vdots & 0 \\ 0 & 0 & 3 & \vdots & 9 \end{bmatrix}$$

$$\tfrac{1}{3}R_3 \rightarrow \begin{bmatrix} 1 & 0 & -1 & \vdots & -1 \\ 0 & 1 & 1 & \vdots & 0 \\ 0 & 0 & 1 & \vdots & 3 \end{bmatrix}$$

$$\begin{matrix} R_3 + R_1 \rightarrow \\ -R_3 + R_2 \rightarrow \end{matrix} \begin{bmatrix} 1 & 0 & 0 & \vdots & 2 \\ 0 & 1 & 0 & \vdots & -3 \\ 0 & 0 & 1 & \vdots & 3 \end{bmatrix}$$

$x = 2, y = -3, z = 3$

Solution: $(2, -3, 3)$

27.
$$\begin{bmatrix} 2 & -1 & 9 & \vdots & -8 \\ -1 & -3 & 4 & \vdots & -15 \\ 5 & 2 & -1 & \vdots & 17 \end{bmatrix}$$

$$R_2 + R_1 \rightarrow \begin{bmatrix} 1 & -4 & 13 & \vdots & -23 \\ -1 & -3 & 4 & \vdots & -15 \\ 5 & 2 & -1 & \vdots & 17 \end{bmatrix}$$

$$\begin{matrix} R_1 + R_2 \rightarrow \\ -5R_1 + R_3 \rightarrow \end{matrix} \begin{bmatrix} 1 & -4 & 13 & \vdots & -23 \\ 0 & -7 & 17 & \vdots & -38 \\ 0 & 22 & -66 & \vdots & 132 \end{bmatrix}$$

$$\begin{matrix} R_3 \\ R_2 \end{matrix} \begin{bmatrix} 1 & -4 & 13 & \vdots & -23 \\ 0 & 22 & -66 & \vdots & 132 \\ 0 & -7 & 17 & \vdots & -38 \end{bmatrix}$$

$$\tfrac{1}{22}R_2 \rightarrow \begin{bmatrix} 1 & -4 & 13 & \vdots & -23 \\ 0 & 1 & -3 & \vdots & 6 \\ 0 & -7 & 17 & \vdots & -38 \end{bmatrix}$$

$$7R_2 + R_3 \rightarrow \begin{bmatrix} 1 & -4 & 13 & \vdots & -23 \\ 0 & 1 & -3 & \vdots & 6 \\ 0 & 0 & -4 & \vdots & 4 \end{bmatrix}$$

$$-\tfrac{1}{4}R_3 \rightarrow \begin{bmatrix} 1 & -4 & 13 & \vdots & -23 \\ 0 & 1 & -3 & \vdots & 6 \\ 0 & 0 & 1 & \vdots & -1 \end{bmatrix}$$

$$4R_2 + R_1 \rightarrow \begin{bmatrix} 1 & 0 & 1 & \vdots & 1 \\ 0 & 1 & -3 & \vdots & 6 \\ 0 & 0 & 1 & \vdots & -1 \end{bmatrix}$$

$$\begin{matrix} -R_3 + R_1 \rightarrow \\ 3R_3 + R_2 \rightarrow \end{matrix} \begin{bmatrix} 1 & 0 & 0 & \vdots & 2 \\ 0 & 1 & 0 & \vdots & 3 \\ 0 & 0 & 1 & \vdots & -1 \end{bmatrix}$$

$x = 2, y = 3, z = -1$

Solution: $(2, 3, -1)$

29. Use the reduced row-echelon form feature of a graphing utility.

$$\begin{bmatrix} 3 & -1 & 5 & -2 & \vdots & -44 \\ 1 & 6 & 4 & -1 & \vdots & 1 \\ 5 & -1 & 1 & 3 & \vdots & -15 \\ 0 & 4 & -1 & -8 & \vdots & 58 \end{bmatrix} \Rightarrow \begin{bmatrix} 1 & 0 & 0 & 0 & \vdots & 2 \\ 0 & 1 & 0 & 0 & \vdots & 6 \\ 0 & 0 & 1 & 0 & \vdots & -10 \\ 0 & 0 & 0 & 1 & \vdots & -3 \end{bmatrix}$$

$x = 2, y = 6, z = -10, w = -3$

Solution: $(2, 6, -10, -3)$

31. $\begin{bmatrix} -1 & x \\ y & 9 \end{bmatrix} = \begin{bmatrix} -1 & 12 \\ -7 & 9 \end{bmatrix} \Rightarrow x = 12$ and $y = -7$

33. $\begin{bmatrix} x + 3 & -4 & 4y \\ 0 & -3 & 2 \\ -2 & y + 5 & 6x \end{bmatrix} = \begin{bmatrix} 5x - 1 & -4 & 44 \\ 0 & -3 & 2 \\ -2 & 16 & 6 \end{bmatrix}$

$\left.\begin{matrix} x + 3 = 5x - 1 \\ 4y = 44 \\ y + 5 = 16 \\ 6x = 6 \end{matrix}\right\} x = 1$ and $y = 11$

35. (a) $A + B = \begin{bmatrix} 2 & -2 \\ 3 & 5 \end{bmatrix} + \begin{bmatrix} -3 & 10 \\ 12 & 8 \end{bmatrix} = \begin{bmatrix} -1 & 8 \\ 15 & 13 \end{bmatrix}$

(b) $A - B = \begin{bmatrix} 2 & -2 \\ 3 & 5 \end{bmatrix} - \begin{bmatrix} -3 & 10 \\ 12 & 8 \end{bmatrix} = \begin{bmatrix} 5 & -12 \\ -9 & -3 \end{bmatrix}$

(c) $4A = 4\begin{bmatrix} 2 & -2 \\ 3 & 5 \end{bmatrix} = \begin{bmatrix} 8 & -8 \\ 12 & 20 \end{bmatrix}$

(d) $A + 3B = \begin{bmatrix} 2 & -2 \\ 3 & 5 \end{bmatrix} + 3\begin{bmatrix} -3 & 10 \\ 12 & 8 \end{bmatrix} = \begin{bmatrix} 2 & -2 \\ 3 & 5 \end{bmatrix} + \begin{bmatrix} -9 & 30 \\ 36 & 24 \end{bmatrix} = \begin{bmatrix} -7 & 28 \\ 39 & 29 \end{bmatrix}$

37. (a) $A + B = \begin{bmatrix} 5 & 4 \\ -7 & 2 \\ 11 & 2 \end{bmatrix} + \begin{bmatrix} 0 & 3 \\ 4 & 12 \\ 20 & 40 \end{bmatrix} = \begin{bmatrix} 5 & 7 \\ -3 & 14 \\ 31 & 42 \end{bmatrix}$

(b) $A - B = \begin{bmatrix} 5 & 4 \\ -7 & 2 \\ 11 & 2 \end{bmatrix} - \begin{bmatrix} 0 & 3 \\ 4 & 12 \\ 20 & 40 \end{bmatrix} = \begin{bmatrix} 5 & 1 \\ -11 & -10 \\ -9 & -38 \end{bmatrix}$

(c) $4A = 4\begin{bmatrix} 5 & 4 \\ -7 & 2 \\ 11 & 2 \end{bmatrix} = \begin{bmatrix} 20 & 16 \\ -28 & 8 \\ 44 & 8 \end{bmatrix}$

(d) $A + 3B = \begin{bmatrix} 5 & 4 \\ -7 & 2 \\ 11 & 2 \end{bmatrix} + 3\begin{bmatrix} 0 & 3 \\ 4 & 12 \\ 20 & 40 \end{bmatrix} = \begin{bmatrix} 5 & 4 \\ -7 & 2 \\ 11 & 2 \end{bmatrix} + \begin{bmatrix} 0 & 9 \\ 12 & 36 \\ 60 & 120 \end{bmatrix} = \begin{bmatrix} 5 & 13 \\ 5 & 38 \\ 71 & 122 \end{bmatrix}$

39. $\begin{bmatrix} 7 & 3 \\ -1 & 5 \end{bmatrix} + \begin{bmatrix} 10 & -20 \\ 14 & -3 \end{bmatrix} = \begin{bmatrix} 7+10 & 3-20 \\ -1+14 & 5-3 \end{bmatrix} = \begin{bmatrix} 17 & -17 \\ 13 & 2 \end{bmatrix}$

41. $-2\begin{bmatrix} 1 & 2 \\ 5 & -4 \\ 6 & 0 \end{bmatrix} + 8\begin{bmatrix} 7 & 1 \\ 1 & 2 \\ 1 & 4 \end{bmatrix} = \begin{bmatrix} -2 & -4 \\ -10 & 8 \\ -12 & 0 \end{bmatrix} + \begin{bmatrix} 56 & 8 \\ 8 & 16 \\ 8 & 32 \end{bmatrix} = \begin{bmatrix} 54 & 4 \\ -2 & 24 \\ -4 & 32 \end{bmatrix}$

43. $3\begin{bmatrix} 8 & -2 & 5 \\ 1 & 3 & -1 \end{bmatrix} + 6\begin{bmatrix} 4 & -2 & -3 \\ 2 & 7 & 6 \end{bmatrix} = \begin{bmatrix} 24 & -6 & 15 \\ 3 & 9 & -3 \end{bmatrix} + \begin{bmatrix} 24 & -12 & -18 \\ 12 & 42 & 36 \end{bmatrix} = \begin{bmatrix} 48 & -18 & -3 \\ 15 & 51 & 33 \end{bmatrix}$

45. $X = 3A - 2B = 3\begin{bmatrix} -4 & 0 \\ 1 & -5 \\ -3 & 2 \end{bmatrix} - 2\begin{bmatrix} 1 & 2 \\ -2 & 1 \\ 4 & 4 \end{bmatrix}$

$\qquad = \begin{bmatrix} -14 & -4 \\ 7 & -17 \\ -17 & -2 \end{bmatrix}$

47. $X = \dfrac{1}{3}[B - 2A] = \dfrac{1}{3}\left(\begin{bmatrix} 1 & 2 \\ -2 & 1 \\ 4 & 4 \end{bmatrix} - 2\begin{bmatrix} -4 & 0 \\ 1 & -5 \\ -3 & 2 \end{bmatrix} \right)$

$\qquad = \dfrac{1}{3}\begin{bmatrix} 9 & 2 \\ -4 & 11 \\ 10 & 0 \end{bmatrix} = \begin{bmatrix} 3 & \frac{2}{3} \\ -\frac{4}{3} & \frac{11}{3} \\ \frac{10}{3} & 0 \end{bmatrix}$

49. A and B are both 2×2 so AB exists.

$AB = \begin{bmatrix} 2 & -2 \\ 3 & 5 \end{bmatrix}\begin{bmatrix} -3 & 10 \\ 12 & 8 \end{bmatrix} = \begin{bmatrix} 2(-3) + (-2)(12) & 2(10) + (-2)(8) \\ 3(-3) + 5(12) & 3(10) + 5(8) \end{bmatrix} = \begin{bmatrix} -30 & 4 \\ 51 & 70 \end{bmatrix}$

51. Since A is 3×2 and B is 2×2, AB exists.

$AB = \begin{bmatrix} 5 & 4 \\ -7 & 2 \\ 11 & 2 \end{bmatrix}\begin{bmatrix} 4 & 12 \\ 20 & 40 \end{bmatrix} = \begin{bmatrix} 5(4) + 4(20) & 5(12) + 4(40) \\ -7(4) + 2(20) & -7(12) + 2(40) \\ 11(4) + 2(20) & 11(12) + 2(40) \end{bmatrix} = \begin{bmatrix} 100 & 220 \\ 12 & -4 \\ 84 & 212 \end{bmatrix}$

53.
$$\begin{bmatrix} 1 & 2 \\ 5 & -4 \\ 6 & 0 \end{bmatrix} \begin{bmatrix} 6 & -2 & 8 \\ 4 & 0 & 0 \end{bmatrix} = \begin{bmatrix} 1(6) + 2(4) & 1(-2) + 2(0) & 1(8) + 2(0) \\ 5(6) + (-4)(4) & 5(-2) + (-4)(0) & 5(8) + (-4)(0) \\ 6(6) + (0)(4) & 6(-2) + (0)(0) & 6(8) + (0)(0) \end{bmatrix}$$

$$= \begin{bmatrix} 14 & -2 & 8 \\ 14 & -10 & 40 \\ 36 & -12 & 48 \end{bmatrix}$$

55.
$$\begin{bmatrix} 1 & 5 & 6 \\ 2 & -4 & 0 \end{bmatrix} \begin{bmatrix} 6 & 4 \\ -2 & 0 \\ 8 & 0 \end{bmatrix} = \begin{bmatrix} 1(6) + 5(-2) + 6(8) & 1(4) + 5(0) + 6(0) \\ 2(6) - 4(-2) + 0(8) & 2(4) - 4(0) + 0(0) \end{bmatrix}$$

$$= \begin{bmatrix} 44 & 4 \\ 20 & 8 \end{bmatrix}$$

57.
$$\begin{bmatrix} 4 \\ 6 \end{bmatrix} \begin{bmatrix} 6 & -2 \end{bmatrix} = \begin{bmatrix} 4(6) & 4(-2) \\ 6(6) & 6(-2) \end{bmatrix} = \begin{bmatrix} 24 & -8 \\ 36 & -12 \end{bmatrix}$$

59.
$$\begin{bmatrix} 2 & 1 \\ 6 & 0 \end{bmatrix} \left(\begin{bmatrix} 4 & 2 \\ -3 & 1 \end{bmatrix} + \begin{bmatrix} -2 & 4 \\ 0 & 4 \end{bmatrix} \right) = \begin{bmatrix} 2 & 1 \\ 6 & 0 \end{bmatrix} \begin{bmatrix} 2 & 6 \\ -3 & 5 \end{bmatrix}$$

$$= \begin{bmatrix} 2(2) + 1(-3) & 2(6) + 1(5) \\ 6(2) + 0 & 6(6) + 0 \end{bmatrix}$$

$$= \begin{bmatrix} 1 & 17 \\ 12 & 36 \end{bmatrix}$$

61.
$$\begin{bmatrix} 4 & 1 \\ 11 & -7 \\ 12 & 3 \end{bmatrix} \begin{bmatrix} 3 & -5 & 6 \\ 2 & -2 & -2 \end{bmatrix} = \begin{bmatrix} 14 & -22 & 22 \\ 19 & -41 & 80 \\ 42 & -66 & 66 \end{bmatrix}$$

63. $0.95A = 0.95 \begin{bmatrix} 80 & 120 & 140 \\ 40 & 100 & 80 \end{bmatrix} = \begin{bmatrix} 76 & 114 & 133 \\ 38 & 95 & 76 \end{bmatrix}$

65. $BA = \begin{bmatrix} 10.25 & 14.50 & 17.75 \end{bmatrix} \begin{bmatrix} 8200 & 7400 \\ 6500 & 9800 \\ 5400 & 4800 \end{bmatrix} = \begin{bmatrix} \$274,150 & \$303,150 \end{bmatrix}$

The merchandise shipped to warehouse 1 is worth \$274,150, and the merchandise shipped to warehouse 2 is worth \$303,150.

67. $AB = \begin{bmatrix} -4 & -1 \\ 7 & 2 \end{bmatrix} \begin{bmatrix} -2 & -1 \\ 7 & 4 \end{bmatrix} = \begin{bmatrix} -4(-2) + (-1)(7) & -4(-1) + (-1)(4) \\ 7(-2) + 2(7) & 7(-1) + 2(4) \end{bmatrix}$

$= \begin{bmatrix} 1 & 0 \\ 0 & 1 \end{bmatrix} = I$

$BA = \begin{bmatrix} -2 & -1 \\ 7 & 4 \end{bmatrix} \begin{bmatrix} -4 & -1 \\ 7 & 2 \end{bmatrix} = \begin{bmatrix} -2(-4) + (-1)(7) & -2(-1) + (-1)(2) \\ 7(-4) + 4(7) & 7(-1) + 4(2) \end{bmatrix}$

$= \begin{bmatrix} 1 & 0 \\ 0 & 1 \end{bmatrix} = I$

69. $AB = \begin{bmatrix} 1 & 1 & 0 \\ 1 & 0 & 1 \\ 6 & 2 & 3 \end{bmatrix} \begin{bmatrix} -2 & -3 & 1 \\ 3 & 3 & -1 \\ 2 & 4 & -1 \end{bmatrix}$

$= \begin{bmatrix} 1(-2) + 1(3) + 0(2) & 1(-3) + 1(3) + 0(4) & 1(1) + 1(-1) + 0(-1) \\ 1(-2) + 0(3) + 1(2) & 1(-3) + 0(3) + 1(4) & 1(1) + 0(-1) + 1(-1) \\ 6(-2) + 2(3) + 3(2) & 6(-3) + 2(3) + 3(4) & 6(1) + 2(-1) + 3(-1) \end{bmatrix}$

$= \begin{bmatrix} 1 & 0 & 0 \\ 0 & 1 & 0 \\ 0 & 0 & 1 \end{bmatrix} = I$

$BA = \begin{bmatrix} -2 & -3 & 1 \\ 3 & 3 & -1 \\ 2 & 4 & -1 \end{bmatrix} \begin{bmatrix} 1 & 1 & 0 \\ 1 & 0 & 1 \\ 6 & 2 & 3 \end{bmatrix}$

$= \begin{bmatrix} -2(1) + (-3)(1) + 1(6) & -2(1) + (-3)(0) + 1(2) & -2(0) + (-3)(1) + 1(3) \\ 3(1) + 3(1) + (-1)(6) & 3(1) + 3(0) + (-1)(2) & 3(0) + 3(1) + (-1)(3) \\ 2(1) + 4(1) + (-1)(6) & 2(1) + 4(0) + (-1)(2) & 2(0) + 4(1) + (-1)(3) \end{bmatrix}$

$= \begin{bmatrix} 1 & 0 & 0 \\ 0 & 1 & 0 \\ 0 & 0 & 1 \end{bmatrix} = I$

71. $[A \;\vdots\; I] = \begin{bmatrix} -6 & 5 & \vdots & 1 & 0 \\ -5 & 4 & \vdots & 0 & 1 \end{bmatrix}$

$-\tfrac{1}{6}R_1 \rightarrow \begin{bmatrix} 1 & -\tfrac{5}{6} & \vdots & -\tfrac{1}{6} & 0 \\ -5 & 4 & \vdots & 0 & 1 \end{bmatrix}$

$5R_1 + R_2 \rightarrow \begin{bmatrix} 1 & -\tfrac{5}{6} & \vdots & -\tfrac{1}{6} & 0 \\ 0 & -\tfrac{1}{6} & \vdots & -\tfrac{5}{6} & 1 \end{bmatrix}$

$-6R_2 \rightarrow \begin{bmatrix} 1 & -\tfrac{5}{6} & \vdots & -\tfrac{1}{6} & 0 \\ 0 & 1 & \vdots & 5 & -6 \end{bmatrix}$

$\tfrac{5}{6}R_2 + R_1 \rightarrow \begin{bmatrix} 1 & 0 & \vdots & 4 & -5 \\ 0 & 1 & \vdots & 5 & -6 \end{bmatrix} = [I \;\vdots\; A^{-1}]$

$A^{-1} = \begin{bmatrix} 4 & -5 \\ 5 & -6 \end{bmatrix}$

73. $[A \;\vdots\; I] = \begin{bmatrix} -1 & -2 & -2 & \vdots & 1 & 0 & 0 \\ 3 & 7 & 9 & \vdots & 0 & 1 & 0 \\ 1 & 4 & 7 & \vdots & 0 & 0 & 1 \end{bmatrix}$

$-R_1 \rightarrow \begin{bmatrix} 1 & 2 & 2 & \vdots & -1 & 0 & 0 \\ 3 & 7 & 9 & \vdots & 0 & 1 & 0 \\ 1 & 4 & 7 & \vdots & 0 & 0 & 1 \end{bmatrix}$

$\begin{matrix} -3R_1 + R_2 \rightarrow \\ -R_1 + R_3 \rightarrow \end{matrix} \begin{bmatrix} 1 & 2 & 2 & \vdots & -1 & 0 & 0 \\ 0 & 1 & 3 & \vdots & 3 & 1 & 0 \\ 0 & 2 & 5 & \vdots & 1 & 0 & 1 \end{bmatrix}$

$\begin{matrix} -2R_2 + R_1 \rightarrow \\ \\ -2R_2 + R_3 \rightarrow \end{matrix} \begin{bmatrix} 1 & 0 & -4 & \vdots & -7 & -2 & 0 \\ 0 & 1 & 3 & \vdots & 3 & 1 & 0 \\ 0 & 0 & -1 & \vdots & -5 & -2 & 1 \end{bmatrix}$

$\begin{matrix} -4R_3 + R_1 \rightarrow \\ 3R_3 + R_2 \rightarrow \\ -R_3 \rightarrow \end{matrix} \begin{bmatrix} 1 & 0 & 0 & \vdots & 13 & 6 & -4 \\ 0 & 1 & 0 & \vdots & -12 & -5 & 3 \\ 0 & 0 & 1 & \vdots & 5 & 2 & -1 \end{bmatrix} = [I \;\vdots\; A^{-1}]$

$A^{-1} = \begin{bmatrix} 13 & 6 & -4 \\ -12 & -5 & 3 \\ 5 & 2 & -1 \end{bmatrix}$

75. $\begin{bmatrix} 2 & 0 & 3 \\ -1 & 1 & 1 \\ 2 & -2 & 1 \end{bmatrix}^{-1} = \begin{bmatrix} \frac{1}{2} & -1 & -\frac{1}{2} \\ \frac{1}{2} & -\frac{2}{3} & -\frac{5}{6} \\ 0 & \frac{2}{3} & \frac{1}{3} \end{bmatrix}$

77. $\begin{bmatrix} 1 & 3 & 1 & 6 \\ 4 & 4 & 2 & 6 \\ 3 & 4 & 1 & 2 \\ -1 & 2 & -1 & -2 \end{bmatrix}^{-1} = \begin{bmatrix} -3 & 6 & -\frac{11}{2} & \frac{7}{2} \\ 1 & -2 & 2 & -1 \\ 7 & -15 & \frac{29}{2} & -\frac{19}{2} \\ -1 & \frac{5}{2} & -\frac{5}{2} & \frac{3}{2} \end{bmatrix}$

$= \begin{bmatrix} -3 & 6 & -5.5 & 3.5 \\ 1 & -2 & 2 & -1 \\ 7 & -15 & 14.5 & -9.5 \\ -1 & 2.5 & -2.5 & 1.5 \end{bmatrix}$

79. $A = \begin{bmatrix} -7 & 2 \\ -8 & 2 \end{bmatrix}$

$A^{-1} = \dfrac{1}{-7(2) - 2(-8)} \begin{bmatrix} 2 & -2 \\ 8 & -7 \end{bmatrix} = \dfrac{1}{2} \begin{bmatrix} 2 & -2 \\ 8 & -7 \end{bmatrix} = \begin{bmatrix} 1 & -1 \\ 4 & -\frac{7}{2} \end{bmatrix}$

81. $A = \begin{bmatrix} -\frac{1}{2} & 20 \\ \frac{3}{10} & -6 \end{bmatrix}$

$A^{-1} = \dfrac{1}{-\frac{1}{2}(-6) - 20\left(\frac{3}{10}\right)} \begin{bmatrix} -6 & -20 \\ -\frac{3}{10} & -\frac{1}{2} \end{bmatrix} = -\dfrac{1}{3} \begin{bmatrix} -6 & -20 \\ -\frac{3}{10} & -\frac{1}{2} \end{bmatrix}$

$= \begin{bmatrix} 2 & \frac{20}{3} \\ \frac{1}{10} & \frac{1}{6} \end{bmatrix}$

83. $\begin{cases} -x + 4y = 8 \\ 2x - 7y = -5 \end{cases}$

$\begin{bmatrix} x \\ y \end{bmatrix} = \begin{bmatrix} -1 & 4 \\ 2 & -7 \end{bmatrix}^{-1} \begin{bmatrix} 8 \\ -5 \end{bmatrix} = \begin{bmatrix} 7 & 4 \\ 2 & 1 \end{bmatrix} \begin{bmatrix} 8 \\ -5 \end{bmatrix}$

$= \begin{bmatrix} 7(8) + 4(-5) \\ 2(8) + 1(-5) \end{bmatrix} = \begin{bmatrix} 36 \\ 11 \end{bmatrix}$

Solution: $(36, 11)$

85. $\begin{cases} -3x + 10y = 8 \\ 5x - 17y = -13 \end{cases}$

$\begin{bmatrix} x \\ y \end{bmatrix} = \begin{bmatrix} -3 & 10 \\ 5 & -17 \end{bmatrix}^{-1} \begin{bmatrix} 8 \\ -13 \end{bmatrix} = \begin{bmatrix} -17 & -10 \\ -5 & -3 \end{bmatrix} \begin{bmatrix} 8 \\ -13 \end{bmatrix}$

$= \begin{bmatrix} -17(8) + (-10)(-13) \\ -5(8) + (-3)(-13) \end{bmatrix} = \begin{bmatrix} -6 \\ -1 \end{bmatrix}$

Solution: $(-6, -1)$

87. $\begin{cases} 3x + 2y - z = 6 \\ x - y + 2z = -1 \\ 5x + y + z = 7 \end{cases}$

$\begin{bmatrix} x \\ y \\ z \end{bmatrix} = \begin{bmatrix} 3 & 2 & -1 \\ 1 & -1 & 2 \\ 5 & 1 & 1 \end{bmatrix}^{-1} \begin{bmatrix} 6 \\ -1 \\ 7 \end{bmatrix} = \begin{bmatrix} -1 & -1 & 1 \\ 3 & \frac{8}{3} & -\frac{7}{3} \\ 2 & \frac{7}{3} & -\frac{5}{3} \end{bmatrix} \begin{bmatrix} 6 \\ -1 \\ 7 \end{bmatrix}$

$= \begin{bmatrix} -1(6) - 1(-1) + 1(7) \\ 3(6) + \frac{8}{3}(-1) - \frac{7}{3}(7) \\ 2(6) + \frac{7}{3}(-1) - \frac{5}{3}(7) \end{bmatrix} = \begin{bmatrix} 2 \\ -1 \\ -2 \end{bmatrix}$

Solution: $(2, -1, -2)$

89. $\begin{cases} -2x + y + 2z = -13 \\ -x - 4y + z = -11 \\ -y - z = 0 \end{cases}$

$\begin{bmatrix} x \\ y \\ z \end{bmatrix} = \begin{bmatrix} -2 & 1 & 2 \\ -1 & -4 & 1 \\ 0 & -1 & -1 \end{bmatrix}^{-1} \begin{bmatrix} -13 \\ -11 \\ 0 \end{bmatrix} = \begin{bmatrix} -\frac{5}{9} & \frac{1}{9} & -1 \\ \frac{1}{9} & -\frac{2}{9} & 0 \\ -\frac{1}{9} & \frac{2}{9} & -1 \end{bmatrix} \begin{bmatrix} -13 \\ -11 \\ 0 \end{bmatrix}$

$= \begin{bmatrix} -\frac{5}{9}(-13) + \frac{1}{9}(-11) - 1(0) \\ \frac{1}{9}(-13) - \frac{2}{9}(-11) + 0(0) \\ -\frac{1}{9}(-13) + \frac{2}{9}(-11) - 1(0) \end{bmatrix} = \begin{bmatrix} 6 \\ 1 \\ -1 \end{bmatrix}$

Solution: $(6, 1, -1)$

91. $\begin{cases} x + 2y = -1 \\ 3x + 4y = -5 \end{cases}$

$\begin{bmatrix} x \\ y \end{bmatrix} = \begin{bmatrix} 1 & 2 \\ 3 & 4 \end{bmatrix}^{-1} \begin{bmatrix} -1 \\ -5 \end{bmatrix} = \begin{bmatrix} -2 & 1 \\ \frac{3}{2} & -\frac{1}{2} \end{bmatrix} \begin{bmatrix} -1 \\ -5 \end{bmatrix} = \begin{bmatrix} -3 \\ 1 \end{bmatrix}$

Solution: $(-3, 1)$

93. $\begin{cases} -3x - 3y - 4z = 2 \\ y + z = -1 \\ 4x + 3y + 4z = -1 \end{cases}$

$\begin{bmatrix} x \\ y \\ z \end{bmatrix} = \begin{bmatrix} -3 & -3 & -4 \\ 0 & 1 & 1 \\ 4 & 3 & 4 \end{bmatrix}^{-1} \begin{bmatrix} 2 \\ -1 \\ -1 \end{bmatrix} = \begin{bmatrix} 1 & 0 & 1 \\ 4 & 4 & 3 \\ -4 & -3 & -3 \end{bmatrix} \begin{bmatrix} 2 \\ -1 \\ -1 \end{bmatrix} = \begin{bmatrix} 1 \\ 1 \\ -2 \end{bmatrix}$

Solution: $(1, 1, -2)$

95. $\begin{vmatrix} 8 & 5 \\ 2 & -4 \end{vmatrix} = 8(-4) - 5(2) = -42$

97. $\begin{vmatrix} 50 & -30 \\ 10 & 5 \end{vmatrix} = 50(5) - (-30)(10) = 550$

99. $\begin{bmatrix} 2 & -1 \\ 7 & 4 \end{bmatrix}$

(a) $M_{11} = 4$ (b) $C_{11} = M_{11} = 4$

$M_{12} = 7$ $C_{12} = -M_{12} = -7$

$M_{21} = -1$ $C_{21} = -M_{21} = 1$

$M_{22} = 2$ $C_{22} = M_{22} = 2$

101. $\begin{bmatrix} 3 & 2 & -1 \\ -2 & 5 & 0 \\ 1 & 8 & 6 \end{bmatrix}$

(a) $M_{11} = \begin{vmatrix} 5 & 0 \\ 8 & 6 \end{vmatrix} = 30$ (b) $C_{11} = M_{11} = 30$

$M_{12} = \begin{vmatrix} -2 & 0 \\ 1 & 6 \end{vmatrix} = -12$ $C_{12} = -M_{12} = 12$

$M_{13} = \begin{vmatrix} -2 & 5 \\ 1 & 8 \end{vmatrix} = -21$ $C_{13} = M_{13} = -21$

$M_{21} = \begin{vmatrix} 2 & -1 \\ 8 & 6 \end{vmatrix} = 20$ $C_{21} = -M_{21} = -20$

$M_{22} = \begin{vmatrix} 3 & -1 \\ 1 & 6 \end{vmatrix} = 19$ $C_{22} = M_{22} = 19$

$M_{23} = \begin{vmatrix} 3 & 2 \\ 1 & 8 \end{vmatrix} = 22$ $C_{23} = -M_{23} = -22$

$M_{31} = \begin{vmatrix} 2 & -1 \\ 5 & 0 \end{vmatrix} = 5$ $C_{31} = M_{31} = 5$

$M_{32} = \begin{vmatrix} 3 & -1 \\ -2 & 0 \end{vmatrix} = -2$ $C_{32} = -M_{32} = 2$

$M_{33} = \begin{vmatrix} 3 & 2 \\ -2 & 5 \end{vmatrix} = 19$ $C_{33} = M_{33} = 19$

103. Expand using Column 2.

$\begin{vmatrix} -2 & 4 & 1 \\ -6 & 0 & 2 \\ 5 & 3 & 4 \end{vmatrix} = -4 \begin{vmatrix} -6 & 2 \\ 5 & 4 \end{vmatrix} - 3 \begin{vmatrix} -2 & 1 \\ -6 & 2 \end{vmatrix}$

$= -4(-34) - 3(2) = 130$

105. Expand along Row 1.

$$\begin{vmatrix} 3 & 0 & -4 & 0 \\ 0 & 8 & 1 & 2 \\ 6 & 1 & 8 & 2 \\ 0 & 3 & -4 & 1 \end{vmatrix} = 3 \begin{vmatrix} 8 & 1 & 2 \\ 1 & 8 & 2 \\ 3 & -4 & 1 \end{vmatrix} + (-4) \begin{vmatrix} 0 & 8 & 2 \\ 6 & 1 & 2 \\ 0 & 3 & 1 \end{vmatrix}$$

$$= 3[8(8 - (-8)) - 1(1 - 6) + 2(-4 - 24)] - 4[0 - 6(8 - 6) + 0]$$

$$= 3[128 + 5 - 56] - 4[-12]$$

$$= 279$$

107. $\begin{cases} 5x - 2y = 6 \\ -11x + 3y = -23 \end{cases}$

$$x = \dfrac{\begin{vmatrix} 6 & -2 \\ -23 & 3 \end{vmatrix}}{\begin{vmatrix} 5 & -2 \\ -11 & 3 \end{vmatrix}} = \dfrac{-28}{-7} = 4, \qquad y = \dfrac{\begin{vmatrix} 5 & 6 \\ -11 & -23 \end{vmatrix}}{\begin{vmatrix} 5 & -2 \\ -11 & 3 \end{vmatrix}} = \dfrac{-49}{-7} = 7$$

Solution: $(4, 7)$

109. $\begin{cases} -2x + 3y - 5z = -11 \\ 4x - y + z = -3 \\ -x - 4y + 6z = 15 \end{cases}$

$$D = \begin{vmatrix} -2 & 3 & -5 \\ 4 & -1 & 1 \\ -1 & -4 & 6 \end{vmatrix} = -2(-1)^2 \begin{vmatrix} -1 & 1 \\ -4 & 6 \end{vmatrix} + 4(-1)^3 \begin{vmatrix} 3 & -5 \\ -4 & 6 \end{vmatrix} - 1(-1)^4 \begin{vmatrix} 3 & -5 \\ -1 & 1 \end{vmatrix}$$

$$= -2(-2) - 4(-2) - (-2) = 14$$

$$x = \dfrac{\begin{vmatrix} -11 & 3 & -5 \\ -3 & -1 & 1 \\ 15 & -4 & 6 \end{vmatrix}}{14} = \dfrac{-11(-1)^2 \begin{vmatrix} -1 & 1 \\ -4 & 6 \end{vmatrix} - 3(-1)^3 \begin{vmatrix} 3 & -5 \\ -4 & 6 \end{vmatrix} + 15(-1)^4 \begin{vmatrix} 3 & -5 \\ -1 & 1 \end{vmatrix}}{14}$$

$$= \dfrac{-11(-2) + 3(-2) + 15(-2)}{14} = \dfrac{-14}{14} = -1$$

$$y = \dfrac{\begin{vmatrix} -2 & -11 & -5 \\ 4 & -3 & 1 \\ -1 & 15 & 6 \end{vmatrix}}{14} = \dfrac{-2(-1)^2 \begin{vmatrix} -3 & 1 \\ 15 & 6 \end{vmatrix} + 4(-1)^3 \begin{vmatrix} -11 & -5 \\ 15 & 6 \end{vmatrix} - 1(-1)^4 \begin{vmatrix} -11 & -5 \\ -3 & 1 \end{vmatrix}}{14}$$

$$= \dfrac{-2(-33) - 4(9) - 1(-26)}{14} = \dfrac{56}{14} = 4$$

$$z = \dfrac{\begin{vmatrix} -2 & 3 & -11 \\ 4 & -1 & -3 \\ -1 & -4 & 15 \end{vmatrix}}{14} = \dfrac{-2(-1)^2 \begin{vmatrix} -1 & -3 \\ -4 & 15 \end{vmatrix} + 4(-1)^3 \begin{vmatrix} 3 & -11 \\ -4 & 15 \end{vmatrix} - 1(-1)^4 \begin{vmatrix} 3 & -11 \\ -1 & -3 \end{vmatrix}}{14}$$

$$= \dfrac{-2(-27) - 4(1) - 1(-20)}{14} = \dfrac{70}{14} = 5$$

Solution: $(-1, 4, 5)$

111. $(1, 0), (5, 0), (5, 8)$

$$\text{Area} = \frac{1}{2} \begin{vmatrix} 1 & 0 & 1 \\ 5 & 0 & 1 \\ 5 & 8 & 1 \end{vmatrix} = \frac{1}{2} \left(1 \begin{vmatrix} 0 & 1 \\ 8 & 1 \end{vmatrix} + 1 \begin{vmatrix} 5 & 0 \\ 5 & 8 \end{vmatrix} \right) = \frac{1}{2}(-8 + 40) = \frac{1}{2}(32) = 16 \text{ square units}$$

113. $(1, -4), (-2, 3), (0, 5)$

$$\text{Area} = -\frac{1}{2}\begin{vmatrix} 1 & -4 & 1 \\ -2 & 3 & 1 \\ 0 & 5 & 1 \end{vmatrix}$$

$$= -\frac{1}{2}\left(-5\begin{vmatrix} 1 & 1 \\ -2 & 1 \end{vmatrix} + \begin{vmatrix} 1 & -4 \\ -2 & 3 \end{vmatrix}\right)$$

$$= -\frac{1}{2}(-5(3) + (-5)) = 10 \text{ square units}$$

115. $(-1, 7), (3, -9), (-3, 15)$

$$\begin{vmatrix} -1 & 7 & 1 \\ 3 & -9 & 1 \\ -3 & 15 & 1 \end{vmatrix} = 0$$

The points are collinear.

117. $(-4, 0), (4, 4)$

$$\begin{vmatrix} x & y & 1 \\ -4 & 0 & 1 \\ 4 & 4 & 1 \end{vmatrix} = 0$$

$$1\begin{vmatrix} -4 & 0 \\ 4 & 4 \end{vmatrix} - 1\begin{vmatrix} x & y \\ 4 & 4 \end{vmatrix} + 1\begin{vmatrix} x & y \\ -4 & 0 \end{vmatrix} = 0$$

$$-16 - (4x - 4y) + 4y = 0$$

$$-4x + 8y - 16 = 0$$

$$x - 2y + 4 = 0$$

119. $\left(-\frac{5}{2}, 3\right), \left(\frac{7}{2}, 1\right)$

$$\begin{vmatrix} x & y & 1 \\ -\frac{5}{2} & 3 & 1 \\ \frac{7}{2} & 1 & 1 \end{vmatrix} = 0$$

$$1\begin{vmatrix} -\frac{5}{2} & 3 \\ \frac{7}{2} & 1 \end{vmatrix} - 1\begin{vmatrix} x & y \\ \frac{7}{2} & 1 \end{vmatrix} + 1\begin{vmatrix} x & y \\ -\frac{5}{2} & 3 \end{vmatrix} = 0$$

$$-13 - \left(x - \frac{7}{2}y\right) + \left(3x + \frac{5}{2}y\right) = 0$$

$$2x + 6y - 13 = 0$$

121. L O O K _ O U T _ B E L O W _
[12 15 15] [11 0 15] [21 20 0] [2 5 12] [15 23 0]

$$A = \begin{bmatrix} 2 & -2 & 0 \\ 3 & 0 & -3 \\ -6 & 2 & 3 \end{bmatrix}$$

$$[12 \quad 15 \quad 15]\begin{bmatrix} 2 & -2 & 0 \\ 3 & 0 & -3 \\ -6 & 2 & 3 \end{bmatrix} = [-21 \quad 6 \quad 0]$$

$$[11 \quad 0 \quad 15]\begin{bmatrix} 2 & -2 & 0 \\ 3 & 0 & -3 \\ -6 & 2 & 3 \end{bmatrix} = [-68 \quad 8 \quad 45]$$

$$[21 \quad 20 \quad 0]\begin{bmatrix} 2 & -2 & 0 \\ 3 & 0 & -3 \\ -6 & 2 & 3 \end{bmatrix} = [102 \quad -42 \quad -60]$$

$$[2 \quad 5 \quad 12]\begin{bmatrix} 2 & -2 & 0 \\ 3 & 0 & -3 \\ -6 & 2 & 3 \end{bmatrix} = [-53 \quad 20 \quad 21]$$

$$[15 \quad 23 \quad 0]\begin{bmatrix} 2 & -2 & 0 \\ 3 & 0 & -3 \\ -6 & 2 & 3 \end{bmatrix} = [99 \quad -30 \quad -69]$$

Cryptogram: $-21 \quad 6 \quad 0 \quad -68 \quad 8 \quad 45 \quad 102$
$-42 \quad -60 \quad -53 \quad 20 \quad 21 \quad 99 \quad -30 \quad -69$

123. $A^{-1} = \begin{bmatrix} -1 & 2 & -3 \\ 2 & 1 & 0 \\ 4 & -2 & 5 \end{bmatrix}$

$[-5 \quad 11 \quad -2] \begin{bmatrix} -1 & 2 & -3 \\ 2 & 1 & 0 \\ 4 & -2 & 5 \end{bmatrix} = [19 \quad 5 \quad 5] \qquad S \quad E \quad E$

$[370 \quad -265 \quad 225] \begin{bmatrix} -1 & 2 & -3 \\ 2 & 1 & 0 \\ 4 & -2 & 5 \end{bmatrix} = [0 \quad 25 \quad 15] \qquad \_ \quad Y \quad O$

$[-57 \quad 48 \quad -33] \begin{bmatrix} -1 & 2 & -3 \\ 2 & 1 & 0 \\ 4 & -2 & 5 \end{bmatrix} = [21 \quad 0 \quad 6] \qquad U \quad \_ \quad F$

$[32 \quad -15 \quad 20] \begin{bmatrix} -1 & 2 & -3 \\ 2 & 1 & 0 \\ 4 & -2 & 5 \end{bmatrix} = [18 \quad 9 \quad 4] \qquad R \quad I \quad D$

$[245 \quad -171 \quad 147] \begin{bmatrix} -1 & 2 & -3 \\ 2 & 1 & 0 \\ 4 & -2 & 5 \end{bmatrix} = [1 \quad 25 \quad 0] \qquad A \quad Y \quad \_$

Message: SEE YOU FRIDAY

125. False. The matrix must be square.

127. The matrix must be square and its determinant nonzero to have an inverse.

129. No. Each matrix is in row-echelon form, but the third matrix cannot be achieved from the first or second matrix with elementary row operations. Also, the first two matrices describe a system of equations with one solution. The third matrix describes a system with infinitely many solutions.

131. $\begin{vmatrix} 2 - \lambda & 5 \\ 3 & -8 - \lambda \end{vmatrix} = 0$

$(2 - \lambda)(-8 - \lambda) - 15 = 0$

$-16 + 6\lambda + \lambda^2 - 15 = 0$

$\lambda^2 + 6\lambda - 31 = 0$

$\lambda = \dfrac{-6 \pm \sqrt{36 - 4(-31)}}{2}$

$\lambda = -3 \pm 2\sqrt{10}$

Problem Solving for Chapter 10

1. $A = \begin{bmatrix} 0 & -1 \\ 1 & 0 \end{bmatrix}$ $T = \begin{bmatrix} 1 & 2 & 3 \\ 1 & 4 & 2 \end{bmatrix}$

(a) $AT = \begin{bmatrix} -1 & -4 & -2 \\ 1 & 2 & 3 \end{bmatrix}$ $AAT = \begin{bmatrix} -1 & -2 & -3 \\ -1 & -4 & -2 \end{bmatrix}$

Original Triangle

AT Triangle

AAT Triangle

The transformation A interchanges the x and y coordinates and then takes the negative of the x coordinate. A represents a counterclockwise rotation by 90°.

(b) $A^{-1}(AAT) = (A^{-1}A)(AT) = (I)(AT) = AT$

$A^{-1}(AT) = (A^{-1}A)T = IT = T$

$A^{-1} = \begin{bmatrix} 0 & 1 \\ -1 & 0 \end{bmatrix}$

A^{-1} represents a clockwise rotation by 90°.

3. (a) $A^2 = \begin{bmatrix} 1 & 0 \\ 0 & 0 \end{bmatrix}\begin{bmatrix} 1 & 0 \\ 0 & 0 \end{bmatrix} = \begin{bmatrix} 1 & 0 \\ 0 & 0 \end{bmatrix} = A$

A *is* idempotent.

(b) $A^2 = \begin{bmatrix} 0 & 1 \\ 1 & 0 \end{bmatrix}\begin{bmatrix} 0 & 1 \\ 1 & 0 \end{bmatrix} = \begin{bmatrix} 1 & 0 \\ 0 & 1 \end{bmatrix} \neq A$

A is *not* idempotent.

(c) $A^2 = \begin{bmatrix} 2 & 3 \\ -1 & -2 \end{bmatrix}\begin{bmatrix} 2 & 3 \\ -1 & -2 \end{bmatrix} = \begin{bmatrix} 1 & 0 \\ 0 & 1 \end{bmatrix} \neq A$

A is *not* idempotent.

(d) $A^2 = \begin{bmatrix} 2 & 3 \\ 1 & 2 \end{bmatrix}\begin{bmatrix} 2 & 3 \\ 1 & 2 \end{bmatrix} = \begin{bmatrix} 7 & 12 \\ 4 & 7 \end{bmatrix} \neq A$

A is *not* idempotent.

5. (a) $\begin{bmatrix} 0.70 & 0.15 & 0.15 \\ 0.20 & 0.80 & 0.15 \\ 0.10 & 0.05 & 0.70 \end{bmatrix}\begin{bmatrix} 25,000 \\ 30,000 \\ 45,000 \end{bmatrix} = \begin{bmatrix} 28,750 \\ 35,750 \\ 35,500 \end{bmatrix}$

Gold Cable Company: 28,750 households

Galaxy Cable Company: 35,750 households

Nonsubscribers: 35,500 households

(c) $\begin{bmatrix} 0.70 & 0.15 & 0.15 \\ 0.20 & 0.80 & 0.15 \\ 0.10 & 0.05 & 0.70 \end{bmatrix}\begin{bmatrix} 30,812.5 \\ 39,675 \\ 29,512.5 \end{bmatrix} \approx \begin{bmatrix} 31,947 \\ 42,329 \\ 25,724 \end{bmatrix}$

Gold Cable Company: 31,947 households

Galaxy Cable Company: 42,329 households

Nonsubscribers: 25,724 households

(b) $\begin{bmatrix} 0.70 & 0.15 & 0.15 \\ 0.20 & 0.80 & 0.15 \\ 0.10 & 0.05 & 0.70 \end{bmatrix}\begin{bmatrix} 28,750 \\ 35,750 \\ 35,500 \end{bmatrix} \approx \begin{bmatrix} 30,813 \\ 39,675 \\ 29,513 \end{bmatrix}$

Gold Cable Company: 30,813 households

Galaxy Cable Company: 39,675 households

Nonsubscribers: 29,513 households

(d) Both cable companies are increasing the number of subscribers, while the number of nonsubscribers is decreasing each year.

7. If $A = \begin{bmatrix} 4 & x \\ -2 & -3 \end{bmatrix}$ is singular then

$$ad - bc = -12 + 2x = 0.$$

Thus, $x = 6$.

9. $(a - b)(b - c)(c - a) = -a^2b + a^2c + ab^2 - ac^2 - b^2c + bc^2$

$$\begin{vmatrix} 1 & 1 & 1 \\ a & b & c \\ a^2 & b^2 & c^2 \end{vmatrix} = \begin{vmatrix} b & c \\ b^2 & c^2 \end{vmatrix} - \begin{vmatrix} a & c \\ a^2 & c^2 \end{vmatrix} + \begin{vmatrix} a & b \\ a^2 & b^2 \end{vmatrix} = bc^2 - b^2c - ac^2 + a^2c + ab^2 - a^2b$$

Thus, $\begin{vmatrix} 1 & 1 & 1 \\ a & b & c \\ a^2 & b^2 & c^2 \end{vmatrix} = (a - b)(b - c)(c - a).$

11. $\begin{vmatrix} x & 0 & c \\ -1 & x & b \\ 0 & -1 & a \end{vmatrix} = x \begin{vmatrix} x & b \\ -1 & a \end{vmatrix} + c \begin{vmatrix} -1 & x \\ 0 & -1 \end{vmatrix} = x(ax + b) + c(1 - 0) = ax^2 + bx + c$

13. $4S + 4N \quad\quad = 184$

$\quad\quad S \quad\quad + 6F = 146$

$\quad\quad\quad 2N + 4F = 104$

$$D = \begin{vmatrix} 4 & 4 & 0 \\ 1 & 0 & 6 \\ 0 & 2 & 4 \end{vmatrix} = -64 \quad\quad N = \dfrac{\begin{vmatrix} 4 & 184 & 0 \\ 1 & 146 & 6 \\ 0 & 104 & 4 \end{vmatrix}}{-64} = \dfrac{-896}{-64} = 14$$

$$S = \dfrac{\begin{vmatrix} 184 & 4 & 0 \\ 146 & 0 & 6 \\ 104 & 2 & 4 \end{vmatrix}}{-64} = \dfrac{-2048}{-64} = 32 \quad\quad F = \dfrac{\begin{vmatrix} 4 & 4 & 184 \\ 1 & 0 & 146 \\ 0 & 2 & 104 \end{vmatrix}}{-64} = \dfrac{-1216}{-64} = 19$$

Element	Atomic mass
Sulfur	32
Nitrogen	14
Fluoride	19

15. $A = \begin{bmatrix} -1 & 1 & -2 \\ 2 & 0 & 1 \end{bmatrix}, \quad\quad B = \begin{bmatrix} -3 & 0 \\ 1 & 2 \\ 1 & -1 \end{bmatrix}$

$A^T = \begin{bmatrix} -1 & 2 \\ 1 & 0 \\ -2 & 1 \end{bmatrix}, \quad\quad B^T = \begin{bmatrix} -3 & 1 & 1 \\ 0 & 2 & -1 \end{bmatrix}$

$AB = \begin{bmatrix} 2 & 4 \\ -5 & -1 \end{bmatrix}, \quad\quad (AB)^T = \begin{bmatrix} 2 & -5 \\ 4 & -1 \end{bmatrix}$

$B^TA^T = \begin{bmatrix} -3 & 1 & 1 \\ 0 & 2 & -1 \end{bmatrix} \begin{bmatrix} -1 & 2 \\ 1 & 0 \\ -2 & 1 \end{bmatrix} = \begin{bmatrix} 2 & -5 \\ 4 & -1 \end{bmatrix}$

Thus, $(AB)^T = B^TA^T$.

17. (a) $[45 \quad -35]\begin{bmatrix} w & x \\ y & z \end{bmatrix} = [10 \quad 15]$

$[38 \quad -30]\begin{bmatrix} w & x \\ y & z \end{bmatrix} = [8 \quad 14]$

$45w - 35y = 10$

$45x - 35z = 15$

$38w - 30y = 8$

$38x - 30z = 14$

$\left.\begin{array}{l} 45w - 35y = 10 \\ 38w - 30y = \ 8 \end{array}\right\} \Rightarrow w = 1, y = 1$

$\left.\begin{array}{l} 45x - 35z = 15 \\ 38x - 30z = 14 \end{array}\right\} \Rightarrow x = -2, z = -3$

$A^{-1} = \begin{bmatrix} 1 & -2 \\ 1 & -3 \end{bmatrix}$

(b) $\begin{bmatrix} 45 & -35 \\ 38 & -30 \\ 18 & -18 \\ 35 & -30 \\ 81 & -60 \\ 42 & -28 \\ 75 & -55 \\ 2 & -2 \\ 22 & -21 \\ 15 & -10 \end{bmatrix} \begin{bmatrix} 1 & -2 \\ 1 & -3 \end{bmatrix} = \begin{bmatrix} 10 & 15 \\ 8 & 14 \\ 0 & 18 \\ 5 & 20 \\ 21 & 18 \\ 14 & 0 \\ 20 & 15 \\ 0 & 2 \\ 1 & 19 \\ 5 & 0 \end{bmatrix}$

J	O
H	N
—	R
E	T
U	R
N	—
T	O
—	B
A	S
E	—

JOHN RETURN TO BASE

19. Let $A = \begin{bmatrix} 3 & -3 \\ 5 & -5 \end{bmatrix}$, then $|A| = 0$.

Let $A = \begin{bmatrix} 2 & 4 & -6 \\ -3 & 1 & 2 \\ 5 & -8 & 3 \end{bmatrix}$, then $|A| = 0$.

Let $A = \begin{bmatrix} 3 & -7 & 5 & -1 \\ -6 & 4 & 0 & 2 \\ 5 & 8 & -6 & -7 \\ 9 & 11 & -4 & -16 \end{bmatrix}$, then $|A| = 0$.

Conjecture: If A is an $n \times n$ matrix, each of whose rows add up to zero, then $|A| = 0$.

Chapter 10   Practice Test

1. Put the matrix in reduced row-echelon form.

$$\begin{bmatrix} 1 & -2 & 4 \\ 3 & -5 & 9 \end{bmatrix}$$

For Exercises 2–4, use matrices to solve the system of equations.

2. $\begin{cases} 3x + 5y = 3 \\ 2x - y = -11 \end{cases}$

3. $\begin{cases} 2x + 3y = -3 \\ 3x + 2y = 8 \\ x + y = 1 \end{cases}$

4. $\begin{cases} x + 3z = -5 \\ 2x + y = 0 \\ 3x + y - z = 3 \end{cases}$

5. Multiply $\begin{bmatrix} 1 & 4 & 5 \\ 2 & 0 & -3 \end{bmatrix} \begin{bmatrix} 1 & 6 \\ 0 & -7 \\ -1 & 2 \end{bmatrix}$.

6. Given $A = \begin{bmatrix} 9 & 1 \\ -4 & 8 \end{bmatrix}$ and $B = \begin{bmatrix} 6 & -2 \\ 3 & 5 \end{bmatrix}$, find $3A - 5B$.

7. Find $f(A)$.

$$f(x) = x^2 - 7x + 8, \ A = \begin{bmatrix} 3 & 0 \\ 7 & 1 \end{bmatrix}$$

8. True or false:

$(A + B)(A + 3B) = A^2 + 4AB + 3B^2$ where A and B are matrices.

(Assume that A^2, AB, and B^2 exist.)

For Exercises 9–10, find the inverse of the matrix, if it exists.

9. $\begin{bmatrix} 1 & 2 \\ 3 & 5 \end{bmatrix}$

10. $\begin{bmatrix} 1 & 1 & 1 \\ 3 & 6 & 5 \\ 6 & 10 & 8 \end{bmatrix}$

11. Use an inverse matrix to solve the systems.

(a) $x + 2y = 4$
    $3x + 5y = 1$

(b) $x + 2y = 3$
    $3x + 5y = -2$

For Exercises 12–14, find the determinant of the matrix.

12. $\begin{bmatrix} 6 & -1 \\ 3 & 4 \end{bmatrix}$

13. $\begin{bmatrix} 1 & 3 & -1 \\ 5 & 9 & 0 \\ 6 & 2 & -5 \end{bmatrix}$

14. $\begin{bmatrix} 1 & 4 & 2 & 3 \\ 0 & 1 & -2 & 0 \\ 3 & 5 & -1 & 1 \\ 2 & 0 & 6 & 1 \end{bmatrix}$

15. Evaluate $\begin{vmatrix} 6 & 4 & 3 & 0 & 6 \\ 0 & 5 & 1 & 4 & 8 \\ 0 & 0 & 2 & 7 & 3 \\ 0 & 0 & 0 & 9 & 2 \\ 0 & 0 & 0 & 0 & 1 \end{vmatrix}$.

16. Use a determinant to find the area of the triangle with vertices $(0, 7)$, $(5, 0)$, and $(3, 9)$.

17. Find the equation of the line through $(2, 7)$ and $(-1, 4)$.

For Exercises 18–20, use Cramer's Rule to find the indicated value.

18. Find x.

$$\begin{cases} 6x - 7y = 4 \\ 2x + 5y = 11 \end{cases}$$

19. Find z.

$$\begin{cases} 3x \quad\ + z = 1 \\ \quad\ y + 4z = 3 \\ x - y \quad\ = 2 \end{cases}$$

20. Find y.

$$\begin{cases} 721.4x - 29.1y = 33.77 \\ 45.9x + 105.6y = 19.85 \end{cases}$$

CHAPTER 11
Sequences, Series, and Probability

CHAPTER 11
Sequences, Series, and Probability

Section 11.1   Sequences and Series

■ Given the general nth term in a sequence, you should be able to find, or list, some of the terms.

■ You should be able to find an expression for the apparent nth term of a sequence.

■ You should be able to use and evaluate factorials.

■ You should be able to use summation notation for a sum.

■ You should know that the sum of the terms of a sequence is a series.

Vocabulary Check

1. infinite sequence

2. terms

3. finite

4. recursively

5. factorial

6. summation notation

7. index; upper; lower

8. series

9. nth partial sum

---

1. $a_n = 3n + 1$

$a_1 = 3(1) + 1 = 4$

$a_2 = 3(2) + 1 = 7$

$a_3 = 3(3) + 1 = 10$

$a_4 = 3(4) + 1 = 13$

$a_5 = 3(5) + 1 = 16$

3. $a_n = 2^n$

$a_1 = 2^1 = 2$

$a_2 = 2^2 = 4$

$a_3 = 2^3 = 8$

$a_4 = 2^4 = 16$

$a_5 = 2^5 = 32$

5. $a_n = (-2)^n$

$a_1 = (-2)^1 = -2$

$a_2 = (-2)^2 = 4$

$a_3 = (-2)^3 = -8$

$a_4 = (-2)^4 = 16$

$a_5 = (-2)^5 = -32$

7. $a_n = \dfrac{n + 2}{n}$

$a_1 = \dfrac{1 + 2}{1} = 3$

$a_2 = \dfrac{4}{2} = 2$

$a_3 = \dfrac{5}{3}$

$a_4 = \dfrac{6}{4} = \dfrac{3}{2}$

$a_5 = \dfrac{7}{5}$

9. $a_n = \dfrac{6n}{3n^2 - 1}$

$a_1 = \dfrac{6(1)}{3(1)^2 - 1} = 3$

$a_2 = \dfrac{6(2)}{3(2)^2 - 1} = \dfrac{12}{11}$

$a_3 = \dfrac{6(3)}{3(3)^2 - 1} = \dfrac{9}{13}$

$a_4 = \dfrac{6(4)}{3(4)^2 - 1} = \dfrac{24}{47}$

$a_5 = \dfrac{6(5)}{3(5)^2 - 1} = \dfrac{15}{37}$

11. $a_n = \dfrac{1 + (-1)^n}{n}$

$a_1 = 0$

$a_2 = \dfrac{2}{2} = 1$

$a_3 = 0$

$a_4 = \dfrac{2}{4} = \dfrac{1}{2}$

$a_5 = 0$

13. $a_n = 2 - \dfrac{1}{3^n}$

$a_1 = 2 - \dfrac{1}{3} = \dfrac{5}{3}$

$a_2 = 2 - \dfrac{1}{9} = \dfrac{17}{9}$

$a_3 = 2 - \dfrac{1}{27} = \dfrac{53}{27}$

$a_4 = 2 - \dfrac{1}{81} = \dfrac{161}{81}$

$a_5 = 2 - \dfrac{1}{243} = \dfrac{485}{243}$

15. $a_n = \dfrac{1}{n^{3/2}}$

$a_1 = \dfrac{1}{1} = 1$

$a_2 = \dfrac{1}{2^{3/2}}$

$a_3 = \dfrac{1}{3^{3/2}}$

$a_4 = \dfrac{1}{4^{3/2}} = \dfrac{1}{8}$

$a_5 = \dfrac{1}{5^{3/2}}$

17. $a_n = \dfrac{(-1)^n}{n^2}$

$a_1 = -\dfrac{1}{1} = -1$

$a_2 = \dfrac{1}{4}$

$a_3 = -\dfrac{1}{9}$

$a_4 = \dfrac{1}{16}$

$a_5 = -\dfrac{1}{25}$

19. $a_n = \dfrac{2}{3}$

$a_1 = \dfrac{2}{3}$

$a_2 = \dfrac{2}{3}$

$a_3 = \dfrac{2}{3}$

$a_4 = \dfrac{2}{3}$

$a_5 = \dfrac{2}{3}$

21. $a_n = n(n-1)(n-2)$

$a_1 = (1)(0)(-1) = 0$

$a_2 = (2)(1)(0) = 0$

$a_3 = (3)(2)(1) = 6$

$a_4 = (4)(3)(2) = 24$

$a_5 = (5)(4)(3) = 60$

23. $a_{25} = (-1)^{25}(3(25) - 2) = -73$

25. $a_{11} = \dfrac{4(11)}{2(11)^2 - 3} = \dfrac{44}{239}$

27. $a_n = \dfrac{3}{4}n$

29. $a_n = 16(-0.5)^{n-1}$

31. $a_n = \dfrac{2n}{n+1}$

33. $a_n = \dfrac{8}{n+1}$

$a_1 = 4,\ a_{10} = \dfrac{8}{11}$

The sequence decreases.

Matches graph (c).

35. $a_n = 4(0.5)^{n-1}$

$a_1 = 4,\ a_{10} = \dfrac{1}{128}$

The sequence decreases.

Matches graph (d).

37. $1, 4, 7, 10, 13, \ldots$

$a_n = 1 + (n-1)3 = 3n - 2$

39. $0, 3, 8, 15, 24, \ldots$

$a_n = n^2 - 1$

41. $-\dfrac{2}{3}, \dfrac{3}{4}, -\dfrac{4}{5}, \dfrac{5}{6}, -\dfrac{6}{7}, \ldots$

$a_n = (-1)^n \left(\dfrac{n+1}{n+2} \right)$

43. $\dfrac{2}{1}, \dfrac{3}{3}, \dfrac{4}{5}, \dfrac{5}{7}, \dfrac{6}{9}, \ldots$

$a_n = \dfrac{n+1}{2n-1}$

45. $1, \dfrac{1}{4}, \dfrac{1}{9}, \dfrac{1}{16}, \dfrac{1}{25}, \ldots$

$a_n = \dfrac{1}{n^2}$

47. $1, -1, 1, -1, 1, \ldots$

$a_n = (-1)^{n+1}$

49. $1 + \frac{1}{1}, 1 + \frac{1}{2}, 1 + \frac{1}{3}, 1 + \frac{1}{4}, 1 + \frac{1}{5}, \ldots$

$a_n = 1 + \frac{1}{n}$

51. $a_1 = 28$ and $a_{k+1} = a_k - 4$

$a_1 = 28$

$a_2 = a_1 - 4 = 28 - 4 = 24$

$a_3 = a_2 - 4 = 24 - 4 = 20$

$a_4 = a_3 - 4 = 20 - 4 = 16$

$a_5 = a_4 - 4 = 16 - 4 = 12$

53. $a_1 = 3$ and $a_{k+1} = 2(a_k - 1)$

$a_1 = 3$

$a_2 = 2(a_1 - 1) = 2(3 - 1) = 4$

$a_3 = 2(a_2 - 1) = 2(4 - 1) = 6$

$a_4 = 2(a_3 - 1) = 2(6 - 1) = 10$

$a_5 = 2(a_4 - 1) = 2(10 - 1) = 18$

55. $a_1 = 6$ and $a_{k+1} = a_k + 2$

$a_1 = 6$

$a_2 = a_1 + 2 = 6 + 2 = 8$

$a_3 = a_2 + 2 = 8 + 2 = 10$

$a_4 = a_3 + 2 = 10 + 2 = 12$

$a_5 = a_4 + 2 = 12 + 2 = 14$

In general, $a_n = 2n + 4$.

57. $a_1 = 81$ and $a_{k+1} = \frac{1}{3}a_k$

$a_1 = 81$

$a_2 = \frac{1}{3}a_1 = \frac{1}{3}(81) = 27$

$a_3 = \frac{1}{3}a_2 = \frac{1}{3}(27) = 9$

$a_4 = \frac{1}{3}a_3 = \frac{1}{3}(9) = 3$

$a_5 = \frac{1}{3}a_4 = \frac{1}{3}(3) = 1$

In general,

$a_n = 81\left(\frac{1}{3}\right)^{n-1} = 81(3)\left(\frac{1}{3}\right)^n = \frac{243}{3^n}$.

59. $a_n = \frac{3^n}{n!}$

$a_0 = \frac{3^0}{0!} = 1$

$a_1 = \frac{3^1}{1!} = 3$

$a_2 = \frac{3^2}{2!} = \frac{9}{2}$

$a_3 = \frac{3^3}{3!} = \frac{27}{6} = \frac{9}{2}$

$a_4 = \frac{3^4}{4!} = \frac{81}{24} = \frac{27}{8}$

61. $a_n = \frac{1}{(n+1)!}$

$a_0 = \frac{1}{1!} = 1$

$a_1 = \frac{1}{2!} = \frac{1}{2}$

$a_2 = \frac{1}{3!} = \frac{1}{6}$

$a_3 = \frac{1}{4!} = \frac{1}{24}$

$a_4 = \frac{1}{5!} = \frac{1}{120}$

63. $a_n = \frac{(-1)^{2n}}{(2n)!} = \frac{1}{(2n)!}$

$a_0 = \frac{1}{0!} = 1$

$a_1 = \frac{1}{2!} = \frac{1}{2}$

$a_2 = \frac{1}{4!} = \frac{1}{24}$

$a_3 = \frac{1}{6!} = \frac{1}{720}$

$a_4 = \frac{1}{8!} = \frac{1}{40,320}$

65. $\frac{4!}{6!} = \frac{1 \cdot 2 \cdot 3 \cdot 4}{1 \cdot 2 \cdot 3 \cdot 4 \cdot 5 \cdot 6} = \frac{1}{5 \cdot 6} = \frac{1}{30}$

67. $\frac{10!}{8!} = \frac{1 \cdot 2 \cdot 3 \cdot 4 \cdot 5 \cdot 6 \cdot 7 \cdot 8 \cdot 9 \cdot 10}{1 \cdot 2 \cdot 3 \cdot 4 \cdot 5 \cdot 6 \cdot 7 \cdot 8} = \frac{9 \cdot 10}{1} = 90$

69. $\frac{(n+1)!}{n!} = \frac{1 \cdot 2 \cdot 3 \cdots n \cdot (n+1)}{1 \cdot 2 \cdot 3 \cdots n} = \frac{n+1}{1}$

$= n + 1$

71. $\frac{(2n-1)!}{(2n+1)!} = \frac{1 \cdot 2 \cdot 3 \cdots (2n-1)}{1 \cdot 2 \cdot 3 \cdots (2n-1) \cdot (2n) \cdot (2n+1)}$

$= \frac{1}{2n(2n+1)}$

73. $\displaystyle\sum_{i=1}^{5} (2i + 1) = (2 + 1) + (4 + 1) + (6 + 1) + (8 + 1) + (10 + 1) = 35$

75. $\displaystyle\sum_{k=1}^{4} 10 = 10 + 10 + 10 + 10 = 40$

77. $\displaystyle\sum_{i=0}^{4} i^2 = 0^2 + 1^2 + 2^2 + 3^2 + 4^2 = 30$

79. $\displaystyle\sum_{k=0}^{3} \frac{1}{k^2+1} = \frac{1}{1} + \frac{1}{1+1} + \frac{1}{4+1} + \frac{1}{9+1} = \frac{9}{5}$

81. $\displaystyle\sum_{k=2}^{5} (k+1)^2(k-3) = (3)^2(-1) + (4)^2(0) + (5)^2(1) + (6)^2(2) = 88$

83. $\displaystyle\sum_{i=1}^{4} 2^i = 2^1 + 2^2 + 2^3 + 2^4 = 30$　　**85.** $\displaystyle\sum_{j=1}^{6}(24-3j) = 81$　　　　**87.** $\displaystyle\sum_{k=0}^{4} \frac{(-1)^k}{k+1} = \frac{47}{60}$

89. $\displaystyle\frac{1}{3(1)} + \frac{1}{3(2)} + \frac{1}{3(3)} + \cdots + \frac{1}{3(9)} = \sum_{i=1}^{9} \frac{1}{3i}$

91. $\displaystyle\left[2\left(\frac{1}{8}\right)+3\right] + \left[2\left(\frac{2}{8}\right)+3\right] + \left[2\left(\frac{3}{8}\right)+3\right] + \cdots + \left[2\left(\frac{8}{8}\right)+3\right] = \sum_{i=1}^{8}\left[2\left(\frac{i}{8}\right)+3\right]$

93. $\displaystyle 3 - 9 + 27 - 81 + 243 - 729 = \sum_{i=1}^{6}(-1)^{i+1}3^i$　　**95.** $\displaystyle\frac{1}{1^2} - \frac{1}{2^2} + \frac{1}{3^2} - \frac{1}{4^2} + \cdots - \frac{1}{20^2} = \sum_{i=1}^{20} \frac{(-1)^{i+1}}{i^2}$

97. $\displaystyle\frac{1}{4} + \frac{3}{8} + \frac{7}{16} + \frac{15}{32} + \frac{31}{64} = \sum_{i=1}^{5} \frac{2^i-1}{2^{i+1}}$　　**99.** $\displaystyle\sum_{i=1}^{4} 5\left(\frac{1}{2}\right)^i = 5\left(\frac{1}{2}\right) + 5\left(\frac{1}{2}\right)^2 + 5\left(\frac{1}{2}\right)^3 + 5\left(\frac{1}{2}\right)^4 = \frac{75}{16}$

101. $\displaystyle\sum_{n=1}^{3} 4\left(-\frac{1}{2}\right)^n = 4\left(-\frac{1}{2}\right) + 4\left(-\frac{1}{2}\right)^2 + 4\left(-\frac{1}{2}\right)^3 = -\frac{3}{2}$　　**103.** $\displaystyle\sum_{i=1}^{\infty} 6\left(\frac{1}{10}\right)^i = 0.6 + 0.06 + 0.006 + 0.0006 + \cdots = \frac{2}{3}$

105. By using a calculator, we have

$\displaystyle\sum_{k=1}^{10} 7\left(\frac{1}{10}\right)^k \approx 0.7777777777$

$\displaystyle\sum_{k=1}^{50} 7\left(\frac{1}{10}\right)^k \approx 0.7777777778$

$\displaystyle\sum_{k=1}^{100} 7\left(\frac{1}{10}\right)^k \approx \frac{7}{9}.$

The terms approach zero as $n \rightarrow \infty$.

Thus, we conclude that $\displaystyle\sum_{k=1}^{\infty} 7\left(\frac{1}{10}\right)^k = \frac{7}{9}.$

107. $A_n = 5000\left(1 + \dfrac{0.08}{4}\right)^n, \ n = 1, 2, 3, \ldots$

(a) $A_1 = \$5100.00$

$A_2 = \$5202.00$

$A_3 = \$5306.04$

$A_4 = \$5412.16$

$A_5 = \$5520.40$

$A_6 = \$5630.81$

$A_7 = \$5743.43$

$A_8 = \$5858.30$

(b) $A_{40} = \$11{,}040.20$

109. (a) Linear model: $a_n \approx 60.57n - 182$

(b) Quadratic model: $a_n \approx 1.61n^2 + 26.8n - 9.5$

(c)

Year	n	Actual Data	Linear Model	Quadratic Model
1998	8	311	303	308
1999	9	357	363	362
2000	10	419	424	420
2001	11	481	484	480
2002	12	548	545	544
2003	13	608	605	611

The quadratic model is a better fit.

(d) For the year 2008 we have the following predictions:

Linear model: 908 stores

Quadratic model: 995 stores

Since the quadratic model is a better fit, the predicted number of stores in 2008 is 995.

111. (a) $a_n = 2.7698n^3 - 61.372n^2 + 600.00n + 3102.9$

$a_0 = \$3102.9$ billion	$a_7 \approx \$5245.7$ billion
$a_1 \approx \$3644.3$ billion	$a_8 \approx \$5393.2$ billion
$a_2 \approx \$4079.6$ billion	$a_9 \approx \$5551.0$ billion
$a_3 \approx \$4425.3$ billion	$a_{10} = \$5735.5$ billion
$a_4 \approx \$4698.2$ billion	$a_{11} \approx \$5963.5$ billion
$a_5 \approx \$4914.8$ billion	$a_{12} \approx \$6251.5$ billion
$a_6 \approx \$5091.8$ billion	$a_{13} \approx \$6616.3$ billion

(b) The federal debt is increasing.

113. True, $\displaystyle\sum_{i=1}^{4}(i^2 + 2i) = \sum_{i=1}^{4}i^2 + 2\sum_{i=1}^{4}i$ by the Properties of Sums.

115. $a_1 = 1, a_2 = 1, a_{k+2} = a_{k+1} + a_k, k \geq 1$

$a_1 = 1$	$b_1 = \frac{1}{1} = 1$
$a_2 = 1$	$b_2 = \frac{2}{1} = 2$
$a_3 = 1 + 1 = 2$	$b_3 = \frac{3}{2}$
$a_4 = 2 + 1 = 3$	$b_4 = \frac{5}{3}$
$a_5 = 3 + 2 = 5$	$b_5 = \frac{8}{5}$
$a_6 = 5 + 3 = 8$	$b_6 = \frac{13}{8}$
$a_7 = 8 + 5 = 13$	$b_7 = \frac{21}{13}$
$a_8 = 13 + 8 = 21$	$b_8 = \frac{34}{21}$
$a_9 = 21 + 13 = 34$	$b_9 = \frac{55}{34}$
$a_{10} = 34 + 21 = 55$	$b_{10} = \frac{89}{55}$
$a_{11} = 55 + 34 = 89$	
$a_{12} = 89 + 55 = 144$	

117. $\dfrac{327.15 + 785.69 + 433.04 + 265.38 + 604.12 + 590.30}{6} \approx \$500.95$

119. $\sum_{i=1}^{n} (x_i - \bar{x}) = \sum_{i=1}^{n} x_i - \sum_{i=1}^{n} \bar{x}$

$\qquad = \left(\sum_{i=1}^{n} x_i \right) - n\bar{x}$

$\qquad = \left(\sum_{i=1}^{n} x_i \right) - n\left(\frac{1}{n}\sum_{i=1}^{n} x_i \right)$

$\qquad = 0$

121. $a_n = \dfrac{x^n}{n!}$

$a_1 = \dfrac{x^1}{1!} = x$

$a_2 = \dfrac{x^2}{2!} = \dfrac{x^2}{2}$

$a_3 = \dfrac{x^3}{3!} = \dfrac{x^3}{6}$

$a_4 = \dfrac{x^4}{4!} = \dfrac{x^4}{24}$

$a_5 = \dfrac{x^5}{5!} = \dfrac{x^5}{120}$

123. $a_n = \dfrac{(-1)^n x^{2n}}{(2n)!}$

$a_1 = \dfrac{-x^2}{2!} = -\dfrac{x^2}{2}$

$a_2 = \dfrac{x^4}{4!} = \dfrac{x^4}{24}$

$a_3 = \dfrac{-x^6}{6!} = -\dfrac{x^6}{720}$

$a_4 = \dfrac{x^8}{8!} = \dfrac{x^8}{40,320}$

$a_5 = \dfrac{-x^{10}}{10!} = -\dfrac{x^{10}}{3,628,800}$

125. $f(x) = 4x - 3$ is one-to-one, so it has an inverse.

$y = 4x - 3$

$x = 4y - 3$

$\dfrac{x + 3}{4} = y$

$f^{-1}(x) = \dfrac{x + 3}{4}$

127. $h(x) = \sqrt{5x + 1}$ is one-to-one, so it has in inverse.

Domain: $x \geq -\dfrac{1}{5}$

Range: $y \geq 0$

$y = \sqrt{5x + 1}, x \geq -\dfrac{1}{5}, y \geq 0$

$x = \sqrt{5y + 1}, x \geq 0, y \geq -\dfrac{1}{5}$

$x^2 = 5y + 1, x \geq 0$

$\dfrac{x^2 - 1}{5} = y, x \geq 0$

$h^{-1}(x) = \dfrac{x^2 - 1}{5} = \dfrac{1}{5}(x^2 - 1), x \geq 0$

129. (a) $A - B = \begin{bmatrix} 6 & 5 \\ 3 & 4 \end{bmatrix} - \begin{bmatrix} -2 & 4 \\ 6 & -3 \end{bmatrix} = \begin{bmatrix} 6 - (-2) & 5 - 4 \\ 3 - 6 & 4 - (-3) \end{bmatrix} = \begin{bmatrix} 8 & 1 \\ -3 & 7 \end{bmatrix}$

(b) $4B - 3A = 4 \begin{bmatrix} -2 & 4 \\ 6 & -3 \end{bmatrix} - 3 \begin{bmatrix} 6 & 5 \\ 3 & 4 \end{bmatrix} = \begin{bmatrix} -8 - 18 & 16 - 15 \\ 24 - 9 & -12 - 12 \end{bmatrix} = \begin{bmatrix} -26 & 1 \\ 15 & -24 \end{bmatrix}$

(c) $AB = \begin{bmatrix} 6 & 5 \\ 3 & 4 \end{bmatrix} \begin{bmatrix} -2 & 4 \\ 6 & -3 \end{bmatrix} = \begin{bmatrix} -12 + 30 & 24 - 15 \\ -6 + 24 & 12 - 12 \end{bmatrix} = \begin{bmatrix} 18 & 9 \\ 18 & 0 \end{bmatrix}$

(d) $BA = \begin{bmatrix} -2 & 4 \\ 6 & -3 \end{bmatrix} \begin{bmatrix} 6 & 5 \\ 3 & 4 \end{bmatrix} = \begin{bmatrix} -12 + 16 & -10 + 12 \\ 36 - 9 & 30 - 12 \end{bmatrix} = \begin{bmatrix} 0 & 6 \\ 27 & 18 \end{bmatrix}$

131. (a) $A - B = \begin{bmatrix} -2 & -3 & 6 \\ 4 & 5 & 7 \\ 1 & 7 & 4 \end{bmatrix} - \begin{bmatrix} 1 & 4 & 2 \\ 0 & 1 & 6 \\ 0 & 3 & 1 \end{bmatrix} = \begin{bmatrix} -2 - 1 & -3 - 4 & 6 - 2 \\ 4 - 0 & 5 - 1 & 7 - 6 \\ 1 - 0 & 7 - 3 & 4 - 1 \end{bmatrix} = \begin{bmatrix} -3 & -7 & 4 \\ 4 & 4 & 1 \\ 1 & 4 & 3 \end{bmatrix}$

(b) $4B - 3A = 4 \begin{bmatrix} 1 & 4 & 2 \\ 0 & 1 & 6 \\ 0 & 3 & 1 \end{bmatrix} - 3 \begin{bmatrix} -2 & -3 & 6 \\ 4 & 5 & 7 \\ 1 & 7 & 4 \end{bmatrix} = \begin{bmatrix} 4 - (-6) & 16 - (-9) & 8 - 18 \\ 0 - 12 & 4 - 15 & 24 - 21 \\ 0 - 3 & 12 - 21 & 4 - 12 \end{bmatrix} = \begin{bmatrix} 10 & 25 & -10 \\ -12 & -11 & 3 \\ -3 & -9 & -8 \end{bmatrix}$

(c) $AB = \begin{bmatrix} -2 & -3 & 6 \\ 4 & 5 & 7 \\ 1 & 7 & 4 \end{bmatrix} \begin{bmatrix} 1 & 4 & 2 \\ 0 & 1 & 6 \\ 0 & 3 & 1 \end{bmatrix} = \begin{bmatrix} -2 + 0 + 0 & -8 - 3 + 18 & -4 - 18 + 6 \\ 4 + 0 + 0 & 16 + 5 + 21 & 8 + 30 + 7 \\ 1 + 0 + 0 & 4 + 7 + 12 & 2 + 42 + 4 \end{bmatrix} = \begin{bmatrix} -2 & 7 & -16 \\ 4 & 42 & 45 \\ 1 & 23 & 48 \end{bmatrix}$

(d) $BA = \begin{bmatrix} 1 & 4 & 2 \\ 0 & 1 & 6 \\ 0 & 3 & 1 \end{bmatrix} \begin{bmatrix} -2 & -3 & 6 \\ 4 & 5 & 7 \\ 1 & 7 & 4 \end{bmatrix} = \begin{bmatrix} -2 + 16 + 2 & -3 + 20 + 14 & 6 + 28 + 8 \\ 0 + 4 + 6 & 0 + 5 + 42 & 0 + 7 + 24 \\ 0 + 12 + 1 & 0 + 15 + 7 & 0 + 21 + 4 \end{bmatrix} = \begin{bmatrix} 16 & 31 & 42 \\ 10 & 47 & 31 \\ 13 & 22 & 25 \end{bmatrix}$

133. $|A| = \begin{vmatrix} 3 & 5 \\ -1 & 7 \end{vmatrix} = 3(7) - 5(-1) = 26$

135. $|A| = \begin{vmatrix} 3 & 4 & 5 \\ 0 & 7 & 3 \\ 4 & 9 & -1 \end{vmatrix} = 3 \begin{vmatrix} 7 & 3 \\ 9 & -1 \end{vmatrix} + 4 \begin{vmatrix} 4 & 5 \\ 7 & 3 \end{vmatrix}$

$= 3[7(-1) - 3(9)] + 4[4(3) - 5(7)] = -194$

Section 11.2   Arithmetic Sequences and Partial Sums

■   You should be able to recognize an arithmetic sequence, find its common difference, and find its nth term.

■   You should be able to find the nth partial sum of an arithmetic sequence by using the formula

$$S_n = \frac{n}{2}(a_1 + a_n).$$

Vocabulary Check

1. arithmetic; common

2. $a_n = dn + c$

3. sum of a finite arithmetic sequence

1. $10, 8, 6, 4, 2, \ldots$

Arithmetic sequence, $d = -2$

3. $1, 2, 4, 8, 16, \ldots$

Not an arithmetic sequence

5. $\frac{9}{4}, 2, \frac{7}{4}, \frac{3}{2}, \frac{5}{4}, \ldots$

Arithmetic sequence, $d = -\frac{1}{4}$

7. $\frac{1}{3}, \frac{2}{3}, 1, \frac{4}{3}, \frac{5}{6}, \ldots$

Not an arithmetic sequence

9. $\ln 1, \ln 2, \ln 3, \ln 4, \ln 5, \ldots$

Not an arithmetic sequence

11. $a_n = 5 + 3n$

8, 11, 14, 17, 20

Arithmetic sequence, $d = 3$

13. $a_n = 3 - 4(n - 2)$

$7, 3, -1, -5, -9$

Arithmetic sequence, $d = -4$

15. $a_n = (-1)^n$

$-1, 1, -1, 1, -1$

Not an arithmetic sequence

17. $a_n = \dfrac{(-1)^n 3}{n}$

$-3, \dfrac{3}{2}, -1, \dfrac{3}{4}, -\dfrac{3}{5}$

Not an arithmetic sequence

19. $a_1 = 1, \ d = 3$

$a_n = a_1 + (n - 1)d = 1 + (n - 1)(3) = 3n - 2$

21. $a_1 = 100, \ d = -8$

$a_n = a_1 + (n - 1)d = 100 + (n - 1)(-8)$

$= -8n + 108$

23. $a_1 = x, \ d = 2x$

$a_n = a_1 + (n - 1)d = x + (n - 1)(2x) = 2xn - x$

25. $4, \frac{3}{2}, -1, -\frac{7}{2}, \ldots$

$d = -\frac{5}{2}$

$a_n = a_1 + (n - 1)d = 4 + (n - 1)\left(-\frac{5}{2}\right) = -\frac{5}{2}n + \frac{13}{2}$

27. $a_1 = 5, \ a_4 = 15$

$a_4 = a_1 + 3d \implies 15 = 5 + 3d \implies d = \frac{10}{3}$

$a_n = a_1 + (n - 1)d = 5 + (n - 1)\left(\frac{10}{3}\right) = \frac{10}{3}n + \frac{5}{3}$

29. $a_3 = 94, \ a_6 = 85$

$a_6 = a_3 + 3d \implies 85 = 94 + 3d \implies d = -3$

$a_1 = a_3 - 2d \implies a_1 = 94 - 2(-3) = 100$

$a_n = a_1 + (n - 1)d = 100 + (n - 1)(-3)$

$= -3n + 103$

31. $a_1 = 5, \ d = 6$

$a_1 = 5$

$a_2 = 5 + 6 = 11$

$a_3 = 11 + 6 = 17$

$a_4 = 17 + 6 = 23$

$a_5 = 23 + 6 = 29$

33. $a_1 = -2.6, \ d = -0.4$

$a_1 = -2.6$

$a_2 = -2.6 + (-0.4) = -3.0$

$a_3 = -3.0 + (-0.4) = -3.4$

$a_4 = -3.4 + (-0.4) = -3.8$

$a_5 = -3.8 + (-0.4) = -4.2$

35. $a_1 = 2, \ a_{12} = 46$

$46 = 2 + (12 - 1)d$

$44 = 11d$

$4 = d$

$a_1 = 2$

$a_2 = 2 + 4 = 6$

$a_3 = 6 + 4 = 10$

$a_4 = 10 + 4 = 14$

$a_5 = 14 + 4 = 18$

37. $a_8 = 26$, $a_{12} = 42$

$a_{12} = a_8 + 4d$

$42 = 26 + 4d \Rightarrow d = 4$

$a_8 = a_1 + 7d$

$26 = a_1 + 28 \Rightarrow a_1 = -2$

$a_1 = -2$

$a_2 = -2 + 4 = 2$

$a_3 = 2 + 4 = 6$

$a_4 = 6 + 4 = 10$

$a_5 = 10 + 4 = 14$

39. $a_1 = 15$, $a_{k+1} = a_k + 4$

$a_2 = 15 + 4 = 19$

$a_3 = 19 + 4 = 23$

$a_4 = 23 + 4 = 27$

$a_5 = 27 + 4 = 31$

$d = 4$

$c = a_1 - d = 15 - 4 = 11$

$a_n = 4n + 11$

41. $a_1 = 200$, $a_{k+1} = a_k - 10$

$a_2 = 200 - 10 = 190$

$a_3 = 190 - 10 = 180$

$a_4 = 180 - 10 = 170$

$a_5 = 170 - 10 = 160$

$d = -10$

$c = a_1 - d = 200 - (-10) = 210$

$a_n = -10n + 210$

43. $a_1 = \frac{5}{8}$, $a_{k+1} = a_k - \frac{1}{8}$

$a_1 = \frac{5}{8}$

$a_2 = \frac{5}{8} - \frac{1}{8} = \frac{1}{2}$

$a_3 = \frac{1}{2} - \frac{1}{8} = \frac{3}{8}$

$a_4 = \frac{3}{8} - \frac{1}{8} = \frac{1}{4}$

$a_5 = \frac{1}{4} - \frac{1}{8} = \frac{1}{8}$

$d = -\frac{1}{8}$

$c = a_1 - d = \frac{5}{8} - \left(-\frac{1}{8}\right) = \frac{3}{4}$

$a_n = -\frac{1}{8}n + \frac{3}{4}$

45. $a_1 = 5$, $a_2 = 11 \Rightarrow d = 11 - 5 = 6$

$a_n = a_1 + (n - 1)d \Rightarrow a_{10} = 5 + 9(6) = 59$

47. $a_1 = 4.2$, $a_2 = 6.6 \Rightarrow d = 6.6 - 4.2 = 2.4$

$a_n = a_1 + (n - 1)d \Rightarrow a_7 = 4.2 + 6(2.4) = 18.6$

49. $a_n = -\frac{3}{4}n + 8$

$d = -\frac{3}{4}$ so the sequence is decreasing and $a_1 = 7\frac{1}{4}$.

Matches (b).

51. $a_n = 2 + \frac{3}{4}n$

$d = \frac{3}{4}$ so the sequence is increasing and $a_1 = 2\frac{3}{4}$.

Matches (c).

53. $a_n = 15 - \frac{3}{2}n$

55. $a_n = 0.2n + 3$

57. 8, 20, 32, 44, . . .

$a_1 = 8$, $d = 12$, $n = 10$

$a_{10} = 8 + 9(12) = 116$

$S_{10} = \frac{10}{2}(8 + 116) = 620$

59. 4.2, 3.7, 3.2, 2.7, . . .

$a_1 = 4.2$, $d = -0.5$, $n = 12$

$a_{12} = 4.2 + 11(-0.5) = -1.3$

$S_{12} = \frac{12}{2}[4.2 + (-1.3)] = 17.4$

61. 40, 37, 34, 31, . . .

$a_1 = 40$, $d = -3$, $n = 10$

$a_{10} = 40 + 9(-3) = 13$

$S_{10} = \frac{10}{2}(40 + 13) = 265$

63. $a_1 = 100$, $a_{25} = 220$, $n = 25$

$S_n = \frac{n}{2}[a_1 + a_n]$

$S_{25} = \frac{25}{2}(100 + 220) = 4000$

65. $a_n = 2n - 1$

$a_1 = 1$, $a_{100} = 199$

$\displaystyle\sum_{n=1}^{100}(2n - 1) = \frac{100}{2}(1 + 199)$

$= 10{,}000$

67. $a_1 = 1$, $a_{50} = 50$, $n = 50$

$\displaystyle\sum_{n=1}^{50} n = \frac{50}{2}(1 + 50) = 1275$

69. $a_{10} = 60, a_{100} = 600, n = 91$

$\sum_{n=10}^{100} 6n = \frac{91}{2}(60 + 600) = 30{,}030$

71. $\sum_{n=11}^{30} n - \sum_{n=1}^{10} n = \frac{20}{2}(11 + 30) - \frac{10}{2}(1 + 10) = 355$

73. $a_1 = 1, a_{400} = 799, n = 400$

$\sum_{n=1}^{400} (2n - 1) = \frac{400}{2}(1 + 799) = 160{,}000$

75. $\sum_{n=1}^{20} (2n + 5) = 520$

77. $\sum_{n=1}^{100} \frac{n + 4}{2} = 2725$

79. $\sum_{i=1}^{60} \left(250 - \frac{8}{3}i \right) = 10{,}120$

81. (a) $a_1 = 32{,}500, \ d = 1500$

   $a_6 = a_1 + 5d = 32{,}500 + 5(1500) = \$40{,}000$

   (b) $S_6 = \frac{6}{2}[32{,}500 + 40{,}000] = \$217{,}500$

83. $a_1 = 20, \ d = 4, \ n = 30$

   $a_{30} = 20 + 29(4) = 136$

   $S_{30} = \frac{30}{2}(20 + 136) = 2340$ seats

85. $a_1 = 14, a_{18} = 31$

   $S_{18} = \frac{18}{2}(14 + 31) = 405$ bricks

87. $4.9, 14.7, 24.5, 34.3, \ldots$

   $d = 9.8$

   $a_{10} = 4.9 + 9(9.8) = 93.1$ meters

   $S_{10} = \frac{10}{2}(4.9 + 93.1) = 490$ meters

89. (a) $a_1 = 200, a_2 = 175 \implies d = -25$

   $c = 200 - (-25) = 225$

   $a_n = -25n + 225$

   (b) $a_8 = -25(8) + 225 = 25$

   $S_8 = \frac{8}{2}(200 + 25) = \$900$

91. $a_n = 1500n + 6500$

   $a_1 = 8000, a_6 = 15{,}500$

   $S_6 = \frac{6}{2}(8000 + 15{,}500) = \$70{,}500$

   The cost of gasoline, labor, equipment, insurance, and maintenance are a few economic factors that could prevent the company from meeting its goals, but the biggest unknown variable is the amount of annual snowfall.

93. (a)

Monthly Payment	Unpaid Balance
$a_1 = 200 + 0.01(2000) = \$220$	\$1800
$a_2 = 200 + 0.01(1800) = \$218$	\$1600
$a_3 = 200 + 0.01(1600) = \$216$	\$1400
$a_4 = 200 + 0.01(1400) = \$214$	\$1200
$a_5 = 200 + 0.01(1200) = \$212$	\$1000
$a_6 = 200 + 0.01(1000) = \$210$	\$800

(b) $a_n = -2n + 222 \implies a_{10} = 202$

   $S_{10} = \frac{10}{2}(220 + 202) = \$2110$

   Interest paid: \$110

95. (a) Using (5, 23,078) and (6, 24,176) we have $d = 1098$ and $c = 23,078 - 5(1098) = 17,588$.

$$a_n \approx 1098n + 17,588$$

(b) $a_n \approx 1114.95n + 17,795.07$

The models are similar.

(c)

(d) For 2004 use $n = 14$: $32,960

    For 2005 use $n = 15$: $34,058

(e) Answers will vary.

97. True; given a_1 and a_2 then $d = a_2 - a_1$ and $a_n = a_1 + (n - 1)d$.

99. A sequence is arithmetic if the differences between consecutive terms are the same.

$$a_{n+1} - a_n = d \text{ for } n \geq 1$$

101. (a) $a_n = 2 + 3n$

(b) $y = 3x + 2$

(c) The graph of $a_n = 2 + 3n$ contains only points at the positive integers. The graph of $y = 3x + 2$ is a solid line which contains these points.

(d) The slope $m = 3$ is equal to the common difference $d = 3$. In general, these should be equal.

103.
$$S_{20} = \frac{20}{2}\{a_1 + [a_1 + (20 - 1)(3)]\} = 650$$
$$10(2a_1 + 57) = 650$$
$$2a_1 + 57 = 65$$
$$2a_1 = 8$$
$$a_1 = 4$$

105. $2x - 4y = 3$

$$y = \frac{1}{2}x - \frac{3}{4}$$

Slope: $m = \frac{1}{2}$

y-intercept: $\left(0, -\frac{3}{4}\right)$

107. $x - 7 = 0$

$x = 7$

Vertical line

No slope

No y-intercept

109. $\begin{cases} 2x - y + 7z = -10 \\ 3x + 2y - 4z = 17 \\ 6x - 5y + z = -20 \end{cases}$ Equation 1
Equation 2
Equation 3

$\begin{cases} x - \frac{1}{2}y + \frac{7}{2}z = -5 \\ 3x + 2y - 4z = 17 \\ 6x - 5y + z = -20 \end{cases}$ $\frac{1}{2}$ Eq.1

$\begin{cases} x - \frac{1}{2}y + \frac{7}{2}z = -5 \\ \frac{7}{2}y - \frac{29}{2}z = 32 \\ -2y - 20z = 10 \end{cases}$ (-3)Eq.1 + Eq.2
(-6)Eq.1 + Eq.3

$\begin{cases} x - \frac{1}{2}y + \frac{7}{2}z = -5 \\ -2y - 20z = 10 \\ \frac{7}{2}y - \frac{29}{2}z = 32 \end{cases}$

$\begin{cases} x - \frac{1}{2}y + \frac{7}{2}z = -5 \\ y + 10z = -5 \\ 7y - 29z = 64 \end{cases}$ $(-\frac{1}{2})$Eq.2
2 Eq.3

$\begin{cases} x + \frac{17}{2}z = -\frac{15}{2} \\ y + 10z = -5 \\ -99z = 99 \end{cases}$ $(\frac{1}{2})$Eq.2 + Eq.1
(-7)Eq.2 + Eq.3

$\begin{cases} x + \frac{17}{2}z = -\frac{15}{2} \\ y + 10z = -5 \\ z = -1 \end{cases}$ $(-\frac{1}{99})$Eq.3

$\begin{cases} x = 1 \\ y = 5 \\ z = -1 \end{cases}$ $(-\frac{17}{2})$ Eq.3 + Eq.1
(-10)Eq.3 + Eq.2

Answer: $x = 1, y = 5, z = -1$

111. Answers will vary.

Section 11.3   Geometric Sequences and Series

- You should be able to identify a geometric sequence, find its common ratio, and find the nth term.
- You should know that the nth term of a geometric sequence with common ratio r is given by $a_n = a_1 r^{n-1}$.
- You should know that the nth partial sum of a geometric sequence with common ratio $r \neq 1$ is given by
$$S_n = a_1\left(\frac{1 - r^n}{1 - r}\right).$$
- You should know that if $|r| < 1$, then
$$\sum_{n=1}^{\infty} a_1 r^{n-1} = \sum_{n=0}^{\infty} a_1 r^n = \frac{a_1}{1 - r}.$$

Vocabulary Check

1. geometric; common

2. $a_n = a_1 r^{n-1}$

3. $S_n = a_1\left(\frac{1 - r^n}{1 - r}\right)$

4. geometric series

5. $S = \frac{a_1}{1 - r}$

1. $5, 15, 45, 135, \ldots$

Geometric sequence, $r = 3$

3. $3, 12, 21, 30, \ldots$

Not a geometric sequence

Note: It is an arithmetic sequence with $d = 9$.

5. $1, -\frac{1}{2}, \frac{1}{4}, -\frac{1}{8}, \ldots$

Geometric sequence, $r = -\frac{1}{2}$

7. $\frac{1}{8}, \frac{1}{4}, \frac{1}{2}, 1, \ldots$

Geometric sequence, $r = 2$

9. $1, \frac{1}{2}, \frac{1}{3}, \frac{1}{4}, \ldots$

Not a geometric sequence

11. $a_1 = 2, r = 3$

$a_1 = 2$

$a_2 = 2(3) = 6$

$a_3 = 6(3) = 18$

$a_4 = 18(3) = 54$

$a_5 = 54(3) = 162$

13. $a_1 = 1$, $r = \frac{1}{2}$

$a_1 = 1$

$a_2 = 1\left(\frac{1}{2}\right) = \frac{1}{2}$

$a_3 = \frac{1}{2}\left(\frac{1}{2}\right) = \frac{1}{4}$

$a_4 = \frac{1}{4}\left(\frac{1}{2}\right) = \frac{1}{8}$

$a_5 = \frac{1}{8}\left(\frac{1}{2}\right) = \frac{1}{16}$

15. $a_1 = 5$, $r = -\frac{1}{10}$

$a_1 = 5$

$a_2 = 5\left(-\frac{1}{10}\right) = -\frac{1}{2}$

$a_3 = \left(-\frac{1}{2}\right)\left(-\frac{1}{10}\right) = \frac{1}{20}$

$a_4 = \frac{1}{20}\left(-\frac{1}{10}\right) = -\frac{1}{200}$

$a_5 = \left(-\frac{1}{200}\right)\left(-\frac{1}{10}\right) = \frac{1}{2000}$

17. $a_1 = 1$, $r = e$

$a_1 = 1$

$a_2 = 1(e) = e$

$a_3 = (e)(e) = e^2$

$a_4 = (e^2)(e) = e^3$

$a_5 = (e^3)(e) = e^4$

19. $a_1 = 2$, $r = \frac{x}{4}$

$a_1 = 2$

$a_2 = 2\left(\frac{x}{4}\right) = \frac{x}{2}$

$a_3 = \left(\frac{x}{2}\right)\left(\frac{x}{4}\right) = \frac{x^2}{8}$

$a_4 = \left(\frac{x^2}{8}\right)\left(\frac{x}{4}\right) = \frac{x^3}{32}$

$a_5 = \left(\frac{x^3}{32}\right)\left(\frac{x}{4}\right) = \frac{x^4}{128}$

21. $a_1 = 64$, $a_{k+1} = \frac{1}{2}a_k$

$a_1 = 64$

$a_2 = \frac{1}{2}(64) = 32$

$a_3 = \frac{1}{2}(32) = 16$

$a_4 = \frac{1}{2}(16) = 8$

$a_5 = \frac{1}{2}(8) = 4$

$r = \frac{1}{2}$

$a_n = 64\left(\frac{1}{2}\right)^{n-1} = 128\left(\frac{1}{2}\right)^n$

23. $a_1 = 7$, $a_{k+1} = 2a_k$

$a_1 = 7$

$a_2 = 2(7) = 14$

$a_3 = 2(14) = 28$

$a_4 = 2(28) = 56$

$a_5 = 2(56) = 112$

$r = 2$

$a_n = 7(2)^{n-1} = \frac{7}{2}(2)^n$

25. $a_1 = 6$, $a_{k+1} = -\frac{3}{2}a_k$

$a_1 = 6$

$a_2 = -\frac{3}{2}(6) = -9$

$a_3 = -\frac{3}{2}(-9) = \frac{27}{2}$

$a_4 = -\frac{3}{2}\left(\frac{27}{2}\right) = -\frac{81}{4}$

$a_5 = -\frac{3}{2}\left(-\frac{81}{4}\right) = \frac{243}{8}$

$r = -\frac{3}{2}$

$a_n = 6\left(-\frac{3}{2}\right)^{n-1}$ or $a_n = -4\left(-\frac{3}{2}\right)^n$

27. $a_1 = 4$, $r = \frac{1}{2}$, $n = 10$

$a_n = a_1 r^{n-1} = 4\left(\frac{1}{2}\right)^{n-1}$

$a_{10} = 4\left(\frac{1}{2}\right)^9 = \left(\frac{1}{2}\right)^7 = \frac{1}{128}$

29. $a_1 = 6$, $r = -\frac{1}{3}$, $n = 12$

$a_n = a_1 r^{n-1} = 6\left(-\frac{1}{3}\right)^{n-1}$

$a_{12} = 6\left(-\frac{1}{3}\right)^{11} = -\frac{2}{3^{10}}$

31. $a_1 = 100$, $r = e^x$, $n = 9$

$a_n = a_1 r^{n-1} = 100(e^x)^{n-1}$

$a_9 = 100(e^x)^8 = 100e^{8x}$

33. $a_1 = 500$, $r = 1.02$, $n = 40$

$a_n = a_1 r^{n-1} = 500(1.02)^{n-1}$

$a_{40} = 500(1.02)^{39} \approx 1082.372$

35. $7, 21, 63, \ldots \implies r = 3$

$a_n = 7(3)^{n-1}$

$a_9 = 7(3)^8 = 45{,}927$

37. $5, 30, 180, \ldots \implies r = 6$

$a_n = 5(6)^{n-1}$

$a_{10} = 5(6)^9 = 50,388,480$

39. $a_1 = 16, a_4 = \dfrac{27}{4}$

$a_4 = a_1 r^3$

$\dfrac{27}{4} = 16 r^3$

$\dfrac{27}{64} = r^3$

$\dfrac{3}{4} = r$

$a_n = 16\left(\dfrac{3}{4}\right)^{n-1}$

$a_3 = 16\left(\dfrac{3}{4}\right)^2 = 9$

41. $a_4 = -18, a_7 = \dfrac{2}{3}$

$a_7 = a_4 r^3$

$\dfrac{2}{3} = -18 r^3$

$-\dfrac{1}{27} = r^3$

$-\dfrac{1}{3} = r$

$a_6 = \dfrac{a_7}{r} = \dfrac{2/3}{-1/3} = -2$

43. $a_n = 18\left(\tfrac{2}{3}\right)^{n-1}$

$a_1 = 18$ and $r = \dfrac{2}{3}$

Since $0 < r < 1$, the sequence is decreasing.

Matches (a).

45. $a_n = 18\left(\tfrac{3}{2}\right)^{n-1}$

$a_1 = 18$ and $r = \dfrac{3}{2} > 1$, so the sequence is increasing.

Matches (b).

47. $a_n = 12(-0.75)^{n-1}$

49. $a_n = 12(-0.4)^{n-1}$

51. $a_n = 2(1.3)^{n-1}$

53. $\displaystyle\sum_{n=1}^{9} 2^{n-1} = 1 + 2^1 + 2^2 + \cdots + 2^8 \implies a_1 = 1, r = 2$

$S_9 = \dfrac{1(1 - 2^9)}{1 - 2} = 511$

55. $\displaystyle\sum_{n=1}^{9} (-2)^{n-1} \implies a_1 = 1, r = -2, n = 9$

$S_9 = 1\left(\dfrac{1 - (-2)^9}{1 - (-2)}\right) = 171$

57. $\displaystyle\sum_{i=1}^{7} 64\left(-\dfrac{1}{2}\right)^{i-1} = 64 + 64\left(-\dfrac{1}{2}\right)^1 + 64\left(-\dfrac{1}{2}\right)^2 + \cdots + 64\left(-\dfrac{1}{2}\right)^6 \implies a_1 = 64, r = -\dfrac{1}{2}$

$S_7 = 64\left[\dfrac{1 - \left(-\frac{1}{2}\right)^7}{1 - \left(-\frac{1}{2}\right)}\right] = \dfrac{128}{3}\left[1 - \left(-\dfrac{1}{2}\right)^7\right] = 43$

59. $\displaystyle\sum_{i=1}^{6} 32\left(\dfrac{1}{4}\right)^{i-1} = 32 + 32\left(\dfrac{1}{4}\right)^1 + 32\left(\dfrac{1}{4}\right)^2 + 32\left(\dfrac{1}{4}\right)^3 + 32\left(\dfrac{1}{4}\right)^4 + 32\left(\dfrac{1}{4}\right)^5 \implies a_1 = 32, r = \dfrac{1}{4}, n = 6$

$S_6 = 32\left(\dfrac{1 - \left(\frac{1}{4}\right)^6}{1 - \frac{1}{4}}\right) = \dfrac{1365}{32}$

61. $\displaystyle\sum_{n=0}^{20} 3\left(\dfrac{3}{2}\right)^n = \sum_{n=1}^{21} 3\left(\dfrac{3}{2}\right)^{n-1} = 3 + 3\left(\dfrac{3}{2}\right)^1 + 3\left(\dfrac{3}{2}\right)^2 + \cdots + 3\left(\dfrac{3}{2}\right)^{20} \implies a_1 = 3, r = \dfrac{3}{2}$

$S_{21} = 3\left[\dfrac{1 - \left(\frac{3}{2}\right)^{21}}{1 - \frac{3}{2}}\right] = -6\left[1 - \left(\dfrac{3}{2}\right)^{21}\right] \approx 29,921.311$

63. $\displaystyle\sum_{n=0}^{15} 2\left(\frac{4}{3}\right)^n = \sum_{n=1}^{16} 2\left(\frac{4}{3}\right)^{n-1} = 2 + 2\left(\frac{4}{3}\right)^1 + 2\left(\frac{4}{3}\right)^2 + \cdots + 2\left(\frac{4}{3}\right)^{15} \Rightarrow a_1 = 2, r = \frac{4}{3}, n = 16$

$S_{16} = 2\left(\dfrac{1 - \left(\frac{4}{3}\right)^{16}}{1 - \frac{4}{3}}\right) \approx 592.647$

65. $\displaystyle\sum_{n=0}^{5} 300(1.06)^n = \sum_{n=1}^{6} 300(1.06)^{n-1}$

$= 300 + 300(1.06)^1 + 300(1.06)^2 + 300(1.06)^3 + 300(1.06)^4 + 300(1.06)^5 \Rightarrow a_1 = 300, r = 1.06$

$S_6 = 300\left[\dfrac{1 - (1.06)^6}{1 - 1.06}\right] \approx 2092.596$

67. $\displaystyle\sum_{n=0}^{40} 2\left(-\frac{1}{4}\right)^n = 2 + 2\left(-\frac{1}{4}\right) + 2\left(-\frac{1}{4}\right)^2 + \cdots + 2\left(-\frac{1}{4}\right)^{40} \Rightarrow a_1 = 2, r = -\frac{1}{4}, n = 41$

$S_{41} = 2\left[\dfrac{1 - \left(-\frac{1}{4}\right)^{41}}{1 - \left(-\frac{1}{4}\right)}\right] = \dfrac{8}{5}\left[1 - \left(-\frac{1}{4}\right)^{41}\right] \approx 1.6 = \dfrac{8}{5}$

69. $\displaystyle\sum_{i=1}^{10} 8\left(-\frac{1}{4}\right)^{i-1} = 8 + 8\left(-\frac{1}{4}\right)^1 + 8\left(-\frac{1}{4}\right)^2 + \cdots + 8\left(-\frac{1}{4}\right)^9 \Rightarrow a_1 = 8, r = -\frac{1}{4}$

$S_{10} = 8\left[\dfrac{1 - \left(-\frac{1}{4}\right)^{10}}{1 - \left(-\frac{1}{4}\right)}\right] = \dfrac{32}{5}\left[1 - \left(-\frac{1}{4}\right)^{10}\right] \approx 6.400$

71. $\displaystyle\sum_{i=1}^{10} 5\left(-\frac{1}{3}\right)^{i-1} = 5 + 5\left(-\frac{1}{3}\right)^1 + 5\left(-\frac{1}{3}\right)^2 + \cdots + 5\left(-\frac{1}{3}\right)^9 \Rightarrow a_1 = 5, r = -\frac{1}{3}, n = 10$

$S_{10} = 5\left(\dfrac{1 - \left(-\frac{1}{3}\right)^{10}}{1 - \left(-\frac{1}{3}\right)}\right) \approx 3.750$

73. $5 + 15 + 45 + \cdots + 3645$

$r = 3$ and $3645 = 5(3)^{n-1}$

$729 = 3^{n-1} \Rightarrow 6 = n - 1 \Rightarrow n = 7$

Thus, the sum can be written as $\displaystyle\sum_{n=1}^{7} 5(3)^{n-1}$.

75. $2 - \dfrac{1}{2} + \dfrac{1}{8} - \cdots + \dfrac{1}{2048}$

$r = -\dfrac{1}{4}$ and $\dfrac{1}{2048} = 2\left(-\dfrac{1}{4}\right)^{n-1}$

By trial and error, we find that $n = 7$.

Thus, the sum can be written as $\displaystyle\sum_{n=1}^{7} 2\left(-\frac{1}{4}\right)^{n-1}$.

77. $0.1 + 0.4 + 1.6 + \cdots + 102.4$

$r = 4$ and $102.4 = 0.1(4)^{n-1}$

$1024 = 4^{n-1} \Rightarrow 5 = n - 1 \Rightarrow n = 6$

Thus, the sum can be written as $\displaystyle\sum_{n=1}^{6} 0.1(4)^{n-1}$.

79. $\displaystyle\sum_{n=0}^{\infty} \left(\frac{1}{2}\right)^n = 1 + \left(\frac{1}{2}\right)^1 + \left(\frac{1}{2}\right)^2 + \cdots$

$a_1 = 1, r = \dfrac{1}{2}$

$\displaystyle\sum_{n=0}^{\infty} \left(\frac{1}{2}\right)^n = \dfrac{a_1}{1-r} = \dfrac{1}{1 - \left(\frac{1}{2}\right)} = 2$

81. $\sum_{n=0}^{\infty}\left(-\frac{1}{2}\right)^n = 1 + \left(-\frac{1}{2}\right)^1 + \left(-\frac{1}{2}\right)^2 + \cdots$

$a_1 = 1,\; r = -\frac{1}{2}$

$\sum_{n=0}^{\infty}\left(-\frac{1}{2}\right)^n = \frac{a_1}{1-r} = \frac{1}{1-\left(-\frac{1}{2}\right)} = \frac{2}{3}$

83. $\sum_{n=0}^{\infty} 4\left(\frac{1}{4}\right)^n = 4 + 4\left(\frac{1}{4}\right)^1 + 4\left(\frac{1}{4}\right)^2 + \cdots$

$a_1 = 4,\; r = \frac{1}{4}$

$\sum_{n=0}^{\infty} 4\left(\frac{1}{4}\right)^n = \frac{a_1}{1-r} = \frac{4}{1-\left(\frac{1}{4}\right)} = \frac{16}{3}$

85. $\sum_{n=0}^{\infty} (0.4)^n = 1 + (0.4)^1 + (0.4)^2 + \cdots$

$a_1 = 1,\; r = 0.4$

$\sum_{n=0}^{\infty} (0.4)^n = \frac{1}{1-0.4} = \frac{5}{3}$

87. $\sum_{n=0}^{\infty} -3(0.9)^n = -3 - 3(0.9)^1 - 3(0.9)^2 - \cdots$

$a_1 = -3,\; r = 0.9$

$\sum_{n=0}^{\infty} -3(0.9)^n = \frac{-3}{1-0.9} = -30$

89. $8 + 6 + \frac{9}{2} + \frac{27}{8} + \cdots = \sum_{n=0}^{\infty} 8\left(\frac{3}{4}\right)^n = \frac{8}{1-\frac{3}{4}} = 32$

91. $\frac{1}{9} - \frac{1}{3} + 1 - 3 + \cdots = \sum_{n=0}^{\infty} \frac{1}{9}(-3)^n$

The sum is undefined because

$|r| = |-3| = 3 > 1.$

93. $0.\overline{36} = \sum_{n=0}^{\infty} 0.36(0.01)^n = \frac{0.36}{1-0.01} = \frac{0.36}{0.99} = \frac{36}{99} = \frac{4}{11}$

95. $0.3\overline{18} = 0.3 + \sum_{n=0}^{\infty} 0.018(0.01)^n = \frac{3}{10} + \frac{0.018}{1-0.01}$

$\quad = \frac{3}{10} + \frac{0.018}{0.99} = \frac{3}{10} + \frac{18}{990} = \frac{3}{10} + \frac{2}{110}$

$\quad = \frac{35}{110} = \frac{7}{22}$

97. $f(x) = 6\left[\dfrac{1-(0.5)^x}{1-(0.5)}\right],\; \sum_{n=0}^{\infty} 6\left(\frac{1}{2}\right)^n = \dfrac{6}{1-\frac{1}{2}} = 12$

The horizontal asymptote of $f(x)$ is $y = 12$.
This corresponds to the sum of the series.

99. (a) $a_n \approx 1190.88(1.006)^n$

(b) The population is growing at a rate of 0.6% per year.

(c) For 2010, let $n = 20$: $a_n = 1190.88(1.006)^{20}$

≈ 1342.2 million

(d) $1190.88(1.006)^n = 1320$

$1.006^n = \dfrac{1320}{1190.88}$

$\ln 1.006^n = \ln\left(\dfrac{1320}{1190.88}\right)$

$n \ln 1.006 = \ln\left(\dfrac{1320}{1190.88}\right)$

$n = \dfrac{\ln\left(\dfrac{1320}{1190.88}\right)}{\ln 1.006} \approx 17.21$

This corresponds with the year 2008.

101. $A = P\left(1 + \dfrac{r}{n}\right)^{nt} = 2500\left(1 + \dfrac{0.02}{n}\right)^{n(20)}$

(a) $n = 1$: $A = 2500\left(1 + \dfrac{0.02}{1}\right)^{(1)(20)} \approx \$3714.87$

(b) $n = 2$: $A = 2500\left(1 + \dfrac{0.02}{2}\right)^{(2)(20)} \approx \$3722.16$

(c) $n = 4$: $A = 2500\left(1 + \dfrac{0.02}{4}\right)^{(4)(20)} \approx \$3725.85$

(d) $n = 12$: $A = 2500\left(1 + \dfrac{0.02}{12}\right)^{(12)(20)} \approx \$3728.32$

(e) $n = 365$: $A = 2500\left(1 + \dfrac{0.02}{365}\right)^{(365)(20)} \approx \$3729.52$

103. $A = \displaystyle\sum_{n=1}^{60} 100\left(1 + \dfrac{0.06}{12}\right)^n$

$\qquad = \displaystyle\sum_{n=1}^{60} 100(1.005)^n$

$\qquad = 100(1.005) \cdot \dfrac{[1 - 1.005^{60}]}{[1 - 1.005]}$

$\qquad \approx \$7011.89$

105. Let $N = 12t$ be the total number of deposits.

$A = P\left(1 + \dfrac{r}{12}\right) + P\left(1 + \dfrac{r}{12}\right)^2 + \cdots + P\left(1 + \dfrac{r}{12}\right)^N$

$\quad = \left(1 + \dfrac{r}{12}\right)\left[P + P\left(1 + \dfrac{r}{12}\right) + \cdots + P\left(1 + \dfrac{r}{12}\right)^{N-1}\right]$

$\quad = P\left(1 + \dfrac{r}{12}\right)\displaystyle\sum_{n=1}^{N}\left(1 + \dfrac{r}{12}\right)^{n-1}$

$\quad = P\left(1 + \dfrac{r}{12}\right)\left[\dfrac{1 - \left(1 + \dfrac{r}{12}\right)^N}{1 - \left(1 + \dfrac{r}{12}\right)}\right]$

$\quad = P\left(1 + \dfrac{r}{12}\right)\left(-\dfrac{12}{r}\right)\left[1 - \left(1 + \dfrac{r}{12}\right)^N\right]$

$\quad = P\left(\dfrac{12}{r} + 1\right)\left[-1 + \left(1 + \dfrac{r}{12}\right)^N\right]$

$\quad = P\left[\left(1 + \dfrac{r}{12}\right)^N - 1\right]\left(1 + \dfrac{12}{r}\right)$

$\quad = P\left[\left(1 + \dfrac{r}{12}\right)^{12t} - 1\right]\left(1 + \dfrac{12}{r}\right)$

107. $P = \$50$, $r = 7\%$, $t = 20$ years

(a) Compounded monthly:

$A = 50\left[\left(1 + \dfrac{0.07}{12}\right)^{12(20)} - 1\right]\left(1 + \dfrac{12}{0.07}\right)$
$\quad \approx \$26{,}198.27$

(b) Compounded continuously:

$A = \dfrac{50e^{0.07/12}(e^{0.07(20)} - 1)}{e^{0.07/12} - 1} \approx \$26{,}263.88$

109. $P = \$100$, $r = 10\%$, $t = 40$ years

(a) Compounded monthly: $A = 100\left[\left(1 + \dfrac{0.10}{12}\right)^{12(40)} - 1\right]\left(1 + \dfrac{12}{0.10}\right) \approx \$637{,}678.02$

(b) Compounded continuously: $A = \dfrac{100e^{0.10/12}(e^{(0.10)(40)} - 1)}{e^{0.10/12} - 1} \approx \$645{,}861.43$

111. $P = W \sum_{n=1}^{12t} \left[\left(1 + \dfrac{r}{12} \right)^{-1} \right]^n$

$= W \left(1 + \dfrac{r}{12} \right)^{-1} \left[\dfrac{1 - \left(1 + \dfrac{r}{12} \right)^{-12t}}{1 - \left(1 + \dfrac{r}{12} \right)^{-1}} \right]$

$= W \left(\dfrac{1}{1 + \dfrac{r}{12}} \right) \dfrac{\left[1 - \left(1 + \dfrac{r}{12} \right)^{-12t} \right]}{1 - \dfrac{1}{\left(1 + \dfrac{r}{12} \right)}}$

$= W \dfrac{\left[1 - \left(1 + \dfrac{r}{12} \right)^{-12t} \right]}{\left(1 + \dfrac{r}{12} \right) - 1}$

$= W \left(\dfrac{12}{r} \right) \left[1 - \left(1 + \dfrac{r}{12} \right)^{-12t} \right]$

113. $\sum_{n=1}^{\infty} 400(0.75)^n = \dfrac{300}{1 - 0.75} = \$1200$

115. $\sum_{n=1}^{\infty} 600(0.725)^n = \dfrac{435}{1 - 0.725} \approx \$1581.82$

117. $64 + 32 + 16 + 8 + 4 + 2 = 126$

Total area of shaded region is approximately 126 square inches.

119. $a_n = 30{,}000(1.05)^{n-1}$

$T = \sum_{n=1}^{40} 30{,}000(1.05)^{n-1} = 30{,}000 \dfrac{(1 - 1.05^{40})}{(1 - 1.05)} \approx \$3{,}623{,}993.23$

121. False. A sequence is geometric if the ratios of consecutive terms are the same.

123. Given a real number r between -1 and 1, as the exponent n increases, r^n approaches zero.

125. $g(x) = x^2 - 1$

$g(x + 1) = (x + 1)^2 - 1$

$= x^2 + 2x + 1 - 1 = x^2 + 2x$

127. $f(x) = 3x + 1, \ g(x) = x^2 - 1$

$f(g(x + 1)) = f(x^2 + 2x)$

$= 3(x^2 + 2x) + 1$

$= 3x^2 + 6x + 1$

129. $9x^3 - 64x = x(9x^2 - 64) = x(3x + 8)(3x - 8)$

131. $6x^2 - 13x - 5 = (3x + 1)(2x - 5)$

133. $\dfrac{3}{x + 3} \cdot \dfrac{x(x + 3)}{x - 3} = \dfrac{3x}{x - 3}, \ x \neq -3$

135. $\dfrac{x}{3} \div \dfrac{3x}{6x + 3} = \dfrac{x}{3} \cdot \dfrac{3(2x + 1)}{3x} = \dfrac{2x + 1}{3}, \ x \neq 0, -\dfrac{1}{2}$

137. $5 + \dfrac{7}{x + 2} + \dfrac{2}{x - 2} = \dfrac{5(x + 2)(x - 2) + 7(x - 2) + 2(x + 2)}{(x + 2)(x - 2)}$

$= \dfrac{5(x^2 - 4) + 7(x - 2) + 2(x + 2)}{(x + 2)(x - 2)}$

$= \dfrac{5x^2 - 20 + 7x - 14 + 2x + 4}{(x + 2)(x - 2)} = \dfrac{5x^2 + 9x - 30}{(x + 2)(x - 2)}$

139. Answers will vary.

Section 11.4   Mathematical Induction

■   You should be sure that you understand the principle of mathematical induction. If P_n is a statement involving the positive integer n, where P_1 is true and the truth of P_k implies the truth of P_{k+1} for every positive k, then P_n is true for all positive integers n.

■   You should be able to verify (by induction) the formulas for the sums of powers of integers and be able to use these formulas.

■   You should be able to calculate the first and second differences of a sequence.

■   You should be able to find the quadratic model for a sequence, when it exists.

Vocabulary Check

1. mathematical induction

2. first

3. arithmetic

4. second

---

1. $P_k = \dfrac{5}{k(k+1)}$

$P_{k+1} = \dfrac{5}{(k+1)[(k+1)+1]} = \dfrac{5}{(k+1)(k+2)}$

3. $P_k = \dfrac{k^2(k+1)^2}{4}$

$P_{k+1} = \dfrac{(k+1)^2[(k+1)+1]^2}{4} = \dfrac{(k+1)^2(k+2)^2}{4}$

5. 1. When $n = 1$, $S_1 = 2 = 1(1+1)$.

2. Assume that

$S_k = 2 + 4 + 6 + 8 + \cdots + 2k = k(k+1)$.

Then,

$S_{k+1} = 2 + 4 + 6 + 8 + \cdots + 2k + 2(k+1)$

$= S_k + 2(k+1) = k(k+1) + 2(k+1) = (k+1)(k+2)$.

Therefore, we conclude that the formula is valid for all positive integer values of n.

7. 1. When $n = 1$, $S_1 = 2 = \dfrac{1}{2}(5(1) - 1)$.

2. Assume that

$S_k = 2 + 7 + 12 + 17 + \cdots + (5k - 3) = \dfrac{k}{2}(5k - 1)$.

Then,

$S_{k+1} = 2 + 7 + 12 + 17 + \cdots + (5k - 3) + [5(k+1) - 3]$

$= S_k + (5k + 5 - 3) = \dfrac{k}{2}(5k - 1) + 5k + 2$

$= \dfrac{5k^2 - k + 10k + 4}{2} = \dfrac{5k^2 + 9k + 4}{2}$

$= \dfrac{(k+1)(5k+4)}{2} = \dfrac{(k+1)}{2}[5(k+1) - 1]$.

Therefore, we conclude that this formula is valid for all positive integer values of n.

9. 1. When $n = 1$, $S_1 = 1 = 2^1 - 1$.

2. Assume that

$$S_k = 1 + 2 + 2^2 + 2^3 + \cdots + 2^{k-1} = 2^k - 1.$$

Then,

$$S_{k+1} = 1 + 2 + 2^2 + 2^3 + \cdots + 2^{k-1} + 2^k$$

$$= S_k + 2^k = 2^k - 1 + 2^k = 2(2^k) - 1 = 2^{k+1} - 1.$$

Therefore, we conclude that this formula is valid for all positive integer values of n.

11. 1. When $n = 1$, $S_1 = 1 = \dfrac{1(1 + 1)}{2}$.

2. Assume that

$$S_k = 1 + 2 + 3 + 4 + \cdots + k = \frac{k(k + 1)}{2}.$$

Then,

$$S_{k+1} = 1 + 2 + 3 + 4 + \cdots + k + (k + 1)$$

$$= S_k + (k + 1) = \frac{k(k + 1)}{2} + \frac{2(k + 1)}{2} = \frac{(k + 1)(k + 2)}{2}.$$

Therefore, we conclude that this formula is valid for all positive integer values of n.

13. 1. When $n = 1$, $S_1 = 1 = \dfrac{(1)^2(1 + 1)^2(2(1)^2 + 2(1) - 1)}{12}$.

2. Assume that

$$S_k = \sum_{i=1}^{k} i^5 = \frac{k^2(k + 1)^2(2k^2 + 2k - 1)}{12}.$$

Then,

$$S_{k+1} = \sum_{i=1}^{k+1} i^5 = \left(\sum_{i=1}^{k} i^5 \right) + (k + 1)^5$$

$$= \frac{k^2(k + 1)^2(2k^2 + 2k - 1)}{12} + \frac{12(k + 1)^5}{12}$$

$$= \frac{(k + 1)^2[k^2(2k^2 + 2k - 1) + 12(k + 1)^3]}{12}$$

$$= \frac{(k + 1)^2[2k^4 + 2k^3 - k^2 + 12(k^3 + 3k^2 + 3k + 1)]}{12}$$

$$= \frac{(k + 1)^2[2k^4 + 14k^3 + 35k^2 + 36k + 12]}{12}$$

$$= \frac{(k + 1)^2(k^2 + 4k + 4)(2k^2 + 6k + 3)}{12}$$

$$= \frac{(k + 1)^2(k + 2)^2[2(k + 1)^2 + 2(k + 1) - 1]}{12}.$$

Therefore, we conclude that this formula is valid for all positive integer values of n.

Note: The easiest way to complete the last two steps is to "work backwards." Start with the desired expression for S_{k+1} and multiply out to show that it is equal to the expression you found for $S_k + (k + 1)^5$.

15. 1. When $n = 1$, $S_1 = 2 = \dfrac{1(2)(3)}{3}$.

2. Assume that

$$S_k = 1(2) + 2(3) + 3(4) + \cdots + k(k + 1) = \dfrac{k(k + 1)(k + 2)}{3}.$$

Then,

$$S_{k+1} = 1(2) + 2(3) + 3(4) + \cdots + k(k + 1) + (k + 1)(k + 2)$$

$$= S_k + (k + 1)(k + 2) = \dfrac{k(k + 1)(k + 2)}{3} + \dfrac{3(k + 1)(k + 2)}{3}$$

$$= \dfrac{(k + 1)(k + 2)(k + 3)}{3}.$$

Therefore, we conclude that this formula is valid for all positive integer values of n.

17. 1. When $n = 4$, $4! = 24$ and $2^4 = 16$, thus $4! > 2^4$.

2. Assume

$k! > 2^k$, $k > 4$.

Then,

$(k + 1)! = k!(k + 1) > 2^k(2)$ since $k! > 2^k$ and $k + 1 > 2$.

Thus, $(k + 1)! > 2^{k+1}$.

Therefore, by extended mathematical induction, the inequality is valid for all integers n such that $n \geq 4$.

19. 1. When $n = 2$, $\dfrac{1}{\sqrt{1}} + \dfrac{1}{\sqrt{2}} \approx 1.707$ and $\sqrt{2} \approx 1.414$, thus $\dfrac{1}{\sqrt{1}} + \dfrac{1}{\sqrt{2}} > \sqrt{2}$.

2. Assume that

$$\dfrac{1}{\sqrt{1}} + \dfrac{1}{\sqrt{2}} + \dfrac{1}{\sqrt{3}} + \cdots + \dfrac{1}{\sqrt{k}} > \sqrt{k}, k > 2.$$

Then,

$$\dfrac{1}{\sqrt{1}} + \dfrac{1}{\sqrt{2}} + \dfrac{1}{\sqrt{3}} + \cdots + \dfrac{1}{\sqrt{k}} + \dfrac{1}{\sqrt{k + 1}} > \sqrt{k} + \dfrac{1}{\sqrt{k + 1}}.$$

Now it is sufficient to show that

$$\sqrt{k} + \dfrac{1}{\sqrt{k + 1}} > \sqrt{k + 1}, \ k > 2,$$

or equivalently $\left(\text{multiplying by } \sqrt{k + 1}\right)$,

$$\sqrt{k}\sqrt{k + 1} + 1 > k + 1.$$

This is true because

$$\sqrt{k}\sqrt{k + 1} + 1 > \sqrt{k}\sqrt{k} + 1 = k + 1.$$

Therefore,

$$\dfrac{1}{\sqrt{1}} + \dfrac{1}{\sqrt{2}} + \dfrac{1}{\sqrt{3}} + \ldots + \dfrac{1}{\sqrt{k}} + \dfrac{1}{\sqrt{k + 1}} > \sqrt{k + 1}.$$

Therefore, by extended mathematical induction, the inequality is valid for all integers n such that $n \geq 2$.

21. $(1 + a)^n \geq na$, $n \geq 1$ and $a > 0$

Since a is positive, then all of the terms in the binomial expansion are positive.

$(1 + a)^n = 1 + na + \cdots + na^{n-1} + a^n > na$

23. 1. When $n = 1$, $(ab)^1 = a^1b^1 = ab$.

2. Assume that $(ab)^k = a^kb^k$.

Then, $(ab)^{k+1} = (ab)^k(ab)$

$$= a^kb^kab$$

$$= a^{k+1}b^{k+1}.$$

Thus, $(ab)^n = a^nb^n$.

25. 1. When $n = 2$, $(x_1x_2)^{-1} = \dfrac{1}{x_1x_2} = \dfrac{1}{x_1} \cdot \dfrac{1}{x_2} = x_1^{-1}x_2^{-1}$.

2. Assume that

$$(x_1x_2x_3 \cdots x_k)^{-1} = x_1^{-1}x_2^{-1}x_3^{-1} \cdots x_k^{-1}.$$

Then,

$$(x_1x_2x_3 \cdots x_kx_{k+1})^{-1} = [(x_1x_2x_3 \cdots x_k)x_{k+1}]^{-1}$$

$$= (x_1x_2x_3 \ldots x_k)^{-1}x_{k+1}^{-1}$$

$$= x_1^{-1}x_2^{-1}x_3^{-1} \cdots x_k^{-1}x_{k+1}^{-1}.$$

Thus, the formula is valid.

27. 1. When $n = 1$, $x(y_1) = xy_1$.

2. Assume that

$$x(y_1 + y_2 + \cdots + y_k) = xy_1 + xy_2 + \cdots + xy_k.$$

Then,

$$xy_1 + xy_2 + \cdots + xy_k + xy_{k+1} = x(y_1 + y_2 + \cdots + y_k) + xy_{k+1}$$

$$= x[(y_1 + y_2 + \cdots + y_k) + y_{k+1}]$$

$$= x(y_1 + y_2 + \cdots + y_k + y_{k+1}).$$

Hence, the formula holds.

29. 1. When $n = 1$, $[1^3 + 3(1)^2 + 2(1)] = 6$ and 3 is a factor.

2. Assume that 3 is a factor of $k^3 + 3k^2 + 2k$.

Then,

$$(k + 1)^3 + 3(k + 1)^2 + 2(k + 1) = k^3 + 3k^2 + 3k + 1 + 3k^2 + 6k + 3 + 2k + 2$$

$$= (k^3 + 3k^2 + 2k) + (3k^2 + 9k + 6)$$

$$= (k^3 + 3k^2 + 2k) + 3(k^2 + 3k + 2).$$

Since 3 is a factor of $(k^3 + 3k^2 + 2k)$, our assumption, and 3 is a factor of $3(k^2 + 3k + 2)$, we conclude that 3 is a factor of the whole sum.

Thus, 3 is a factor of $(n^3 + 3n^2 + 2n)$ for every positive integer n.

31. A factor of $n^4 - n + 4$ is 2.

1. When $n = 1$, $1^4 - 1 + 4 = 4$ and 2 is a factor.

2. Assume that 2 is a factor of $k^4 - k + 4$.

Then,

$$(k + 1)^4 - (k + 1) + 4 = k^4 + 4k^3 + 6k^2 + 4k + 1 - k - 1 + 4$$

$$= (k^4 - k + 4) + (4k^3 + 6k^2 + 4k)$$

$$= (k^4 - k + 4) + 2(2k^3 + 3k^2 + 2k).$$

Since 2 is a factor of $k^4 - k + 4$, our assumption, and 2 is a factor of $2(2k^3 + 3k^2 + 2k)$, we conclude that 2 is a factor of the entire expression.

Thus, 2 is a factor of $n^4 - n + 4$ for every positive integer n.

33. A factor of $2^{4n-2} + 1$ is 5.

1. When $n = 1$,

   $2^{4(1)-2} + 1 = 5$ and 5 is a factor.

2. Assume that 5 is a factor of $2^{4k-2} + 1$.

   Then,

   $$2^{4(k+1)-2} + 1 = 2^{4k+4-2} + 1$$
   $$= 2^{4k-2} \cdot 2^4 + 1$$
   $$= 2^{4k-2} \cdot 16 + 1$$
   $$= (2^{4k-2} + 1) + 15 \cdot 2^{4k-2}.$$

   Since 5 is a factor of $2^{4k-2} + 1$, our assumption, and 5 is a factor of $15 \cdot 2^{4k-2}$, we conclude that 5 is a factor of the entire expression.

   Thus, 5 is a factor of $2^{4n-2} + 1$ for every positive integer n.

35. $S_n = 1 + 5 + 9 + 13 + \cdots + (4n - 3)$

$S_1 = 1 = 1 \cdot 1$

$S_2 = 1 + 5 = 6 = 2 \cdot 3$

$S_3 = 1 + 5 + 9 = 15 = 3 \cdot 5$

$S_4 = 1 + 5 + 9 + 13 = 28 = 4 \cdot 7$

From this sequence, it appears that $S_n = n(2n - 1)$. This can be verified by mathematical induction. The formula has already been verified for $n = 1$. Assume that the formula is valid for $n = k$. Then,

$$S_{k+1} = [1 + 5 + 9 + 13 + \cdots + (4k - 3)] + [4(k + 1) - 3]$$
$$= k(2k - 1) + (4k + 1)$$
$$= 2k^2 + 3k + 1$$
$$= (k + 1)(2k + 1)$$
$$= (k + 1)[2(k + 1) - 1].$$

Thus, the formula is valid.

37. $S_n = 1 + \dfrac{9}{10} + \dfrac{81}{100} + \dfrac{729}{1000} + \cdots + \left(\dfrac{9}{10}\right)^{n-1}$

Since this series is geometric, we have

$$S_n = \sum_{i=1}^{n} \left(\frac{9}{10}\right)^{i-1} = \frac{1 - \left(\frac{9}{10}\right)^n}{1 - \frac{9}{10}} = 10\left[1 - \left(\frac{9}{10}\right)^n\right]$$

$$= 10 - 10\left(\frac{9}{10}\right)^n.$$

39. $S_n = \dfrac{1}{4} + \dfrac{1}{12} + \dfrac{1}{24} + \dfrac{1}{40} + \cdots + \dfrac{1}{2n(n+1)}$

$S_1 = \dfrac{1}{4} = \dfrac{1}{2(2)}$

$S_2 = \dfrac{1}{4} + \dfrac{1}{12} = \dfrac{4}{12} = \dfrac{2}{6} = \dfrac{2}{2(3)}$

$S_3 = \dfrac{1}{4} + \dfrac{1}{12} + \dfrac{1}{24} = \dfrac{9}{24} = \dfrac{3}{8} = \dfrac{3}{2(4)}$

$S_4 = \dfrac{1}{4} + \dfrac{1}{12} + \dfrac{1}{24} + \dfrac{1}{40} = \dfrac{16}{40} = \dfrac{4}{10} = \dfrac{4}{2(5)}$

From this sequence, it appears that

$S_n = \dfrac{n}{2(n+1)}.$

This can be verified by mathematical induction. The formula has already been verified for $n = 1$.
Assume that the formula is valid for $n = k$. Then,

$S_{k+1} = \left[\dfrac{1}{4} + \dfrac{1}{12} + \dfrac{1}{40} + \cdots + \dfrac{1}{2k(k+1)}\right] + \dfrac{1}{2(k+1)(k+2)}$

$= \dfrac{k}{2(k+1)} + \dfrac{1}{2(k+1)(k+2)}$

$= \dfrac{k(k+2) + 1}{2(k+1)(k+2)}$

$= \dfrac{k^2 + 2k + 1}{2(k+1)(k+2)}$

$= \dfrac{(k+1)^2}{2(k+1)(k+2)}$

$= \dfrac{k+1}{2(k+2)}.$

Thus, the formula is valid.

41. $\displaystyle\sum_{n=1}^{15} n = \dfrac{15(15+1)}{2} = 120$

43. $\displaystyle\sum_{n=1}^{6} n^2 = \dfrac{6(6+1)[2(6)+1]}{6} = 91$

45. $\displaystyle\sum_{n=1}^{5} n^4 = \dfrac{5(5+1)[2(5)+1][3(5)^2 + 3(5) - 1]}{30} = 979$

47. $\displaystyle\sum_{n=1}^{6} (n^2 - n) = \sum_{n=1}^{6} n^2 - \sum_{n=1}^{6} n$

$= \dfrac{6(6+1)[2(6)+1]}{6} - \dfrac{6(6+1)}{2}$

$= 91 - 21 = 70$

49. $\displaystyle\sum_{i=1}^{6} (6i - 8i^3) = 6\sum_{i=1}^{6} i - 8\sum_{i=1}^{6} i^3 = 6\left[\dfrac{6(6+1)}{2}\right] - 8\left[\dfrac{(6)^2(6+1)^2}{4}\right] = 6(21) - 8(441) = -3402$

51. $a_1 = 0, a_n = a_{n-1} + 3$

$a_1 = a_1 = 0$

$a_2 = a_1 + 3 = 0 + 3 = 3$

$a_3 = a_2 + 3 = 3 + 3 = 6$

$a_4 = a_3 + 3 = 6 + 3 = 9$

$a_5 = a_4 + 3 = 9 + 3 = 12$

$a_6 = a_5 + 3 = 12 + 3 = 15$

a_n: 0   3   6   9   12   15

First differences: 3   3   3   3   3

Second differences: 0   0   0   0

Since the first differences are equal, the sequence has a linear model.

53. $a_1 = 3, a_n = a_{n-1} - n$

$a_1 = a_1 = 3$

$a_2 = a_1 - 2 = 3 - 2 = 1$

$a_3 = a_2 - 3 = 1 - 3 = -2$

$a_4 = a_3 - 4 = -2 - 4 = -6$

$a_5 = a_4 - 5 = -6 - 5 = -11$

$a_6 = a_5 - 6 = -11 - 6 = -17$

a_n: 3   1   -2   -6   -11   -17

First differences: -2   -3   -4   -5   -6

Second differences: -1   -1   -1   -1

Since the second differences are all the same, the sequence has a quadratic model.

55. $a_0 = 2, a_n = (a_{n-1})^2$

$a_0 = 2$

$a_1 = a_0^2 = 2^2 = 4$

$a_2 = a_1^2 = 4^2 = 16$

$a_3 = a_2^2 = 16^2 = 256$

$a_4 = a_3^2 = 256^2 = 65,536$

$a_5 = a_4^2 = 65,536^2 = 4,294,967,296$

a_n: 2   4   16   256   65,536   4,294,967,296

First differences: 2   12   240   65,280   4,294,901,760

Second differences: 10   228   65,040   4,294,836,480

Since neither the first differences nor the second differences are equal, the sequence does not have a linear or quadratic model.

57. $a_0 = 3, a_1 = 3, a_4 = 15$

Let $a_n = an^2 + bn + c$.

Thus: $a_0 = a(0)^2 + b(0) + c = 3 \implies c = 3$

$a_1 = a(1)^2 + b(1) + c = 3 \implies a + b + c = 3$

$a + b = 0$

$a_4 = a(4)^2 + b(4) + c = 15 \implies 16a + 4b + c = 15$

$16a + 4b = 12$

$4a + b = 3$

By elimination: $-a - b = 0$

$\underline{4a + b = 3}$

$3a \quad = 3$

$a = 1 \implies b = -1$

Thus, $a_n = n^2 - n + 3$.

59. $a_0 = -3, a_2 = 1, a_4 = 9$

Let $a_n = an^2 + bn + c$.

Then: $a_0 = a(0)^2 + b(0) + c = -3 \implies c = -3$

$a_2 = a(2)^2 + b(2) + c = 1 \implies 4a + 2b + c = 1$

$4a + 2b = 4$

$2a + b = 2$

$a_4 = a(4)^2 + b(4) + c = 9 \implies 16a + 4b + c = 9$

$16a + 4b = 12$

$4a + b = 3$

By elimination: $-2a - b = -2$

$\underline{4a + b = \quad 3}$

$2a \quad = 1$

$a = \frac{1}{2} \implies b = 1$

Thus, $a_n = \frac{1}{2}n^2 + n - 3$.

61. (a)

First differences: 2.2   2.4   2.2   2.3   0.9

(b) The first differences are not equal, but are fairly close to each other, so a linear model can be used. If we let $m = 2.2$, then $b = 120.3 - 2.2(8) = 102.7$

$a_n \approx 2.2n + 102.7$

(c) $a_n \approx 2.08n + 103.9$ is obtained by using the regression feature of a graphing utility.

(d) For 2008, let $n = 18$.

$a_n \approx 2.2(18) + 102.7 = 142.3$

$a_n = 2.08(18) + 103.9 = 141.34$

These are very similar.

63. True. P_7 may be false.

65. True. If the second differences are all zero, then the first differences are all the same, so the sequence is arithmetic.

67. $(2x^2 - 1)^2 = (2x^2 - 1)(2x^2 - 1) = 4x^4 - 4x^2 + 1$

69. $(5 - 4x)^3 = -64x^3 + 240x^2 - 300x + 125$

71. $f(x) = \dfrac{x}{x + 3}$

(a) Domain: All real numbers x except $x = -3$

(b) Intercept: $(0, 0)$

(c) Vertical asymptote: $x = -3$

Horizontal asymptote: $y = 1$

(d)

x	-5	-4	-2	-1	1
$f(x)$	$\frac{5}{2}$	4	-2	$-\frac{1}{2}$	$\frac{1}{4}$

73. $h(t) = \dfrac{t - 7}{t}$

(a) Domain: All real numbers t except $t = 0$

(b) Intercept: $(7, 0)$

(c) Vertical asymptote: $t = 0$

Horizontal asymptote: $y = 1$

(d)

t	-2	-1	1	2	3
$h(t)$	$\frac{9}{2}$	8	-6	$-\frac{5}{2}$	$-\frac{4}{3}$

Section 11.5   The Binomial Theorem

- You should be able to use the formula

$$(x + y)^n = x^n + nx^{n-1}y + \frac{n(n-1)}{2!}x^{n-2}y^2 + \cdots + {}_nC_r x^{n-r}y^r + \cdots + y^n$$

where ${}_nC_r = \dfrac{n!}{(n-r)!r!}$, to expand $(x + y)^n$. Also, ${}_nC_r = \begin{pmatrix} n \\ r \end{pmatrix}$.

- You should be able to use Pascal's Triangle in binomial expansion.

Vocabulary Check

1. binomial coefficients

2. Binomial Theorem/Pascal's Triangle

3. $\begin{pmatrix} n \\ r \end{pmatrix}$ or ${}_nC_r$

4. expanding a binomial

1. ${}_5C_3 = \dfrac{5!}{3!2!} = \dfrac{5 \cdot 4}{2 \cdot 1} = 10$

3. ${}_{12}C_0 = \dfrac{12!}{0!12!} = 1$

5. ${}_{20}C_{15} = \dfrac{20!}{15!5!} = \dfrac{20 \cdot 19 \cdot 18 \cdot 17 \cdot 16}{5 \cdot 4 \cdot 3 \cdot 2 \cdot 1} = 15{,}504$

7. $\begin{pmatrix} 10 \\ 4 \end{pmatrix} = \dfrac{10!}{6!4!} = \dfrac{10 \cdot 9 \cdot 8 \cdot 7 \cdot 6!}{6!(24)} = 210$

9. $\dbinom{100}{98} = \dfrac{100!}{2!98!} = \dfrac{100 \cdot 99}{2 \cdot 1} = 4950$

11.
```
              1
            1   1
          1   2   1
        1   3   3   1
      1   4   6   4   1
    1   5  10  10   5   1
  1   6  15  20  15   6   1
 1  7  21  35  35  21   7   1
1  8  28  56  70 (56) 28   8   1
```

$\dbinom{8}{5} = 56$, the 6th entry in the 8th row.

13.
```
              1
            1   1
          1   2   1
        1   3   3   1
      1   4   6   4   1
    1   5  10  10   5   1
  1   6  15  20  15   6   1
 1  7  21  35 (35) 21   7   1
```

$_7C_4 = 35$, the 5th entry in the 7th row.

15. $(x + 1)^4 = {_4C_0}x^4 + {_4C_1}x^3(1) + {_4C_2}x^2(1)^2 + {_4C_3}x(1)^3 + {_4C_4}(1)^4$

$\qquad = x^4 + 4x^3 + 6x^2 + 4x + 1$

17. $(a + 6)^4 = {_4C_0}a^4 + {_4C_1}a^3(6) + {_4C_2}a^2(6)^2 + {_4C_3}a(6)^3 + {_4C_4}(6)^4$

$\qquad = 1a^4 + 4a^3(6) + 6a^2(6)^2 + 4a(6)^3 + 1(6)^4$

$\qquad = a^4 + 24a^3 + 216a^2 + 864a + 1296$

19. $(y - 4)^3 = {_3C_0}y^3 - {_3C_1}y^2(4) + {_3C_2}y(4)^2 - {_3C_3}(4)^3$

$\qquad = 1y^3 - 3y^2(4) + 3y(4)^2 - 1(4)^3$

$\qquad = y^3 - 12y^2 + 48y - 64$

21. $(x + y)^5 = {_5C_0}x^5 + {_5C_1}x^4y + {_5C_2}x^3y^2 + {_5C_3}x^2y^3 + {_5C_4}xy^4 + {_5C_5}y^5$

$\qquad = x^5 + 5x^4y + 10x^3y^2 + 10x^2y^3 + 5xy^4 + y^5$

23. $(r + 3s)^6 = {_6C_0}r^6 + {_6C_1}r^5(3s) + {_6C_2}r^4(3s)^2 + {_6C_3}r^3(3s)^3 + {_6C_4}r^2(3s)^4 + {_6C_5}r(3s)^5 + {_6C_6}(3s)^6$

$\qquad = 1r^6 + 6r^5(3s) + 15r^4(3s)^2 + 20r^3(3s)^3 + 15r^2(3s)^4 + 6r(3s)^5 + 1(3s)^6$

$\qquad = r^6 + 18r^5s + 135r^4s^2 + 540r^3s^3 + 1215r^2s^4 + 1458rs^5 + 729s^6$

25. $(3a - 4b)^5 = {_5C_0}(3a)^5 - {_5C_1}(3a)^4(4b) + {_5C_2}(3a)^3(4b)^2 - {_5C_3}(3a)^2(4b)^3 + {_5C_4}(3a)(4b)^4 - {_5C_5}(4b)^5$

$\qquad = (1)(243a^5) - 5(81a^4)(4b) + 10(27a^3)(16b^2) - 10(9a^2)(64b^3) + 5(3a)(256b^4) - (1)(1024b^5)$

$\qquad = 243a^5 - 1620a^4b + 4320a^3b^2 - 5760a^2b^3 + 3840ab^4 - 1024b^5$

27. $(2x + y)^3 = {_3C_0}(2x)^3 + {_3C_1}(2x)^2(y) + {_3C_2}(2x)(y^2) + {_3C_3}(y^3)$

$\qquad = (1)(8x^3) + (3)(4x^2)(y) + (3)(2x)(y^2) + (1)(y^3)$

$\qquad = 8x^3 + 12x^2y + 6xy^2 + y^3$

29. $(x^2 + y^2)^4 = {_4C_0}(x^2)^4 + {_4C_1}(x^2)^3(y^2) + {_4C_2}(x^2)^2(y^2)^2 + {_4C_3}(x^2)(y^2)^3 + {_4C_4}(y^2)^4$

$\qquad = (1)(x^8) + (4)(x^6y^2) + (6)(x^4y^4) + (4)(x^2y^6) + (1)(y^8)$

$\qquad = x^8 + 4x^6y^2 + 6x^4y^4 + 4x^2y^6 + y^8$

31. $\left(\dfrac{1}{x} + y\right)^5 = {}_5C_0\left(\dfrac{1}{x}\right)^5 + {}_5C_1\left(\dfrac{1}{x}\right)^4 y + {}_5C_2\left(\dfrac{1}{x}\right)^3 y^2 + {}_5C_3\left(\dfrac{1}{x}\right)^2 y^3 + {}_5C_4\left(\dfrac{1}{x}\right)y^4 + {}_5C_5 y^5$

$\qquad = \dfrac{1}{x^5} + \dfrac{5y}{x^4} + \dfrac{10y^2}{x^3} + \dfrac{10y^3}{x^2} + \dfrac{5y^4}{x} + y^5$

33. $2(x - 3)^4 + 5(x - 3)^2 = 2[x^4 - 4(x^3)(3) + 6(x^2)(3^2) - 4(x)(3^3) + 3^4] + 5[x^2 - 2(x)(3) + 3^2]$

$\qquad = 2(x^4 - 12x^3 + 54x^2 - 108x + 81) + 5(x^2 - 6x + 9)$

$\qquad = 2x^4 - 24x^3 + 113x^2 - 246x + 207$

35. 5th Row of Pascal's Triangle: 1   5   10   10   5   1

$(2t - s)^5 = 1(2t)^5 - 5(2t)^4(s) + 10(2t)^3(s)^2 - 10(2t)^2(s)^3 + 5(2t)(s)^4 - 1(s)^5$

$\qquad = 32t^5 - 80t^4 s + 80t^3 s^2 - 40t^2 s^3 + 10t s^4 - s^5$

37. 5th Row of Pascal's Triangle: 1   5   10   10   5   1

$(x + 2y)^5 = 1x^5 + 5x^4(2y) + 10x^3(2y)^2 + 10x^2(2y)^3 + 5x(2y)^4 + 1(2y)^5$

$\qquad = x^5 + 10x^4 y + 40x^3 y^2 + 80x^2 y^3 + 80xy^4 + 32y^5$

39. The 4th term in the expansion of $(x + y)^{10}$ is

$\qquad {}_{10}C_3 x^{10-3} y^3 = 120x^7 y^3$.

41. The 3rd term in the expansion of $(x - 6y)^5$ is

$\qquad {}_5C_2 x^{5-2}(-6y)^2 = 10x^3(36y^2) = 360x^3 y^2$.

43. The 8th term in the expansion of $(4x + 3y)^9$ is

$\qquad {}_9C_7(4x)^{9-7}(3y)^7 = 36(16x^2)(2187y^7)$

$\qquad\qquad = 1{,}259{,}712x^2 y^7$.

45. The 9th term in the expansion of $(10x - 3y)^{12}$ is

$\qquad {}_{12}C_8(10x)^{12-8}(-3y)^8 = 495(10{,}000x^4)(6561y^8)$

$\qquad\qquad = 32{,}476{,}950{,}000x^4 y^8$.

47. The term involving x^5 in the expansion of $(x + 3)^{12}$ is

$\qquad {}_{12}C_7 x^5(3)^7 = \dfrac{12!}{7!5!} \cdot 3^7 x^5 = 1{,}732{,}104x^5$.

The coefficient is 1,732,104.

49. The term involving $x^8 y^2$ in the expansion of $(x - 2y)^{10}$ is

$\qquad {}_{10}C_2 x^8(-2y)^2 = \dfrac{10!}{2!8!} \cdot 4x^8 y^2 = 180x^8 y^2$.

The coefficient is 180.

51. The term involving $x^4 y^5$ in the expansion of $(3x - 2y)^9$ is

$\qquad {}_9C_5(3x)^4(-2y)^5 = \dfrac{9!}{5!4!}(81x^4)(-32y^5) = -326{,}592x^4 y^5$.

The coefficient is $-326{,}592$.

53. The term involving $x^8 y^6 = (x^2)^4 y^6$ in the expansion of

$(x^2 + y)^{10}$ is ${}_{10}C_6(x^2)^4 y^6 = \dfrac{10!}{4!6!}(x^2)^4 y^6 = 210x^8 y^6$.

The coefficient is 210.

55. $\left(\sqrt{x} + 3\right)^4 = \left(\sqrt{x}\right)^4 + 4\left(\sqrt{x}\right)^3(3) + 6\left(\sqrt{x}\right)^2(3)^2 + 4\left(\sqrt{x}\right)(3)^3 + (3)^4$

$\qquad = x^2 + 12x\sqrt{x} + 54x + 108\sqrt{x} + 81$

$\qquad = x^2 + 12x^{3/2} + 54x + 108x^{1/2} + 81$

57. $(x^{2/3} - y^{1/3})^3 = (x^{2/3})^3 - 3(x^{2/3})^2 (y^{1/3}) + 3(x^{2/3}) (y^{1/3})^2 - (y^{1/3})^3$

$\qquad = x^2 - 3x^{4/3} y^{1/3} + 3x^{2/3} y^{2/3} - y$

59. $\dfrac{f(x+h)-f(x)}{h} = \dfrac{(x+h)^3 - x^3}{h}$

$\qquad\qquad = \dfrac{x^3 + 3x^2h + 3xh^2 + h^3 - x^3}{h}$

$\qquad\qquad = \dfrac{h(3x^2 + 3xh + h^2)}{h}$

$\qquad\qquad = 3x^2 + 3xh + h^2, h \neq 0$

61. $\dfrac{f(x+h)-f(x)}{h} = \dfrac{\sqrt{x+h}-\sqrt{x}}{h}$

$\qquad\qquad = \dfrac{\sqrt{x+h}-\sqrt{x}}{h} \cdot \dfrac{\sqrt{x+h}+\sqrt{x}}{\sqrt{x+h}+\sqrt{x}}$

$\qquad\qquad = \dfrac{(x+h)-x}{h(\sqrt{x+h}+\sqrt{x})}$

$\qquad\qquad = \dfrac{1}{\sqrt{x+h}+\sqrt{x}}, h \neq 0$

63. $(1+i)^4 = {}_4C_0(1)^4 + {}_4C_1(1)^3i + {}_4C_2(1)^2i^2 + {}_4C_3(1)i^3 + {}_4C_4i^4$

$\qquad\quad = 1 + 4i - 6 - 4i + 1$

$\qquad\quad = -4$

65. $(2-3i)^6 = {}_6C_0 2^6 - {}_6C_1 2^5(3i) + {}_6C_2 2^4(3i)^2 - {}_6C_3 2^3(3i)^3 + {}_6C_4 2^2(3i)^4 - {}_6C_5 2(3i)^5 + {}_6C_6(3i)^6$

$\qquad\qquad = (1)(64) - (6)(32)(3i) + 15(16)(-9) - 20(8)(-27i) + 15(4)(81) - 6(2)(243i) + (1)(-729)$

$\qquad\qquad = 64 - 576i - 2160 + 4320i + 4860 - 2916i - 729$

$\qquad\qquad = 2035 + 828i$

67. $\left(-\dfrac{1}{2} + \dfrac{\sqrt{3}}{2}i\right)^3 = \dfrac{1}{8}\left[(-1)^3 + 3(-1)^2(\sqrt{3}i) + 3(-1)(\sqrt{3}i)^2 + (\sqrt{3}i)^3\right]$

$\qquad\qquad\qquad = \dfrac{1}{8}\left[-1 + 3\sqrt{3}i + 9 - 3\sqrt{3}i\right]$

$\qquad\qquad\qquad = 1$

69. $(1.02)^8 = (1 + 0.02)^8$

$\qquad\quad = 1 + 8(0.02) + 28(0.02)^2 + 56(0.02)^3 + 70(0.02)^4 + 56(0.02)^5 + 28(0.02)^6 + 8(0.02)^7 + (0.02)^8$

$\qquad\quad = 1 + 0.16 + 0.0112 + 0.000448 + \cdots \approx 1.172$

71. $(2.99)^{12} = (3 - 0.01)^{12}$

$\qquad\quad = 3^{12} - 12(3)^{11}(0.01) + 66(3)^{10}(0.01)^2 - 220(3)^9(0.01)^3 + 495(3)^8(0.01)^4$

$\qquad\qquad - 792(3)^7(0.01)^5 + 924(3)^6(0.01)^6 - 792(3)^5(0.01)^7 + 495(3)^4(0.01)^8$

$\qquad\qquad - 220(3)^3(0.01)^9 + 66(3)^2(0.01)^{10} - 12(3)(0.01)^{11} + (0.01)^{12}$

$\qquad\quad \approx 531,441 - 21,257.64 + 389.7234 - 4.3303 + 0.0325 - 0.0002 + \cdots \approx 510,568.785$

73. $f(x) = x^3 - 4x$

$\quad g(x) = f(x + 4)$

$\qquad = (x+4)^3 - 4(x+4)$

$\qquad = x^3 + 3x^2(4) + 3x(4)^2 + (4)^3 - 4x - 16$

$\qquad = x^3 + 12x^2 + 48x + 64 - 4x - 16$

$\qquad = x^3 + 12x^2 + 44x + 48$

The graph of g is the same as the graph of f shifted four units to the left.

75. $_7C_4\left(\dfrac{1}{2}\right)^4\left(\dfrac{1}{2}\right)^3 = \dfrac{7!}{3!4!}\left(\dfrac{1}{16}\right)\left(\dfrac{1}{8}\right) = 35\left(\dfrac{1}{16}\right)\left(\dfrac{1}{8}\right) \approx 0.273$

77. $_8C_4\left(\dfrac{1}{3}\right)^4\left(\dfrac{2}{3}\right)^4 = \dfrac{8!}{4!4!}\left(\dfrac{1}{81}\right)\left(\dfrac{16}{81}\right) = 70\left(\dfrac{1}{81}\right)\left(\dfrac{16}{81}\right) \approx 0.171$

79. (a) $f(t) \approx 0.0025t^3 - 0.015t^2 + 0.88t + 7.7$

(b)

(c) $g(t) = f(t + 10) = 0.0025(t + 10)^3 - 0.015(t + 10)^2$

$\qquad\qquad\qquad\qquad + 0.88(t + 10) + 7.7$

$\qquad\quad = 0.0025(t^3 + 30t^2 + 300t + 1000)$

$\qquad\qquad - 0.015(t^2 + 20t + 100) + 0.88(t + 10) + 7.7$

$\qquad\quad = 0.0025t^3 + 0.06t^2 + 1.33t + 17.5$

(d)

(e) For 2008 use $t = 18$ in $f(t)$ and $t = 8$ in $g(t)$.

$f(18) = 33.26$ gallons

$g(8) = 33.26$ gallons

Both models yield the same answer.

(f) The trend is for the per capita consumption of bottled water to increase. This may be due to the increasing concern with contaminants in tap water.

81. True. The coefficients from the Binomial Theorem can be used to find the numbers in Pascal's Triangle.

83. False.

The coefficient of the x^{10}-term is $_{12}C_7(3)^7 = 1,732,104$.

The coefficient of the x^{14}-term is $_{12}C_5(3)^5 = 192,456$.

85.

```
                        1
                     1     1
                  1     2     1
               1     3     3     1
            1     4     6     4     1
         1     5    10    10     5     1
      1     6    15    20    15     6     1
   1     7    21    35    35    21     7     1
1     8    28    56    70    56    28     8     1
1  9   36   84   126   126   84   36   9   1
1  10  45  120  210  252  210  120  45  10  1
```

87. The signs of the terms in the expansion of $(x - y)^n$ alternate from positive to negative.

89. $_nC_{n-r} = \dfrac{n!}{(n - (n - r))!(n - r)!}$

$\qquad\quad = \dfrac{n!}{r!(n - r)!}$

$\qquad\quad = \dfrac{n!}{(n - r)!r!}$

$\qquad\quad = _nC_r$

91. $_nC_r + {_nC_{r-1}} = \dfrac{n!}{(n - r)!r!} + \dfrac{n!}{(n - r + 1)!(r - 1)!}$

$\qquad\qquad = \dfrac{n!(n - r + 1)!(r - 1)! + n!(n - r)!r!}{(n - r)!r!(n - r + 1)!(r - 1)!}$

$\qquad\qquad = \dfrac{n![(n - r + 1)!(r - 1)! + r!(n - r)!]}{(n - r)!r!(n - r + 1)!(r - 1)!}$

$\qquad\qquad = \dfrac{n!(r - 1)![(n - r + 1)! + r(n - r)!]}{(n - r)!r!(n - r + 1)!(r - 1)!}$

$\qquad\qquad = \dfrac{n!(n - r)![(n - r + 1) + r]}{(n - r)!r!(n - r + 1)!}$

$\qquad\qquad = \dfrac{n![n + 1]}{r!(n - r + 1)!}$

$\qquad\qquad = \dfrac{(n + 1)!}{[(n + 1) - r]!r!}$

$\qquad\qquad = _{n+1}C_r$

93. The graph of $f(x) = x^2$ is shifted three units to the right. Thus, $g(x) = (x - 3)^2$.

95. The graph of $f(x) = \sqrt{x}$ is shifted two units to the left and shifted one unit upward. Thus, $g(x) = \sqrt{x + 2} + 1$.

97. $A^{-1} = \dfrac{1}{(-6)(4) - (5)(-5)} \begin{bmatrix} 4 & -5 \\ 5 & -6 \end{bmatrix} = \begin{bmatrix} 4 & -5 \\ 5 & -6 \end{bmatrix}$

Section 11.6   Counting Principles

- You should know The Fundamental Counting Principle.

- $_nP_r = \dfrac{n!}{(n - r)!}$ is the number of permutations of n elements taken r at a time.

- Given a set of n objects that has n_1 of one kind, n_2 of a second kind, and so on, the number of distinguishable permutations is

$$\frac{n!}{n_1! n_2! \ldots n_k!}.$$

- $_nC_r = \dfrac{n!}{(n - r)! r!}$ is the number of combinations of n elements taken r at a time.

Vocabulary Check

1. Fundamental Counting Principle

2. permutation

3. $_nP_r = \dfrac{n!}{(n - r)!}$

4. distinguishable permutations

5. combinations

1. Odd integers: 1, 3, 5, 7, 9, 11

6 ways

3. Prime integers: 2, 3, 5, 7, 11

5 ways

5. Divisible by 4: 4, 8, 12

3 ways

7. Sum is 9: $1 + 8, 2 + 7, 3 + 6, 4 + 5, 5 + 4,$

$6 + 3, 7 + 2, 8 + 1$

8 ways

9. Amplifiers: 3 choices

Compact disc players: 2 choices

Speakers: 5 choices

Total: $3 \cdot 2 \cdot 5 = 30$ ways

11. Math courses: 2

Science courses: 3

Social sciences and humanities courses: 5

Total: $2 \cdot 3 \cdot 5 = 30$ schedules

13. $2^6 = 64$

15. $26 \cdot 26 \cdot 26 \cdot 10 \cdot 10 \cdot 10 \cdot 10 = 175,760,000$
distinct license plate numbers

17. (a) $9 \cdot 10 \cdot 10 = 900$

(b) $9 \cdot 9 \cdot 8 = 648$

(c) $9 \cdot 10 \cdot 2 = 180$

(d) $6 \cdot 10 \cdot 10 = 600$

19. $40^3 = 64,000$

21. (a) $8 \cdot 7 \cdot 6 \cdot 5 \cdot 4 \cdot 3 \cdot 2 \cdot 1 = 40,320$

(b) $8 \cdot 1 \cdot 6 \cdot 1 \cdot 4 \cdot 1 \cdot 2 \cdot 1 = 384$

23. $_nP_r = \dfrac{n!}{(n-r)!}$

So, $_4P_4 = \dfrac{4!}{0!} = 4! = 24.$

25. $_8P_3 = \dfrac{8!}{5!} = 8 \cdot 7 \cdot 6 = 336$

27. $_5P_4 = \dfrac{5!}{1!} = 120$

29. $14 \cdot {}_nP_3 = {}_{n+2}P_4$ **Note:** $n \geq 3$ for this to be defined.

$$14\left(\frac{n!}{(n-3)!}\right) = \frac{(n+2)!}{(n-2)!}$$

$14n(n-1)(n-2) = (n+2)(n+1)n(n-1)$ (We can divide here by $n(n-1)$ since $n \neq 0, n \neq 1$.)

$14(n-2) = (n+2)(n+1)$

$14n - 28 = n^2 + 3n + 2$

$0 = n^2 - 11n + 30$

$0 = (n-5)(n-6)$

$n = 5$ or $n = 6$

31. $_{20}P_5 = 1,860,480$

33. $_{100}P_3 = 970,200$

35. $_{20}C_5 = 15,504$

37. $5! = 120$ ways

39. $_{12}P_4 = \dfrac{12!}{8!} = 12 \cdot 11 \cdot 10 \cdot 9 = 11,880$ ways

41. $\dfrac{7!}{2!1!3!1!} = \dfrac{7!}{2!3!} = 420$

43. $\dfrac{7!}{2!1!1!1!1!1!} = \dfrac{7!}{2!} = 7 \cdot 6 \cdot 5 \cdot 4 \cdot 3 = 2520$

45.

ABCD	BACD	CABD	DABC
ABDC	BADC	CADB	DACB
ACBD	BCAD	CBAD	DBAC
ACDB	BCDA	CBDA	DBCA
ADBC	BDAC	CDAB	DCAB
ADCB	BDCA	CDBA	DCBA

47. $_{15}P_9 = \dfrac{15!}{6!} = 1,816,214,400$
different batting orders

49. $_{40}C_{12} = \dfrac{40!}{28!12!} = 5,586,853,480$ ways

51. $_6C_2 = 15$

The 15 ways are listed below.

AB, AC, AD, AE, AF, BC, BD, BE,

BF, CD, CE, CF, DE, DF, EF

53. $_{35}C_5 = \dfrac{35!}{30!5!} = 324{,}632$ ways

55. There are 7 good units and 3 defective units.

(a) $_7C_4 = \dfrac{7!}{3!4!} = 35$ ways

(b) $_7C_2 \cdot {_3C_2} = \dfrac{7!}{5!2!} \cdot \dfrac{3!}{1!2!} = 21 \cdot 3 = 63$ ways

(c) $_7C_4 + {_7C_3} \cdot {_3C_1} + {_7C_2} \cdot {_3C_2} = \dfrac{7!}{3!4!} + \dfrac{7!}{4!3!} \cdot \dfrac{3!}{2!1!} + \dfrac{7!}{5!2!} \cdot \dfrac{3!}{1!2!}$

$$= 35 + 35 \cdot 3 + 21 \cdot 3$$

$$= 203 \text{ ways}$$

57. (a) Select type of card for three of a kind: $_{13}C_1$

Select three of four cards for three of a kind: $_4C_3$

Select type of card for pair: $_{12}C_1$

Select two of four cards for pair: $_4C_2$

$$_{13}C_1 \cdot {_4C_3} \cdot {_{12}C_1} \cdot {_4C_2} = \dfrac{13!}{(13-1)!1!} \cdot \dfrac{4!}{(4-3)!3!} \cdot \dfrac{12!}{(12-1)!1!} \cdot \dfrac{4!}{(4-2)!2!} = 3744$$

(b) Select two jacks: $_4C_2$

Select three aces: $_4C_3$

$$_4C_2 \cdot {_4C_3} = \dfrac{4!}{(4-2)!2!} \cdot \dfrac{4!}{(4-3)!3!} = 24$$

59. $_7C_1 \cdot {_{12}C_3} \cdot {_{20}C_2} = \dfrac{7!}{(7-1)!1!} \cdot \dfrac{12!}{(12-3)!3!} \cdot \dfrac{20!}{(20-2)!2!} = 292{,}600$

61. $_5C_2 - 5 = 10 - 5 = 5$ diagonals

63. $_8C_2 - 8 = 28 - 8 = 20$ diagonals

65. (a) $_{53}C_5 \cdot (42) = 120{,}526{,}770$

(b) 1. If the jackpot is won, then there is only one winning number.

(c) There are 22,957,480 possible winning numbers in the state lottery, which is less than the possible number of winning Powerball numbers.

67. False.

It is an example of a combination.

69. $_nC_r = {_nC_{n-r}}$ They are the same.

71. $_nP_{n-1} = \dfrac{n!}{(n-(n-1))!} = \dfrac{n!}{1!} = \dfrac{n!}{0!} = {_nP_n}$

73. $_nC_{n-1} = \dfrac{n!}{(n-(n-1))!(n-1)!} = \dfrac{n!}{(1)!(n-1)!}$

$$= \dfrac{n!}{(n-1)!1!} = {_nC_1}$$

75. $_{100}P_{80} \approx 3.836 \times 10^{139}$

This number is too large for some calculators to evaluate.

77. $f(x) = 3x^2 + 8$

    (a) $f(3) = 3(3)^2 + 8 = 35$

    (b) $f(0) = 3(0)^2 + 8 = 8$

    (c) $f(-5) = 3(-5)^2 + 8 = 83$

79. $f(x) = -|x - 5| + 6$

    (a) $f(-5) = -|-5 - 5| + 6 = -10 + 6 = -4$

    (b) $f(-1) = -|-1 - 5| + 6 = -6 + 6 = 0$

    (c) $f(11) = -|11 - 5| + 6 = -6 + 6 = 0$

81. $\sqrt{x - 3} = x - 6$

$\left(\sqrt{x - 3}\right)^2 = (x - 6)^2$

$x - 3 = x^2 - 12x + 36$

$0 = x^2 - 13x + 39$

By the Quadratic Formula we have: $x = \dfrac{13 \pm \sqrt{13}}{2}$

$x = \dfrac{13 - \sqrt{13}}{2}$ is extraneous.

The only valid solution is $x = \dfrac{13 + \sqrt{13}}{2} \approx 8.30$.

83. $\log_2(x - 3) = 5$

$x - 3 = 2^5$

$x - 3 = 32$

$x = 35$

Section 11.7   Probability

You should know the following basic principles of probability.

■ If an event E has $n(E)$ equally likely outcomes and its sample space has $n(S)$ equally likely outcomes, then the probability of event E is

$$P(E) = \frac{n(E)}{n(S)}, \text{ where } 0 \leq P(E) \leq 1.$$

■ If A and B are mutually exclusive events, then $P(A \cup B) = P(A) + P(B)$.

    If A and B are not mutually exclusive events, then $P(A \cup B) = P(A) + P(B) - P(A \cap B)$.

■ If A and B are independent events, then the probability that both A and B will occur is $P(A)P(B)$.

■ The complement of an event A is denoted by A' and its probability is $P(A') = 1 - P(A)$.

Vocabulary Check

1. experiment; outcomes

2. sample space

3. probability

4. impossible; certain

5. mutually exclusive

6. independent

7. complement

8. (a) iii   (b) i   (c) iv   (d) ii

---

1. $\{(H, 1), (H, 2), (H, 3), (H, 4), (H, 5), (H, 6),$
    $(T, 1), (T, 2), (T, 3), (T, 4), (T, 5), (T, 6)\}$

3. $\{ABC, ACB, BAC, BCA, CAB, CBA\}$

5. $\{AB, AC, AD, AE, BC, BD, BE, CD, CE, DE\}$

7. $E = \{HHT, HTH, THH\}$

$$P(E) = \frac{n(E)}{n(S)} = \frac{3}{8}$$

9. $E = \{HHH, HHT, HTH, HTT, THH, THT, TTH\}$

$$P(E) = \frac{n(E)}{n(S)} = \frac{7}{8}$$

11. $E = \{K\clubsuit, K\diamondsuit, K\heartsuit, K\spadesuit, Q\clubsuit, Q\diamondsuit, Q\heartsuit, Q\spadesuit, J\clubsuit, J\diamondsuit, J\heartsuit, J\spadesuit\}$

$$P(E) = \frac{n(E)}{n(S)} = \frac{12}{52} = \frac{3}{13}$$

13. $E = \{K\diamondsuit, K\heartsuit, Q\diamondsuit, Q\heartsuit, J\diamondsuit, J\heartsuit\}$

$$P(E) = \frac{n(E)}{n(S)} = \frac{6}{52} = \frac{3}{26}$$

15. $E = \{(1, 3), (2, 2), (3, 1)\}$

$$P(E) = \frac{n(E)}{n(S)} = \frac{3}{36} = \frac{1}{12}$$

17. Use the complement.

$$E' = \{(5, 6), (6, 5), (6, 6)\}$$

$$P(E') = \frac{n(E')}{n(S)} = \frac{3}{36} = \frac{1}{12}$$

$$P(E) = 1 - P(E') = 1 - \frac{1}{12} = \frac{11}{12}$$

19. $E_3 = \{(1, 2), (2, 1)\},\ n(E_3) = 2$

$E_5 = \{(1, 4), (2, 3), (3, 2), (4, 1)\},\ n(E_5) = 4$

$E_7 = \{(1, 6), (2, 5), (3, 4), (4, 3), (5, 2), (6, 1)\},\ n(E_7) = 6$

$E = E_3 \cup E_5 \cup E_7$

$n(E) = 2 + 4 + 6 = 12$

$$P(E) = \frac{n(E)}{n(S)} = \frac{12}{36} = \frac{1}{3}$$

21. $P(E) = \dfrac{_3C_2}{_6C_2} = \dfrac{3}{15} = \dfrac{1}{5}$

23. $P(E) = \dfrac{_4C_2}{_6C_2} = \dfrac{6}{15} = \dfrac{2}{5}$

25. $P(E') = 1 - P(E) = 1 - 0.7 = 0.3$

27. $P(E') = 1 - P(E) = 1 - \frac{1}{4} = \frac{3}{4}$

29. $P(E) = 1 - P(E')$
$= 1 - 0.14 = 0.86$

31. $P(E) = 1 - P(E') = 1 - \frac{17}{35} = \frac{18}{35}$

33. (a) $\dfrac{290}{500} = 0.58 = 58\%$

(b) $\dfrac{478}{500} = 0.956 = 95.6\%$

(c) $\dfrac{2}{500} = 0.004 = 0.4\%$

35. (a) $0.24(1011) \approx 243$ adults

(b) $2\% = \dfrac{1}{50}$

(c) $52\% + 12\% = 64\% = \dfrac{16}{25}$

37. (a) $\dfrac{672}{1254} = \dfrac{112}{209}$

(b) $\dfrac{582}{1254} = \dfrac{97}{209}$

(c) $\dfrac{672 - 124}{1254} = \dfrac{548}{1254} = \dfrac{274}{627}$

39. $p + p + 2p = 1$

$p = 0.25$

Taylor: $0.50 = \frac{1}{2}$, Moore: $0.25 = \frac{1}{4}$, Jenkins: $0.25 = \frac{1}{4}$

41. (a) $\dfrac{_{15}C_{10}}{_{20}C_{10}} = \dfrac{3003}{184{,}756} = \dfrac{21}{1292} \approx 0.016$

(b) $\dfrac{_{15}C_8 \cdot {_5}C_2}{_{20}C_{10}} = \dfrac{64{,}350}{184{,}756} = \dfrac{225}{646} \approx 0.348$

(c) $\dfrac{_{15}C_9 \cdot {_5}C_1}{_{20}C_{10}} + \dfrac{_{15}C_{10}}{_{20}C_{10}} = \dfrac{25{,}025 + 3003}{184{,}756} = \dfrac{28{,}028}{184{,}756} = \dfrac{49}{323} \approx 0.152$

43. (a) $\dfrac{1}{_5P_5} = \dfrac{1}{120}$

(b) $\dfrac{1}{_4P_4} = \dfrac{1}{24}$

45. (a) $\dfrac{20}{52} = \dfrac{5}{13}$

(b) $\dfrac{26}{52} = \dfrac{1}{2}$

(c) $\dfrac{16}{52} = \dfrac{4}{13}$

47. (a) $\dfrac{_9C_4}{_{12}C_4} = \dfrac{126}{495} = \dfrac{14}{55}$ \quad (4 good units)

(b) $\dfrac{_9C_2 \cdot _3C_2}{_{12}C_4} = \dfrac{108}{495} = \dfrac{12}{55}$ \quad (2 good units)

(c) $\dfrac{_9C_3 \cdot _3C_1}{_{12}C_4} = \dfrac{252}{495} = \dfrac{28}{55}$ \quad (3 good units)

At least 2 good units: $\dfrac{12}{55} + \dfrac{28}{55} + \dfrac{14}{55} = \dfrac{54}{55}$

49. $(0.78)^3 \approx 0.4746$

51. (a) $P(SS) = (0.985)^2 \approx 0.9702$

(b) $P(S) = 1 - P(FF) = 1 - (0.015)^2 \approx 0.9998$

(c) $P(FF) = (0.015)^2 \approx 0.0002$

53. (a) $P(BBBB) = \left(\frac{1}{2}\right)^4 = \frac{1}{16}$

(b) $P(BBBB) + P(GGGG) = \left(\frac{1}{2}\right)^4 + \left(\frac{1}{2}\right)^4 = \frac{1}{8}$

(c) $P(\text{at least one boy}) = 1 - P(\text{no boys})$
$= 1 - P(GGGG) = 1 - \frac{1}{16} = \frac{15}{16}$

55. $1 - \dfrac{(45)^2}{(60)^2} = 1 - \left(\dfrac{45}{60}\right)^2 = 1 - \left(\dfrac{3}{4}\right)^2 = 1 - \dfrac{9}{16} = \dfrac{7}{16}$

57. True. Two events are independent if the occurance of one has no effect on the occurance of the other.

59. (a) As you consider successive people with distinct birthdays, the probabilities must decrease to take into account the birth dates already used. Because the birth dates of people are independent events, multiply the respective probabilities of distinct birthdays.

(b) $\dfrac{365}{365} \cdot \dfrac{364}{365} \cdot \dfrac{363}{365} \cdot \dfrac{362}{365}$

(c) $P_1 = \dfrac{365}{365} = 1$

$P_2 = \dfrac{365}{365} \cdot \dfrac{364}{365} = \dfrac{364}{365}P_1 = \dfrac{365 - (2-1)}{365}P_1$

$P_3 = \dfrac{365}{365} \cdot \dfrac{364}{365} \cdot \dfrac{363}{365} = \dfrac{363}{365}P_2 = \dfrac{365 - (3-1)}{365}P_2$

$P_n = \dfrac{365}{365} \cdot \dfrac{364}{365} \cdot \dfrac{363}{365} \cdots \dfrac{365 - (n-1)}{365} = \dfrac{365 - (n-1)}{365}P_{n-1}$

(d) Q_n is the probability that the birthdays are not distinct which is equivalent to at least two people having the same birthday.

(e)

n	10	15	20	23	30	40	50
P_n	0.88	0.75	0.59	0.49	0.29	0.11	0.03
Q_n	0.12	0.25	0.41	(0.51)	0.71	0.89	0.97

(f) 23, see the chart above.

61. $6x^2 + 8 = 0$

$6x^2 = -8$

$x^2 = -\dfrac{4}{3}$

No real solution

63. $x^3 - x^2 - 3x = 0$

$x(x^2 - x - 3) = 0$

$x = 0$ or $x^2 - x - 3 = 0$

$x = \dfrac{1 \pm \sqrt{1 - 4(1)(-3)}}{2(1)} = \dfrac{1 \pm \sqrt{13}}{2}$

65. $\dfrac{12}{x} = -3$

$12 = -3x$

$-4 = x$

67. $\dfrac{2}{x-5} = 4$

$2 = 4(x-5)$

$2 = 4x - 20$

$22 = 4x$

$\dfrac{11}{2} = x$

69. $\dfrac{3}{x-2} + \dfrac{x}{x+2} = 1$

$3(x+2) + x(x-2) = 1(x-2)(x+2)$

$3x + 6 + x^2 - 2x = x^2 - 4$

$x^2 + x + 6 = x^2 - 4$

$x + 6 = -4$

$x = -10$

71. $\begin{cases} y \geq -3 \\ x \geq -1 \\ -x - y \geq -8 \end{cases}$

73. $\begin{cases} x^2 + y \geq -2 \\ y \geq x - 4 \end{cases}$

Review Exercises for Chapter 11

1. $a_n = 2 + \dfrac{6}{n}$

$a_1 = 2 + \dfrac{6}{1} = 8$

$a_2 = 2 + \dfrac{6}{2} = 5$

$a_3 = 2 + \dfrac{6}{3} = 4$

$a_4 = 2 + \dfrac{6}{4} = \dfrac{7}{2}$

$a_5 = 2 + \dfrac{6}{5} = \dfrac{16}{5}$

3. $a_n = \dfrac{72}{n!}$

$a_1 = \dfrac{72}{1!} = 72$

$a_2 = \dfrac{72}{2!} = 36$

$a_3 = \dfrac{72}{3!} = 12$

$a_4 = \dfrac{72}{4!} = 3$

$a_5 = \dfrac{72}{5!} = \dfrac{3}{5}$

5. $-2, 2, -2, 2, -2, \ldots$

$a_n = 2(-1)^n$

7. $4, 2, \dfrac{4}{3}, 1, \dfrac{4}{5}, \ldots$

$a_n = \dfrac{4}{n}$

9. $5! = 5 \cdot 4 \cdot 3 \cdot 2 \cdot 1 = 120$

11. $\dfrac{3! \, 5!}{6!} = \dfrac{(3 \cdot 2 \cdot 1)5!}{6 \cdot 5!} = 1$

13. $\displaystyle\sum_{i=1}^{6} 5 = 6(5) = 30$

15. $\displaystyle\sum_{j=1}^{4} \frac{6}{j^2} = \frac{6}{1^2} + \frac{6}{2^2} + \frac{6}{3^2} + \frac{6}{4^2} = 6 + \frac{3}{2} + \frac{2}{3} + \frac{3}{8} = \frac{205}{24}$

17. $\displaystyle\sum_{k=1}^{10} 2k^3 = 2(1)^3 + 2(2)^3 + 2(3)^3 + \cdots + 2(10)^3 = 6050$

19. $\displaystyle\frac{1}{2(1)} + \frac{1}{2(2)} + \frac{1}{2(3)} + \cdots + \frac{1}{2(20)} = \sum_{k=1}^{20} \frac{1}{2k}$

21. $\displaystyle\sum_{i=1}^{\infty} \frac{5}{10^i} = 0.5 + 0.05 + 0.005 + 0.0005 + \cdots = 0.5555\cdots = \frac{5}{9}$

23. $\displaystyle\sum_{k=1}^{\infty} \frac{2}{100^k} = 0.02 + 0.0002 + 0.000002 + \cdots = 0.020202\cdots = \frac{2}{99}$

25. $A_n = 10{,}000\left(1 + \dfrac{0.08}{12}\right)^n$

    (a) $A_1 \approx \$10{,}066.67$ $A_6 \approx \$10{,}406.73$

        $A_2 \approx \$10{,}133.78$ $A_7 \approx \$10{,}476.10$

        $A_3 \approx \$10{,}201.34$ $A_8 \approx \$10{,}545.95$

        $A_4 \approx \$10{,}269.35$ $A_9 \approx \$10{,}616.25$

        $A_5 \approx \$10{,}337.81$ $A_{10} \approx \$10{,}687.03$

    (b) $A_{120} \approx \$22{,}196.40$

27. $5, 3, 1, -1, -3, \ldots$

    Arithmetic sequence, $d = -2$

29. $\frac{1}{2}, 1, \frac{3}{2}, 2, \frac{5}{2}, \ldots$

    Arithmetic sequence, $d = \frac{1}{2}$

31. $a_1 = 4, \; d = 3$

    $a_1 = 4$

    $a_2 = 4 + 3 = 7$

    $a_3 = 7 + 3 = 10$

    $a_4 = 10 + 3 = 13$

    $a_5 = 13 + 3 = 16$

33. $a_1 = 25, \; a_{k+1} = a_k + 3$

    $a_1 = 25$

    $a_2 = 25 + 3 = 28$

    $a_3 = 28 + 3 = 31$

    $a_4 = 31 + 3 = 34$

    $a_5 = 34 + 3 = 37$

35. $a_1 = 7, \; d = 12$

    $a_n = 7 + (n-1)12$

    $ = 7 + 12n - 12$

    $ = 12n - 5$

37. $a_1 = y, \; d = 3y$

    $a_n = y + (n-1)3y$

    $ = y + 3ny - 3y$

    $ = 3ny - 2y$

39. $a_2 = 93, \; a_6 = 65$

    $a_6 = a_2 + 4d \implies 65 = 93 + 4d \implies -28 = 4d \implies d = -7$

    $a_1 = a_2 - d \implies a_1 = 93 - (-7) = 100$

    $a_n = a_1 + (n-1)d = 100 + (n-1)(-7) = -7n + 107$

41. $\displaystyle\sum_{j=1}^{10} (2j - 3)$ is arithmetic. Therefore, $a_1 = -1, \; a_{10} = 17, \; S_{10} = \frac{10}{2}[-1 + 17] = 80.$

43. $\displaystyle\sum_{k=1}^{11} \left(\frac{2}{3}k + 4\right)$ is arithmetic. Therefore, $a_1 = \frac{14}{3}, \; a_{11} = \frac{34}{3}, \; S_{11} = \frac{11}{2}\left[\frac{14}{3} + \frac{34}{3}\right] = 88.$

45. $\sum_{k=1}^{100} 5k$ is arithmetic. Therefore, $a_1 = 5$, $a_{100} = 500$, $S_{500} = \frac{100}{2}(5 + 500) = 25,250$.

47. $a_n = 34,000 + (n - 1)(2250)$

    (a) $a_5 = 34,000 + 4(2250) = \$43,000$

    (b) $S_5 = \frac{5}{2}(34,000 + 43,000) = \$192,500$

49. $5, 10, 20, 40, \ldots$

    The sequence *is* geometric, $r = 2$

51. $\frac{1}{3}, -\frac{2}{3}, \frac{4}{3}, -\frac{8}{3}, \ldots$

    The sequence *is* geometric, $r = -2$

53. $a_1 = 4$, $r = -\frac{1}{4}$

    $a_1 = 4$

    $a_2 = 4\left(-\frac{1}{4}\right) = -1$

    $a_3 = -1\left(-\frac{1}{4}\right) = \frac{1}{4}$

    $a_4 = \frac{1}{4}\left(-\frac{1}{4}\right) = -\frac{1}{16}$

    $a_5 = -\frac{1}{16}\left(-\frac{1}{4}\right) = \frac{1}{64}$

55. $a_1 = 9$, $a_3 = 4$

    $a_3 = a_1 r^2$

    $4 = 9r^2$

    $\frac{4}{9} = r^2 \implies r = \pm\frac{2}{3}$

    $a_1 = 9 \qquad\qquad a_1 = 9$

    $a_2 = 9\left(\frac{2}{3}\right) = 6 \qquad a_2 = 9\left(-\frac{2}{3}\right) = -6$

    $a_3 = 6\left(\frac{2}{3}\right) = 4 \quad$ or $\quad a_3 = -6\left(-\frac{2}{3}\right) = 4$

    $a_4 = 4\left(\frac{2}{3}\right) = \frac{8}{3} \qquad a_4 = 4\left(-\frac{2}{3}\right) = -\frac{8}{3}$

    $a_5 = \frac{8}{3}\left(\frac{2}{3}\right) = \frac{16}{9} \qquad a_5 = -\frac{8}{3}\left(-\frac{2}{3}\right) = \frac{16}{9}$

57. $a_1 = 16$, $a_2 = -8$

    $a_2 = a_1 r \implies -8 = 16r \implies r = -\frac{1}{2}$

    $a_n = 16\left(-\frac{1}{2}\right)^{n-1}$

    $a_{20} = 16\left(-\frac{1}{2}\right)^{19} \approx -3.052 \times 10^{-5}$

59. $a_1 = 100$, $r = 1.05$

    $a_n = 100(1.05)^{n-1}$

    $a_{20} = 100(1.05)^{19} \approx 252.695$

61. $\sum_{i=1}^{7} 2^{i-1} = \frac{1 - 2^7}{1 - 2} = 127$

63. $\sum_{i=1}^{4} \left(\frac{1}{2}\right)^i = \frac{1}{2} + \frac{1}{4} + \frac{1}{8} + \frac{1}{16} = \frac{15}{16}$

65. $\sum_{i=1}^{5} (2)^{i-1} = 1 + 2 + 4 + 8 + 16 = 31$

67. $\sum_{i=1}^{10} 10\left(\frac{3}{5}\right)^{i-1} \approx 24.85$

69. $\sum_{i=1}^{25} 100(1.06)^{i-1} \approx 5486.45$

71. $\sum_{i=1}^{\infty} \left(\frac{7}{8}\right)^{i-1} = \frac{1}{1 - \frac{7}{8}} = 8$

73. $\sum_{i=1}^{\infty} (0.1)^{i-1} = \frac{1}{1 - 0.1} = \frac{10}{9}$

75. $\sum_{k=1}^{\infty} 4\left(\frac{2}{3}\right)^{k-1} = \frac{4}{1 - \frac{2}{3}} = 12$

77. (a) $a_t = 120,000(0.7)^t$

    (b) $a_5 = 120,000(0.7)^5$

            $= \$20,168.40$

79. 1. When $n = 1, 3 = 1(1 + 2)$.

    2. Assume that $S_k = 3 + 5 + 7 + \cdots + (2k + 1) = k(k + 2)$.

       Then, $S_{k+1} = 3 + 5 + 7 + \cdots + (2k + 1) + [2(k + 1) + 1] = S_k + (2k + 3)$

$$= k(k + 2) + 2k + 3$$
$$= k^2 + 4k + 3$$
$$= (k + 1)(k + 3)$$
$$= (k + 1)[(k + 1) + 2].$$

Therefore, by mathematical induction, the formula is valid for all positive integer values of n.

81. 1. When $n = 1, a = a\left(\dfrac{1 - r}{1 - r}\right)$.

    2. Assume that $S_k = \displaystyle\sum_{i=0}^{k-1} ar^i = \dfrac{a(1 - r^k)}{1 - r}$.

       Then, $S_{k+1} = \displaystyle\sum_{i=0}^{k} ar^i = \left(\sum_{i=0}^{k-1} ar^i\right) + ar^k = \dfrac{a(1 - r^k)}{1 - r} + ar^k$

$$= \dfrac{a(1 - r^k + r^k - r^{k+1})}{1 - r} = \dfrac{a(1 - r^{k+1})}{1 - r}.$$

Therefore, by mathematical induction, the formula is valid for all positive integer values of n.

83. $S_1 = 9 = 1(9) = 1[2(1) + 7]$

$S_2 = 9 + 13 = 22 = 2(11) = 2[2(2) + 7]$

$S_3 = 9 + 13 + 17 = 39 = 3(13) = 3[2(3) + 7]$

$S_4 = 9 + 13 + 17 + 21 = 60 = 4(15) = 4[2(4) + 7]$

$S_n = n(2n + 7)$

85. $S_1 = 1$

$S_2 = 1 + \dfrac{3}{5} = \dfrac{8}{5}$

$S_3 = 1 + \dfrac{3}{5} + \dfrac{9}{25} = \dfrac{49}{25}$

$S_4 = 1 + \dfrac{3}{5} + \dfrac{9}{25} + \dfrac{27}{125} = \dfrac{272}{125}$

Since the series is geometric,

$$S_n = \dfrac{1 - \left(\frac{3}{5}\right)^n}{1 - \frac{3}{5}} = \dfrac{5}{2}\left[1 - \left(\dfrac{3}{5}\right)^n\right].$$

87. $\displaystyle\sum_{n=1}^{30} n = \dfrac{30(31)}{2} = 465$

89. $\displaystyle\sum_{n=1}^{7} (n^4 - n) = \sum_{n=1}^{7} n^4 - \sum_{n=1}^{7} n = \dfrac{(7)(8)(15)[(3)(49) + 21 - 1]}{30} - \dfrac{(7)(8)}{2}$

$$= \dfrac{(7)(8)(15)(167)}{30} - \dfrac{(7)(8)}{2}$$

$$= 4676 - 28 = 4648$$

91. $a_1 = f(1) = 5,\quad a_n = a_{n-1} + 5$

$a_1 = 5$

$a_2 = 5 + 5 = 10$

$a_3 = 10 + 5 = 15$

$a_4 = 15 + 5 = 20$

$a_5 = 20 + 5 = 25$

n:	1	2	3	4	5
a_n:	5	10	15	20	25

First differences: 5   5   5   5

Second differences: 0   0   0

The sequence has a linear model.

93. $a_1 = f(1) = 16,\quad a_n = a_{n-1} - 1$

$a_1 = 16$

$a_2 = 16 - 1 = 15$

$a_3 = 15 - 1 = 14$

$a_4 = 14 - 1 = 13$

$a_5 = 13 - 1 = 12$

n:	1	2	3	4	5
a_n:	16	15	14	13	12

First differences: -1 -1 -1 -1

Second differences: 0   0   0

The sequence has a linear model.

95. $_6C_4 = \dfrac{6!}{2!4!} = 15$

97. $_8C_5 = \dfrac{8!}{3!5!} = 56$

99. $\dbinom{7}{3} = 35$

```
            1
          1   1
        1   2   1
      1   3   3   1
    1   4   6   4   1
  1   5   10  10  5   1
1   6   15  20  15  6   1
1   7   21  35 (35) 21  7   1
```

$\dbinom{7}{3} = 35$, the 5th entry in the 7th row

101. $\dbinom{8}{6} = 28$

```
              1
            1   1
          1   2   1
        1   3   3   1
      1   4   6   4   1
    1   5   10  10  5   1
  1   6   15  20  15  6   1
1   7   21  35  35  21  7   1
1  8  28  56  70  56 (28) 8   1
```

$\dbinom{8}{6} = 28$, the 7th entry in the 8th row

103. $(x + 4)^4 = x^4 + 4x^3(4) + 6x^2(4)^2 + 4x(4)^3 + 4^4$

$\qquad\qquad = x^4 + 16x^3 + 96x^2 + 256x + 256$

105. $(a - 3b)^5 = a^5 - 5a^4(3b) + 10a^3(3b)^2 - 10a^2(3b)^3 + 5a(3b)^4 - (3b)^5$

$\qquad\qquad\quad = a^5 - 15a^4b + 90a^3b^2 - 270a^2b^3 + 405ab^4 - 243b^5$

107. $(5 + 2i)^4 = (5)^4 + 4(5)^3(2i) + 6(5)^2(2i)^2 + 4(5)(2i)^3 + (2i)^4$

$\qquad\qquad\quad = 625 + 1000i + 600i^2 + 160i^3 + 16i^4$

$\qquad\qquad\quad = 625 + 1000i - 600 - 160i + 16 = 41 + 840i$

109.

First number:	1	2	3	4	5	6	7	8	9	10	11
Second number:	11	10	9	8	7	6	5	4	3	2	1

From this list, you can see that a total of 12 occurs 11 different ways.

111. $(10)(10)(10)(10) = 10{,}000$ different telephone numbers

113. $_{10}P_3 = \dfrac{10!}{7!} = \dfrac{10 \cdot 9 \cdot 8 \cdot 7!}{7!}$

$\qquad\qquad = 10 \cdot 9 \cdot 8 = 720$ different ways

115. $_8C_3 = \dfrac{8!}{5!3!} = 56$

117. $(1)\left(\dfrac{1}{9}\right) = \dfrac{1}{9}$

119. (a) $25\% + 18\% = 43\%$

(b) $100\% - 18\% = 82\%$

121. $\left(\dfrac{1}{6}\right)\left(\dfrac{1}{6}\right)\left(\dfrac{1}{6}\right) = \dfrac{1}{216}$

123. $1 - \dfrac{13}{52} = 1 - \dfrac{1}{4} = \dfrac{3}{4}$

125. True. $\dfrac{(n+2)!}{n!} = \dfrac{(n+2)(n+1)n!}{n!} = (n+2)(n+1)$

127. True. $\displaystyle\sum_{k=1}^{8} 3k = 3\sum_{k=1}^{8} k$ by the Properties of Sums.

129. False. If $r = 0$ or $r = 1$, then $_nP_r = {}_nC_r$.

131. (a) Odd-numbered terms are negative.

(b) Even-numbered terms are negative.

133. Each term of the sequence is defined in terms of preceding terms.

135. $a_n = 4\left(\dfrac{1}{2}\right)^{n-1}$

$a_1 = 4, a_2 = 2, a_{10} = \dfrac{1}{128}$

The sequence is geometric and is decreasing.

Matches graph (d).

137. $a_n = \displaystyle\sum_{k=1}^{n} 4\left(\dfrac{1}{2}\right)^{k-1}$

$a_1 = 4$ and $a_n \rightarrow 8$ as $n \rightarrow \infty$

Matches graph (b).

139. $S_6 = S_5 + S_4 + S_3 = 130 + 70 + 40 = 240$

$S_7 = S_6 + S_5 + S_4 = 240 + 130 + 70 = 440$

$S_8 = S_7 + S_6 + S_5 = 440 + 240 + 130 = 810$

$S_9 = S_8 + S_7 + S_6 = 810 + 440 + 240 = 1490$

$S_{10} = S_9 + S_8 + S_7 = 1490 + 810 + 440 = 2740$

Problem Solving for Chapter 11

1. $x_0 = 1$ and $x_n = \dfrac{1}{2}x_{n-1} + \dfrac{1}{x_{n-1}}, n = 1, 2, \ldots$

$x_0 = 1$

$x_1 = \dfrac{1}{2}(1) + \dfrac{1}{1} = \dfrac{3}{2} = 1.5$

$x_2 = \dfrac{1}{2}\left(\dfrac{3}{2}\right) + \dfrac{1}{3/2} = \dfrac{17}{12} = 1.41\overline{6}$

$x_3 = \dfrac{1}{2}\left(\dfrac{17}{12}\right) + \dfrac{1}{17/12} = \dfrac{577}{408} \approx 1.414215686$

$x_4 = \dfrac{1}{2}\left(\dfrac{577}{408}\right) + \dfrac{1}{577/408} \approx 1.414213562$

$x_5 = \dfrac{1}{2}x_4 + \dfrac{1}{x_4} \approx 1.414213562$

$x_6 \approx x_7 \approx x_8 \approx x_9 \approx 1.414213562$

Conjecture: $x_n \rightarrow \sqrt{2}$ as $n \rightarrow \infty$

3. $a_n = 3 + (-1)^n$

(a)

(b) $a_n = \begin{cases} 2, & \text{if } n \text{ is odd} \\ 4, & \text{if } n \text{ is even} \end{cases}$

(c)

n	1	10	101	1000	10,001
a_n	2	4	2	4	2

(d) As $n \rightarrow \infty$, a_n oscillates between 2 and 4 and does not approach a fixed value.

5. (a)

$$1 \quad 4 \quad 9 \quad 16 \quad 25 \quad 36 \quad 49 \quad 64 \quad 81$$

First differences: $\quad 3 \quad 5 \quad 7 \quad 9 \quad 11 \quad 13 \quad 15 \quad 17$

In general, $b_n = 2n + 1$ for the first differences.

(b) Find the second differences of the perfect cubes.

(c)

$$1 \quad 8 \quad 27 \quad 64 \quad 125 \quad 216 \quad 343 \quad 512 \quad 729$$

First differences: $\quad 7 \quad 19 \quad 37 \quad 61 \quad 91 \quad 127 \quad 169 \quad 217$

Second differences: $\quad 12 \quad 18 \quad 24 \quad 30 \quad 36 \quad 42 \quad 48$

In general, $c_n = 6(n + 1) = 6n + 6$ for the second differences.

(d) Find the third differences of the perfect fourth powers.

(e)

$$1 \quad 16 \quad 81 \quad 256 \quad 625 \quad 1296 \quad 2401 \quad 4096 \quad 6561$$

First differences: $\quad 15 \quad 65 \quad 175 \quad 369 \quad 671 \quad 1105 \quad 1695 \quad 2465$

Second differences: $\quad 50 \quad 110 \quad 194 \quad 302 \quad 434 \quad 590 \quad 770$

Third differences: $\quad 60 \quad 84 \quad 108 \quad 132 \quad 156 \quad 180$

In general, $d_n = 24n + 36$ for the third differences.

7. Side lengths: $1, \dfrac{1}{2}, \dfrac{1}{4}, \dfrac{1}{8}, \ldots$

$$S_n = \left(\frac{1}{2}\right)^{n-1} \text{ for } n \geq 1$$

Areas: $\dfrac{\sqrt{3}}{4}, \dfrac{\sqrt{3}}{4}\left(\dfrac{1}{2}\right)^2, \dfrac{\sqrt{3}}{4}\left(\dfrac{1}{4}\right)^2, \dfrac{\sqrt{3}}{4}\left(\dfrac{1}{8}\right)^2, \ldots$

$$A_n = \frac{\sqrt{3}}{4}\left[\left(\frac{1}{2}\right)^{n-1}\right]^2 = \frac{\sqrt{3}}{4}\left(\frac{1}{2}\right)^{2n-2} = \frac{\sqrt{3}}{4}S_n^{\,2}$$

9. The numbers $1, 5, 12, 22, 35, 51, \ldots$ can be written recursively as $P_n = P_{n-1} + (3n - 2)$. Show that $P_n = n(3n - 1)/2$.

1. For $n = 1$: $1 = \dfrac{1(3 - 1)}{2}$

2. Assume $P_k = \dfrac{k(3k - 1)}{2}$.

Then, $P_{k+1} = P_k + [3(k + 1) - 2]$

$$= \frac{k(3k - 1)}{2} + (3k + 1) = \frac{k(3k - 1) + 2(3k + 1)}{2}$$

$$= \frac{3k^2 + 5k + 2}{2} = \frac{(k + 1)(3k + 2)}{2}$$

$$= \frac{(k + 1)[3(k + 1) - 1]}{2}.$$

Therefore, by mathematical induction, the formula is valid for all integers $n \geq 1$.

11. (a) The Fibonacci sequence is defined as follows: $f_1 = 1, f_2 = 1, f_n = f_{n-2} + f_{n-1}$ for $n \geq 3$.

By this definition $f_3 = f_1 + f_2 = 2, f_4 = f_2 + f_3 = 3, f_5 = f_4 + f_3 = 5, f_6 = f_5 + f_4 = 8, \ldots$

1. For $n = 2$: $f_1 + f_2 = 2$ and $f_4 - 1 = 2$

2. Assume $f_1 + f_2 + \ldots + f_k = f_{k+2} - 1$.

Then, $f_1 + f_2 + f_3 + \ldots + f_k + f_{k+1} = f_{k+2} - 1 + f_{k+1} = (f_{k+2} + f_{k+1}) - 1 = f_{k+3} - 1 = f_{(k+1)+2} - 1.$

Therefore, by mathematical induction, the formula is valid for all integers $n \geq 2$.

(b) $S_{20} = f_{22} - 1 = 17,711 - 1 = 17,710$

13. $\dfrac{1}{3}$

15. (a) $V = \left(\dfrac{1}{{}_{47}C_5(27)}\right)(12,000,000) + \left(1 - \dfrac{1}{{}_{47}C_5(27)}\right)(-1)$

$\approx -\$0.71$

(b) $V = \dfrac{1}{36}(1) + \dfrac{1}{36}(4) + \dfrac{1}{36}(9) + \dfrac{1}{36}(16) + \dfrac{1}{36}(25) + \dfrac{1}{36}(36) + \dfrac{30}{36}(0) \approx 2.53$

$\dfrac{60}{2.53} \approx 24$ turns

Chapter 11   Practice Test

1. Write out the first five terms of the sequence $a_n = \dfrac{2n}{(n+2)!}$.

2. Write an expression for the nth term of the sequence $\frac{4}{3}, \frac{5}{9}, \frac{6}{27}, \frac{7}{81}, \frac{8}{243}, \ldots$

3. Find the sum $\displaystyle\sum_{i=1}^{6}(2i-1)$.

4. Write out the first five terms of the arithmetic sequence where $a_1 = 23$ and $d = -2$.

5. Find a_n for the arithmetic sequence with $a_1 = 12$, $d = 3$, and $n = 50$.

6. Find the sum of the first 200 positive integers.

7. Write out the first five terms of the geometric sequence with $a_1 = 7$ and $r = 2$.

8. Evaluate $\displaystyle\sum_{n=1}^{10} 6\left(\frac{2}{3}\right)^{n-1}$.

9. Evaluate $\displaystyle\sum_{n=0}^{\infty}(0.03)^n$.

10. Use mathematical induction to prove that $1 + 2 + 3 + 4 + \cdots + n = \dfrac{n(n+1)}{2}$.

11. Use mathematical induction to prove that $n! > 2^n$, $n \geq 4$.

12. Evaluate $_{13}C_4$.

13. Expand $(x+3)^5$.

14. Find the term involving x^7 in $(x-2)^{12}$.

15. Evaluate $_{30}P_4$.

16. How many ways can six people sit at a table with six chairs?

17. Twelve cars run in a race. How many different ways can they come in first, second, and third place? (Assume that there are no ties.)

18. Two six-sided dice are tossed. Find the probability that the total of the two dice is less than 5.

19. Two cards are selected at random form a deck of 52 playing cards without replacement. Find the probability that the first card is a King and the second card is a black ten.

20. A manufacturer has determined that for every 1000 units it produces, 3 will be faulty. What is the probability that an order of 50 units will have one or more faulty units?

Chapter P   Practice Test Solutions

1. $\dfrac{|-42| - 20}{15 - |-4|} = \dfrac{42 - 20}{15 - 4} = \dfrac{22}{11} = 2$

2. $\dfrac{x}{z} - \dfrac{z}{y} = \dfrac{x}{z} \cdot \dfrac{y}{y} - \dfrac{z}{y} \cdot \dfrac{z}{z} = \dfrac{xy - z^2}{yz}$

3. $|x - 7| \le 4$

4. $10(-5)^3 = 10(-125) = -1250$

5. $(-4x^3)(-2x^{-5})\left(\dfrac{1}{16}x\right) = (-4)(-2)\left(\dfrac{1}{16}\right)x^{3+(-5)+1} = \dfrac{8}{16}x^{-1} = \dfrac{1}{2x}$

6. $0.0000412 = 4.12 \times 10^{-5}$

7. $125^{2/3} = \left(\sqrt[3]{125}\right)^2 = (5)^2 = 25$

8. $\sqrt[4]{64x^7y^9} = \sqrt[4]{16 \cdot 4x^4x^3y^8y} = 2xy^2\sqrt[4]{4x^3y}$

9. $\dfrac{6}{\sqrt{12}} = \dfrac{6}{2\sqrt{3}} \cdot \dfrac{\sqrt{3}}{\sqrt{3}} = \dfrac{6\sqrt{3}}{6} = \sqrt{3}$

10. $3\sqrt{80} - 7\sqrt{500} = 3\left(4\sqrt{5}\right) - 7\left(10\sqrt{5}\right) = 12\sqrt{5} - 70\sqrt{5} = -58\sqrt{5}$

11. $(8x^4 - 9x^2 + 2x - 1) - (3x^3 + 5x + 4) = 8x^4 - 3x^3 - 9x^2 - 3x - 5$

12. $(x - 3)(x^2 + x - 7) = x^3 + x^2 - 7x - 3x^2 - 3x + 21 = x^3 - 2x^2 - 10x + 21$

13. $[(x - 2) - y]^2 = (x - 2)^2 - 2y(x - 2) + y^2$
$$= x^2 - 4x + 4 - 2xy + 4y + y^2 = x^2 + y^2 - 2xy - 4x + 4y + 4$$

14. $16x^4 - 1 = (4x^2 + 1)(4x^2 - 1) = (4x^2 + 1)(2x + 1)(2x - 1)$

15. $6x^2 + 5x - 4 = (2x - 1)(3x + 4)$

16. $x^3 - 64 = x^3 - 4^3 = (x - 4)(x^2 + 4x + 16)$

17. $-\dfrac{3}{x} + \dfrac{x}{x^2 + 2} = \dfrac{-3(x^2 + 2) + x^2}{x(x^2 + 2)} = \dfrac{-2x^2 - 6}{x(x^2 + 2)} = -\dfrac{2(x^2 + 3)}{x(x^2 + 2)}$

18. $\dfrac{x - 3}{4x} \div \dfrac{x^2 - 9}{x^2} = \dfrac{x - 3}{4x} \cdot \dfrac{x^2}{(x + 3)(x - 3)} = \dfrac{x}{4(x + 3)},\ x \ne 0, 3$

19. $\dfrac{1 - \dfrac{1}{x}}{1 - \dfrac{1}{1 - (1/x)}} = \dfrac{\dfrac{x - 1}{x}}{1 - \dfrac{1}{(x - 1)/x}} = \dfrac{\dfrac{x - 1}{x}}{1 - \dfrac{x}{x - 1}} = \dfrac{\dfrac{x - 1}{x}}{\dfrac{-1}{x - 1}} = \dfrac{x - 1}{x} \cdot \dfrac{x - 1}{-1} = \dfrac{-(x - 1)^2}{x},\ x \ne 1$

20. (a)

(b) $d = \sqrt{[5 - (-3)]^2 + (-1 - 7)^2}$
$$= \sqrt{(8)^2 + (-8)^2}$$
$$= \sqrt{64 + 64}$$
$$= \sqrt{128}$$
$$= 8\sqrt{2}$$

(c) $\left(\dfrac{-3 + 5}{2}, \dfrac{7 + (-1)}{2}\right)$
$$= (1, 3)$$

Chapter 1   Practice Test Solutions

1. $3x - 5y = 15$

Line

x-intercept: $(5, 0)$

y-intercept: $(0, -3)$

2. $y = \sqrt{9 - x}$

Domain: $(-\infty, 9]$

x-intercept: $(9, 0)$

y-intercept: $(0, 3)$

3. $5x + 4 = 7x - 8$

$4 + 8 = 7x - 5x$

$12 = 2x$

$x = 6$

4. $\dfrac{x}{3} - 5 = \dfrac{x}{5} + 1$

$15\left(\dfrac{x}{3} - 5\right) = 15\left(\dfrac{x}{5} + 1\right)$

$5x - 75 = 3x + 15$

$2x = 90$

$x = 45$

5. $\dfrac{3x + 1}{6x - 7} = \dfrac{2}{5}$

$5(3x + 1) = 2(6x - 7)$

$15x + 5 = 12x - 14$

$3x = -19$

$x = -\dfrac{19}{3}$

6. $(x - 3)^2 + 4 = (x + 1)^2$

$x^2 - 6x + 9 + 4 = x^2 + 2x + 1$

$-8x = -12$

$x = \dfrac{-12}{-8}$

$x = \dfrac{3}{2}$

7. $A = \dfrac{1}{2}(a + b)h$

$2A = ah + bh$

$2A - bh = ah$

$\dfrac{2A - bh}{h} = a$

8. Percent $= \dfrac{301}{4300} = 0.07 = 7\%$

9. Let x = number of quarters.

Then $53 - x$ = number of nickels.

$25x + 5(53 - x) = 605$

$20x + 265 = 605$

$20x = 340$

$x = 17$ quarters

$53 - x = 36$ nickels

10. Let x = amount in $9\frac{1}{2}\%$ fund.

Then $15,000 - x$ = amount in 11% fund.

$0.095x + 0.11(15,000 - x) = 1582.50$

$-0.015x + 1650 = 1582.50$

$-0.015x = -67.5$

$x = \$4500$ at $9\frac{1}{2}\%$

$15,000 - x = \$10,500$ at 11%

11. $28 + 5x - 3x^2 = 0$

$(4 - x)(7 + 3x) = 0$

$4 - x = 0 \Rightarrow x = 4$

$7 + 3x = 0 \Rightarrow x = -\dfrac{7}{3}$

12. $(x - 2)^2 = 24$

$x - 2 = \pm\sqrt{24}$

$x - 2 = \pm 2\sqrt{6}$

$x = 2 \pm 2\sqrt{6}$

13. $x^2 - 4x - 9 = 0$

$x^2 - 4x + 2^2 = 9 + 2^2$

$(x - 2)^2 = 13$

$x - 2 = \pm\sqrt{13}$

$x = 2 \pm \sqrt{13}$

14. $x^2 + 5x - 1 = 0$

$a = 1, \ b = 5, \ c = -1$

$$x = \frac{-5 \pm \sqrt{(5)^2 - 4(1)(-1)}}{2(1)}$$

$$= \frac{-5 \pm \sqrt{25 + 4}}{2}$$

$$= \frac{-5 \pm \sqrt{29}}{2}$$

15. $3x^2 - 2x + 4 = 0$

$a = 3, \ b = -2, \ c = 4$

$$x = \frac{-(-2) \pm \sqrt{(-2)^2 - 4(3)(4)}}{2(3)}$$

$$= \frac{2 \pm \sqrt{4 - 48}}{6}$$

$$= \frac{2 \pm \sqrt{-44}}{6}$$

$$= \frac{2 \pm 2i\sqrt{11}}{6}$$

$$= \frac{1 \pm i\sqrt{11}}{3} = \frac{1}{3} \pm \frac{\sqrt{11}}{3}i$$

16.

$$60,000 = xy$$

$$y = \frac{60,000}{x}$$

$$2x + 2y = 1100$$

$$2x + 2\left(\frac{60,000}{x}\right) = 1100$$

$$x + \frac{60,000}{x} = 550$$

$$x^2 + 60,000 = 550x$$

$$x^2 - 550x + 60,000 = 0$$

$$(x - 150)(x - 400) = 0$$

$$x = 150 \ \text{ or } \ x = 400$$

$$y = 400 \qquad y = 150$$

Length: 400 feet; width: 150 feet

17.

$$x(x + 2) = 624$$

$$x^2 + 2x - 624 = 0$$

$$(x - 24)(x + 26) = 0$$

$$x = 24 \ \text{ or } \ x = -26, \text{ (extraneous solution)}$$

$$x + 2 = 26$$

The integers are 24 and 26.

18. $x^3 - 10x^2 + 24x = 0$

$$x(x^2 - 10x + 24) = 0$$

$$x(x - 4)(x - 6) = 0$$

$$x = 0, \ x = 4, \ x = 6$$

19. $\sqrt[3]{6 - x} = 4$

$$6 - x = 64$$

$$-x = 58$$

$$x = -58$$

20. $(x^2 - 8)^{2/5} = 4$

$$x^2 - 8 = \pm 4^{5/2}$$

$$x^2 - 8 = 32 \ \text{ or } \ x^2 - 8 = -32$$

$$x^2 = 40 \qquad\qquad x^2 = -24$$

$$x = \pm\sqrt{40} \qquad\quad x = \pm\sqrt{-24}$$

$$x = \pm 2\sqrt{10} \qquad\quad x = \pm 2\sqrt{6}i$$

21. $x^4 - x^2 - 12 = 0$

$(x^2 - 4)(x^2 + 3) = 0$

$x^2 = 4$ or $x^2 = -3$

$x^2 = \pm 2$ $x = \pm\sqrt{3}i$

22. $4 - 3x > 16$

$-3x > 12$

$x < -4$

23. $\left|\dfrac{x - 3}{2}\right| < 5$

$-5 < \dfrac{x - 3}{2} < 5$

$-10 < x - 3 < 10$

$-7 < x < 13$

24. $\dfrac{x + 1}{x - 3} < 2$

$\dfrac{x + 1}{x - 3} - 2 < 0$

$\dfrac{x + 1 - 2(x - 3)}{x - 3} < 0$

$\dfrac{7 - x}{x - 3} < 0$

Critical numbers: $x = 7$ and $x = 3$

Test intervals: $(-\infty, 3), (3, 7), (7, \infty)$

Test: Is $\dfrac{7 - x}{x - 3} < 0$?

Solution intervals: $(-\infty, 3) \cup (7, \infty)$

25. $|3x - 4| \geq 9$

$3x - 4 \leq -9$ or $3x - 4 \geq 9$

$3x \leq -5$ $3x \geq 13$

$x \leq -\dfrac{5}{3}$ $x \geq \dfrac{13}{3}$

Chapter 2   Practice Test Solutions

1. $m = \dfrac{-1 - 4}{3 - 2} = -5$

$y - 4 = -5(x - 2)$

$y - 4 = -5x + 10$

$y = -5x + 14$

2. $y = \dfrac{4}{3}x - 3$

3. $2x + 3y = 0$

$y = -\dfrac{2}{3}x$

$m_1 = -\dfrac{2}{3}$

$\perp m_2 = \dfrac{3}{2}$ through $(4, 1)$

$y - 1 = \dfrac{3}{2}(x - 4)$

$y - 1 = \dfrac{3}{2}x - 6$

$y = \dfrac{3}{2}x - 5$

4. $(5, 32)$ and $(9, 44)$

$m = \dfrac{44 - 32}{9 - 5} = \dfrac{12}{4} = 3$

$y - 32 = 3(x - 5)$

$y - 32 = 3x - 15$

$y = 3x + 17$

When $x = 20$, $y = 3(20) + 17$

$y = \$77.$

5. $f(x - 3) = (x - 3)^2 - 2(x - 3) + 1$ 0

$\qquad = x^2 - 6x + 9 - 2x + 6 + 1$

$\qquad = x^2 - 8x + 16$

6. $\qquad f(3) = 12 - 11 = 1$

$\dfrac{f(x) - f(3)}{x - 3} = \dfrac{(4x - 11) - 1}{x - 3}$

$\qquad = \dfrac{4x - 12}{x - 3}$

$\qquad = \dfrac{4(x - 3)}{x - 3} = 4, x \neq 3$

7. $f(x) = \sqrt{36 - x^2} = \sqrt{(6 + x)(6 - x)}$

Domain: $[-6, 6]$

Range: $[0, 6]$, because

$(6 + x)(6 - x) \geq 0$ on this interval

8. (a) $6x - 5y + 4 = 0$

$\qquad y = \dfrac{6x + 4}{5}$ is a function of x.

(b) $x^2 + y^2 = 9$

$\qquad y = \pm\sqrt{9 - x^2}$ is not a function of x.

(c) $y^3 = x^2 + 6$

$\qquad y = \sqrt[3]{x^2 + 6}$ is a function of x.

9. Parabola

Vertex: $(0, -5)$

Intercepts:
$(0, -5), \left(\pm\sqrt{5}, 0\right)$

y-axis symmetry

10. Intercepts: $(0, 3), (-3, 0)$

x	-4	-3	-2	-1	0	1	2
y	1	0	1	2	3	4	5

11.

x	0	1	2	3
y	1	3	5	7

x	-1	-2	-3
y	2	6	12

12. (a) $f(x + 2)$

Horizontal shift two units to the left

(b) $-f(x) + 2$

Reflection in the x-axis and a vertical shift two units upward

13. (a) $(g - f)(x) = g(x) - f(x)$

$\qquad = (2x^2 - 5) - (3x + 7)$

$\qquad = 2x^2 - 3x - 12$

(b) $(fg)(x) = f(x)g(x)$

$\qquad = (3x + 7)(2x^2 - 5)$

$\qquad = 6x^3 + 14x^2 - 15x - 35$

14. $f(g(x)) = f(2x + 3)$

$\qquad = (2x + 3)^2 - 2(2x + 3) + 16$

$\qquad = 4x^2 + 12x + 9 - 4x - 6 + 16$

$\qquad = 4x^2 + 8x + 19$

15. $f(x) = x^3 + 7$

$\qquad y = x^3 + 7$

$\qquad x = y^3 + 7$

$\qquad x - 7 = y^3$

$\qquad \sqrt[3]{x - 7} = y$

$\qquad f^{-1}(x) = \sqrt[3]{x - 7}$

16. (a) $f(x) = |x - 6|$ does not have an inverse. Its graph does not pass the Horizontal Line Test.

(b) $f(x) = ax + b, a \neq 0$ does have an inverse.

$\qquad y = ax + b$

$\qquad x = ay + b$

$\qquad \dfrac{x - b}{a} = y$

$\qquad f^{-1}(x) = \dfrac{x - b}{a}$

(c) $\quad f(x) = x^3 - 19$ does have an inverse.

$\qquad y = x^3 - 19$

$\qquad x = y^3 - 19$

$\qquad x + 19 = y^3$

$\qquad \sqrt[3]{x + 19} = y$

$\qquad f^{-1}(x) = \sqrt[3]{x + 19}$

17. $\quad f(x) = \sqrt{\dfrac{3 - x}{x}},\ 0 < x \leq 3, y \geq 0$

$\qquad y = \sqrt{\dfrac{3 - x}{x}}$

$\qquad x = \sqrt{\dfrac{3 - y}{y}},\ 0 < y \leq 3, x \geq 0$

$\qquad x^2 = \dfrac{3 - y}{y}$

$\qquad x^2 y = 3 - y$

$\qquad x^2 y + y = 3$

$\qquad y(x^2 + 1) = 3$

$\qquad y = \dfrac{3}{x^2 + 1}$

$\qquad f^{-1}(x) = \dfrac{3}{x^2 + 1},\ x \geq 0$

18. False. The slopes of 3 and $\frac{1}{3}$ are not **negative** reciprocals.

19. True. Let $y = (f \circ g)(x)$. Then $x = (f \circ g)^{-1}(y)$. Also,

$\quad (f \circ g)(x) = y$

$\qquad f(g(x)) = y$

$\qquad g(x) = f^{-1}(y)$

$\qquad x = g^{-1}(f^{-1}(y))$

$\qquad x = (g^{-1} \circ f^{-1})(y)$

Since $x = x$, we have $(f \circ g)^{-1}(y) = (g^{-1} \circ f^{-1})(y)$.

20. True. It must pass the Vertical Line Test to be a function and it must pass the Horizontal Line Test to have an inverse.

Chapter 3   Practice Test Solutions

1. x-intercepts: $(1, 0), (5, 0)$

y-intercept: $(0, 5)$

Vertex: $(3, -4)$

2. $a = 0.01, b = -90$

$$\frac{-b}{2a} = \frac{90}{2(0.01)} = 4500 \text{ units}$$

3. Vertex $(1, 7)$ opening downward through $(2, 5)$

$y = a(x - 1)^2 + 7$ Standard form

$5 = a(2 - 1)^2 + 7$

$5 = a + 7$

$a = -2$

$y = -2(x - 1)^2 + 7$

$\quad = -2(x^2 - 2x + 1) + 7$

$\quad = -2x^2 + 4x + 5$

4. $y = \pm a(x - 2)(3x - 4)$ where a is any real nonzero number.

$y = \pm(3x^2 - 10x + 8)$

5. Leading coefficient: -3

Degree: 5 (odd)

Falls to the right.

Rises to the left.

6. $0 = x^5 - 5x^3 + 4x$

$\quad = x(x^4 - 5x^2 + 4)$

$\quad = x(x^2 - 1)(x^2 - 4)$

$\quad = x(x + 1)(x - 1)(x + 2)(x - 2)$

$x = 0, x = \pm 1, x = \pm 2$

7. $f(x) = x(x - 3)(x + 2)$

$\quad = x(x^2 - x - 6)$

$\quad = x^3 - x^2 - 6x$

8. $f(x) = x^3 - 12x$

Intercepts: $(0, 0), \left(\pm 2\sqrt{3}, 0\right)$

Rises to the right.

Falls to the left.

x	-2	-1	0	1	2
y	16	11	0	-11	-16

9.

$$
\begin{array}{r}
3x^3 + 9x^2 + 20x + 62 \\
x - 3 \overline{\smash{)}\ 3x^4 + 0x^3 - 7x^2 + 2x - 10} \\
\underline{3x^4 - 9x^3} \\
9x^3 - 7x^2 \\
\underline{9x^3 - 27x^2} \\
20x^2 + 2x \\
\underline{20x^2 - 60x} \\
62x - 10 \\
\underline{62x - 186} \\
176
\end{array}
$$

$$\frac{3x^4 - 7x^2 + 2x - 10}{x - 3} = 3x^3 + 9x^2 + 20x + 6 + \frac{176}{x - 3}$$

10.

$$
\begin{array}{r}
x - 2 \\
x^2 + 2x - 1 \overline{\smash{\big)}\ x^3 + 0x^2 + 0x - 11} \\
\underline{x^3 + 2x^2 -\ x} \\
-2x^2 +\ x - 11 \\
\underline{-2x^2 - 4x +\ 2} \\
5x - 13
\end{array}
$$

$$\frac{x^3 - 11}{x^2 + 2x - 1} = x - 2 + \frac{5x - 13}{x^2 + 2x - 1}$$

11.

$$
\begin{array}{r|rrrrrr}
-5 & 3 & 13 & 0 & 0 & 12 & -1 \\
   &   & -15 & 10 & -50 & 250 & -1310 \\
\hline
   & 3 & -2 & 10 & -50 & 262 & -1311
\end{array}
$$

$$\frac{3x^5 + 13x^4 + 12x - 1}{x + 5} = 3x^4 - 2x^3 + 10x^2 -$$

$$50x + 262 - \frac{1311}{x + 5}$$

12.

$$
\begin{array}{r|rrrr}
-6 & 7 & 40 & -12 & 15 \\
   &   & -42 & 12 & 0 \\
\hline
   & 7 & -2 & 0 & 15
\end{array}
$$

$$f(-6) = 15$$

13. $0 = x^3 - 19x - 30$

Possible rational zeros: $\pm 1, \pm 2, \pm 3, \pm 5, \pm 6, \pm 10, \pm 15, \pm 30$

$$
\begin{array}{r|rrrr}
-2 & 1 & 0 & -19 & -30 \\
   &   & -2 & 4 & 30 \\
\hline
   & 1 & -2 & -15 & 0
\end{array}
$$

$$0 = (x + 2)(x^2 - 2x - 15)$$

$$0 = (x + 2)(x + 3)(x - 5)$$

Zeros: $x = -2, x = -3, x = 5$

14. $0 = x^4 + x^3 - 8x^2 - 9x - 9$

Possible rational zeros: $\pm 1, \pm 3, \pm 9$

$$
\begin{array}{r|rrrrr}
3 & 1 & 1 & -8 & -9 & -9 \\
  &   & 3 & 12 & 12 & 9 \\
\hline
  & 1 & 4 & 4 & 3 & 0
\end{array}
$$

$$0 = (x - 3)(x^3 + 4x^2 + 4x + 3)$$

Possible rational zeros of $x^3 + 4x^2 + 4x + 3$: $\pm 1, \pm 3$

$$
\begin{array}{r|rrrr}
-3 & 1 & 4 & 4 & 3 \\
   &   & -3 & -3 & -3 \\
\hline
   & 1 & 1 & 1 & 0
\end{array}
$$

$$0 = (x - 3)(x + 3)(x^2 + x + 1)$$

Use the Quadratic Formula. The zeros of $x^2 + x + 1$ are $x = \dfrac{-1 \pm \sqrt{3}i}{2}$.

Zeros: $x = 3, x = -3, x = -\dfrac{1}{2} + \dfrac{\sqrt{3}}{2}i, x = -\dfrac{1}{2} - \dfrac{\sqrt{3}}{2}i$

The real zeros are $x = 3$ and $x = -3$.

15. $0 = 6x^3 - 5x^2 + 4x - 15$

Possible rational zeros:
$\pm 1, \pm 3, \pm 5, \pm 15, \pm \frac{1}{2}, \pm \frac{3}{2}, \pm \frac{5}{2}, \pm \frac{15}{2}, \pm \frac{1}{3}, \pm \frac{5}{3}, \pm \frac{1}{6}, \pm \frac{5}{6}$

16. $0 = x^3 - \frac{20}{3}x^2 + 9x - \frac{10}{3}$

$$0 = 3x^3 - 20x^2 + 27x - 10$$

Possible rational zeros: $\pm 1, \pm 2, \pm 5, \pm 10, \pm \frac{1}{3}, \pm \frac{2}{3}, \pm \frac{5}{3}, \pm \frac{10}{3}$

$$
\begin{array}{r|rrrr}
1 & 3 & -20 & 27 & -10 \\
  &   & 3 & -17 & 10 \\
\hline
  & 3 & -17 & 10 & 0
\end{array}
$$

$$0 = (x - 1)(3x^2 - 17x + 10)$$

$$0 = (x - 1)(3x - 2)(x - 5)$$

Zeros: $x = 1, x = \frac{2}{3}, x = 5$

17. Possible rational zeros: $\pm 1, \pm 2, \pm 5, \pm 10$

$$
\begin{array}{r|rrrrr}
1 & 1 & 1 & 3 & 5 & -10 \\
  &   & 1 & 2 & 5 & 10 \\
\hline
  & 1 & 2 & 5 & 10 & 0 \\
\end{array}
$$

$$
\begin{array}{r|rrrr}
-2 & 1 & 2 & 5 & 10 \\
   &   & -2 & 0 & -10 \\
\hline
   & 1 & 0 & 5 & 0 \\
\end{array}
$$

$$
\begin{aligned}
f(x) &= (x-1)(x+2)(x^2+5) \\
&= (x-1)(x+2)\left(x+\sqrt{5}i\right)\left(x-\sqrt{5}i\right)
\end{aligned}
$$

18.
$$
\begin{aligned}
f(x) &= (x-2)[x-(3+i)][x-(3-i)] \\
&= (x-2)[(x-3)-i][(x-3)+i] \\
&= (x-2)[(x-3)^2-(i)^2] \\
&= (x-2)[x^2-6x+10] \\
&= x^3-8x^2+22x-20
\end{aligned}
$$

19.
$$
\begin{array}{r|rrrr}
3i & 1 & 4 & 9 & 36 \\
   &   & 3i & 12i-9 & -36 \\
\hline
   & 1 & 4+3i & 12i & 0 \\
\end{array}
$$

Thus, $f(3i) = 0$.

20. $z = \dfrac{kx^2}{\sqrt{y}}$

Chapter 4   Practice Test Solutions

1. Vertical asymptote: $x = 0$

Horizontal asymptote: $y = \dfrac{1}{2}$

x-intercept: $(1, 0)$

2. Vertical asymptote: $x = 0$

Slant asymptote: $y = 3x$

x-intercepts: $\left(\pm\dfrac{2}{\sqrt{3}}, 0\right)$

3. $y = 8$ is a horizontal asymptote since the degree of the numerator equals the degree of the denominator. There are no vertical asymptotes.

4. $x = 1$ is a vertical asymptote.

$$
\frac{4x^2 - 2x + 7}{x - 1} = 4x + 2 + \frac{9}{x - 1}
$$

so $y = 4x + 2$ is a slant asymptote.

5. $f(x) = \dfrac{x-5}{(x-5)^2} = \dfrac{1}{x-5}$

Vertical asymptote: $x = 5$

Horizontal asymptote: $y = 0$

y-intercept: $\left(0, -\dfrac{1}{5}\right)$

6. $(x-0)^2 = 4(5)(y-0)$

Vertex: $(0, 0)$

Focus: $(0, 5)$

Directrix: $y = -5$

7. $(y - 0)^2 = 4(7)(x - 0)$

$$y^2 = 28x$$

8. $a = 12, b = 5, h = k = 0,$

$c = \sqrt{144 - 25} = \sqrt{119}$

Center: $(0, 0)$

Foci: $\left(\pm\sqrt{119}, 0\right)$

Vertices: $(\pm 12, 0)$

9. Center: $(0, 0)$

$c = 4, 2b = 6 \implies b = 3,$

$a = \sqrt{16 + 9} = 5$

$$\frac{x^2}{25} + \frac{y^2}{9} = 1$$

10. $a = 12, b = 13, c = \sqrt{144 + 169} = \sqrt{313}$

Center: $(0, 0)$

Foci: $\left(0, \pm\sqrt{313}\right)$

Vertices: $(0, \pm 12)$

Asymptotes: $y = \pm\frac{12}{13}x$

11. Center: $(0, 0)$

$a = 4, \pm\frac{1}{2} = \pm\frac{b}{4} \implies b = 2$

$$\frac{x^2}{16} - \frac{y^2}{4} = 1$$

12. $p = 4$

$(x - 6)^2 = 4(4)(y + 1)$

$(x - 6)^2 = 16(y + 1)$

13. $16x^2 - 96x + 9y^2 + 36y = -36$

$16(x^2 - 6x + 9) + 9(y^2 + 4y + 4) = -36 + 144 + 36$

$16(x - 3)^2 + 9(y + 2)^2 = 144$

$$\frac{(x - 3)^2}{9} + \frac{(y + 2)^2}{16} = 1$$

$a = 4, b = 3, c = \sqrt{16 - 9} = \sqrt{7}$

Center: $(3, -2)$

Foci: $\left(3, -2 \pm \sqrt{7}\right)$

Vertices: $(3, -2 \pm 4)$ OR $(3, 2)$ and $(3, -6)$

14. Center: $(3, 1)$

$a = 4, 2b = 2 \implies b = 1$

$$\frac{(x - 3)^2}{16} + \frac{(y - 1)^2}{1} = 1$$

15. $\dfrac{(x + 3)^2}{1/4} - \dfrac{(y - 1)^2}{1/9} = 1$

$a = \frac{1}{2}, b = \frac{1}{3}, c = \sqrt{\frac{1}{4} + \frac{1}{9}} = \frac{\sqrt{13}}{6}$

Center: $(-3, 1)$

Vertices: $\left(-3 \pm \frac{1}{2}, 1\right)$ OR $\left(-\frac{5}{2}, 1\right)$ and $\left(-\frac{7}{2}, 1\right)$

Foci: $\left(-3 \pm \frac{\sqrt{13}}{6}, 1\right)$

Asymptotes: $y = \pm\frac{1/3}{1/2}(x + 3) + 1 = \pm\frac{2}{3}(x + 3) + 1$

16. Center: $(3, 0)$

$a = 4, c = 7, b = \sqrt{49 - 16} = \sqrt{33}$

$$\frac{y^2}{16} - \frac{(x - 3)^2}{33} = 1$$

Chapter 5   Practice Test Solutions

1. $x^{3/5} = 8$

$\qquad x = 8^{5/3} = \left(\sqrt[3]{8}\right)^5 = 2^5 = 32$

2. $3^{x-1} = \frac{1}{81}$

$\qquad 3^{x-1} = 3^{-4}$

$\qquad x - 1 = -4$

$\qquad x = -3$

3. $f(x) = 2^{-x} = \left(\frac{1}{2}\right)^x$

x	-2	-1	0	1	2
$f(x)$	4	2	1	$\frac{1}{2}$	$\frac{1}{4}$

4. $g(x) = e^x + 1$

x	-2	-1	0	1	2
$g(x)$	1.14	1.37	2	3.72	8.39

5. (a) $A = P\left(1 + \dfrac{r}{n}\right)^{nt}$

$\qquad A = 5000\left(1 + \dfrac{0.04}{12}\right)^{12(3)} \approx \$5636.36$

(b) $A = P\left(1 + \dfrac{r}{n}\right)^{nt}$

$\qquad A = 5000\left(1 + \dfrac{0.04}{4}\right)^{4(3)} \approx \$5634.13$

(c) $A = Pe^{rt}$

$\qquad A = 5000e^{(0.04)(3)} \approx \$5637.48$

6. $\quad 7^{-2} = \dfrac{1}{49}$

$\quad \log_7 \dfrac{1}{49} = -2$

7. $x - 4 = \log_2 \dfrac{1}{64}$

$\qquad 2^{x-4} = \dfrac{1}{64}$

$\qquad 2^{x-4} = 2^{-6}$

$\qquad x - 4 = -6$

$\qquad x = -2$

8. $\ln\left(\dfrac{x^2\sqrt{y}}{z^4}\right) = \ln x^2\sqrt{y} - \ln z^4$

$\qquad\qquad\quad = \ln x^2 + \ln\sqrt{y} - \ln z^4$

$\qquad\qquad\quad = 2\ln x + \dfrac{1}{2}\ln y - 4\ln z$

9. $5\ln x - \dfrac{1}{2}\ln y + 6\ln z = \ln x^5 - \ln\sqrt{y} + \ln z^6 = \ln\left(\dfrac{x^5 z^6}{\sqrt{y}}\right),\ z > 0$

10. $\log_9 28 = \dfrac{\log 28}{\log 9} \approx 1.5166$

11. $\log N = 0.6646$

$\qquad N = 10^{0.6646} \approx 4.62$

12.

13. Domain:

$$x^2 - 9 > 0$$

$$(x + 3)(x - 3) > 0$$

$$x < -3 \text{ or } x > 3$$

14.

15. False. $\dfrac{\ln x}{\ln y} \neq \ln(x - y)$ since $\dfrac{\ln x}{\ln y} = \log_y x$.

16. $5^3 = 41$

$$x = \log_5 41 = \frac{\ln 41}{\ln 5} \approx 2.3074$$

17. $x - x^2 = \log_5 \frac{1}{25}$

$$5^{x - x^2} = \frac{1}{25}$$

$$5^{x - x^2} = 5^{-2}$$

$$x - x^2 = -2$$

$$0 = x^2 - x - 2$$

$$0 = (x + 1)(x - 2)$$

$$x = -1 \text{ or } x = 2$$

18. $\log_2 x + \log_2(x - 3) = 2$

$$\log_2[x(x - 3)] = 2$$

$$x(x - 3) = 2^2$$

$$x^2 - 3x = 4$$

$$x^2 - 3x - 4 = 0$$

$$(x + 1)(x - 4) = 0$$

$$x = 4$$

$$x = -1 \text{ (extraneous)}$$

$x = 4$ is the only solution.

19. $y = ae^{bx}$

$$7 = ae^{b(0)} \Longrightarrow a = 7$$

$$3 = 7e^{b(5)} \Longrightarrow \frac{3}{7} = e^{5b} \Longrightarrow \ln\left(\frac{3}{7}\right) = 5b \Longrightarrow b = \frac{\ln(3/7)}{5}$$

$$\approx -0.1695$$

Thus, $y = 7e^{-0.1695x}$.

20. $A = Pe^{et}$

$$12{,}000 = 6000e^{0.06t}$$

$$2 = e^{0.06t}$$

$$0.06t = \ln 2$$

$$t = \frac{\ln 2}{0.06}$$

$$t \approx 11.552 \text{ years}$$

Chapter 6   Practice Test Solutions

1. $350° = 350\left(\dfrac{\pi}{180}\right) = \dfrac{35\pi}{18}$

2. $\dfrac{5\pi}{9} = \dfrac{5\pi}{9} \cdot \dfrac{180}{\pi} = 100°$

3. $135°14'12'' = \left(135 + \dfrac{14}{60} + \dfrac{12}{3600}\right)°$

$$\approx 135.2367°$$

4. $-22.569° = -(22° + 0.569(60)')$

$$= -22°34.14'$$

$$= -(22°34' + 0.14(60)'')$$

$$\approx -22°34'8''$$

5. $\cos \theta = \dfrac{2}{3}$

$x = 2, r = 3, y = \pm \sqrt{9 - 4} = \pm \sqrt{5}$

$\tan \theta = \dfrac{y}{x} = \pm \dfrac{\sqrt{5}}{2}$

6. $\sin \theta = 0.9063$

$\theta = \arcsin(0.9063)$

$\theta \approx 65° = \dfrac{13\pi}{36}$ or $\theta \approx 180° - 65° = 115° = \dfrac{23\pi}{36}$

7. $\tan 20° = \dfrac{35}{x}$

$x = \dfrac{35}{\tan 20°} \approx 96.1617$

8. $\theta = \dfrac{6\pi}{5}$, θ is in Quadrant III.

Reference angle: $\dfrac{6\pi}{5} - \pi = \dfrac{\pi}{5}$ or $36°$

9. $\csc 3.92 = \dfrac{1}{\sin 3.92} \approx -1.4242$

10. $\tan \theta = 6 = \dfrac{6}{1}$, θ lies in Quadrant III.

$y = -6, x = -1, r = \sqrt{36 + 1} = \sqrt{37}$,

so $\sec \theta = \dfrac{\sqrt{37}}{-1} \approx -6.0828$.

11. Period: 4π

Amplitude: 3

12. Period: 2π

Amplitude: 2

13. Period: $\dfrac{\pi}{2}$

14. Period: 2π

15.

16.

17. $\theta = \arcsin 1$

$\sin \theta = 1$

$\theta = \dfrac{\pi}{2} = 90°$

18. $\theta = \arctan(-3)$

$\tan \theta = -3$

$\theta \approx -1.249 \approx -71.565°$

19. $\sin\left(\arccos\dfrac{4}{\sqrt{35}}\right)$

$\sin\theta = \dfrac{\sqrt{19}}{\sqrt{35}} \approx 0.7368$

$x = \sqrt{35-16}$
$= \sqrt{19}$

20. $\cos\left(\arcsin\dfrac{x}{4}\right)$

$\cos\theta = \dfrac{\sqrt{16-x^2}}{4}$

21. Given $A = 40°$, $c = 12$

$B = 90° - 40° = 50°$

$\sin 40° = \dfrac{a}{12}$

$a = 12\sin 40° \approx 7.713$

$\cos 40° = \dfrac{b}{12}$

$b = 12\cos 40° \approx 9.193$

22. Given $B = 6.84°$, $a = 21.3$

$A = 90° - 6.84° = 83.16°$

$\sin 83.16° = \dfrac{21.3}{c}$

$c = \dfrac{21.3}{\sin 83.16°} \approx 21.453$

$\tan 83.16° = \dfrac{21.3}{b}$

$b = \dfrac{21.3}{\tan 83.16°} \approx 2.555$

23. Given $a = 5$, $b = 9$

$c = \sqrt{25 + 81} = \sqrt{106} \approx 10.296$

$\tan A = \dfrac{5}{9}$

$A = \arctan\dfrac{5}{9} \approx 29.055°$

$B \approx 90° - 29.055° = 60.945°$

24. $\sin 67° = \dfrac{x}{20}$

$x = 20\sin 67° \approx 18.41$ feet

25. $\tan 5° = \dfrac{250}{x}$

$x = \dfrac{250}{\tan 5°}$

≈ 2857.513 feet

≈ 0.541 mi

Chapter 7   Practice Test Solutions

1. $\tan x = \dfrac{4}{11}$, $\sec x < 0 \implies x$ is in Quadrant III.

$y = -4$, $x = -11$, $r = \sqrt{16 + 121} = \sqrt{137}$

$\sin x = -\dfrac{4}{\sqrt{137}} = -\dfrac{4\sqrt{137}}{137}$ $\csc x = -\dfrac{\sqrt{137}}{4}$

$\cos x = -\dfrac{11}{\sqrt{137}} = -\dfrac{11\sqrt{137}}{137}$ $\sec x = -\dfrac{\sqrt{137}}{11}$

$\tan x = \dfrac{4}{11}$ $\cot x = \dfrac{11}{4}$

2. $\dfrac{\sec^2 x + \csc^2 x}{\csc^2 x(1 + \tan^2 x)} = \dfrac{\sec^2 x + \csc^2 x}{\csc^2 x + (\csc^2 x)\tan^2 x} = \dfrac{\sec^2 x + \csc^2 x}{\csc^2 x + \dfrac{1}{\sin^2 x}\cdot\dfrac{\sin^2 x}{\cos^2 x}}$

$= \dfrac{\sec^2 x + \csc^2 x}{\csc^2 x + \dfrac{1}{\cos^2 x}} = \dfrac{\sec^2 x + \csc^2 x}{\csc^2 x + \sec^2 x} = 1$

3. $\ln|\tan \theta| - \ln|\cot \theta| = \ln\left|\dfrac{\tan \theta}{\cot \theta}\right| = \ln\left|\dfrac{\sin \theta/\cos\theta}{\cos \theta/\sin\theta}\right| = \ln\left|\dfrac{\sin^2 \theta}{\cos^2 \theta}\right| = \ln|\tan^2 \theta| = 2\ln|\tan \theta|$

4. $\cos\left(\dfrac{\pi}{2} - x\right) = \dfrac{1}{\csc x}$ is true since $\cos\left(\dfrac{\pi}{2} - x\right) = \sin x = \dfrac{1}{\csc x}$.

5. $\sin^4 x + (\sin^2 x)\cos^2 x = \sin^2 x(\sin^2 x + \cos^2 x) = \sin^2 x(1) = \sin^2 x$

6. $(\csc x + 1)(\csc x - 1) = \csc^2 x - 1 = \cot^2 x$

7. $\dfrac{\cos^2 x}{1 - \sin x} \cdot \dfrac{1 + \sin x}{1 + \sin x} = \dfrac{\cos^2 x(1 + \sin x)}{1 - \sin^2 x} = \dfrac{\cos^2 x(1 + \sin x)}{\cos^2 x} = 1 + \sin x$

8. $\dfrac{1 + \cos \theta}{\sin \theta} + \dfrac{\sin \theta}{1 + \cos \theta} = \dfrac{(1 + \cos \theta)^2 + \sin^2 \theta}{\sin \theta(1 + \cos \theta)}$

$$= \dfrac{1 + 2\cos \theta + \cos^2 \theta + \sin^2 \theta}{\sin \theta(1 + \cos \theta)} = \dfrac{2 + 2\cos \theta}{\sin \theta(1 + \cos \theta)} = \dfrac{2}{\sin \theta} = 2\csc \theta$$

9. $\tan^4 x + 2\tan^2 x + 1 = (\tan^2 x + 1)^2 = (\sec^2 x)^2 = \sec^4 x$

10. (a) $\sin 105° = \sin(60° + 45°) = \sin 60° \cos 45° + \cos 60° \sin 45°$

$$= \dfrac{\sqrt{3}}{2} \cdot \dfrac{\sqrt{2}}{2} + \dfrac{1}{2} \cdot \dfrac{\sqrt{2}}{2} = \dfrac{\sqrt{2}}{4}\left(\sqrt{3} + 1\right)$$

(b) $\tan 15° = \tan(60° - 45°) = \dfrac{\tan 60° - \tan 45°}{1 + \tan 60° \tan 45°}$

$$= \dfrac{\sqrt{3} - 1}{1 + \sqrt{3}} \cdot \dfrac{1 - \sqrt{3}}{1 - \sqrt{3}} = \dfrac{2\sqrt{3} - 1 - 3}{1 - 3} = \dfrac{2\sqrt{3} - 4}{-2} = 2 - \sqrt{3}$$

11. $(\sin 42°)\cos 38° - (\cos 42°)\sin 38° = \sin(42° - 38°) = \sin 4°$

12. $\tan\left(\theta + \dfrac{\pi}{4}\right) = \dfrac{\tan \theta + \tan\left(\dfrac{\pi}{4}\right)}{1 - (\tan \theta)\tan\left(\dfrac{\pi}{4}\right)} = \dfrac{\tan \theta + 1}{1 - \tan \theta(1)} = \dfrac{1 + \tan \theta}{1 - \tan \theta}$

13. $\sin(\arcsin x - \arccos x) = \sin(\arcsin x)\cos(\arccos x) - \cos(\arcsin x)\sin(\arccos x)$

$$= (x)(x) - \left(\sqrt{1 - x^2}\right)\left(\sqrt{1 - x^2}\right) = x^2 - (1 - x^2) = 2x^2 - 1$$

14. (a) $\cos(120°) = \cos[2(60°)] = 2\cos^2 60° - 1 = 2\left(\dfrac{1}{2}\right)^2 - 1 = -\dfrac{1}{2}$

(b) $\tan(300°) = \tan[2(150°)] = \dfrac{2\tan 150°}{1 - \tan^2 150°} = \dfrac{-\dfrac{2\sqrt{3}}{3}}{1 - \left(\dfrac{1}{3}\right)} = -\sqrt{3}$

15. (a) $\sin 22.5° = \sin \dfrac{45°}{2} = \sqrt{\dfrac{1 - \cos 45°}{2}} = \sqrt{\dfrac{1 - \frac{\sqrt{2}}{2}}{2}} = \dfrac{\sqrt{2 - \sqrt{2}}}{2}$

(b) $\tan \dfrac{\pi}{12} = \tan \dfrac{\frac{\pi}{6}}{2} = \dfrac{\sin \frac{\pi}{6}}{1 + \cos\left(\frac{\pi}{6}\right)} = \dfrac{\frac{1}{2}}{1 + \frac{\sqrt{3}}{2}} = \dfrac{1}{2 + \sqrt{3}} = 2 - \sqrt{3}$

16. $\sin \theta = \dfrac{4}{5}$, θ lies in Quadrant II $\implies \cos \theta = -\dfrac{3}{5}$.

$\cos \dfrac{\theta}{2} = \sqrt{\dfrac{1 + \cos \theta}{2}} = \sqrt{\dfrac{1 - \frac{3}{5}}{2}} = \sqrt{\dfrac{2}{10}} = \dfrac{1}{\sqrt{5}} = \dfrac{\sqrt{5}}{5}$

17. $(\sin^2 x)\cos^2 x = \dfrac{1 - \cos 2x}{2} \cdot \dfrac{1 + \cos 2x}{2} = \dfrac{1}{4}[1 - \cos^2 2x] = \dfrac{1}{4}\left[1 - \dfrac{1 + \cos 4x}{2}\right]$

$= \dfrac{1}{8}[2 - (1 + \cos 4x)] = \dfrac{1}{8}[1 - \cos 4x]$

18. $6(\sin 5\theta)\cos 2\theta = 6\left\{\dfrac{1}{2}[\sin(5\theta + 2\theta) + \sin(5\theta - 2\theta)]\right\} = 3[\sin 7\theta + \sin 3\theta]$

19. $\sin(x + \pi) + \sin(x - \pi) = 2\left(\sin\dfrac{[(x + \pi) + (x - \pi)]}{2}\right)\cos\dfrac{[(x + \pi) - (x - \pi)]}{2}$

$= 2\sin x \cos \pi = -2\sin x$

20. $\dfrac{\sin 9x + \sin 5x}{\cos 9x - \cos 5x} = \dfrac{2\sin 7x \cos 2x}{-2\sin 7x \sin 2x} = -\dfrac{\cos 2x}{\sin 2x} = -\cot 2x$

21. $\dfrac{1}{2}[\sin(u + v) - \sin(u - v)] = \dfrac{1}{2}\{(\sin u)\cos v + (\cos u)\sin v - [(\sin u)\cos v - (\cos u)\sin v]\}$

$= \dfrac{1}{2}[2(\cos u)\sin v] = (\cos u)\sin v$

22. $4\sin^2 x = 1$

$\sin^2 x = \dfrac{1}{4}$

$\sin x = \pm\dfrac{1}{2}$

$\sin x = \dfrac{1}{2}$ or $\sin x = -\dfrac{1}{2}$

$x = \dfrac{\pi}{6}$ or $\dfrac{5\pi}{6}$ $x = \dfrac{7\pi}{6}$ or $\dfrac{11\pi}{6}$

23. $\tan^2 \theta + \left(\sqrt{3} - 1\right)\tan \theta - \sqrt{3} = 0$

$(\tan\theta - 1)(\tan\theta + \sqrt{3}) = 0$

$\tan \theta = 1$ or $\tan \theta = -\sqrt{3}$

$\theta = \dfrac{\pi}{4}$ or $\dfrac{5\pi}{4}$ $\theta = \dfrac{2\pi}{3}$ or $\dfrac{5\pi}{3}$

24. $\sin 2x = \cos x$

$2(\sin x)\cos x - \cos x = 0$

$\cos x(2\sin x - 1) = 0$

$\cos x = 0$ or $\sin x = \dfrac{1}{2}$

$x = \dfrac{\pi}{2}$ or $\dfrac{3\pi}{2}$ $x = \dfrac{\pi}{6}$ or $\dfrac{5\pi}{6}$

25. $\tan^2 x - 6\tan x + 4 = 0$

$\tan x = \dfrac{-(-6) \pm \sqrt{(-6)^2 - 4(1)(4)}}{2(1)}$

$\tan x = \dfrac{6 \pm \sqrt{20}}{2} = 3 \pm \sqrt{5}$

$\tan x = 3 + \sqrt{5}$ or $\tan x = 3 - \sqrt{5}$

$x \approx 1.3821$ or 4.5237 $x = 0.6524$ or 3.7940

Chapter 8   Practice Test Solutions

1. $C = 180° - (40° + 12°) = 128°$

$$a = \sin 40° \left(\frac{100}{\sin 12°} \right) \approx 309.164$$

$$c = \sin 128° \left(\frac{100}{\sin 12°} \right) \approx 379.012$$

2. $\sin A = 5 \left(\dfrac{\sin 150°}{20} \right) = 0.125$

$A \approx 7.181°$

$B \approx 180° - (150° + 7.181°) = 22.819°$

$b = \sin 22.819° \left(\dfrac{20}{\sin 150°} \right) \approx 15.513$

3. Area $= \frac{1}{2}ab \sin C$

$\quad = \frac{1}{2}(3)(6)\sin 130°$

$\quad \approx 6.894$ square units

4. $h = b \sin A$

$\quad = 35 \sin 22.5°$

$\quad \approx 13.394$

$a = 10$

Since $a < h$ and A is acute, the triangle has no solution.

5. $\cos A = \dfrac{(53)^2 + (38)^2 - (49)^2}{2(53)(38)} \approx 0.4598$

$A \approx 62.627°$

$\cos B = \dfrac{(49)^2 + (38)^2 - (53)^2}{2(49)(38)} \approx 0.2782$

$B \approx 73.847°$

$C \approx 180° - (62.627° + 73.847°)$

$\quad = 43.526°$

6. $c^2 = (100)^2 + (300)^2 - 2(100)(300)\cos 29°$

$\quad \approx 47522.8176$

$c \approx 218$

$\cos A = \dfrac{(300)^2 + (218)^2 - (100)^2}{2(300)(218)} \approx 0.97495$

$A \approx 12.85°$

$B \approx 180° - (12.85° + 29°) = 138.15°$

7. $s = \dfrac{a + b + c}{2} = \dfrac{4.1 + 6.8 + 5.5}{2} = 8.2$

Area $= \sqrt{s(s - a)(s - b)(s - c)}$

$\quad = \sqrt{8.2(8.2 - 4.1)(8.2 - 6.8)(8.2 - 5.5)}$

$\quad \approx 11.273$ square units

8. $x^2 = (40)^2 + (70)^2 - 2(40)(70)\cos 168°$

$\quad \approx 11977.6266$

$x \approx 190.442$ miles

9. $\mathbf{w} = 4(3\mathbf{i} + \mathbf{j}) - 7(-\mathbf{i} + 2\mathbf{j})$

$\quad = 19\mathbf{i} - 10\mathbf{j}$

10. $\dfrac{\mathbf{v}}{\|\mathbf{v}\|} = \dfrac{5\mathbf{i} - 3\mathbf{j}}{\sqrt{25 + 9}} = \dfrac{5}{\sqrt{34}}\mathbf{i} - \dfrac{3}{\sqrt{34}}\mathbf{j}$

$\quad = \dfrac{5\sqrt{34}}{34}\mathbf{i} - \dfrac{3\sqrt{34}}{34}\mathbf{j}$

11. $\mathbf{u} = 6\mathbf{i} + 5\mathbf{j} \quad \mathbf{v} = 2\mathbf{i} - 3\mathbf{j}$

$\mathbf{u} \cdot \mathbf{v} = 6(2) + 5(-3) = -3$

$\|\mathbf{u}\| = \sqrt{61} \quad \|\mathbf{v}\| = \sqrt{13}$

$\cos \theta = \dfrac{-3}{\sqrt{61}\sqrt{13}}$

$\theta \approx 96.116°$

12. $4(\mathbf{i} \cos 30° + \mathbf{j} \sin 30°)$

$\quad = 4\left(\dfrac{\sqrt{3}}{2}\mathbf{i} + \dfrac{1}{2}\mathbf{j} \right)$

$\quad = \langle 2\sqrt{3}, 2 \rangle$

13. $\text{proj}_\mathbf{v}\mathbf{u} = \left(\dfrac{\mathbf{u}\cdot\mathbf{v}}{\|\mathbf{v}\|^2}\right)\mathbf{v} = \dfrac{-10}{20}\langle -2, 4\rangle = \langle 1, -2\rangle$

14. $r = \sqrt{25 + 25} = \sqrt{50} = 5\sqrt{2}$

$\tan\theta = \dfrac{-5}{5} = -1$

Since z is in Quadrant IV,

$\theta = 315°$

$z = 5\sqrt{2}(\cos 315° + i\sin 315°).$

15. $\cos 225° = -\dfrac{\sqrt{2}}{2},\ \ \sin 225° = -\dfrac{\sqrt{2}}{2}$

$z = 6\left(-\dfrac{\sqrt{2}}{2} - i\dfrac{\sqrt{2}}{2}\right)$

$= -3\sqrt{2} - 3\sqrt{2}i$

16. $[7(\cos 23° + i\sin 23°)][4(\cos 7° + i\sin 7°)] = 7(4)[\cos(23° + 7°) + i\sin(23° + 7°)]$

$= 28(\cos 30° + i\sin 30°)$

17. $\dfrac{9\left(\cos\dfrac{5\pi}{4} + i\sin\dfrac{5\pi}{4}\right)}{3(\cos\pi + i\sin\pi)} = \dfrac{9}{3}\left[\cos\left(\dfrac{5\pi}{4} - \pi\right) + i\sin\left(\dfrac{5\pi}{4} - \pi\right)\right] = 3\left(\cos\dfrac{\pi}{4} + i\sin\dfrac{\pi}{4}\right)$

18. $(2 + 2i)^8 = [2\sqrt{2}(\cos 45° + i\sin 45°)]^8 = \left(2\sqrt{2}\right)^8[\cos(8)(45°) + i\sin(8)(45°)]$

$= 4096[\cos 360° + i\sin 360°] = 4096$

19. $z = 8\left(\cos\dfrac{\pi}{3} + i\sin\dfrac{\pi}{3}\right),\ n = 3$

The cube roots of z are: $\sqrt[3]{8}\left[\cos\dfrac{\dfrac{\pi}{3} + 2\pi k}{3} + i\sin\dfrac{\dfrac{\pi}{3} + 2\pi k}{3}\right], k = 0, 1, 2$

For $k = 0$, $\sqrt[3]{8}\left[\cos\dfrac{\dfrac{\pi}{3}}{3} + i\sin\dfrac{\dfrac{\pi}{3}}{3}\right] = 2\left(\cos\dfrac{\pi}{9} + i\sin\dfrac{\pi}{9}\right)$

For $k = 1$, $\sqrt[3]{8}\left[\cos\dfrac{\left(\dfrac{\pi}{3}\right) + 2\pi}{3} + i\sin\dfrac{\left(\dfrac{\pi}{3}\right) + 2\pi}{3}\right] = 2\left(\cos\dfrac{7\pi}{9} + i\sin\dfrac{7\pi}{9}\right)$

For $k = 2$, $\sqrt[3]{8}\left[\cos\dfrac{\dfrac{\pi}{3} + 4\pi}{3} + i\sin\dfrac{\dfrac{\pi}{3} + 4\pi}{3}\right] = 2\left(\cos\dfrac{13\pi}{9} + i\sin\dfrac{13\pi}{9}\right)$

20. $x^4 = -i = 1\left(\cos\dfrac{3\pi}{2} + i\sin\dfrac{3\pi}{2}\right)$

The fourth roots are: $\sqrt[4]{1}\left[\cos\dfrac{\left(\dfrac{3\pi}{2}\right)+2\pi k}{4} + i\sin\dfrac{\left(\dfrac{3\pi}{2}\right)+2\pi k}{4}\right], k = 0, 1, 2, 3$

For $k = 0$, $\cos\dfrac{\dfrac{3\pi}{2}}{4} + i\sin\dfrac{\dfrac{3\pi}{2}}{4} = \cos\dfrac{3\pi}{8} + i\sin\dfrac{3\pi}{8}$

For $k = 1$, $\cos\dfrac{\dfrac{3\pi}{2}+2\pi}{4} + i\sin\dfrac{\dfrac{3\pi}{2}+2\pi}{4} = \cos\dfrac{7\pi}{8} + i\sin\dfrac{7\pi}{8}$

For $k = 2$, $\cos\dfrac{\dfrac{3\pi}{2}+4\pi}{4} + i\sin\dfrac{\dfrac{3\pi}{2}+4\pi}{4} = \cos\dfrac{11\pi}{8} + i\sin\dfrac{11\pi}{8}$

For $k = 3$, $\cos\dfrac{\dfrac{3\pi}{2}+6\pi}{4} + i\sin\dfrac{\dfrac{3\pi}{2}+6\pi}{4} = \cos\dfrac{15\pi}{8} + i\sin\dfrac{15\pi}{8}$

Chapter 9   Practice Test Solutions

1. $\begin{cases} x + y = 1 \\ 3x - y = 15 \end{cases} \implies y = 3x - 15$

$x + (3x - 15) = 1$

$4x = 16$

$x = 4$

$y = -3$

Solution: $(4, -3)$

2. $\begin{cases} x - 3y = -3 \implies x = 3y - 3 \\ x^2 + 6y = 5 \end{cases}$

$(3y - 3)^2 + 6y = 5$

$9y^2 - 18y + 9 + 6y = 5$

$9y^2 - 12y + 4 = 0$

$(3y - 2)^2 = 0$

$y = \tfrac{2}{3}$

$x = -1$

Solution: $\left(-1, \tfrac{2}{3}\right)$

3. $\begin{cases} x + y + z = 6 \implies z = 6 - x - y \\ 2x - y + 3z = 0 \implies 2x - y + 3(6 - x - y) = 0 \implies -x - 4y = -18 \implies x = 18 - 4y \\ 5x + 2y - z = -3 \implies 5x + 2y - (6 - x - y) = -3 \implies 6x + 3y = 3 \end{cases}$

$6(18 - 4y) + 3y = 3$

$-21y = -105$

$y = 5$

$x = 18 - 4y = -2$

$z = 6 - x - y = 3$

Solution: $(-2, 5, 3)$

4.
$$x + y = 110 \implies y = 110 - x$$
$$xy = 2800$$
$$x(110 - x) = 2800$$
$$0 = x^2 - 110x + 2800$$
$$0 = (x - 40)(x - 70)$$
$$x = 40 \quad \text{or} \quad x = 70$$
$$y = 70 \qquad y = 40$$

Solution: The two numbers are 40 and 70.

5.
$$2x + 2y = 170 \implies y = \frac{170 - 2x}{2} = 85 - x$$
$$xy = 1500$$
$$x(85 - x) = 1500$$
$$0 = x^2 - 85x + 1500$$
$$0 = (x - 25)(x - 60)$$
$$x = 25 \quad \text{or} \quad x = 60$$
$$y = 60 \qquad y = 25$$

Dimensions: $60 \text{ ft} \times 25 \text{ ft}$

6.
$$\begin{cases} 2x + 15y = 4 \\ x - 3y = 23 \end{cases} \implies \begin{aligned} 2x + 15y &= 4 \\ \underline{5x - 15y} &= \underline{115} \\ 7x &= 119 \\ x &= 17 \end{aligned}$$
$$y = \frac{x - 23}{3} = -2$$

Solution: $(17, -2)$

7.
$$\begin{cases} x + y = 2 \\ 38x - 19y = 7 \end{cases} \implies \begin{aligned} 19x + 19y &= 38 \\ \underline{38x - 19y} &= \underline{7} \\ 57x &= 45 \end{aligned}$$
$$x = \frac{45}{57} = \frac{15}{19}$$
$$y = 2 - x = \frac{38}{19} - \frac{15}{19} = \frac{23}{19}$$

Solution: $\left(\dfrac{15}{19}, \dfrac{23}{19} \right)$

8.
$$\begin{cases} 0.4x + 0.5y = 0.112 \\ 0.3x - 0.7y = -0.131 \end{cases} \implies \begin{aligned} 0.28x + 0.35y &= 0.0784 \\ \underline{0.15x - 0.35y} &= \underline{-0.0655} \\ 0.43x &= 0.0129 \end{aligned}$$
$$x = \frac{0.0129}{0.43} = 0.03$$
$$y = \frac{0.112 - 0.4x}{0.5} = 0.20$$

Solution: $(0.03, 0.20)$

9. Let x = amount in 11% fund and y = amount in 13% fund.
$$x + y = 17000 \implies y = 17000 - x$$
$$0.11x + 0.13y = 2080$$
$$0.11x + 0.13(17{,}000 - x) = 2080$$
$$-0.02x = -130$$
$$x = \$6500 \text{ at } 11\%$$
$$y = \$10{,}500 \text{ at } 13\%$$

10. $(4, 3), (1, 1), (-1, -2), (-2, -1)$

Use a calculator.
$$y = ax + b = \tfrac{11}{14}x - \tfrac{1}{7}$$

11. $\begin{cases} x + y & = -2 \\ 2x - y + z & = 11 \\ 4y - 3z & = -20 \end{cases}$

$\begin{cases} x + y & = -2 \\ -3y + z & = 15 \\ 4y - 3z & = -20 \end{cases}$ $-2\text{Eq.1} + \text{Eq.2}$

$\begin{cases} x + y & = -2 \\ y - 2z & = -5 \\ 4y - 3z & = -20 \end{cases}$ $\text{Eq.3} + \text{Eq.2}$

$\begin{cases} x + y & = -2 \\ y - 2z & = -5 \\ 5z & = 0 \end{cases}$ $-4\text{Eq.2} + \text{Eq.3}$

$\begin{cases} x + y & = -2 \\ y - 2z & = -5 \\ z & = 0 \end{cases}$

$y - 2(0) = -5 \implies y = -5$

$x + (-5) = -2 \implies x = 3$

Solution: $(3, -5, 0)$

12. $\begin{cases} 3x + 2y - z = 5 \\ 6x - y + 5z = 2 \end{cases}$

$\begin{cases} 3x + 2y - z = 5 \\ -5y + 7z = -8 \end{cases}$ $-2\text{Eq.1} + \text{Eq.2}$

$\begin{cases} x + \frac{2}{3}y - \frac{1}{3}z = \frac{5}{3} & \frac{1}{3}\text{Eq.1} \\ y - \frac{7}{5}z = \frac{8}{5} & -\frac{1}{5}\text{Eq.2} \end{cases}$

Let $a = z$.

Then $y = \frac{7}{5}a + \frac{8}{5}$, and

$x + \frac{2}{3}\left(\frac{7}{5}a + \frac{8}{5}\right) - \frac{1}{3}a = \frac{5}{3}$

$x + \frac{3}{5}a = \frac{3}{5}$

$x = -\frac{3}{5}a + \frac{3}{5}.$

Solution: $\left(-\frac{3}{5}a + \frac{3}{5}, \frac{7}{5}a + \frac{8}{5}, a\right)$, where a is any real number.

13. $y = ax^2 + bx + c$ passes through $(0, -1)$, $(1, 4)$, and $(2, 13)$.

At $(0, -1)$: $-1 = a(0)^2 + b(0) + c \implies c = -1$

At $(1, 4)$: $4 = a(1)^2 + b(1) - 1 \implies 5 = a + b \implies 5 = a + b$

At $(2, 13)$: $13 = a(2)^2 + b(2) - 1 \implies 14 = 4a + 2b \implies \underline{-7 = -2a - b}$

$-2 = -a$

$a = 2$

$b = 3$

Thus, the equation of the parabola is $y = 2x^2 + 3x - 1$.

14. $\dfrac{10x - 17}{x^2 - 7x - 8} = \dfrac{10x - 17}{(x + 1)(x - 8)} = \dfrac{A}{x + 1} + \dfrac{B}{x - 8}$

$10x - 17 = A(x - 8) + B(x + 1)$

Let $x = -1$: $-27 = -9A \implies A = 3$

Let $x = 8$: $63 = 9B \implies B = 7$

$\dfrac{10x - 17}{x^2 - 7x - 8} = \dfrac{3}{x + 1} + \dfrac{7}{x - 8}$

15. $\dfrac{x^2 + 4}{x^4 + x^2} = \dfrac{x^2 + 4}{x^2(x^2 + 1)} = \dfrac{A}{x} + \dfrac{B}{x^2} + \dfrac{Cx + D}{x^2 + 1}$

$x^2 + 4 = Ax(x^2 + 1) + B(x^2 + 1) + (Cx + D)x^2$

$= Ax^3 + Ax + Bx^2 + B + Cx^3 + Dx^2$

$= (A + C)x^3 + (B + D)x^2 + Ax + B$

Equate the coefficients of like terms:

$0 = A + C$

$1 = B + D$

$0 = A$

$4 = B$

Thus, $A = 0, B = 4, C = 0, D = -3$.

$\dfrac{x^2 + 4}{x^4 + x^2} = \dfrac{4}{x^2} - \dfrac{3}{x^2 + 1}$

16. $x^2 + y^2 \geq 9$

17. $\begin{cases} x + y \leq 6 \\ x \geq 2 \\ y \geq 0 \end{cases}$

18. Line through $(0, 0)$ and $(0, 7)$:

$x = 0$

Line through $(0, 0)$ and $(2, 3)$:

$y = \frac{3}{2}x$ or $3x - 2y = 0$

Line through $(0, 7)$ and $(2, 3)$:

$y = -2x + 7$ or $2x + y = 7$

Inequalities: $\begin{cases} x \geq 0 \\ 3x - 2y \leq 0 \\ 2x + y \leq 7 \end{cases}$

19. Vertices: $(0, 0), (0, 7), (6, 0), (3, 5)$

$z = 30x + 26y$

At $(0, 0)$: $z = 0$

At $(0, 7)$: $z = 182$

At $(6, 0)$: $z = 180$

At $(3, 5)$: $z = 220$

The optimal value of z occurs at $(3, 5)$ and is 220.

20. $\quad x^2 + y^2 \leq 4$

$(x - 2)^2 + y^2 \geq 4$

Chapter 10   Practice Test Solutions

1. $\begin{bmatrix} 1 & -2 & 4 \\ 3 & -5 & 9 \end{bmatrix}$

$-3R_1 + R_2 \rightarrow \begin{bmatrix} 1 & -2 & 4 \\ 0 & 1 & -3 \end{bmatrix}$

$2R_2 + R_1 \rightarrow \begin{bmatrix} 1 & 0 & -2 \\ 0 & 1 & -3 \end{bmatrix}$

2. $\begin{cases} 3x + 5y = 3 \\ 2x - y = -11 \end{cases}$

$\begin{bmatrix} 3 & 5 & \vdots & 3 \\ 2 & -1 & \vdots & -11 \end{bmatrix}$

$-R_2 + R_1 \rightarrow \begin{bmatrix} 1 & 6 & \vdots & 14 \\ 2 & -1 & \vdots & -11 \end{bmatrix}$

$-2R_1 + R_2 \rightarrow \begin{bmatrix} 1 & 6 & \vdots & 14 \\ 0 & -13 & \vdots & -39 \end{bmatrix}$

$-\frac{1}{13}R_2 \rightarrow \begin{bmatrix} 1 & 6 & \vdots & 14 \\ 0 & 1 & \vdots & 3 \end{bmatrix}$

$-6R_2 + R_1 \rightarrow \begin{bmatrix} 1 & 0 & \vdots & -4 \\ 0 & 1 & \vdots & 3 \end{bmatrix}$

$x = -4, y = 3$

Solution: $(-4, 3)$

3. $\begin{cases} 2x + 3y = -3 \\ 3x - 2y = 8 \\ x + y = 1 \end{cases}$

$$\begin{bmatrix} 2 & 3 & \vdots & -3 \\ 3 & 2 & \vdots & 8 \\ 1 & 1 & \vdots & 1 \end{bmatrix}$$

$$\begin{matrix} R_3 \rightarrow \\ \\ R_1 \rightarrow \end{matrix} \begin{bmatrix} 1 & 1 & \vdots & 1 \\ 3 & 2 & \vdots & 8 \\ 2 & 3 & \vdots & -3 \end{bmatrix}$$

$$\begin{matrix} \\ -3R_1 + R_2 \rightarrow \\ -2R_1 + R_3 \rightarrow \end{matrix} \begin{bmatrix} 1 & 1 & \vdots & 1 \\ 0 & -1 & \vdots & 5 \\ 0 & 1 & \vdots & -5 \end{bmatrix}$$

$$\begin{matrix} \\ -R_2 \rightarrow \\ \\ \end{matrix} \begin{bmatrix} 1 & 1 & \vdots & 1 \\ 0 & 1 & \vdots & -5 \\ 0 & 1 & \vdots & -5 \end{bmatrix}$$

$$\begin{matrix} -R_2 + R_1 \rightarrow \\ \\ -R_2 + R_3 \rightarrow \end{matrix} \begin{bmatrix} 1 & 0 & \vdots & 6 \\ 0 & 1 & \vdots & -5 \\ 0 & 0 & \vdots & 0 \end{bmatrix}$$

$x = 6, y = -5$

Solution: $(6, -5)$

4. $\begin{cases} x \quad + 3z = -5 \\ 2x + y \quad = 0 \\ 3x + y - z = 3 \end{cases}$

$$\begin{bmatrix} 1 & 0 & 3 & \vdots & -5 \\ 2 & 1 & 0 & \vdots & 0 \\ 3 & 1 & -1 & \vdots & 3 \end{bmatrix}$$

$$\begin{matrix} \\ -2R_1 + R_2 \rightarrow \\ -3R_1 + R_3 \rightarrow \end{matrix} \begin{bmatrix} 1 & 0 & 3 & \vdots & -5 \\ 0 & 1 & -6 & \vdots & 10 \\ 0 & 1 & -10 & \vdots & 18 \end{bmatrix}$$

$$\begin{matrix} \\ \\ -R_2 + R_3 \rightarrow \end{matrix} \begin{bmatrix} 1 & 0 & 3 & \vdots & -5 \\ 0 & 1 & -6 & \vdots & 10 \\ 0 & 0 & -4 & \vdots & 8 \end{bmatrix}$$

$$\begin{matrix} \\ \\ -\frac{1}{4}R_3 \rightarrow \end{matrix} \begin{bmatrix} 1 & 0 & 3 & \vdots & -5 \\ 0 & 1 & -6 & \vdots & 10 \\ 0 & 0 & 1 & \vdots & -2 \end{bmatrix}$$

$$\begin{matrix} -3R_3 + R_1 \rightarrow \\ 6R_3 + R_2 \rightarrow \\ \end{matrix} \begin{bmatrix} 1 & 0 & 0 & \vdots & 1 \\ 0 & 1 & 0 & \vdots & -2 \\ 0 & 0 & 1 & \vdots & -2 \end{bmatrix}$$

$x = 1, y = -2, z = -2$

Solution: $(1, -2, -2)$

5. $\begin{bmatrix} 1 & 4 & 5 \\ 2 & 0 & -3 \end{bmatrix} \begin{bmatrix} 1 & 6 \\ 0 & -7 \\ -1 & 2 \end{bmatrix} = \begin{bmatrix} (1)(1) + (4)(0) + (5)(-1) & (1)(6) + (4)(-7) + (5)(2) \\ (2)(1) + (0)(0) + (-3)(-1) & (2)(6) + (0)(-7) + (-3)(2) \end{bmatrix} = \begin{bmatrix} -4 & -12 \\ 5 & 6 \end{bmatrix}$

6. $3A - 5B = 3\begin{bmatrix} 9 & 1 \\ -4 & 8 \end{bmatrix} - 5\begin{bmatrix} 6 & -2 \\ 3 & 5 \end{bmatrix}$

$= \begin{bmatrix} 27 & 3 \\ -12 & 24 \end{bmatrix} - \begin{bmatrix} 30 & -10 \\ 15 & 25 \end{bmatrix}$

$= \begin{bmatrix} -3 & 13 \\ -27 & -1 \end{bmatrix}$

7. $f(A) = \begin{bmatrix} 3 & 0 \\ 7 & 1 \end{bmatrix}^2 - 7\begin{bmatrix} 3 & 0 \\ 7 & 1 \end{bmatrix} + 8\begin{bmatrix} 1 & 0 \\ 0 & 1 \end{bmatrix}$

$= \begin{bmatrix} 3 & 0 \\ 7 & 1 \end{bmatrix}\begin{bmatrix} 3 & 0 \\ 7 & 1 \end{bmatrix} - \begin{bmatrix} 21 & 0 \\ 49 & 7 \end{bmatrix} + \begin{bmatrix} 8 & 0 \\ 0 & 8 \end{bmatrix}$

$= \begin{bmatrix} 9 & 0 \\ 28 & 1 \end{bmatrix} - \begin{bmatrix} 21 & 0 \\ 49 & 7 \end{bmatrix} + \begin{bmatrix} 8 & 0 \\ 0 & 8 \end{bmatrix}$

$= \begin{bmatrix} -4 & 0 \\ -21 & 2 \end{bmatrix}$

8. False since

$(A + B)(A + 3B) = A(A + 3B) + B(A + 3B)$

$\qquad\qquad = A^2 + 3AB + BA + 3B^2$

and, in general, $AB \neq BA$.

9.

$$\begin{bmatrix} 1 & 2 & \vdots & 1 & 0 \\ 3 & 5 & \vdots & 0 & 1 \end{bmatrix}$$

$$-3R_1 + R_2 \rightarrow \begin{bmatrix} 1 & 2 & \vdots & 1 & 0 \\ 0 & -1 & \vdots & -3 & 1 \end{bmatrix}$$

$$2R_2 + R_1 \rightarrow \begin{bmatrix} 1 & 0 & \vdots & -5 & 2 \\ 0 & -1 & \vdots & -3 & 1 \end{bmatrix}$$

$$-R_2 \rightarrow \begin{bmatrix} 1 & 0 & \vdots & -5 & 2 \\ 0 & 1 & \vdots & 3 & -1 \end{bmatrix}$$

$A^{-1} = \begin{bmatrix} -5 & 2 \\ 3 & -1 \end{bmatrix}$

10.
$$\begin{bmatrix} 1 & 1 & 1 & \vdots & 1 & 0 & 0 \\ 3 & 6 & 5 & \vdots & 0 & 1 & 0 \\ 6 & 10 & 8 & \vdots & 0 & 0 & 1 \end{bmatrix}$$

$$\begin{array}{c} \\ -3R_1 + R_2 \rightarrow \\ -6R_1 + R_3 \rightarrow \end{array} \begin{bmatrix} 1 & 1 & 1 & \vdots & 1 & 0 & 0 \\ 0 & 3 & 2 & \vdots & -3 & 1 & 0 \\ 0 & 4 & 2 & \vdots & -6 & 0 & 1 \end{bmatrix}$$

$$\begin{array}{c} \\ -R_3 + R_2 \rightarrow \\ \\ \end{array} \begin{bmatrix} 1 & 1 & 1 & \vdots & 1 & 0 & 0 \\ 0 & -1 & 0 & \vdots & 3 & 1 & -1 \\ 0 & 4 & 2 & \vdots & -6 & 0 & 1 \end{bmatrix}$$

$$\begin{array}{c} R_2 + R_1 \rightarrow \\ \\ 4R_2 + R_3 \rightarrow \end{array} \begin{bmatrix} 1 & 0 & 1 & \vdots & 4 & 1 & -1 \\ 0 & -1 & 0 & \vdots & 3 & 1 & -1 \\ 0 & 0 & 2 & \vdots & 6 & 4 & -3 \end{bmatrix}$$

$$\begin{array}{c} \\ -R_2 \rightarrow \\ \tfrac{1}{2}R_3 \rightarrow \end{array} \begin{bmatrix} 1 & 0 & 1 & \vdots & 4 & 1 & -1 \\ 0 & 1 & 0 & \vdots & -3 & -1 & 1 \\ 0 & 0 & 1 & \vdots & 3 & 2 & -\tfrac{3}{2} \end{bmatrix}$$

$$\begin{array}{c} -R_3 + R_1 \rightarrow \\ \\ \\ \end{array} \begin{bmatrix} 1 & 0 & 0 & \vdots & 1 & -1 & \tfrac{1}{2} \\ 0 & 1 & 0 & \vdots & -3 & -1 & 1 \\ 0 & 0 & 1 & \vdots & 3 & 2 & -\tfrac{3}{2} \end{bmatrix}$$

$$A^{-1} = \begin{bmatrix} 1 & -1 & \tfrac{1}{2} \\ -3 & -1 & 1 \\ 3 & 2 & -\tfrac{3}{2} \end{bmatrix}$$

11. (a) $\begin{cases} x + 2y = 4 \\ 3x + 5y = 1 \end{cases}$

$$A = \begin{bmatrix} 1 & 2 \\ 3 & 5 \end{bmatrix}$$

$$A^{-1} = \frac{1}{5-6}\begin{bmatrix} 5 & -2 \\ -3 & 1 \end{bmatrix} = \begin{bmatrix} -5 & 2 \\ 3 & -1 \end{bmatrix}$$

$$\begin{bmatrix} x \\ y \end{bmatrix} = A^{-1}B = \begin{bmatrix} -5 & 2 \\ 3 & -1 \end{bmatrix}\begin{bmatrix} 4 \\ 1 \end{bmatrix} = \begin{bmatrix} -18 \\ 11 \end{bmatrix}$$

$x = -18, y = 11$

Solution: $(-18, 11)$

(b) $\begin{cases} x + 2y = 3 \\ 3x + 5y = -2 \end{cases}$

Again, $A^{-1} = \begin{bmatrix} -5 & 2 \\ 3 & -1 \end{bmatrix}$.

$$\begin{bmatrix} x \\ y \end{bmatrix} = A^{-1}B = \begin{bmatrix} -5 & 2 \\ 3 & -1 \end{bmatrix}\begin{bmatrix} 3 \\ -2 \end{bmatrix} = \begin{bmatrix} -19 \\ 11 \end{bmatrix}$$

$x = -19, y = 11$

Solution: $(-19, 11)$

12. $\begin{vmatrix} 6 & -1 \\ 3 & 4 \end{vmatrix} = 24 - (-3) = 27$

13. $\begin{vmatrix} 1 & 3 & -1 \\ 5 & 9 & 0 \\ 6 & 2 & -5 \end{vmatrix} = -1\begin{vmatrix} 5 & 9 \\ 6 & 2 \end{vmatrix} - 5\begin{vmatrix} 1 & 3 \\ 5 & 9 \end{vmatrix} = -(-44) - 5(-6) = 74$

14. Expand along Row 2.

$$\begin{vmatrix} 1 & 4 & 2 & 3 \\ 0 & 1 & -2 & 0 \\ 3 & 5 & -1 & 1 \\ 2 & 0 & 6 & 1 \end{vmatrix} = \begin{vmatrix} 1 & 2 & 3 \\ 3 & -1 & 1 \\ 2 & 6 & 1 \end{vmatrix} + 2\begin{vmatrix} 1 & 4 & 3 \\ 3 & 5 & 1 \\ 2 & 0 & 1 \end{vmatrix}$$

$$= 51 + 2(-29) = -7$$

15. $\begin{vmatrix} 6 & 4 & 3 & 0 & 6 \\ 0 & 5 & 1 & 4 & 8 \\ 0 & 0 & 2 & 7 & 3 \\ 0 & 0 & 0 & 9 & 2 \\ 0 & 0 & 0 & 0 & 1 \end{vmatrix} = 6\begin{vmatrix} 5 & 1 & 4 & 8 \\ 0 & 2 & 7 & 3 \\ 0 & 0 & 9 & 2 \\ 0 & 0 & 0 & 1 \end{vmatrix} = 6(5)\begin{vmatrix} 2 & 7 & 3 \\ 0 & 9 & 2 \\ 0 & 0 & 1 \end{vmatrix} = 6(5)(2)\begin{vmatrix} 9 & 2 \\ 0 & 1 \end{vmatrix} = 6(5)(2)(9) = 540$

16. Area $= \frac{1}{2}\begin{vmatrix} 0 & 7 & 1 \\ 5 & 0 & 1 \\ 3 & 9 & 1 \end{vmatrix} = \frac{1}{2}(31) = \frac{31}{2}$

17. $\begin{vmatrix} x & y & 1 \\ 2 & 7 & 1 \\ -1 & 4 & 1 \end{vmatrix} = 3x - 3y + 15 = 0$

or equivalently, $x - y + 5 = 0$

18. $x = \dfrac{\begin{vmatrix} 4 & -7 \\ 11 & 5 \end{vmatrix}}{\begin{vmatrix} 6 & -7 \\ 2 & 5 \end{vmatrix}} = \dfrac{97}{44}$

20. $y = \dfrac{\begin{vmatrix} 721.4 & 33.77 \\ 45.9 & 19.85 \end{vmatrix}}{\begin{vmatrix} 721.4 & -29.1 \\ 45.9 & 105.6 \end{vmatrix}} = \dfrac{12{,}769.747}{77{,}515.530} \approx 0.1647$

19. $z = \dfrac{\begin{vmatrix} 3 & 0 & 1 \\ 0 & 1 & 3 \\ 1 & -1 & 2 \end{vmatrix}}{\begin{vmatrix} 3 & 0 & 1 \\ 0 & 1 & 4 \\ 1 & -1 & 0 \end{vmatrix}} = \dfrac{14}{11}$

Chapter 11   Practice Test Solutions

1. $a_n = \dfrac{2n}{(n + 2)!}$

$a_1 = \dfrac{2(1)}{3!} = \dfrac{2}{6} = \dfrac{1}{3}$

$a_2 = \dfrac{2(2)}{4!} = \dfrac{4}{24} = \dfrac{1}{6}$

$a_3 = \dfrac{2(3)}{5!} = \dfrac{6}{120} = \dfrac{1}{20}$

$a_4 = \dfrac{2(4)}{6!} = \dfrac{8}{720} = \dfrac{1}{90}$

$a_5 = \dfrac{2(5)}{7!} = \dfrac{10}{5040} = \dfrac{1}{504}$

Terms: $\dfrac{1}{3}, \dfrac{1}{6}, \dfrac{1}{20}, \dfrac{1}{90}, \dfrac{1}{504}$

2. $a_n = \dfrac{n + 3}{3^n}$

3. $\displaystyle\sum_{i=1}^{6} (2i - 1) = 1 + 3 + 5 + 7 + 9 + 11 = 36$

4. $a_1 = 23, \; d = -2$

$a_2 = 23 + (-2) = 21$

$a_3 = 21 + (-2) = 19$

$a_4 = 19 + (-2) = 17$

$a_5 = 17 + (-2) = 15$

Terms: 23, 21, 19, 17, 15

5. $a_1 = 12, d = 3, n = 50$

$a_n = a_1 + (n - 1)d$

$a_{50} = 12 + (50 - 1)3 = 159$

6. $a_1 = 1$

$a_{200} = 200$

$S_n = \dfrac{n}{2}(a_1 + a_n)$

$S_{200} = \dfrac{200}{2}(1 + 200) = 20{,}100$

7. $a_1 = 7, \; r = 2$

$a_2 = 7(2) = 14$

$a_3 = 7(2)^2 = 28$

$a_4 = 7(2)^3 = 56$

$a_5 = 7(2)^4 = 112$

Terms: 7, 14, 28, 56, 112

8. $\sum\limits_{n=1}^{10} 6\left(\frac{2}{3}\right)^{n-1}, a_1 = 6, r = \frac{2}{3}, n = 10$

$S_n = \dfrac{a_1(1 - r^n)}{1 - r} = \dfrac{6\left[1 - \left(\frac{2}{3}\right)^{10}\right]}{1 - \frac{2}{3}} = 18\left(1 - \dfrac{1024}{59,049}\right) = \dfrac{116,050}{6561} \approx 17.6879$

9. $\sum\limits_{n=0}^{\infty} (0.03)^n = \sum\limits_{n=1}^{\infty} (0.03)^{n-1}, a_1 = 1, r = 0.03$

$S = \dfrac{a_1}{1 - r} = \dfrac{1}{1 - 0.03} = \dfrac{1}{0.97} = \dfrac{100}{97} \approx 1.0309$

10. For $n = 1, 1 = \dfrac{1(1 + 1)}{2}$.

Assume that $S_k = 1 + 2 + 3 + 4 + \cdots + k = \dfrac{k(k + 1)}{2}$.

Then $S_{k+1} = 1 + 2 + 3 + 4 + \cdots + k + (k + 1) = \dfrac{k(k + 1)}{2} + k + 1$

$$= \dfrac{k(k + 1)}{2} + \dfrac{2(k + 1)}{2}$$

$$= \dfrac{(k + 1)(k + 2)}{2}.$$

Thus, by the principle of mathematical induction, $1 + 2 + 3 + 4 + \cdots + n = \dfrac{n(n + 1)}{2}$ for all integers $n \geq 1$.

11. For $n = 4, 4! > 2^4$. Assume that $k! > 2^k$.

Then $(k + 1)! = (k + 1)(k!) > (k + 1)2^k > 2 \cdot 2^k = 2^{k+1}$.

Thus, by the extended principle of mathematical induction, $n! > 2^n$ for all integers $n \geq 4$.

12. $_{13}C_4 = \dfrac{13!}{(13 - 4)!4!} = 715$

13. $(x + 3)^5 = x^5 + 5x^4(3) + 10x^3(3)^2 + 10x^2(3)^3 + 5x(3)^4 + (3)^5$

$\qquad\quad = x^5 + 15x^4 + 90x^3 + 270x^2 + 405x + 243$

14. $_{12}C_5 x^7(2)^5 = 25,344x^7$

15. $_{30}P_4 = \dfrac{30!}{(30 - 4)!} = 657,720$

16. $6! = 720$ ways

17. $_{12}P_3 = 1320$

18. $P(2) + P(3) + P(4) = \dfrac{1}{36} + \dfrac{2}{36} + \dfrac{3}{36}$

$$= \dfrac{6}{36} = \dfrac{1}{6}$$

19. $P(K, B10) = \dfrac{4}{52} \cdot \dfrac{2}{51} = \dfrac{2}{663}$

20. Let A = probability of no faulty units.

$P(A) = \left(\dfrac{997}{1000}\right)^{50} \approx 0.8605$

$P(A') = 1 - P(A) \approx 0.1395$

PART II

Chapter P   Chapter Test Solutions

1. $-\frac{10}{3} = -3\frac{1}{3}$

$-|-4| = -4$

$-\frac{10}{3} > -|-4|$

2. $\left| -5.4 - 3\frac{3}{4} \right| = 9.15$

3. $(5 - x) + 0 = 5 - x$

Additive Identity Property

4. (a) $27\left(-\frac{2}{3}\right) = -18$

(b) $\dfrac{5}{18} \div \dfrac{5}{8} = \dfrac{5}{18} \cdot \dfrac{8}{5} = \dfrac{4}{9}$

(c) $\left(-\dfrac{3}{5}\right)^3 = -\dfrac{27}{125}$

(d) $\left(\dfrac{3^2}{2}\right)^{-3} = \left(\dfrac{2}{9}\right)^3 = \dfrac{8}{729}$

5. (a) $\sqrt{5} \cdot \sqrt{125} = \sqrt{625} = 25$

(b) $\dfrac{\sqrt{27}}{\sqrt{2}} = \dfrac{3\sqrt{3}}{\sqrt{2}} \cdot \dfrac{\sqrt{2}}{\sqrt{2}} = \dfrac{3\sqrt{6}}{2}$

(c) $\dfrac{5.4 \times 10^8}{3 \times 10^3} = \dfrac{5.4}{3} \times 10^{8-3} = 1.8 \times 10^5$

(d) $(3 \times 10^4)^3 = 27 \times 10^{12} = 2.7 \times 10^{13}$

6. (a) $3z^2(2z^3)^2 = 3z^2(4z^6) = 12z^8$

(b) $(u - 2)^{-4}(u - 2)^{-3} = (u - 2)^{-7} = \dfrac{1}{(u - 2)^7}$

(c) $\left(\dfrac{x^{-2}y^2}{3}\right)^{-1} = \dfrac{x^2y^{-2}}{3^{-1}} = \dfrac{3x^2}{y^2}$

7. (a) $9z\sqrt{8z} - 3\sqrt{2z^3} = 18z\sqrt{2z} - 3z\sqrt{2z} = 15z\sqrt{2z}$

Since $\sqrt{8z}$ appears in the expression, we may assume that $z \geq 0$. It is not necessary to use an absolute value when simplifying $\sqrt{2z^3}$.

(b) $(4x^{3/5})(x^{1/3}) = 4x^{(3/5) + (1/3)} = 4x^{14/15}$

(c) $\sqrt[3]{\dfrac{16}{v^5}} = \sqrt[3]{\dfrac{8}{v^6} \cdot 2v} = \dfrac{2}{v^2}\sqrt[3]{2v}$

8. Standard form: $4x^4 - 3x^2 + x - 5$

Degree: 4

Leading coefficient: 4

9. $(x^2 + 3) - [3x + (8 - x^2)] = x^2 + 3 - 3x - 8 + x^2$

$= 2x^2 - 3x - 5$

10. $\left(x + \sqrt{5}\right)\left(x - \sqrt{5}\right) = x^2 - \left(\sqrt{5}\right)^2 = x^2 - 5$

11. $\dfrac{8x}{x - 3} + \dfrac{24}{3 - x} = \dfrac{8x}{x - 3} - \dfrac{24}{x - 3} = \dfrac{8x - 24}{x - 3}$

$= \dfrac{8(x - 3)}{x - 3} = 8, \quad x \neq 3$

12. $\dfrac{\left(\dfrac{2}{x} - \dfrac{2}{x + 1}\right)}{\left(\dfrac{4}{x^2 - 1}\right)} = \dfrac{2(x + 1) - 2x}{x(x + 1)} \cdot \dfrac{x^2 - 1}{4}$

$= \dfrac{2}{x(x + 1)} \cdot \dfrac{(x + 1)(x - 1)}{4}$

$= \dfrac{x - 1}{2x}, x \neq \pm 1$

13. (a) $2x^4 - 3x^3 - 2x^2 = x^2(2x^2 - 3x - 2)$

$= x^2(2x + 1)(x - 2)$

(b) $x^3 + 2x^2 - 4x - 8 = x^2(x + 2) - 4(x + 2)$

$= (x + 2)(x^2 - 4)$

$= (x + 2)(x + 2)(x - 2)$

$= (x + 2)^2(x - 2)$

14. (a) $\dfrac{16}{\sqrt[3]{16}} = \dfrac{16}{\sqrt[3]{16}} \cdot \dfrac{\sqrt[3]{4}}{\sqrt[3]{4}} = \dfrac{16\sqrt[3]{4}}{\sqrt[3]{64}} = \dfrac{16\sqrt[3]{4}}{4} = 4\sqrt[3]{4}$

(b) $\dfrac{6}{1 - \sqrt{3}} = \dfrac{6}{1 - \sqrt{3}} \cdot \dfrac{1 + \sqrt{3}}{1 + \sqrt{3}} = \dfrac{6\left(1 + \sqrt{3}\right)}{1 - 3} = -3\left(1 + \sqrt{3}\right)$

15. The domain of $\dfrac{1 - x}{4 - x}$ is all real numbers except $x = 4$.

16. $\dfrac{y^2 + 8y + 16}{2y - 4} \cdot \dfrac{8y - 16}{(y + 4)^3} = \dfrac{(y + 4)^2}{2(y - 2)} \cdot \dfrac{8(y - 2)}{(y + 4)^3}$

$$= \dfrac{4}{y + 4}, \; y \neq 2$$

17. $P = R - C$

$\quad = 15x - (1480 + 6x)$

$\quad = 9x - 1480$

When $x = 225$,

$P = 9(225) - 1480$

$\quad = \$545$

18.

Midpoint: $\left(\dfrac{-2 + 6}{2}, \dfrac{5 + 0}{2}\right) = \left(2, \dfrac{5}{2}\right)$

Distance: $d = \sqrt{(-2 - 6)^2 + (5 - 0)^2}$

$\quad = \sqrt{64 + 25}$

$\quad = \sqrt{89}$

19. Area = Area of large triangle − Area of small triangle

$$A = \dfrac{1}{2}(3x)\left(\sqrt{3}x\right) - \dfrac{1}{2}(2x)\left(\dfrac{2}{3}\sqrt{3}x\right)$$

$$= \dfrac{3\sqrt{3}x^2}{2} - \dfrac{2\sqrt{3}x^2}{3}$$

$$= \dfrac{9\sqrt{3}x^2 - 4\sqrt{3}x^2}{6}$$

$$= \dfrac{5\sqrt{3}x^2}{6}$$

$$= \dfrac{5}{6}\sqrt{3}x^2$$

Chapter 1 · Chapter Test Solutions

1. $y = 4 - \frac{3}{4}x$

No symmetry

x-intercept: $\left(\frac{16}{3}, 0\right)$

y-intercept: $(0, 4)$

2. $y = 4 - \frac{3}{4}|x|$

y-axis symmetry

x-intercepts: $\left(\pm\frac{16}{3}, 0\right)$

y-intercept: $(0, 4)$

3. $y = 4 - (x - 2)^2$

Parabola; vertex: $(2, 4)$

No x-axis, y-axis, or origin symmetry

x-intercepts: $(0, 0)$ and $(4, 0)$

$$0 = 4 - (x - 2)^2$$

$$(x - 2)^2 = 4$$

$$x - 2 = \pm 2$$

$$x = 2 \pm 2$$

$$x = 4 \quad \text{or} \quad x = 0$$

y-intercept: $(0, 0)$

4. $y = x - x^3$

Origin symmetry

x-intercepts: $(0, 0), (1, 0), (-1, 0)$

$$0 = x - x^3$$

$$0 = x(1 + x)(1 - x)$$

$$x = 0, x = \pm 1$$

y-intercept: $(0, 0)$

5. $y = \sqrt{3 - x}$

Domain: $x \leq 3$

No symmetry

x-intercept: $(3, 0)$

y-intercept: $\left(0, \sqrt{3}\right)$

6. $(x - 3)^2 + y^2 = 9$

Circle

Center: $(3, 0)$

Radius: 3

x-axis symmetry

x-intercepts: $(0, 0)$ and $(6, 0)$

y-intercept: $(0, 0)$

7. $\frac{2}{3}(x - 1) + \frac{1}{4}x = 10$

$$12\left[\frac{2}{3}(x - 1) + \frac{1}{4}x\right] = 12(10)$$

$$8(x - 1) + 3x = 120$$

$$8x - 8 + 3x = 120$$

$$11x = 128$$

$$x = \frac{128}{11}$$

8. $(x - 3)(x + 2) = 14$

$$x^2 - x - 6 = 14$$

$$x^2 - x - 20 = 0$$

$$(x + 4)(x - 5) = 0$$

$$x = -4 \quad \text{or} \quad x = 5$$

9. $\dfrac{x - 2}{x + 2} + \dfrac{4}{x + 2} + 4 = 0, \quad x \neq -2$

$$\frac{x + 2}{x + 2} = -4$$

$$1 \neq -4 \implies \text{No solution}$$

10. $x^4 + x^2 - 6 = 0$

$$(x^2 - 2)(x^2 + 3) = 0$$

$$x^2 = 2 \implies x = \pm\sqrt{2}$$

$$x^2 = -3 \implies x = \pm\sqrt{3}i$$

11. $2\sqrt{x} - \sqrt{2x+1} = 1$

$-\sqrt{2x+1} = 1 - 2\sqrt{x}$

$\left(-\sqrt{2x+1}\right)^2 = \left(1 - 2\sqrt{x}\right)^2$

$2x+1 = 1 - 4\sqrt{x} + 4x$

$-2x = -4\sqrt{x}$

$x = 2\sqrt{x}$

$x^2 = 4x$

$x^2 - 4x = 0$

$x(x-4) = 0$

$x = 0 \ \text{ or } \ x = 4$

Only $x = 4$ is a solution to the original equation.
$x = 0$ is extraneous.

12. $|3x - 1| = 7$

$3x - 1 = 7 \ \text{ or } \ 3x - 1 = -7$

$3x = 8 \qquad\qquad 3x = -6$

$x = \frac{8}{3} \qquad\qquad x = -2$

13. $-3 \le 2(x+4) < 14$

$-3 \le 2x + 8 < 14$

$-11 \le 2x < 6$

$-\frac{11}{2} \le x < 3$

14. $\dfrac{2}{x} > \dfrac{5}{x+6}$

$\dfrac{2}{x} - \dfrac{5}{x+6} > 0$

$\dfrac{2(x+6) - 5x}{x(x+6)} > 0$

$\dfrac{-3x + 12}{x(x+6)} > 0$

$\dfrac{-3(x-4)}{x(x+6)} > 0$

Critical numbers: $x = 4, x = 0, x = -6$

Test intervals: $(-\infty, -6), (-6, 0), (0, 4), (4, \infty)$

Test: Is $\dfrac{-3(x-4)}{x(x+6)} > 0$?

Solution set: $(-\infty, -6) \cup (0, 4)$

In inequality notation: $x < -6 \ \text{ or } \ 0 < x < 4$

15. $2x^2 + 5x > 12$

$2x^2 + 5x - 12 > 0$

$(2x - 3)(x + 4) > 0$

Critical numbers: $x = \frac{3}{2}, x = -4$

Test intervals: $(-\infty, -4), \left(-4, \frac{3}{2}\right), \left(\frac{3}{2}, \infty\right)$

Test: Is $(2x - 3)(x + 4) > 0$?

Solution set: $(-\infty, -4) \cup \left(\frac{3}{2}, \infty\right)$

In inequality notation: $x < -4 \ \text{or} \ x > \dfrac{3}{2}$

16. $|x - 15| \ge 5$

$x - 15 \le -5 \ \text{ or } \ x - 15 \ge 5$

$x \le 10 \qquad\qquad x \ge 20$

17. (a) $10i - \left(3 + \sqrt{-25}\right) = 10i - (3 + 5i) = -3 + 5i$

(b) $\left(2 + \sqrt{3}i\right)\left(2 - \sqrt{3}i\right) = 4 - 3i^2 = 4 + 3 = 7$

18. $\dfrac{5}{2+i} = \dfrac{5}{2+i} \cdot \dfrac{2-i}{2-i} = \dfrac{5(2-i)}{4+1} = 2-i$

19. $y = 4.45t - 16.6, \ 4 \le t \le 13$

(a)

(b) For 2008, let $t = 18$: $y \approx \$63.5$ billion dollars

(c) $y = 4.45(18) - 16.6 = 63.5$

20. $V = \dfrac{4}{3}\pi r^3$

$\dfrac{4}{3}\pi r^3 = 455.9$

$r = \sqrt[3]{\dfrac{455.9(3)}{4\pi}} \approx 4.774$ inches

21. $(100 \text{ km/hr})(2\frac{1}{4} \text{ hr}) + (x \text{ km/hr})(1\frac{1}{3} \text{ hr}) = 350 \text{ km}$

$225 + \frac{4}{3}x = 350$

$\frac{4}{3}x = 125$

$x = \frac{375}{4} = 93\frac{3}{4} \text{ km/hr}$

22. $a + b = 100 \Longrightarrow b = 100 - a$

Area of ellipse = Area of circle

$\pi ab = \pi(40)^2$

$a(100 - a) = 1600$

$0 = a^2 - 100a + 1600$

$0 = (a - 80)(a - 20)$

$a = 80 \Longrightarrow b = 20$

OR

$a = 20 \Longrightarrow b = 80$

Since $a > b$, we choose $a = 80$ and $b = 20$.

Chapter 2   Chapter Test Solutions

1. $(2, -3)$ and $(-4, 9)$

$m = \dfrac{9 - (-3)}{-4 - 2} = -2$

$y - (-3) = -2(x - 2)$

$y + 3 = -2x + 4$

$y = -2x + 1$

$2x + y - 1 = 0$

2. $(3, 0.8)$ and $(7, -6)$

$m = \dfrac{-6 - 0.8}{7 - 3} = -1.7$

$y - (-6) = -1.7(x - 7)$

$y + 6 = -1.7x + 11.9$

$y = -1.7x + 5.9$

$10y = -17x + 59$

$17x + 10y - 59 = 0$

3. $-4x + 7y = -5$

$$7y = 4x - 5$$

$$y = \tfrac{4}{7}x - \tfrac{5}{7} \implies m_1 = \tfrac{4}{7}$$

(a) Parallel line: $m_2 = \tfrac{4}{7}$

$$y - 8 = \tfrac{4}{7}(x - 3)$$

$$7y - 56 = 4x - 12$$

$$-4x + 7y = 44$$

$$4x - 7y + 44 = 0$$

(b) Perpendicular line: $m_2 = -\tfrac{7}{4}$

$$y - 8 = -\tfrac{7}{4}(x - 3)$$

$$4y - 32 = -7x + 21$$

$$7x + 4y - 53 = 0$$

4. $f(x) = |x + 2| - 15$

(a) $f(-8) = -9$

(b) $f(14) = 1$

(c) $f(x - 6) = |x - 4| - 15$

5. $f(x) = \dfrac{\sqrt{x + 9}}{x^2 - 81}$

(a) $f(7) = \dfrac{4}{-32} = -\dfrac{1}{8}$

(b) $f(-5) = \dfrac{2}{-56} = -\dfrac{1}{28}$

(c) $f(x - 9) = \dfrac{\sqrt{x}}{(x - 9)^2 - 81} = \dfrac{\sqrt{x}}{x^2 - 18x}$

6. $f(x) = \sqrt{100 - x^2}$

Domain: $100 - x^2 \geq 0 \implies -10 \leq x \leq 10$ or $[-10, 10]$

7. $f(x) = |-x + 6| + 2$

Domain: All real numbers or $(-\infty, \infty)$

8. $f(x) = 2x^6 + 5x^4 - x^2$

(a)

(b) Increasing on $(-0.31, 0)$, $(0.31, \infty)$

Decreasing on $(-\infty, -0.31)$, $(0, 0.31)$

(c) y-axis symmetry $\implies$ the function is even.

9. $f(x) = 4x\sqrt{3 - x}$

(a)

(b) Increasing on $(-\infty, 2)$,

Decreasing on $(2, 3)$

(c) The function is neither odd nor even.

10. $f(x) = |x + 5|$

(a)

(b) Increasing on $(-5, \infty)$, Decreasing on $(-\infty, -5)$

(c) The function is neither odd nor even.

11. $f(x) = -x^3 + 2x - 1$

Relative minimum: $(-0.816, -2.089)$

Relative maximum: $(0.816, 0.089)$

12. $f(x) = -2x^2 + 5x - 3$

$$\frac{f(3) - f(1)}{3 - 1} = \frac{-6 - 0}{2} = -3$$

The average rate of change of f from $x_1 = 1$ to $x_2 = 3$ is -3.

13. $f(x) = \begin{cases} 3x + 7, & x \le -3 \\ 4x^2 - 1, & x > -3 \end{cases}$

14. $h(x) = -[\![x]\!]$

(a) Common function: $f(x) = [\![x]\!]$

(b) Transformation: Reflection in the x-axis

(c)

15. $h(x) = -\sqrt{x + 5} + 8$

(a) Common function: $f(x) = \sqrt{x}$

(b) Transformation: Reflection in the x-axis, a horizontal shift 5 units to the left, and a vertical shift 8 units upward

(c)

16. $h(x) = \frac{1}{4}|x + 1| - 3$

(a) Common fraction: $f(x) = |x|$

(b) Transformation: Vertical shrink, horizontal shift 1 unit to the left, vertical shift 3 units down

(c)

17. $f(x) = 3x^2 - 7$, $g(x) = -x^2 - 4x + 5$

(a) $(f + g)(x) = (3x^2 - 7) + (-x^2 - 4x + 5) = 2x^2 - 4x - 2$

(b) $(f - g)(x) = (3x^2 - 7) - (-x^2 - 4x + 5) = 4x^2 + 4x - 12$

(c) $(fg)(x) = (3x^2 - 7)(-x^2 - 4x + 5) = -3x^4 - 12x^3 + 22x^2 + 28x - 35$

(d) $\left(\dfrac{f}{g}\right)(x) = \dfrac{3x^2 - 7}{-x^2 - 4x + 5}$, $x \ne -5, 1$

(e) $(f \circ g)(x) = f(g(x)) = f(-x^2 - 4x + 5) = 3(-x^2 - 4x + 5)^2 - 7 = 3x^4 + 24x^3 + 18x^2 - 120x + 68$

(f) $(g \circ f)(x) = g(f(x)) = g(3x^2 - 7) = -(3x^2 - 7)^2 - 4(3x^2 - 7) + 5 = -9x^4 + 30x^2 - 16$

18. $f(x) = \dfrac{1}{x}$, $g(x) = 2\sqrt{x}$

(a) $(f + g)(x) = \dfrac{1}{x} + 2\sqrt{x} = \dfrac{1 + 2x^{3/2}}{x}$, $x > 0$

(b) $(f - g)(x) = \dfrac{1}{x} - 2\sqrt{x} = \dfrac{1 - 2x^{3/2}}{x}$, $x > 0$

(c) $(fg)(x) = \left(\dfrac{1}{x}\right)(2\sqrt{x}) = \dfrac{2\sqrt{x}}{x}$, $x > 0$

(d) $\left(\dfrac{f}{g}\right)(x) = \dfrac{1/x}{2\sqrt{x}} = \dfrac{1}{2x\sqrt{x}} = \dfrac{1}{2x^{3/2}}$, $x > 0$

(e) $(f \circ g)(x) = f(g(x)) = f(2\sqrt{x}) = \dfrac{1}{2\sqrt{x}} = \dfrac{\sqrt{x}}{2x}$, $x > 0$

(f) $(g \circ f)(x) = g(f(x)) = g\left(\dfrac{1}{x}\right) = 2\sqrt{\dfrac{1}{x}} = \dfrac{2}{\sqrt{x}} = \dfrac{2\sqrt{x}}{x}$, $x > 0$

19. $f(x) = x^3 + 8$

Since f is one-to-one, f has an inverse.

$$y = x^3 + 8$$
$$x = y^3 + 8$$
$$x - 8 = y^3$$
$$\sqrt[3]{x - 8} = y$$
$$f^{-1}(x) = \sqrt[3]{x - 8}$$

20. $f(x) = |x^2 - 3| + 6$

Since f is not one-to-one, f does not have an inverse.

21. $f(x) = \dfrac{3x\sqrt{x}}{8} = \dfrac{3}{8}x^{3/2}$, $x \geq 0$

Since f is one-to-one, f has an inverse.

$$y = \dfrac{3}{8}x^{3/2}$$
$$x = \dfrac{3}{8}y^{3/2}$$
$$\dfrac{8}{3}x = y^{3/2}$$
$$\left(\dfrac{8}{3}x\right)^{2/3} = y, \ x \geq 0$$
$$f^{-1}(x) = \left(\dfrac{8}{3}x\right)^{2/3}, \ x \geq 0$$

22. $(6, 58)$ and $(10, 78)$

$$m = \dfrac{78 - 58}{10 - 6} = 5$$
$$C - 58 = 5(x - 6)$$
$$C = 5x + 28$$

When $x = 25$: $C = 5(25) + 28 = \$153$

Chapters P–2   Cumulative Test Solutions

1. $\dfrac{8x^2y^{-3}}{30x^{-1}y^2} = \dfrac{8x^2x}{30y^2y^3} = \dfrac{4x^3}{15y^5}$, $x \neq 0$

2. $\sqrt{24x^4y^3} = \sqrt{4x^4y^2 6y} = 2x^2y\sqrt{6y}$

3. $4x - [2x + 3(2 - x)] = 4x - [2x + 6 - 3x]$
$$= 4x - [-x + 6]$$
$$= 5x - 6$$

4. $(x - 2)(x^2 + x - 3) = x^3 + x^2 - 3x - 2x^2 - 2x + 6$
$$= x^3 - x^2 - 5x + 6$$

5. $\dfrac{2}{s + 3} - \dfrac{1}{s + 1} = \dfrac{2(s + 1) - (s + 3)}{(s + 3)(s + 1)}$
$$= \dfrac{2s + 2 - s - 3}{(s + 3)(s + 1)}$$
$$= \dfrac{s - 1}{(s + 3)(s + 1)}$$

6. $25 - (x - 2)^2 = [5 + (x - 2)][5 - (x - 2)]$
$$= (3 + x)(7 - x)$$

7. $x - 5x^2 - 6x^3 = x(1 - 5x - 6x^2)$

$ = x(1 + x)(1 - 6x)$

8. $54 - 16x^3 = 2(27 - 8x^3)$

$ = 2(3 - 2x)(9 + 6x + 4x^2)$

9. $x(2x + 4) + 2x(x + 4) = 2x^2 + 4x + 2x^2 + 8x$

$ = 4x^2 + 12x$

10. $\frac{1}{2}(x + 5)[(x - 1) + 2(x + 1)] = \frac{1}{2}(x + 5)[x - 1 + 2x + 2]$

$\phantom{\frac{1}{2}(x + 5)[(x - 1) + 2(x + 1)]} = \frac{1}{2}(x + 5)(3x + 1)$

$\phantom{\frac{1}{2}(x + 5)[(x - 1) + 2(x + 1)]} = \frac{3}{2}x^2 + 8x + \frac{5}{2}$

11. $x - 3y + 12 = 0$

Line

x-intercept: $(-12, 0)$

y-intercept: $(0, 4)$

12. $y = x^2 - 9$

Parabola

x-intercepts: $(\pm 3, 0)$

y-intercept: $(0, -9)$

13. $y = \sqrt{4 - x}$

Domain: $x \leq 4$

x-intercept: $(4, 0)$

y-intercept: $(0, 2)$

14. $2x - 4 = 7x + 5$

$-5x = 9$

$x = -\frac{9}{5} = -1.8$

15. $-(x + 3) = 14(x - 6)$

$-x - 3 = 14x - 84$

$-15x = -81$

$x = \dfrac{-81}{-15} = \dfrac{27}{5} = 5.4$

16. $\dfrac{1}{x - 2} = \dfrac{10}{4x + 3}$

$4x + 3 = 10(x - 2)$

$4x + 3 = 10x - 20$

$-6x = -23$

$x = \dfrac{23}{6} \approx 3.833$

17. Factoring

$x^2 - 4x + 3 = 0$

$(x - 1)(x - 3) = 0$

$x - 1 = 0 \implies x = 1$

$x - 3 = 0 \implies x = 3$

18. Completing the Square

$-2x^2 + 8x + 12 = 0$

$-2(x^2 - 4x - 6) = 0$

$x^2 - 4x - 6 = 0$

$x^2 - 4x = 6$

$x^2 - 4x + 4 = 6 + 4$

$(x - 2)^2 = 10$

$x - 2 = \pm\sqrt{10}$

$x = 2 \pm \sqrt{10}$

19. Extracting Square Roots

$\frac{3}{4}x^2 = 12$

$x^2 = \frac{4}{3}(12)$

$x^2 = 16$

$x = \pm\sqrt{16}$

$x = \pm 4$

20. Quadratic Formula

$3x^2 + 5x - 6 = 0$

$a = 3, b = 5, c = -6$

$x = \dfrac{-5 \pm \sqrt{5^2 - 4(3)(-6)}}{2(3)} = \dfrac{-5 \pm \sqrt{25 + 72}}{6} = \dfrac{-5 \pm \sqrt{97}}{6}$

21. Quadratic Formula

$3x^2 + 9x + 1 = 0$

$a = 3, b = 9, c = 1$

$x = \dfrac{-9 \pm \sqrt{9^2 - 4(3)(1)}}{2(3)}$

$= \dfrac{-9 \pm \sqrt{81 - 12}}{6}$

$= \dfrac{-9 \pm \sqrt{69}}{6}$

$= -\dfrac{3}{2} \pm \dfrac{\sqrt{69}}{6}$

22. Extracting Square Roots

$\dfrac{1}{2}x^2 - 7 = 25$

$\dfrac{1}{2}x^2 = 32$

$x^2 = 64$

$x = \pm\sqrt{64}$

$x = \pm 8$

23. $x^4 + 12x^3 + 4x^2 + 48x = 0$

$x^3(x + 12) + 4x(x + 12) = 0$

$(x^3 + 4x)(x + 12) = 0$

$x(x^2 + 4)(x + 12) = 0$

$x = 0$

$x^2 + 4 = 0 \implies x = \pm 2i$

$x + 12 = 0 \implies x = -12$

24. $8x^3 - 48x^2 + 72x = 0$

$8x(x^2 - 6x + 9) = 0$

$8x(x - 3)^2 = 0$

$8x = 0 \implies x = 0$

$x - 3 = 0 \implies x = 3$

25. $x^{2/3} + 13 = 17$

$\sqrt[3]{x^2} = 4$

$x^2 = 4^3$

$x = \pm\sqrt{64}$

$x = \pm 8$

26. $\sqrt{x + 10} = x - 2$

$x + 10 = x^2 - 4x + 4$

$0 = x^2 - 5x - 6$

$0 = (x - 6)(x + 1)$

$x = 6 \quad \text{or} \quad x = -1$

Only $x = 6$ is a solution to the original equation. $x = -1$ is extraneous.

27. $|4(x - 2)| = 28$

$4(x - 2) = -28 \quad \text{or} \quad 4(x - 2) = 28$

$x - 2 = -7 \qquad\qquad x - 2 = 7$

$x = -5 \qquad\qquad\quad x = 9$

28. $|x - 12| = -2$

No solution. The absolute value of a number cannot be negative.

29. $4x + 2 > 7$

(a) $4(-1) + 2 \not> 7$

$x = -1$ *is not* a solution.

(b) $4\left(\frac{1}{2}\right) + 2 \not> 7$

$x = \frac{1}{2}$ *is not* a solution.

(c) $4\left(\frac{3}{2}\right) + 2 > 7$

$x = \frac{3}{2}$ *is* a solution.

(d) $4(2) + 2 > 7$

$x = 2$ *is* a solution.

30. $|5x - 1| < 4$

(a) $|5(-1) - 1| \not< 4$

$x = -1$ is *not* a solution.

(b) $\left|5\left(-\frac{1}{2}\right) - 1\right| < 4$

$x = -\frac{1}{2}$ *is* a solution.

(c) $|5(1) - 1| \not< 4$

$x = 1$ is *not* a solution.

(d) $|5(2) - 1| \not< 4$

$x = 2$ *is not* a solution.

31. $|x + 1| \le 6$

$-6 \le x + 1 \le 6$

$-7 \le \quad x \quad \le 5$

32. $|7 + 8x| > 5$

$7 + 8x < -5 \quad \text{or} \quad 7 + 8x > 5$

$8x < -12 \qquad\qquad 8x > -2$

$x < -\dfrac{3}{2} \qquad\qquad x > -\dfrac{1}{4}$

33. $5x^2 + 12x + 7 \geq 0$

$(5x + 7)(x + 1) \geq 0$

Critical numbers: $x = -\dfrac{7}{5}, -1$

Test intervals: $\left(-\infty, -\dfrac{7}{5}\right), \left(-\dfrac{7}{5}, -1\right), (-1, \infty)$

Test: Is $5x^2 + 12x + 7 \geq 0$?

Solution: $x \leq -\dfrac{7}{5}, x \geq -1$

34. $-x^2 + x + 4 < 0$

$x^2 - x - 4 > 0$

Critical numbers: $x = \dfrac{1 \pm \sqrt{17}}{2}$ (by the Quadratic Formula)

Test intervals: $\left(-\infty, \dfrac{1 - \sqrt{17}}{2}\right), \left(\dfrac{1 - \sqrt{17}}{2}, \dfrac{1 + \sqrt{17}}{2}\right),$

$\left(\dfrac{1 + \sqrt{17}}{2}, \infty\right)$

Test: Is $-x^2 + x + 4 < 0$?

Solution: $x < \dfrac{1 - \sqrt{17}}{2}, x > \dfrac{1 + \sqrt{17}}{2}$

35. $\left(-\dfrac{1}{2}, 1\right)$ and $(3, 8)$

$$m = \frac{8 - 1}{3 - (-1/2)} = \frac{7}{7/2} = 2$$

$y - 8 = 2(x - 3)$

$y - 8 = 2x - 6$

$0 = 2x - y + 2$

36. It fails the Vertical Line Test. For some values of x there correspond two values of y.

37. $f(x) = \dfrac{x}{x - 2}$

(a) $f(6) = \dfrac{6}{4} = \dfrac{3}{2}$

(b) $f(2)$ is undefined because division by zero is undefined.

(c) $f(s + 2) = \dfrac{s + 2}{(s + 2) - 2} = \dfrac{s + 2}{s}$

38. $f(x) = 10 - \sqrt{3 - x}$

$f(-x) = 10 - \sqrt{3 - (-x)} = 10 - \sqrt{3 + x}$

$f(-x) \neq f(x)$ and $f(-x) \neq -f(x)$

The function is neither even nor odd.

39. $f(x) = x^5 - x^3 + 2$

$f(-x) = (-x)^5 - (-x)^3 + 2 = -x^5 + x^3 + 2$

$f(-x) \neq f(x)$ and $f(-x) \neq -f(x)$

The function is neither even nor odd.

40. $f(x) = 2x^4 - 4$

$f(-x) = 2(-x)^4 - 4 = 2x^4 - 4 = f(x)$

The function is even.

41. $y = \sqrt[3]{x}$

(a) $r(x) = \dfrac{1}{2}\sqrt[3]{x}$ is a vertical shrink by a factor of $\dfrac{1}{2}$.

(b) $h(x) = \sqrt[3]{x} + 2$ is a vertical shift two units upward.

(c) $g(x) = \sqrt[3]{x + 2}$ is a horizontal shift two units to the left.

42. $f(x) = x - 3, g(x) = 4x + 1$

(a) $(f + g)(x) = f(x) + g(x)$

$= (x - 3) + (4x + 1)$

$= 5x - 2$

(b) $(f - g)(x) = f(x) - g(x)$

$= (x - 3) - (4x + 1)$

$= -3x - 4$

(c) $(fg)(x) = f(x)g(x)$

$= (x - 3)(4x + 1)$

$= 4x^2 - 11x - 3$

(d) $\left(\dfrac{f}{g}\right)(x) = \dfrac{f(x)}{g(x)} = \dfrac{x - 3}{4x + 1}$

Domain: all real numbers except $x = -\frac{1}{4}$

43. $f(x) = \sqrt{x - 1}, g(x) = x^2 + 1$

(a) $(f + g)(x) = f(x) + g(x)$

$= \sqrt{x - 1} + x^2 + 1$

(b) $(f - g)(x) = f(x) - g(x)$

$= \sqrt{x - 1} - x^2 - 1$

(c) $(fg)(x) = f(x)g(x)$

$= \sqrt{x - 1}(x^2 + 1) = x^2\sqrt{x - 1} + \sqrt{x - 1}$

(d) $\left(\dfrac{f}{g}\right)(x) = \dfrac{f(x)}{g(x)} = \dfrac{\sqrt{x - 1}}{x^2 + 1}$

Domain: $x \geq 1$

44. $f(x) = 2x^2, g(x) = \sqrt{x + 6}$

(a) $(f \circ g)(x) = f(g(x))$

$= f\left(\sqrt{x + 6}\right)$

$= 2\left(\sqrt{x + 6}\right)^2$

$= 2(x + 6)$

$= 2x + 12, \text{Domain: } x \geq -6$

(b) $(g \circ f)(x) = g(f(x))$

$= g(2x^2)$

$= \sqrt{2x^2 + 6}, \text{Domain: all real numbers}$

45. $f(x) = x - 2, g(x) = |x|$

(a) $(f \circ g)(x) = f(g(x))$

$= f(|x|)$

$= |x| - 2, \text{Domain: all real numbers}$

(b) $(g \circ f)(x) = g(f(x))$

$= g(x - 2)$

$= |x - 2|, \text{Domain: all real numbers}$

46. The graph of h is a one-to-one line so h has an inverse.

$h(x) = 5x - 2$

$y = 5x - 2$

$x = 5y - 2$

$x + 2 = 5y$

$\frac{1}{5}(x + 2) = y$

$h^{-1}(x) = \frac{1}{5}(x + 2)$

47. Cost per person: $\dfrac{36{,}000}{n}$

If three additional people join the group, the cost

per person is $\dfrac{36{,}000}{n + 3}$.

$\dfrac{36{,}000}{n} = \dfrac{36{,}000}{n + 3} + 1000$

$36{,}000\,(n + 3) = 36{,}000n + 1000n(n + 3)$

$36(n + 3) = 36n + n(n + 3)$

$36n + 108 = 36n + n^2 + 3n$

$0 = n^2 + 3n - 108$

$0 = (n + 12)(n - 9)$

Choosing the positive value, we have $n = 9$ people.

48. Rate $= 8.00 - 0.05(n - 80), n \geq 80$

(a) Revenue $=$ (number of people)(rate per person)

$= n[8.00 - 0.05(n - 80)]$

$= 8.00n - 0.05n(n - 80)$

$= 8.00n - 0.05n^2 + 4.00n$

$= -0.05n^2 + 12n, n \geq 80$

(b)

The revenue is maximum when $n = 120$ passengers.

49. $s(t) = -16t^2 + 64t + 5$

$$\frac{s(2) - s(0)}{2 - 0} = \frac{69 - 5}{2} = 32$$

The average rate of change in the height of the object from $t = 0$ to $t = 2$ seconds is 32 feet per second.

Chapter 3   Chapter Test Solutions

1. $f(x) = x^2$

(a) $g(x) = 2 - x^2$

Reflection in the x-axis followed by a vertical translation two units upward

(b) $g(x) = \left(x - \frac{3}{2}\right)^2$

Horizontal translation $\frac{3}{2}$ units to the right

2. $y = x^2 + 4x + 3$

$= x^2 + 4x + 4 - 4 + 3$

$= (x + 2)^2 - 1$

Vertex: $(-2, -1)$

x-intercepts: $0 = x^2 + 4x + 3$

$0 = (x + 3)(x + 1)$

$x = -3$ or $x = -1$

$(-3, 0)$ or $(-1, 0)$

y-intercept: $(0, 3)$

3. Vertex: $(3, -6)$

$y = a(x - 3)^2 - 6$

Point on the graph: $(0, 3)$

$3 = a(0 - 3)^2 - 6$

$9 = 9a \Rightarrow a = 1$

Thus, $y = (x - 3)^2 - 6$.

4. (a) $y = -\frac{1}{20}x^2 + 3x + 5$

$= -\frac{1}{20}(x^2 - 60x + 900 - 900) + 5$

$= -\frac{1}{20}[(x - 30)^2 - 900] + 5$

$= -\frac{1}{20}(x - 30)^2 + 50$

Vertex: $(30, 50)$

The maximum height is 50 feet.

(b) The constant term, $c = 5$, determines the height at which the ball was thrown. Changing this constant results in a vertical translation of the graph, and, therefore, changes the maximum height.

5. $h(t) = -\frac{3}{4}t^5 + 2t^2$

The degree is odd and the leading coefficient is negative. The graph rises to the left and falls to the right.

6.

$$\begin{array}{r} 3x + \dfrac{x - 1}{x^2 + 1} \\ x^2 + 0x + 1 \overline{\smash{)}\ 3x^3 + 0x^2 + 4x - 1} \\ \underline{3x^3 + 0x^2 + 3x} \\ x - 1 \end{array}$$

Thus, $\dfrac{3x^3 + 4x - 1}{x^2 + 1} = 3x + \dfrac{x - 1}{x^2 + 1}$.

7. 2 | 2   0   −5   0   −3
          4   8   6   12
     ———————————————
        2   4   3   6   9

Thus, $\dfrac{2x^4 - 5x^2 - 3}{x - 2} = 2x^3 + 4x^2 + 3x + 6 + \dfrac{9}{x - 2}$.

8. $\sqrt{3}$ | 4   −1   −12   3
              $4\sqrt{3}$ $12 - \sqrt{3}$ −3
     ————————————————————————
        4    $4\sqrt{3} - 1$ $-\sqrt{3}$ 0

$-\sqrt{3}$ | 4 $4\sqrt{3} - 1$ $-\sqrt{3}$
                   $-4\sqrt{3}$ $\sqrt{3}$
     ————————————————————
        4   −1   0

$4x^3 - x^2 - 12x + 3 = (x - \sqrt{3})(x + \sqrt{3})(4x - 1)$

The real solutions are $x = \pm\sqrt{3}$ and $x = \frac{1}{4}$.

9. $g(t) = 2t^4 - 3t^3 + 16t - 24$

Possible rational zeros:
$\pm1, \pm2, \pm3, \pm4, \pm6, \pm8, \pm12, \pm24, \pm\frac{1}{2}, \pm\frac{3}{2}$

From the graph, we have $t = -2$ and $t = \frac{3}{2}$.

10. $h(x) = 3x^5 + 2x^4 - 3x - 2$

Possible rational zeros: $\pm1, \pm2, \pm\frac{1}{3}, \pm\frac{2}{3}$

From the graph, we have $x = \pm1$ and $x = -\frac{2}{3}$.

11. $f(x) = x(x - 3)[x - (3 + i)][x - (3 - i)]$

$= (x^2 - 3x)[(x - 3) - i][(x - 3) + i]$

$= (x^2 - 3x)[(x - 3)^2 - i^2]$

$= (x^2 - 3x)(x^2 - 6x + 10)$

$= x^4 - 9x^3 + 28x^2 - 30x$

12. $f(x) = [x - (1 + \sqrt{3}i)][x - (1 - \sqrt{3}i)](x - 2)(x - 2)$

$= [(x - 1) - \sqrt{3}i][(x - 1) + \sqrt{3}i](x^2 - 4x + 4)$

$= [(x - 1)^2 - 3i^2](x^2 - 4x + 4)$

$= (x^2 - 2x + 4)(x^2 - 4x + 4)$

$= x^4 - 6x^3 + 16x^2 - 24x + 16$

13. $f(x) = x^3 + 2x^2 + 5x + 10$

$= x^2(x + 2) + 5(x + 2)$

$= (x + 2)(x^2 + 5)$

$= (x + 2)(x + \sqrt{5}i)(x - \sqrt{5}i)$

Zeros: $x = -2, \pm\sqrt{5}i$

14. $f(x) = x^4 - 9x^2 - 22x - 24$

Possible rational zeros: $\pm1, \pm2, \pm3, \pm4, \pm6, \pm8, \pm12, \pm24$

−2 | 1   0   −9   −22   −24
         −2   4   10   24
    ————————————————————
    1   −2   −5   −12   0

4 | 1   −2   −5   −12
        4   8   12
   ————————————————
   1   2   3   0

$f(x) = (x + 2)(x - 4)(x^2 + 2x + 3)$

By the Quadratic Formula the zeros of $x^2 + 2x + 3$ are
$x = -1 \pm \sqrt{2}i$. The zeros of f are: $x = -2, 4, -1 \pm \sqrt{2}i$.

15. $v = k\sqrt{s}$

$24 = k\sqrt{16}$

$6 = k$

$v = 6\sqrt{s}$

16. $A = kxy$

$500 = k(15)(8)$

$500 = k(120)$

$\dfrac{25}{6} = k$

$A = \dfrac{25}{6}xy$

17. $b = \dfrac{k}{a}$

$32 = \dfrac{k}{1.5}$

$48 = k$

$b = \dfrac{48}{a}$

18. $y = 476.1t + 625.5$

Year, t	Salary, S	Model, y
1	1038	1101.6
2	1600	1577.7
3	2200	2053.8
4	2425	2529.9

From the table (and since $r \approx 0.98$), we see that the model is a fairly good fit to the actual data.

Chapter 4   Chapter Test Solutions

1. $y = \dfrac{2}{4 - x}$

Domain: all real numbers except $x = 4$

Vertical asymptote: $x = 4$

Horizontal asymptote: $y = 0$

2. $f(x) = \dfrac{3 - x^2}{3 + x^2} = \dfrac{-x^2 + 3}{x^2 + 3}$

Domain: all real numbers

Vertical asymptote: None

Horizontal asymptote: $y = \dfrac{-1}{1} = -1$

3. $g(x) = \dfrac{x^2 + 2x - 3}{x - 2} = x + 4 + \dfrac{5}{x - 2}$

Domain: all real numbers except $x = 2$

Vertical asymptote: $x = 2$

Slant asymptote: $y = x + 4$

4. $h(x) = \dfrac{4}{x^2} - 1$

$= \dfrac{4 - x^2}{x^2}$

$= \dfrac{(2 - x)(2 + x)}{x^2}$

Vertical asymptote: $x = 0$

Horizontal asymptote: $y = -1$

x-intercepts: $(\pm 2, 0)$

5. $g(x) = \dfrac{x^2 + 2}{x - 1} = x + 1 + \dfrac{3}{x - 1}$

Vertical asymptote: $x = 1$

Slant asymptote: $y = x + 1$

y-intercept: $(0, -2)$

6. $f(x) = \dfrac{x + 1}{x^2 + x - 12} = \dfrac{x + 1}{(x + 4)(x - 3)}$

Vertical asymptotes: $x = -4$ and $x = 3$

Horizontal asymptote: $y = 0$

x-intercept: $(-1, 0)$

y-intercept: $\left(0, -\dfrac{1}{12}\right)$

7. $f(x) = \dfrac{2x^2 - 5x - 12}{x^2 - 16}$

$= \dfrac{(2x + 3)(x - 4)}{(x + 4)(x - 4)} = \dfrac{2x + 3}{x + 4}, \; x \neq 4$

x-intercept: $\left(-\dfrac{3}{2}, 0\right)$

y-intercept: $\left(0, \dfrac{3}{4}\right)$

Vertical asymptote: $x = -4$

Horizontal asymptote: $y = 2$

8. $f(x) = \dfrac{x^2 - 9}{2x - 6}$

$= \dfrac{(x + 3)(x - 3)}{2(x - 3)} = \dfrac{x + 3}{2}, \; x \neq 3$

x-intercept: $(-3, 0)$

y-intercept: $\left(0, \dfrac{3}{2}\right)$

No horizontal, vertical, or slant asymptotes

9. $g(x) = \dfrac{2x^3 - 7x^2 + 4x + 4}{x^2 - x - 2}$

$= 2x - 5 + \dfrac{3(x - 2)}{(x - 2)(x + 1)}$

$= 2x - 5 + \dfrac{3}{x + 1}$

$= \dfrac{2x^2 - 3x - 2}{x + 1}$

$= \dfrac{(2x + 1)(x - 2)}{x + 1}, \; x \neq 2$

x-intercept: $\left(-\dfrac{1}{2}, 0\right)$

y-intercept: $(0, -2)$

Vertical asymptote: $x = -1$

Slant asymptote: $y = 2x - 5$

10.

Minimize $A = xy$.

Given: $(x - 4)(y - 2) = 36$

$y = \dfrac{36}{x - 4} + 2$

$A = x\left(\dfrac{36}{x - 4} + 2\right)$

$= x\left(\dfrac{2x + 28}{x - 4}\right)$

$= \dfrac{2x(x + 14)}{x - 4}$

Domain: $x > 4$

From the graph of A we see that the minimum occurs when $x \approx 12.49$ inches and

$y = \dfrac{36}{x - 4} + 2 \approx 6.24$ inches.

Note: The exact values are $x = 4 + 6\sqrt{2}$ and $y = 2 + 3\sqrt{2}$.

11. (a) Equate the slopes.

$\dfrac{y - 1}{0 - 2} = \dfrac{1 - 0}{2 - x}$

$\dfrac{y - 1}{-2} = \dfrac{1}{2 - x}$

$y - 1 = -2\left(\dfrac{1}{2 - x}\right)$

$y = 1 + \dfrac{2}{x - 2}$

(b) $A = \dfrac{1}{2}xy = \dfrac{1}{2}x\left[1 + \dfrac{2}{x - 2}\right]$

$= \dfrac{x}{2} + \dfrac{x}{x - 2} = \dfrac{x^2}{2(x - 2)}$

In context, we have $x > 2$ for the domain.

(c)

The minimum area occurs at $x = 4$ and is $A = 4$.

12. $y^2 - 8x = 0$

$$y^2 = 4(2)x$$

Parabola, opens to the right

Vertex: $(0, 0)$

Focus: $(2, 0)$

13.
$$x^2 + y^2 - 10x + 4y + 4 = 0$$
$$(x^2 - 10x) + (y^2 + 4y) = -4$$
$$(x^2 - 10x + 25) + (y^2 + 4y + 4) = -4 + 25 + 4$$
$$(x - 5)^2 + (y + 2)^2 = 25$$

Circle

Center: $(5, -2)$

Radius: 5

14. $x^2 - 10x - 2y + 19 = 0$

$$x^2 - 10x = 2y - 19$$
$$x^2 - 10x + 25 = 2y - 19 + 25$$
$$(x - 5)^2 = 2(y + 3)$$

Parabola

Vertex: $(5, -3)$

Focus: $\left(5, -\dfrac{5}{2}\right)$

15. $\dfrac{x^2}{1} - \dfrac{y^2}{4} = 1$

Hyperbola

Center: $(0, 0)$

$a = 1, b = 2, c = \sqrt{5}$

Horizontal transverse axis

Vertices: $(\pm 1, 0)$

Foci: $\left(\pm \sqrt{5}, 0\right)$

Asymptotes: $y = \pm 2x$

16.
$$x^2 - 4y^2 - 4x = 0$$
$$(x^2 - 4x + 4) - 4y^2 = 0 + 4$$
$$\dfrac{(x - 2)^2}{4} - \dfrac{y^2}{1} = 1$$

Hyperbola

Center: $(2, 0)$

$a = 2, b = 1, c = \sqrt{5}$

Horizontal transverse axis

Vertices: $(0, 0)$ and $(4, 0)$

Foci: $\left(2 \pm \sqrt{5}, 0\right)$

Asymptotes: $y = \pm \dfrac{1}{2}(x - 2)$

17.
$$x^2 + 3y^2 - 2x + 36y + 100 = 0$$
$$(x^2 - 2x) + 3(y^2 + 12y) = -100$$
$$(x^2 - 2x + 1) + 3(y^2 + 12y + 36) = -100 + 1 + 108$$
$$(x - 1)^2 + 3(y + 6)^2 = 9$$
$$\dfrac{(x - 1)^2}{9} + \dfrac{(y + 6)^2}{3} = 1$$

Ellipse

Center: $(1, -6)$

$a = 3, b = \sqrt{3}, c = \sqrt{6}$

Horizontal major axis

Vertices: $(-2, -6), (4, -6)$

Foci: $\left(1 \pm \sqrt{6}, -6\right)$

18. Ellipse

Vertices: $(0, 2)$ and $(8, 2)$

Center: $(4, 2)$

Horizontal major axis: $a = 4$

Minor axis of length 4: $2b = 4 \implies b = 2$

$$\frac{(x - h)^2}{a^2} + \frac{(y - k)^2}{b^2} = 1$$

$$\frac{(x - 4)^2}{16} + \frac{(y - 2)^2}{4} = 1$$

19. Hyperbola

Vertices: $(0, \pm 3)$

Center: $(0, 0)$

Vertical transverse axis: $a = 3$

Asymptotes: $y = \pm\frac{3}{2}x$

$$\pm\frac{a}{b} = \pm\frac{3}{2} \implies b = 2$$

$$\frac{(y - k)^2}{a^2} - \frac{(x - h)^2}{b^2} = 1$$

$$\frac{y^2}{9} - \frac{x^2}{4} = 1$$

20.

$$x^2 = 4p(y - 16)$$

$$36 = 4p(14 - 16)$$

$$-\frac{9}{2} = p$$

$$x^2 = -18(y - 16)$$

When $y = 0$: $x^2 = -18(-16) \implies x \approx 17 \implies 2x \approx 34$

At ground level the archway is approximately 34 meters.

21. $a = \frac{1}{2}(768{,}800) = 384{,}400$

$b = \frac{1}{2}(767{,}640) = 383{,}820$

$c = \sqrt{384{,}400^2 - 383{,}820^2} \approx 21{,}108$

Perigee: $a - c \approx 363{,}292$ kilometers

Apogee: $a + c \approx 405{,}508$ kilometers

Chapter 5   Chapter Test Solutions

1. $12.4^{2.79} \approx 1123.690$

2. $4^{3\pi/2} \approx 687.291$

3. $e^{-7/10} \approx 0.497$

4. $e^{3.1} \approx 22.198$

5. $f(x) = 10^{-x}$

x	-1	$-\frac{1}{2}$	0	$\frac{1}{2}$	1
$f(x)$	10	3.162	1	0.316	0.1

Horizontal asymptote: $y = 0$

6. $f(x) = -6^{x-2}$

x	-1	0	1	2	3
$f(x)$	-0.005	-0.028	-0.167	-1	-6

Horizontal asymptote: $y = 0$

7. $f(x) = 1 - e^{2x}$

x	-1	$-\frac{1}{2}$	0	$\frac{1}{2}$	1
$f(x)$	0.865	0.632	0	-1.718	-6.389

Horizontal asymptote: $y = 1$

8. (a) $\log_7 7^{-0.89} = -0.89$

(b) $4.6 \ln e^2 = 4.6(2) = 9.2$

9. $f(x) = -\log x - 6$

x	$\frac{1}{2}$	1	$\frac{3}{2}$	2	4
$f(x)$	-5.699	-6	-6.176	-6.301	-6.602

Vertical asymptote: $x = 0$

10. $f(x) = \ln(x - 4)$

x	5	7	9	11	13
$f(x)$	0	1.099	1.609	1.946	2.197

Vertical asymptote: $x = 4$

11. $f(x) = 1 + \ln(x + 6)$

x	-5	-3	-1	0	1
$f(x)$	1	2.099	2.609	2.792	2.946

Vertical asymptote: $x = -6$

12. $\log_7 44 = \dfrac{\ln 44}{\ln 7} = \dfrac{\log 44}{\log 7} \approx 1.945$

13. $\log_{2/5} 0.9 = \dfrac{\ln 0.9}{\ln(2/5)} = \dfrac{\log 0.9}{\log(2/5)} \approx 0.115$

14. $\log_{24} 68 = \dfrac{\ln 68}{\ln 24} = \dfrac{\log 68}{\log 24} \approx 1.328$

15. $\log_2 3a^4 = \log_2 3 + \log_2 a^4 = \log_2 3 + 4 \log_2 |a|$

16. $\ln \dfrac{5\sqrt{x}}{6} = \ln(5\sqrt{x}) - \ln 6 = \ln 5 + \ln \sqrt{x} - \ln 6 = \ln 5 + \dfrac{1}{2} \ln x - \ln 6$

17. $\log\left(\dfrac{7x^2}{yz^3}\right) = \log 7x^2 - \log yz^3 = \log 7 + \log x^2 - (\log y + \log z^3) = \log 7 + 2 \log x - \log y - 3 \log z$

18. $\log_3 13 + \log_3 y = \log_3 13y$

19. $4 \ln x - 4 \ln y = \ln x^4 - \ln y^4 = \ln\left(\dfrac{x^4}{y^4}\right), \ x > 0, y > 0$

20. $2 \ln x + \ln(x - 5) - 3 \ln y = \ln x^2 + \ln(x - 5) - \ln y^3$
$$= \ln x^2(x - 5) - \ln y^3$$
$$= \ln \dfrac{x^2(x - 5)}{y^3}$$

21. $5^x = \dfrac{1}{25}$
$$5^x = 5^{-2}$$
$$x = -2$$

22. $3e^{-5x} = 132$
$$e^{-5x} = 44$$
$$-5x = \ln 44$$
$$x = \dfrac{\ln 44}{-5} \approx -0.757$$

23. $\dfrac{1025}{8 + e^{4x}} = 5$
$$1025 = 5(8 + e^{4x})$$
$$205 = 8 + e^{4x}$$
$$197 = e^{4x}$$
$$\ln 197 = 4x$$
$$\dfrac{\ln 197}{4} = x$$
$$x \approx 1.321$$

24. $\ln x = \frac{1}{2}$
$$x = e^{1/2} \approx 1.649$$

25. $18 + 4 \ln x = 7$
$$4 \ln x = -11$$
$$\ln x = -\tfrac{11}{4}$$
$$x = e^{-11/4} \approx 0.064$$

26. $\log x - \log(8 - 5x) = 2$
$$\log \dfrac{x}{8 - 5x} = 2$$
$$\dfrac{x}{8 - 5x} = 10^2$$
$$x = 100(8 - 5x)$$
$$x = 800 - 500x$$
$$501x = 800$$
$$x = \dfrac{800}{501} \approx 1.597$$

27. $y = ae^{bt}$

$(0, 2745)$: $2745 = ae^{b(0)} \Longrightarrow a = 2745$
$$y = 2745e^{bt}$$

$(9, 11{,}277)$: $11{,}277 = 2745e^{b(9)}$
$$\dfrac{11{,}277}{2745} = e^{9b}$$
$$\ln\left(\dfrac{11{,}277}{2745}\right) = 9b$$
$$\dfrac{1}{9} \ln\left(\dfrac{11{,}277}{2745}\right) = b \Longrightarrow b \approx 0.1570$$

Thus, $y = 2745e^{0.1570t}$.

28. $y = ae^{bt}$

$\dfrac{1}{2}a = ae^{b(21.77)}$

$\dfrac{1}{2} = e^{21.77b}$

$\ln\left(\dfrac{1}{2}\right) = 21.77b$

$b = \dfrac{\ln(1/2)}{21.77} \approx -0.0318$

$y = ae^{-0.0318t}$

When $t = 19$: $y = ae^{-0.0318(19)} \approx 0.55a$

Thus, 55% will remain after 19 years.

29. $H = 70.228 + 5.104x + 9.222 \ln x, \ \dfrac{1}{4} \le x \le 6$

(a)

x	H (cm)
$\frac{1}{4}$	58.720
$\frac{1}{2}$	66.388
1	75.332
2	86.828
3	95.671
4	103.43
5	110.59
6	117.38

(b) When $x = 4$, $H \approx 103.43$ cm.

Chapters 3–5   Cumulative Test Solutions

1. Vertex $(-8, 5)$

Point $(-4, -7)$

$y - k = a(x - h)^2$

$y - 5 = a(x + 8)^2$

$-7 - 5 = a(-4 + 8)^2$

$-12 = 16a$

$-\dfrac{3}{4} = a$

$y = -\dfrac{3}{4}(x + 8)^2 + 5$

2. $h(x) = -(x^2 + 4x)$

$= -(x^2 + 4x + 4 - 4)$

$= -(x + 2)^2 + 4$

Parabola

Vertex: $(-2, 4)$

Intercepts: $(-4, 0), (0, 0)$

3. $f(t) = \frac{1}{4}t(t - 2)^2$

Cubic

Falls to the left

Rises to the right

Intercepts: $(0, 0), (2, 0)$

4. $g(s) = s^2 + 4s + 10$

$= (s^2 + 4s + 4) - 4 + 10$

$= (s + 2)^2 + 6$

Parabola

Vertex: $(-2, 6)$

Intercept: $(0, 10)$

5. $f(x) = x^3 + 2x^2 + 4x + 8$

$ = x^2(x + 2) + 4(x + 2)$

$ = (x + 2)(x^2 + 4)$

$x + 2 = 0 \Longrightarrow x = -2$

$x^2 + 4 = 0 \Longrightarrow x = \pm 2i$

The zeros of $f(x)$ are -2 and $\pm 2i$.

6. $f(x) = x^4 + 4x^3 - 21x^2$

$ = x^2(x^2 + 4x - 21)$

$ = x^2(x + 7)(x - 3)$

The zeros of $f(x)$ are 0, -7, and 3.

7.
$$2x^2 + 0x + 1 \overline{)\,6x^3 - 4x^2 + 0x + 0} \qquad 3x - 2 + \dfrac{-3x + 2}{2x^2 + 1}$$

$$\underline{6x^3 + 0x^2 + 3x}$$
$$-4x^2 - 3x + 0$$
$$\underline{-4x^2 + 0x - 2}$$
$$-3x + 2$$

Thus, $\dfrac{6x^3 - 4x^2}{2x^2 + 1} = 3x - 2 - \dfrac{3x - 2}{2x^2 + 1}$.

8.

$$-2 \,\bigg|\begin{array}{rrrrr} 2 & 3 & 0 & -6 & 5 \\ & -4 & 2 & -4 & 20 \\ \hline 2 & -1 & 2 & -10 & 25 \end{array}$$

Thus, $\dfrac{2x^4 + 3x^3 - 6x + 5}{x + 2} = 2x^3 - x^2 + 2x - 10 + \dfrac{25}{x + 2}$.

9. $g(x) = x^3 + 3x^2 - 6$

$x \approx 1.20$

10. Since $2 + \sqrt{3}i$ is a zero, so is $2 - \sqrt{3}i$.

$f(x) = (x + 5)(x + 2)\big[x - \big(2 + \sqrt{3}i\big)\big]\big[x - \big(2 - \sqrt{3}i\big)\big]$

$ = (x^2 + 7x + 10)\big[(x - 2) - \sqrt{3}i\big]\big[(x - 2) + \sqrt{3}i\big]$

$ = (x^2 + 7x + 10)\big[(x - 2)^2 + 3\big]$

$ = (x^2 + 7x + 10)(x^2 - 4x + 7)$

$ = x^4 + 3x^3 - 11x^2 + 9x + 70$

11. $f(x) = \dfrac{2x}{x - 3}$

Domain: all real numbers x except $x = 3$

Vertical asymptote: $x = 3$

Horizontal asymptote: $y = 2$

Intercept: $(0, 0)$

12. $f(x) = \dfrac{4x^2}{x - 5} = 4x + 20 + \dfrac{100}{x - 5}$

Domain: all real numbers x except $x = 5$

Vertical asymptote: $x = 5$

Slant asymptote: $y = 4x + 20$

Intercept: $(0, 0)$

13. $f(x) = \dfrac{2x}{x^2 - 9}$

Vertical asymptotes: $x = \pm 3$

Horizontal asymptote: $y = 0$

Intercept: $(0, 0)$

14. $f(x) = \dfrac{x^2 - 4x + 3}{x^2 - 2x - 3}$

$= \dfrac{(x - 1)(x - 3)}{(x + 1)(x - 3)}$

$= \dfrac{x - 1}{x + 1}, \quad x \neq 3$

Vertical asymptote: $x = -1$

Horizontal asymptote: $y = 1$

x-intercept: $(1, 0)$

y-intercept: $(0, -1)$

15. $f(x) = \dfrac{x^3 + 3x^2 - 4x - 12}{x^2 - x - 2}$

$= \dfrac{(x + 2)(x - 2)(x + 3)}{(x + 1)(x - 2)}$

$= \dfrac{(x + 2)(x + 3)}{x + 1}$

$= \dfrac{x^2 + 5x + 6}{x + 1}$

$= x + 4 + \dfrac{2}{x + 1}, \quad x \neq 2$

Vertical asymptote: $x = -1$

Slant asymptote: $y = x + 4$

x-intercepts: $(-2, 0), (-3, 0)$

y-intercept: $(0, 6)$

16. $\dfrac{(x + 3)^2}{16} - \dfrac{(y + 4)^2}{25} = 1$

Hyperbola

Center: $(-3, -4)$

Vertices: $(-7, -4)$ and $(1, -4)$

17. $\dfrac{(x - 2)^2}{4} + \dfrac{(y + 1)^2}{9} = 1$

Ellipse

Center: $(2, -1)$

Vertices: $(2, -4)$ and $(2, 2)$

18. Parabola

Vertex: $(3, -2) \Longrightarrow y = a(x - 3)^2 - 2$

Point: $(0, 4) \Longrightarrow 4 = a(0 - 3)^2 - 2$

$6 = 9a \Longrightarrow a = \dfrac{2}{3}$

Equation: $\quad y = \dfrac{2}{3}(x - 3)^2 - 2$

$y + 2 = \dfrac{2}{3}(x - 3)^2$

$\dfrac{3}{2}(y + 2) = (x - 3)^2$

$(x - 3)^2 = \dfrac{3}{2}(y + 2)$

19. Hyperbola

Foci: $(0, 0)$ and $(0, 4) \Longrightarrow$ Center: $(0, 2)$ and vertical transverse axis

Asymptotes: $y = \pm\dfrac{1}{2}x + 2 \Longrightarrow \dfrac{a}{b} = \dfrac{1}{2} \Longrightarrow 2a = b$

$c^2 = a^2 + b^2 \Longrightarrow 4 = a^2 + 4a^2 \Longrightarrow a^2 = \dfrac{4}{5}$ and $b^2 = \dfrac{16}{5}$

Equation: $\dfrac{(y - 2)^2}{4/5} - \dfrac{x^2}{16/5} = 1$

20. $f(x) = \left(\frac{2}{5}\right)^x$

$g(x) = -\left(\frac{2}{5}\right)^{-x+3}$

g is a reflection in the x-axis, a reflection in the y-axis, and a horizontal shift three units to the right of the graph of f.

21. $f(x) = 2.2^x$

$g(x) = -2.2^x + 4$

g is a reflection in the x-axis, and a vertical shift four units upward of the graph of f.

22. $\log 98 \approx 1.991$

23. $\log\left(\frac{6}{7}\right) \approx -0.067$

24. $\ln\sqrt{31} \approx 1.717$

25. $\ln\left(\sqrt{40} - 5\right) \approx 0.281$

26. $\log_7 1.8 = \dfrac{\log_{10} 1.8}{\log_{10} 7} = \dfrac{\ln 1.8}{\ln 7} \approx 0.302$

27. $\log_3 0.149 = \dfrac{\log_{10} 0.149}{\log_{10} 3} = \dfrac{\ln 0.149}{\ln 3} \approx -1.733$

28. $\log_{1/2} 17 = \dfrac{\log_{10} 17}{\log_{10}(1/2)} = \dfrac{\ln 17}{\ln(1/2)} \approx -4.087$

29. $\ln\left(\dfrac{x^2 - 16}{x^4}\right) = \ln(x^2 - 16) - \ln x^4$

$= \ln(x + 4)(x - 4) - 4\ln x$

$= \ln(x + 4) + \ln(x - 4) - 4\ln x,\ x > 4$

30. $2\ln x - \dfrac{1}{2}\ln(x + 5) = \ln x^2 - \ln\sqrt{x + 5}$

$= \ln\dfrac{x^2}{\sqrt{x + 5}},\ x > 0$

31. $6e^{2x} = 72$

$e^{2x} = 12$

$2x = \ln 12$

$x = \dfrac{\ln 12}{2} \approx 1.242$

32. $4^{x-5} + 21 = 30$

$4^{x-5} = 9$

$x - 5 = \log_4 9$

$x = 5 + \log_4 9$

$x = 5 + \dfrac{\ln 9}{\ln 4}$

$x \approx 6.585$

33. $e^{2x} - 11e^x + 24 = 0$

$(e^x - 3)(e^x - 8) = 0$

$e^x - 3 = 0 \Longrightarrow e^x = 3 \Longrightarrow x = \ln 3 \approx 1.099$

$e^x - 8 = 0 \Longrightarrow e^x = 8 \Longrightarrow x = \ln 8 \approx 2.079$

34. $\log_2 x + \log_2 5 = 6$

$\log_2 5x = 6$

$5x = 2^6$

$x = \frac{64}{5} = 12.8$

35. $\ln 4x - \ln 2 = 8$

$$\ln \frac{4x}{2} = 8$$

$$\ln 2x = 8$$

$$2x = e^8$$

$$x = \frac{e^8}{2}$$

$$x \approx 1490.479$$

36. $\ln \sqrt{x + 2} = 3$

$$\frac{1}{2}\ln(x + 2) = 3$$

$$\ln(x + 2) = 6$$

$$x + 2 = e^6$$

$$x = e^6 - 2 \approx 401.429$$

37. $f(x) = \dfrac{1000}{1 + 4e^{-0.2x}}$

Horizontal asymptotes: $y = 0$ and $y = 1000$

38. $P = -\dfrac{1}{2}x^2 + 20x + 230$

$$= -\frac{1}{2}(x^2 - 40x + 400 - 400) + 230$$

$$= -\frac{1}{2}(x - 20)^2 + 430$$

The maximum occurs at the vertex when $x = 20$, which corresponds to spending $(20)(100) = \$2000$ on advertising.

39. (a)

(b) $S \approx 0.274t^2 - 4.08t + 50.6$

(c)

The model is a good fit to the actual data.

(d) For 2008, use $t = 18$: $S(18) \approx \$65.9$ billion

Yes, this seems reasonable, but since the model goes to infinity as t increases, it cannot be used for predictions much beyond this value.

40. $A = 2500e^{(0.075)(25)} \approx \$16,302.05$

41.
$$N = 175e^{kt}$$

$$420 = 175e^{k(8)}$$

$$2.4 = e^{8k}$$

$$\ln 2.4 = 8k$$

$$\frac{\ln 2.4}{8} = k$$

$$k \approx 0.1094$$

$$N = 175e^{0.1094t}$$

$$350 = 175e^{0.1094t}$$

$$2 = e^{0.1094t}$$

$$\ln 2 = 0.1094t$$

$$t = \frac{\ln 2}{0.1094} \approx 6.3 \text{ hours to double}$$

42.
$$P = 16,989e^{0.0207t}$$

$$24 \text{ million} = 24,000 \text{ thousand}$$

$$24,000 = 16,989e^{0.0207t}$$

$$\frac{24,000}{16,989} = e^{0.0207t}$$

$$\ln\left(\frac{24,000}{16,989}\right) = 0.0207t$$

$$t = \frac{\ln(24,000/16,989)}{0.0207} \approx 16.69$$

According to the model, the population will reach 24 million during the year 2006.

43. $p = \dfrac{1200}{1 + 3e^{-t/5}}$

(a) $p(0) = \dfrac{1200}{1 + 3e^0} = \dfrac{1200}{4} = 300$ birds

(b) $p(5) = \dfrac{1200}{1 + 3e^{-1}} \approx 570$ birds

(c)
$$800 = \dfrac{1200}{1 + 3e^{-t/5}}$$
$$800(1 + 3e^{-t/5}) = 1200$$
$$1 + 3e^{-t/5} = 1.5$$
$$3e^{-t/5} = 0.5$$
$$e^{-t/5} = \dfrac{1}{6}$$
$$-\dfrac{t}{5} = \ln\!\left(\dfrac{1}{6}\right)$$
$$t = -5\ln\!\left(\dfrac{1}{6}\right) \approx 9 \text{ years}$$

Chapter 6   Chapter Test Solutions

1. $\theta = \dfrac{5\pi}{4}$

(a)

(b) $\dfrac{5\pi}{4} + 2\pi = \dfrac{13\pi}{4}$

$\dfrac{5\pi}{4} - 2\pi = -\dfrac{3\pi}{4}$

(c) $\dfrac{5\pi}{4}\left(\dfrac{180°}{\pi}\right) = 225°$

2. $90\dfrac{\text{km}}{\text{hr}} \times \dfrac{1 \text{ hr}}{60 \text{ min}} \times \dfrac{1000 \text{ m}}{1 \text{ km}} = 1500$ meters per minute

$\dfrac{\text{Revolutions}}{\text{minute}} = \dfrac{1500}{\pi}$

Circumference $= 2\pi\left(\dfrac{1}{2}\right) = \pi = \pi$ meters

Angular speed $= \left(\dfrac{1500 \text{ revolutions}}{\pi \text{ minute}}\right)\left(\dfrac{2\pi \text{ radians}}{\text{revolution}}\right)$

$= 3000$ radians per minute

3. $130° = \dfrac{130\pi}{180} = \dfrac{13\pi}{18}$ radians

$A = \dfrac{1}{2}r^2\theta = \dfrac{1}{2}(25)^2\left(\dfrac{13\pi}{18}\right) \approx 709.04$ square feet

4. $x = -2,\ y = 6$

$r = \sqrt{(-2)^2 + (6)^2} = 2\sqrt{10}$

$\sin\theta = \dfrac{y}{r} = \dfrac{6}{2\sqrt{10}} = \dfrac{3}{\sqrt{10}} = \dfrac{3\sqrt{10}}{10}$

$\cos\theta = \dfrac{x}{r} = \dfrac{-2}{2\sqrt{10}} = -\dfrac{1}{\sqrt{10}} = -\dfrac{\sqrt{10}}{10}$

$\tan\theta = \dfrac{y}{x} = \dfrac{6}{-2} = -3$

$\csc\theta = \dfrac{r}{y} = \dfrac{2\sqrt{10}}{6} = \dfrac{\sqrt{10}}{3}$

$\sec\theta = \dfrac{r}{x} = \dfrac{2\sqrt{10}}{-2} = -\sqrt{10}$

$\cot\theta = \dfrac{x}{y} = \dfrac{-2}{6} = -\dfrac{1}{3}$

5.

For $0 \le \theta < \dfrac{\pi}{2}$, we have:

$$\sin \theta = \frac{\text{opp}}{\text{hyp}} = \frac{3}{\sqrt{13}} = \frac{3\sqrt{13}}{13}$$

$$\cos \theta = \frac{\text{adj}}{\text{hyp}} = \frac{2}{\sqrt{13}} = \frac{2\sqrt{13}}{13}$$

$$\csc \theta = \frac{\text{hyp}}{\text{opp}} = \frac{\sqrt{13}}{3}$$

$$\sec \theta = \frac{\text{hyp}}{\text{adj}} = \frac{\sqrt{13}}{2}$$

$$\cot \theta = \frac{\text{adj}}{\text{opp}} = \frac{2}{3}$$

For $\pi \le \theta < \dfrac{3\pi}{2}$, we have:

$$\sin \theta = -\frac{3\sqrt{13}}{13}$$

$$\cos \theta = -\frac{2\sqrt{13}}{13}$$

$$\csc \theta = -\frac{\sqrt{13}}{3}$$

$$\sec \theta = -\frac{\sqrt{13}}{2}$$

$$\cot \theta = \frac{2}{3}$$

6. $\theta = 290°$

$\theta' = 360° - 290° = 70°$

7. $\sec \theta < 0$ and $\tan \theta > 0$

$$\frac{r}{x} < 0 \text{ and } \frac{y}{x} > 0$$

Quandrant III

8. $\cos \theta = -\dfrac{\sqrt{3}}{2}$

Reference angle is $30°$ and θ is in Quandrant II or III.

$\theta = 150°$ or $210°$

9. $\csc \theta = 1.030$

$$\frac{1}{\sin \theta} = 1.030$$

$$\sin \theta = \frac{1}{1.030}$$

$$\theta = \arcsin \frac{1}{1.030}$$

$$\theta \approx 1.33 \text{ and } \pi - 1.33 \approx 1.81$$

10. $\cos \theta = \frac{3}{5}$, $\tan \theta < 0 \implies \theta$ lies in Quadrant IV.

Let $x = 3$, $r = 5 \implies y = -4$.

$\sin \theta = -\frac{4}{5}$	$\csc \theta = -\frac{5}{4}$
$\cos \theta = \frac{3}{5}$	$\sec \theta = \frac{5}{3}$
$\tan \theta = -\frac{4}{3}$	$\cot \theta = -\frac{3}{4}$

11. $\sec \theta = -\frac{17}{8}$, $\sin \theta > 0 \implies \theta$ lies in Quadrant II.

Let $r = 17$, $x = -8 \implies y = 15$.

$\sin \theta = \frac{15}{17}$	$\csc \theta = \frac{17}{15}$
$\cos \theta = -\frac{8}{17}$	$\sec \theta = -\frac{17}{8}$
$\tan \theta = -\frac{15}{8}$	$\cot \theta = -\frac{8}{15}$

12. $g(x) = -2 \sin\left(x - \dfrac{\pi}{4}\right)$

Period: 2π

Amplitude: $|-2| = 2$

Shifted to the right by $\dfrac{\pi}{4}$ units
and reflected in the x-axis.

x	0	$\dfrac{\pi}{4}$	$\dfrac{3\pi}{4}$	$\dfrac{5\pi}{4}$	$\dfrac{7\pi}{4}$
y	$\sqrt{2}$	0	-2	0	2

13. $f(\alpha) = \dfrac{1}{2} \tan 2\alpha$

Period: $\dfrac{\pi}{2}$

Asymptotes:

$x = -\dfrac{\pi}{4}, x = \dfrac{\pi}{4}$

α	$-\dfrac{\pi}{8}$	0	$\dfrac{\pi}{8}$
$f(\alpha)$	$-\dfrac{1}{2}$	0	$\dfrac{1}{2}$

14. $y = \sin 2\pi x + 2 \cos \pi x$

Periodic: period $= 2$

15. $y = 6e^{-0.12t} \cos(0.25t), 0 \le t \le 32$

Not periodic

16. $f(x) = a \sin(bx + c)$

Amplitude: $2 \implies |a| = 2$

Reflected in the x-axis: $a = -2$

Period: $4\pi = \dfrac{2\pi}{b} \implies b = \dfrac{1}{2}$

Phase shift: $\dfrac{c}{b} = -\dfrac{\pi}{2} \implies c = -\dfrac{\pi}{4}$

$f(x) = -2 \sin\left(\dfrac{x}{2} - \dfrac{\pi}{4}\right)$

17. Let $u = \arccos \dfrac{2}{3}$,

$\cos u = \dfrac{2}{3}$.

$\tan\left(\arccos \dfrac{2}{3}\right) = \tan u = \dfrac{\sqrt{5}}{2}$

18. $f(x) = 2 \arcsin\left(\tfrac{1}{2}x\right)$

Domain: $[-2, 2]$

Range: $[-\pi, \pi]$

19.

$\tan \theta = -\dfrac{80}{95} \implies \theta \approx -40.1°$

Bearing: $360° - (90° - 40.1°) = 310.1°$

20. $d = a \cos bt$

$a = -6$

$\dfrac{2\pi}{b} = 2 \implies b = \pi$

$d = -6 \cos \pi t$

Chapter 7   Chapter Test Solutions

1. $\tan \theta = \dfrac{3}{2}$ and $\cos \theta < 0$

θ is in Quadrant III.

$\sec \theta = -\sqrt{1 + \tan^2 \theta} = -\sqrt{1 + \left(\dfrac{3}{2}\right)^2} = -\dfrac{\sqrt{13}}{2}$

$\cos \theta = \dfrac{1}{\sec \theta} = -\dfrac{2}{\sqrt{13}} = -\dfrac{2\sqrt{13}}{13}$

$\sin \theta = \tan \theta \cos \theta = \left(\dfrac{3}{2}\right)\left(-\dfrac{2}{\sqrt{13}}\right) = -\dfrac{3}{\sqrt{13}} = -\dfrac{3\sqrt{13}}{13}$

$\csc \theta = \dfrac{1}{\sin \theta} = -\dfrac{\sqrt{13}}{3}$

$\cot \theta = \dfrac{1}{\tan \theta} = \dfrac{2}{3}$

2. $\csc^2 \beta (1 - \cos^2 \beta) = \dfrac{1}{\sin^2 \beta}(\sin^2 \beta) = 1$

3. $\dfrac{\sec^4 x - \tan^4 x}{\sec^2 x + \tan^2 x} = \dfrac{(\sec^2 x + \tan^2 x)(\sec^2 x - \tan^2 x)}{\sec^2 x + \tan^2 x}$

$\qquad\qquad = \sec^2 x - \tan^2 x = 1$

4. $\dfrac{\cos \theta}{\sin \theta} + \dfrac{\sin \theta}{\cos \theta} = \dfrac{\cos^2 \theta + \sin^2 \theta}{\sin \theta \cos \theta} = \dfrac{1}{\sin \theta \cos \theta}$

$\qquad\qquad = \csc \theta \sec \theta$

5. $y = \tan \theta,\ y = -\sqrt{\sec^2 \theta - 1}$

$\tan \theta = -\sqrt{\sec^2 \theta - 1}$ on

$\theta = 0, \dfrac{\pi}{2} < \theta \le \pi, \dfrac{3\pi}{2} < \theta < 2\pi.$

6. $y_1 = \cos x + \sin x \tan x,\ y_2 = \sec x$

It appears that $y_1 = y_2$.

$\cos x + \sin x \tan x = \cos\ + \sin x \dfrac{\sin x}{\cos x}$

$\qquad\qquad = \cos\ + \dfrac{\sin^2 x}{\cos x}$

$\qquad\qquad = \dfrac{\cos^2 x + \sin^2 x}{\cos x}$

$\qquad\qquad = \dfrac{1}{\cos x} = \sec x$

7. $\sin \theta \sec \theta = \sin \theta \dfrac{1}{\cos \theta} = \dfrac{\sin \theta}{\cos \theta} = \tan \theta$

8. $\sec^2 x \tan^2 x + \sec^2 x = \sec^2 x(\sec^2 x - 1) + \sec^2 x$

$\qquad\qquad = \sec^4 x - \sec^2 x + \sec^2 x$

$\qquad\qquad = \sec^4 x$

9. $\dfrac{\csc \alpha + \sec \alpha}{\sin \alpha + \cos \alpha} = \dfrac{\dfrac{1}{\sin \alpha} + \dfrac{1}{\cos \alpha}}{\sin \alpha + \cos \alpha} = \dfrac{\dfrac{\cos \alpha + \sin \alpha}{\sin \alpha \cos \alpha}}{\sin \alpha + \cos \alpha} = \dfrac{1}{\sin \alpha \cos \alpha}$

$\qquad = \dfrac{\cos^2 \alpha + \sin^2 \alpha}{\sin \alpha \cos \alpha} = \dfrac{\cos^2 \alpha}{\sin \alpha \cos \alpha} + \dfrac{\sin^2 \alpha}{\sin \alpha \cos \alpha}$

$\qquad = \dfrac{\cos \alpha}{\sin \alpha} + \dfrac{\sin \alpha}{\cos \alpha} = \cot \alpha + \tan \alpha$

10. $\cos\left(x + \dfrac{\pi}{2}\right) = \cos\left(\dfrac{\pi}{2} - (-x)\right) = \sin(-x) = -\sin x$

11. $\sin(n\pi + \theta) = (-1)^n \sin \theta$, n is an integer.

For n odd: $\sin(n\pi + \theta) = \sin n\pi \cos \theta + \cos n\pi \sin \theta$

$\qquad\qquad\qquad\qquad = (0) \cos \theta + (-1) \sin \theta = -\sin \theta$

For n even: $\sin(n\pi + \theta) = \sin n\pi \cos \theta + \cos n\pi \sin \theta$

$\qquad\qquad\qquad\qquad = (0) \cos \theta + (1) \sin \theta = \sin \theta$

When n is odd, $(-1)^n = -1$. When n is even $(-1)^n = 1$.

Thus, $\sin(n\pi + \theta) = (-1)^n \sin \theta$ for any integer n.

12. $(\sin x + \cos x)^2 = \sin^2 x + 2 \sin x \cos x + \cos^2 x$

$\qquad\qquad\qquad\quad = 1 + 2 \sin x \cos x$

$\qquad\qquad\qquad\quad = 1 + \sin 2x$

13. $\sin^4 x \tan^2 x = \sin^4 x \left(\dfrac{\sin^2 x}{\cos^2 x}\right) = \dfrac{\sin^6 x}{\cos^2 x} = \dfrac{(\sin^2 x)^3}{\cos^2 x}$

$\qquad = \dfrac{\left(\dfrac{1 - \cos 2x}{2}\right)^3}{\dfrac{1 + \cos 2x}{2}}$

$\qquad = \dfrac{\dfrac{1 - 3\cos 2x + 3\cos^2 2x - \cos^3 2x}{8}}{\dfrac{1 + \cos 2x}{2}}$

$\qquad = \dfrac{\dfrac{1}{4}\left[1 - 3\cos 2x + 3\left(\dfrac{1 + \cos 4x}{2}\right) - \cos 2x\left(\dfrac{1 + \cos 4x}{2}\right)\right]}{1 + \cos 2x}$

$\qquad = \dfrac{\dfrac{1}{8}[2 - 6\cos 2x + 3 + 3\cos 4x - \cos 2x - \cos 2x \cos 4x]}{1 + \cos 2x}$

$\qquad = \dfrac{1}{8}\left[\dfrac{5 - 7\cos 2x + 3\cos 4x - \dfrac{1}{2}(\cos(-2x) + \cos(6x))}{1 + \cos 2x}\right]$

$\qquad = \dfrac{1}{16}\left[\dfrac{10 - 14\cos 2x + 6\cos 4x - \cos 2x - \cos 6x}{1 + \cos 2x}\right]$

$\qquad = \dfrac{1}{16}\left[\dfrac{10 - 15\cos 2x + 6\cos 4x - \cos 6x}{1 + \cos 2x}\right]$

14. $\dfrac{\sin 4\theta}{1 + \cos 4\theta} = \tan \dfrac{4\theta}{2} = \tan 2\theta$

15. $4 \cos 2\theta \sin 4\theta = 4\left(\dfrac{1}{2}\right)[\sin(2\theta + 4\theta) - \sin(2\theta - 4\theta)]$

$\qquad\qquad\qquad\quad = 2[\sin 6\theta - \sin(-2\theta)]$

$\qquad\qquad\qquad\quad = 2(\sin 6\theta + \sin 2\theta)$

16. $\sin 3\theta - \sin 4\theta = 2\cos\left(\dfrac{3\theta + 4\theta}{2}\right)\sin\left(\dfrac{3\theta - 4\theta}{2}\right)$

$$= 2\cos\frac{7\theta}{2}\sin\left(\frac{-\theta}{2}\right)$$

$$= -2\cos\frac{7\theta}{2}\sin\frac{\theta}{2}$$

17. $\tan^2 x + \tan x = 0$

$\tan x(\tan x + 1) = 0$

$\tan x = 0 \quad\text{or}\quad \tan x + 1 = 0$

$x = 0, \pi \qquad \tan x = -1$

$$x = \frac{3\pi}{4}, \frac{7\pi}{4}$$

18. $\sin 2\alpha - \cos \alpha = 0$

$2\sin\alpha\cos\alpha - \cos\alpha = 0$

$\cos\alpha(2\sin\alpha - 1) = 0$

$\cos\alpha = 0 \quad\text{or}\quad 2\sin\alpha - 1 = 0$

$\alpha = \dfrac{\pi}{2}, \dfrac{3\pi}{2} \qquad \sin\alpha = \dfrac{1}{2}$

$$\alpha = \frac{\pi}{6}, \frac{5\pi}{6}$$

19. $4\cos^2 x - 3 = 0$

$\cos^2 x = \dfrac{3}{4}$

$\cos x = \pm\sqrt{\dfrac{3}{4}} = \pm\dfrac{\sqrt{3}}{2}$

$$x = \frac{\pi}{6}, \frac{5\pi}{6}, \frac{7\pi}{6}, \frac{11\pi}{6}$$

20. $\csc^2 x - \csc x - 2 = 0$

$(\csc x - 2)(\csc x + 1) = 0$

$\csc x - 2 = 0 \quad\text{or}\quad \csc x + 1 = 0$

$\csc x = 2 \qquad\qquad \csc = -1$

$\dfrac{1}{\sin x} = 2 \qquad\qquad \dfrac{1}{\sin x} = -1$

$\sin x = \dfrac{1}{2} \qquad\qquad \sin x = -1$

$x = \dfrac{\pi}{6}, \dfrac{5\pi}{6} \qquad\qquad x = \dfrac{3\pi}{2}$

21. $3\cos x - x = 0$

$x \approx -2.938, -2.663, 1.170$

22. $105° = 135° - 30°$

$\cos 105° = \cos(135° - 30°)$

$\quad = \cos 135°\cos 30° + \sin 135°\sin 30°$

$\quad = -\cos 45°\cos 30° + \sin 45°\sin 30°$

$\quad = \left(-\dfrac{\sqrt{2}}{2}\right)\left(\dfrac{\sqrt{3}}{2}\right) + \left(\dfrac{\sqrt{2}}{2}\right)\left(\dfrac{1}{2}\right)$

$\quad = \dfrac{-\sqrt{6} + \sqrt{2}}{4} = \dfrac{\sqrt{2} - \sqrt{6}}{4}$

23. $x = 1, y = 2, r = \sqrt{5}$

$\sin 2u = 2\sin u\cos u$

$\quad = 2\left(\dfrac{2}{\sqrt{5}}\right)\left(\dfrac{1}{\sqrt{5}}\right) = \dfrac{4}{5}$

$\cos 2u = \cos^2 u - \sin^2 u$

$\quad = \left(\dfrac{1}{\sqrt{5}}\right)^2 - \left(\dfrac{2}{\sqrt{5}}\right)^2 = \dfrac{1}{5} - \dfrac{4}{5} = -\dfrac{3}{5}$

$\tan 2u = \dfrac{2\tan u}{1 - \tan^2 u} = \dfrac{2(2)}{1 - (2)^2} = \dfrac{4}{-3} = -\dfrac{4}{3}$

24. Let $y_1 = 31 \sin\left(\dfrac{2\pi t}{365} - 1.4\right)$ and $y_2 = 20$.

The points of intersection occur when $t \approx 123$ and $t \approx 223$. The number of days that $D > 20°$ is 100, from day 123 to day 223.

25. $28 \cos 10t + 38 = 28 \cos\left[10\left(t - \dfrac{\pi}{6}\right)\right] + 38$

$$\cos 10t = \cos\left[10\left(t - \dfrac{\pi}{6}\right)\right]$$

$$0 = \cos\left[10\left(t - \dfrac{\pi}{6}\right)\right] - \cos 10t$$

$$= -2 \sin\left(\dfrac{10(t - (\pi/6)) + 10t}{2}\right) \sin\left(\dfrac{10(t - (\pi/6)) - 10t}{2}\right)$$

$$= -2 \sin\left(10t - \dfrac{5\pi}{6}\right) \sin\left(-\dfrac{5\pi}{6}\right)$$

$$= -2 \sin\left(10t - \dfrac{5\pi}{6}\right)\left(-\dfrac{1}{2}\right)$$

$$= \sin\left(10t - \dfrac{5\pi}{6}\right)$$

$10t - \dfrac{5\pi}{6} = n\pi$ where n is any integer.

$t = \dfrac{n\pi}{10} + \dfrac{\pi}{12}$ where n is any integer.

The first six times the two people are at the same height are:

0.26 minutes, 0.58 minutes, 0.89 minutes, 1.20 minutes, 1.52 minutes, 1.83 minutes.

Chapter 8   Chapter Test Solutions

1. $A = 24°, B = 68°, a = 12.2$

$C = 180° - 24° - 68° = 88°$

$b = \dfrac{a \sin B}{\sin A} = \dfrac{12.2 \sin 68°}{\sin 24°} \approx 27.81$

$c = \dfrac{a \sin C}{\sin A} = \dfrac{12.2 \sin 88°}{\sin 24°} \approx 29.98$

2. $B = 104°, C = 33°, a = 18.1$

$A = 180° - 104° - 33° = 43°$

$b = \dfrac{a \sin B}{\sin A} = \dfrac{18.1 \sin 104°}{\sin 43°} \approx 25.75$

$c = \dfrac{a \sin C}{\sin A} = \dfrac{18.1 \sin 33°}{\sin 43°} \approx 14.45$

3. $A = 24°$, $a = 11.2$, $b = 13.4$

$$\sin B = \frac{b \sin A}{a} = \frac{13.4 \sin 24°}{11.2} \approx 0.4866$$

Two Solutions

$B \approx 29.12°$ or $B \approx 150.88°$

$C \approx 126.88°$ $C \approx 5.12°$

$$c = \frac{a \sin C}{\sin A} = \frac{11.2 \sin 126.88°}{\sin 24°} \qquad c = \frac{11.2 \sin 5.12°}{\sin 24°}$$

$c \approx 22.03$ $c \approx 2.46$

4. $a = 4.0$, $b = 7.3$, $c = 12.4$

$$\cos C = \frac{a^2 + b^2 - c^2}{2ab} = \frac{4^2 + 7.3^2 - 12.4^2}{2(4)(7.3)} \approx -1.4464 < -1$$

No solution

5. $B = 100°$, $a = 15$, $b = 23$

$$\sin A = \frac{a \sin B}{b} = \frac{15 \sin 100°}{23} \Rightarrow A \approx 39.96°$$

$$C \approx 180° - 100° - 39.96° = 40.04°$$

$$c \approx \frac{b \sin C}{\sin B} = \frac{23 \sin 40.04°}{\sin 100°} \approx 15.02$$

6. $C = 123°$, $a = 41$, $b = 57$

$$c^2 = 41^2 + 57^2 - 2(41)(57)\cos 123° \Rightarrow c \approx 86.46$$

$$\sin A = \frac{a \sin C}{c} = \frac{41 \sin 123°}{86.46} \Rightarrow A \approx 23.43°$$

$$B \approx 180° - 23.43° - 123° = 33.57°$$

7. $a = 60$, $b = 70$, $c = 82$

$$s = \frac{60 + 70 + 82}{2} = 106$$

$$\text{Area} = \sqrt{106(46)(36)(24)} \approx 2052.5 \text{ square meters}$$

8.

Not drawn to scale

$$b^2 = 370^2 + 240^2 - 2(370)(240)\cos 167°$$

$$b \approx 606.3 \text{ miles}$$

$$\sin A = \frac{a \sin B}{b} = \frac{240 \sin 167°}{606.3}$$

$$A \approx 5.1°$$

Bearing: $24° + 5.1° = 29.1°$

9. Initial Point: $(-3, 7)$

Terminal Point: $(11, -16)$

$\mathbf{v} = \langle 11 - (-3), -16 - 7 \rangle = \langle 14, -23 \rangle$

10. $\mathbf{v} = 12\left(\dfrac{\mathbf{u}}{\|\mathbf{u}\|}\right) = 12\left(\dfrac{\langle 3, -5 \rangle}{\sqrt{3^2 + (-5)^2}}\right) = \dfrac{12}{\sqrt{34}}\langle 3, -5 \rangle$

$$= \frac{6\sqrt{34}}{17}\langle 3, -5 \rangle = \left\langle \frac{18\sqrt{34}}{17}, -\frac{30\sqrt{34}}{17} \right\rangle$$

11. $\mathbf{u} + \mathbf{v} = \langle 3, 5 \rangle + \langle -7, 1 \rangle = \langle -4, 6 \rangle$

12. $\mathbf{u} - \mathbf{v} = \langle 3, 5 \rangle - \langle -7, 1 \rangle = \langle 10, 4 \rangle$

13. $5\mathbf{u} - 3\mathbf{v} = 5\langle 3, 5 \rangle - 3\langle -7, 1 \rangle = \langle 15, 25 \rangle + \langle 21, -3 \rangle$
$$= \langle 36, 22 \rangle$$

14. $\dfrac{\mathbf{u}}{\|\mathbf{u}\|} = \dfrac{\langle 4, -3 \rangle}{\sqrt{4^2 + (-3)^2}} = \dfrac{1}{5}\langle 4, -3 \rangle = \left\langle \dfrac{4}{5}, -\dfrac{3}{5} \right\rangle$

15. $\mathbf{u} = 250(\cos 45° \, \mathbf{i} + \sin 45° \, \mathbf{j})$

$\mathbf{v} = 130(\cos(-60°)\mathbf{i} + \sin(-60°)\mathbf{j})$

$\mathbf{R} = \mathbf{u} + \mathbf{v} \approx 241.7767 \, \mathbf{i} + 64.1934 \, \mathbf{j}$

$\|\mathbf{R}\| \approx \sqrt{241.7767^2 + 64.1934^2} \approx 250.15$ pounds

$\tan \theta \approx \dfrac{64.1934}{241.7767} \implies \theta \approx 14.9°$

16. $\mathbf{u} = \langle -1, 5 \rangle, \mathbf{v} = \langle 3, -2 \rangle$

$\cos \theta = \dfrac{\mathbf{u} \cdot \mathbf{v}}{\|\mathbf{u}\| \|\mathbf{v}\|} = \dfrac{-13}{\sqrt{26}\sqrt{13}} \implies \theta = 135°$

17. $\mathbf{u} = \langle 6, 10 \rangle, \mathbf{v} = \langle 2, 3 \rangle$

$\mathbf{u} \cdot \mathbf{v} = 42 \neq 0 \implies \mathbf{u}$ and $\mathbf{v}$ are not orthogonal.

18. $\mathbf{u} = \langle 6, 7 \rangle, \mathbf{v} = \langle -5, -1 \rangle$

$\mathbf{w}_1 = \text{proj}_{\mathbf{v}} \, \mathbf{u} = \left(\dfrac{\mathbf{u} \cdot \mathbf{v}}{\|\mathbf{v}\|^2} \right)\mathbf{v} = -\dfrac{37}{26}\langle -5, -1 \rangle = \dfrac{37}{26}\langle 5, 1 \rangle$

$\mathbf{w}_2 = \mathbf{u} - \mathbf{w}_1 = \langle 6, 7 \rangle - \dfrac{37}{26}\langle 5, 1 \rangle$

$= \left\langle -\dfrac{29}{26}, \dfrac{145}{26} \right\rangle$

$= \dfrac{29}{26}\langle -1, 5 \rangle$

$\mathbf{u} = \mathbf{w}_1 + \mathbf{w}_2 = \dfrac{37}{26}\langle 5, 1 \rangle + \dfrac{29}{26}\langle -1, 5 \rangle$

19. $\mathbf{F} = -500\mathbf{j}, \mathbf{v} = (\cos 12°)\mathbf{i} + (\sin 12°)\mathbf{j}$

$\mathbf{w}_1 = \text{proj}_{\mathbf{v}} \, \mathbf{F} = \left(\dfrac{\mathbf{F} \cdot \mathbf{v}}{\|\mathbf{v}\|^2} \right)\mathbf{v} = (\mathbf{F} \cdot \mathbf{v})\mathbf{v}$

$= (-500 \sin 12°)\mathbf{v}$

The magnitude of the force is $500 \sin 12° \approx 104$ pounds.

20. $z = 5 - 5i$

$|z| = \sqrt{5^2 + (-5)^2} = \sqrt{50} = 5\sqrt{2}$

$\tan \theta = \dfrac{-5}{5} = -1$ and θ is in Quadrant IV $\implies \theta = \dfrac{7\pi}{4}$

$z = 5\sqrt{2}\left(\cos \dfrac{7\pi}{4} + i \sin \dfrac{7\pi}{4}\right)$

21. $z = 6(\cos 120° + i \sin 120°)$

$= 6\left(-\dfrac{1}{2} + \dfrac{\sqrt{3}}{2}i\right) = -3 + 3\sqrt{3}i$

22. $\left[3\left(\cos \dfrac{7\pi}{6} + i \sin \dfrac{7\pi}{6}\right)\right]^8 = 3^8\left(\cos \dfrac{28\pi}{3} + i \sin \dfrac{28\pi}{3}\right)$

$= 6561\left(-\dfrac{1}{2} - \dfrac{\sqrt{3}}{2}i\right)$

$= -\dfrac{6561}{2} - \dfrac{6561\sqrt{3}}{2}i$

23. $(3 - 3i)^6 = \left[3\sqrt{2}\left(\cos \dfrac{7\pi}{4} + i \sin \dfrac{7\pi}{4}\right)\right]^6$

$= \left(3\sqrt{2}\right)^6\left(\cos \dfrac{21\pi}{2} + i \sin \dfrac{21\pi}{2}\right)$

$= 5832(0 + i)$

$= 5832i$

24. $z = 256\left(1 + \sqrt{3}i\right)$

$|z| = 256\sqrt{1^2 + \left(\sqrt{3}\right)^2} = 256\sqrt{4} = 512$

$\tan \theta = \dfrac{\sqrt{3}}{1} \implies \theta = \dfrac{\pi}{3}$

$z = 512\left(\cos \dfrac{\pi}{3} + i \sin \dfrac{\pi}{3}\right)$

Fourth roots of $z = \sqrt[4]{512}\left[\cos \dfrac{\dfrac{\pi}{3} + 2\pi k}{4} + i \sin \dfrac{\dfrac{\pi}{3} + 2\pi k}{4}\right]$, $k = 0, 1, 2, 3$

$k = 0$: $4\sqrt[4]{2}\left(\cos \dfrac{\pi}{12} + i \sin \dfrac{\pi}{12}\right)$

$k = 1$: $4\sqrt[4]{2}\left(\cos \dfrac{7\pi}{12} + i \sin \dfrac{7\pi}{12}\right)$

$k = 2$: $4\sqrt[4]{2}\left(\cos \dfrac{13\pi}{12} + i \sin \dfrac{13\pi}{12}\right)$

$k = 3$: $4\sqrt[4]{2}\left(\cos \dfrac{19\pi}{12} + i \sin \dfrac{19\pi}{12}\right)$

25. $x^3 - 27i = 0 \implies x = 27i$

The solutions to the equation are the cube roots of $27i = 27\left(\cos \dfrac{\pi}{2} + i \sin \dfrac{\pi}{2}\right)$.

Cube roots: $\sqrt[3]{27}\left[\cos \dfrac{\dfrac{\pi}{2} + 2\pi k}{3} + i \sin \dfrac{\dfrac{\pi}{2} + 2\pi k}{3}\right]$, $k = 0, 1, 2$

$k = 0$: $3\left(\cos \dfrac{\pi}{6} + i \sin \dfrac{\pi}{6}\right) = 3\left(\dfrac{\sqrt{3}}{2} + \dfrac{1}{2}i\right) = \dfrac{3\sqrt{3}}{2} + \dfrac{3}{2}i$

$k = 1$: $3\left(\cos \dfrac{5\pi}{6} + i \sin \dfrac{5\pi}{6}\right) = 3\left(-\dfrac{\sqrt{3}}{2} + \dfrac{1}{2}i\right) = -\dfrac{3\sqrt{3}}{2} + \dfrac{3}{2}i$

$k = 2$: $3\left(\cos \dfrac{3\pi}{2} + i \sin \dfrac{3\pi}{2}\right) = 3(0 - i) = -3i$

Chapters 6–8   Cumulative Test Solutions

1. (a)

(b) $-120° + 360° = 240°$

(c) $-120\left(\dfrac{\pi}{180°}\right) = -\dfrac{2\pi}{3}$

(d) $-120° + 360° = 240°$

   $\theta' = 240° - 180° = 60°$

2. $2.35\left(\dfrac{180°}{\pi}\right) \approx 134.6°$

4. $f(x) = 3 - 2\sin \pi x$

Period: $\dfrac{2\pi}{\pi} = 2$

Amplitude: $|a| = |-2| = 2$

Upward shift of 3 units (reflected in x-axis prior to shift)

(e) $\sin(-120°) = -\sin 60° = -\dfrac{\sqrt{3}}{2}$

   $\cos(-120°) = -\cos 60° = -\dfrac{1}{2}$

   $\tan(-120°) = \tan 60° = \sqrt{3}$

   $\csc(-120°) = \dfrac{1}{-\sin 60°} = -\dfrac{2\sqrt{3}}{3}$

   $\sec(-120°) = \dfrac{1}{-\cos 60°} = -2$

   $\cot(-120°) = \dfrac{1}{\tan 60°} = \dfrac{\sqrt{3}}{3}$

3. $\tan \theta = \dfrac{y}{x} = -\dfrac{4}{3} \implies r = 5$

Since $\sin \theta < 0$ θ is in Quadrant IV, $\implies x = 3$.

   $\cos \theta = \dfrac{x}{r} = \dfrac{3}{5}$

5. $g(x) = \dfrac{1}{2}\tan\left(x - \dfrac{\pi}{2}\right)$

Period: π

Asymptotes: $x = 0, x = \pi$

6. $h(x) = -\sec(x + \pi)$

Graph $y = -\cos(x + \pi)$ first.

Period: 2π

Amplitude: 1

Set $x + \pi = 0$ and $x + \pi = 2\pi$ for one cycle

$\qquad x = -\pi \qquad\qquad x = \pi$

The asymptotes of $h(x)$ corresponds to the x-intercepts of

$y = -\cos(x + \pi)$

$x + \pi = \dfrac{(2n + 1)\pi}{2}$

$\quad x = \dfrac{(2n - 1)\pi}{2}$ where n is any integer

7. $h(x) = a \cos(bx + c)$

Graph is reflected in x-axis.

Amplitude: $a = -3$

Period: $2 = \dfrac{2\pi}{\pi} \Rightarrow b = \pi$

No phase shift: $c = 0$

$h(x) = -3\cos(\pi x)$

8. $f(x) = \dfrac{x}{2}\sin x,\ -3\pi \le x \le 3\pi$

$\qquad -\dfrac{x}{2} \le f(x) \le \dfrac{x}{2}$

9. $\tan(\arctan 6.7) = 6.7$

10. $\tan\left(\arcsin\dfrac{3}{5}\right) = \dfrac{3}{4}$

11. $\qquad y = \arccos(2x)$

$\sin y = \sin(\arccos(2x)) = \sqrt{1 - 4x^2}$

12. $\cos\left(\dfrac{\pi}{2} - x\right)\csc x = \sin x\left(\dfrac{1}{\sin x}\right) = 1$

13. $\dfrac{\sin\theta - 1}{\cos\theta} - \dfrac{\cos\theta}{\sin\theta - 1} = \dfrac{\sin\theta - 1}{\cos\theta} - \dfrac{\cos\theta(\sin\theta + 1)}{\sin^2\theta - 1}$

$\qquad\qquad = \dfrac{\sin\theta - 1}{\cos\theta} + \dfrac{\cos\theta(\sin\theta + 1)}{\cos^2\theta} = \dfrac{\sin\theta - 1}{\cos\theta} + \dfrac{\sin\theta + 1}{\cos\theta} = \dfrac{2\sin\theta}{\cos\theta} = 2\tan\theta$

14. $\cot^2\alpha(\sec^2\alpha - 1) = \cot^2\alpha\tan^2\alpha = 1$

15. $\sin(x + y)\sin(x - y) = \dfrac{1}{2}[\cos(x + y - (x - y)) - \cos(x + y + x - y)]$

$$= \dfrac{1}{2}[\cos 2y - \cos 2x] = \dfrac{1}{2}[1 - 2\sin^2 y - (1 - 2\sin^2 x)] = \sin^2 x - \sin^2 y$$

16. $\sin^2 x \cos^2 x = \left(\dfrac{1 - \cos 2x}{2}\right)\left(\dfrac{1 + \cos 2x}{2}\right)$

$$= \dfrac{1}{4}(1 - \cos 2x)(1 + \cos 2x)$$

$$= \dfrac{1}{4}(1 - \cos^2 2x)$$

$$= \dfrac{1}{4}\left(1 - \dfrac{1 + \cos 4x}{2}\right)$$

$$= \dfrac{1}{8}(2 - (1 + \cos 4x))$$

$$= \dfrac{1}{8}(1 - \cos 4x)$$

17. $2\cos^2 \beta - \cos \beta = 0$

$\cos \beta(2\cos \beta - 1) = 0$

$\cos \beta = 0 \qquad 2\cos \beta - 1 = 0$

$\beta = \dfrac{\pi}{2}, \dfrac{3\pi}{2} \qquad \cos \beta = \dfrac{1}{2}$

$$\beta = \dfrac{\pi}{3}, \dfrac{5\pi}{3}$$

Answer: $\dfrac{\pi}{3}, \dfrac{\pi}{2}, \dfrac{3\pi}{2}, \dfrac{5\pi}{3}$

18. $3\tan \theta - \cot \theta = 0$

$3\tan \theta - \dfrac{1}{\tan \theta} = 0$

$\dfrac{3\tan^2 \theta - 1}{\tan \theta} = 0$

$3\tan^2 \theta - 1 = 0$

$\tan^2 \theta = \dfrac{1}{3}$

$\tan \theta = \pm\dfrac{\sqrt{3}}{3}$

$\theta = \dfrac{\pi}{6}, \dfrac{5\pi}{6}, \dfrac{7\pi}{6}, \dfrac{11\pi}{6}$

19. $\sin^2 x + 2\sin x + 1 = 0$

$(\sin x + 1)(\sin x + 1) = 0$

$\sin x + 1 = 0$

$\sin x = -1$

$x = \dfrac{3\pi}{2}$

20. $\sin u = \dfrac{12}{13} \Rightarrow \cos u = \dfrac{5}{13}$ and $\tan u = \dfrac{12}{5}$ since u is in Quadrant I.

$\cos v = \dfrac{3}{5} \Rightarrow \sin v = \dfrac{4}{5}$ and $\tan v = \dfrac{4}{3}$ since v is in Quadrant I.

$$\tan(u - v) = \dfrac{\tan u - \tan v}{1 + \tan u \tan v} = \dfrac{\dfrac{12}{5} - \dfrac{4}{3}}{1 + \left(\dfrac{12}{5}\right)\left(\dfrac{4}{3}\right)} = \dfrac{16}{63}$$

21. $\tan \theta = \dfrac{1}{2}$

$$\tan 2\theta = \dfrac{2\tan \theta}{1 - \tan^2 \theta} = \dfrac{2\left(\frac{1}{2}\right)}{1 - \left(\frac{1}{2}\right)^2} = \dfrac{4}{3}$$

22. $\tan \theta = \dfrac{4}{3} \Rightarrow \cos \theta = \pm\dfrac{3}{5}$

$$\sin \dfrac{\theta}{2} = \sqrt{\dfrac{1 - \cos \theta}{2}} = \sqrt{\dfrac{1 - \frac{3}{5}}{2}} = \dfrac{\sqrt{5}}{5}$$

$$\text{or} = \sqrt{\dfrac{1 + \frac{3}{5}}{2}} = \dfrac{2\sqrt{5}}{5}$$

23. $5 \sin \dfrac{3\pi}{4} \cos \dfrac{7\pi}{4} = \dfrac{5}{2}\left[\sin\left(\dfrac{3\pi}{4} + \dfrac{7\pi}{4} \right) + \sin\left(\dfrac{3\pi}{4} - \dfrac{7\pi}{4} \right) \right]$

$\qquad\qquad\qquad\quad = \dfrac{5}{2}\left[\sin \dfrac{5\pi}{2} + \sin(-\pi) \right]$

$\qquad\qquad\qquad\quad = \dfrac{5}{2}\left(\sin \dfrac{5\pi}{2} - \sin \pi \right)$

24. $\cos 8x + \cos 4x = 2 \cos\left(\dfrac{8x + 4x}{2} \right) \cos\left(\dfrac{8x - 4x}{2} \right)$

$\qquad\qquad\qquad\quad = 2 \cos 6x \cos 2x$

25. Given: $A = 30°, a = 9, b = 8$

$\qquad \dfrac{\sin B}{8} = \dfrac{\sin 30°}{9}$

$\qquad \sin B = \dfrac{8}{9}\left(\dfrac{1}{2} \right)$

$\qquad\qquad B = \arcsin\left(\dfrac{4}{9} \right)$

$\qquad\qquad B \approx 26.39°$

$\qquad\qquad C = 180° - A - B \approx 123.61°$

$\dfrac{c}{\sin 123.61°} = \dfrac{9}{\sin 30°}$

$\qquad\qquad c \approx 14.99$

26. Given: $A = 30°, b = 8, c = 10$

$\qquad a^2 = 8^2 + 10^2 - 2(8)(10)\cos 30°$

$\qquad a^2 \approx 25.4359$

$\qquad a \approx 5.04$

$\qquad \cos B = \dfrac{5.04^2 + 10^2 - 8^2}{2(5.04)(10)}$

$\qquad \cos B \approx 0.6091$

$\qquad B \approx 52.48°$

$\qquad C = 180° - A - B \approx 97.52°$

27. Given: $A = 30°, C = 90°, b = 10$

$\qquad B = 180° - 30° - 90° = 60°$

$\qquad \tan 30° = \dfrac{a}{10} \implies a = 10 \tan 30° \approx 5.77$

$\qquad \cos 30° = \dfrac{10}{c} \implies c = \dfrac{10}{\cos 30°} \approx 11.55$

28. $a = 4, b = 8, c = 9$

$\qquad \cos C = \dfrac{4^2 + 8^2 - 9^2}{2(4)(8)} = \dfrac{-1}{64} \implies C \approx 90.90°$

$\qquad \sin A \approx \dfrac{4 \sin 90.90°}{9} \implies A \approx 26.38°$

$\qquad B \approx 180° - 26.38° - 90.90° = 62.72°$

29. Area $= \dfrac{1}{2}(7)(12) \sin 60° \approx 36.4$ square inches

30. $s = \dfrac{11 + 16 + 17}{2} = 22$

Area $= \sqrt{22(11)(6)(5)} \approx 85.2$ square inches

31. $\mathbf{u} = \langle 3, 5 \rangle = 3\mathbf{i} + 5\mathbf{j}$

32. $\mathbf{v} = \mathbf{i} + \mathbf{j}$

$\qquad \|\mathbf{v}\| = \sqrt{1^2 + 1^2} = \sqrt{2}$

$\qquad \mathbf{u} = \dfrac{\mathbf{v}}{\|\mathbf{v}\|} = \dfrac{1}{\sqrt{2}}(\mathbf{i} + \mathbf{j}) = \dfrac{\sqrt{2}}{2}(\mathbf{i} + \mathbf{j})$

33. $\mathbf{u} = 3\mathbf{i} + 4\mathbf{j}, \mathbf{v} = \mathbf{i} - 2\mathbf{j}$

$\qquad \mathbf{u} \cdot \mathbf{v} = 3(1) + 4(-2) = -5$

34. $\mathbf{u} = \langle 8, -2 \rangle, \mathbf{v} = \langle 1, 5 \rangle$

$\mathbf{w}_1 = \text{proj}_{\mathbf{v}}\,\mathbf{u} = \left(\dfrac{\mathbf{u}\cdot\mathbf{v}}{\|\mathbf{v}\|^2}\right)\mathbf{v} = \dfrac{-2}{26}\langle 1, 5 \rangle = -\dfrac{1}{13}\langle 1, 5 \rangle$

$\mathbf{w}_2 = \mathbf{u} - \mathbf{w}_1 = \langle 8, -2 \rangle - \left\langle -\dfrac{1}{13}, -\dfrac{5}{13} \right\rangle = \left\langle \dfrac{105}{13}, -\dfrac{21}{13} \right\rangle$

$\quad = \dfrac{21}{13}\langle 5, -1 \rangle$

$\mathbf{u} = \mathbf{w}_1 + \mathbf{w}_2 = -\dfrac{1}{13}\langle 1, 5 \rangle + \dfrac{21}{13}\langle 5, -1 \rangle$

35. $r = |-2 + 2i| = \sqrt{(-2)^2 + (2)^2} = 2\sqrt{2}$

$\tan\theta = \dfrac{2}{-2} = -1$

Since $\tan\theta = -1$ and $-2 + 2i$ lies in Quadrant II,

$\theta = \dfrac{3\pi}{4}$. Thus, $-2 + 2i = 2\sqrt{2}\left(\cos\dfrac{3\pi}{4} + i\sin\dfrac{3\pi}{4}\right)$.

36. $[4(\cos 30° + i\sin 30°)][6(\cos 120° + i\sin 120°)] = (4)(6)[\cos(30° + 120°) + i\sin(30° + 120°)]$

$= 24(\cos 150° + i\sin 150°)$

$= 24\left(-\dfrac{\sqrt{3}}{2} + \dfrac{1}{2}i\right)$

$= -12\sqrt{3} + 12i$

37. $1 = 1(\cos 0 + i\sin 0)$

$\sqrt[3]{1} = \sqrt[3]{1}\left[\cos\left(\dfrac{0 + 2\pi k}{3}\right) + i\sin\left(\dfrac{0 + 2\pi k}{3}\right)\right], k = 0, 1, 2$

$k = 0: \sqrt[3]{1}\left[\left(\cos\left(\dfrac{0 + 2\pi(0)}{3}\right) + i\sin\left(\dfrac{0 + 2\pi(0)}{3}\right)\right)\right] = \cos 0 + i\sin 0 = 1$

$k = 1: \sqrt[3]{1}\left[\left(\cos\left(\dfrac{0 + 2\pi(1)}{3}\right) + i\sin\left(\dfrac{0 + 2\pi(1)}{3}\right)\right)\right] = \cos\dfrac{2\pi}{3} + i\sin\dfrac{2\pi}{3} = -\dfrac{1}{2} + \dfrac{\sqrt{3}}{2}i$

$k = 2: \sqrt[3]{1}\left[\left(\cos\left(\dfrac{0 + 2\pi(2)}{3}\right) + i\sin\left(\dfrac{0 + 2\pi(2)}{3}\right)\right)\right] = \cos\dfrac{4\pi}{3} + i\sin\dfrac{4\pi}{3} = -\dfrac{1}{2} - \dfrac{\sqrt{3}}{2}i$

38. $x^5 + 243 = 0 \Rightarrow x^5 = -243$

The solutions to the equation are the fifth roots of

$-243 = 243(\cos\pi + i\sin\pi)$, which are:

$\sqrt[5]{243}\left[\cos\left(\dfrac{\pi + 2\pi k}{5} + i\sin\dfrac{\pi + 2\pi k}{5}\right)\right], k = 0, 1, 2, 3, 4$

$k = 0: 3\left(\cos\dfrac{\pi}{5} + i\sin\dfrac{\pi}{5}\right)$

$k = 1: 3\left(\cos\dfrac{3\pi}{5} + i\sin\dfrac{3\pi}{5}\right)$

$k = 2: 3(\cos\pi + i\sin\pi)$

$k = 3: 3\left(\cos\dfrac{7\pi}{5} + i\sin\dfrac{7\pi}{5}\right)$

$k = 4: 3\left(\cos\dfrac{9\pi}{5} + i\sin\dfrac{9\pi}{5}\right)$

39. Angular speed $= \dfrac{\theta}{t} = \dfrac{2\pi(63)}{1} \approx 395.8$ radians per minute

Linear speed $= \dfrac{s}{t} = \dfrac{42\pi(63)}{1} \approx 8312.7$ inches per minute

40. Area $= \dfrac{\theta r^2}{2} = \dfrac{(114°)\left(\dfrac{\pi}{180°}\right)(8)^2}{2} \approx 20.267\pi \approx 63.67$ square yards

41. Height of smaller triangle:

$$\tan 16° 45' = \frac{h_1}{200}$$

$$h_1 = 200 \tan 16.75° \approx 60.2 \text{ feet}$$

Height of larger triangle:

$$\tan 18° = \frac{h_2}{200}$$

$$h_2 = 200 \tan 18° \approx 65.0 \text{ feet}$$

Height of flag:

$$h_2 - h_1 = 65.0 - 60.2 \approx 5 \text{ feet}$$

Not drawn to scale

42. $\tan \theta = \dfrac{5}{12} \implies \theta \approx 22.6°$

43. $d = a \cos bt$

$$|a| = 4 \implies a = 4$$

$$\frac{2\pi}{b} = 8 \implies b = \frac{\pi}{4}$$

$$d = 4 \cos \frac{\pi}{4}t$$

44. $\mathbf{v}_1 = 500\langle \cos 60°, \sin 60° \rangle = \langle 250, 250\sqrt{3} \rangle$

$\mathbf{v}_2 = 50\langle \cos 30°, \sin 30° \rangle = \langle 25\sqrt{3}, 25 \rangle$

$\mathbf{v} = \mathbf{v}_1 + \mathbf{v}_2 = \langle 250 + 25\sqrt{3}, 250\sqrt{3} + 25 \rangle$

$\approx \langle 293.3, 458.0 \rangle$

$\|\mathbf{v}\| = \sqrt{(293.3)^2 + (458.0)^2} \approx 543.9$

$\tan \theta = \dfrac{458.0}{293.3} \approx 1.56 \implies \theta \approx 57.4°$

Bearing: $90° - 57.4° = 32.6°$

The plane is traveling on a bearing of 32.6° at 543.9 kilometers per hour.

45. $\mathbf{w} = (85)(10)\cos 60° = 425 \text{ foot-pounds}$

Chapter 9   Chapter Test Solutions

1. $\begin{cases} x - y = -7 \implies y = x + 7 \\ 4x + 5y = 8 \implies 4x + 5(x + 7) = 8 \end{cases}$

$$9x + 35 = 8$$
$$9x = -27$$
$$x = -3 \implies y = 4$$

Solution: $(-3, 4)$

2. $\begin{cases} y = x - 1 \\ y = (x - 1)^3 \end{cases}$

$$x - 1 = (x - 1)^3$$
$$x - 1 = x^3 - 3x^2 + 3x - 1$$
$$0 = x^3 - 3x^2 + 2x$$
$$0 = x(x - 1)(x - 2)$$
$$x = 0, \quad x = 1, \quad x = 2$$
$$y = -1, \quad y = 0, \quad y = 1$$

Solutions: $(0, -1), (1, 0), (2, 1)$

3. $\begin{cases} x - y = 4 \implies x = y + 4 \\ 2x - y^2 = 0 \implies 2(y + 4) - y^2 = 0 \end{cases}$

$$0 = y^2 - 2y - 8$$
$$0 = (y + 2)(y - 4)$$
$$y = -2 \text{ or } y = 4$$
$$x = 2 \qquad x = 8$$

Solutions: $(2, -2), (8, 4)$

4. $\begin{cases} 2x - 3y = 0 \\ 2x + 3y = 12 \end{cases}$

Solution: $(3, 2)$

5. $\begin{cases} y = 9 - x^2 \\ y = x + 3 \end{cases}$

Solutions: $(-3, 0), (2, 5)$

6. $\begin{cases} y - \ln x = 12 \implies y = 12 + \ln x \\ 7x - 2y + 11 = -6 \implies y = \frac{7}{2}x + \frac{17}{2} \end{cases}$

Solutions: $(1, 12), (0.034, 8.619)$

7. $\begin{cases} 2x + 3y = 17 \implies 8x + 12y = 68 \\ 5x - 4y = -15 \implies 15x - 12y = -45 \end{cases}$

$$\overline{23x = 23}$$
$$x = 1 \implies y = 5$$

Solution: $(1, 5)$

8. $\begin{cases} 2.5x - y = 6 \implies 10x - 4y = 24 \\ 3x + 4y = 2 \implies 3x + 4y = 2 \end{cases}$

$$\overline{13x = 26}$$
$$x = 2 \implies y = -1$$

Solution: $(2, -1)$

9. $\begin{cases} x - 2y + 3z = 11 \\ 2x \quad\;\; - \;\; z = 3 \\ \quad\;\; 3y + z = -8 \end{cases}$

$\begin{cases} x - 2y + 3z = 11 \\ 4y - 7z = -19 \quad -2\text{Eq.1} + \text{Eq.2} \\ 3y + z = -8 \end{cases}$

$\begin{cases} x - 2y + 3z = 11 \\ y - 8z = -11 \quad -\text{Eq.3} + \text{Eq.2} \\ 3y + z = -8 \end{cases}$

$\begin{cases} x - 2y + 3z = 11 \\ y - 8z = -11 \\ 25z = 25 \quad -3\text{Eq.2} + \text{Eq.3} \end{cases}$

$\begin{cases} x - 2y + 3z = 11 \\ y - 8z = -11 \\ z = 1 \quad \frac{1}{25}\text{Eq.3} \end{cases}$

$y - 8(1) = -11 \implies y = -3$

$x - 2(-3) + 3(1) = 11 \implies x = 2$

Solution: $(2, -3, 1)$

10. $\begin{cases} 3x + 2y + z = 17 \quad \text{Equation 1} \\ -x + y + z = 4 \quad \text{Equation 2} \\ x - y - z = 3 \quad \text{Equation 3} \end{cases}$

Interchange Equations 1 and 3

$\begin{cases} x - y - z = 3 \\ -x + y + z = 4 \\ 3x + 2y + z = 17 \end{cases}$

$\begin{cases} x - y - z = 3 \\ 0 \neq 7 \quad \text{Eq. 1} + \text{Eq. 2} \\ 3x + 2y + z = 17 \end{cases}$

Inconsistent
No solution

11. $\dfrac{2x + 5}{x^2 - x - 2} = \dfrac{2x + 5}{(x - 2)(x + 1)} = \dfrac{A}{x - 2} + \dfrac{B}{x + 1}$

$2x + 5 = A(x + 1) + B(x - 2)$

Let $x = 2$: $9 = 3A \implies A = 3$

Let $x = -1$: $3 = -3B \implies B = -1$

$\dfrac{2x + 5}{x^2 - x - 2} = \dfrac{3}{x - 2} - \dfrac{1}{x + 1}$

12. $\dfrac{3x^2 - 2x + 4}{x^2(2 - x)} = \dfrac{A}{x} + \dfrac{B}{x^2} + \dfrac{C}{2 - x}$

$3x^2 - 2x + 4 = Ax(2 - x) + B(2 - x) + Cx^2$

Let $x = 0$: $4 = 2B \implies B = 2$

Let $x = 2$: $12 = 4C \implies C = 3$

Let $x = 1$: $5 = A + B + C = A + 2 + 3 \implies A = 0$

$\dfrac{3x^2 - 2x + 4}{x^2(2 - x)} = \dfrac{2}{x^2} + \dfrac{3}{2 - x}$

13. $\dfrac{x^2 + 5}{x^3 - x} = \dfrac{x^2 + 5}{x(x + 1)(x - 1)} = \dfrac{A}{x} + \dfrac{B}{x + 1} + \dfrac{C}{x - 1}$

$x^2 + 5 = A(x + 1)(x - 1) + Bx(x - 1) + Cx(x + 1)$

Let $x = 0$: $5 = -A \implies A = -5$

Let $x = -1$: $6 = 2B \implies B = 3$

Let $x = 1$: $6 = 2C \implies C = 3$

$\dfrac{x^2 + 5}{x^3 - x} = -\dfrac{5}{x} + \dfrac{3}{x + 1} + \dfrac{3}{x - 1}$

14. $\dfrac{x^2 - 4}{x^3 + 2x} = \dfrac{x^2 - 4}{x(x^2 + 2)} = \dfrac{A}{x} + \dfrac{Bx + C}{x^2 + 2}$

$x^2 - 4 = A(x^2 + 2) + (Bx + C)x$

$= Ax^2 + 2A + Bx^2 + Cx$

$= (A + B)x^2 + Cx + 2A$

Equate the coefficients of like terms:

$1 = A + B, 0 = C, -4 = 2A$

Thus, $A = -2, B = 3, C = 0$.

$\dfrac{x^2 - 4}{x^3 + 2x} = -\dfrac{2}{x} + \dfrac{3x}{x^2 + 2}$

15. $2x + y \le 4$

$2x - y \ge 0$

$x \ge 0$

16. $y < -x^2 + x + 4$

$y > 4x$

17. $x^2 + y^2 \le 16$

$x \ge 1$

$y \ge -3$

18. Maximize $z = 20x + 12y$ subject to:

$$\begin{cases} x \ge 0, y \ge 0 \\ x + 4y \le 32 \\ 3x + 2y \le 36 \end{cases}$$

At $(0, 0)$ we have $z = 0$.

At $(0, 8)$ we have $z = 96$.

At $(8, 6)$ we have $z = 232$.

At $(12, 0)$ we have $z = 240$.

The maximum value, $z = 240$, occurs at $(12, 0)$.

The minimum value, $z = 0$ occurs at $(0, 0)$.

19. Let $x =$ amount in 8% fund.

Let $y =$ amount in 8.5% fund.

$x + y = 50,000 \implies y = 50,000 - x$

$0.08x + 0.085y = 4150$

Use substitution

$0.08x + 0.085(50,000 - x) = 4150$

$0.08x + 4250 - 0.085x = 4150$

$100 = 0.005x$

$20,000 = x \implies y = 30,000$

\$20,000 is invested at 8%.

\$30,000 is invested at 8.5%.

20. $y = ax^2 + bx + c$

$(0, 6)$: $6 = c$

$(-2, 2)$: $2 = 4a - 2b + c$

$\left(3, \frac{9}{2}\right)$: $\frac{9}{2} = 9a + 3b + c$

Solving this system yields: $a = -\frac{1}{2}$, $b = 1$, and $c = 6$.

Thus, $y = -\frac{1}{2}x^2 + x + 6$.

21. Optimize $P = 30x + 40y$ subject to:

$$\begin{cases} x \ge 0, y \ge 0 \\ 0.5x + 0.75y \le 4000 \\ 2.0x + 1.5y \le 8950 \\ 0.5x + 0.5y \le 2650 \end{cases}$$

At $(0, 0)$: $P = 0$

At $(0, 5300)$: $P = 212,000$

At $(2000, 3300)$: $P = 192,000$

At $(4475, 0)$: $P = 134,250$

The manufacturer should produce 5300 units of Model II and not produce any of Model I to realize an optimal profit of \$212,000.

Chapter 10   Chapter Test Solutions

1.
$$\begin{bmatrix} 1 & -1 & 5 \\ 6 & 2 & 3 \\ 5 & 3 & -3 \end{bmatrix}$$

$$\begin{matrix} -6R_1 + R_2 \to \\ -5R_1 + R_3 \to \end{matrix} \begin{bmatrix} 1 & -1 & 5 \\ 0 & 8 & -27 \\ 0 & 8 & -28 \end{bmatrix}$$

$$-R_2 + R_3 \to \begin{bmatrix} 1 & -1 & 5 \\ 0 & 8 & -27 \\ 0 & 0 & -1 \end{bmatrix}$$

$$\begin{matrix} \frac{1}{8}R_2 \to \\ -R_3 \to \end{matrix} \begin{bmatrix} 1 & -1 & 5 \\ 0 & 1 & -\frac{27}{8} \\ 0 & 0 & 1 \end{bmatrix}$$

$$R_2 + R_1 \to \begin{bmatrix} 1 & 0 & \frac{13}{8} \\ 0 & 1 & -\frac{27}{8} \\ 0 & 0 & 1 \end{bmatrix}$$

$$\begin{matrix} -\frac{13}{8}R_3 + R_1 \to \\ \frac{27}{8}R_3 + R_2 \to \end{matrix} \begin{bmatrix} 1 & 0 & 0 \\ 0 & 1 & 0 \\ 0 & 0 & 1 \end{bmatrix}$$

2.
$$\begin{bmatrix} 1 & 0 & -1 & 2 \\ -1 & 1 & 1 & -3 \\ 1 & 1 & -1 & 1 \\ 3 & 2 & -3 & 4 \end{bmatrix}$$

$$\begin{matrix} R_1 + R_2 \to \\ -R_1 + R_3 \to \\ -3R_1 + R_4 \to \end{matrix} \begin{bmatrix} 1 & 0 & -1 & 2 \\ 0 & 1 & 0 & -1 \\ 0 & 1 & 0 & -1 \\ 0 & 2 & 0 & -2 \end{bmatrix}$$

$$\begin{matrix} -R_2 + R_3 \to \\ -2R_2 + R_4 \to \end{matrix} \begin{bmatrix} 1 & 0 & -1 & 2 \\ 0 & 1 & 0 & -1 \\ 0 & 0 & 0 & 0 \\ 0 & 0 & 0 & 0 \end{bmatrix}$$

3.
$$\begin{bmatrix} 4 & 3 & -2 & \vdots & 14 \\ -1 & -1 & 2 & \vdots & -5 \\ 3 & 1 & -4 & \vdots & 8 \end{bmatrix}$$

$$3R_2 + R_1 \to \begin{bmatrix} 1 & 0 & 4 & \vdots & -1 \\ -1 & -1 & 2 & \vdots & -5 \\ 3 & 1 & -4 & \vdots & 8 \end{bmatrix}$$

$$\begin{matrix} R_1 + R_2 \to \\ -3R_1 + R_3 \to \end{matrix} \begin{bmatrix} 1 & 0 & 4 & \vdots & -1 \\ 0 & -1 & 6 & \vdots & -6 \\ 0 & 1 & -16 & \vdots & 11 \end{bmatrix}$$

$$R_2 + R_3 \to \begin{bmatrix} 1 & 0 & 4 & \vdots & -1 \\ 0 & -1 & 6 & \vdots & -6 \\ 0 & 0 & -10 & \vdots & 5 \end{bmatrix}$$

$$\begin{matrix} -R_2 \to \\ -\frac{1}{10}R_3 \to \end{matrix} \begin{bmatrix} 1 & 0 & 4 & \vdots & -1 \\ 0 & 1 & -6 & \vdots & 6 \\ 0 & 0 & 1 & \vdots & -\frac{1}{2} \end{bmatrix}$$

$$\begin{matrix} -4R_3 + R_1 \to \\ 6R_3 + R_2 \to \end{matrix} \begin{bmatrix} 1 & 0 & 0 & \vdots & 1 \\ 0 & 1 & 0 & \vdots & 3 \\ 0 & 0 & 1 & \vdots & -\frac{1}{2} \end{bmatrix}$$

Solution: $\left(1, 3, -\frac{1}{2}\right)$

4. (a) $A - B = \begin{bmatrix} 5 & 4 \\ -4 & -4 \end{bmatrix} - \begin{bmatrix} 4 & -1 \\ -4 & 0 \end{bmatrix}$

$$= \begin{bmatrix} 1 & 5 \\ 0 & -4 \end{bmatrix}$$

(b) $3A = 3\begin{bmatrix} 5 & 4 \\ -4 & -4 \end{bmatrix} = \begin{bmatrix} 15 & 12 \\ -12 & -12 \end{bmatrix}$

(c) $3A - 2B = 3\begin{bmatrix} 5 & 4 \\ -4 & -4 \end{bmatrix} - 2\begin{bmatrix} 4 & -1 \\ -4 & 0 \end{bmatrix}$

$$= \begin{bmatrix} 15 & 12 \\ -12 & -12 \end{bmatrix} - \begin{bmatrix} 8 & -2 \\ -8 & 0 \end{bmatrix}$$

$$= \begin{bmatrix} 7 & 14 \\ -4 & -12 \end{bmatrix}$$

(d) $AB = \begin{bmatrix} 5 & 4 \\ -4 & -4 \end{bmatrix}\begin{bmatrix} 4 & -1 \\ -4 & 0 \end{bmatrix}$

$$= \begin{bmatrix} (5)(4) + (4)(-4) & (5)(-1) + (4)(0) \\ (-4)(4) + (-4)(-4) & (-4)(-1) + (-4)(0) \end{bmatrix}$$

$$= \begin{bmatrix} 4 & -5 \\ 0 & 4 \end{bmatrix}$$

5. $\begin{bmatrix} -6 & 4 \\ 10 & -5 \end{bmatrix}^{-1} = \dfrac{1}{(-6)(-5) - (4)(10)}\begin{bmatrix} -5 & -4 \\ -10 & -6 \end{bmatrix} = \begin{bmatrix} \frac{1}{2} & \frac{2}{5} \\ 1 & \frac{3}{5} \end{bmatrix}$

6.

$$\begin{bmatrix} -2 & 4 & -6 & \vdots & 1 & 0 & 0 \\ 2 & 1 & 0 & \vdots & 0 & 1 & 0 \\ 4 & -2 & 5 & \vdots & 0 & 0 & 1 \end{bmatrix}$$

$$\begin{matrix} R_1 + R_2 \rightarrow \\ 2R_1 + R_3 \rightarrow \end{matrix} \begin{bmatrix} -2 & 4 & -6 & \vdots & 1 & 0 & 0 \\ 0 & 5 & -6 & \vdots & 1 & 1 & 0 \\ 0 & 6 & -7 & \vdots & 2 & 0 & 1 \end{bmatrix}$$

$$\begin{matrix} -\frac{1}{2}R_1 \rightarrow \\ -R_3 + R_2 \rightarrow \\ {} \end{matrix} \begin{bmatrix} 1 & -2 & 3 & \vdots & -\frac{1}{2} & 0 & 0 \\ 0 & -1 & 1 & \vdots & -1 & 1 & -1 \\ 0 & 6 & -7 & \vdots & 2 & 0 & 1 \end{bmatrix}$$

$$\begin{matrix} -2R_2 + R_1 \rightarrow \\ {} \\ 6R_2 + R_3 \rightarrow \end{matrix} \begin{bmatrix} 1 & 0 & 1 & \vdots & \frac{3}{2} & -2 & 2 \\ 0 & -1 & 1 & \vdots & -1 & 1 & -1 \\ 0 & 0 & -1 & \vdots & -4 & 6 & -5 \end{bmatrix}$$

$$\begin{matrix} {} \\ -R_2 \rightarrow \\ -R_3 \rightarrow \end{matrix} \begin{bmatrix} 1 & 0 & 1 & \vdots & \frac{3}{2} & -2 & 2 \\ 0 & 1 & -1 & \vdots & 1 & -1 & 1 \\ 0 & 0 & 1 & \vdots & 4 & -6 & 5 \end{bmatrix}$$

$$\begin{matrix} -R_3 + R_1 \rightarrow \\ R_3 + R_2 \rightarrow \\ {} \end{matrix} \begin{bmatrix} 1 & 0 & 0 & \vdots & -\frac{5}{2} & 4 & -3 \\ 0 & 1 & 0 & \vdots & 5 & -7 & 6 \\ 0 & 0 & 1 & \vdots & 4 & -6 & 5 \end{bmatrix}$$

$$A^{-1} = \begin{bmatrix} -\frac{5}{2} & 4 & -3 \\ 5 & -7 & 6 \\ 4 & -6 & 5 \end{bmatrix}$$

7. $\begin{cases} -6x + 4y = 10 \\ 10x - 5y = 20 \end{cases}$

$$\begin{bmatrix} x \\ y \end{bmatrix} = \begin{bmatrix} \frac{1}{2} & \frac{2}{5} \\ 1 & \frac{3}{5} \end{bmatrix} \begin{bmatrix} 10 \\ 20 \end{bmatrix} = \begin{bmatrix} \frac{1}{2}(10) + \frac{2}{5}(20) \\ 1(10) + \frac{3}{5}(20) \end{bmatrix} = \begin{bmatrix} 13 \\ 22 \end{bmatrix}$$

Solution: $(13, 22)$

8. $\begin{vmatrix} -9 & 4 \\ 13 & 16 \end{vmatrix} = (-9)(16) - (4)(13) = -196$

9. $\begin{vmatrix} \frac{5}{2} & \frac{13}{4} \\ -8 & \frac{6}{5} \end{vmatrix} = \left(\frac{5}{2}\right)\left(\frac{6}{5}\right) - \left(\frac{13}{4}\right)(-8) = 29$

10. $\begin{vmatrix} 6 & -7 & 2 \\ 3 & -2 & 0 \\ 1 & 5 & 1 \end{vmatrix} = 2\begin{vmatrix} 3 & -2 \\ 1 & 5 \end{vmatrix} + \begin{vmatrix} 6 & -7 \\ 3 & -2 \end{vmatrix} = 2(17) + 9 = 43$

Expand along column 3.

11. $\begin{cases} 7x + 6y = 9 \\ -2x - 11y = -49 \end{cases}$ $D = \begin{vmatrix} 7 & 6 \\ -2 & -11 \end{vmatrix} = -65$

$$x = \frac{\begin{vmatrix} 9 & 6 \\ -49 & -11 \end{vmatrix}}{-65} = \frac{195}{-65} = -3$$

$$y = \frac{\begin{vmatrix} 7 & 9 \\ -2 & -49 \end{vmatrix}}{-65} = \frac{-325}{-65} = 5$$

Solution: $(-3, 5)$

12. $\begin{cases} 6x - y + 2z = -4 \\ -2x + 3y - z = 10 \\ 4x - 4y + z = -18 \end{cases}$ $D = \begin{vmatrix} 6 & -1 & 2 \\ -2 & 3 & -1 \\ 4 & -4 & 1 \end{vmatrix} = -12$

$$x = \frac{\begin{vmatrix} -4 & -1 & 2 \\ 10 & 3 & -1 \\ -18 & -4 & 1 \end{vmatrix}}{-12} = \frac{24}{-12} = -2$$

$$y = \frac{\begin{vmatrix} 6 & -4 & 2 \\ -2 & 10 & -1 \\ 4 & -18 & 1 \end{vmatrix}}{-12} = \frac{-48}{-12} = 4$$

$$z = \frac{\begin{vmatrix} 6 & -1 & -4 \\ -2 & 3 & 10 \\ 4 & -4 & -18 \end{vmatrix}}{-12} = \frac{-72}{-12} = 6$$

Solution: $(-2, 4, 6)$

13. $A = -\frac{1}{2}\begin{vmatrix} -5 & 0 & 1 \\ 4 & 4 & 1 \\ 3 & 2 & 1 \end{vmatrix} = -\frac{1}{2}(-14) = 7$

14.
$$\begin{matrix} K & N & O \\ C & K & - \\ O & N & - \\ W & O & O \\ D & - & - \end{matrix} \begin{bmatrix} 11 & 14 & 15 \\ 3 & 11 & 0 \\ 15 & 14 & 0 \\ 23 & 15 & 15 \\ 4 & 0 & 0 \end{bmatrix} \begin{bmatrix} 1 & -1 & 0 \\ 1 & 0 & -1 \\ 6 & -2 & -3 \end{bmatrix} = \begin{bmatrix} 115 & -41 & -59 \\ 14 & -3 & -11 \\ 29 & -15 & -14 \\ 128 & -53 & -60 \\ 4 & -4 & 0 \end{bmatrix}$$

Message: $[11 \ 14 \ 15], [3 \ 11 \ 0], [15 \ 14 \ 0], [23 \ 15 \ 15], [4 \ 0 \ 0]$

Encoded Message: $115 \ -41 \ -59 \ 14 \ -3 \ -11 \ 29 \ -15 \ -14 \ 128 \ -53 \ -60 \ 4 \ -4 \ 0$

15. Let $x =$ amount of 60% solution and $y =$ amount of 20% solution.

$$\begin{cases} x + y = 100 \implies y = 100 - x \\ 0.60x + 0.20y = 0.50(100) \implies 6x + 2y = 500 \end{cases}$$

By substitution, we have

$$6x + 2(100 - x) = 500$$
$$6x + 200 - 2x = 500$$
$$4x = 300$$
$$x = 75$$
$$y = 100 - x = 25$$

Answer: 75 liters of 60% solution and 25 liters of 20% solution.

Chapter 11   Chapter Test Solutions

1. $a_n = \dfrac{(-1)^n}{3n + 2}$

$a_1 = -\dfrac{1}{5}$

$a_2 = \dfrac{1}{8}$

$a_3 = -\dfrac{1}{11}$

$a_4 = \dfrac{1}{14}$

$a_5 = -\dfrac{1}{17}$

2. $\dfrac{3}{1!}, \dfrac{4}{2!}, \dfrac{5}{3!}, \dfrac{6}{4!}, \dfrac{7}{5!}, \cdots$

$a_n = \dfrac{n + 2}{n!}$

3. $6 + 17 + 28 + 39 + \cdots$

$a_n = 11n - 5$

$a_5 = 50, a_6 = 61, a_7 = 72$

$S_5 = 6 + 17 + 28 + 39 + 50$

$= 140$

4. $a_5 = 5.4, a_{12} = 11.0$

$a_{12} = a_5 + 7d$

$11.0 = 5.4 + 7d$

$5.6 = 7d$

$0.8 = d$

$a_1 = a_5 - 4d$

$a_1 = 5.4 - 4(0.8)$

$= 2.2$

$a_n = a_1 + (n - 1)d$

$= 2.2 + (n - 1)(0.8)$

$= 0.8n + 1.4$

5. $a_n = 5(2)^{n-1}$

$a_1 = 5$

$a_2 = 10$

$a_3 = 20$

$a_4 = 40$

$a_5 = 80$

6. $\displaystyle\sum_{i=1}^{50}(2i^2+5)=2\sum_{i=1}^{50}i^2+\sum_{i=1}^{50}5$

$\qquad = 2\left[\dfrac{50(51)(101)}{6}\right]+50(5)$

$\qquad = 86{,}100$

7. $\displaystyle\sum_{n=1}^{7}(8n-5)=8\sum_{n=1}^{7}n-\sum_{n=1}^{7}5$

$\qquad = 8\left[\dfrac{(7)(8)}{2}\right]-7(5)$

$\qquad = 189$

8. $\displaystyle\sum_{i=1}^{\infty}4\left(\dfrac{1}{2}\right)^i=\dfrac{2}{1-\frac{1}{2}}=4$

9. $5+10+15+\cdots+5n=\dfrac{5n(n+1)}{2}$

When $n=1$, $S_1=5=\dfrac{5(1)(2)}{2}$, so the formula is valid.

Assume that $S_k=5+10+15+\cdots+5k=\dfrac{5k(k+1)}{2}$, then

$\quad S_{k+1}=S_k+a_{k+1}$

$\qquad = \dfrac{5k(k+1)}{2}+5(k+1)$

$\qquad = \dfrac{5k(k+1)}{2}+\dfrac{10(k+1)}{2}$

$\qquad = \dfrac{5k(k+1)+10(k+1)}{2}$

$\qquad = \dfrac{5(k+1)(k+2)}{2}$

$\qquad = \dfrac{5(k+1)[(k+1)+1]}{2}.$

Thus, the formula is valid for all integers $n\ge 1$.

10. $(x+2y)^4=x^4+4x^3(2y)+6x^2(2y)^2+4x(2y)^3+(2y)^4$

$\qquad = x^4+8x^3y+24x^2y^2+32xy^3+16y^4$

11. $_nC_r\,x^{n-r}y^r={}_8C_5(2a)^3(-3b)^5$

$\qquad = 56(8a^3)(-243b^5)$

$\qquad = -108{,}864a^3b^5$

So, the coefficient of a^3b^5 is $-108{,}864$.

12. (a) $_9P_2=\dfrac{9!}{7!}=72$

(b) $_{70}P_3=\dfrac{70!}{67!}=328{,}440$

13. (a) $_{11}C_4=\dfrac{11!}{7!4!}=330$

(b) $_{66}C_4=\dfrac{66!}{62!4!}=720{,}720$

14. $(26)(10)(10)(10)=26{,}000$ distinct license plates

15. $\underbrace{(1)}_{\text{owner}}\cdot\underbrace{(3)(2)}_{\substack{\text{bow}\\\text{seats}}}\cdot\underbrace{(5)(4)(3)(2)(1)}_{\substack{\text{remaining}\\\text{seats}}}=720$ seating arrangements

16. $\dfrac{20}{300}=\dfrac{1}{15}\approx 0.0667$

17. $\dfrac{1}{_{60}C_8}\approx 3.908\times10^{-10}$

18. $P(E')=1-P(E)$

$\qquad = 1-0.75$

$\qquad = 0.25$ or 25%

Chapters 9–11   Cumulative Test Solutions

1. $\begin{cases} y = 3 - x^2 \\ 2(y - 2) = x - 1 \end{cases} \Rightarrow 2(3 - x^2 - 2) = x - 1$

$$2(1 - x^2) = x - 1$$
$$2 - 2x^2 = x - 1$$
$$0 = 2x^2 + x - 3$$
$$0 = (2x + 3)(x - 1)$$
$$x = -\tfrac{3}{2} \quad \text{or} \quad x = 1$$
$$y = \tfrac{3}{4} \qquad \quad y = 2$$

Solutions: $\left(-\tfrac{3}{2}, \tfrac{3}{4}\right), (1, 2)$

2. $\begin{cases} x + 3y = -1 \implies 4x + 12y = -4 \\ 2x + 4y = 0 \implies -6x - 12y = 0 \end{cases}$
$$-2x = -4$$
$$x = 2 \implies y = -1$$

Solution: $(2, -1)$

3. $\begin{cases} -2x + 4y - z = 3 \\ x - 2y + 2z = -6 \\ x - 3y - z = 1 \end{cases}$

Interchange equations.

$\begin{cases} x - 2y + 2z = -6 & \text{Eq.1} \\ -2x + 4y - z = 3 & \text{Eq.2} \\ x - 3y - z = 1 & \text{Eq.3} \end{cases}$

$\begin{cases} x - 2y + 2z = -6 & \\ 3z = -9 & 2\text{Eq.1} + \text{Eq.2} \\ {-y} - 3z = 7 & -\text{Eq.1} + \text{Eq.3} \end{cases}$

From Equation 2 we have $z = -3$. Substituting this into Equation 3 yields $y = 2$. Using these in Equation 1 yields $x = 4$.

Solution: $(4, 2, -3)$

4. $\begin{cases} x + 3y - 2z = -7 \\ -2x + y - z = -5 \\ 4x + y + z = 3 \end{cases}$

$\begin{cases} x + 3y - 2z = -7 & \\ 7y - 5z = -19 & 2\text{Eq.1} + \text{Eq.2} \\ -11y + 9z = 31 & -4\text{Eq.1} + \text{Eq.3} \end{cases}$

$\begin{cases} x + 3y - 2z = -7 & \\ y - \tfrac{5}{7}z = -\tfrac{19}{7} & \tfrac{1}{7}\text{Eq.2} \\ -11y + 9z = 31 & \end{cases}$

$\begin{cases} x + \tfrac{1}{7}z = \tfrac{8}{7} & -3\text{Eq.2} + \text{Eq.1} \\ y - \tfrac{5}{7}z = -\tfrac{19}{7} & \\ \tfrac{8}{7}z = \tfrac{8}{7} & 11\text{Eq.2} + \text{Eq.3} \end{cases}$

$\begin{cases} x + \tfrac{1}{7}z = \tfrac{8}{7} & \\ y - \tfrac{5}{7}z = -\tfrac{19}{7} & \\ z = 1 & \tfrac{7}{8}\text{Eq.3} \end{cases}$

$\begin{cases} x = 1 & -\tfrac{1}{7}\text{Eq.3} + \text{Eq.1} \\ y = -2 & \tfrac{5}{7}\text{Eq.3} + \text{Eq.2} \\ z = 1 & \end{cases}$

Solution: $(1, -2, 1)$

5. $\begin{cases} 2x + y \geq -3 \\ x - 3y \leq 2 \end{cases}$

6. $\begin{cases} x - y > 6 \\ 5x + 2y < 10 \end{cases}$

7. Objective function: $z = 3x + 2y$

Subject to: $x + 4y \leq 20$

$\qquad\quad 2x + y \leq 12$

$\qquad\quad x \geq 0, y \geq 0$

At $(0, 0)$: $z = 0$

At $(0, 5)$: $z = 10$

At $(4, 4)$: $z = 20$

At $(6, 0)$: $z = 18$

Minimum of $z = 0$ at $(0, 0)$

Maximum of $z = 20$ at $(4, 4)$

8. $\begin{cases} x + y = 200 \implies y = 200 - x \\ 0.75x + 1.25y = 0.95(200) \end{cases}$

$0.75x + 1.25(200 - x) = 190$

$0.75x + 250 - 1.25x = 190$

$\qquad\qquad -0.50x = -60$

$\qquad\qquad\qquad x = 120$

$\qquad\qquad y = 200 - x = 80$

Answer: 120 pounds of \$0.75 seed and 80 pounds of \$1.25 seed.

9. $y = ax^2 + bx + c$

$(0, 4)$: $4 = a(0)^2 + b(0) + c \implies c = 4$

$(3, 1)$: $1 = a(3)^2 + b(3) + 4 \implies 9a + 3b = -3$

$\qquad\qquad\qquad\qquad\qquad\qquad 3a + b = -1$

$(6, 4)$: $4 = a(6)^2 + b(6) + 4 \implies 36a + 6b = 0$

$\qquad\qquad\qquad\qquad\qquad\qquad 6a + b = 0$

Solving the system:

$\begin{cases} 3a + b = -1 \\ 6a + b = 0 \end{cases}$ yields $a = \frac{1}{3}$ and $b = -2$.

Thus, the equation of the parabola is $y = \frac{1}{3}x^2 - 2x + 4$.

10. $\begin{cases} -x + 2y - z = 9 \\ 2x - y + 2z = -9 \\ 3x + 3y - 4z = 7 \end{cases}$ $\quad \begin{bmatrix} -1 & 2 & -1 & \vdots & 9 \\ 2 & -1 & 2 & \vdots & -9 \\ 3 & 3 & -4 & \vdots & 7 \end{bmatrix}$

11. $\begin{bmatrix} -1 & 2 & -1 & \vdots & 9 \\ 2 & -1 & 2 & \vdots & -9 \\ 3 & 3 & -4 & \vdots & 7 \end{bmatrix}$

$\begin{matrix} \\ 2R_1 + R_2 \to \\ 3R_1 + R_3 \to \end{matrix} \begin{bmatrix} -1 & 2 & -1 & \vdots & 9 \\ 0 & 3 & 0 & \vdots & 9 \\ 0 & 9 & -7 & \vdots & 34 \end{bmatrix}$

$\begin{matrix} -R_1 \to \\ \\ -3R_2 + R_3 \to \end{matrix} \begin{bmatrix} 1 & -2 & 1 & & -9 \\ 0 & 3 & 0 & & 3 \\ 0 & 0 & -7 & & 7 \end{bmatrix}$

$\begin{matrix} \\ \frac{1}{3}R_2 \to \\ -\frac{1}{7}R_3 \to \end{matrix} \begin{bmatrix} 1 & -2 & 1 & \vdots & -9 \\ 0 & 1 & 0 & \vdots & 3 \\ 0 & 0 & 1 & \vdots & -1 \end{bmatrix}$

$\begin{matrix} 2R_2 + R_1 \to \\ \\ \end{matrix} \begin{bmatrix} 1 & 0 & 1 & \vdots & -3 \\ 0 & 1 & 0 & \vdots & 3 \\ 0 & 0 & 1 & \vdots & -1 \end{bmatrix}$

$\begin{matrix} -R_3 + R_1 \to \\ \\ \end{matrix} \begin{bmatrix} 1 & 0 & 0 & \vdots & -2 \\ 0 & 1 & 0 & \vdots & 3 \\ 0 & 0 & 1 & \vdots & -1 \end{bmatrix}$

Solution: $(-2, 3, -1)$

12. $A + B = \begin{bmatrix} 4 & 0 \\ -1 & 2 \end{bmatrix} + \begin{bmatrix} -1 & 3 \\ 1 & 0 \end{bmatrix} = \begin{bmatrix} 3 & 3 \\ 0 & 2 \end{bmatrix}$

13. $-2B = -2\begin{bmatrix} -1 & 3 \\ 1 & 0 \end{bmatrix} = \begin{bmatrix} 2 & -6 \\ -2 & 0 \end{bmatrix}$

14. Use the result of Exercise 13.

$A - 2B = A + (-2B) = \begin{bmatrix} 4 & 0 \\ -1 & 2 \end{bmatrix} + \begin{bmatrix} 2 & -6 \\ -2 & 0 \end{bmatrix} = \begin{bmatrix} 6 & -6 \\ -3 & 2 \end{bmatrix}$

15. $AB = \begin{bmatrix} 4 & 0 \\ -1 & 2 \end{bmatrix}\begin{bmatrix} -1 & 3 \\ 1 & 0 \end{bmatrix} = \begin{bmatrix} (4)(-1) + (0)(1) & (4)(3) + (0)(0) \\ (-1)(-1) + 2(1) & (-1)(3) + (2)(0) \end{bmatrix} = \begin{bmatrix} -4 & 12 \\ 3 & -3 \end{bmatrix}$

16. $\begin{vmatrix} 8 & 0 & -5 \\ 1 & 3 & -1 \\ -2 & 6 & 4 \end{vmatrix} = 8\begin{vmatrix} 3 & -1 \\ 6 & 4 \end{vmatrix} - 5\begin{vmatrix} 1 & 3 \\ -2 & 6 \end{vmatrix}$

$= 8(18) - 5(12)$

$= 84$

Expand along Row 1.

17.

$\begin{bmatrix} 1 & 2 & -1 & \vdots & 1 & 0 & 0 \\ 3 & 7 & -10 & \vdots & 0 & 1 & 0 \\ -5 & -7 & -15 & \vdots & 0 & 0 & 1 \end{bmatrix}$

$\begin{matrix} -3R_1 + R_2 \rightarrow \\ 5R_1 + R_3 \rightarrow \end{matrix} \begin{bmatrix} 1 & 2 & -1 & \vdots & 1 & 0 & 0 \\ 0 & 1 & -7 & \vdots & -3 & 1 & 0 \\ 0 & 3 & -20 & \vdots & 5 & 0 & 1 \end{bmatrix}$

$\begin{matrix} -2R_2 + R_1 \rightarrow \\ \\ -3R_2 + R_3 \rightarrow \end{matrix} \begin{bmatrix} 1 & 0 & 13 & \vdots & 7 & -2 & 0 \\ 0 & 1 & -7 & \vdots & -3 & 1 & 0 \\ 0 & 0 & 1 & \vdots & 14 & -3 & 1 \end{bmatrix}$

$\begin{matrix} -13R_3 + R_1 \rightarrow \\ 7R_3 + R_2 \rightarrow \end{matrix} \begin{bmatrix} 1 & 0 & 0 & \vdots & -175 & 37 & -13 \\ 0 & 1 & 0 & \vdots & 95 & -20 & 7 \\ 0 & 0 & 1 & \vdots & 14 & -3 & 1 \end{bmatrix}$

$\begin{bmatrix} 1 & 2 & -1 \\ 3 & 7 & -10 \\ -5 & -7 & -15 \end{bmatrix}^{-1} = \begin{bmatrix} -175 & 37 & -13 \\ 95 & -20 & 7 \\ 14 & -3 & 1 \end{bmatrix}$

18. Let x = total sales of gym shoes (in millions),

y = total sales of jogging shoes (in millions),

z = total sales of walking shoes (in millions).

$\begin{bmatrix} 0.09 & 0.09 & 0.03 \\ 0.06 & 0.10 & 0.05 \\ 0.12 & 0.25 & 0.12 \end{bmatrix}\begin{bmatrix} x \\ y \\ z \end{bmatrix} = \begin{bmatrix} 442.20 \\ 466.57 \\ 1088.09 \end{bmatrix}$

$\begin{bmatrix} x \\ y \\ z \end{bmatrix} = \begin{bmatrix} 0.09 & 0.09 & 0.03 \\ 0.06 & 0.10 & 0.05 \\ 0.12 & 0.25 & 0.12 \end{bmatrix}^{-1}\begin{bmatrix} 442.20 \\ 466.57 \\ 1088.09 \end{bmatrix}$

$\approx \begin{bmatrix} 2042 \\ 1733 \\ 3415 \end{bmatrix}$

Thus, sales for each type of shoe amounted to:

Gym shoes: $2042 million

Jogging shoes: $1733 million

Walking shoes: $3415 million

19. $\begin{cases} 8x - 3y = -52 \\ 3x + 5y = 5 \end{cases}$, $\quad D = \begin{vmatrix} 8 & -3 \\ 3 & 5 \end{vmatrix} = 49$

$x = \dfrac{\begin{vmatrix} -52 & -3 \\ 5 & 5 \end{vmatrix}}{49} = \dfrac{-245}{49} = -5$

$y = \dfrac{\begin{vmatrix} 8 & -52 \\ 3 & 5 \end{vmatrix}}{49} = \dfrac{196}{49} = 4$

Solution: $(-5, 4)$

20. $\begin{cases} 5x + 4y + 3z = 7 \\ -3x - 8y + 7z = -9, \\ 7x - 5y - 6z = -53 \end{cases}$ $D = \begin{vmatrix} 5 & 4 & 3 \\ -3 & -8 & 7 \\ 7 & -5 & -6 \end{vmatrix} = 752$

21. $A = \pm\dfrac{1}{2}\begin{vmatrix} -2 & 3 & 1 \\ 1 & 5 & 1 \\ 4 & 1 & 1 \end{vmatrix} = -\dfrac{1}{2}(-18) = 9$

$x = \dfrac{\begin{vmatrix} 7 & 4 & 3 \\ -9 & -8 & 7 \\ -53 & -5 & -6 \end{vmatrix}}{752} = \dfrac{-2256}{752} = -3$

$y = \dfrac{\begin{vmatrix} 5 & 7 & 3 \\ -3 & -9 & 7 \\ 7 & -53 & -6 \end{vmatrix}}{752} = \dfrac{3008}{752} = 4$

$z = \dfrac{\begin{vmatrix} 5 & 4 & 7 \\ -3 & -8 & -9 \\ 7 & -5 & -53 \end{vmatrix}}{752} = \dfrac{1504}{752} = 2$

Solution: $(-3, 4, 2)$

22. $a_n = \dfrac{(-1)^{n+1}}{2n + 3}$

$a_1 = \dfrac{1}{5}$

$a_2 = -\dfrac{1}{7}$

$a_3 = \dfrac{1}{9}$

$a_4 = -\dfrac{1}{11}$

$a_5 = \dfrac{1}{13}$

23. $\dfrac{2!}{4}, \dfrac{3!}{5}, \dfrac{4!}{6}, \dfrac{5!}{7}, \dfrac{6!}{8}, \ldots$

$a_n = \dfrac{(n + 1)!}{n + 3}$

24. $8, 12, 16, 20, \ldots$

$a_n = 4n + 4$

$a_1 = 8, \; a_{20} = 84$

$S_{20} = \dfrac{20}{2}(8 + 84) = 920$

25. (a) $a_6 = 20.6$

$a_9 = 30.2$

$a_9 = a_6 + 3d$

$30.2 = 20.6 + 3d$

$9.6 = 3d$

$3.2 = d$

$a_{20} = a_9 + 11d = 30.2 + 11(3.2) = 65.4$

(b) $a_1 = a_6 - 5d$

$a_1 = 20.6 - 5(3.2)$

$= 4.6$

$a_n = a_1 + (n - 1)d$

$= 4.6 + (n - 1)(3.2)$

$= 3.2n + 1.4$

26. $a_n = 3(2)^{n-1}$

$a_1 = 3$

$a_2 = 6$

$a_3 = 12$

$a_4 = 24$

$a_5 = 48$

27. $\displaystyle\sum_{i=6}^{\infty} 1.3\left(\dfrac{1}{10}\right)^{i-1} = \dfrac{1.3}{1 - \frac{1}{10}} = 1.3\left(\dfrac{10}{9}\right) = \dfrac{13}{9}$

28. $S_1 = 3 = 1[2(1) + 1]$

Assume that $S_k = 3 + 7 + 11 + 15 + \cdots + (4k - 1) = k(2k + 1)$.

Then, $S_{k+1} = 3 + 7 + 11 + 15 + \cdots + (4k - 1) + [4(k + 1) - 1]$

$$= S_k + (4k + 3)$$

$$= k(2k + 1) + (4k + 3)$$

$$= 2k^2 + 5k + 3$$

$$= (k + 1)(2k + 3)$$

$$= (k + 1)[2(k + 1) + 1].$$

Therefore, the formula is valid for all integers $n \geq 1$.

29. $(z - 3)^4 = z^4 - 4z^3(3) + 6z^2(3)^2 - 4z(3)^3 + (3)^4$

$$= z^4 - 12z^3 + 54z^2 - 108z + 81$$

30. $_7P_3 = \dfrac{7!}{(7 - 3)!} = \dfrac{7!}{4!} = 210$

31. $_{25}P_2 = \dfrac{25!}{(25 - 2)!} = \dfrac{25!}{23!} = 600$

32. $\dbinom{8}{4} = {}_8C_4 = \dfrac{8!}{(8 - 4)!4!} = \dfrac{8!}{4!4!} = 70$

33. $_{10}C_3 = \dfrac{10!}{(10 - 3)!3!} = \dfrac{10!}{7!3!} = 120$

34. B A S K E T B A L L

$$\dfrac{10!}{2!2!2!1!1!1!1!1!} = 453{,}600 \text{ distinguishable permutations}$$

35. A N T A R C T I C A

$$\dfrac{10!}{3!2!2!1!1!1!1!} = 151{,}200 \text{ distinguishable permutations}$$

36. $_{10}P_3 = \dfrac{10!}{(10 - 3)!} = \dfrac{10!}{7!} = 720$

37. The first digit is 4 or 5, so the probability of picking it correctly is $\frac{1}{2}$.

Then there are two numbers left for the second digit so its probability is also $\frac{1}{2}$.

If these two are correct, then the third digit must be the remaining number.

The probability of winning is:

$$\left(\tfrac{1}{2}\right)\left(\tfrac{1}{2}\right)(1) = \tfrac{1}{4}$$